Studies in Advanced Mathematics

KENNETH KUTTLER
Michigan Technological University

Modern Analysis

CRC PRESS
Boca Raton   Boston   New York   Washington   London

Library of Congress Cataloging-in-Publication Data

Kuttler, Kenneth.
   Modern analysis / Kenneth Kuttler.
        p. cm. – – (Studies in advanced mathematics)
       Includes bibliographical references and index.
       ISBN 0-8493-7166-X (alk. paper)
       1. Mathematical analysis. I. Title. II. Series.
   QR300.K883 1997
   515—dc21   97-35735
                                                     CIP

Preface

This book is on real and abstract analysis. It is intended to give a reasonable introduction to functional analysis and to give a thorough treatment of Lebesgue integration as well as topics related to it.

The book will use the theorems and definitions about determinants and linear algebra. In particular, it is assumed the reader knows about normed vector space, sometimes referred to as a linear space, along with the basic theorems on linear algebra. In particular, theorems about determinants such as $\det(AB) = \det(A)\det(B)$ are assumed. It will also use the theorem from linear algebra which says that if A is a linear transformation mapping $\mathbb{R}^n$ to $\mathbb{R}^n$ and $A = A^*$, then there is an orthonormal basis of eigenvectors of A. The books [29] and [36] contain enough on the topic of determinants and provide the necessary background in linear algebra. The reader is also assumed to know the meaning of the Riemann integral,

$$\int_{-\infty}^{\infty} f(x)\, dx$$

for f a continuous function defined on $\mathbb{R}$ which is zero outside of some finite interval. The books [4], [5], [45] contain more than enough on this subject. The reader should also know basic set theory and the notation used in this subject. Other than these things, the book is essentially self contained.

There are at least two routes through the first 16 chapters, the order listed in the table of contents being one. This order involves placing all the abstract analysis before the real analysis and is similar to the organization found in [33]. Another order is the following: 1, 2, 7, 8, 9, 10, 11, 12, 3, 4, 13, 14, 6,16,15, 17. This alternate order involves spreading the abstract analysis out, introducing it as needed. Once these chapters have been studied, the following diagram shows the logical dependence of the remaining chapters.

$$
\begin{array}{ccccc}
& 6 & & & \\
& \downarrow & & & \\
& 18 & & & \\
13 & \downarrow & 16,\,6 & 6,\,18,\,20,\,23 & \\
\downarrow & 19 & \downarrow & \downarrow & \\
23 & \downarrow & 22 & 24 & \\
& 20 & & & \\
& \downarrow & & & \\
& 21 & & &
\end{array}
$$

The final chapters in the book are appendices and contain topics which are of general interest.

Many of the exercises extend the theorems and supply examples to illustrate the theorems proved in the book. When an exercise is dependent on an earlier exercise, not necessarily the preceding one, a small vertical arrow appears at the beginning.

I am grateful to CRC Press for publishing this book and to many people who have helped in its preparation, especially Yuri Dimitrov who found numerous errors in the first part of the book; Ivo Dinov who helped me in my attempts to write Chapter 23; Prof. Pinelis who found several ommissions and errors in Chapter 17; and Dan Hayden whose Master's Degree project helped me prepare Chapters 14 and 15. I am also grateful to my wife for her help typing.

Contents

Chapter 1
Set Theory and General Topology

1.1   Set theory

To begin with we review some basic set theory. There are axioms of set theory which may be found in the book by Halmos, *Naive Set Theory* [23], which seem very obvious. Of course much can be said about these axioms and we refer to the above book for more discussion.

1. Two sets are equal if and only if they have the same elements.

2. To every set, A, and to every condition $S(x)$ there corresponds a set, B, whose elements are exactly those elements x of A for which $S(x)$ holds.

3. For every collection of sets there exists a set that contains all the elements that belong to at least one set of the given collection.

4. The Cartesian product of a nonempty family of nonempty sets is nonempty.

These axioms are referred to as the axiom of extension, axiom of specification, axiom of unions, and axiom of choice respectively.

The reader is assumed to be familiar with the usual notions and notation of set theory such as unions, intersections, etc. The following theorem will be used to prove the very important Schroder Bernstein theorem but it is also very interesting for its own sake.

Theorem 1.1 *Let $f : X \to Y$ and $g : Y \to X$ be two mappings. Then there exist sets A, B, C, D, such that*

$$A \cup B = X, \ C \cup D = Y, \ A \cap B = \emptyset, \ C \cap D = \emptyset,$$

$$f(A) = C, \ g(D) = B.$$

Proof: We will say $A \subseteq X$ satisfies $\mathcal{P}$ if whenever $y \in Y \backslash f(A)$, $g(y) \notin A$. Note $\emptyset$ satisfies $\mathcal{P}$.

$$\mathcal{A} \equiv \{A_0 \subseteq X : A_0 \text{ satisfies } \mathcal{P}\}.$$

Let $A = \cup \mathcal{A}$. If $y \in Y \setminus f(A)$, then for each $A_0 \in \mathcal{A}$, $y \in Y \setminus f(A_0)$ and so $g(y) \notin A_0$. Since $g(y) \notin A_0$ for all $A_0 \in \mathcal{A}$, it follows $g(y) \notin A$. Hence

A satisfies $\mathcal{P}$ and is the largest subset of A which does so. Define

$$C \equiv f(A), \ D \equiv Y \setminus C, \ B \equiv X \setminus A.$$

Thus all conditions of the theorem are satisfied except for $g(D) = B$ and we verify this condition now.

Let $x \in B = X \setminus A$. Then $A \cup \{x\}$ does not satisfy $\mathcal{P}$ because this set is larger than A. Therefore there exists

$$y \in Y \setminus f(A \cup \{x\}) \subseteq Y \setminus f(A) \equiv D$$

such that $g(y) \in A \cup \{x\}$. But $g(y) \notin A$ because $y \in Y \setminus f(A)$ and A satisfies $\mathcal{P}$. Hence $g(y) = x$ and this proves the theorem.

Theorem 1.2 *(Schroder Bernstein) If $f : X \to Y$ and $g : Y \to X$ are one to one, then there exists $h : X \to Y$ which is one to one and onto.*

Proof: Let A, B, C, D be the sets of Theorem 1.1 and define

$$h(x) \equiv \begin{cases} f(x) & \text{if } x \in A, \\ g^{-1}(x) & \text{if } x \in B. \end{cases}$$

It is clear h is one to one and onto.

Definition 1.3 Let I be a set and let X_i be a set for each $i \in I$. We say that f is a choice function and write

$$f \in \prod_{i \in I} X_i$$

if $f(i) \in X_i$ for each $i \in I$.

The axiom of choice says that if $X_i \neq \emptyset$ for each $i \in I$, for I a set, then

$$\prod_{i \in I} X_i \neq \emptyset.$$

The symbol above denotes the collection of all choice functions. Using the axiom of choice, we can obtain the following interesting corollary to the Schroder Bernstein theorem.

Corollary 1.4 *If $f : X \to Y$ is onto and $g : Y \to X$ is onto, then there exists $h : X \to Y$ which is one to one and onto.*

Proof: For each $y \in Y$, let $f_0^{-1}(y) \in f^{-1}(y)$ and similarly let $g_0^{-1}(x) \in g^{-1}(x)$. We used the axiom of choice to pick a single element, $f_0^{-1}(y)$ in $f^{-1}(y)$ and similarly for $g^{-1}(x)$. Then f_0^{-1} and g_0^{-1} are one to one; so, by the Schroder Bernstein theorem, there exists $h : X \to Y$ which is one to one and onto.

Definition 1.5 We say a set, S, is finite if there exists a natural number n and a map θ which maps $\{1, \cdots, n\}$ one to one and onto S. We say S is infinite if it is not finite. A set, S, is called countable if there exists a map θ mapping $\mathbb{N}$ one to one and onto S. (When θ maps a set A to a set B, we will write $\theta : A \to B$ in the future.) Here $\mathbb{N} \equiv \{1, 2, \cdots\}$, the natural numbers. If there exists a map $\theta : \mathbb{N} \to S$ which is onto, we say that S is at most countable.

In the literature, the property of being at most countable is often referred to as being countable. When this is done, there is usually no harm incurred from this sloppiness because the question of interest is normally whether one can list all elements of the set, designating a first, second, third etc. in such a way as to exhaust the entire set. The possibility that a single element of the set may occur more than once in the list is often not important.

Theorem 1.6 *If X and Y are both at most countable, then $X \times Y$ is also at most countable.*

Proof: We know there exists a mapping $\eta : \mathbb{N} \to X$ which is onto. If we define $\eta(i) \equiv x_i$ we may consider X as the sequence $\{x_i\}_{i=1}^{\infty}$, written in the traditional way. Similarly, we may consider Y as the sequence $\{y_j\}_{j=1}^{\infty}$. It follows we can represent all elements of $X \times Y$ by the following infinite rectangular array.

$$
\begin{array}{cccc}
(x_1, y_1) & (x_1, y_2) & (x_1, y_3) & \cdots \\
(x_2, y_1) & (x_2, y_2) & (x_2, y_3) & \cdots \\
(x_3, y_1) & (x_3, y_2) & (x_3, y_3) & \cdots \cdot \\
\vdots & \vdots & \vdots &
\end{array}
$$

We follow a path through this array as follows.

$$
\begin{array}{ccc}
(x_1, y_1) & \to & (x_1, y_2) & & (x_1, y_3) & \to \\
& \swarrow & & \nearrow & \\
(x_2, y_1) & & (x_2, y_2) & \\
\downarrow & \nearrow & \\
(x_3, y_1) & &
\end{array}
$$

Thus the first element of $X \times Y$ is (x_1, y_1), the second element of $X \times Y$ is (x_1, y_2), the third element of $X \times Y$ is (x_2, y_1) etc. In this way we see that we can assign a number from $\mathbb{N}$ to each element of $X \times Y$. In other words there exists a mapping from $\mathbb{N}$ onto $X \times Y$. This proves the theorem.

3

Corollary 1.7 *If either X or Y is countable, then $X \times Y$ is also countable.*

Proof: By Theorem 1.6, there exists a mapping $\theta : \mathbb{N} \to X \times Y$ which is onto. Suppose without loss of generality that X is countable. Then there exists $\alpha : \mathbb{N} \to X$ which is one to one and onto. Let $\beta : X \times Y \to \mathbb{N}$ be defined by $\beta((x, y)) \equiv \alpha^{-1}(x)$. Then by Corollary 1.4, there is a one to one and onto mapping from $X \times Y$ to $\mathbb{N}$. This proves the corollary.

Theorem 1.8 *If X and Y are at most countable, then $X \cup Y$ is at most countable.*

Proof: Let $X = \{x_i\}_{i=1}^{\infty}, Y = \{y_j\}_{j=1}^{\infty}$ and consider the following array consisting of $X \cup Y$ and path through it.

$$
\begin{array}{ccccc}
x_1 & \to & x_2 & & x_3 & \to \\
 & \swarrow & & \nearrow & & \\
y_1 & \to & y_2 & & &
\end{array}
$$

Thus the first element of $X \cup Y$ is x_1, the second is x_2, the third is y_1, the fourth is y_2, etc. This proves the theorem.

Corollary 1.9 *If either X or Y are countable, then $X \cup Y$ is countable.*

Proof: There is a map from $\mathbb{N}$ onto $X \times Y$. Suppose without loss of generality that X is countable and $\alpha : \mathbb{N} \to X$ is one to one and onto. Then define $\beta(y) \equiv 1$, for all $y \in Y$, and $\beta(x) \equiv \alpha^{-1}(x)$. Thus, β maps $X \times Y$ onto $\mathbb{N}$ and applying Corollary 1.4 yields the conclusion and proves the corollary.

1.2 General topology

This section is a brief introduction to some basic general topology. These theorems and definitions will be referred to in the rest of the book. The best way to understand these things is to prove them yourself and this is the approach taken here. Nevertheless, there are many good books on the subject such as [40] in which proofs of these and many other theorems may be found. Topological spaces consist of a set and a subset of the set of all subsets of this set called the open sets or topology which satisfy certain axioms. Like other areas in mathematics the abstraction inherent in this approach is an attempt to unify many different useful examples into one general theory.

For example, consider $\mathbb{R}^n$ with the usual norm given by

$$
||\mathbf{x}|| \equiv \left(\sum_{i=1}^{n} |x_i|^2 \right)^{1/2}
$$

We say a set U in $\mathbb{R}^n$ is an open set if every point of U is an "interior" point which means that if $\mathbf{x} \in U$, there exists $\delta > 0$ such that if $||\mathbf{y} - \mathbf{x}|| < \delta$, then $\mathbf{y} \in U$. It is easy to see that with this definition of open sets, the axioms 1 - 2 given below are satisfied if τ is the collection of open sets as just described. There are many other sets of interest besides $\mathbb{R}^n$ however, and the appropriate definition of the concept of "open set" may be very different and yet the collection of open sets may still satisfy these axioms. By abstracting the concept of open sets, we can unify many different examples. Here is the definition of a general topological space.

Let X be a set and let τ be a collection of subsets of X satisfying

$$\emptyset \in \tau, \ X \in \tau, \tag{1}$$

$$\text{If } \mathcal{C} \subseteq \tau, \text{ then } \cup \mathcal{C} \in \tau$$

$$\text{If } A, B \in \tau, \text{ then } A \cap B \in \tau. \tag{2}$$

Definition 1.10 A set X together with such a collection of its subsets satisfying 1-2 is called a topological space. τ is called the topology or set of open sets of X. Note $\tau \subseteq \mathcal{P}(X)$, the set of all subsets of X, also called the power set.

Definition 1.11 A subset $\mathcal{B}$ of τ is called a basis for τ if whenever $p \in U \in \tau$, there exists a set $B \in \mathcal{B}$ such that $p \in B \subseteq U$. The elements of $\mathcal{B}$ are called basic open sets.

The preceding definition implies that every open set (element of τ) may be written as a union of basic open sets (elements of $\mathcal{B}$). This brings up an interesting and important question. If a collection of subsets $\mathcal{B}$ of a set X is specified, does there exist a topology τ for X satisfying 1-2 such that $\mathcal{B}$ is a basis for τ?

Theorem 1.12 *Let X be a set and let $\mathcal{B}$ be a set of subsets of X. Then $\mathcal{B}$ is a basis for a topology τ if and only if whenever $p \in B \cap C$ for $B, C \in \mathcal{B}$, there exists $D \in \mathcal{B}$ such that $p \in D \subseteq C \cap B$ and $\cup \mathcal{B} = X$. In this case τ consists of all unions of subsets of $\mathcal{B}$.*

Proof: The only if part is left to the reader. Let τ consist of all unions of sets of $\mathcal{B}$ and suppose $\mathcal{B}$ satisfies the conditions of the proposition. Then $\emptyset \in \tau$ because $\emptyset \subseteq \mathcal{B}$. $X \in \tau$ because $\cup \mathcal{B} = X$ by assumption. If $\mathcal{C} \subseteq \tau$ then clearly $\cup \mathcal{C} \in \tau$. Now suppose $A, B \in \tau$, $A = \cup \mathcal{S}$, $B = \cup \mathcal{R}$, $\mathcal{S}, \mathcal{R} \subseteq \mathcal{B}$. We need to show $A \cap B \in \tau$. If $A \cap B = \emptyset$, we are done. Suppose $p \in A \cap B$. Then $p \in S \cap R$ where $S \in \mathcal{S}$, $R \in \mathcal{R}$. Hence there exists $U \in \mathcal{B}$ such that $p \in U \subseteq S \cap R$. It follows, since $p \in A \cap B$ was arbitrary, that $A \cap B = $ union of sets of $\mathcal{B}$. Thus $A \cap B \in \tau$. Hence τ satisfies 1-2.

Definition 1.13 A topological space is said to be Hausdorff if whenever p and q are distinct points of X, there exist disjoint open sets U, V such that $p \in U, q \in V$.

Hausdorff

Definition 1.14 A subset of a topological space is said to be closed if its complement is open. Let p be a point of X and let $E \subseteq X$. Then p is said to be a limit point of E if every open set containing p contains a point of E distinct from p.

Theorem 1.15 *A subset, E, of X is closed if and only if it contains all its limit points.*

Theorem 1.16 *If (X, τ) is a Hausdorff space and if $p \in X$, then $\{p\}$ is a closed set.*

Definition 1.17 A topological space (X, τ) is said to be regular if whenever C is a closed set and p is a point not in C, then there exist disjoint open sets U and V such that $p \in U$, $C \subseteq V$. The topological space, (X, τ) is said to be normal if whenever C and K are disjoint closed sets, there exist disjoint open sets U and V such that $C \subseteq U$, $K \subseteq V$.

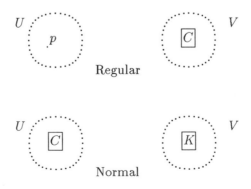

Regular

Normal

Definition 1.18 Let E be a subset of X. $\bar{E}$ is defined to be the smallest closed set containing E. Note that this is well defined since X is closed and the intersection of any collection of closed sets is closed.

Theorem 1.19 $\bar{E} = E \cup \{limit\ points\ of\ E\}$.

Definition 1.20 Let X be a set and let $d : X \times X \rightarrow [0, \infty)$ satisfy

$$d(x, y) = d(y, x), \tag{3}$$

$$d(x, y) + d(y, z) \geq d(x, z),\ \text{(triangle inequality)}$$

$$d(x, y) = 0 \text{ if and only if } x = y. \tag{4}$$

Such a function is called a metric. For $r \in [0, \infty)$ and $x \in X$, define

$$B(x, r) = \{y \in X : d(x, y) < r\}$$

This may also be denoted by $N(x, r)$.

Definition 1.21 A topological space (X, τ) is called a metric space if there exists a metric, d, such that the sets $\{B(x, r), x \in X,\ r > 0\}$ form a basis for τ. We write (X, d) for the metric space.

Theorem 1.22 *Suppose X is a set and d satisfies 3-4. Then the sets $\{B(x, r) : r > 0,\ x \in X\}$ form a basis for a topology on X.*

Theorem 1.23 *If (X, τ) is a metric space, then (X, τ) is Hausdorff, regular, and normal.*

Definition 1.24 A metric space is said to be separable if there is a countable dense subset of the space. This means there exists $D = \{p_i\}_{i=1}^{\infty}$ such that for all x and $r > 0$, $B(x, r) \cap D \neq \emptyset$.

Definition 1.25 A topological space is said to be completely separable if it has a countable basis for the topology.

Theorem 1.26 *A metric space is separable if and only if it is completely separable.*

Definition 1.27 A sequence $\{p_n\}_{n=1}^{\infty}$ in a metric space is called a Cauchy sequence if for every $\varepsilon > 0$ there exists N such that $d(p_n, p_m) < \varepsilon$ whenever $n, m > N$. A metric space is called complete if every Cauchy sequence converges to some element of the metric space.

Example 1.28 $\mathbb{R}^n$ and $\mathbb{C}^n$ are complete metric spaces for the metric defined by $d(\mathbf{x}, \mathbf{y}) \equiv ||\mathbf{x} - \mathbf{y}|| \equiv (\sum_{i=1}^n |x_i - y_i|^2)^{1/2}$.

Definition 1.29 Let (X, τ) and (Y, η) be two topological spaces and let $f : X \to Y$. We say f is continuous at $x \in X$ if whenever V is an open set of Y containing $f(x)$, there exists an open set $U \in \tau$ such that $x \in U$ and $f(U) \subseteq V$. We say that f is continuous if $f^{-1}(V) \in \tau$ whenever $V \in \eta$.

Definition 1.30 Let (X, τ) and (Y, η) be two topological spaces. $X \times Y$ is the Cartesian product. $(X \times Y = \{(x, y) : x \in X, y \in Y\})$. We can define a product topology as follows. Let $\mathcal{B} = \{(A \times B) : A \in \tau, B \in \eta\}$. $\mathcal{B}$ is a basis for the product topology.

Theorem 1.31 $\mathcal{B}$ defined above is a basis satisfying the conditions of Theorem 1.12.

More generally we have the following definition which considers any finite Cartesian product of topological spaces.

Definition 1.32 If (X_i, τ_i) is a topological space, we make $\prod_{i=1}^n X_i$ into a topological space by letting a basis be $\prod_{i=1}^n A_i$ where $A_i \in \tau_i$.

Theorem 1.33 Definition 1.32 yields a basis for a topology.

Definition 1.34 A subset, E, of a topological space (X, τ) is said to be compact if whenever $\mathcal{C} \subseteq \tau$ and $E \subseteq \cup \mathcal{C}$, there exists a finite subset of $\mathcal{C}, \{U_1 \cdots U_n\}$, such that $E \subseteq \cup_{i=1}^n U_i$. (Every open covering admits a finite subcovering.) A topological space is called Locally Compact if it has a basis $\mathcal{B}$, with the property that $\bar{B}$ is compact for each $B \in \mathcal{B}$.

Lemma 1.35 Let (X, τ) be a topological space and let $\mathcal{B}$ be a basis for τ. Then K is compact if and only if every open cover of basic open sets admits a finite subcover.

The proof follows directly from the definition and is left to the reader.

Theorem 1.36 Let X and Y be topological spaces, and K_1, K_2 be compact sets in X and Y respectively. Then $K_1 \times K_2$ is compact in the topological space $X \times Y$.

Proof: Let $\mathcal{C}$ be an open cover of $K_1 \times K_2$ of sets $A \times B$ where A and B are open sets. Thus $\mathcal{C}$ is a open cover of basic open sets. For $y \in Y$, define

$$\mathcal{C}_y = \{A \times B \in \mathcal{C} : y \in B\}, \quad \mathcal{D}_y = \{A : A \times B \in \mathcal{C}_y\}$$

Claim: $\mathcal{D}_y$ covers K_1.

Proof: Let $x \in K_1$. Then $(x, y) \in K_1 \times K_2$ so $(x, y) \in A \times B \in \mathcal{C}$. Therefore $A \times B \in \mathcal{C}_y$ and so $x \in A \in \mathcal{D}_y$.

Since K_1 is compact,

$$\{A_1, \cdots, A_{n(y)}\} \subseteq \mathcal{D}_y$$

covers K_1. Let

$$B_y = \cap_{i=1}^{n(y)} B_i$$

Thus $\{A_1, \cdots, A_{n(y)}\}$ covers K_1 and $A_i \times B_y \subseteq A_i \times B_i \in \mathcal{C}_y$.

Since K_2 is compact, there is a finite list of elements of K_2, $y_1, \cdots, y_r$ such that

$$\{B_{y_1}, \cdots, B_{y_r}\}$$

covers K_2. Consider

$$\{A_i \times B_{y_l}\}_{i=1\ l=1}^{n(y_l)\ r}.$$

If $(x, y) \in K_1 \times K_2$, then $y \in B_{y_j}$ for some $j \in \{1, \cdots, r\}$. Then $x \in A_i$ for some $i \in \{1, \cdots, n(y_j)\}$. Hence $(x, y) \in A_i \times B_{y_j}$. Each of the sets $A_i \times B_{y_j}$ is contained in some set of $\mathcal{C}$ and so this proves the theorem.

Definition 1.37 $\mathcal{S} \subseteq \tau$ is called a subbasis for the topology τ if the set $\mathcal{B}$ of finite intersections of sets of $\mathcal{S}$ is a basis for the topology, τ. If (X_α, τ_α) is a topological space for each $\alpha \in \Lambda$, an index set which is possibly infinite, we use the axiom of choice to consider $\prod_{\alpha \in \Lambda} X_\alpha$. A subbasis for the product topology consists of all sets of the form $\prod_{\alpha \in \Lambda} A_\alpha$ where $A_\alpha = X_\alpha$ except for a single α, say $\alpha = \beta$, and $A_\beta \in \tau_\beta$.

1.3   Urysohn's lemma

Urysohn's lemma which characterizes normal spaces is a very important result which is useful in general topology and in the construction of measures. Because it is somewhat technical we give a proof of the part of it we will need later here.

Theorem 1.38 *(Urysohn) Let (X, τ) be normal and let $H \subseteq U$ where H is closed and U is open. Then there exists $g : X \to [0, 1]$ such that g is continuous, $g(x) = 1$ on H and $g(x) = 0$ if $x \notin U$.*

Proof: Let $D \equiv \{r_n\}_{n=1}^{\infty}$ be the rational numbers in $(0, 1)$. Choose V_{r_1} an open set such that

$$H \subseteq V_{r_1} \subseteq \overline{V}_{r_1} \subseteq U.$$

We can do this by applying the assumption that X is normal to the disjoint closed sets, H and U^C, to obtain open sets V and W with

$$H \subseteq V, \ U^C \subseteq W, \text{and } V \cap W = \emptyset.$$

Then

$$H \subseteq V \subseteq \overline{V}, \ \overline{V} \cap U^C = \emptyset$$

and so we let $V_{r_1} = V$.

Suppose $V_{r_1}, \cdots, V_{r_k}$ have been chosen and list the rational numbers $r_1, \cdots, r_k$ in order,

$$r_{l_1} < r_{l_2} < \cdots < r_{l_k} \text{ for } \{l_1, \cdots, l_k\} = \{1, \cdots, k\}.$$

If $r_{k+1} > r_{l_k}$ then letting $p = r_{l_k}$, we let

$$\overline{V}_p \subseteq V_{r_{k+1}} \subseteq \overline{V}_{r_{k+1}} \subseteq U.$$

If $r_{k+1} \in (r_{l_i}, r_{l_{i+1}})$, let $p = r_{l_i}$ and let $q = r_{l_{i+1}}$. Then let $V_{r_{k+1}}$ satisfy

$$\overline{V}_p \subseteq V_{r_{k+1}} \subseteq \overline{V}_{r_{k+1}} \subseteq V_q.$$

If $r_{k+1} < r_{l_1}$, let $p = r_{l_1}$ and let $V_{r_{k+1}}$ satisfy

$$H \subseteq V_{r_{k+1}} \subseteq \overline{V}_{r_{k+1}} \subseteq V_p.$$

Thus there exist open sets V_r for each $r \in \mathbb{Q} \cap (0, 1)$ with the property that if $r < s$,

$$H \subseteq V_r \subseteq \overline{V}_r \subseteq V_s \subseteq \overline{V}_s \subseteq U.$$

Now let

$$f(x) = \inf\{t \in D : x \in V_t\}, \ f(x) \equiv 1 \text{ if } x \notin \bigcup_{t \in D} V_t.$$

We show f is continuous.

$$f^{-1}([0, a)) = \cup\{V_t : t < a, t \in D\}.$$

Next consider $f^{-1}([0, a])$. If $t > a$, then $x \in V_t$ because if not, then

$$\inf\{t \in D : x \in V_t\} > a.$$

Thus

$$f^{-1}([0, a]) = \cap\{V_t : t > a\} = \cap\{\overline{V}_t : t > a\}$$

which is a closed set. If $a = 1$, $f^{-1}([0, 1]) = f^{-1}([0, a]) = X$. Therefore,

$$f^{-1}((a, 1]) = X \setminus f^{-1}([0, a]) = \text{open set}.$$

It follows f is continuous. Clearly $f(x) = 0$ on H. If $x \in U^C$, then $x \notin V_t$ for any $t \in D$ so $f(x) = 1$ on U^C. Let $g(x) = 1 - f(x)$. This proves the theorem.

Theorem 1.39 *Every compact Hausdorff space is normal.*

Proof: First we show the compact Hausdorff space, X, is regular. Let H be a closed set and let $p \notin H$. Then for each $h \in H$, there exists an open set U_h containing p and an open set V_h containing h such that $U_h \cap V_h = \emptyset$. Since H must be compact, it follows there are finitely many of the sets $V_h, V_{h_1} \cdots V_{h_n}$ such that $H \subseteq \cup_{i=1}^n V_{h_i}$. Then letting $U = \cap_{i=1}^n U_{h_i}$ and $V = \cup_{i=1}^n V_{h_i}$, it follows that $p \in U$, $H \in V$ and $U \cap V = \emptyset$. Thus X is regular as claimed.

Next let K and H be disjoint nonempty closed sets. Using regularity of X, for every $k \in K$, there exists an open set U_k containing k and an open set V_k containing H such that these two open sets have empty intersection. Thus $H \cap \overline{U}_k = \emptyset$. Finitely many of the U_k, $U_{k_1}, \cdots, U_{k_p}$ cover K and so $\cup_{i=1}^p \overline{U}_{k_i}$ is a closed set which has empty intersection with H. Therefore, $K \subseteq \cup_{i=1}^p U_{k_i}$ and $H \subseteq \left(\cup_{i=1}^p \overline{U}_{k_i} \right)^C$. This proves the theorem.

A useful construction when dealing with locally compact Hausdorff spaces is the notion of the one point compactification of the space. Suppose (X, τ) is a locally compact Hausdorff space. Then let $\widetilde{X} \equiv X \cup \{\infty\}$ where ∞ is just the name of some point which is not in X which we call the point at infinity. A basis for the topology $\widetilde{\tau}$ for $\widetilde{X}$ is $\tau \cup \{K^C$ where K is a compact subset of $X\}$. The reason this is called a compactification is contained in the next lemma.

Lemma 1.40 *If (X, τ) is a locally compact Hausdorff space, then $\left(\widetilde{X}, \widetilde{\tau} \right)$ is a compact Hausdorff space.*

Proof: Since (X, τ) is a locally compact Hausdorff space, it is easy to see $\left(\widetilde{X}, \widetilde{\tau} \right)$ is a Hausdorff topological space. (See Problem 14 of the exercises in this chapter.) Now let $\mathcal{C}$ be an open cover of $\widetilde{X}$ with sets from $\widetilde{\tau}$. Then ∞ must be in some set, U_∞ from $\mathcal{C}$, which must contain a set of the form K^C where K is a compact subset of X. Then there exist sets from $\mathcal{C}$, $U_1, \cdots, U_r$ which cover K. Therefore, a finite subcover of $\widetilde{X}$ is $U_1, \cdots, U_r, U_\infty$.

Theorem 1.41 *Let X be a locally compact Hausdorff space, and let K be a compact subset of the open set V. Then there exists a continuous function, $f : X \to [0,1]$, such that f equals 1 on K and $\overline{\{x : f(x) \neq 0\}}$ is a compact subset of V.*

Proof: Let $\tilde{X}$ be the space just described. Then K and V are respectively closed and open in $\tilde{\tau}$. By Theorem 1.39 there exist open sets in $\tilde{\tau}$, U, and W such that $K \subseteq U, \infty \in V^C \subseteq W$, and $U \cap W = U \cap (W \setminus \{\infty\}) = \emptyset$. Now note $W \setminus \{\infty\}$ is an open set in τ because it is a union of open sets, K^C for K compact and open sets from τ. Thus, $(W \setminus \{\infty\})^C \subseteq V$ and $(W \setminus \{\infty\})^C$ is a closed set in τ which contains U. Now $\infty \in V^C$ and so $\infty \in K_1^C \subseteq W$ for some K_1 compact in τ. It follows that $U \cap K_1^C = \emptyset$ and so

$$K \subseteq U \subseteq K_1 \cap (W \setminus \{\infty\})^C \subseteq V$$

and it follows $\overline{U}$ is a compact subset of V. By Theorem 1.39 and Urysohn's lemma, there exists $f : \overline{U} \to [0,1]$ with $f = 1$ on K, f continuous with respect to τ, and $\{x : f(x) \neq 0\} \subseteq U$. If we extend f to equal 0 off $\overline{U}$, the resulting function satisfies the conclusion of the theorem.

1.4   Exercises

1. Prove Theorem 1.15.

2. Prove Theorem 1.16

3. Prove Theorem 1.19

4. Prove Theorem 1.22.

5. Prove Theorem 1.23.

6. Prove Theorem 1.26.

7. Prove Lemma 1.35.

8. Prove the definition of distance in $\mathbb{R}^n$ or $\mathbb{C}^n$ satisfies 3-4. In addition to this, prove that $||\cdot||$ given by $||\mathbf{x}|| = (\sum_{i=1}^n |x_i|^2)^{1/2}$ is a norm. This means it satisfies the following.

$$||\mathbf{x}|| \geq 0, \ ||\mathbf{x}|| = 0 \text{ if and only if } \mathbf{x} = 0.$$

$$||\alpha\mathbf{x}|| = |\alpha|||\mathbf{x}|| \text{ for } \alpha \text{ a number.}$$

$$||\mathbf{x} + \mathbf{y}|| \leq ||\mathbf{x}|| + ||\mathbf{y}||.$$

9. Completeness of $\mathbb{R}$ is an axiom. Using this, show $\mathbb{R}^n$ and $\mathbb{C}^n$ are complete metric spaces with respect to the distance given by the usual norm.

10. Prove Urysohn's lemma. A Hausdorff space, X, is normal if and only if whenever K and H are disjoint nonempty closed sets, there exists a continuous function $f : X \to [0,1]$ such that $f(k) = 0$ for all $k \in K$ and $f(h) = 1$ for all $h \in H$. All that is left is the if part of the implication.

11. Prove that $f : X \to Y$ is continuous if and only if f is continuous at every point of X.

12. Suppose (X, d), and (Y, ρ) are metric spaces and let $f : X \to Y$. Show f is continuous at $x \in X$ if and only if whenever $x_n \to x$, $f(x_n) \to f(x)$. (Recall that $x_n \to x$ means that for all $\epsilon > 0$, there exists n_ϵ such that $d(x_n, x) < \epsilon$ whenever $n > n_\epsilon$.)

13. If (X, d) is a metric space, give an easy proof independent of Problem 10 that whenever K, H are disjoint non empty closed sets, there exists $f : X \to [0,1]$ such that f is continuous, $f(K) = \{0\}$, and $f(H) = \{1\}$.

14. ↑ Let (Ω, τ) be a locally compact Hausdorff space with $\mathcal{B}$ the basis of sets whose closures are compact and consider the set $\Omega \cup \{\infty\}$ where ∞ is called the point at ∞ and is an element of H^C for all sets, H, compact in (Ω, τ). (This is called the one point compactification of a locally compact Hausdorff space.) Thus ∞ is a new point not in Ω. Show $\mathcal{B} \cup \{H^C : H$ is a compact set in $(\Omega, \tau)\}$ is a basis for a topology which makes $\Omega \cup \{\infty\}$ into a compact Hausdorff space and is therefore a normal space. Use this to show that if K is compact in (Ω, τ) and V is open in (Ω, τ) containing K, then there exists an open set W whose closure is compact such that

$$K \subseteq W \subseteq \overline{W} \subseteq V.$$

15. Let $\mathcal{C}$ be a set whose elements are compact subsets of a Hausdorff space and suppose every finite subset of $\mathcal{C}$ has nonempty intersection, the finite intersection property. Show that then $\cap \mathcal{C} \neq \emptyset$.

16. ↑ This problem gives a different approach to Problem 14 which does not involve the idea of the one point compactification. Let (Ω, τ) be a locally compact Hausdorff space and let $K \subseteq U$ where K is a compact set and U is an open set. Show there exists an open set V

whose closure is compact and

$$K \subseteq V \subseteq \overline{V} \subseteq U.$$

Hint: First obtain an open set W whose closure is compact which contains K and which is contained in U. If $U \neq \Omega$, for each $p \in U^C$ show there exists an open set W_p, containing K, whose closure is compact and such that $p \notin \overline{W_p}$. Show

$$\cap_{p \in U^C} U^C \cap \overline{W} \cap \overline{W_p} = \emptyset$$

and then use compactness of these sets and the finite intersection property described in 15 to conclude some finite set of them has empty intersection. Thus, $U^C \cap \left(\cap_{i=1}^m \overline{W} \cap \overline{W_{p_i}} \right) = \emptyset$. Consider

$$V = \cap_{i=1}^m W \cap W_{p_i}.$$

17. Let (X, τ) (Y, η) be topological spaces with (X, τ) compact and let $f : X \to Y$ be continuous. Show $f(X)$ is compact.

18. (An example) Let $X = [-\infty, \infty]$ and consider $\mathcal{B}$ defined by sets of the form (a, b), $[-\infty, b)$, and $(a, \infty]$. Show $\mathcal{B}$ is the basis for a topology on X.

19. ↑ Show (X, τ) defined in Problem 18 is a compact Hausdorff space.

20. ↑ Show (X, τ) defined in Problem 18 is completely separable.

21. ↑ In Problem 18, show sets of the form $[-\infty, b)$ and $(a, \infty]$ form a subbasis for the topology described in Problem 18.

22. Let (X, τ) and (Y, η) be topological spaces and let $f : X \to Y$. Also let $\mathcal{S}$ be a subbasis for η. Show f is continuous if and only if $f^{-1}(V) \in \tau$ for all $V \in \mathcal{S}$. Thus, it suffices to check inverse images of subbasic sets in checking for continuity.

23. Show the usual topology of $\mathbb{R}^n$ is the same as the product topology of

$$\prod_{i=1}^n \mathbb{R} \equiv \mathbb{R} \times \mathbb{R} \times \cdots \times \mathbb{R}.$$

Do the same for $\mathbb{C}^n$.

24. If M is a separable metric space and $T \subseteq M$, then T is separable also.

25. Show the rational numbers, $\mathbb{Q}$, are countable.

26. Let α be an n dimensional multi-index. This means

$$\alpha = (\alpha_1, \cdots, \alpha_n)$$

where each α_i is a natural number or zero. Also, we let

$$|\alpha| \equiv \sum_{i=1}^{n} |\alpha_i|$$

When we write $\mathbf{x}^\alpha$, we mean

$$\mathbf{x}^\alpha \equiv x_1^{\alpha_1} x_2^{\alpha_2} \cdots x_3^{\alpha_n}.$$

An n dimensional polynomial of degree m is a function of the form

$$\sum_{|\alpha| \le m} d_\alpha \mathbf{x}^\alpha.$$

Let $\mathcal{R}$ be all n dimensional polynomials whose coefficients d_α come from the rational numbers, $\mathbb{Q}$. Show $\mathcal{R}$ is countable.

Chapter 2
Compactness and Continuous Functions

2.1   Compactness in metric space

In any metric space, we say a set E is totally bounded if for every $\epsilon > 0$ there exists a finite set of points $\{x_1, \cdots, x_n\}$ such that

$$E \subseteq \cup_{i=1}^n B(x_i, \epsilon).$$

This finite set of points is called an ϵ net.

Many existence theorems in analysis depend on some set being compact. Therefore, it is important to be able to identify compact sets. The following proposition tells which sets in a metric space are compact.

Proposition 2.1 *Let (X, d) be a metric space. Then the following are equivalent.*

$$(X, d) \text{ is compact,} \tag{1}$$

$$(X, d) \text{ is sequentially compact,} \tag{2}$$

$$(X, d) \text{ is complete and totally bounded.} \tag{3}$$

Sequentially compact means every sequence has a convergent subsequence.

Proof: Suppose 1 and let $\{x_k\}$ be a sequence. Suppose $\{x_k\}$ has no convergent subsequence. If this is so, then $\{x_k\}$ has no limit point and no value of the sequence is repeated more than finitely many times. Thus the set

$$C_n = \cup\{x_k : k \geq n\}$$

is a closed set and if

$$U_n = C_n^C,$$

then

$$X = \cup_{n=1}^\infty U_n$$

but there is no finite subcovering, contradicting compactness of (X, d).

Now suppose Formula 2 and let $\{x_n\}$ be a Cauchy sequence. Then $x_{n_k} \to x$ for some subsequence. Let $\epsilon > 0$ be given. Let n_0 be such that if $m, n \geq n_0$, then $d(x_n, x_m) < \frac{\epsilon}{2}$ and let l be such that if $k \geq l$ then $d(x_{n_k}, x) < \frac{\epsilon}{2}$. Let $n_1 > \max(n_l, n_0)$. If $n > n_1$, let $k > l$ and $n_k > n_0$.

$$
\begin{aligned}
d(x_n, x) &\leq d(x_n, x_{n_k}) + d(x_{n_k}, x) \\
&< \frac{\epsilon}{2} + \frac{\epsilon}{2} = \epsilon.
\end{aligned}
$$

Thus $\{x_n\}$ converges to x and this shows (X, d) is complete. If (X, d) is not totally bounded, then there exists $\epsilon > 0$ for which there is no ϵ net. Hence there exists a sequence $\{x_k\}$ with $d(x_k, x_l) \geq \epsilon$ for all $l \neq k$. This contradicts Formula 2 because this is a sequence having no convergent subsequence. This shows Formula 2 implies Formula 3.

Now suppose Formula 3. We show this implies Formula 2. Let $\{p_n\}$ be a sequence and let $\{x_i^n\}_{i=1}^{m_n}$ be a 2^{-n} net for $n = 1, 2, \cdots$. Let

$$B_n \equiv B\left(x_{i_n}^n, 2^{-n}\right)$$

be such that B_n contains p_k for infinitely many values of k and $B_n \cap B_{n+1} \neq \emptyset$. Let p_{n_k} be a subsequence having

$$p_{n_k} \in B_k.$$

Then if $k \geq l$,

$$
\begin{aligned}
d\left(p_{n_k}, p_{n_l}\right) &\leq \sum_{i=l}^{k-1} d\left(p_{n_{i+1}}, p_{n_i}\right) \\
&< \sum_{i=l}^{k-1} 2^{-(i-1)} < 2^{-(l-2)}.
\end{aligned}
$$

Consequently $\{p_{n_k}\}$ is a Cauchy sequence. Hence it converges. This proves Formula 2.

Now suppose Formula 2 and Formula 3. Let D_n be a n^{-1} net for $n = 1, 2, \cdots$ and let

$$D = \cup_{n=1}^{\infty} D_n.$$

Thus D is a countable dense subset of (X, d). The set of balls

$$\mathcal{B} = \{B(q, r) : q \in D, \ r \in Q \cap (0, \infty)\}$$

is a countable basis for (X, d). To see this, let $p \in B(x, \epsilon)$ and choose $r \in Q \cap (0, \infty)$ such that

$$\epsilon - d(p, x) > 2r.$$

Let $q \in B(p, r) \cap D$. If $y \in B(q, r)$, then

$$
\begin{aligned}
d(y, x) &\leq d(y, q) + d(q, p) + d(p, x) \\
&< r + r + \epsilon - 2r = \epsilon.
\end{aligned}
$$

Hence $p \in B(q, r) \subseteq B(x, \epsilon)$ and this shows each ball is the union of balls of $\mathcal{B}$. Now suppose $\mathcal{C}$ is any open cover of X. Let $\tilde{\mathcal{B}}$ denote the balls of $\mathcal{B}$ which are contained in some set of $\mathcal{C}$. Thus

$$\cup \tilde{\mathcal{B}} = X.$$

For each $B \in \widetilde{\mathcal{B}}$, pick $U \in \mathcal{C}$ such that $U \supseteq B$. Let $\widetilde{\mathcal{C}}$ be the resulting countable collection of sets. Then $\widetilde{\mathcal{C}}$ is a countable open cover of X. Say $\widetilde{\mathcal{C}} = \{U_n\}_{n=1}^{\infty}$. If $\mathcal{C}$ admits no finite subcover, then neither does $\widetilde{\mathcal{C}}$ and we can pick $p_n \in X \setminus \cup_{k=1}^n U_k$. Then since X is sequentially compact, there is a subsequence $\{p_{n_k}\}$ such that $\{p_{n_k}\}$ converges. Say

$$p = \lim_{k \to \infty} p_{n_k}.$$

All but finitely many points of $\{p_{n_k}\}$ are in $X \setminus \cup_{k=1}^n U_k$. Therefore $p \in X \setminus \cup_{k=1}^n U_k$ for each n. Hence

$$p \notin \cup_{k=1}^{\infty} U_k$$

contradicting the construction of $\{U_n\}_{n=1}^{\infty}$. Hence X is compact. This proves the proposition.

Next we apply this very general result to a familiar example, $\mathbb{R}^n$. In this setting totally bounded and bounded are the same. This will imply the familiar Heine Borel theorem from advanced calculus.

Lemma 2.2 *A subset of $\mathbb{R}^n$ is totally bounded if and only if it is bounded.*

Proof: Let A be totally bounded. We need to show it is bounded. Let $\mathbf{x}_1, \cdots, \mathbf{x}_p$ be a 1 net for A. Now consider the ball $B(\mathbf{0}, r+1)$ where $r > \max(\|\mathbf{x}_i\| : i = 1, \cdots, p)$. If $\mathbf{z} \in A$, then $\mathbf{z} \in B(\mathbf{x}_j, 1)$ for some j and so by the triangle inequality,

$$\|\mathbf{z} - \mathbf{0}\| \le \|\mathbf{z} - \mathbf{x}_j\| + \|\mathbf{x}_j\| < 1 + r.$$

Thus $A \subseteq B(\mathbf{0}, r+1)$ and so A is bounded.

Now suppose A is bounded and suppose A is not totally bounded. Then there exists $\epsilon > 0$ such that there is no ϵ net for A. Therefore, there exists a sequence of points $\{a_i\}$ with $\|a_i - a_j\| \ge \epsilon$ if $i \ne j$. Since A is bounded, there exists $r > 0$ such that

$$A \subseteq [-r, r)^n.$$

($\mathbf{x} \in [-r, r)^n$ means $x_i \in [-r, r)$ for each i.) Now define $\mathcal{S}$ to be all cubes of the form

$$\prod_{k=1}^{n} [a_k, b_k)$$

where

$$a_k = -r + i2^{-p}r, \quad b_k = -r + (i+1)2^{-p}r,$$

for $i \in \{0, 1, \cdots, 2^{p+1} - 1\}$. Thus $\mathcal{S}$ is a collection of $(2^{p+1})^n$ nonoverlapping cubes whose union equals $[-r, r)^n$ and whose diameters are all equal to $2^{-p}r\sqrt{n}$. Now choose p large enough that the diameter of these cubes is less than ϵ. This yields a contradiction because one of the cubes must contain infinitely many points of $\{a_i\}$. This proves the lemma.

The next theorem is called the Heine Borel theorem and it characterizes the compact sets in $\mathbb{R}^n$.

Theorem 2.3 *A subset of $\mathbb{R}^n$ is compact if and only if it is closed and bounded.*

Proof: Since a set in $\mathbb{R}^n$ is totally bounded if and only if it is bounded, this theorem follows from Proposition 2.1 and the observation that a subset of $\mathbb{R}^n$ is closed if and only if it is complete. This proves the theorem.

The following corollary is an important existence theorem which depends on compactness.

Corollary 2.4 *Let (X, τ) be a compact topological space and let $f : X \to \mathbb{R}$ be continuous. Then $\max\{f(x) : x \in X\}$ and $\min\{f(x) : x \in X\}$ both exist.*

Proof: By Problem 17 of Chapter 1, it follows $f(X)$ is compact. From Theorem 2.3 $f(X)$ is closed and bounded. This implies it has a largest and a smallest value. This proves the corollary.

Definition 2.5 Let $(X, d), (Y, \rho)$ be metric spaces and let $f : X \to Y$ be a function. We say f is uniformly continuous on X if for every $\epsilon > 0$, there exists a $\delta > 0$ such that if

$$d(x, y) < \delta,$$

then

$$\rho(f(x), f(y)) < \epsilon.$$

The difference between this and ordinary continuity is that in this case the δ depends only on ϵ and not on the point being considered. For example, consider the function

$$f(x) = 1/x$$

defined on $(0, \infty)$. This function is continuous but not uniformly continuous because as x becomes close to 0, the δ needed for a given ϵ must be made very small to satisfy the definition of continuity. However, in the case when (X, d) is a compact metric space, continuity does imply uniform continuity. This is the content of the next theorem.

Recall from problem 12 of Chapter 1 that a function mapping a metric space to a metric space is continuous at a point x if and only if $x_n \to x$ implies $f(x_n) \to f(x)$. Also consider the expression

$$|d(x_1, y_1) - d(x, y)|.$$

If $d(x_1, y_1) > d(x, y)$, this expression is no smaller than

$$
\begin{aligned}
d(x_1, x) + d(x, y_1) - d(x, y) &\leq d(x_1, x) + d(x, y) + d(y, y_1) - d(x, y) \\
&= d(x, x_1) + d(y, y_1).
\end{aligned}
$$

A similar argument yields the same inequality in the case where

$$
d(x_1, y_1) < d(x, y).
$$

Thus

$$
|d(x_1, y_1) - d(x, y)| \leq d(x, x_1) + d(y, y_1) \tag{4}
$$

Theorem 2.6 *Let $(X, d), (Y, \rho)$ be metric spaces and suppose (X, d) is compact. Then if $f : X \to Y$ is continuous, it follows that f is uniformly continuous.*

Proof: If this is not so, there exists an $\epsilon > 0$ and points $x_n, y_n \in X$ such that

$$
d(x_n, y_n) \leq n^{-1}, \quad \text{but } \rho(f(x_n), f(y_n)) \geq \epsilon.
$$

By Proposition 2.1, there exists a subsequence x_{n_k} which converges to a point $x \in X$. But this requires y_{n_k} to also converge to x because $d(x_n, y_n) \leq n^{-1}$. Thus

$$
f(x_{n_k}) \to f(x), \ f(y_{n_k}) \to f(x),
$$

and by Formula 4,

$$
0 = \rho(f(x), f(x)) = \lim_{k \to \infty} \rho(f(x_{n_k}), f(y_{n_k})) \geq \epsilon
$$

a contradiction. This proves the theorem.

2.2 Compactness in spaces of continuous functions

Let (X, τ) be a compact space and let $C(X; \mathbb{R}^n)$ denote the space of continuous $\mathbb{R}^n$ valued functions. For $f \in C(X; \mathbb{R}^n)$ let

$$
\|f\|_\infty \equiv \sup\{|f(x)| : x \in X\}
$$

where the norm in the parenthesis refers to the usual norm in $\mathbb{R}^n$.

Definition 2.7 A complete normed linear space is called a Banach space.

The following proposition shows that $C(X; \mathbb{R}^n)$ is an example of a Banach space.

Proposition 2.8 $(C(X;\mathbb{R}^n), ||\ ||_\infty)$ *is a Banach space.*

Proof: It is obvious $||\ ||_\infty$ is a norm because (X,τ) is compact. Also it is clear that $C(X;\mathbb{R}^n)$ is a linear space. Suppose $\{f_r\}$ is a Cauchy sequence in $C(X;\mathbb{R}^n)$. Then for each $x \in X$, $\{f_r(x)\}$ is a Cauchy sequence in $\mathbb{R}^n$. Let

$$f(x) \equiv \lim_{k \to \infty} f_k(x).$$

Therefore,

$$\sup_{x \in X} |f(x) - f_k(x)| = \sup_{x \in X} \lim_{m \to \infty} |f_m(x) - f_k(x)|$$

$$\leq \lim_{m \to \infty} \sup ||f_m - f_k||_\infty < \epsilon$$

for all k large enough. Thus,

$$\lim_{k \to \infty} \sup_{x \in X} |f(x) - f_k(x)| = 0.$$

It only remains to show that f is continuous. Let

$$\sup_{x \in X} |f(x) - f_k(x)| < \epsilon/3$$

whenever $k \geq k_0$ and pick $k \geq k_0$.

$$|f(x) - f(y)| \leq |f(x) - f_k(x)| + |f_k(x) - f_k(y)| + |f_k(y) - f(y)|$$
$$< 2\epsilon/3 + |f_k(x) - f_k(y)|$$

Now f_k is continuous and so there exists U an open set containing x such that if $y \in U$, then

$$|f_k(x) - f_k(y)| < \epsilon/3.$$

Thus, for all $y \in U$, $|f(x) - f(y)| < \epsilon$ and this shows that f is continuous and proves the proposition.

This space is an example of an infinite dimensional normed linear space. As such, it is a metric space. The next task is to find the compact subsets of this metric space. We know these are the subsets which are complete and totally bounded by Proposition 2.1, but which sets are those? We need another way to identify them which is more convenient. This is the extremely important Ascoli Arzela theorem which is the next big theorem.

Definition 2.9 We say $\mathcal{F} \subseteq C(X;\mathbb{R}^n)$ is equicontinuous at x_0 if for all $\epsilon > 0$ there exists $U \in \tau$, $x_0 \in U$, such that if $x \in U$, then for all $f \in \mathcal{F}$,

$$|f(x) - f(x_0)| < \epsilon.$$

If $\mathcal{F}$ is equicontinuous at every point of X, we say $\mathcal{F}$ is equicontinuous. We say $\mathcal{F}$ is bounded if there exists a constant, M, such that $||f||_\infty < M$ for all $f \in \mathcal{F}$.

Lemma 2.10 *Let $\mathcal{F} \subseteq C(X;\mathbb{R}^n)$ be equicontinuous and bounded and let $\epsilon > 0$ be given. Then if $\{f_r\} \subseteq \mathcal{F}$, there exists a subsequence $\{g_k\}$, depending on ϵ, such that*

$$||g_k - g_m||_\infty < \epsilon$$

whenever k, m are large enough.

Proof: If $x \in X$ there exists an open set U_x containing x such that for all $f \in \mathcal{F}$ and $y \in U_x$,

$$|f(x) - f(y)| < \epsilon/4. \tag{5}$$

Since X is compact, finitely many of these sets, $U_{x_1}, \cdots, U_{x_p}$, cover X. Let $\{f_{1k}\}$ be a subsequence of $\{f_k\}$ such that $\{f_{1k}(x_1)\}$ converges. Such a subsequence exists because $\mathcal{F}$ is bounded. Let $\{f_{2k}\}$ be a subsequence of $\{f_{1k}\}$ such that $\{f_{2k}(x_i)\}$ converges for $i = 1, 2$. Continue in this way and let $\{g_k\} = \{f_{pk}\}$. Thus $\{g_k(x_i)\}$ converges for each x_i. Now if $y \in X$, then $y \in U_{x_i}$ for some x_i. Denote this x_i by x_y. Then using Formula 5,

$$\limsup_{k \to \infty} ||g_k - g_m||_\infty = \limsup_{k \to \infty} \max_{y \in X} |g_k(y) - g_m(y)|$$

$$\leq \limsup_{k \to \infty} \max_{y \in X} (|g_k(y) - g_k(x_y)| + |g_k(x_y) - g_m(x_y)| + |g_m(x_y) - g_m(y)|)$$

$$\leq \epsilon/2 + \limsup_{k \to \infty} \max\{|g_k(x_i) - g_m(x_i)|, \ i = 1, \cdots, p\} < \epsilon$$

whenever m is large enough. This yields the conclusion of the lemma because it implies that whenever m is large enough, say $m \geq m_0$, then for l large enough, $l \geq m_0$,

$$\sup\{||g_k - g_m||_\infty : k \geq l\} < \epsilon.$$

Thus if $m, k > l$,

$$||g_k - g_m||_\infty < \epsilon.$$

Theorem 2.11 *(Ascoli Arzela) Let $\mathcal{F} \subseteq C(X;\mathbb{R}^n)$. Then $\mathcal{F}$ is compact if and only if $\mathcal{F}$ is closed, bounded, and equicontinuous.*

Proof: Suppose $\mathcal{F}$ is closed, bounded, and equicontinuous. We will show this implies $\mathcal{F}$ is totally bounded. Then since $\mathcal{F}$ is closed, it follows that $\mathcal{F}$ is complete and will therefore be compact by Proposition 2.1.

Suppose $\mathcal{F}$ is not totally bounded. Then there exists $\epsilon > 0$ such that there is no ϵ net. Hence there exists a sequence $\{f_k\} \subseteq \mathcal{F}$ such that

$$\|f_k - f_l\| \geq \epsilon$$

for all $k \neq l$. This contradicts Lemma 2.10. Thus $\mathcal{F}$ must be totally bounded and this proves half of the theorem.

Now suppose $\mathcal{F}$ is compact. Then it must be closed and totally bounded. This implies $\mathcal{F}$ is bounded. It remains to show $\mathcal{F}$ is equicontinuous. Suppose not. Then there exists $x \in X$ such that $\mathcal{F}$ is not equicontinuous at x. Thus there exists $\epsilon > 0$ such that for every open U containing x, there exists $f \in \mathcal{F}$ such that $|f(x) - f(y)| \geq \epsilon$ for some $y \in U$.

Let $\{h_1, \cdots, h_p\}$ be an $\epsilon/4$ net for $\mathcal{F}$. For each x, let U_x be an open set containing x such that for all $y \in U_x$,

$$|h_i(x) - h_i(y)| < \epsilon/8$$

for all $i = 1, \cdots, p$. Let $U_{x_1}, \cdots, U_{x_m}$ cover X. Then $x \in U_{x_i}$ for some x_i and so, for some $y \in U_{x_i}$, $|f(x) - f(y)| \geq \epsilon$. Hence, since $\{h_1, \cdots, h_p\}$ is an $\epsilon/4$ net,

$$\epsilon \leq |f(x) - f(y)| \leq |f(x) - h_i(x)| + |h_i(x) - h_i(y)| +$$

$$|h_i(y) - f(y)| \leq \epsilon/2 + |h_i(x) - h_i(y)| \leq \epsilon/2 +$$

$$|h_i(x) - h_i(x_i)| + |h_i(x_i) - h_i(y)| \leq 3\epsilon/4,$$

a contradiction. This proves the theorem.

2.3   Stone Weierstrass theorem

In this section we give a proof of the important approximation theorem of Weierstrass and its generalization by Stone. This theorem is about approximating an arbitrary continuous function uniformly by a polynomial or some other such function.

Definition 2.12 We say $\mathcal{A}$ is an algebra of functions if $\mathcal{A}$ is a vector space and if whenever $f, g \in \mathcal{A}$ then $fg \in \mathcal{A}$.

We will assume that the field of scalars is $\mathbb{R}$ in this section unless otherwise indicated. The approach to the Stone Weierstrass depends on the following estimate which may look familiar to someone who has taken a probability class. The left side of the following estimate is the variance of a binomial distribution. However, it is not necessary to know anything about probability to follow the proof below although what is being done is an application of the moment generating function technique to find the variance.

Lemma 2.13 *The following estimate holds for $x \in [0, 1]$.*

$$\sum_{k=0}^{n} \binom{n}{k} (k - nx)^2 \, x^k \, (1 - x)^{n-k} \leq 2n$$

Proof: By the Binomial theorem,

$$\sum_{k=0}^{n} \binom{n}{k} (e^t x)^k (1 - x)^{n-k} = (1 - x + e^t x)^n.$$

Differentiating both sides with respect to t and then evaluating at $t = 0$ yields

$$\sum_{k=0}^{n} \binom{n}{k} k x^k (1 - x)^{n-k} = nx.$$

Now doing two derivatives with respect to t yields

$$\sum_{k=0}^{n} \binom{n}{k} k^2 (e^t x)^k (1 - x)^{n-k} = n(n - 1)(1 - x + e^t x)^{n-2} e^{2t} x^2$$

$$+ n(1 - x + e^t x)^{n-1} x e^t.$$

Evaluating this at $t = 0$,

$$\sum_{k=0}^{n} \binom{n}{k} k^2 (x)^k (1 - x)^{n-k} = n(n - 1) x^2 + nx.$$

Therefore,

$$\sum_{k=0}^{n} \binom{n}{k} (k - nx)^2 \, x^k \, (1 - x)^{n-k} = n(n - 1) x^2 + nx - 2n^2 x^2 + n^2 x^2$$

$$= n(x - x^2) \leq 2n.$$

This proves the lemma.

Definition 2.14 Let $f \in C([0, 1])$. Then the following polynomials are known as the Bernstein polynomials.

$$p_n(x) \equiv \sum_{k=0}^{n} \binom{n}{k} f\left(\frac{k}{n}\right) x^k (1 - x)^{n-k}.$$

Theorem 2.15 *Let $f \in C([0,1])$ and let p_n be given in Definition 2.14. Then*

$$\lim_{n \to \infty} ||f - p_n||_\infty = 0.$$

Proof: Since f is continuous on the compact $[0, 1]$, it follows f is uniformly continuous there and so if $\epsilon > 0$ is given, there exists $\delta > 0$ such that if

$$|y - x| \leq \delta,$$

then

$$|f(x) - f(y)| < \epsilon/2.$$

By the Binomial theorem,

$$f(x) = \sum_{k=0}^{n} \binom{n}{k} f(x) x^k (1-x)^{n-k}$$

and so

$$|p_n(x) - f(x)| \leq \sum_{k=0}^{n} \binom{n}{k} \left| f\left(\frac{k}{n}\right) - f(x) \right| x^k (1-x)^{n-k}$$

$$\leq \sum_{|k/n-x|>\delta} \binom{n}{k} \left| f\left(\frac{k}{n}\right) - f(x) \right| x^k (1-x)^{n-k} +$$

$$\sum_{|k/n-x|\leq\delta} \binom{n}{k} \left| f\left(\frac{k}{n}\right) - f(x) \right| x^k (1-x)^{n-k}$$

$$< \epsilon/2 + 2||f||_\infty \sum_{(k-nx)^2 > n^2\delta^2} \binom{n}{k} x^k (1-x)^{n-k}$$

$$\leq \frac{2||f||_\infty}{n^2\delta^2} \sum_{k=0}^{n} \binom{n}{k} (k-nx)^2 x^k (1-x)^{n-k} + \epsilon/2.$$

By the lemma,

$$\leq \frac{4||f||_\infty}{\delta^2 n} + \epsilon/2 < \epsilon$$

whenever n is large enough. This proves the theorem.

The next corollary is called the Weierstrass approximation theorem.

Corollary 2.16 *The polynomials are dense in $C([a,b])$.*

Proof: Let $f \in C([a,b])$ and let $h : [0,1] \to [a,b]$ be linear and onto. Then

$$|f(h(t)) - p_n(t)| < \epsilon$$

for all $t \in [0, 1]$. Therefore for all $x \in [a, b]$,

$$\left| f(x) - p_n\left(h^{-1}(x)\right) \right| < \epsilon.$$

Since h is linear $p_n \circ h^{-1}$ is a polynomial. This proves the theorem.

The next result is the key to the profound generalization of the Weierstrass theorem due to Stone in which an interval will be replaced by a compact or locally compact set and polynomials will be replaced with elements of an algebra satisfying certain axioms.

Corollary 2.17 *On the interval $[-M, M]$, there exist polynomials p_n such that*

$$p_n(0) = 0$$

and

$$\lim_{n \to \infty} \left\| p_n - |\cdot| \right\|_\infty = 0.$$

Proof: Let $\tilde{p}_n \to |\cdot|$ uniformly and let

$$p_n \equiv \tilde{p}_n - \tilde{p}_n(0).$$

This proves the corollary.

The following generalization is known as the Stone Weierstrass approximation theorem. First, we say an algebra of functions, $\mathcal{A}$ defined on A, annihilates no point of A if for all $x \in A$, there exists $g \in \mathcal{A}$ such that $g(x) \neq 0$. We say the algebra separates points if whenever $x_1 \neq x_2$, then there exists $g \in \mathcal{A}$ such that $g(x_1) \neq g(x_2)$.

Theorem 2.18 *Let A be a compact topological space and let $\mathcal{A} \subseteq C(A; \mathbb{R})$ be an algebra of functions which separates points and annihilates no point. Then $\mathcal{A}$ is dense in $C(A; \mathbb{R})$.*

Proof: We begin by proving a simple lemma.

Lemma 2.19 *Let c_1 and c_2 be two real numbers and let $x_1 \neq x_2$ be two points of A. Then there exists a function $f_{x_1 x_2}$ such that*

$$f_{x_1 x_2}(x_1) = c_1, \ f_{x_1 x_2}(x_2) = c_2.$$

Proof of the lemma: Let $g \in \mathcal{A}$ satisfy

$$g(x_1) \neq g(x_2).$$

Such a g exists because the algebra separates points. Since the algebra annihilates no point, there exist functions h and k such that

$$h(x_1) \neq 0, \ k(x_2) \neq 0.$$

Then let
$$u \equiv gh - g(x_2) h, \quad v \equiv gk - g(x_1) k.$$
It follows that $u(x_1) \neq 0$ and $u(x_2) = 0$ while $v(x_2) \neq 0$ and $v(x_1) = 0$. Let
$$f_{x_1 x_2} \equiv \frac{c_1 u}{u(x_1)} + \frac{c_2 v}{v(x_2)}.$$
This proves the lemma. Now we continue with the proof of the theorem.

First note that $\overline{A}$ satisfies the same axioms as A but in addition to these axioms, $\overline{A}$ is closed. Suppose $f \in \overline{A}$ and suppose M is large enough that
$$\|f\|_\infty < M.$$
Using Corollary 2.17, let p_n be a sequence of polynomials such that
$$\|p_n - |\cdot|\|_\infty \to 0, \ p_n(0) = 0.$$
It follows that $p_n \circ f \in \overline{A}$ and so $|f| \in \overline{A}$ whenever $f \in \overline{A}$. Also note that
$$\max(f, g) = \frac{|f - g| + (f + g)}{2}$$
$$\min(f, g) = \frac{(f + g) - |f - g|}{2}.$$
Therefore, this shows that if $f, g \in \overline{A}$ then
$$\max(f, g), \ \min(f, g) \in \overline{A}.$$
By induction, if $f_i, i = 1, 2, \cdots, m$ are in $\overline{A}$ then
$$\max(f_i, i = 1, 2, \cdots, m), \ \min(f_i, i = 1, 2, \cdots, m) \in \overline{A}.$$

Now let $h \in C(A; \mathbb{R})$ and use Lemma 2.19 to obtain f_{xy}, a function of $\overline{A}$ which agrees with h at x and y. Let $\epsilon > 0$ and let $x \in A$. Then there exists an open set $U(y)$ containing y such that
$$f_{xy}(z) > h(z) - \epsilon \text{ if } z \in U(y).$$
Since A is compact, let $U(y_1), \cdots, U(y_l)$ cover A. Let
$$f_x \equiv \max(f_{xy_1}, f_{xy_2}, \cdots, f_{xy_l}).$$
Then $f_x \in \overline{A}$ and
$$f_x(z) > h(z) - \epsilon$$
for all $z \in A$ and $f_x(x) = h(x)$. Then for each $x \in A$ there exists an open set $V(x)$ containing x such that for $z \in V(x)$,
$$f_x(z) < h(z) + \epsilon.$$

Let $V(x_1), \cdots, V(x_m)$ cover A and let

$$f \equiv \min(f_{x_1}, \cdots, f_{x_m}).$$

Therefore,

$$f(z) < h(z) + \epsilon$$

for all $z \in A$ and since each

$$f_x(z) > h(z) - \epsilon,$$

it follows

$$f(z) > h(z) - \epsilon$$

also and so

$$|f(z) - h(z)| < \epsilon$$

for all z. Since ϵ is arbitrary, this shows $h \in \overline{\mathcal{A}}$ and proves $\overline{\mathcal{A}} = C(A; \mathbb{R})$. This proves the theorem.

2.4 Exercises

1. Let $(X, \tau), (Y, \eta)$ be topological spaces and let $A \subseteq X$ be compact. Then if $f : X \rightarrow Y$ is continuous, show that $f(A)$ is also compact.

2. ↑ In the context of Problem 1, suppose $\mathbb{R} = Y$ where the usual topology is placed on $\mathbb{R}$. Show f achieves its maximum and minimum on A.

3. Let V be an open set in $\mathbb{R}^n$. Show there is an increasing sequence of compact sets, K_m, such that $V = \cup_{m=1}^{\infty} K_m$. **Hint:** Let

$$C_m \equiv \left\{ \mathbf{x} \in \mathbb{R}^n : dist\left(\mathbf{x}, V^C\right) \geq \frac{1}{m} \right\}$$

where

$$dist(\mathbf{x}, S) \equiv \inf \{|\mathbf{y} - \mathbf{x}| \text{ such that } \mathbf{y} \in S\}.$$

Consider $K_m \equiv C_m \cap \overline{B(\mathbf{0}, m)}$.

4. Let $B(X; \mathbb{R}^n)$ be the space of functions $\mathbf{f}$, mapping X to $\mathbb{R}^n$ such that

$$\sup\{|\mathbf{f}(\mathbf{x})| : \mathbf{x} \in X\} < \infty.$$

Show $B(X; \mathbb{R}^n)$ is a complete normed linear space if

$$\|\mathbf{f}\| \equiv \sup\{|\mathbf{f}(\mathbf{x})| : \mathbf{x} \in X\}.$$

5. Let $\alpha \in [0, 1]$. We define, for X a compact subset of $\mathbb{R}^p$,

$$C^\alpha (X; \mathbb{R}^n) \equiv \{\mathbf{f} \in C (X; \mathbb{R}^n) : \rho_\alpha (\mathbf{f}) + ||\mathbf{f}|| \equiv ||\mathbf{f}||_\alpha < \infty\}$$

where

$$||\mathbf{f}|| \equiv \sup\{|\mathbf{f} (\mathbf{x})| : \mathbf{x} \in X\}$$

and

$$\rho_\alpha (\mathbf{f}) \equiv \sup\{\frac{|\mathbf{f} (\mathbf{x}) - \mathbf{f} (\mathbf{y})|}{|\mathbf{x} - \mathbf{y}|^\alpha} : \mathbf{x}, \mathbf{y} \in X, \ \mathbf{x} \neq \mathbf{y}\}.$$

Show that $(C^\alpha (X; \mathbb{R}^n), ||\cdot||_\alpha)$ is a complete normed linear space.

6. Let $\{\mathbf{f}_n\}_{n=1}^\infty \subseteq C^\alpha (X; \mathbb{R}^n)$ where X is a compact subset of $\mathbb{R}^p$ and suppose

$$||\mathbf{f}_n||_\alpha \leq M$$

for all n. Show there exists a subsequence, n_k, such that $\mathbf{f}_{n_k}$ converges in $C (X; \mathbb{R}^n)$. We say the given sequence is precompact when this happens. (This also shows the embedding of $C^\alpha (X; \mathbb{R}^n)$ into $C (X; \mathbb{R}^n)$ is a compact embedding.)

7. Show the set of polynomials $\mathcal{R}$ described in Problem 26 of Chapter 1 is dense in the space $C (A; \mathbb{R})$ when A is a compact subset of $\mathbb{R}^n$. Conclude from this other problem that $C (A; \mathbb{R})$ is separable.

8. Let H and K be disjoint closed sets in a metric space, (X, d), and let

$$g (x) \equiv \frac{2}{3} h (x) - \frac{1}{3}$$

where

$$h (x) \equiv \frac{dist (x, H)}{dist (x, H) + dist (x, K)}.$$

Show $g (x) \in [-\frac{1}{3}, \frac{1}{3}]$ for all $x \in X$, g is continuous, and g equals $\frac{-1}{3}$ on H while g equals $\frac{1}{3}$ on K. Is it necessary to be in a metric space to do this?

9. ↑ Suppose H is a closed set in X where X is the metric space of problem 8 and suppose $f : H \rightarrow [-1, 1]$ is continuous. Show there exists $g : X \rightarrow [-1, 1]$ such that g is continuous and $g = f$ on H. **Hint:** Show there exists

$$g_1 \in C (X), \ g_1 (x) \in \left[\frac{-1}{3}, \frac{1}{3}\right],$$

and $|f(x) - g_1(x)| \leq \frac{2}{3}$ for all $x \in H$. To do this, consider the disjoint closed sets

$$H \equiv f^{-1}\left(\left[-1, \frac{-1}{3}\right]\right), \quad K \equiv f^{-1}\left(\left[\frac{1}{3}, 1\right]\right)$$

and use Problem 8 if the two sets are nonempty. When this has been done, let

$$\frac{3}{2}(f(x) - g_1(x))$$

play the role of f and let g_2 be like g_1. Obtain

$$\left| f(x) - \sum_{i=1}^{n} \left(\frac{2}{3}\right)^{i-1} g_i(x) \right| \leq \left(\frac{2}{3}\right)^n$$

and consider

$$g(x) \equiv \sum_{i=1}^{\infty} \left(\frac{2}{3}\right)^{i-1} g_i(x).$$

Is it necessary to be in a metric space to do this?

10. ↑ Let H be a closed set in a metric space (X, d) and suppose $f \in C(H)$. Show there exists $g \in C(X)$ such that $g(x) = f(x)$ for all $x \in H$ and if $f(H) \subseteq [a, b]$, then $g(X) \subseteq [a, b]$. This is a version of the Tietze extension theorem. Is it necessary to be in a metric space for this to work?

11. Let X be a compact topological space and suppose $\{f_n\}$ is a sequence of functions continuous on X having values in $\mathbb{R}^n$. Show there exists a countable dense subset of X, $\{x_i\}$ and a subsequence of $\{f_n\}$, $\{f_{n_k}\}$, such that $\{f_{n_k}(x_i)\}$ converges for each x_i. **Hint:** First get a subsequence which converges at x_1, then a subsequence of this subsequence which converges at x_2 and a subsequence of this one which converges at x_3 and so forth. Thus the second of these subsequences converges at both x_1 and x_2 while the third converges at these two points and also at x_3 and so forth. List them so the second is under the first and the third is under the second and so forth thus obtaining an infinite matrix of entries. Now consider the diagonal sequence and argue it is ultimately a subsequence of every one of these subsequences described earlier and so it must converge at each x_i. This procedure is called the Cantor diagonal process.

12. ↑ Use the Cantor diagonal process to give a different proof of the Ascoli Arzela theorem than that presented in this chapter. **Hint:** Start with a sequence of functions in $C(X; \mathbb{R}^n)$ and use the Cantor diagonal process to produce a subsequence which converges at each

point of a countable dense subset of X. Then show this sequence is a Cauchy sequence in $C(X;\mathbb{R}^n)$.

13. Let (X, τ) be a locally compact Hausdorff space and let $C_0(X)$ denote the space of continuous functions defined on X with the property that if $f \in C_0(X)$, then for each $\epsilon > 0$ there exists a compact set K such that $|f(x)| < \epsilon$ for all $x \notin K$. Define

$$\|f\|_\infty = \sup\{|f(x)| : x \in X\}.$$

Show that with this norm, $C_0(X)$ is a Banach space. Now let $\mathcal{A}$ be an algebra of functions in $C_0(X)$ which separates the points and annihilates no point. Show $\mathcal{A}$ is dense in $C_0(X)$. **Hint:** Let $\left(\widetilde{X}, \widetilde{\tau}\right)$ be the one point compactification as described in Lemma 1.40 and let $\widetilde{\mathcal{A}}$ denote all finite linear combinations of the form

$$\left\{\sum_{i=1}^n c_i \widetilde{f} + c_0 : f \in \mathcal{A}, \ c_i \in \mathbb{R}\right\}.$$

Here,

$$\widetilde{f}(x) \equiv \begin{cases} f(x) \text{ if } x \in X, \\ 0 \text{ if } x = \infty. \end{cases}$$

Show $\widetilde{\mathcal{A}}$ is an algebra of functions on $C\left(\widetilde{X}\right)$ which separates the points and annihilates no point. Apply the Stone Weierstrass theorem and argue that $c_0 = 0$.

14. What about the case where $C_0(X)$ consists of complex valued functions and the field of scalars is $\mathbb{C}$ rather than $\mathbb{R}$? In this case, suppose $\mathcal{A}$ is an algebra of functions in $C_0(X)$ which separates the points, annihilates no point, and has the property that if $f \in \mathcal{A}$, then $\overline{f} \in \mathcal{A}$. Show that $\mathcal{A}$ is dense in $C_0(X)$. **Hint:** Let $\operatorname{Re}\mathcal{A} \equiv \{\operatorname{Re} f : f \in \mathcal{A}\}$, $\operatorname{Im}\mathcal{A} \equiv \{\operatorname{Re} f : f \in \mathcal{A}\}$. Show $\mathcal{A} = \operatorname{Re}\mathcal{A} + i\operatorname{Im}\mathcal{A} = \operatorname{Im}\mathcal{A} + i\operatorname{Re}\mathcal{A}$. Then argue that both $\operatorname{Re}\mathcal{A}$ and $\operatorname{Im}\mathcal{A}$ are real algebras which annihilate no point of X and separate the points of X. Apply the Stone Weierstrass theorem to approximate $\operatorname{Re} f$ and $\operatorname{Im} f$ with functions from these real algebras.

Chapter 3
Banach Spaces

3.1 Baire category theorem

Functional analysis is the study of various types of vector spaces which are also topological spaces and the linear operators defined on these spaces. As such, it is really a generalization of linear algebra and calculus. The vector spaces which are of interest in this subject include the usual spaces $\mathbb{R}^n$ and $\mathbb{C}^n$ but also many which are infinite dimensional such as the space $C(X; \mathbb{R}^n)$ discussed in Chapter 2 in which we think of a function as a point or a vector. When the topology comes from a norm, the vector space is called a normed linear space and this is the case of interest here. A normed linear space is called real if the field of scalars is $\mathbb{R}$ and complex if the field of scalars is $\mathbb{C}$. We will assume a linear space is complex unless stated otherwise. A normed linear space may be considered as a vector space if we define $d(x, y) \equiv ||x - y||$. As usual, if every Cauchy sequence converges, the metric space is called complete.

Definition 3.1 A complete normed linear space is called a Banach space.

The purpose of this chapter is to prove some of the most important theorems about Banach spaces. The next theorem is called the Baire category theorem and it will be used in the proofs of many of the other theorems.

Theorem 3.2 *Let (X, d) be a complete metric space and let $\{U_n\}_{n=1}^{\infty}$ be a sequence of open subsets of X satisfying $\overline{U_n} = X$ (U_n is dense). Then $D \equiv \cap_{n=1}^{\infty} U_n$ is a dense subset of X.*

Proof: Let $p \in X$ and let $r_0 > 0$. We need to show $D \cap B(p, r_0) \neq \emptyset$. Since U_1 is dense, there exists $p_1 \in U_1 \cap B(p, r_0)$, an open set. Let $p_1 \in B(p_1, r_1) \subseteq \overline{B(p_1, r_1)} \subseteq U_1 \cap B(p, r_0)$ and $r_1 < 2^{-1}$. We are using Theorem 1.23.

There exists $p_2 \in U_2 \cap B(p_1, r_1)$ because U_2 is dense. Let

$$p_2 \in B(p_2, r_2) \subseteq \overline{B(p_2, r_2)} \subseteq U_2 \cap B(p_1, r_1) \subseteq U_1 \cap U_2 \cap B(p, r_0).$$

and let $r_2 < 2^{-2}$. Continue in this way. Thus

$$r_n < 2^{-n},$$

$$\overline{B(p_n, r_n)} \subseteq U_1 \cap U_2 \cap ... \cap U_n \cap B(p, r_0),$$

$$\overline{B(p_n, r_n)} \subseteq B(p_{n-1}, r_{n-1}).$$

Consider the Cauchy sequence, $\{p_n\}$. Since X is complete, let

$$\lim_{n \to \infty} p_n = p_\infty.$$

Since all but finitely many terms of $\{p_n\}$ are in $\overline{B(p_m, r_m)}$, it follows that $p_\infty \in \overline{B(p_m, r_m)}$. Since this holds for every m,

$$p_\infty \in \cap_{m=1}^\infty \overline{B(p_m, r_m)} \subseteq \cap_{i=1}^\infty U_i \cap B(p, r_0).$$

This proves the theorem.

Corollary 3.3 *Let X be a complete metric space and suppose $X = \cup_{i=1}^\infty F_i$ where each F_i is a closed set. Then for some i, interior $F_i \neq \emptyset$.*

The set D of Theorem 3.2 is called a G_δ set because it is the countable intersection of open sets. Thus D is a dense G_δ set.

Recall that a norm satisfies:

a.) $||x|| \geq 0$, $||x|| = 0$ if and only if $x = 0$.

b.) $||x + y|| \leq ||x|| + ||y||$.

c.) $||cx|| = |c| \, ||x||$ if c is a scalar and $x \in X$.

We also recall the following lemma which gives a simple way to tell if a function mapping a metric space to a metric space is continuous.

Lemma 3.4 *If $(X, d), (Y, p)$ are metric spaces, f is continuous at x if and only if*

$$\lim_{n \to \infty} x_n = x$$

implies

$$\lim_{n \to \infty} f(x_n) = f(x).$$

The proof is left to the reader and follows quickly from the definition of continuity. See Problem 12 of Chapter 1. For the sake of simplicity, we will write $x_n \to x$ sometimes instead of $\lim_{n \to \infty} x_n = x$.

Theorem 3.5 *Let X and Y be two normed linear spaces and let $L : X \to Y$ be linear $(L(ax + by) = aL(x) + bL(y)$ for a, b scalars and $x, y \in X)$. The following are equivalent*

a.) L is continuous at 0

b.) L is continuous

c.) There exists $K > 0$ such that $||Lx||_Y \leq K \, ||x||_X$ for all $x \in X$ (L is bounded).

Proof: a.)$\Rightarrow$b.) Let $x_n \to x$. Then $(x_n - x) \to 0$. It follows $Lx_n - Lx \to 0$ so $Lx_n \to Lx$. b.)$\Rightarrow$c.) Since L is continuous, L is continuous at 0. Hence $||Lx||_Y < 1$ whenever $||x||_X \leq \delta$ for some δ. Therefore, suppressing the subscript on the $|| \, ||$,

$$||L \left(\frac{\delta x}{||x||} \right) || \leq 1.$$

Hence

$$||Lx|| \leq \frac{1}{\delta} ||x||.$$

c.)$\Rightarrow$a.) is obvious.

Definition 3.6 Let $L : X \to Y$ be linear and continuous where X and Y are normed linear spaces. We denote the set of all such continuous linear maps by $\mathcal{L}(X, Y)$ and define

$$||L|| = \sup\{||Lx|| : ||x|| \leq 1\}. \tag{1}$$

The proof of the next lemma is left to the reader.

Lemma 3.7 *With $||L||$ defined in Formula 1, $\mathcal{L}(X, Y)$ is a normed linear space. Also $||Lx|| \leq ||L|| \, ||x||$.*

For example, we could consider the space of linear transformations defined on $\mathbb{R}^n$ having values in $\mathbb{R}^m$, and the above gives a way to measure the distance between two linear transformations. In this case, the linear transformations are all continuous because if L is such a linear transformation, and $\{e_k\}_{k=1}^n$ and $\{e_i\}_{i=1}^m$ are the standard basis vectors in $\mathbb{R}^n$ and $\mathbb{R}^m$ respectively, there are scalars l_{ik} such that

$$L(e_k) = \sum_{i=1}^{m} l_{ik} e_i.$$

Thus, letting $\mathbf{a} = \sum_{k=1}^{n} a_k e_k$,

$$L(\mathbf{a}) = L \left(\sum_{k=1}^{n} a_k e_k \right) = \sum_{k=1}^{n} a_k L(e_k) = \sum_{k=1}^{n} \sum_{i=1}^{m} a_k l_{ik} e_i.$$

Consequently, letting $K \geq |l_{ik}|$ for all i, k,

$$
\begin{aligned}
||L\mathbf{a}|| & \leq \left(\sum_{i=1}^{m} \left| \sum_{k=1}^{n} a_k l_{ik} \right|^2 \right)^{1/2} \leq K m^{1/2} n \left(\max\{|a_k|, k = 1, \cdots, n\} \right) \\
& \leq K m^{1/2} n \left(\sum_{k=1}^{n} a_k^2 \right)^{1/2} = K m^{1/2} n \, ||\mathbf{a}||.
\end{aligned}
$$

This type of thing occurs whenever one is dealing with a linear transformation between finite dimensional normed linear spaces. This will be shown later. Thus, in finite dimensions the algebraic condition that an operator is linear is sufficient to imply the topological condition that the operator is continuous. The situation is not so simple in infinite dimensional spaces such as $C(X; \mathbb{R}^n)$. This is why we impose the topological condition of continuity as a criterion for membership in $\mathcal{L}(X, Y)$ in addition to the algebraic condition of linearity.

Theorem 3.8 *If Y is a Banach space, then $\mathcal{L}(X, Y)$ is also a Banach space.*

Proof: Let $\{L_n\}$ be a Cauchy sequence in $\mathcal{L}(X, Y)$ and let $x \in X$.

$$||L_n x - L_m x|| \le ||x|| \, ||L_n - L_m||.$$

Thus $\{L_n x\}$ is a Cauchy sequence. Let

$$Lx = \lim_{n \to 0} L_n x.$$

Then, clearly, L is linear. Also L is continuous. To see this, note that $\{||L_n||\}$ is a Cauchy sequence of real numbers because $|\,||L_n|| - ||L_m||\,| \le ||L_n - L_m||$. Hence there exists $K > \sup\{||L_n|| : n \in \mathbb{N}\}$. Thus, if $x \in X$,

$$||Lx|| = \lim_{n \to 0} ||L_n x|| \le K ||x||.$$

This proves the theorem.

3.2 Uniform boundedness closed graph and open mapping theorems

The next big result is sometimes called the Uniform Boundedness theorem, or the Banach-Steinhaus theorem. This is a very surprising theorem which implies that for a collection of bounded linear operators, if they are bounded pointwise, then they are also bounded uniformly. As an example of a situation in which pointwise bounded does not imply uniformly bounded, consider the functions $f_\alpha(x) \equiv \mathcal{X}_{(\alpha, 1)}(x) x^{-1}$ for $\alpha \in (0, 1)$. Clearly each function is bounded and the collection of functions is bounded at each point of $(0, 1)$, but there is no bound for all the functions taken together.

Theorem 3.9 *Let X be a Banach space and let Y be a normed linear space. Let $\{L_\alpha\}_{\alpha \in \Lambda}$ be a collection of elements of $\mathcal{L}(X, Y)$. Then one of the following happens.*

a.) $\sup\{\|L_\alpha\| \mid \alpha \in \Lambda\} < \infty$

b.) There exists a dense G_δ set, D, such that for all $x \in D$,

$$\sup\{\|L_\alpha x\| \mid \alpha \in \Lambda\} = \infty.$$

Proof: For each $n \in \mathbb{N}$, define

$$U_n = \{x \in X : \sup\{\|L_\alpha x\| : \alpha \in \Lambda\} > n\}.$$

Then U_n is an open set. Case b.) is obtained from Theorem 3.2 if each U_n is dense. The other case is that for some n, U_n is not dense. If this occurs, there exists x_0 and $r > 0$ such that for all $x \in B(x_0, r)$, $\|L_\alpha x\| \leq n$ for all α. Now if $y \in B(0, r)$, $x_0 + y \in B(x_0, r)$. Consequently, for all such y, $\|L_\alpha(x_0 + y)\| \leq n$. This implies that for such y and all α,

$$\|L_\alpha y\| \leq n + \|L_\alpha(x_0)\| \leq 2n.$$

Hence $\|L_\alpha\| \leq \frac{2n}{r}$ for all α, and we obtain case a.).

The next theorem is called the Open Mapping theorem. Unlike Theorem 3.9 it requires both X and Y to be Banach spaces.

Theorem 3.10 *Let X and Y be Banach spaces, let $L \in \mathcal{L}(X, Y)$, and suppose L is onto. Then L maps open sets onto open sets.*

To aid in the proof of this important theorem, we give a lemma.

Lemma 3.11 *Let a and b be positive constants and suppose*

$$B(0, a) \subseteq \overline{L(B(0, b))}.$$

Then

$$\overline{L(B(0, b))} \subseteq L(B(0, 2b)).$$

Proof of Lemma 3.11: Let $y \in \overline{L(B(0, b))}$. Pick $x_1 \in B(0, b)$ such that $\|y - Lx_1\| < \frac{a}{2}$. Now

$$2y - 2Lx_1 \in B(0, a) \subseteq \overline{L(B(0, b))}.$$

Then choose $x_2 \in B(0, b)$ such that $\|2y - 2Lx_1 - Lx_2\| < a/2$. Thus $\|y - Lx_1 - L\left(\frac{x_2}{2}\right)\| < a/2^2$. Continuing in this way, we pick $x_3, x_4, \ldots$ in $B(0, b)$ such that

$$\left\|y - \sum_{i=1}^{n} 2^{-(i-1)}L(x_i)\right\| = \left\|y - L\sum_{i=1}^{n} 2^{-(i-1)}x_i\right\| < a2^{-n}. \qquad (2)$$

Let $x = \sum_{i=1}^{\infty} 2^{-(i-1)}x_i$. The series converges because X is complete and $\|\sum_{i=m}^{n} 2^{-(i-1)}x_i\| \leq b \sum_{i=m}^{\infty} 2^{-(i-1)} = b\, 2^{-m+2}$. Thus the sequence of partial sums is Cauchy. Letting $n \to \infty$ in Formula 2 yields $||y - Lx|| = 0$. Now

$$||x|| = \lim_{n \to \infty} || \sum_{i=1}^{n} 2^{-(i-1)} x_i ||$$

$$\leq \lim_{n \to \infty} \sum_{i=1}^{n} 2^{-(i-1)} ||x_i|| < \lim_{n \to \infty} \sum_{i=1}^{n} 2^{-(i-1)} b = 2b.$$

This proves the lemma.

Proof of Theorem 3.10: $Y = \cup_{n=1}^{\infty} \overline{L(B(0,n))}$. By Corollary 3.3, $\overline{L(B(0,n_0))}$ has nonempty interior for some n_0. Thus $B(y,r) \subseteq \overline{L(B(0,n_0))}$ for some y and some $r > 0$. Since L is linear $B(-y,r) \subseteq \overline{L(B(0,n_0))}$ also (why?). Therefore

$$
\begin{aligned}
B(0,r) &\subseteq B(y,r) + B(-y,r) \\
&\equiv \{x + z : x \in B(y,r) \text{ and } z \in B(-y,r)\} \\
&\subseteq \overline{L(B(0,2n_0))}
\end{aligned}
$$

By Lemma 3.11, $\overline{L(B(0,2n_0))} \subseteq L(B(0,4n_0))$. Letting $a = r(4n_0)^{-1}$, it follows, since L is linear, that $B(0,a) \subseteq L(B(0,1))$.

Now let U be open in X and let $x + B(0,r) = B(x,r) \subseteq U$. Then

$$L(U) \supseteq L(x + B(0,r))$$

$$= Lx + L(B(0,r)) \supseteq Lx + B(0,ar) = B(Lx, ar)$$

$(L(B(0,r)) \supseteq B(0,ar)$ because $L(B(0,1)) \supseteq B(0,a)$ and L is linear). Hence

$$Lx \in B(Lx, ar) \subseteq L(U).$$

This shows that every point, $Lx \in LU$, is an interior point of LU and so LU is open. This proves the theorem.

This theorem is surprising because it implies that if $|\cdot|$ and $||\cdot||$ are two norms with respect to which a vector space X is a Banach space such that $|\cdot| \leq K ||\cdot||$, then there exists a constant k, such that $||\cdot|| \leq k |\cdot|$. This can be useful because sometimes it is not clear how to compute k when all that is needed is its existence. To see the open mapping theorem implies this, consider the identity map $ix = x$. Then $i : (X, ||\cdot||) \to (X, |\cdot|)$ is continuous and onto. Hence i is an open map which implies i^{-1} is continuous. This gives the existence of the constant k. Of course there are many other situations where this theorem will be of use.

Definition 3.12 Let $f : D \to E$. The set of all ordered pairs of the form $\{(x, f(x)) : x \in D\}$ is called the graph of f.

Definition 3.13 If X and Y are normed linear spaces, we make $X \times Y$ into a normed linear space by using the norm $||(x, y)|| = ||x|| + ||y||$ along with component-wise addition and scalar multiplication. Thus $a(x, y) + b(z, w) \equiv (ax + bz, ay + bw)$.

There are other ways to give a norm for $X \times Y$. See Problem 5 for some alternatives.

Lemma 3.14 *The norm defined in Formula 3.13 on $X \times Y$ along with the definition of addition and scalar multiplication given there make $X \times Y$ into a normed linear space. Furthermore, the topology induced by this norm is identical to the product topology defined in Chapter 1.*

Lemma 3.15 *If X and Y are Banach spaces, then $X \times Y$ with the norm and vector space operations defined in Definition 3.13 is also a Banach space.*

Lemma 3.16 *Every closed subspace of a Banach space is a Banach space.*

Definition 3.17 Let X and Y be Banach spaces and let $D \subseteq X$ be a subspace. A linear map $L : D \rightarrow Y$ is said to be closed if its graph is a closed subspace of $X \times Y$. Equivalently, L is closed if $x_n \rightarrow x$ and $Lx_n \rightarrow y$ implies $x \in D$ and $y = Lx$.

Note the distinction between closed and continuous. If the operator is closed the assertion that $y = Lx$ only follows if it is known that the sequence $\{Lx_n\}$ converges. In the case of a continuous operator, the convergence of $\{Lx_n\}$ follows from the assumption that $x_n \rightarrow x$. It is not always the case that a mapping which is closed is necessarily continuous. Consider the function $f(x) = \tan(x)$ if x is not an odd multiple of $\frac{\pi}{2}$ and $f(x) \equiv 0$ at every odd multiple of $\frac{\pi}{2}$. Then the graph is closed and the function is defined on $\mathbb{R}$ but it clearly fails to be continuous. The next theorem, the closed graph theorem, gives conditions under which closed implies continuous.

Theorem 3.18 *Let X and Y be Banach spaces and suppose $L : X \rightarrow Y$ is closed and linear. Then L is continuous.*

Proof: Let G be the graph of L. $G = \{(x, Lx) : x \in X\}$. Define $P : G \rightarrow X$ by $P(x, Lx) = x$. P maps the Banach space G onto the Banach space X and is continuous and linear. By the open mapping theorem, P maps open sets onto open sets. Since P is also 1-1, this says that P^{-1} is continuous. Thus $||P^{-1}x|| \leq K||x||$. Hence

$$||x|| + ||Lx|| \leq K||x||$$

and so $||Lx|| \leq (K-1)||x||$. This shows L is continuous and proves the theorem.

3.3   Hahn Banach theorem

The closed graph, open mapping, and uniform boundedness theorems are the three major topological theorems in functional analysis. The other major theorem is the Hahn-Banach theorem which has nothing to do with topology. Before presenting this theorem, we need some preliminaries.

Definition 3.19 Let $\mathcal{F}$ be a nonempty set. $\mathcal{F}$ is called a partially ordered set if there is a relation, denoted here by $\leq$, such that

$$x \leq x \text{ for all } x \in \mathcal{F}.$$

If $x \leq y$ and $y \leq z$ then $x \leq z$.

$\mathcal{C} \subseteq \mathcal{F}$ is said to be a chain if every two elements of $\mathcal{C}$ are related. By this we mean that if $x, y \in \mathcal{C}$, then either $x \leq y$ or $y \leq x$. Sometimes we call a chain a totally ordered set. $\mathcal{C}$ is said to be a maximal chain if whenever $\mathcal{D}$ is a chain containing $\mathcal{C}$, $\mathcal{D} = \mathcal{C}$.

The most common example of a partially ordered set is the power set of a given set with $\subseteq$ being the relation. The following theorem is equivalent to the axiom of choice. For a discussion of this, see Chapter 1.

Theorem 3.20 *(Hausdorff Maximal Principle) Let $\mathcal{F}$ be a nonempty partially ordered set. Then there exists a maximal chain.*

Definition 3.21 Let X be a real vector space $\rho : X \to \mathbb{R}$ is called a gauge function if

$$\rho(x+y) \leq \rho(x) + \rho(y),$$

$$\rho(ax) = a\rho(x) \text{ if } a \geq 0. \tag{3}$$

Suppose M is a subspace of X and $z \notin M$. Suppose also that f is a linear real-valued function having the property that $f(x) \leq \rho(x)$ for all $x \in M$. We want to consider the problem of extending f to $M \oplus \mathbb{R}z$ such that if F is the extended function, $F(y) \leq \rho(y)$ for all $y \in M \oplus \mathbb{R}z$ and F is linear. Since F is to be linear, we see that we only need to determine how to define $F(z)$. Letting $a > 0$, we need to have the following hold for all $x, y \in M$.

$$F(x+az) \leq \rho(x+az), \ F(y-az) \leq \rho(y-az).$$

40

Multiplying by a^{-1} using the fact that M is a subspace, and Formula 3, we see this is the same as

$$f(x) + F(z) \leq \rho(x + z), \; f(y) - \rho(y - z) \leq F(z)$$

for all $x, y \in M$. Hence we need to have $F(z)$ such that for all $x, y \in M$

$$f(y) - \rho(y - z) \leq F(z) \leq \rho(x + z) - f(x). \tag{4}$$

Is there any such number between $f(y) - \rho(y - z)$ and $\rho(x + z) - f(x)$ for every pair $x, y \in M$? This is where we use that $f(x) \leq \rho(x)$ on M. For $x, y \in M$,

$$\rho(x + z) - f(x) - [f(y) - \rho(y - z)] = \rho(x + z) + \rho(y - z) - (f(x) + f(y))$$

$$\geq \rho(x + y) - f(x + y) \geq 0.$$

Therefore there exists a number between $\sup \{f(y) - \rho(y - z) : y \in M\}$ and $\inf \{\rho(x + z) - f(x) : x \in M\}$. We choose $F(z)$ to satisfy Formula 4. With this preparation, we state a simple lemma which will be used to prove the Hahn Banach theorem.

Lemma 3.22 *Let M be a subspace of X, a real linear space, and let ρ be a gauge function on X. Suppose $f : M \to \mathbb{R}$ is linear and $z \notin M$. Then f can be extended to $M \oplus \mathbb{R}z$ such that, if F is the extended function, F is linear and $F(x) \leq \rho(x)$ for all $x \in M \oplus \mathbb{R}z$.*

Proof: Let $f(y) - \rho(y - z) \leq F(z) \leq \rho(x + z) - f(x)$ for all $x, y \in M$ and let $F(x + az) = f(x) + aF(z)$ whenever $x \in M$, $a \in \mathbb{R}$. If $a > 0$

$$
\begin{aligned}
F(x + az) &= f(x) + aF(z) \\
&\leq f(x) + a\left[\rho\left(\frac{x}{a} + z\right) - f\left(\frac{x}{a}\right)\right] \\
&= \rho(x + az).
\end{aligned}
$$

If $a < 0$,

$$
\begin{aligned}
F(x + az) &= f(x) + aF(z) \\
&\leq f(x) + a\left[f\left(\frac{-x}{a}\right) - \rho\left(\frac{-x}{a} - z\right)\right] \\
&= f(x) - f(x) + \rho(x + az) = \rho(x + az).
\end{aligned}
$$

This proves the lemma.

Theorem 3.23 *(Hahn Banach theorem) Let X be a real vector space, let M be a subspace of X, let $f : M \to \mathbb{R}$ be linear, let ρ be a gauge function* on X, and suppose $f(x) \leq \rho(x)$ for all $x \in M$. Then there exists a linear function, $F : X \to \mathbb{R}$, such that

a.) $F(x) = f(x)$ for all $x \in M$
b.) $F(x) \leq \rho(x)$ for all $x \in X$.

Proof: Let $\mathcal{F} = \{(V, g) : V \supseteq M,\ V$ is a subspace of X, $g : V \to \mathbb{R}$ is linear, $g(x) = f(x)$ for all $x \in M$, and $g(x) \leq \rho(x)\}$. Then $(M, f) \in \mathcal{F}$ so $\mathcal{F} \neq \emptyset$. Define a partial order by the following rule.

$$(V, g) \leq (W, h)$$

means

$$V \subseteq W \text{ and } h(x) = g(x) \text{ if } x \in V.$$

Let $\mathcal{C}$ be a maximal chain in $\mathcal{F}$ (Hausdorff Maximal theorem). Let $Y = \cup\{V : (V, g) \in \mathcal{C}\}$. Let $h : Y \to \mathbb{R}$ be defined by $h(x) = g(x)$ where $x \in V$ and $(V, g) \in \mathcal{C}$. This is well defined since $\mathcal{C}$ is a chain. Also h is clearly linear and $h(x) \leq \rho(x)$ for all $x \in Y$. We want to argue that $Y = X$. If not, there exists $z \in X \setminus Y$ and we can extend h to $Y \oplus \mathbb{R}z$ using Lemma 3.22. But this will contradict the maximality of $\mathcal{C}$. Indeed, $\mathcal{C} \cup \{Y \oplus \mathbb{R}z,\ \overline{h}\}$ would be a longer chain where $\overline{h}$ is the extended h. This proves the Hahn Banach theorem.

This is the original version of the theorem. There is also a version of this theorem for complex vector spaces which is based on a trick.

Corollary 3.24 *(Hahn Banach) Let M be a subspace of a complex normed linear space, X, and suppose $f : M \to \mathbb{C}$ is linear and satisfies $|f(x)| \leq K||x||$ for all $x \in M$. Then there exists a linear function, F, defined on all of X such that $F(x) = f(x)$ for all $x \in M$ and $|F(x)| \leq K||x||$ for all x.*

Proof: First note $f(x) = \operatorname{Re} f(x) + i \operatorname{Im} f(x)$ and so

$$\operatorname{Re} f(ix) + i\ \operatorname{Im} f(ix) = f(ix) = i\,f(x) = i\ \operatorname{Re} f(x) - \operatorname{Im} f(x).$$

Therefore, $\operatorname{Im} f(x) = -\operatorname{Re} f(ix)$, and we may write

$$f(x) = \operatorname{Re} f(x) - i \operatorname{Re} f(ix).$$

If c is a real scalar

$$\operatorname{Re} f(cx) - i\ \operatorname{Re} f(icx) = cf(x) = c \operatorname{Re} f(x) - i\,c \operatorname{Re} f(ix).$$

Thus $\operatorname{Re} f(cx) = c \operatorname{Re} f(x)$. It is also clear that $\operatorname{Re} f(x + y) \doteq \operatorname{Re} f(x) + \operatorname{Re} f(y)$. Consider X as a real vector space and let $\rho(x) = K||x||$. Then for all $x \in M$,

$$|\operatorname{Re} f(x)| \leq K||x|| = \rho(x).$$

From Theorem 3.23, Re f may be extended to a function, h which satisfies

$$\begin{aligned} h(ax + by) &= ah(x) + bh(y) \text{ if } a, b \in \mathbb{R} \\ |h(x)| &\leq K||x|| \text{ for all } x \in X. \end{aligned}$$

Let

$$F(x) \equiv h(x) - i\, h(ix).$$

It is routine to show F is linear. Now $wF(x) = |F(x)|$ for some $|w| = 1$. Therefore

$$\begin{aligned} |F(x)| &= wF(x) = h(wx) - i\, h(iwx) = h(wx) \\ &= |h(wx)| \leq K||x||. \end{aligned}$$

This proves the corollary.

Definition 3.25 Let X be a Banach space. We denote by X' the space $\mathcal{L}(X, \mathbb{C})$. By Theorem 3.8, X' is a Banach space. Remember

$$||f|| = \sup\{|f(x)| : ||x|| \leq 1\}$$

for $f \in X'$. We call X' the dual space.

Definition 3.26 Let X and Y be Banach spaces and suppose $L \in \mathcal{L}(X, Y)$. Then we define the adjoint map in $\mathcal{L}(Y', X')$, denoted by L^*, by

$$L^* y^* (x) \equiv y^* (Lx)$$

for all $y^* \in Y'$.

$$\begin{array}{ccc} & L^* & \\ X' & \leftarrow & Y' \\ X & \underset{L}{\rightarrow} & Y \end{array}$$

In terms of linear algebra, this adjoint map is algebraically very similar to, and is in fact a generalization of, the transpose of a matrix considered as a map on $\mathbb{R}^n$. Recall that if A is such a matrix, A^T satisfies $A^T \mathbf{x} \cdot \mathbf{y} = \mathbf{x} \cdot A\mathbf{y}$. In the case of $\mathbb{C}^n$ the adjoint is similar to the conjugate transpose of the matrix and it behaves the same with respect to the complex inner product on $\mathbb{C}^n$. What is being done here is to generalize this algebraic concept to arbitrary Banach spaces.

Theorem 3.27 Let $L \in \mathcal{L}(X, Y)$ where X and Y are Banach spaces. Then
  a.) $L^* \in \mathcal{L}(Y', X')$ as claimed and $||L^*|| \leq ||L||$.
  b.) If L is 1-1 onto a closed subspace of Y, then L^* is onto.

c.) If L is onto a dense subset of Y, then L* is 1-1.

Proof: Clearly L^*y^* is linear and L^* is also a linear map.

$$\begin{aligned}
||L^*y^*|| &= \sup\{|L^*y^*(x)| : ||x|| \leq 1\} = \sup\{|y^*(Lx)| : ||x|| \leq 1\} \\
&\leq ||y^*||\,||L||.
\end{aligned}$$

Hence, $||L^*|| \leq ||L||$ and this shows part a.).

If L is 1-1 and onto a closed subset of Y, then we can apply the Open Mapping theorem to conclude that $L^{-1} : L(X) \to X$ is continuous. Hence

$$||x|| = ||L^{-1}Lx|| \leq K||Lx||$$

for some K. Now let $x^* \in X'$ be given. Define $f \in \mathcal{L}(L(X), \mathbb{C})$ by $f(Lx) = x^*(x)$. Since L is 1-1, it follows that f is linear and well defined. Also

$$|f(Lx)| = |x^*(x)| \leq ||x^*||\,||x|| \leq K||x^*||\,||Lx||.$$

By the Hahn Banach theorem, we can extend f to an element $y^* \in Y'$ such that $||y^*|| \leq K||x^*||$. Then

$$L^*y^*(x) = y^*(Lx) = f(Lx) = x^*(x)$$

so $L^*y^* = x^*$ and we have shown L^* is onto. This shows b.).

Now suppose LX is dense in Y. If $L^*y^* = 0$, then $y^*(Lx) = 0$ for all x. Since LX is dense, this can only happen if $y^* = 0$. Hence L^* is 1-1.

Corollary 3.28 *Suppose X and Y are Banach spaces, $L \in \mathcal{L}(X,Y)$, and L is 1-1 and onto. Then L^* is also 1-1 and onto.*

There exists a natural mapping from a normed linear space, X, to the dual of the dual space.

Definition 3.29 Define $J : X \to X''$ by $J(x)(x^*) = x^*(x)$. This map is called the James map.

Theorem 3.30 *The map, J, has the following properties.*
a.) J is 1-1 and linear.
b.) $||Jx|| = ||x||$ and $||J|| = 1$.
c.) $J(X)$ is a closed subspace of X'' if X is complete.
Also if $x^* \in X'$,

$$||x^*|| = \sup\left\{|x^{**}(x^*)| : ||x^{**}|| \leq 1,\ x^{**} \in X''\right\}.$$

Proof: To prove this, we will use a simple but useful lemma which depends on the Hahn Banach theorem.

Lemma 3.31 *Let X be a normed linear space and let $x \in X$. Then there exists $x^* \in X'$ such that $||x^*|| = 1$ and $x^*(x) = ||x||$.*

Proof: Let $f : \mathbb{C}x \to \mathbb{C}$ be defined by $f(\alpha x) = \alpha ||x||$. Then for $y \in \mathbb{C}x$, $|f(y)| \leq ||y||$. By the Hahn Banach theorem, there exists $x^* \in X'$ such that $x^*(\alpha x) = f(\alpha x)$ and $||x^*|| \leq 1$. Since $x^*(x) = ||x||$ it follows that $||x^*|| = 1$. This proves the lemma.

Now we prove the theorem. It is obvious that J is linear. If $Jx = 0$, then let $x^*(x) = ||x||$ with $||x^*|| = 1$.

$$0 = J(x)(x^*) = x^*(x) = ||x||.$$

This shows a.). To show b.), let $x \in X$ and $x^*(x) = ||x||$ with $||x^*|| = 1$. Then

$$
\begin{aligned}
||x|| &\geq \sup\{|y^*(x)| : ||y^*|| \leq 1\} = \sup\{|J(x)(y^*)| : ||y^*|| \leq 1\} = ||Jx|| \\
&\geq |J(x)(x^*)| = |x^*(x)| = ||x||
\end{aligned}
$$

$$||J|| = \sup\{||Jx|| : ||x|| \leq 1\} = \sup\{||x|| : ||x|| \leq 1\} = 1.$$

This shows b.). To verify c.), use b.). If $Jx_n \to y^{**} \in X''$ then by b.), x_n is a Cauchy sequence converging to some $x \in X$. Then $Jx = \lim_{n \to \infty} Jx_n = y^{**}$.

Finally, to show the assertion about the norm of x^*, use what was just shown applied to the James map from X' to X'''. More specifically,

$$||x^*|| = \sup\{|x^*(x)| : ||x|| \leq 1\} = \sup\{|J(x)(x^*)| : ||Jx|| \leq 1\}$$

$$\leq \sup\{|x^{***}(x^*)| : ||x^{**}|| \leq 1\} = \{J(x^*)(x^{**}) : ||x^{**}|| \leq 1\}$$

$$= ||Jx^*|| = ||x^*||.$$

This proves the theorem.

Definition 3.32 When J maps X onto X'', we say that X is Reflexive.

Later on we will give examples of reflexive spaces. In particular, it will be shown that the space of square integrable and *pth* power integrable functions are reflexive.

3.4   Exercises

1. Show that no countable dense subset of $\mathbb{R}$ is a G_δ set. In particular, the rational numbers are not a G_δ set.

2. ↑ Let $f : \mathbb{R} \to \mathbb{C}$ be a function. Let $\omega_r f(x) = \sup\{|f(z) - f(y)| : y, z \in B(x, r)\}$. Let $\omega f(x) = \lim_{r \to 0} \omega_r f(x)$. Show f is continuous at x if and only if $\omega f(x) = 0$. Then show the set of points where f is continuous is a G_δ set (try $U_n = \{x : \omega f(x) < \frac{1}{n}\}$). Does there exist a function continuous at only the rational numbers? Does there exist a function continuous at every irrational and discontinuous elsewhere?

3. Let $f \in C([0, 1])$ and suppose $f'(x)$ exists. Show there exists a constant, K, such that $|f(x) - f(y)| \leq K|x - y|$ for all $y \in [0, 1]$. Let $U_n = \{f \in C([0, 1])$ such that for each $x \in [0, 1]$ there exists $y \in [0, 1]$ such that $|f(x) - f(y)| > n|x - y|\}$. Show that U_n is open and dense in $C([0, 1])$ where for $f \in C([0, 1])$,

    $$\|f\| \equiv \sup\{|f(x)| : x \in [0, 1]\}.$$

    Show that if $f \in C([0, 1])$, there exists $g \in C([0, 1])$ such that $\|g - f\| < \varepsilon$ but $g'(x)$ does not exist for any $x \in [0, 1]$.

4. Let X be a normed linear space and suppose $A \subseteq X$ is "weakly bounded". This means that for each $x^* \in X'$, $\sup\{|x^*(x)| : x \in A\} < \infty$. Show A is bounded. That is, show $\sup\{\|x\| : x \in A\} < \infty$.

5. Let X and Y be two Banach spaces. Define the norm

    $$\|\|(x, y)\|\| \equiv \max(\|x\|_X, \|y\|_Y).$$

    Show this is a norm on $X \times Y$ which is equivalent to the norm given in the chapter for $X \times Y$. Can you do the same for the norm defined by

    $$|(x, y)| \equiv \left(\|x\|_X^2 + \|y\|_Y^2\right)^{1/2}?$$

6. Prove Lemmas 3.14 - 3.16.

7. Let $f : \mathbb{R} \to \mathbb{C}$ be continuous and periodic with period 2π. That is $f(x + 2\pi) = f(x)$ for all x. Then it is not unreasonable to try to write $f(x) = \sum_{n=-\infty}^{\infty} a_n e^{inx}$. Find what a_n should be. **Hint:** Multiply both sides by e^{-imx} and do $\int_{-\pi}^{\pi}$. Pretend there is no problem writing $\int \sum = \sum \int$ (such a series is called a Fourier series).

8. ↑ If you did 7 correctly, you found

    $$a_n = \left(\int_{-\pi}^{\pi} f(x) e^{-inx} dx\right)(2\pi)^{-1}.$$

    The nth partial sum will be denoted by $S_n f$ and defined by $S_n f(x) =$

$\sum_{k=-n}^{n} a_k e^{ikx}$. Show $S_n f(x) = \int_{-\pi}^{\pi} f(y) D_n(x-y) dy$ where

$$D_n(t) = \frac{\sin((n+\frac{1}{2})t)}{2\pi \sin(\frac{t}{2})}.$$

This is called the Dirichlet kernel.

9. ↑ Let $Y = \{f$ such that f is continuous, defined on $\mathbb{R}$, and 2π periodic$\}$. Define $||f||_Y = \sup\{|f(x)| : x \in [-\pi, \pi]\}$. Show that $(Y, || \ ||_Y)$ is a Banach space. Let $x \in \mathbb{R}$ and define $L_n(f) = S_n f(x)$. Show $L_n \in Y'$ but $\lim_{n\to\infty} ||L_n|| = \infty$. **Hint:** Let $f(y)$ approximate $\text{sign}(D_n(x-y))$.

10. ↑ Show there exists a dense G_δ subset of Y such that for f in this set, $|S_n f(x)|$ is unbounded. Show there is a dense G_δ subset of Y having the property that $|S_n f(x)|$ is unbounded on a dense G_δ subset of $\mathbb{R}$. This shows Fourier series can fail to converge pointwise to continuous periodic functions in a fairly spectacular way.

11. Let X be a normed linear space and let M be a convex open set containing 0. Define

$$\rho(x) = \inf\{t > 0 : \frac{x}{t} \in M\}.$$

Show ρ is a gauge function defined on X. This particular example is called a Minkowski functional. Recall a set, M, is convex if $\lambda x + (1-\lambda)y \in M$ whenever $\lambda \in [0,1]$ and $x, y \in M$.

12. ↑ This problem explores the use of the Hahn Banach theorem in establishing separation theorems. Let M be an open convex set containing 0. Let $x \notin M$. Show there exists $x^* \in X'$ such that $\operatorname{Re} x^*(x) \geq 1 > \operatorname{Re} x^*(y)$ for all $y \in M$. **Hint:** If $y \in M, \rho(y) < 1$. Show this. If $x \notin M$, $\rho(x) \geq 1$. Try $f(\alpha x) = \alpha \rho(x)$ for $\alpha \in \mathbb{R}$. Then extend f to F, show F is continuous, then fix it so F is the real part of $x^* \in X'$.

13. A Banach space is said to be strictly convex if whenever $||x|| = ||y||$ and $x \neq y$, then

$$\left\| \frac{x+y}{2} \right\| < ||x||.$$

$F : X \to X'$ is said to be a duality map if it satisfies the following: a.) $F(x) = ||x||$. b.) $F(x)(x) = ||x||^2$. Show that if X' is strictly convex, then such a duality map exists. **Hint:** Let $f(\alpha x) = \alpha ||x||^2$ and use Hahn Banach theorem, then strict convexity.

14. Suppose $D \subseteq X$, a Banach space, and $L : D \to Y$ is a closed operator. D might not be a Banach space with respect to the norm on X. Define a new norm on D by $||x||_D = ||x||_X + ||Lx||_Y$. Show $(D, || \; ||_D)$ is a Banach space.

15. Prove the following theorem which is an improved version of the open mapping theorem, [15]. Let X and Y be Banach spaces and let $A \in \mathcal{L}(X, Y)$. Then the following are equivalent.

$$AX = Y,$$

A is an open map.

There exists a constant M such that for every $y \in Y$, there exists $x \in X$ with $y = Ax$ and

$$||x|| \leq M \, ||y||.$$

16. Here is an example of a closed unbounded operator. Let $X = Y = C([0, 1])$ and let

$$D = \{f \in C^1([0, 1]) : f(0) = 0\}.$$

$L : D \to C([0, 1])$ is defined by $Lf = f'$. Show L is closed.

17. Suppose $D \subseteq X$ and D is dense in X. Suppose $L : D \to Y$ is linear and $||Lx|| \leq K||x||$ for all $x \in D$. Show there is a unique extension of L, $\tilde{L}$, defined on all of X with $||\tilde{L}x|| \leq K||x||$ and $\tilde{L}$ is linear.

18. ↑ A Banach space is uniformly convex if whenever $||x_n||, \; ||y_n|| \leq 1$ and $||x_n + y_n|| \to 2$, it follows that $||x_n - y_n|| \to 0$. Show uniform convexity implies strict convexity.

19. We say that x_n converges weakly to x if for every $x^* \in X'$, $x^*(x_n) \to x^*(x)$. We write $x_n \rightharpoonup x$ to denote weak convergence. Show that if $||x_n - x|| \to 0$, then $x_n \rightharpoonup x$.

20. ↑ Show that if X is uniformly convex, then $x_n \rightharpoonup x$ and $||x_n|| \to ||x||$ implies $||x_n - x|| \to 0$. **Hint:** Use Lemma 3.31 to obtain $f \in X'$ with $||f|| = 1$ and $f(x) = ||x||$. See Problem 18 for the definition of uniform convexity.

21. Suppose $L \in \mathcal{L}(X, Y)$ and $M \in \mathcal{L}(Y, Z)$. Show $ML \in \mathcal{L}(X, Z)$ and that $(ML)^* = L^*M^*$.

Chapter 4
Hilbert Spaces

Let X be a vector space. An inner product is a mapping from $X \times X$ to $\mathbb{C}$ if X is complex and from $X \times X$ to $\mathbb{R}$ if X is real, denoted by (x, y) which satisfies the following.

$$(x, x) \geq 0, \ (x, x) = 0 \ \text{ if and only if } x = 0, \tag{1}$$

$$(x, y) = \overline{(y, x)}. \tag{2}$$

For $a, b \in \mathbb{C}$ and $x, y, z \in X$,

$$(ax + by, z) = a(x, z) + b(y, z). \tag{3}$$

Note that Formula 2 and Formula 3 imply $(x, ay + bz) = \overline{a}(x, y) + \overline{b}(x, z)$.

We will show that if $(\cdot, \cdot)$ is an inner product, then $(x, x)^{1/2}$ defines a norm and we say that a normed linear space is an inner product space if $||x|| = (x, x)^{1/2}$.

Definition 4.1 A normed linear space in which the norm comes from an inner product as just described is called an inner product space. A Hilbert space is a complete inner product space.

Thus a Hilbert space is a Banach space whose norm comes from an inner product as just described.

Example 4.2 *Let* $X = \mathbb{C}^n$ *with the inner product given by* $(\mathbf{x}, \mathbf{y}) \equiv \sum_{i=1}^{n} x_i \overline{y}_i$. *This is a complex Hilbert space.*

Example 4.3 *Let* $X = \mathbb{R}^n$, $(\mathbf{x}, \mathbf{y}) = \mathbf{x} \cdot \mathbf{y}$. *This is a real Hilbert space.*

Theorem 4.4 *(Cauchy Schwarz) In any inner product space*

$$|(x, y)| \leq ||x|| \, ||y||.$$

Proof: Let $\omega \in \mathbb{C}$, $|\omega| = 1$, and $\overline{\omega}(x, y) = |(x, y)| = \text{Re}(x, y\omega)$. Let

$$F(t) = (x + ty\omega, x + t\omega y).$$

If $y = 0$ there is nothing to prove because

$$(x, 0) = (x, 0 + 0) = (x, 0) + (x, 0)$$

and so $(x, 0) = 0$. Thus, we may assume $y \neq 0$. Then from the axioms of the inner product, Formula 1,

$$F(t) = ||x||^2 + 2t \, \text{Re}(x, \omega y) + t^2 ||y||^2 \geq 0.$$

This yields

$$||x||^2 + 2t|(x, y)| + t^2||y||^2 \geq 0.$$

Since this inequality holds for all $t \in \mathbb{R}$, it follows from the quadratic formula that

$$4|(x, y)|^2 - 4||x||^2||y||^2 \leq 0.$$

This yields the conclusion and proves the theorem.

Earlier it was claimed that the inner product defines a norm. In this next proposition this claim is proved.

Proposition 4.5 *For an inner product space, $||x|| \equiv (x, x)^{1/2}$ does specify a norm.*

Proof: All the axioms are obvious except the triangle inequality. To verify this,

$$\begin{aligned} ||x + y||^2 &\equiv (x + y, x + y) \equiv ||x||^2 + ||y||^2 + 2\operatorname{Re}(x, y) \\ &\leq ||x||^2 + ||y||^2 + 2|(x, y)| \\ &\leq ||x||^2 + ||y||^2 + 2||x||\,||y|| = (||x|| + ||y||)^2. \end{aligned}$$

The following lemma is called the parallelogram identity.

Lemma 4.6 *In an inner product space,*

$$||x + y||^2 + ||x - y||^2 = 2||x||^2 + 2||y||^2.$$

The proof, a straightforward application of the inner product axioms, is left to the reader. See Problem 3.

One of the best things about Hilbert space is the theorem about projection onto a closed convex set. Recall that a set, K, is convex if whenever $\lambda \in [0, 1]$ and $x, y \in K$, $\lambda x + (1 - \lambda)y \in K$.

Theorem 4.7 *Let K be a closed convex nonempty subset of a Hilbert space, H, and let $x \in H$. Then there exists a unique point $Px \in K$ such that $||Px - x|| \leq ||y - x||$ for all $y \in K$.*

Proof: First we show uniqueness. Suppose $||z_i - x|| \leq ||y - x||$ for all $y \in K$. Then using the parallelogram identity and

$$||z_1 - x|| \leq ||y - x||$$

for all $y \in K$,

$$\begin{aligned} ||z_1 - x||^2 &\leq ||\frac{z_1 + z_2}{2} - x||^2 = ||\frac{z_1 - x}{2} + \frac{z_2 - x}{2}||^2 \\ &= 2(||\frac{z_1 - x}{2}||^2 + ||\frac{z_2 - x}{2}||^2) - ||\frac{z_1 - z_2}{2}||^2 \\ &\leq ||z_1 - x||^2 - ||\frac{z_1 - z_2}{2}||^2, \end{aligned}$$

where the last inequality holds because

$$||z_2 - x|| \leq ||z_1 - x||.$$

Hence $z_1 = z_2$ and this shows uniqueness.

Now let $\lambda = \inf\{||x-y|| : y \in K\}$ and let y_n be a minimizing sequence. Thus $\lim_{n \to \infty} ||x - y_n|| = \lambda$, $y_n \in K$. Then by the parallelogram identity,

$$\begin{aligned}
||y_n - y_m||^2 &= 2(||y_n - x||^2 + ||y_m - x||^2) - 4(||\frac{y_n + y_m}{2} - x||^2) \\
&\leq 2(||y_n - x||^2 + ||y_m - x||^2) - 4\lambda^2.
\end{aligned}$$

Since $||x - y_n|| \to \lambda$, this shows $\{y_n\}$ is a Cauchy sequence. Since H is complete, $y_n \to y$ for some $y \in H$ which must be in K because K is closed. Therefore $||x - y|| = \lambda$ and we let $Px = y$.

Corollary 4.8 *Let K be a closed, convex, nonempty subset of a Hilbert space, H, and let $x \notin K$. Then for $z \in K$, $z = Px$ if and only if*

$$\text{Re}(x - z, y - z) \leq 0 \tag{4}$$

for all $y \in K$.

Before proving this, consider what it says in the case where the Hilbert space is $\mathbb{R}^n$.

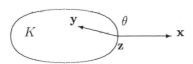

Condition 4 says the angle, θ, shown in the diagram is always obtuse. Remember, the sign of $\mathbf{x} \cdot \mathbf{y}$ is the same as the sign of the cosine of their included angle.

The inequality 4 is an example of a variational inequality and this corollary characterizes the projection of x onto K as the solution of this variational inequality.

Proof of Corollary: Let $z \in K$. Since K is convex, every point of K is in the form $z + t(y - z)$ where $t \in [0, 1]$ and $y \in K$. Therefore, $z = Px$ if and only if for all $y \in K$ and $t \in [0, 1]$,

$$||x - (z + t(y - z))||^2 = ||(x - z) - t(y - z)||^2 \geq ||x - z||^2$$

for all $t \in [0, 1]$ if and only if for all $t \in [0, 1]$ and $y \in K$

$$||x - z||^2 + t^2 ||y - z||^2 - 2t \, \text{Re} \, (x - z, y - z) \geq ||x - z||^2$$

51

which is equivalent to Formula 4. This proves the corollary.

The case where the closed convex set is a closed subspace is of special importance and in this case the above corollary implies the following.

Corollary 4.9 *Let K be a closed subspace of a Hilbert space, H, and let $x \notin K$. Then for $z \in K$, $z = Px$ if and only if*

$$(x - z, y) = 0$$

for all $y \in K$.

Proof: Since K is a subspace, the condition 4 implies

$$\text{Re}(x - z, y) \leq 0$$

for all $y \in K$. But this implies this inequality holds with $\leq$ replaced with $=$. To see this, replace y with $-y$. Now let $|\alpha| = 1$ and

$$\alpha(x - z, y) = |(x - z, y)|.$$

Since $\overline{\alpha}y \in K$ for all $y \in K$,

$$0 = \text{Re}(x - z, \overline{\alpha}y) = (x - z, \overline{\alpha}y) = \alpha(x - z, y) = |(x - z, y)|.$$

This proves the corollary.

The following theorem is called the Riesz representation theorem for the dual of a Hilbert space. If $z \in H$ then we may define an element $f \in H'$ by the rule

$$(x, z) \equiv f(x).$$

It follows from the Cauchy Schwartz inequality and the properties of the inner product that $f \in H'$. The Riesz representation theorem says that all elements of H' are of this form.

Theorem 4.10 *Let H be a Hilbert space and let $f \in H'$. Then there exists a unique $z \in H$ such that $f(x) = (x, z)$ for all $x \in H$.*

Proof: If $f = 0$, there is nothing to prove so assume without loss of generality that $f \neq 0$. Let $M = \{x \in H : f(x) = 0\}$. Thus M is a closed proper subspace of H. Let $y \notin M$. Then $y - Py \equiv w$ has the property that $(x, w) = 0$ for all $x \in M$ by Corollary 4.9. Let $x \in H$ be arbitrary. Then

$$xf(w) - f(x)w \in M$$

so

$$0 = (f(w)x - f(x)w, w) = f(w)(x, w) - f(x)||w||^2.$$

Thus

$$f(x) = (x, \overline{\frac{f(w)w}{||w||^2}})$$

and so we let

$$z = \overline{\frac{f(w)w}{||w||^2}}.$$

This proves the existence of z. If $f(x) = (x, z_i)$ $i = 1, 2$, for all $x \in H$, then for all $x \in H$,

$$(x, z_1 - z_2) = 0.$$

Let $x = z_1 - z_2$ to conclude uniqueness. This proves the theorem.

4.1 Finite dimensional normed linear space

If X is a normed linear space, the norm specifies a topology for X. There may be other norms which specify the same topology but which also make X into a Hilbert space. We will show that this can be done for every finite dimensional normed linear space.

Let X be a finite dimensional normed linear space with norm $||\cdot||$ where the field of scalars is denoted by $\mathbb{F}$ and is understood to be either $\mathbb{R}$ or $\mathbb{C}$. Let $\{v_1, \cdots, v_n\}$ be a basis for X. If $\mathbf{x} \in X$, we will denote by x_i the *ith* component of $\mathbf{x}$ with respect to this basis. Thus

$$\mathbf{x} = \sum_{i=1}^{n} x_i \mathbf{v}_i.$$

Definition 4.11 For $\mathbf{x} \in X$ and $\{v_1, \cdots, v_n\}$ a basis, we define a new norm on X by

$$|\mathbf{x}| \equiv \left(\sum_{i=1}^{n} |x_i|^2 \right)^{1/2}.$$

The first thing we will show is that these two norms, $||\cdot||$ and $|\cdot|$, are equivalent. This means the conclusion of the following theorem holds.

Theorem 4.12 *Let $(X, ||\cdot||)$ be a finite dimensional normed linear space and let $|\cdot|$ be described above relative to a given basis, $\{v_1, \cdots, v_n\}$. Then $|\cdot|$ is a norm and there exist constants $\delta, \Delta > 0$ independent of $\mathbf{x}$ such that*

$$\delta ||\mathbf{x}|| \leq |\mathbf{x}| \leq \Delta ||\mathbf{x}||.$$

Proof: All of the properties of a norm are obvious except the triangle inequality. To establish this inequality, we use the Cauchy Schwartz inequality for the Hilbert space $\mathbb{F}^n$ to write

$$\begin{aligned}
|\mathbf{x} + \mathbf{y}|^2 &\equiv \sum_{i=1}^{n} |x_i + y_i|^2 \le \sum_{i=1}^{n} |x_i|^2 + \sum_{i=1}^{n} |y_i|^2 + 2\,\mathrm{Re} \sum_{i=1}^{n} x_i \bar{y}_i \\
&\le |\mathbf{x}|^2 + |\mathbf{y}|^2 + 2 \left(\sum_{i=1}^{n} |x_i|^2 \right)^{1/2} \left(\sum_{i=1}^{n} |y_i|^2 \right)^{1/2} \\
&= |\mathbf{x}|^2 + |\mathbf{y}|^2 + 2\,|\mathbf{x}|\,|\mathbf{y}| = (|\mathbf{x}| + |\mathbf{y}|)^2
\end{aligned}$$

and this proves the triangle inequality.

It remains to show the equivalence of the two norms. By the Cauchy Schwartz inequality again,

$$\begin{aligned}
||\mathbf{x}|| &\equiv \left\| \sum_{i=1}^{n} x_i \mathbf{v}_i \right\| \le \sum_{i=1}^{n} |x_i|\,||\mathbf{v}_i|| \le |\mathbf{x}| \left(\sum_{i=1}^{n} ||\mathbf{v}_i||^2 \right)^{1/2} \\
&\equiv \delta^{-1}\,|\mathbf{x}|.
\end{aligned}$$

This proves the first half of the inequality for the equivalence of the norms.

Suppose the second half of the inequality is not valid. Then there exists a sequence $\mathbf{x}^k \in X$ such that

$$\left| \mathbf{x}^k \right| > k\,||\mathbf{x}^k||, \quad k = 1, 2, \cdots.$$

Then define

$$\mathbf{y}^k \equiv \frac{\mathbf{x}^k}{|\mathbf{x}^k|}.$$

It follows

$$\left| \mathbf{y}^k \right| = 1, \quad \left| \mathbf{y}^k \right| > k\,||\mathbf{y}^k||. \tag{5}$$

Letting y_i^k be the components of $\mathbf{y}^k$ with respect to the given basis, it follows the vector

$$\left(y_1^k, \cdots, y_n^k \right)$$

is a unit vector in $\mathbb{F}^n$. By the Heine Borel theorem, there exists a subsequence, still denoted by k, such that

$$\left(y_1^k, \cdots, y_n^k \right) \to (y_1, \cdots, y_n),$$

where $(y_1, \cdots, y_n)$ is a unit vector in $\mathbb{F}^n$. It follows from Formula 5 that $\mathbf{y}^k \to 0$ and so

$$0 = \lim_{k \to \infty} \mathbf{y}^k = \lim_{k \to \infty} \sum_{i=1}^{n} y_i^k \mathbf{v}_i = \sum_{i=1}^{n} y_i \mathbf{v}_i,$$

contradicting the linear independence of $\{\mathbf{v}_1, \cdots, \mathbf{v}_n\}$ and proving the second half of the inequality.

This has the following immediate corollary.

Corollary 4.13 *If $(X, ||\cdot||)$ is a finite dimensional normed linear space with the field of scalars $\mathbb{F} = \mathbb{C}$ or $\mathbb{R}$, then X is complete. Furthermore any two norms on a finite dimensional vector space are equivalent.*

Proof: Let $\{\mathbf{x}^k\}$ be a Cauchy sequence. Then letting the components of $\mathbf{x}^k$ with respect to the given basis be

$$x_1^k, \cdots, x_n^k,$$

it follows from Theorem 4.12 that

$$\left(x_1^k, \cdots, x_n^k\right)$$

is a Cauchy sequence in $\mathbb{F}^n$ and so

$$\left(x_1^k, \cdots, x_n^k\right) \rightarrow (x_1, \cdots, x_n) \in \mathbb{F}^n.$$

Thus,

$$\mathbf{x}^k = \sum_{i=1}^n x_i^k \mathbf{v}_i \rightarrow \sum_{i=1}^n x_i \mathbf{v}_i \in X.$$

Let $\{\mathbf{v}_1, \cdots, \mathbf{v}_n\}$ be a basis, let $|\cdot|$ be the norm described above, and let $||\cdot||$ and $|||\cdot|||$ be two norms on X. Then Theorem 4.12 implies that both norms are equivalent to $|\cdot|$ which implies there are constants δ and Δ such that $\delta \, |||\mathbf{x}||| \leq ||\mathbf{x}|| \leq \Delta \, |||\mathbf{x}|||$. This proves the corollary.

The norm $|\cdot|$ described above makes X into a Hilbert space with the inner product

$$(\mathbf{x}, \mathbf{y}) = \sum_{i=1}^n x_i \overline{y_i}$$

where $\mathbf{x} = \sum_{i=1}^n x_i \mathbf{v}_i$, $\mathbf{y} = \sum_{i=1}^n y_i \mathbf{v}_i$.

4.2 Uniformly convex Banach spaces

As the above section shows, the difference between a Banach space and a Hilbert space is not explained by topological considerations alone since the same topology can be realized in terms of two different norms although not both norms give a Hilbert space. It can be shown that a Banach space is a Hilbert space if and only if the parallelogram identity holds. Now the parallelogram identity implies a more general notion known as uniform convexity and it turns out that many Banach spaces important in applications are uniformly convex. It also turns out that some of the best features of Hilbert space are preserved in this more general class of Banach space.

We will use some of these ideas later in an alternate explanation of some representation theorems, but other than this, the section is not essential to the logical development of the rest of the book.

Definition 4.14 We say a Banach space $(X, ||\cdot||)$ is uniformly convex if whenever $||x_n||, ||y_n|| \leq 1$ and $||x_n + y_n|| \to 2$ then $||x_n - y_n|| \to 0$.

Lemma 4.15 *Let $0 \in M$, an open convex subset of X, a complex Banach space. Suppose $x_0 \notin M$. Then there exists $L \in X'$ such that*

$$\operatorname{Re} L(x_0) > \operatorname{Re} L(x)$$

for all $x \in M$.

Proof: We introduce the Minkowski functional, ρ, defined by

$$\rho(x) = \inf\{t > 0 : \frac{x}{t} \in M\}.$$

Then one can show ρ is a gauge function. This means

$$\rho(x + y) \leq \rho(x) + \rho(y), \quad \rho(\alpha x) = \alpha\rho(x) \text{ if } \alpha > 0.$$

Now define $f : \mathbb{R}x_0 \to \mathbb{R}$ by

$$f(\alpha x_0) = \alpha\rho(x_0).$$

Then $f(\alpha x_0) \leq \rho(\alpha x_0)$ for all $\alpha \in \mathbb{R}$. By the Hahn Banach theorem f can be extended to a linear function $F : X \to \mathbb{R}$ such that

$$F = f \text{ on } \mathbb{R}x_0, \quad F(x) \leq \rho(x) \quad \text{all } x.$$

Thus $F(x_0) = f(x_0) = \rho(x_0) \geq 1$. If $x \in M$, $\rho(x) < 1$. We see easily that F is continuous. (If $x_n \to 0$, then $\frac{x_n}{t} \in M$ for small t. Hence $\rho(x_n) \to 0$ and so $\lim \sup_{n \to \infty} F(x_n) \leq 0$. Now consider $-x_n$.)

Now let $Lx = F(x) - iF(ix)$. L is linear, continuous, and $\operatorname{Re} L(x_0) = F(x_0) \geq 1$ while if $x \in M$, $\operatorname{Re} L(x) = F(x) \leq \rho(x) < 1$.

Lemma 4.16 *(Helly) Let X be a complex Banach Space and suppose $\gamma > 0$ and $\{f_1, \cdots, f_n\} \subseteq X'$. Suppose $\alpha_1, \cdots, \alpha_n$ are given numbers. Then for each $\varepsilon > 0$, there exists x_ε satisfying*

$$\begin{aligned} a.) \quad & ||x_\varepsilon|| \leq \gamma + \varepsilon \\ b.) \quad & f_i(x_\varepsilon) = \alpha_i \ i = 1, \cdots, n \end{aligned} \tag{6}$$

if and only if

$$\left|\sum_{i=1}^{n}\beta_i\alpha_i\right| \le \gamma \left\|\sum_{i=1}^{n}\beta_i f_i\right\| \tag{7}$$

for all $(\beta_1,\cdots,\beta_n) = \overrightarrow{\beta} \in \mathbb{C}^n$.

Proof: If Formula 6 holds, then

$$|\sum_{i=1}^{n}\beta_i\alpha_i| = |(\sum_{i=1}^{n}\beta_i f_i)(x_\varepsilon)| \le (\gamma+\varepsilon)\|\sum_{i=1}^{n}\beta_i f_i\|.$$

Since this holds for each $\varepsilon > 0$, Formula 7 follows.

Now suppose Formula 7. To begin with, suppose $\{f_1,\cdots,f_n\}$ is linearly independent. Define $\phi : X \to \mathbb{C}^n$ by $\phi(x) = (f_1(x),\cdots,f_n(x))^T$. Thus $\phi(X)$ is a finite dimensional closed subspace of $\mathbb{C}^n$. If ϕ is not onto, then there exists $\overrightarrow{\sigma} \in \mathbb{C}^n, \overrightarrow{\sigma} \ne \mathbf{0}$, such that $\left(\overrightarrow{\beta},\overrightarrow{\sigma}\right) = 0$ for all $\overrightarrow{\beta} \in \phi(X)$. Thus $(\overrightarrow{\sigma},\phi(x)) = 0$ for all x. This says $\sum_{i=1}^{n}\sigma_i f_i = 0$ so $\{f_1,\cdots,f_n\}$ is not linearly independent after all. Therefore, we can conclude ϕ is onto.

Let $B_\varepsilon = B(0,\gamma+\varepsilon)$. Then $\phi(B_\varepsilon)$ is an open (by Open Map theorem) and convex neighborhood of $\mathbf{0}$. We want to show that $\overrightarrow{\alpha} \in \phi(B_\varepsilon)$. If not, we apply Lemma 4.15 and the Riesz representation theorem, Theorem 4.10, to conclude there exists $\overrightarrow{\beta} \in \mathbb{C}^n$ such that

$$\text{Re}(\overrightarrow{\alpha},\overrightarrow{\beta}) > \text{Re}(\overrightarrow{\zeta},\overrightarrow{\beta})$$

for all $\overrightarrow{\zeta} \in \phi(B_\varepsilon)$. Thus, for all $x \in B_\varepsilon$

$$\text{Re}\left(\sum_{i=1}^{n}\alpha_i\overline{\beta}_i\right) > \text{Re}\left(\sum_{i=1}^{n}f_i(x)\overline{\beta}_i\right).$$

Let $|w| = 1$ and $w(\sum_{i=1}^{n}f_i(x)\overline{\beta}_i) = |\sum_{i=1}^{n}f_i(x)\overline{\beta}_i|$. Then $wx \in B_\varepsilon$ and for this $x \in B_\varepsilon$

$$\begin{aligned}|\sum_{i=1}^{n}f_i(x)\overline{\beta}_i| &= \sum_{i=1}^{n}f_i(wx)\overline{\beta}_i = \text{Re}(\sum_{i=1}^{n}f_i(wx)\overline{\beta}_i)\\ &< \text{Re}(\sum_{i=1}^{n}\alpha_i\overline{\beta}_i) \le |\sum_{i=1}^{n}\alpha_i\overline{\beta}_i|.\end{aligned}$$

Since x is arbitrary, we can take the sup over all such $x \in B_\varepsilon$, and write the following using Formula 7.

$$(\gamma+\varepsilon)\left\|\sum_{i=1}^{n}f_i\overline{\beta}_i\right\| = \sup\left\{\left|\left(\sum_{i=1}^{n}f_i\overline{\beta}_i\right)(x)\right| : \|x\| < \gamma+\varepsilon\right\}$$

$$\leq |\sum_{i=1}^{n} \alpha_i \overline{\beta}_i| \leq \gamma \|\sum_{i=1}^{n} f_i \overline{\beta}_i\|,$$

a contradiction because $\{f_1, \cdots, f_n\}$ is linearly independent. Thus $\overrightarrow{\alpha} \in \phi(B_\varepsilon)$ as desired. Hence a.) and b.) are obtained.

Now suppose $f_1, \cdots, f_m$ are linearly dependent. Then by renumbering the functions if necessary, we may obtain r such that $f_1, \cdots, f_r$ are linearly independent and for each $p = r+1, \cdots, m$, $f_p = \sum_{i=1}^{r} \beta_i^p f_i$ for some scalars $\beta_1^p, \cdots, \beta_r^p$. Suppose also Formula 7 holds for n replaced with m. Then condition 7 holds for each $p \leq m$ also. From the first part of the proof, there exists x_ε, $\|x_\varepsilon\| \leq \gamma + \varepsilon$ and $f_i(x_\varepsilon) = \alpha_i$ for $i = 1, \cdots, r$. Then $f_p(x_\varepsilon) = \alpha_p$ for $p = r+1, \cdots, m$ since

$$f_p(x_\varepsilon) = \sum_{i=1}^{r} \beta_i^p f_i(x_\varepsilon) = \sum_{i=1}^{r} \beta_i^p \alpha_i = \alpha_p.$$

This follows because by Formula 7 for p,

$$|\sum_{i=1}^{r} \beta_i^p \alpha_i - \alpha_p| \leq \gamma \|\sum_{i=1}^{r} \beta_i^p f_i - f_p\| = 0.$$

This proves the lemma.

Lemma 4.17 *(Corollary to Helly's Lemma) Let $\{f_1, \cdots, f_n\} \subseteq X'$ and let $x_0^{**} \in X''$ with $\|x_0^{**}\| = 1$. Then there exists $x \in X$ such that*
a.) $\|x\| \leq 1 + \varepsilon$
b.) $f_i(x) = x_0^{**}(f_i)$ $i = 1, \cdots, n$.

Proof: Let $\beta \in \mathbb{C}^n$. Using Theorem 3.30 of Chapter 3,

$$\|\sum_{i=1}^{n} \beta_i f_i\| = \sup\{|\sum_{i=1}^{n} \beta_i f_i(x)| : \|x\| \leq 1\} =$$

$$= \sup\{|\sum_{i=1}^{n} \beta_i x^{**}(f_i)| : \|x^{**}\| \leq 1\} \geq |\sum_{i=1}^{n} \beta_i x_0^{**}(f_i)|.$$

By Helly's Lemma with $\gamma = 1$, there exists an x satisfying a.) and b.).

Theorem 4.18 *(Milman) If X is uniformly convex, then X is reflexive.*

Proof: We need to show the James map, J, is onto. Let $x_0^{**} \in X''$, $\|x_0^{**}\| = 1$. (It is enough to show that all such elements of X'' are in the range of J.) Choose $f_n \in X'$, $n = 1, 2, \cdots$ such that

$$\|f_n\| = 1, \quad x_0^{**}(f_n) \to \|x_0^{**}\| = 1, \quad x_0^{**}(f_n) > 0. \tag{8}$$

Let f_0 be any element of X' and use Lemma 4.17 to obtain $x_n \in X$ with

$$||x_n|| \leq 1 + \frac{1}{n}, \quad f_i(x_n) = x_0^{**}(f_i) \quad i = 0, 1, \cdots, n. \tag{9}$$

Claim For all $\varepsilon > 0$ there exists N such that if $n, m \geq N$, then $||x_n + x_m|| > 2 - \varepsilon$.

Proof of Claim: If not, there exists $\alpha > 0$ and arbitrarily large values of m, n such that $||x_n + x_m|| \leq 2 - \alpha$. Pick N such that if $n > N$, then

$$x_0^{**}(f_n) > 1 - \eta$$

where $\eta > 0$ is arbitrary. Pick m, n, with $N < m \leq n$ and

$$||x_n + x_m|| \leq 2 - \alpha.$$

Then

$$2 - \alpha \geq ||x_n + x_m|| \geq f_m(x_n + x_m) = x_0^{**}(f_n) + x_0^{**}(f_m) \geq 2 - 2\eta,$$

a contradiction if $\eta < \frac{\alpha}{2}$. This proves the claim.

By the claim and uniform convexity, $\{x_n(1 + \frac{1}{n})^{-1}\}$ is a Cauchy sequence converging to $x \in X$ where $||x|| = 1$. Thus $x_n \to x$. From Formula 9, $f_i(x) = x_0^{**}(f_i)$ for all i.

Claim: There is at most one solution to

$$||x|| = 1, \quad f_i(x) = x_0^{**}(f_i) \quad i = 1, 2 \cdots. \tag{10}$$

Proof of Claim: Say x and $\hat{x}$ both solve Formula 10, then

$$2x_0^{**}(f_i) = f_i(x + \hat{x}) \leq ||x + \hat{x}|| < 2||x|| = 2.$$

Now let $i \to \infty$ and use Formula 8 to conclude

$$2 \leq ||x + \hat{x}|| < 2,$$

a contradiction. The strict inequality comes from the uniform convexity which implies that if $||x|| = ||\hat{x}|| = 1$ and if $x \neq \hat{x}$, then $||\frac{x + \hat{x}}{2}|| < 1$.

This claim shows that x does not depend on f_0 which was an arbitrary element of X'. Hence $f_0(x) = x_0^{**}(f_0)$ for all $f_0 \in X'$ so $x_0^{**} = Jx$. This proves Milman's theorem.

4.3   Exercises

1. Prove Examples 4.2 and 4.3 are Hilbert spaces. For $f, g \in C([0,1])$ let $(f, g) = \int_0^1 f(x) \overline{g(x)} dx$. Show this is an inner product. What does the Cauchy Schwarz inequality say in this context?

2. Suppose, in the definition of inner product, Condition 1 is weakened to read only $(x, x) \geq 0$. Thus the requirement that $(x, x) = 0$ if and only if $x = 0$ has been dropped. Show that then $|(x, y)| \leq |(x, x)|^{1/2} |(y, y)|^{1/2}$. This generalizes the usual Cauchy Schwarz inequality.

3. Prove the parallelogram identity. Next suppose $(X, \|\ \|)$ is a real normed linear space. Show that if the parallelogram identity holds, then $(X, \|\ \|)$ is actually an inner product space. That is, there exists an inner product $(\cdot, \cdot)$ such that $\|x\| = (x, x)^{1/2}$.

4. Show that every inner product space is uniformly convex.

5. Let H be a Hilbert space. $S \subseteq H$ is called an orthonormal set if $\|x\| = 1$ for all $x \in S$ and $(x, y) = 0$ if $x, y \in S$ and $x \neq y$. Show that every Hilbert space has a maximal orthonormal set. **Hint:** Use the Hausdorff maximal principle.

6. Let H be separable and let S be an orthonormal set. Show S is countable.

7. Suppose $\{x_1, \cdots, x_m\}$ is a linearly independent set of vectors in a normed linear space. Show $span(x_1, \cdots, x_m)$ is a closed subspace.

8. ↑ Prove Parseval's inequality, which says that if $\{x_n\}_{n=1}^{\infty}$ is an orthonormal set in H, then for all $x \in H$, $\|x\|^2 \geq \sum_{k=1}^{\infty} |(x, x_k)|^2$. **Hint:** Show that if $M = span(x_1, \cdots, x_n)$, then $Px = \sum_{k=1}^{n} x_k(x, x_k)$. Then observe $\|x\|^2 = \|x - Px\|^2 + \|Px\|^2$.

9. ↑ Show S is a maximal orthonormal set if and only if

$$span(S) \equiv \{\text{all finite linear combinations of elements of } S\}$$

is dense in H.

60

10. ↑ Suppose $\{x_n\}_{n=1}^{\infty}$ is a maximal orthonormal set. Show that

$$x = \sum_{n=1}^{\infty}(x, x_n)x_n \equiv \lim_{N \to \infty}\sum_{n=1}^{N}(x, x_n)x_n$$

and $||x||^2 = \sum_{i=1}^{\infty}|(x, x_i)|^2$. Also show $(x, y) = \sum_{n=1}^{\infty}(x, x_n)(\overline{y, x_n})$.

11. Let $S = \{e^{inx}(2\pi)^{-\frac{1}{2}}\}_{n=-\infty}^{\infty}$. Show S is an orthonormal set if the inner product is given by $(f, g) = \int_{-\pi}^{\pi} f(x)\overline{g(x)}dx$.

12. The nth Ceasaro mean, $\sigma_n f(x)$, is defined for a 2π periodic function f, $f(x + 2\pi) = f(x)$, as $\sigma_n f(x) = \frac{1}{n+1}\sum_{k=0}^{n}S_k f(x)$ (see Problems 7 and 8 in Chapter 3). Show that $\sigma_n f(x) = \int_{-\pi}^{\pi} f(x - y)F_n(y)dy$ where F_n, the Fejer kernel, satisfies $F_n(y) \geq 0$, $\int_{-\pi}^{\pi} F_n(y)dy = 1$, and if $|x| \geq \delta$, then $F_n(x) \to 0$ as $n \to \infty$. Show that if f is continuous and 2π periodic then $\sigma_n f$ converges uniformly to f for all x. Contrast this with Fourier series which can fail to even converge (see Problem 10 in Chapter 3).

13. ↑ Suppose H is a Hilbert space and suppose that $\{\phi_n\}_{n=1}^{\infty}$ is an orthonormal set in H. Let $S \subseteq H$ have the property that finite linear combinations of elements of S are dense in H. Suppose also that for each $h \in S$,

$$||h||^2 = \sum_{n=1}^{\infty}|(h, \phi_n)|^2.$$

Can we conclude that for all $g \in H$,

$$||g||^2 = \sum_{n=1}^{\infty}|(g, \phi_n)|^2?$$

Hint: You may want to consider Problem 10 at some point.

14. ↑ Show that if Parseval's equation,

$$||y||^2 = \sum_{n=1}^{\infty}|(y, \phi_n)|^2,$$

holds for all $y \in H$ where $\{\phi_n\}_{n=1}^{\infty}$ is an orthonormal set, then $\{\phi_n\}_{n=1}^{\infty}$ is a maximal orthonormal set and

$$\lim_{N \to \infty}\left|\left|y - \sum_{n=1}^{N}(y, \phi_n)\phi_n\right|\right| = 0.$$

15. Suppose X is an infinite dimensional Banach space and suppose

$$\{x_1 \cdots x_n\}$$

are linearly independent with $||x_i|| = 1$. Show $span\,(x_1 \cdots x_n) \equiv X_n$ is a closed linear subspace of X. Now let $z \notin X_n$ and pick $y \in X_n$ such that $||z - y|| \leq 2\,dist\,(z, X_n)$ and let

$$x_{n+1} = \frac{z - y}{||z - y||}.$$

Show the sequence $\{x_k\}$ satisfies $||x_n - x_k|| \geq 1/2$ whenever $k < n$.
Hint:

$$\left\| \frac{z - y}{||z - y||} - x_k \right\| = \left\| \frac{z - y - x_k\,||z - y||}{||z - y||} \right\|.$$

Now show the unit ball $\{x \in X : ||x|| \leq 1\}$ is compact if and only if X is finite dimensional.

Chapter 5
Calculus in Banach Space

5.1   The derivative

If f is a function of one variable, we say it has a derivative at x, $f'(x)$, if and only if

$$\lim_{v \to 0} \frac{|f(x+v) - f(x) - f'(x)v|}{|v|} = 0.$$

It is this idea which gives the correct generalization for the derivative of a function of more than one variable. More generally, we may consider a function f defined on an open subset of a Banach space X and ask for its derivative. In this chapter, X and Y will be Banach spaces whose field of scalars will be either $\mathbb{C}$ or $\mathbb{R}$, denoted by $\mathbb{F}$, and U will be an open set on which a function f mapping X to Y is defined. The main objective is to discuss the derivative and give a proof of the inverse and implicit function theorems. The inverse function theorem will be used later on in the book.

Definition 5.1 We say a function g is $o(v)$ if

$$\lim_{||v|| \to 0} \frac{g(v)}{||v||} = 0. \tag{1}$$

We say a function $f : U \to Y$ is differentiable at $x \in U$ if there exists a linear transformation $L \in \mathcal{L}(X, Y)$ such that

$$f(x+v) = f(x) + Lv + o(v).$$

Here the symbol $o(v)$ is used both as the name of that which is left over after approximating $f(x+v)$ with $f(x) + Lv$ and as a term which is descriptive of the behaviour of this function. This linear transformation L is the definition of $Df(x)$, the derivative.

Thus the definition means that the error,

$$f(x+v) - f(x) - Lv,$$

converges to 0 faster than $||v||$. The term $o(v)$ is notation that is descriptive of the behavior in Formula 1 and it is only this behavior that concerns us. Thus,

$$o(v) = o(v) + o(v), \; o(tv) = o(v), \; ko(v) = o(v)$$

and other similar observations hold. This notation is both sloppy and useful because it neglects details which are not important.

Theorem 5.2 *The derivative is well defined.*

  Proof: Suppose both L_1 and L_2 work in the above definition. Then let v be any vector and let t be a real scalar which is chosen small enough that $tv + x \in U$. Then

$$f(x + tv) = f(x) + L_1 tv + o(tv),$$

$$f(x + tv) = f(x) + L_2 tv + o(tv).$$

Therefore, subtracting these two yields

$$(L_2 - L_1)(tv) = o(t).$$

Note that $o(tv) = o(t)$ for fixed v. Therefore, dividing by t yields

$$(L_2 - L_1)(v) = \frac{o(t)}{t}.$$

Now let $t \to 0$ to conclude that $(L_2 - L_1)(v) = 0$. This proves the theorem.

Lemma 5.3 *Let f be differentiable at x. Then f is continuous at x and, in fact, there exists $K > 0$ such that whenever $\|v\|$ is small enough,*

$$\|f(x + v) - f(x)\| \leq K \|v\|.$$

  Proof:

$$f(x + v) - f(x) = Df(x)v + o(v).$$

Let $\delta > 0$ be small enough that if $\|v\| < \delta$, then

$$\|o(v)\| \leq \|v\|.$$

Then for $\|v\| < \delta$,

$$
\begin{aligned}
\|f(x + v) - f(x)\| &\leq \|Df(x)v\| + \|v\| \\
&\leq (\|Df(x)\| + 1)\|v\|.
\end{aligned}
$$

This proves the lemma.

Theorem 5.4 *(The chain rule) Let X, Y, and Z be Banach spaces, $U \subseteq X$ be an open set, and let $V \subseteq Y$ also be an open set. Suppose $f : U \to V$ is differentiable at x and suppose $g : V \to Z$ is differentiable at $f(x)$. Then $g \circ f$ is differentiable at x and*

$$D(g \circ f)(x) = D(g(f(x)))D(f(x)).$$

Proof: This follows from a computation. Let $B(x,r) \subseteq U$ and let r also be small enough that for

$$\|v\| \le r,$$

$f(x+v) \in V$. For such v, using the definition of differentiability of g and f,

$$g(f(x+v)) - g(f(x))$$

$$= Dg(f(x))(f(x+v) - f(x)) + o(f(x+v) - f(x))$$

$$= Dg(f(x))[Df(x)v + o(v)] + o(f(x+v) - f(x))$$

$$= D(g(f(x)))D(f(x))v + o(v) + o(f(x+v) - f(x)). \qquad (2)$$

Now by Lemma 5.3, letting $\epsilon > 0$, it follows that for v small enough,

$$\|o(f(x+v) - f(x))\| \le \epsilon \|f(x+v) - f(x)\| \le \epsilon K \|v\|.$$

Since $\epsilon > 0$ is arbitrary, this shows $o(f(x+v) - f(x)) = o(v)$. By Formula 2, this shows

$$g(f(x+v)) - g(f(x)) = D(g(f(x)))D(f(x))v + o(v)$$

which proves the theorem.

5.2 Finite dimensions

We have defined the derivative as a linear transformation. This means that in the case when X and Y are finite dimensional, we can consider the matrix of the linear transformation with respect to various bases on X and Y. In the case where $X = \mathbb{R}^n$ and $Y = \mathbb{R}^m$, we shall denote the matrix taken with respect to the standard basis vectors e_i, the vector with a 1 in the ith slot and zeros elsewhere, by $J\mathbf{f}(\mathbf{x})$. Thus, if the components of $\mathbf{v}$ with respect to the standard basis vectors are v_i,

$$J\mathbf{f}(\mathbf{x})_{ij} v_j = \pi_i(D\mathbf{f}(\mathbf{x})\mathbf{v}) \qquad (3)$$

where π_i is the projection onto the ith component of a vector in $Y = \mathbb{R}^m$. What are the entries of $J\mathbf{f}(x)$? Letting

$$\mathbf{f}(\mathbf{x}) = \sum_{i=1}^{m} f_i(\mathbf{x}) e_i,$$

$$f_i(\mathbf{x} + \mathbf{v}) - f_i(\mathbf{x}) = \pi_i(D\mathbf{f}(\mathbf{x})\mathbf{v}) + o(\mathbf{v}).$$

Thus, letting t be a small scalar,

$$f_i(\mathbf{x} + t e_j) - f_i(\mathbf{x}) = t\pi_i(D\mathbf{f}(\mathbf{x})e_j) + o(t).$$

Dividing by t, and letting $t \to 0$,

$$\frac{\partial f_i (\mathbf{x})}{\partial x_j} = \pi_i \left(D\mathbf{f} (\mathbf{x}) \, \mathbf{e}_j \right).$$

Thus, from Formula 3,

$$J\mathbf{f} (\mathbf{x})_{ij} = \frac{\partial f_i (\mathbf{x})}{\partial x_j}. \tag{4}$$

This proves the following theorem.

Theorem 5.5 *In the case where $X = \mathbb{R}^n$ and $Y = \mathbb{R}^m$, if $\mathbf{f}$ is differentiable at $\mathbf{x}$ then all the partial derivatives*

$$\frac{\partial f_i (\mathbf{x})}{\partial x_j}$$

exist and if $J\mathbf{f} (\mathbf{x})$ is the matrix of the linear transformation with respect to the standard basis vectors, then the ijth entry is given by Formula 4.

What if all the partial derivatives of $\mathbf{f}$ exist? Does it follow that $\mathbf{f}$ is differentiable? Consider the following function. $f : \mathbb{R}^2 \to \mathbb{R}$.

$$f (x, y) = \begin{cases} \frac{xy}{x^2 + y^2} & \text{if } (x, y) \neq (0, 0), \\ 0 & \text{if } (x, y) = (0, 0). \end{cases}$$

Then from the definition of partial derivatives, this function has both partial derivatives at $(0, 0)$. However f is not even continuous at $(0, 0)$ which may be seen by considering the behavior of the function along the line $y = x$ and along the line $x = 0$. By Lemma 5.3 this implies f is not differentiable.

Lemma 5.6 *Suppose $X = \mathbb{R}^n$, $f : U \to \mathbb{R}$ and all the partial derivatives of f exist and are continuous in U. Then f is differentiable in U.*

Proof: Let $B (\mathbf{x}, r) \subseteq U$ and let $||\mathbf{v}|| < r$. Then,

$$f (\mathbf{x} + \mathbf{v}) - f (\mathbf{x}) = \sum_{i=1}^{n} \left(f \left(\mathbf{x} + \sum_{j=1}^{i} v_j \mathbf{e}_j \right) - f \left(\mathbf{x} + \sum_{j=1}^{i-1} v_j \mathbf{e}_j \right) \right)$$

where

$$\sum_{i=1}^{0} v_j \mathbf{e}_j \equiv \mathbf{0}.$$

By the one variable mean value theorem,

$$f (\mathbf{x} + \mathbf{v}) - f (\mathbf{x}) = \sum_{i=1}^{n} \frac{\partial f \left(\mathbf{x} + \sum_{j=1}^{i-1} v_j \mathbf{e}_j + \theta_i v_i \mathbf{e}_i \right)}{\partial x_i} v_i$$

where $\theta_j \in [0,1]$. Therefore,

$$f(\mathbf{x}+\mathbf{v}) - f(\mathbf{x}) = \sum_{i=1}^{n} \frac{\partial f(\mathbf{x})}{\partial x_i} v_i +$$

$$\sum_{i=1}^{n} \left(\frac{\partial f\left(\mathbf{x}+\sum_{j=1}^{i-1} v_j \mathbf{e}_j + \theta_i v_i \mathbf{e}_i\right)}{\partial x_i} - \frac{\partial f(\mathbf{x})}{\partial x_i} \right) v_i.$$

Consider the last term.

$$\left| \sum_{i=1}^{n} \left(\frac{\partial f\left(\mathbf{x}+\sum_{j=1}^{i-1} v_j \mathbf{e}_j + \theta_j v_j \mathbf{e}_j\right)}{\partial x_i} - \frac{\partial f(\mathbf{x})}{\partial x_i} \right) v_i \right| \leq \left(\sum_{i=1}^{n} |v_i|^2 \right)^{1/2} \cdot$$

$$\left(\sum_{i=1}^{n} \left| \left(\frac{\partial f\left(\mathbf{x}+\sum_{j=1}^{i-1} v_j \mathbf{e}_j + \theta_j v_j \mathbf{e}_j\right)}{\partial x_i} - \frac{\partial f(\mathbf{x})}{\partial x_i} \right) \right|^2 \right)^{1/2}$$

and so it follows from continuity of the partial derivatives that this last term is $o(\mathbf{v})$. Therefore, we define

$$L\mathbf{v} \equiv \sum_{i=1}^{n} \frac{\partial f(\mathbf{x})}{\partial x_i} v_i$$

where

$$\mathbf{v} = \sum_{i=1}^{n} v_i \mathbf{e}_i.$$

Then L is a linear transformation which satisfies the conditions needed for it to equal $Df(\mathbf{x})$ and this proves the lemma.

Theorem 5.7 *Suppose $X = \mathbb{R}^n$, $Y = \mathbb{R}^m$ and $\mathbf{f} : U \to Y$ and suppose the partial derivatives*

$$\frac{\partial f_i}{\partial x_j}$$

all exist and are continuous in U. Then $\mathbf{f}$ is differentiable in U.

Proof: From Lemma 5.6,

$$f_i(\mathbf{x}+\mathbf{v}) - f_i(\mathbf{x}) = Df_i(\mathbf{x})\mathbf{v} + o(\mathbf{v}).$$

Letting

$$(D\mathbf{f}(\mathbf{x})\mathbf{v})_i \equiv Df_i(\mathbf{x})\mathbf{v},$$

we see that

$$\mathbf{f}(\mathbf{x}+\mathbf{v}) - \mathbf{f}(\mathbf{x}) = D\mathbf{f}(\mathbf{x})\mathbf{v} + o(\mathbf{v})$$

and this proves the theorem.

When all the partial derivatives exist and are continuous we say the function is a C^1 function. More generally, we give the following definition.

Definition 5.8 In the case where X and Y are Banach spaces, and $U \subseteq X$ is an open set, we say $\mathbf{f} : U \to Y$ is $C^1(U)$ if $\mathbf{f}$ is differentiable and the mapping

$$\mathbf{x} \to D\mathbf{f}(\mathbf{x}),$$

is continuous as a function from U to $\mathcal{L}(X, Y)$.

5.3 Higher order derivatives

If $f : U \to Y$, then

$$x \to Df(x)$$

is a mapping from U to $\mathcal{L}(X, Y)$, a normed linear space. The following is the definition of the second derivative. Simply stated, it is just the derivative of the first derivative.

Definition 5.9 Define

$$D^2 f(x) \equiv D(Df(x)).$$

Thus,

$$Df(x+v) - Df(x) = D^2 f(x) v + o(v).$$

This implies

$$D^2 f(x) \in \mathcal{L}(X, \mathcal{L}(X, Y)), \ D^2 f(x)(u)(v) \in Y,$$

and the map

$$(u,v) \to D^2 f(x)(u)(v)$$

is a bilinear map having values in Y. The same pattern applies to taking higher order derivatives. Thus,

$$D^3 f(x) \equiv D(D^2 f(x))$$

and we can consider $D^3 f(x)$ as a trilinear map. Also, instead of writing

$$D^2 f(x)(u)(v),$$

we sometimes write

$$D^2 f(x)(u,v).$$

We say f is $C^k(U)$ if f and its first k derivatives are all continuous. For example, for f to be $C^2(U)$,

$$x \to D^2 f(x)$$

would have to be continuous as a map from U to $\mathcal{L}(X, \mathcal{L}(X, Y))$. The following theorem deals with the question of symmetry of the map $D^2 f$.

Theorem 5.10 *Suppose $D^2 f(x)$ exists for all $x \in U$ and $D^2 f$ is continuous at $x \in U$. Then*

$$D^2 f(x)(u)(v) = D^2 f(x)(v)(u).$$

Proof: Let $B(x, r) \subseteq U$ and let $t, s \in [0, r/2]$. Now let $L \in Y'$ and define

$$\Delta(s, t) \equiv \frac{\operatorname{Re} L}{st} \{f(x + tu + sv) - f(x + tu) - (f(x + sv) - f(x))\}. \quad (5)$$

Let $h(t) = \operatorname{Re} L(f(x + sv + tu) - f(x + tu))$. Then by the mean value theorem,

$$\begin{aligned}
\Delta(s, t) &= \frac{1}{st}(h(t) - h(0)) = \frac{1}{st}h'(\alpha t)t \\
&= \frac{1}{s}(\operatorname{Re} LDf(x + sv + \alpha tu)u - \operatorname{Re} LDf(x + \alpha tu)u).
\end{aligned}$$

Applying the mean value theorem again,

$$\Delta(s, t) = \operatorname{Re} LD^2 f(x + \beta sv + \alpha tu)(v)(u)$$

where $\alpha, \beta \in (0, 1)$. If the terms $f(x + tu)$ and $f(x + sv)$ are interchanged in Formula 5, $\Delta(s, t)$ is also unchanged and the above argument shows there exist $\gamma, \delta \in (0, 1)$ such that

$$\Delta(s, t) = \operatorname{Re} LD^2 f(x + \gamma sv + \delta tu)(u)(v).$$

Letting $(s, t) \to (0, 0)$ and using the continuity of $D^2 f$ at x,

$$\lim_{(s,t) \to (0,0)} \Delta(s, t) = \operatorname{Re} LD^2 f(x)(u)(v) = \operatorname{Re} LD^2 f(x)(v)(u).$$

By Lemma 3.31 of Chapter 3, there exists $L \in Y'$ such that

$$L\left(D^2 f(x)(u)(v) - D^2 f(x)(v)(u)\right)$$

$$= \left\|D^2 f(x)(u)(v) - D^2 f(x)(v)(u)\right\|$$

For this L,

$$0 = \operatorname{Re} L\left(D^2 f(x)(u)(v) - D^2 f(x)(v)(u)\right)$$

$$= L\left(D^2 f\left(x\right)\left(u\right)\left(v\right) - D^2 f\left(x\right)\left(v\right)\left(u\right)\right)$$
$$= \left\|D^2 f\left(x\right)\left(u\right)\left(v\right) - D^2 f\left(x\right)\left(v\right)\left(u\right)\right\|$$

and this proves the theorem.

Consider the important special case when $X = \mathbb{R}^n$ and $Y = \mathbb{R}$. If e_i are the standard basis vectors, what is

$$D^2 f\left(\mathbf{x}\right)\left(e_i\right)\left(e_j\right)?$$

To see what this is, use the definition to write

$$D^2 f\left(\mathbf{x}\right)\left(e_i\right)\left(e_j\right) = t^{-1} s^{-1} D^2 f\left(\mathbf{x}\right)\left(te_i\right)\left(se_j\right)$$
$$= t^{-1} s^{-1} \left(Df\left(\mathbf{x}+te_i\right) - Df\left(\mathbf{x}\right) + o\left(t\right)\right)\left(se_j\right)$$
$$= t^{-1} s^{-1} \left(f\left(\mathbf{x}+te_i + se_j\right) - f\left(\mathbf{x}+te_i\right)\right.$$
$$\left. +o\left(s\right) - \left(f\left(\mathbf{x}+se_j\right) - f\left(\mathbf{x}\right) + o\left(s\right)\right) + o\left(t\right)\right).$$

First let $s \to 0$ to get

$$t^{-1}\left(\frac{\partial f}{\partial x_j}\left(\mathbf{x}+te_i\right) - \frac{\partial f}{\partial x_j}\left(\mathbf{x}\right) + o\left(t\right)\right)$$

and then let $t \to 0$ to obtain

$$D^2 f\left(\mathbf{x}\right)\left(e_i\right)\left(e_j\right) = \frac{\partial^2 f}{\partial x_i \partial x_j}\left(\mathbf{x}\right).$$

Thus the theorem asserts that in this special case the mixed partial derivatives are equal at $\mathbf{x}$ if they are defined near $\mathbf{x}$ and continuous at $\mathbf{x}$.

5.4 Inverse function theorem

To begin with we give a simple definition.

Definition 5.11 Let $T : X \times Y \to Z$, and denote an element of $X \times Y$ by (x, y) where $x \in X$ and $y \in Y$. Then the map $x \to T\left(x, y\right)$ is a function from X to Z. When this map is differentiable, we denote its derivative by

$$D_x T\left(x, y\right).$$

Thus,

$$T\left(x + v, y\right) - T\left(x, y\right) = D_x T\left(x, y\right) v + o\left(v\right).$$

A similar definition holds for the symbol $D_y T$.

The following lemma is very useful. It is a generalization of the familiar formula for the sum of a geometric series.

70

Lemma 5.12 *Let $A \in \mathcal{L}(X, X)$ and suppose $\|A\| \leq r < 1$. Then*

$$(I - A)^{-1} \quad \text{exists}$$

and

$$\left\| (I - A)^{-1} \right\| \leq (1 - r)^{-1}.$$

Proof: Consider

$$B_k \equiv \sum_{i=0}^{k} A^i.$$

Then if $N < l < k$,

$$\|B_k - B_l\| \leq \sum_{i=N}^{k} \|A^i\| \leq \sum_{i=N}^{k} \|A\|^i \leq \frac{r^N}{1 - r}.$$

It follows B_k is a Cauchy sequence and so it converges to $B \in \mathcal{L}(X, X)$. Also,

$$(I - A) B_k = I - A^{k+1} = B_k (I - A)$$

and so

$$I = \lim_{k \to \infty} (I - A) B_k = (I - A) B, \ I = \lim_{k \to \infty} B_k (I - A) = B (I - A).$$

Thus

$$(I - A)^{-1} = B = \sum_{i=0}^{\infty} A^i.$$

It follows

$$\left\| (I - A)^{-1} \right\| \leq \sum_{i=1}^{\infty} \|A^i\| \leq \sum_{i=0}^{\infty} \|A\|^i \leq \frac{1}{1 - r}.$$

This proves the lemma.

The inverse function theorem gives conditions under which a function has a local inverse which is as smooth as the original function. It is one of the most important theorems in analysis.

Theorem 5.13 *(Inverse Function theorem) Let $x_0 \in U$, an open set in X, and let $f : U \longrightarrow Y$. Suppose*

$$f \text{ is } C^1(U), \quad \text{and} \quad Df(x_0)^{-1} \in \mathcal{L}(Y, X).$$

Then there exist open sets, W and V, such that

$$x_0 \in W \subseteq U,$$

$$f : W \longrightarrow V \text{ is } 1 - 1 \text{ and onto,}$$

71

$$f^{-1} \text{ is } C^1,$$

$$Df^{-1}(y) = \left(Df\left(f^{-1}(y) \right) \right)^{-1}. \tag{6}$$

Proof: Let

$$T(x, y) = x - Df(x_0)^{-1}(f(x) - y).$$

Thus $T(x, y) = x$ if and only if $f(x) = y$.

Lemma 5.14 *There exists $\delta > 0$ such that if $||x - x_0|| < \delta$, then*

$$||D_x T(x, y)|| < 1/2, \tag{7}$$

$$(Df(x))^{-1} \text{ exists }, \ ||Df(x)^{-1}|| \le 2||Df(x_0)^{-1}||, \tag{8}$$

$$||Df(x_1)^{-1} - Df(x_2)^{-1}|| \le 4||Df(x_0)^{-1}||^2||Df(x_1) - Df(x_2)|| \tag{9}$$

for $||x_i - x_0|| < \delta$, $i = 1, 2$.

Proof: $D_x T(x, y) = I - Df(x_0)^{-1}Df(x)$. Thus $D_x T(x_0, y) = 0$. By continuity of Df, there exists $\delta > 0$ such that if $||x - x_0|| < \delta$, then Formula 7 holds and also

$$0 < \delta < \text{distance}(x_0, U^C).$$

Now for $||x_0 - x|| < \delta$,

$$Df(x) = Df(x_0)(I - D_x T(x, y)).$$

Since $||D_x T(x, y)|| < 1/2$, Lemma 5.12 implies

$$(I - D_x T(x, y))^{-1} = \sum_{n=0}^{\infty} (D_x T(x, y))^n$$

and

$$||(I - D_x T(x, y))^{-1}|| \le 2. \tag{10}$$

Thus, $(Df(x))^{-1}$ exists and

$$Df(x)^{-1} = (I - D_x T(x, y))^{-1} Df(x_0)^{-1}$$

$$||Df(x)^{-1}|| \le 2||Df(x_0)^{-1}||. \tag{11}$$

This establishes Formula 8. To establish Formula 9, observe that if $A, B \in \mathcal{L}(X, X)$,

$$||A||, ||B|| < 1/2,$$

then

$$(I - A)^{-1} - (I - B)^{-1} = (I - A)^{-1}(A - B)(I - B)^{-1}.$$

Consequently Formulas 10 - 11 imply

$$||Df(x_1)^{-1} - Df(x_2)^{-1}|| \leq$$

$$||(I - D_xT(x_1, y))^{-1} - (I - D_xT(x_2, y))^{-1}|| \, ||| Df(x_0)^{-1}||$$
$$\leq \; 4||D_xT(x_1, y) - D_xT(x_2, y)|| \, ||| Df(x_0)^{-1}||$$
$$\leq \; 4||Df(x_1) - Df(x_2)|| \, ||| Df(x_0)^{-1}||^2.$$

This proves the lemma.

Up until now, there is no restriction placed on y. We will place a restriction on y now. Let $y_0 \equiv f(x_0)$, let

$$|y - y_0| < \delta(4||Df(x_0)^{-1}||)^{-1}, \tag{12}$$

and let

$$T_y(x) \equiv T(x, y).$$

Lemma 5.15 *For all $n = \{1, 2, \cdots\}$,*

$$|T_y^n(x_0) - x_0| \leq \frac{\delta}{2}, \tag{13}$$

$$|T_y^{n+1}(x_0) - T_y^n(x_0)| \leq \frac{1}{2^n}|T_y(x_0) - x_0| \leq \frac{1}{2^n}(\frac{\delta}{4}), \tag{14}$$

$$|T_y(x_1) - T_y(x_2)| \leq \frac{1}{2}|x_1 - x_2| \tag{15}$$

for every pair $x_1, x_2 \in B(x_0, \delta)$.

Proof: To obtain Formula 15, let $x^* \in X'$ and define

$$h(t) \equiv x^* \left(T(x_1 + t(x_2 - x_1), y) \right).$$

Thus

$$h'(t) = x^* \left(D_xT(x_1 + t(x_2 - x_1), y)(x_2 - x_1) \right)$$

and so for J the James map,

$$|J(T_y(x_1) - T_y(x_2))(x^*)| = |x^* (T_y(x_1) - T_y(x_2))|$$

$$= |h(1) - h(0)| \leq \int_0^1 |x^* (D_xT(x_1 + t(x_2 - x_1), y)(x_2 - x_1))| \, dt$$

$$\leq 2^{-1} ||x_2 - x_1|| \, ||x^*||.$$

Thus

$$||J(T_y(x_1) - T_y(x_2))|| \leq 2^{-1} ||x_2 - x_1||.$$

By Theorem 3.30 of Chapter 3, it follows that

$$\|T_y(x_1) - T_y(x_2)\| \leq 2^{-1}\|x_2 - x_1\|.$$

Now the first inequality in Formula 14 is obvious from iterating Formula 15. The second follows from the definition of T and the assumption 12 on $\|y - y_0\|$. To get Formula 13,

$$
\begin{aligned}
|T_y^n(x_0) - x_0| &\leq \sum_{k=1}^{n} |T_y^k(x_0) - T_y^{k-1}(x_0)| \\
&\leq \sum_{k=1}^{n} 2^{-k+1}\frac{\delta}{4} \leq \frac{\delta}{2}.
\end{aligned}
$$

This proves the Lemma.

Lemma 5.16   *Let*

$$W = f^{-1}(B(y_0, \delta(4\|Df(x_0)^{-1}\|)^{-1})) \cap B(x_0, \delta)$$

and let

$$V = B(y_0,\ \delta(4\|Df(x_0)^{-1}\|)^{-1}).$$

Then $f(W) = V$, and f is 1-1 on W.

Proof: Consider $\{T_y^n(x_0)\}_{n=1}^{\infty}$ for $y \in V$. By the first two parts of Lemma 5.15, $\{T_y^n(x_0)\}$ is a Cauchy sequence converging to a point $x \in \overline{B(x_0, \delta)}$. (In fact, $x \in \overline{B(x_0, \frac{\delta}{2})}$.) Then

$$T_y x = T_y \lim_{n\to\infty} T_y^n(x_0) = \lim_{n\to\infty} T_y^{n+1}(x_0) = x.$$

From the definition of T_y,

$$f(x) = y.$$

This shows f maps W onto V. If $f(x_1) = y = f(x_2)$, then from Lemma 5.15,

$$\|x_1 - x_2\| = \|T_y x_1 - T_y x_2\| \leq \frac{1}{2}\|x_1 - x_2\|.$$

Thus $x_1 = x_2$ so f is 1-1 also.

It remains to verify that f^{-1} is C^1. In order to do this, we first verify f^{-1} is Lipschitz.

Lemma 5.17   *f^{-1} is Lipschitz continuous on V.*

Proof: Let $y_1, y_2 \in V$.

$$\|f^{-1}(y_1) - f^{-1}(y_2)\| = \|T(f^{-1}(y_1), y_1) - T(f^{-1}(y_2), y_2)\|$$

$$\leq \|T(f^{-1}(y_1), y_1) - T(f^{-1}(y_2), y_1)\|$$
$$+\|T(f^{-1}(y_2), y_1) - T(f^{-1}(y_2), y_2)\|$$
$$\leq \frac{1}{2}\|f^{-1}(y_1) - f^{-1}(y_2)\| + \|(Df(x_0))^{-1}\| \, \|y_1 - y_2\|.$$

It follows

$$\|f^{-1}(y_1) - f^{-1}(y_2)\| \leq 2\|(Df(x_0))^{-1}\| \, \|y_1 - y_2\|.$$

This proves the lemma.

All that remains is to prove f^{-1} is C^1. Let $y, y + z \in V$.

$$f(f^{-1}(y+z)) - f(f^{-1}(y)) \quad = \quad Df(f^{-1}(y))(f^{-1}(y+z) - f^{-1}(y)) +$$
$$+o(f^{-1}(y+z) - f^{-1}(y)).$$

Thus, since f^{-1} is Lipschitz,

$$z = Df(f^{-1}(y))(f^{-1}(y+z) - f^{-1}(y)) + o(z).$$

By Lemma 5.14, and Formula 8,

$$f^{-1}(y+z) - f^{-1}(z) \quad = \quad (Df(f^{-1}(y))^{-1}(z) + Df(f^{-1}(y))^{-1}o(z)$$
$$= \quad (Df(f^{-1}(y))^{-1}(z) + o(z).$$

Thus

$$Df^{-1}(y) = (Df(f^{-1}(y)))^{-1}$$

which verifies Formula 6. By Lemma 5.14 again,

$$\|(Df^{-1}(y_1) - (Df^{-1}(y_2)\| = \|(Df(f^{-1}(y_1))^{-1} - (Df(f^{-1}(y_2))^{-1}\|$$
$$\leq 4\|Df(x_0)^{-1}\|^2\|Df(f^{-1}(y_1)) - Df(f^{-1}(y_2))\|.$$

Since f is $C^1(U)$ and f^{-1} is continuous, this shows f^{-1} is $C^1(V)$ as claimed. This proves the inverse function theorem.

Next we show that f^{-1} inherits higher order derivatives from f. To do this, it is necessary to consider the differentiability of the map which takes a linear operator to its inverse.

Lemma 5.18 *Let*

$$O \equiv \{A \in \mathcal{L}(X,Y) : A^{-1} \in \mathcal{L}(Y,X)\}$$

and let

$$\mathfrak{I}: O \to \mathcal{L}(Y,X), \ \mathfrak{I}A \equiv A^{-1}.$$

Then O is open and $\mathfrak{I}$ is in $C^m(O)$ for all $m = 1, 2, \cdots$. Also

$$D\mathfrak{I}(A)(B) = -\mathfrak{I}(A)(B)\mathfrak{I}(A). \tag{16}$$

75

Proof: Let $A \in O$ and let $B \in \mathcal{L}(X, Y)$ with

$$\|B\| \leq \frac{1}{2} \|A^{-1}\|^{-1}.$$

Then

$$\|A^{-1}B\| \leq \|A^{-1}\| \|B\| \leq \frac{1}{2}$$

and so by Lemma 5.12,

$$\left(I + A^{-1}B\right)^{-1} \in \mathcal{L}(X, X).$$

Thus

$$(A + B)^{-1} = \left(I + A^{-1}B\right)^{-1} A^{-1} = \sum_{n=0}^{\infty} (-1)^n \left(A^{-1}B\right)^n A^{-1}$$

$$= \left[I - A^{-1}B + o(B)\right] A^{-1}$$

which shows that O is open and, also,

$$\begin{aligned}
\Im(A + B) - \Im(A) &= \sum_{n=0}^{\infty} (-1)^n \left(A^{-1}B\right)^n A^{-1} - A^{-1} \\
&= -A^{-1}BA^{-1} + o(B) \\
&= -\Im(A)(B)\Im(A) + o(B)
\end{aligned}$$

which demonstrates Formula 16. It follows from this that we can continue taking derivatives of $\Im$. For $\|B_1\|$ small,

$$-\left[D\Im(A + B_1)(B) - D\Im(A)(B)\right] =$$

$$\Im(A + B_1)(B)\Im(A + B_1) - \Im(A)(B)\Im(A)$$

$$= \quad \Im(A + B_1)(B)\Im(A + B_1) - \Im(A)(B)\Im(A + B_1) + \\
\Im(A)(B)\Im(A + B_1) - \Im(A)(B)\Im(A)$$

$$= \left[\Im(A)(B_1)\Im(A) + o(B_1)\right](B)\Im(A + B_1) + \\
\Im(A)(B)\left[\Im(A)(B_1)\Im(A) + o(B_1)\right]$$

$$= \quad \left[\Im(A)(B_1)\Im(A) + o(B_1)\right](B)\left[A^{-1} - A^{-1}B_1A^{-1}\right] + \\
\Im(A)(B)\left[\Im(A)(B_1)\Im(A) + o(B_1)\right]$$

$$= \Im(A)(B_1)\Im(A)(B)\Im(A) + \Im(A)(B)\Im(A)(B_1)\Im(A) + o(B_1)$$

and so

$$D^2\Im(A)(B_1)(B) = \Im(A)(B_1)\Im(A)(B)\Im(A) + \Im(A)(B)\Im(A)(B_1)\Im(A)$$

which shows $\mathfrak{I}$ is $C^2(O)$. Clearly we can continue in this way which shows $\mathfrak{I}$ is in $C^m(O)$ for all $m = 1, 2, \cdots$.

Corollary 5.19 *In the inverse function theorem, assume*

$$f \in C^m(U), m \geq 1.$$

Then

$$f^{-1} \in C^m(V).$$

Proof:
$$Df^{-1}(y) = \mathfrak{I}\left(Df\left(f^{-1}(y)\right)\right).$$

Now by Lemma 5.18, and the chain rule,

$$D^2 f^{-1}(y)(B) = -\mathfrak{I}\left(Df\left(f^{-1}(y)\right)\right)(B)\mathfrak{I}\left(Df\left(f^{-1}(y)\right)\right) \cdot$$
$$D^2 f\left(f^{-1}(y)\right) Df^{-1}(y)$$
$$= -\mathfrak{I}\left(Df\left(f^{-1}(y)\right)\right)(B)\mathfrak{I}\left(Df\left(f^{-1}(y)\right)\right) D^2 f\left(f^{-1}(y)\right) \cdot$$
$$\mathfrak{I}\left(Df\left(f^{-1}(y)\right)\right).$$

Continuing in this way we see that it is possible to continue taking derivatives up to order m. This proves the corollary.

The next theorem is called the implicit function theorem. It considers the question of existence and smoothness of implicitly defined functions. First note that if X, Y are normed linear spaces we can make $X \times Y$ into a normed linear space by defining the norm to be

$$\|(x,y)\| \equiv \max\left(\|x\|, \|y\|\right). \tag{17}$$

Theorem 5.20 *Let X, Y, Z be Banach spaces, let U be an open set in X and V be an open set in Y. Suppose $f : U \times V \to Z$ is C^m and $f(x_0, y_0) = 0$ for some $(x_0, y_0) \in U \times V$. Suppose $(D_x f(x_0, y_0))^{-1} \in \mathcal{L}(Z, X)$. Then there exists an open set A in Y and an open set B in $X \times Y$, $(x_0, y_0) \in B$ and a mapping $\theta : A \to X$ such that*

$$y_0 \in A$$

$$\theta \in C^m(A)$$

$$f(\theta(y), y) = 0, \quad y \in A$$

$$If \quad f(x, y) = 0 \ for \ (x, y) \in B, \ then \ x = \theta(y).$$

Proof: Consider $\mathbf{F}(x, y) = (f(x, y), y)^T$. Then it is routine to see that

$$\mathbf{F} : U \times V \to Z \times Y,$$

$$\mathbf{F} \in C^m \left(U \times V \right),$$

and

$$DF(x, y) = \begin{pmatrix} D_x f(x, y) & D_y f(x, y) \\ 0 & I \end{pmatrix}.$$

It follows from the assumptions on $D_x f(x_0, y_0)$ that $DF(x_0, y_0)$ is one to one, onto, and continuous. By the open mapping theorem,

$$\left(DF(x_0, y_0) \right)^{-1} \in \mathcal{L} \left(Z \times Y, X \times Y \right)$$

and we may apply the inverse function theorem. By this theorem, there exists an open set in $Z \times Y$, $\widetilde{W}$ containing $(0, y_0)$ and an open set $\tilde{B}$ in $X \times Y$ containing (x_0, y_0) such that $\mathbf{F}^{-1}$ is a C^m map from $\widetilde{W}$ to $\tilde{B}$. Let

$$B \left((0, y_0), r \right) \subseteq \widetilde{W}$$

where the norm is defined as in Formula 17 and let

$$B \equiv \mathbf{F}^{-1} \left(B \left((0, y_0), r \right) \right).$$

Then from the definition of the norm in $Z \times Y$, there are open sets $W \subseteq Z$ containing 0 and an open set $A \subseteq Y$ such that $y_0 \in A$ and

$$B \left((0, y_0), r \right) = W \times A.$$

Let $\theta \left(y \right) = \pi_1 \left(\mathbf{F}^{-1} \left(0, y \right) \right)$ where $\pi_1 \left(x, y \right) \equiv x$. Thus

$$\begin{pmatrix} f \left(\theta \left(y \right), y \right) \\ y \end{pmatrix} = \begin{pmatrix} 0 \\ y \end{pmatrix}$$

and θ is a C^m map. If $f(x, y) = 0$ for $(x, y) \in B$, then

$$\mathbf{F} \left(x, y \right) = \begin{pmatrix} f \left(x, y \right) \\ y \end{pmatrix} = \begin{pmatrix} 0 \\ y \end{pmatrix} = \mathbf{F} \left(\theta(y), y \right)$$

which shows that $\theta \left(y \right) = x$ since $\mathbf{F}$ is one to one on B. This proves the theorem.

Thus the relation $f(x, y) = 0$ defines x in terms of y near (x_0, y_0), $x = \theta(y)$ and θ is C^m.

5.5 Ordinary differential equations

In the theory of differential equations, it is important to consider questions of existence, uniqueness and continuous dependence of solutions on initial data and parameters. In this section we give an application of the implicit function theorem to these questions. To do so, we define a few Banach spaces. Let Z be a Banach space and let $C^k\left([-\alpha, \alpha]; Z\right)$ be the vector space of functions which have k continuous derivatives on $[-\alpha, \alpha]$ where the derivative is the right or left derivative as appropriate at the end points of the interval. Then we leave as an exercise to show that $C^k\left([-\alpha, \alpha]; Z\right)$ is a Banach space if the norm is defined as

$$||x||_k \equiv \sum_{i=1}^{k} \left|\left|x^{(i)}\right|\right|_\infty$$

where $x^{(i)} \equiv D^i x$ and

$$||y||_\infty \equiv \max\left\{||y(t)||_Z : t \in [-\alpha, \alpha]\right\}.$$

Theorem 5.21 *Let Λ and Z be Banach spaces and let W be an open subset of $\mathbb{R} \times Z \times \Lambda$ containing $(0, y_0, \lambda)$. Also let $f : W \rightarrow Z$ be a C^k function. Then there exists a unique solution, $y = y(y_0, \lambda)$, to the initial value problem*

$$y' = f(t, y, \lambda), \; y(0) = y_0, \tag{18}$$

valid for $t \in [-\alpha, \alpha]$ where $\alpha > 0$ depends on y_0 and λ. Furthermore, the map

$$(t, y_0, \lambda) \rightarrow y(y_0, \lambda)(t) \tag{19}$$

is C^k on W and for fixed $t \in [-\alpha, \alpha]$,

$$(y_0, \lambda) \rightarrow y(y_0, \lambda)^{(i)}(t) \tag{20}$$

is a C^k function for each $i = 1, \cdots, k+1$.

Proof: Let $\alpha s = t$ and define $\phi(s) \equiv y(t) - y_0$. Then y is a solution to Formula 18 for $t \in [-\alpha, \alpha]$ if and only if ϕ is a solution for $s \in [-1, 1]$ to the equations

$$\phi'(s) = \alpha f(\alpha s, \phi(s) + y_0, \lambda), \; \phi(0) = 0. \tag{21}$$

Let $0 \leq l \leq k$ and define

$$X \equiv \left\{\phi \in C^{l+1}\left([-1, 1]; Z\right) : \phi(0) = 0\right\}, \; Y \equiv C^l\left([-1, 1]; Z\right).$$

Now define

$$\widetilde{W} \equiv \left\{(\alpha, \widehat{y_0}, \mu, \phi) \in \mathbb{R} \times Z \times \Lambda \times X : \right.$$

$$\left. \text{for } s \in [-1, 1], (s\alpha, \widehat{y_0} + \phi(s), \mu) \in W\right\}.$$

For a given $(\alpha, \widehat{y_0}, \mu, \phi) \in \widetilde{W}$,

$$\left\{(s\alpha, \widehat{y_0} + \phi(s), \mu) : s \in [-1, 1]\right\}$$

is a compact subset of W. Consequently, the distance from this set to W^C is positive and so if $(\beta, y_0, \lambda, \psi)$ is sufficiently close to $(\alpha, \widehat{y_0}, \mu, \phi)$ in $\mathbb{R} \times Z \times \Lambda \times X$ it follows $(\beta, y_0, \lambda, \psi)$ is also in $\widetilde{W}$. This shows $\widetilde{W}$ is an open subset of $\mathbb{R} \times Z \times \Lambda \times X$. Now define $F : \widetilde{W} \to Y$ by

$$F(\alpha, \widehat{y_0}, \mu, \phi)(s) \equiv \phi'(s) - \alpha f(\alpha s, \phi(s) + \widehat{y_0}, \mu).$$

Then

$$F(0, y_0, \lambda, 0) = 0,$$

and F is C^k. Also

$$D_4 F(0, y_0, \lambda, 0) \psi = \psi'$$

and so $D_4 F(0, y_0, \lambda, 0) \in \mathcal{L}(X, Y)$, is one to one, onto and continuous. By the open mapping theorem, its inverse is also continuous. Therefore, the conditions of the implicit function theorem are satisfied and so there exists $r > 0$ such that if

$$|\alpha| + ||\mu - \lambda|| + ||\widehat{y_0} - y_0|| < r,$$

then there exists a unique $\phi \in X$ such that

$$F(\alpha, \widehat{y_0}, \mu, \phi) = 0,$$

and ϕ is a C^k function of $(\alpha, \widehat{y_0}, \mu)$. Letting $l = 0$, and fixing $\alpha < r$, this proves uniqueness and existence since every solution of Formula 21 must be a C^1 function. Letting $l = k$, and restricting r to apply in both cases $l = 0$ and $l = k$, then taking $\alpha < r$, it follows

$$(\widehat{y_0}, \mu) \to y(\widehat{y_0}, \mu)$$

is C^k. It follows that for $t \in [-\alpha, \alpha]$,

$$(\widehat{y_0}, \mu) \to y(\widehat{y_0}, \mu)(t)$$

is C^k and $t \to y(\widehat{y_0}, \mu)(t)$ is C^{k+1}. This implies Formula 19 and Formula 20 which proves the theorem.

5.6   Exercises

1. Suppose $L \in \mathcal{L}(X, Y)$ where X and Y are two finite dimensional vector spaces and suppose L is one to one. Show there exists $r > 0$ such that for all $x \in X$,

$$|Lx| \geq r |x|.$$

Hint: Define $|x|_1 \equiv |Lx|$, observe that $|\cdot|_1$ is a norm and then use the theorem proved earlier that all norms are equivalent in a finite dimensional normed linear space.

2. Let X and Y be Banach spaces and let U be an open subset of X. Let $f : U \to Y$ be $C^1(U)$, let $x_0 \in U$, and $\delta > 0$ be given. Show there exists $\epsilon > 0$ such that if $x_1, x_2 \in B(x_0, \epsilon)$, then

$$|f(x_1) - f(x_2) - Df(x_0)(x_1 - x_2)| \leq \delta |x_1 - x_2|.$$

3. ↑ Let U be an open subset of X, $f : U \to Y$ where X, Y are finite dimensional normed linear spaces and suppose $f \in C^1(U)$ and $Df(x_0)$ is one to one. Then show f is one to one near x_0. **Hint:** Use Problem 1 and Problem 2.

4. Suppose U is an open set in a Banach space X, and V is an open set in a Banach space V. Also suppose $f : U \to V$ is one to one, onto and both f and f^{-1} are C^1. Show this implies that for each $x \in U$, $Df(x)$ is one to one and onto. **Hint:** Consider $f \circ f^{-1}$ and $f^{-1} \circ f$ and use the chain rule.

5. Suppose $M \in \mathcal{L}(X, Y)$ where X and Y are finite dimensional linear spaces and suppose M is onto. Show there exists $L \in \mathcal{L}(Y, X)$ such that

$$LMx = Px$$

for all $x \in P(X)$ where $P \in \mathcal{L}(X, X)$, and $P^2 = P$. **Hint:** Let $\{y_1 \cdots y_n\}$ be a basis of Y and let $Mx_i = y_i$. Then define

$$Ly = \sum_{i=1}^{n} \alpha_i x_i \text{ where } y = \sum_{i=1}^{n} \alpha_i y_i.$$

Show $\{x_1 \cdots x_n\}$ is a linearly independent set and let $\{x_1 \cdots x_n \cdots x_m\}$ be a basis for X. Then let

$$Px \equiv \sum_{i=1}^{n} \alpha_i x_i$$

where

$$x = \sum_{i=1}^{m} \alpha_i x_i.$$

6. ↑ Let X and Y be finite dimensional normed linear spaces and let U be an open subset of X, and suppose $f : U \to Y$, $f \in C^1(U)$, and $Df(x_1)$ is onto. Show there exists $\delta, \epsilon > 0$ such that $f(B(x_1, \delta)) \supseteq B(f(x_1), \epsilon)$. **Hint:** Let

$$L \in \mathcal{L}(Y, X), \ LDf(x_1)x = Px,$$

for all $x \in X_1 \equiv PX$ where $P^2 = P$, $x_1 \in X_1$, and let $U_1 \equiv X_1 \cap U$. Now apply the inverse function theorem to f restricted to X_1 or else repeat part of the proof of the inverse function theorem.

7. The following is Graves theorem similar to that stated in [15]. It gives a more general version of the above problem. To see this, see Problem 15 of Chapter 3 and Problem 2 of this chapter.

Let $f : B(0, \epsilon) \subseteq X \to Y$ be $C^1(B(x_0, \epsilon))$ where X and Y are Banach spaces. Suppose $A \in \mathcal{L}(X, Y)$ is onto and let M be such that for all $y \in Y$, there exists $x \in A^{-1}(y)$ such that

$$||x|| \leq M ||y||.$$

Suppose also for all $x_1, x_2 \in B(x_0, \epsilon)$,

$$||f(x_1) - f(x_2) - A(x_1 - x_2)|| \leq \delta ||x_1 - x_2||, \ \delta < M^{-1}. \tag{22}$$

Then if $c \equiv \frac{1 - M\delta}{2}$,

$$\overline{B(f(x_0), c\epsilon)} \subseteq f(B(0, \epsilon)).$$

Hint: The proof involves a sequence similar to the one used in the proof of the inverse function theorem. Let $y \in \overline{B(f(x_0), c\epsilon)}$ and if $x_0, \cdots, x_n$ have been obtained with $x_n \in B(x_0, \epsilon)$, let x_{n+1} be such that

$$x_{n+1} - x_n \in A^{-1}(f(x_n) - y), \ ||x_n - x_{n+1}|| \leq M ||f(x_n) - y||.$$

Show by induction that

$$||x_{k-1} - x_k|| \leq M (M\delta)^{k-1} ||y - f(x_0)|| \text{ and } x_k \in \overline{B\left(x_0, \frac{\epsilon}{2}\right)}.$$

The following will aid in the inductive step. From 22

$$||f(x_{n-1}) - y - (f(x_n) - y) - A(x_{n-1} - x_n)|| < \delta ||x_{n-1} - x_n||$$

and from the construction of the sequence,

$$A(x_{n-1} - x_n) = f(x_{n-1}) - y.$$

Therefore,

$$||f(x_n) - y|| < \delta ||x_{n-1} - x_n|| \leq \delta M (M\delta)^{n-1} ||y - f(x_0)||. \tag{23}$$

By construction,

$$||x_n - x_{n+1}|| \leq M ||f(x_n) - y|| < M (\delta M)^n ||y - f(x_0)||.$$

Conclude $\{x_k\}$ is Cauchy and use Formula 23.

8. ↑ Let $f : U \to Y$, f is C^1, and $Df(x)$ is onto for each $x \in U$. Then show f maps open subsets of U onto open sets in Y. This is called the open mapping theorem in [4].

9. Suppose $U \subseteq \mathbb{R}^2$ is an open set and $\mathbf{f} : U \to \mathbb{R}^3$ is C^1. Suppose $D\mathbf{f}(s_0, t_0)$ has rank two and

$$\mathbf{f}(s_0, t_0) = \begin{pmatrix} x_0 \\ y_0 \\ z_0 \end{pmatrix}.$$

Show that for (s, t) near (s_0, t_0), the points $\mathbf{f}(s, t)$ may be realized in one of the following forms.

$$\{(x, y, \phi(x, y)) : (x, y) \text{ near } (x_0, y_0)\},$$

$$\{(\phi(y, z) \, y, z) : (y, z) \text{ near } (y_0, z_0)\},$$

or

$$\{(x, \phi(x, z), z,) : (x, z) \text{ near } (x_0, z_0)\}.$$

10. Suppose for some $\eta > 0$,

$$h : [0, 1 + \eta) \to \mathbb{R}$$

is continuous, $h(0) = 0$, and

$$\lim_{\delta \to 0+} \sup \frac{h(t + \delta) - h(t)}{\delta} \leq \alpha$$

for all $t \in [0, 1]$. Then show $h(t) \leq \alpha t$ for all $t \in [0, 1]$. **Hint:** Let

$$S \equiv \{t \in [0, 1] : h(s) \leq (\alpha + \epsilon) s \text{ for all } s \in [0, t]\}$$

and let $r \equiv \sup S$. If $r < 1$, argue

$$h(r) = (\alpha + \epsilon) r$$

and there exists $\delta > 0$ such that

$$\delta^{-1} (h(r + \delta) - h(r)) < \alpha + \epsilon/2, \ h(r + \delta) > (\alpha + \epsilon)(r + \delta).$$

Then

$$\alpha + \epsilon/2 > \delta^{-1} (h(r + \delta) - h(r))$$

$$> \delta^{-1} ((\alpha + \epsilon)(r + \delta) - (\alpha + \epsilon) r)$$

$$= \alpha + \epsilon.$$

11. ↑ Suppose B is an open ball in X and $f : B \to Y$ is differentiable. Suppose also there exists $L \in \mathcal{L}(X, Y)$ such that

$$\|Df(x) - L\| < k$$

for all $x \in B$. Show that if $x_1, x_2 \in B$,

$$|f(x_1) - f(x_2) - L(x_1 - x_2)| \le k|x_1 - x_2|.$$

Hint: Let

$$h(t) \equiv |f(x_1 + t(x_2 - x_1)) - f(x_1) - tL(x_2 - x_1)|$$

and show

$$\limsup_{\delta \to 0+} \frac{h(t + \delta) - h(t)}{\delta} \le k|x_2 - x_1|.$$

Now use Problem 10.

12. ↑ Let $f : U \to Y$, $Df(x)$ exists for all $x \in U$, $B(x_0, \delta) \subseteq U$, and there exists $L \in \mathcal{L}(X, Y)$, such that $L^{-1} \in \mathcal{L}(Y, X)$, and for all $x \in B(x_0, \delta)$

$$\|Df(x) - L\| < \frac{r}{\|L^{-1}\|}, \quad r < 1.$$

Show that there exists $\epsilon > 0$ and an open subset of $B(x_0, \delta)$, V, such that $f : V \to B(f(x_0), \epsilon)$ is one to one and onto. Also $Df^{-1}(y)$ exists for each $y \in B(f(x_0), \epsilon)$ and is given by the formula

$$Df^{-1}(y) = \left[Df\left(f^{-1}(y)\right)\right]^{-1}.$$

Hint: Let

$$T_y(x) \equiv T(x, y) \equiv x - L^{-1}(f(x) - y)$$

for $|y - f(x_0)| < \frac{(1-r)\delta}{2\|L^{-1}\|}$, consider $\{T_y^n(x_0)\}$, and follow the other steps in the proof of the inverse function theorem using Problem 11 instead of the arguments presented earlier. This is a version of the inverse function theorem for f only differentiable, not C^1.

13. Let $\mathbf{f} : \mathbb{R} \times \mathbb{R}^n \to \mathbb{R}^n$ be continuous and bounded and let $\mathbf{x}_0 \in \mathbb{R}^n$. If

$$\mathbf{x} : [0, T] \to \mathbb{R}^n$$

and $h > 0$, let

$$\tau_h \mathbf{x}(s) \equiv \begin{cases} \mathbf{x}_0 & \text{if } s \le h, \\ \mathbf{x}(s - h), & \text{if } s > h. \end{cases}$$

84

For $t \in [0, T]$, let

$$\mathbf{x}_h(t) = \mathbf{x}_0 + \int_0^t \mathbf{f}(s, \tau_h \mathbf{x}_h(s)) \, ds.$$

Show using the Ascoli Arzela theorem that there exists a sequence $h \to 0$ such that

$$\mathbf{x}_h \to \mathbf{x}$$

in $C([0, T]; \mathbb{R}^n)$. Next argue

$$\mathbf{x}(t) = \mathbf{x}_0 + \int_0^t \mathbf{f}(s, \mathbf{x}(s)) \, ds$$

and conclude the following theorem. If $\mathbf{f} : \mathbb{R} \times \mathbb{R}^n \to \mathbb{R}^n$ is continuous and bounded, and if $\mathbf{x}_0 \in \mathbb{R}^n$ is given, there exists a solution to the following initial value problem.

$$\begin{aligned} \mathbf{x}' &= \mathbf{f}(t, \mathbf{x}), \quad t \in [0, T] \\ \mathbf{x}(0) &= \mathbf{x}_0. \end{aligned}$$

This is the Peano existence theorem for ordinary differential equations.

14. Consider the function

$$f(x, y) = \begin{cases} \frac{(x^2 - y^4)^2}{(x^2 + y^4)^2} & \text{if } (x, y) \neq (0, 0), \\ 1 & \text{if } (x, y) = (0, 0). \end{cases}$$

Find the directional derivative of f at $(0, 0)$ in the direction (v_1, v_2) where (v_1, v_2) is a unit vector. This is defined as

$$\lim_{t \to 0} \frac{f(x + tv_1, y + tv_2) - f(x, y)}{t}.$$

Show this limit exists. Is the function continuous at $(0, 0)$? What should the general definition of a directional derivative be? Why is this not the way the derivative is defined?

Chapter 6
Locally Convex Topological Vector Spaces

In this chapter we introduce a generalization of the concept of normed linear space. Let X be a vector space and let Ψ be a collection of functions defined on X such that if $\rho \in \Psi$,

$$\rho(x + y) \leq \rho(x) + \rho(y),$$

$$\rho(ax) = |a| \rho(x) \text{ if } a \in \mathbb{F},$$

$$\rho(x) \geq 0,$$

where $\mathbb{F}$ denotes the field of scalars, either $\mathbb{R}$ or $\mathbb{C}$, assumed to be $\mathbb{C}$ unless otherwise specified. These functions are called seminorms because we cannot conclude $x = 0$ when $\rho(x) = 0$. We define a basis, $\mathcal{B}$, for a topology on X as follows. For A a finite subset of Ψ and $r > 0$,

$$B_A(x, r) \equiv \{y \in X : \rho(x - y) < r \text{ for all } \rho \in A\}.$$

Then $\mathcal{B} \equiv \{B_A(x, r) : x \in X, r > 0, \text{ and } A \subseteq \Psi, A \text{ finite}\}$.

Theorem 6.1 $\mathcal{B}$ *is the basis for a topology.*

Proof: From Theorem 1.12, we need to show that if $B_A(x, r_1)$ and $B_B(y, r_2)$ are two elements of $\mathcal{B}$ and if $z \in B_A(x, r_1) \cap B_B(y, r_2)$, then there exists $U \in \mathcal{B}$ such that

$$z \in U \subseteq B_A(x, r_1) \cap B_B(y, r_2).$$

Let

$$r = \min\left(\min\{(r_1 - \rho(z - x)) : \rho \in A\},\right.$$

$$\left.\min\{(r_2 - \rho(z - y)) : \rho \in B\}\right)$$

and consider $B_{A \cup B}(z, r)$. If w belongs to this set, then for $\rho \in A$,

$$\rho(w - z) < r_1 - \rho(z - x).$$

Hence

$$\rho(w - x) \leq \rho(w - z) + \rho(z - x) < r_1$$

for each $\rho \in A$ and so $B_{A \cup B}(z, r) \subseteq B_A(x, r_1)$. Similarly, $B_{A \cup B}(z, r) \subseteq B_B(y, r_2)$. This proves the theorem.

Let τ be the topology consisting of unions of all subsets of $\mathcal{B}$. We call (X, τ) a locally convex topological vector space.

Theorem 6.2 *The vector space operations of addition and scalar multiplication are continuous. More precisely,*

$$+ : X \times X \to X, \quad \cdot : \mathbb{F} \times X \to X$$

are continuous.

Proof: It suffices to show $+^{-1}(B)$ is open in $X \times X$ and $\cdot^{-1}(B)$ is open in $\mathbb{F} \times X$ if B is of the form

$$B = \{y \in X : \rho(y - x) < r\}$$

because finite intersections of such sets form the basis $\mathcal{B}$. (This collection of sets is a subbasis.) Suppose $u + v \in B$ where B is described above. Then

$$\rho(u + v - x) < \lambda r$$

for some $\lambda < 1$. Consider

$$B_\rho(u, \delta) \times B_\rho(v, \delta).$$

If (u_1, v_1) is in this set, then

$$
\begin{aligned}
\rho(u_1 + v_1 - x) &\leq \rho(u + v - x) + \rho(u_1 - u) + \rho(v_1 - v) \\
&< \lambda r + 2\delta.
\end{aligned}
$$

Let δ be positive but small enough that

$$2\delta + \lambda r < r.$$

Thus this choice of δ shows that $+^{-1}(B)$ is open and this shows $+$ is continuous.

Now suppose $\alpha z \in B$. Then

$$\rho(\alpha z - x) < \lambda r < r$$

for some $\lambda \in (0, 1)$. Let $\delta > 0$ be small enough that $\delta < 1$ and also

$$\lambda r + \delta(\rho(z) + 1) + \delta|\alpha| < r.$$

Then consider $(\beta, w) \in B(\alpha, \delta) \times B_\rho(z, \delta)$.

$$
\begin{aligned}
\rho(\beta w - x) - \rho(\alpha z - x) &\leq \rho(\beta w - \alpha z) \\
&\leq |\beta - \alpha|\rho(w) + \rho(w - z)|\alpha| \\
&\leq |\beta - \alpha|(\rho(z) + 1) + \rho(w - z)|\alpha| \\
&< \delta(\rho(z) + 1) + \delta|\alpha|.
\end{aligned}
$$

Hence
$$\rho\left(\beta w - x\right) < \lambda r + \delta\left(\rho\left(z\right) + 1\right) + \delta\left|\alpha\right| < r$$
and so
$$B\left(\alpha, \delta\right) \times B_\rho\left(z, \delta\right) \subseteq \cdot^{-1}(B).$$

This proves the theorem.

Theorem 6.3 *Let x be given and let $f_x(y) = x + y$. Then f_x is $1-1$, onto, and continuous. If $\alpha \neq 0$ and $g_\alpha(x) = \alpha x$, then g_α is also $1-1$ onto and continuous.*

Proof: The assertions about $1-1$ and onto are obvious. We need to show f_x and g_α are continuous. Let $B = B_\rho\left(z, r\right)$ and consider $f_x^{-1}(B)$. Then it is easy to see that
$$f_x^{-1}(B) = B_\rho\left(z - x, r\right)$$
and so f_x is continuous. To see that g_α is continuous, note that
$$g_\alpha^{-1}(B) = B_\rho\left(\frac{z}{\alpha}, \frac{r}{|\alpha|}\right).$$

This proves the theorem.

As in the case of a normed linear space, we consider the dual space X' of continuous linear functionals. Define, for A a finite subset of Ψ,
$$\rho_A\left(x\right) = \max\{\rho\left(x\right) : \rho \in A\}.$$

Theorem 6.4 *The following are equivalent for f, a linear function mapping X to $\mathbb{F}$.*

$$f \text{ is continuous at } 0. \tag{1}$$

For some $A \subseteq \Psi$, A finite,

$$|f\left(x\right)| \leq C\rho_A\left(x\right) \tag{2}$$

for all $x \in X$.

$$f \text{ is continuous at } x \tag{3}$$

for all x.

Proof: Clearly Formula 3 implies Formula 1. Suppose Formula 1. Then
$$0 = f\left(0\right) \in B\left(0, 1\right) \subseteq \mathbb{F}.$$
Since f is continuous at 0, $0 \in f^{-1}\left(B\left(0, 1\right)\right)$ and there exists an open set $V \in \tau$ such that
$$0 \in V \subseteq f^{-1}\left(B\left(0, 1\right)\right).$$

89

Then $0 \in B_A(0, r) \subseteq V$ for some r and some $A \subseteq \Psi$, A finite. Hence

$$|f(y)| < 1 \text{ if } \rho_A(y) < r.$$

Since f is linear

$$|f(x)| \leq \frac{2}{r} \rho_A(x).$$

To see this, note that if $x \neq 0$, then

$$\frac{rx}{2\rho_A(x)} \in B_A(0, r)$$

and so

$$\frac{|f(rx)|}{2\rho_A(x)} \leq 1$$

which shows that Formula 1 implies Formula 2. Now suppose Formula 2 and suppose $f(x) \in V$, an open set in $\mathbb{F}$. Then

$$f(x) \in B(f(x), r) \subseteq V$$

for some $r > 0$. Suppose $\rho_A(x - y) < r(C_A + 1)^{-1}$. Then

$$|f(x) - f(y)| = |f(x - y)| \leq C_A \rho_A(y - x) < r.$$

Hence

$$f\left(B_A\left(x, r(C_A + 1)^{-1}\right)\right) \subseteq B(f(x), r) \subseteq V.$$

Thus f is continuous at x. This proves the theorem.

What are some examples of locally convex topological vector spaces? It is obvious that any normed linear space is such an example. More generally, we give the following theorem.

Theorem 6.5 *Let X be a vector space and let Y be a vector space of linear functionals defined on X. For each $y \in Y$, define*

$$\rho_y(x) \equiv |y(x)|.$$

Then the collection of seminorms $\{\rho_y\}_{y \in Y}$ defined on X makes X into a locally convex topological vector space and $Y = X'$.

Proof: Clearly $\{\rho_y\}_{y \in Y}$ is a collection of seminorms defined on X; so, X supplied with the topology induced by this collection of seminorms is a locally convex topological vector space. We need to show $Y = X'$.

Let $y \in Y$, let $U \subseteq \mathbb{F}$ be open and let $x \in y^{-1}(U)$. Then $B(y(x), r) \subseteq U$ for some $r > 0$. Letting $A = \{y\}$, it is easy to see from the definition that $B_A(x, r) \subseteq y^{-1}(U)$ and so $y^{-1}(U)$ is an open set as desired. Thus, $Y \subseteq X'$.

Now suppose that $z \in X'$. Then by Formula 2, there exists a finite subset of Y, $A \equiv \{y_1, \cdots, y_n\}$, such that

$$|z(x)| \leq C\rho_A(x).$$

Let

$$\pi(x) \equiv (y_1(x), \cdots, y_n(x))$$

and let f be a linear map from $\pi(X)$ to $\mathbb{F}$ defined by

$$f(\pi x) \equiv z(x).$$

(This is well defined because if $\pi(x) = \pi(x_1)$, then $y_i(x) = y_i(x_1)$ for $i = 1, \cdots, n$ and so

$$\rho_A(x - x_1) = 0.)$$

Thus,

$$|z(x_1) - z(x)| = |z(x_1 - x)| \leq C\rho_A(x - x_1) = 0.$$

Extend f to all of $\mathbb{F}^n$ and denote the resulting linear map by F. Then there exists a vector

$$\alpha = (\alpha_1, \cdots, \alpha_n) \in \mathbb{F}^n$$

with $\alpha_i = F(\mathbf{e}_i)$ such that

$$F(\beta) = \alpha \cdot \beta.$$

Hence for each $x \in X$,

$$z(x) = f(\pi x) = F(\pi x) = \sum_{i=1}^{n} \alpha_i y_i(x)$$

and so

$$z = \sum_{i=1}^{n} \alpha_i y_i \in Y.$$

This proves the theorem.

6.1 Separation theorems

A set, K, is said to be convex if whenever $x, y \in K$,

$$\lambda x + (1 - \lambda) y \in K$$

for all $\lambda \in [0, 1]$. Let U be an open convex set containing 0 and define

$$m(x) \equiv \inf\{t > 0 : x/t \in U\}.$$

Locally Convex Topological Vector Spaces

This is called a Minkowski functional.

Proposition 6.6 m *is defined on* X *and satisfies*

$$m(x+y) \leq m(x) + m(y) \tag{4}$$

$$m(\lambda x) = \lambda m(x) \quad \text{if } \lambda > 0. \tag{5}$$

Thus, m *is a gauge function on* X.

Proof: Let $x \in X$ be arbitrary. There exists $A \subseteq \Psi$ such that

$$0 \in B_A(0, r) \subseteq U.$$

Then

$$\frac{rx}{2\rho_A(x)} \in B_A(0, r) \subseteq U$$

which implies

$$\frac{2\rho_A(x)}{r} \geq m(x). \tag{6}$$

Thus $m(x)$ is defined on X.

Let $x/t \in U$, $y/s \in U$. Then since U is convex,

$$\frac{x+y}{t+s} = \left(\frac{t}{t+s}\right)\left(\frac{x}{t}\right) + \left(\frac{s}{t+s}\right)\left(\frac{y}{s}\right) \in U.$$

It follows that

$$m(x+y) \leq t + s.$$

Choosing s, t such that $t - \epsilon < m(x)$ and $s - \epsilon < m(y)$,

$$m(x+y) \leq m(x) + m(y) + 2\epsilon.$$

Since ϵ is arbitrary, this shows Formula 4. It remains to show Formula 5. Let $x/t \in U$. Then if $\lambda > 0$,

$$\frac{\lambda x}{\lambda t} \in U$$

and so $m(\lambda x) \leq \lambda t$. Thus $m(\lambda x) \leq \lambda m(x)$ for all $\lambda > 0$. Hence

$$m(x) = m\left(\lambda^{-1}\lambda x\right) \leq \lambda^{-1} m(\lambda x) \leq \lambda^{-1}\lambda m(x) = m(x)$$

and so

$$\lambda m(x) = m(\lambda x).$$

This proves the proposition.

Lemma 6.7 *Let U be an open convex set containing 0 and let $q \notin U$. Then there exists $f \in X'$ such that*

$$\operatorname{Re} f(q) > \operatorname{Re} f(x)$$

for all $x \in U$.

Proof: Let m be the Minkowski functional just defined and let

$$F(cq) = cm(q)$$

for $c \in \mathbb{R}$. If $c > 0$ then
$$F(cq) = m(cq)$$

while if $c \leq 0$,
$$F(cq) = cm(q) \leq 0 \leq m(cq).$$

By the Hahn Banach theorem, F has an extension, g, defined on all of X satisfying

$$g(x + y) = g(x) + g(y), \ g(cx) = cg(x)$$

for all $c \in \mathbb{R}$, and

$$g(x) \leq m(x).$$

Thus, $g(-x) \leq m(-x)$ and so

$$-m(-x) \leq g(x) \leq m(x).$$

It follows as in Formula 6 that for some $A \subseteq \Psi$, A finite, and $r > 0$,

$$|g(x)| \leq m(x) + m(-x)$$

$$\leq \frac{2}{r}\rho_A(x) + \frac{2}{r}\rho_A(-x) = \frac{4}{r}\rho_A(x)$$

because

$$\rho_A(-x) = |-1|\rho_A(x) = \rho_A(x).$$

Hence g is continuous by Theorem 6.4. Now define

$$f(x) \equiv g(x) - ig(ix).$$

Thus f is linear and continuous so $f \in X'$ and $\operatorname{Re} f(x) = g(x)$. But for $x \in U$, Theorem 6.2 implies that $x/t \in U$ for some $t < 1$ and so $m(x) < 1$. Since U is convex and $0 \in U$, it follows $q/t \notin U$ if $t < 1$ because if it were,

$$q = t\left(\frac{q}{t}\right) + (1 - t)0 \in U.$$

Therefore, $m(q) \geq 1$ and for $x \in U$,

$$\operatorname{Re} f(x) = g(x) \leq m(x) < 1 \leq m(q) = g(q) = \operatorname{Re} f(q)$$

and this proves the lemma.

Theorem 6.8 *Let K be closed and convex and let $p \notin K$. Then there exists a real number, c, and $f \in X'$ such that*

$$\operatorname{Re} f(p) > c > \operatorname{Re} f(k)$$

for all $k \in K$.

Proof: Since K is closed, and $p \notin K$, there exists a finite subset of Ψ, A, and a positive $r > 0$ such that

$$K \cap B_A(p, 2r) = \emptyset.$$

Pick $k_0 \in K$ and let

$$U = K + B_A(0, r) - k_0, \quad q = p - k_0.$$

It follows that U is an open convex set containing 0 and $q \notin U$. Therefore, by Lemma 6.7, there exists $f \in X'$ such that

$$\operatorname{Re} f(p - k_0) = \operatorname{Re} f(q) > \operatorname{Re} f(k + e - k_0) \tag{7}$$

for all $k \in K$ and $e \in B_A(0, r)$. If $\operatorname{Re} f(e) = 0$ for all $e \in B_A(0, r)$, then $\operatorname{Re} f = 0$ and 7 could not hold. Therefore, $\operatorname{Re} f(e) > 0$ for some $e \in B_A(0, r)$ and so,

$$\operatorname{Re} f(p) > \operatorname{Re} f(k) + \operatorname{Re} f(e)$$

for all $k \in K$. Let $c_1 \equiv \sup\{\operatorname{Re} f(k) : k \in K\}$. Then for all $k \in K$,

$$\operatorname{Re} f(p) \geq c_1 + \operatorname{Re} f(e) > c_1 + \frac{\operatorname{Re} f(e)}{2} > \operatorname{Re} f(k).$$

Let $c = c_1 + \frac{\operatorname{Re} f(e)}{2}$. This proves the theorem.

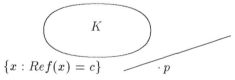

$$\{x : Ref(x) = c\} \qquad \cdot p$$

There are other separation theorems which can be proved in a similar way. The next theorem considers the separation of an open convex set from a convex set.

Theorem 6.9 *Let A and B be disjoint, convex and nonempty sets with B open. Then there exists $f \in X'$ such that*

$$\operatorname{Re} f(a) < \operatorname{Re} f(b)$$

for all $a \in A$ and $b \in B$.

Proof: Let $b_0 \in B, a_0 \in A$. Then the set

$$B - A + a_0 - b_0$$

is open, convex, contains 0, and does not contain $a_0 - b_0$. By Lemma 6.7 there exists $f \in X'$ such that

$$\operatorname{Re} f(a_0 - b_0) > \operatorname{Re} f(b - a + a_0 - b_0)$$

for all $a \in A$ and $b \in B$. Therefore, for all $a \in A, b \in B$,

$$\operatorname{Re} f(b) > \operatorname{Re} f(a).$$

Before giving another separation theorem, we need a lemma.

Lemma 6.10 *If B is convex, then $int(B) \equiv$ union of all open sets contained in B is convex. Also, if $int(B) \neq \emptyset$, then $B \subseteq \overline{int(B)}$.*

Proof: Suppose $x, y \in int(B)$. Then there exists $r > 0$ and a finite set $A \subseteq \Psi$ such that

$$B_A(x, r), B_A(y, r) \subseteq B.$$

Let

$$V \equiv \cup_{\lambda \in [0,1]} \lambda B_A(x, r) + (1 - \lambda) B_A(y, r).$$

Then V is open, $V \subseteq B$, and if $\lambda \in [0, 1]$, then

$$\lambda x + (1 - \lambda) y \in V \subseteq B.$$

Therefore, $int(B)$ is convex as claimed.

Now let $y \in B$ and $x \in int(B)$. Let

$$x \in B_A(x, r) \subseteq int(B)$$

and let $x_\lambda \equiv (1 - \lambda) x + \lambda y$. Define

$$C \equiv \cup_{\lambda \in [0,1]} B_A(x_\lambda, (1 - \lambda) r).$$

This is an open set and we claim $C \subseteq B$. To see this, let

$$z \in B_A(x_\lambda, (1 - \lambda) r), \lambda \in (0, 1).$$

95

Then

$$\rho_A \left(z - x_\lambda \right) < \left(1 - \lambda \right) r$$

and so

$$\rho_A \left(\frac{z}{1-\lambda} - x - \frac{\lambda y}{1-\lambda} \right) < r.$$

Therefore,

$$\frac{z}{1-\lambda} - \frac{\lambda y}{1-\lambda} \in B_A \left(x, r \right) \subseteq B.$$

It follows

$$\left(1 - \lambda \right) \left(\frac{z}{1-\lambda} - \frac{\lambda y}{1-\lambda} \right) + \lambda y = z \in B$$

and so $C \subseteq B$ as claimed. Now this shows $x_\lambda \in int \left(B \right)$ and $\lim_{\lambda \to 1} x_\lambda = y$. Thus, $y \in \overline{int \left(B \right)}$ and this proves the lemma.

Corollary 6.11 *Let A, B be convex, nonempty sets. Suppose $int \left(B \right) \neq \emptyset$ and $A \cap int \left(B \right) = \emptyset$. Then there exists $f \in X'$, $f \neq 0$, such that for all $a \in A$ and $b \in B$,*

$$\operatorname{Re} f \left(b \right) \geq \operatorname{Re} f \left(a \right).$$

Proof: By Theorem 6.9, there exists $f \in X'$ such that for all $b \in int \left(B \right)$, and $a \in A$,

$$\operatorname{Re} f \left(b \right) > \operatorname{Re} f \left(a \right).$$

Thus, in particular, $f \neq 0$. By Lemma 6.10, if $b \in B$ and $a \in A$,

$$\operatorname{Re} f \left(b \right) \geq \operatorname{Re} f \left(a \right).$$

This proves the theorem.

Lemma 6.12 *If X is a topological Hausdorff space then compact implies closed.*

Proof: Let K be compact and suppose K^C is not open. Then there exists $p \in K^C$ such that

$$V_p \cap K \neq \emptyset$$

for all open sets V_p containing p. Let

$$\mathcal{C} = \{ \left(\overline{V_p} \right)^C : V_p \text{ is an open set containing } p \}.$$

Then $\mathcal{C}$ is an open cover of K because if $q \in K$, there exist disjoint open sets V_p and V_q containing p and q respectively. Thus $q \in \left(\overline{V_p} \right)^C$. This is an example of an open cover of K which has no finite subcover, contradicting the assumption that K is compact. This proves the lemma.

Lemma 6.13 *If X is a locally convex topological vector space, and if every point is a closed set, then the seminorms and X' separate the points. By this we mean that if $x \neq y$, then for some $\rho \in \Psi$,*

$$\rho(x - y) \neq 0$$

and for some $f \in X'$,

$$f(x) \neq f(y).$$

In this case, X is a Hausdorff space.

Proof: Let $x \neq y$. Then by Theorem 6.8, there exists $f \in X'$ such that $f(x) \neq f(y)$. Thus X' separates the points. Since $f \in X'$, Theorem 6.4 implies

$$|f(z)| \leq C\rho_A(z)$$

for some A a finite subset of Ψ. Thus

$$0 < |f(x - y)| \leq C\rho_A(x - y)$$

and so $\rho(x - y) \neq 0$ for some $\rho \in A \subseteq \Psi$. Now to show X is Hausdorff, let

$$0 < r < \rho(x - y)2^{-1}.$$

Then the two disjoint open sets containing x and y respectively are

$$B_\rho(x, r) \text{ and } B_\rho(y, r).$$

This proves the lemma.

6.2 The weak and weak* topologies

The weak and weak* topologies are examples which make the underlying vector space into a topological vector space. This section gives a description of these topologies. Unless otherwise specified, X is a locally convex topological vector space. For G a finite subset of X' define $\delta_G : X \to [0, \infty)$ by

$$\delta_G(x) = \max\{|f(x)| : f \in G\}.$$

Lemma 6.14 *The functions δ_G for G a finite subset of X' are seminorms and the sets*

$$B_G(x, r) \equiv \{y \in X : \delta_G(x - y) < r\}$$

form a basis for a topology on X. Furthermore, X with this topology is a locally convex topological vector space. If each point in X is a closed set, then the same is true of X with respect to this new topology.

Proof: It is obvious that the functions δ_G are seminorms and therefore the proof that the sets $B_G(x, r)$ form a basis for a topology is the same as in Theorem 6.1. To see every point is a closed set in this new topology, assuming this is true for X with the original topology, we use Lemma 6.13 to assert that X' separates the points. Let $x \in X$ and let $y \neq x$. There exists $f \in X'$ such that $f(x) \neq f(y)$. Let $G = \{f\}$ and consider

$$B_G(y, |f(x - y)|/2).$$

Then this open set does not contain x. Thus $\{x\}^C$ is open and so $\{x\}$ is closed. This proves the Lemma.

This topology for X is called the weak topology for X. For F a finite subset of X, define $\gamma_F : X' \to [0, \infty)$ by

$$\gamma_F(f) = \max\{|f(x)| : x \in F\}.$$

Lemma 6.15 *The functions γ_F for F a finite subset of X are seminorms and the sets*

$$B_F(f, r) \equiv \{g \in X' : \gamma_F(f - g) < r\}$$

form a basis for a topology on X'. Furthermore, X' with this topology is a locally convex topological vector space having the property that every point is a closed set.

Proof: The proof is similar to that of Lemma 6.14 but there is a difference in the part where every point is shown to be a closed set. Let $f \in X'$ and let $g \neq f$. Thus there exists $x \in X$ such that $f(x) \neq g(x)$. Let $F = \{x\}$. Then

$$B_F(g, |(f - g)(x)|/2)$$

contains g but not f. Thus $\{f\}^C$ is open and so $\{f\}$ is closed. This proves the lemma. Note that we did not need to assume points in X are closed sets to get this.

The topology for X' just described is called the weak* topology. In terms of Theorem 6.5 the weak topology is obtained by letting $Y = X'$ in that theorem while the weak* topology is obtained by letting $Y = X$ with the understanding that X is a vector space of linear functionals on X' defined by

$$x(x^*) \equiv x^*(x).$$

The next theorem is a famous theorem in topology known as the Alexander subbasis theorem. It will be used to develop ideas of weak and weak* compactness.

Theorem 6.16 *Let (X, τ) be a topological space and let $\mathcal{S} \subseteq \tau$ be a subbasis for τ. (Recall this means that finite intersections of sets of $\mathcal{S}$ form* *a basis for τ.) Then if $H \subseteq X$, H is compact if and only if every open cover of H consisting entirely of sets of S admits a finite subcover.*

Proof: The only if part is obvious. To prove the other implication, first note that if every basic open cover, an open cover composed of basic open sets, admits a finite subcover, then H is compact. Now suppose that every subbasic open cover of H admits a finite subcover but H is not compact. This implies that there exists a basic open cover of H, $\mathcal{O}$, which admits no finite subcover. Let $\mathcal{F}$ be defined as

$\{\mathcal{O} : \mathcal{O}$ is a basic open cover of H which admits no finite subcover$\}$.

Partially order $\mathcal{F}$ by set inclusion and use the Hausdorff maximal principle to obtain a maximal chain, $\mathcal{C}$, of such open covers. Let

$$\mathcal{D} = \cup \mathcal{C}.$$

Then it follows that $\mathcal{D}$ is an open cover of H which is maximal with respect to the property of being a basic open cover having no finite subcover of H. (If $\mathcal{D}$ admits a finite subcover, then since $\mathcal{C}$ is a chain and the finite subcover has only finitely many sets, some element of $\mathcal{C}$ would also admit a finite subcover, contrary to the definition of $\mathcal{F}$.) Thus if $\mathcal{D}' \supsetneq \mathcal{D}$ and $\mathcal{D}'$ is a basic open cover of H, then $\mathcal{D}'$ has a finite subcover of H. One of the sets of $\mathcal{D}$, U, has the property that

$$U = \cap_{i=1}^m B_i, \ B_i \in S$$

and no B_i is in $\mathcal{D}$. If not, we could replace each set in $\mathcal{D}$ with a subbasic set also in $\mathcal{D}$ containing it and thereby obtain a subbasic cover which would, by assumption, admit a finite subcover, contrary to the properties of $\mathcal{D}$. Thus $\mathcal{D} \cup \{\mathcal{B}_i\}$ admits a finite subcover,

$$V_1^i, \cdots, V_{m_i}^i, B_i$$

for each $i = 1, \cdots, m$. Consider

$$\{U, V_j^i, \ j = 1, \cdots, m_i, \ i = 1, \cdots, m\}.$$

If $p \in H \setminus \cup \{V_j^i\}$, then $p \in B_i$ for each i and so $p \in U$. This is therefore a finite subcover of $\mathcal{D}$ contradicting the properties of $\mathcal{D}$. This proves the theorem.

Let I be a set and suppose for each $i \in I$, (X_i, τ_i) is a nonempty topological space. The Cartesian product of the X_i, denoted by

$$\prod_{i \in I} X_i,$$

consists of the set of all choice functions defined on I which select a single element of each X_i. Thus

$$f \in \prod_{i \in I} X_i$$

means for every $i \in I$, $f(i) \in X_i$. The axiom of choice says $\prod_{i \in I} X_i$ is nonempty. Let

$$P_j(A) = \prod_{i \in I} B_i$$

where $B_i = X_i$ if $i \neq j$ and $B_j = A$. A subbasis for a topology on the product space consists of all sets $P_j(A)$ where $A \in \tau_j$. (These sets have an open set in the jth slot and the whole space in the other slots.) Thus a basis consists of finite intersections of these sets. It is easy to see that finite intersections do form a basis for a topology. This topology is called the product topology and we will denote it by $\prod \tau_i$. Next we use the Alexander subbasis theorem to prove the Tychonoff theorem.

Theorem 6.17 *If* $(X_i \tau_i)$ *is compact, then so is* $\left(\prod_{i \in I} X_i, \prod \tau_i\right)$.

Proof: By the Alexander subbasis theorem, we will establish compactness of the product space if we show every subbasic open cover admits a finite subcover. Therefore, let $\mathcal{O}$ be a subbasic open cover of $\prod_{i \in I} X_i$. Let

$$\mathcal{O}_j = \{Q \in \mathcal{O} : Q = P_j(A) \text{ for some } A \in \tau_j\}.$$

Let

$$\pi_j \mathcal{O}_j = \{A : P_j(A) \in \mathcal{O}_j\}.$$

If no $\pi_j \mathcal{O}_j$ covers X_j, then pick

$$f \in \prod_{i \in I} X_i \setminus \cup \pi_i \mathcal{O}_i$$

so $f(j) \notin \cup \pi_j \mathcal{O}_j$ and so $f \notin \cup \mathcal{O}$ contradicting $\mathcal{O}$ is an open cover. Hence, for some j,

$$X_j = \cup \theta_j \mathcal{O}_j$$

and so there exist $A_1, \cdots, A_m$, sets in τ_j such that

$$X_j \subseteq \cup_{i=1}^{m} A_i$$

and $P_j(A_i) \in \mathcal{O}$. Therefore, $\{P_j(A_i)\}_{i=1}^{m}$ covers $\prod_{i \in I} X_i$.

The next lemma will be used to prove the very important theorem of Alaoglu sometimes called the Banach Alaoglu theorem.

Lemma 6.18 *Let* $V = B_A(0, r)$ *where A is a finite subset of Ψ and let*

$$K = \{f \in X' : |f(x)| \leq 1 \text{ for all } x \in V\}.$$

The weak and weak* topologies

Then K is compact in (X', τ) where τ is the weak topology.*

Proof: Recall that $\prod_{x \in X} \mathbb{F}$ is the set of mappings of X into $\mathbb{F}$. Thus

$$X' \subseteq \prod_{x \in X} \mathbb{F}.$$

Also the weak* topology on X' is just the intersection of the open sets in $\prod_{x \in X} \mathbb{F}$ with X'. For all $x \in X$,

$$x \in \frac{2}{r} \rho_A(x) V.$$

Therefore, for all $f \in K$,

$$|f(x)| \leq \frac{2}{r} \rho_A(x).$$

Now let α be a limit point of K in the product topology of $\prod_{x \in X} \mathbb{F}$. For a given $x \in X$, it follows from the definition of product topology that there exist $f_n \in K$ such that $|f_n(x) - \alpha(x)| < 1/n$ for $n = 1, 2, \cdots$. Therefore

$$|\alpha(x)| < 1/n + |f_n(x)| \leq \frac{2}{r} \rho_A(x) + 1/n.$$

It follows that

$$|\alpha(x)| \leq \frac{2}{r} \rho_A(x).$$

If α is linear, this will show $\alpha \in X'$. Next we show that α is linear.

Let $a, b \in \mathbb{F}$ and let $x, y \in X$. By definition of product topology there exists $f_n \in K$ such that

$$|f_n(x) - \alpha(x)| < 1/n, \quad |f_n(y) - \alpha(y)| < 1/n,$$

$$|f_n(ax + by) - \alpha(ax + by)| < 1/n.$$

Therefore,

$$|\alpha(ax + by) - (a\alpha(x) + b\alpha(y))|$$
$$\leq |\alpha(ax + by) - f_n(ax + by)|$$
$$+ |a\alpha(x) - af_n(x)| + |b\alpha(y) - bf_n(y)| +$$
$$|f_n(ax + by) - af_n(x) - bf_n(y)| \leq (|a| + |b| + 1) n^{-1}.$$

Hence, since this holds for all n,

$$\alpha(ax + by) - (a\alpha(x) + b\alpha(y)) = 0.$$

This shows $\alpha \in X'$ and so $\alpha \in K$. Thus K is a closed subset of

$$\prod_{x \in X} \overline{B\left(0, \frac{2}{r} \rho_A(x)\right)}.$$

101

Therefore K is compact in the product topology which implies K is compact in the weak$*$ topology because, as already noted, the weak$*$ topology is just the set of intersections of open sets in the product topology with X'. This proves the lemma.

The next theorem replaces $B_A(0, r)$ with an arbitrary open set containing 0.

Theorem 6.19 *Let $0 \in V$, for V an open set and let*

$$K = \{f \in X' : |f(x)| \leq c \text{ for all } x \in V\}.$$

Then K is compact in (X', τ) where τ is the weak$*$ topology.

Proof: First note that

$$K = \{f \in X' : |f(x)| \leq 1 \text{ for } x \in c^{-1}V\}.$$

Now let $0 \in B_A(0, r) \subseteq c^{-1}V$, and let

$$K_0 = \{f \in X' : |f(x)| \leq 1 \text{ for } x \in B_A(0, r)\}.$$

Then K is a closed subset of K_0 and K_0 is compact by Lemma 6.18. Thus K is also compact. This proves the theorem.

Now suppose X is separable. This means there exists a countable set $D = \{x_n\}$ such that whenever $V \neq \emptyset$ is open, $V \cap D \neq \emptyset$.

Theorem 6.20 *If $K \subseteq X'$ is compact and X is separable then there exists a metric, d, on K such that if τ_d is the topology on K induced by d and if τ is the topology on K induced by the weak$*$ topology of X', then $\tau = \tau_d$. Thus we can consider K with the weak$*$ topology as a metric space.*

Proof: Let $D = \{x_n\}$ be the dense countable subset and define for $f \in X'$,

$$\gamma_n(f) = |f(x_n)|.$$

The metric is

$$d(f, g) \equiv \sum_{n=1}^{\infty} 2^{-n} \frac{\gamma_n(f - g)}{1 + \gamma_n(f - g)}.$$

Clearly $d(f, g) = d(g, f) \geq 0$. If $d(f, g) = 0$, then this requires $f(x_n) = g(x_n)$ for all $x_n \in D$. Since f and g are continuous and D is dense, this requires that $f(x) = g(x)$ for all x. (To see this, note that

$$V \equiv \{x : |f(x) - g(x)| > 0\}$$

is an open set. If $V \neq \emptyset$ then it contains some x_n. Hence $|f(x_n) - g(x_n)| > 0$ contradicting $f = g$ on D. Thus $V = \emptyset$ and $f(x) = g(x)$ for all x.)

It is routine to verify the triangle inequality from the easy-to-establish inequality,

$$\frac{x}{1+x} + \frac{y}{1+y} \geq \frac{x+y}{1+x+y},$$

valid whenever $x, y \geq 0$. Therefore this is a metric. Now for each n

$$g \to \frac{\gamma_n (f-g)}{1 + \gamma_n (f-g)}$$

is a continuous function from (K, τ) to $[0, \infty)$ and also the above sum defining d converges uniformly. It follows

$$g \to d(f, g)$$

is continuous. Therefore, the ball with respect to d,

$$B_d(f, r) \equiv \{g \in K : d(g, f) < r\}$$

is open. This implies $\tau_d \subseteq \tau$.

Now suppose $U \in \tau$. Then $K \setminus U$ is closed. Hence, $K \setminus U$ is compact in τ because it is a closed subset of the compact set K. It follows that $K \setminus U$ is compact with respect to τ_d because $\tau_d \subseteq \tau$. But (K, τ_d) is a Hausdorff space and so $K \setminus U$ must be closed with respect to τ_d. This implies $U \in \tau_d$. Thus $\tau \subseteq \tau_d$ and this proves $\tau = \tau_d$.

The fact that this set with the weak* topology can be considered a metric space is very significant because if a point is a limit point in a metric space, one can extract a convergent sequence.

Corollary 6.21 *If X is separable and $K \subseteq X'$ is compact in the weak* topology, then K is sequentially compact. That is, if $\{f_n\}_{n=1}^{\infty} \subseteq K$, then there exists a subsequence f_{n_k} and $f \in K$ such that for all $x \in X$,*

$$\lim_{k \to \infty} f_{n_k}(x) = f(x).$$

Proof: By Theorem 6.20, K is a metric space for the metric described there and it is compact. Therefore by the characterization of compact metric spaces, K is sequentially compact. This proves the corollary.

What does this say about Banach spaces? If X is a normed linear space, it is a locally convex topological vector space in which Ψ consists of only one seminorm, the norm of the space. Consider the ball in X',

$$\{f \in X' : ||f||_{X'} \leq c\}.$$

This is, by definition, the set of $f \in X'$ such that $|f(x)| \leq c$ for all $||x|| < 1$. Therefore by Theorem 6.19, this ball is weak* compact. If X is separable, this ball is sequentially weak* compact.

What about the weak topology on a reflexive Banach space?

Lemma 6.22 *Let $J : X \to X''$ be the James map*

$$Jx(f) \equiv f(x)$$

and let X be reflexive so that J is onto. Then J is a homeomorphism of $(X,$ weak topology$)$ and $(X'',$ weak* topology$)$. This means J is one to one, onto, and both J and J^{-1} are continuous.

Proof: Let $f \in X'$ and let

$$B_f(x, r) \equiv \{y : |f(x) - f(y)| < r\}.$$

Thus $B_f(x, r)$ is a subbasic set for the weak topology on X. Now by the definition of J,

$$y \in B_f(x, r) \text{ if and only if } |Jy(f) - Jx(f)| < r$$

$$\text{if and only if } Jy \in B_f(Jx, r) \equiv$$

$$\{y^{**} \in X'' : |y^{**}(f) - J(x)(f)| < r\},$$

a subbasic set for the weak* topology on X''. Since J^{-1} and J are one to one and onto and map subbasic sets to subbasic sets, it follows that J is a homeomorphism. This proves the Lemma.

The following is an easy corollary.

Corollary 6.23 *If X is a reflexive Banach space, then the closed unit ball is weakly compact.*

Proof: Let B be the closed unit ball. Then $B = J^{-1}(B^{**})$ where B^{**} is the unit ball in X'' which is compact in the weak* topology. Therefore B is weakly compact because J^{-1} is continuous.

Next we discuss the Eberlein Smulian theorem which states that a Banach space is reflexive if and only if the closed unit ball is weakly sequentially compact. Actually, only half the theorem is proved here, the more useful only if part. The book by Yoshida [56] has the complete theorem discussed.

Theorem 6.24 (*Eberlein Smulian*) *The closed unit ball in a reflexive Banach space X, is weakly sequentially compact.*

Proof: Let $\{x_n\} \subseteq B \equiv \overline{B(0, 1)}$. Let Y be the closure of the linear span of $\{x_n\}$. Thus Y is a separable, hence weakly separable Banach space. It is reflexive because it is a closed subspace of a reflexive space. (See Problem 7 of this chapter.) By the Alaoglu theorem, the closed unit ball B^* in Y' is weak* compact. Also by Theorem 6.20, B^* is a metric space with a suitable metric. Thus B^* is complete and totally bounded with respect to this metric and it follows that B^* with the weak* topology is separable. Since Y is reflexive, B^* is also weakly compact and separable. By Corollary 6.21 B^{**}, the closed unit ball in Y'' is weak* sequentially compact. Then by Lemma 6.22 B, the unit ball in Y, is weakly sequentially compact. It follows there exists a subsequence x_{n_k}, of the sequence $\{x_n\}$ and a point $x \in Y$, such that for all $f \in Y'$,

$$f(x_{n_k}) \to f(x).$$

Now if $x^* \in X'$,

$$x^*(x_{n_k}) = i^* x^*(x_{n_k}) \to i^* x^*(x) = x^*(x).$$

which shows x_{n_k} converges weakly and this shows X is weakly sequentially compact.

6.3   The Tychonoff fixed point theorem

The objective in this section is to present a proof of the Tychonoff fixed point theorem, one of the most useful of the generalizations of the Brouwer fixed point theorem. In this section, X will be a locally convex topological vector space in which every point is a closed set. Let $\mathcal{B}$ be the basis described earlier and let $\mathcal{B}_0$ consist of all sets of $\mathcal{B}$ which are of the form $B_A(0, r)$ where A is a finite subset of Ψ as described earlier. Note that for $U \in \mathcal{B}_0$, $U = -U$ and U is convex. Also, if $U \in \mathcal{B}_0$, there exists $V \in \mathcal{B}_0$ such that

$$V + V \subseteq U$$

where

$$V + V \equiv \{v_1 + v_2 : v_i \in V\}.$$

To see this, note

$$B_A(0, r/2) + B_A(0, r/2) \subseteq B(0, r).$$

We let K be a closed convex subset of X and let

$$f \text{ be continuous}, f : K \to K, \text{ and } \overline{f(K)} \text{ is compact}.$$

Lemma 6.25 *For each $U \in \mathcal{B}_0$, there exists a finite set of points*

$$\{y_1 \cdots y_n\} \subseteq \overline{f(K)}$$

and continuous functions ψ_i defined on $\overline{f(K)}$ such that for $x \in \overline{f(K)}$,

$$\sum_{i=1}^{n} \psi_i(x) = 1, \tag{8}$$

$$\psi_i(x) = 0 \text{ if } x \notin y_i + U, \ \psi_i(x) > 0 \text{ if } x \in y_i + U.$$

If

$$f_U(x) \equiv \sum_{i=1}^{n} y_i \psi_i(f(x)), \tag{9}$$

then whenever $x \in K$,

$$f(x) - f_U(x) \in U.$$

Proof: Let $U = B_A(0, r)$. Using the compactness of $\overline{f(K)}$, there exists

$$\{y_1 \cdots y_n\} \subseteq \overline{f(K)}$$

such that

$$\{y_i + U\}_{i=1}^{n}$$

covers $\overline{f(K)}$. Let

$$\phi_i(y) \equiv (r - \rho_A(y - y_i))^{+}.$$

Thus $\phi_i(y) > 0$ if $y \in y_i + U$ and $\phi_i(y) = 0$ if $y \notin y_i + U$. For $x \in \overline{f(K)}$, let

$$\psi_i(x) \equiv \phi_i(x) \left(\sum_{j=1}^{n} \phi_j(x) \right)^{-1}.$$

Then Formula 8 is satisfied. Now let f_U be given by Formula 9 for $x \in f(K)$. For such x,

$$f(x) - f_U(x) = \sum_{\{i:f(x)-y_i \in U\}} (f(x) - y_i)\, \psi_i(f(x))$$

$$+ \sum_{\{i:f(x)-y_i \notin U\}} (f(x) - y_i)\, \psi_i(f(x))$$

$$= \sum_{\{i:f(x)-y_i \in U\}} (f(x) - y_i)\, \psi_i(f(x)) =$$

$$\sum_{\{i:f(x)-y_i \in U\}} (f(x) - y_i)\, \psi_i(f(x)) + \sum_{\{i:f(x)-y_i \notin U\}} 0 \psi_i(f(x)) \in U$$

because $0 \in U$, U is convex, and Formula 8. This proves the lemma. We think of f_U as an approximation to f.

The Tychonoff fixed point theorem

Lemma 6.26 *For each $U \in \mathcal{B}_0$, there exists $x_U \in$ convex hull of $\overline{f(K)}$ such that*

$$f_U(x_U) = x_U.$$

Proof: If $f_U(x_U) = x_U$ and

$$x_U = \sum_{i=1}^{n} a_i y_i$$

for $\sum_{i=1}^{n} a_i = 1$, we need

$$\sum_{j=1}^{n} y_j \psi_j \left(f \left(\sum_{i=1}^{n} a_i y_i \right) \right) = \sum_{j=1}^{n} a_j y_j.$$

This will be satisfied if for each $j = 1, \cdots, n$,

$$a_j = \psi_j \left(f \left(\sum_{i=1}^{n} a_i y_i \right) \right); \tag{10}$$

so, let

$$\Sigma_{n-1} \equiv \left\{ \mathbf{a} \in \mathbb{R}^n : \sum_{i=1}^{n} a_i = 1, \ a_i \geq 0 \right\}$$

and let $h : \Sigma_{n-1} \to \Sigma_{n-1}$ be given by

$$h(\mathbf{a})_j \equiv \psi_j \left(f \left(\sum_{i=1}^{n} a_i y_i \right) \right).$$

Since h is continuous, the Brouwer fixed point theorem applies and we see there exists a fixed point for h which is a solution to Formula 10. This proves the lemma.

For a short proof of the Brouwer fixed point theorem based on integration theory, see the appendix on the subject. The next theorem is the Tychonoff fixed point theorem.

Theorem 6.27 *Let K be a closed and convex subset of X, a locally convex topological vector space in which every point is closed. Let $f : K \to K$ be continuous and suppose $\overline{f(K)}$ is compact. Then f has a fixed point.*

Proof: First we establish the following claim which will yield a candidate for the fixed point.

Claim: There exists $x \in \overline{f(K)}$ with the property that if $V \in \mathcal{B}_0$, there exists $U \subseteq V$, $U \in \mathcal{B}_0$, such that

$$f(x_U) \in x + V.$$

Locally Convex Topological Vector Spaces

Proof of the claim: If no such x exists, then for each $x \in \overline{f(K)}$, there exists $V_x \in \mathcal{B}_0$ such that whenever $U \subseteq V_x$, with $U \in \mathcal{B}_0$,

$$f(x_U) \notin x + V_x.$$

Since $\overline{f(K)}$ is compact, there exist $x_1, \cdots, x_n \in \overline{f(K)}$ such that

$$\{x_i + V_{x_i}\}_{i=1}^n$$

cover $\overline{f(K)}$. Let

$$U \in \mathcal{B}_0, \; U \subseteq \cap_{i=1}^n V_{x_i}$$

and consider x_U.

$$f(x_U) \in x_i + V_{x_i}$$

for some i but $U \subseteq V_{x_i}$, a contradiction. This shows the claim.

Now we show x is the desired fixed point. Let $W \in \mathcal{B}_0$ and let $V \in \mathcal{B}_0$ with

$$V + V + V \subseteq W.$$

Since f is continuous at x, there exists $V_0 \in \mathcal{B}_0$ such that

$$V_0 + V_0 \subseteq V$$

and if

$$y - x \in V_0 + V_0,$$

then

$$f(x) - f(y) \in V.$$

Using the claim, let $U \in \mathcal{B}_0$, $U \subseteq V_0$, such that

$$f(x_U) \in x + V_0.$$

Then

$$x - x_U = x - f(x_U) + f(x_U) - f_U(x_U) \in V_0 + U$$
$$\subseteq V_0 + V_0 \subseteq V$$

and so

$$\begin{aligned} f(x) - x &= f(x) - f(x_U) + f(x_U) - f_U(x_U) + f_U(x_U) - x \\ &= f(x) - f(x_U) + f(x_U) - f_U(x_U) + x_U - x \\ &\subseteq V + U + V \subseteq W. \end{aligned}$$

Since $W \in \mathcal{B}_0$ is arbitrary, it follows from Lemma 6.13 that $f(x) - x = 0$. This proves the theorem.

In the case where X is normed linear space, this fixed point theorem is called the Schauder fixed point theorem. As an example of the usefulness

The Tychonoff fixed point theorem

of this fixed point theorem, consider the following application to the theory of ordinary differential equations. In the context of this theorem, $X = C(0, T; \mathbb{R}^n)$, a Banach space with norm given by

$$\|\mathbf{x}\| \equiv \max\{|\mathbf{x}(t)| : t \in [0, T]\}.$$

Theorem 6.28 *Let* $\mathbf{f} : [0, T] \times \mathbb{R}^n \to \mathbb{R}^n$ *be continuous and suppose there exists* $L > 0$ *such that for all* $\lambda \in (0, 1)$, *if*

$$\mathbf{x}' = \lambda \mathbf{f}(t, \mathbf{x}), \ \mathbf{x}(0) = \mathbf{x}_0 \tag{11}$$

for all $t \in [0, T]$, *then* $\|\mathbf{x}\| < L$. *Then there exists a solution to*

$$\mathbf{x}' = \mathbf{f}(t, \mathbf{x}), \ \mathbf{x}(0) = \mathbf{x}_0 \tag{12}$$

for $t \in [0, T]$.

Proof: Let

$$N\mathbf{x}(t) \equiv \int_0^t \mathbf{f}(s, \mathbf{x}(s)) \, ds.$$

Thus a solution to the initial value problem is obtained if we can show the existence of a solution to

$$\mathbf{x}_0 + N(\mathbf{x}) = \mathbf{x}.$$

Let

$$m \equiv \max\left\{|\mathbf{f}(t, \mathbf{x})| : (t, \mathbf{x}) \in [0, T] \times \overline{B(0, L)}\right\}, \ M \equiv |\mathbf{x}_0| + mT$$

and let

$$K \equiv \{\mathbf{x} \in C(0, T; \mathbb{R}^n) \text{ such that } \mathbf{x}(0) = \mathbf{x}_0 \text{ and } \|\mathbf{x}\| \leq M\}.$$

Now define

$$A\mathbf{x} \equiv \begin{cases} \mathbf{x}_0 + N\mathbf{x} \text{ if } \|N\mathbf{x}\| \leq M - |\mathbf{x}_0|, \\ \mathbf{x}_0 + \frac{(M - |\mathbf{x}_0|)N\mathbf{x}}{\|N\mathbf{x}\|} \text{ if } \|N\mathbf{x}\| > M - |\mathbf{x}_0|. \end{cases}$$

Then A is continuous and maps X to K. Also $A(K)$ is equicontinuous because

$$\mathbf{x}_0 + N\mathbf{x}(t) - (\mathbf{x}_0 + N\mathbf{x}(t_1)) = \int_{t_1}^t \mathbf{f}(s, \mathbf{x}(s)) \, ds$$

and the integrand is bounded. Thus $\overline{A(K)}$ is a compact set in X by the Ascoli Arzela theorem. By the Schauder fixed point theorem, A has a fixed point, $\mathbf{x} \in K$.

If $\|N(\mathbf{x})\| > M - |\mathbf{x}_0|$, then

$$\mathbf{x}_0 + \lambda N(\mathbf{x}) = \mathbf{x}$$

where

$$\lambda = \frac{(M - |\mathbf{x}_0|)}{\|N\mathbf{x}\|} < 1$$

and so Formula 11 holds. Therefore, by the assumed estimate on the solutions to Formula 11, it follows that

$$\|\mathbf{x}\| < L$$

and so $\|N\mathbf{x}\| \le mT = M - |\mathbf{x}_0|$, a contradiction. Therefore, it must be the case that

$$\|N(\mathbf{x})\| \le M - |\mathbf{x}_0|$$

which implies that

$$\mathbf{x}_0 + N(\mathbf{x}) = \mathbf{x}.$$

Since this is equivalent to Formula 12, this proves the theorem.

Note that we do not assume existence for solutions to Formula 11, only estimates of possible solutions to this initial value problem. These estimates are called *a-priori* estimates. Also note this is a global existence theorem, not a local one for a solution defined on only a small interval.

6.4   Set-valued maps

In the abstract theory of partial differential equations and variational inequalities, it is important to consider set-valued maps from a Banach space to the power set of its dual. In this section we give an introduction to this theory by proving a general result on surjectivity for a class of such operators.

To begin with, if $A : X \to \mathcal{P}(Y)$ is a set-valued map, we define the graph of A by

$$G(A) \equiv \{(x, y) : y \in Ax\}.$$

We will first consider a map A which maps $\mathbb{C}^n$ to $\mathcal{P}(\mathbb{C}^n)$ which satisfies

$$Ax \text{ is compact and convex.} \tag{13}$$

If O is open and $O \supseteq Ax$, then there exists $\delta > 0$ such that if

$$\mathbf{y} \in B(0, \delta), \text{ then } Ay \subseteq O. \tag{14}$$

Lemma 6.29 *Let A satisfy Formulas 13 and 14. Then AK is a subset of a compact set whenever K is compact. Also the graph of A is closed.*

Proof: Let $\mathbf{x} \in K$. Then $A\mathbf{x}$ is compact and contained in some open set whose closure is compact, $U_{\mathbf{x}}$. By assumption Formula 14 there exists an open set $V_{\mathbf{x}}$ containing $\mathbf{x}$ such that if $\mathbf{y} \in V_{\mathbf{x}}$, then $A\mathbf{y} \subseteq U_{\mathbf{x}}$. Let $V_{\mathbf{x}_1}, \cdots, V_{\mathbf{x}_m}$ cover K. Then $AK \subseteq \cup_{k=1}^m \overline{U}_{\mathbf{x}_k}$, a compact set. To see the graph of A is closed, let $\mathbf{x}_k \to \mathbf{x}, \mathbf{y}_k \to \mathbf{y}$ where $\mathbf{y}_k \in A\mathbf{x}_k$. Then letting $O = A\mathbf{x} + B(0, r)$ it follows from Formula 14 that $\mathbf{y}_k \in A\mathbf{x}_k \subseteq O$ for all k large enough. Therefore, $\mathbf{y} \in A\mathbf{x} + B(0, 2r)$ and since $r > 0$ is arbitrary and $A\mathbf{x}$ is closed it follows $\mathbf{y} \in A\mathbf{x}$.

The next lemma is an application of the Brouwer fixed point theorem. First we define an n simplex, denoted by $[\mathbf{x}_0, \cdots, \mathbf{x}_n]$, to be the convex hull of the $n+1$ points, $\{\mathbf{x}_0, \cdots, \mathbf{x}_n\}$. Thus

$$[\mathbf{x}_0, \cdots, \mathbf{x}_n] \equiv \left\{ \sum_{i=1}^n t_i \mathbf{x}_i : \sum_{i=1}^n t_i = 1, \ t_i \geq 0 \right\}.$$

If $n = 2$, the simplex is a triangle, line segment, or point. If $n = 3$, it is a tetrahedron, triangle, line segment or point. We also say a collection of simplicies is a tiling of $\mathbb{R}^n$ if $\mathbb{R}^n$ is contained in their union and if S_1, S_2 are two simplicies in the tiling, with

$$S_j = \left[\mathbf{x}_0^j, \cdots, \mathbf{x}_n^j \right],$$

then

$$S_1 \cap S_2 = [\mathbf{x}_{k_0}, \cdots, \mathbf{x}_{k_r}]$$

where

$$\{\mathbf{x}_{k_0}, \cdots, \mathbf{x}_{k_r}\} \subseteq \{\mathbf{x}_0^1, \cdots, \mathbf{x}_n^1\} \cap \{\mathbf{x}_0^2, \cdots, \mathbf{x}_n^2\}$$

or else the two simplicies do not intersect. The collection of simplicies is said to be locally finite if, for every point, there exists a ball containing that point which also intersects only finitely many of the simplicies in the collection. We leave it to the reader to verify that for each $\epsilon > 0$, there exists a locally finite tiling of $\mathbb{R}^n$ which is composed of simplicies which have diameters less than ϵ.

Lemma 6.30 *Suppose $A : \mathbb{C}^n \to \mathcal{P}(\mathbb{C}^n)$ satisfies Formulas 13 and 14 and K is a nonempty closed convex set in $\mathbb{C}^n$. Then if $\mathbf{y} \in \mathbb{C}^n$ there exists $(\mathbf{x}, \mathbf{w}) \in G(A)$ such that $\mathbf{x} \in K$ and*

$$\mathrm{Re}\,(\mathbf{y} - \mathbf{w}, \mathbf{z} - \mathbf{x}) \leq 0$$

for all $\mathbf{z} \in K$.

Proof: Tile $\mathbb{C}^n$ with $2n$ simplicies such that the collection is locally finite and each simplex has diameter less than $\epsilon < 1$. This collection of simplicies is determined by a countable collection of vertices. For each vertex, $\mathbf{x}$, pick $A_\epsilon \mathbf{x} \in A\mathbf{x}$ and define A_ϵ on all of $\mathbb{C}^n$ by the following rule. If

$$\mathbf{x} \in [\mathbf{x}_0, \cdots, \mathbf{x}_{2n}],$$

so $\mathbf{x} = \sum_{i=0}^{2n} t_i \mathbf{x}_i$, then

$$A_\epsilon \mathbf{x} \equiv \sum_{k=0}^{2n} t_k A_\epsilon \mathbf{x}_k.$$

Thus A_ϵ is a continuous map defined on $\mathbb{C}^n$ thanks to the local finiteness of the collection of simplices. Let P_K denote the projection on the convex set K. By the Brouwer fixed point theorem, there exists a fixed point, $\mathbf{x}_\epsilon \in K$ such that

$$P_K \left(\mathbf{y} - A_\epsilon \mathbf{x}_\epsilon + \mathbf{x}_\epsilon \right) = \mathbf{x}_\epsilon.$$

By Corollary 4.8 this requires

$$\mathrm{Re} \left(\mathbf{y} - A_\epsilon \mathbf{x}_\epsilon, \mathbf{z} - \mathbf{x}_\epsilon \right) \le 0$$

for all $\mathbf{z} \in K$.

Suppose $\mathbf{x}_\epsilon \in [\mathbf{x}_0^\epsilon, \cdots, \mathbf{x}_{2n}^\epsilon]$ so $\mathbf{x}_\epsilon = \sum_{k=0}^{2n} t_k^\epsilon \mathbf{x}_k^\epsilon$. Then since $\mathbf{x}_\epsilon$ is contained in K, a compact set, and diameter of each simplex is less than 1, it follows that $A_\epsilon \mathbf{x}_k^\epsilon$ is contained in $A\overline{(K + B(0,1))}$, which is contained in a compact set thanks to Lemma 6.29. Taking a subsequence, we may obtain from the Heine Borel theorem that $\epsilon \to 0$ and

$$t_\epsilon \to 0, \mathbf{x}_\epsilon \to \mathbf{x}, A\mathbf{x}_k^\epsilon \to \mathbf{y}_k$$

for $k = 0, \cdots, 2n$. Since the diameter of the simplex containing $\mathbf{x}_\epsilon$ converges to 0, it follows

$$\mathbf{x}_k^\epsilon \to \mathbf{x}, \, A_\epsilon \mathbf{x}_k^\epsilon \to \mathbf{y}_k.$$

Since the graph of A is closed and $A_\epsilon \mathbf{x}_k^\epsilon \in A\mathbf{x}_k^\epsilon$, this implies $\mathbf{y}_k \in A\mathbf{x}$. Since $A\mathbf{x}$ is convex, this means

$$\sum_{k=1}^{2n} t_k \mathbf{y}_k \in A\mathbf{x}.$$

Hence for all $\mathbf{z} \in K$,

$$\mathrm{Re} \left(\mathbf{y} - \sum_{k=1}^{2n} t_k \mathbf{y}_k, \mathbf{z} - \mathbf{x}_\epsilon \right) = \lim_{\epsilon \to 0} \mathrm{Re} \left(\mathbf{y} - \sum_{k=1}^{2n} t_k^\epsilon A_\epsilon \mathbf{x}_k^\epsilon, \mathbf{z} - \mathbf{x}_\epsilon \right)$$

$$= \lim_{\epsilon \to 0} \mathrm{Re} \left(\mathbf{y} - A_\epsilon \mathbf{x}_\epsilon, \mathbf{z} - \mathbf{x}_\epsilon \right) \le 0.$$

Let $\mathbf{w} = \sum_{k=1}^{2n} t_k \mathbf{y}_k$. This proves the lemma.

112

Lemma 6.31 *Suppose in addition to Formulas 13 and 14, A is coercive,*

$$\lim_{|\mathbf{x}| \to \infty} \left\{ \frac{|(\mathbf{y}, \mathbf{x})|}{|\mathbf{x}|} : \mathbf{y} \in A\mathbf{x} \right\} = \infty.$$

Then A is onto.

Proof: Let $\mathbf{y} \in \mathbb{C}^n$ and let $K_r \equiv \overline{B(\mathbf{0},r)}$. By Lemma 6.30 there exists $\mathbf{x}_r \in K_r$ and $\mathbf{w}_r \in A\mathbf{x}_r$ such that

$$\text{Re}\,(\mathbf{y} - \mathbf{w}_r, \mathbf{z} - \mathbf{x}_r) \leq 0 \tag{15}$$

for all $\mathbf{z} \in K_r$. Letting $\mathbf{z} = \mathbf{0}$,

$$\text{Re}\,(\mathbf{w}_r, \mathbf{x}_r) \leq \text{Re}\,(\mathbf{y}, \mathbf{x}_r).$$

It follows from the assumption of coercivity that $|\mathbf{x}_r|$ is bounded independent of r. Therefore, picking r strictly larger than this bound, Formula 15 implies

$$\text{Re}\,(\mathbf{y} - \mathbf{w}_r, \mathbf{v}) = 0$$

for all $\mathbf{v} \in \mathbb{C}^n$ and so $\mathbf{y} = \mathbf{w}_r \in A\mathbf{x}_r$. This proves the lemma.

Lemma 6.32 *Let F be a finite dimensional Banach space of dimension n, and let T be a mapping from F to $\mathcal{P}(F')$ such that Formula 13 and Formula 14 both hold. Then if T is also coercive,*

$$\lim_{||\mathbf{x}|| \to \infty} \left\{ \frac{|\mathbf{y}^*(\mathbf{x})|}{||\mathbf{x}||} : \mathbf{y}^* \in T\mathbf{x} \right\} = \infty, \tag{16}$$

it follows T is onto.

Proof: Let $|\cdot|$ be an equivalent norm for F such that there is an isometry of $\mathbb{C}^n$ and F, θ. Now define $A : \mathbb{C}^n \to \mathbb{C}^n$ by $\mathbf{y} \in A\mathbf{x}$ if and only if

$$(\mathbf{y}, \mathbf{z}) \equiv (\theta^* T\theta\mathbf{x}, \mathbf{z}).$$

Then A satisfies the conditions of Lemma 6.31 and so A is onto. Consequently T is also onto.

With these lemmas, it is possible to prove a very useful result about a class of mappings which map a reflexive Banach space to the power set of its dual space. For more theorems about these mappings and their applications, see [39]. In the discussion below, we will use the symbol, $\rightharpoonup$, to denote weak convergence.

Definition 6.33 We say $T : V \to \mathcal{P}(V)$ is pseudomonotone   if the following conditions hold.

$$Tu \text{ is closed, nonempty, convex, and bounded.} \tag{17}$$

If F is finite dimensional, then if $u \in F$ and $W \supseteq Tu$ for W a weakly open set in V', then there exists $\delta > 0$ such that

$$v \in B(u, \delta) \cap F \text{ implies } Tv \subseteq W. \tag{18}$$

If $u_k \rightharpoonup u$ and if $u_k^* \in Tu_k$ is such that

$$\limsup_{k \to \infty} \operatorname{Re} u_k^*(u_k - u) \le 0,$$

then for all $v \in V$, there exists $u^*(v) \in Tu$ such that

$$\liminf_{k \to \infty} \operatorname{Re} u_k^*(u_k - v) \ge \operatorname{Re} u^*(v)(u - v). \tag{19}$$

We say T is coercive if Formula 16 holds.

Theorem 6.34 *Let V be a reflexive Banach space and let $T : V \to \mathcal{P}(V')$ be pseudomonotone and coercive. Then T is onto.*

Proof: Let $\mathcal{F}$ be the set of finite subspaces of V and let $F \in \mathcal{F}$. Then define T_F as

$$T_F \equiv i_F^* T i_F.$$

Then T_F satisfies the conditions of Lemma 6.32 and so T_F is onto $\mathcal{P}(F')$. Let $w^* \in V'$. Then since T_F is onto, there exists $u_F \in F$ such that

$$i_F^* w^* \in i_F^* T i_F u_F.$$

Thus for each finite dimensional subspace, F, there exists $u_F \in F$ such that for all $v \in F$,

$$w^*(v) = u_F^*(v), \ u_F^* \in Tu_F. \tag{20}$$

Replacing v with u_F, in Formula 20,

$$\frac{|u_F^*(u_F)|}{||u_F||} = \frac{|w^*(u_F)|}{||u_F||} \le ||w^*||.$$

Therefore, the assumption that T is coercive implies $\{u_F : F \in \mathcal{F}\}$ is bounded in V. Now define

$$W_F \equiv \cup \{u_{F'} : F' \supseteq F\}.$$

Then W_F is bounded and if $\overline{W_F} \equiv$ weak closure of W_F, then

$$\{\overline{W_F} : F \in \mathcal{F}\}$$

is a collection of nonempty weakly compact (since V is reflexive) sets having the finite intersection property because $W_F \ne \emptyset$ for each F. Thus there exists

$$u \in \cap \{\overline{W_F} : F \in \mathcal{F}\}.$$

We will show $w^* \in Tu$. If $w^* \notin Tu$, a closed convex set, there exists $v \in V$ such that

$$\operatorname{Re} w^* (u - v) < \operatorname{Re} u^* (u - v) \tag{21}$$

for all $u^* \in Tu$. This follows from the separation theorems. (These theorems imply there exists $z \in V$ such that

$$\operatorname{Re} w^* (z) < \operatorname{Re} u^* (z)$$

for all $u^* \in Tu$. Define $u - v \equiv z$.) Now let $F \supseteq \{u, v\}$. Since $u \in \overline{W_F}$, a weakly sequentially compact set, there exists a sequence, $\{u_k\}$, such that

$$u_k \rightharpoonup u, \ u_k \in W_F.$$

Then since $F \supseteq \{u, v\}$, there exists $u_k^* \in Tu_k$ such that

$$u_k^* (u_k - u) = w^* (u_k - u).$$

Therefore,

$$\lim \sup_{k \to \infty} \operatorname{Re} u_k^* (u_k - u) = \lim \sup_{k \to \infty} \operatorname{Re} w^* (u_k - u) = 0.$$

It follows by the assumption T is pseudomonotone; the following holds for the v defined above in Formula 21.

$$\lim \inf_{k \to \infty} \operatorname{Re} u_k^* (u_k - v) \geq \operatorname{Re} u^* (v) (u - v), \ u^* (v) \in Tu.$$

But since $v \in F, \operatorname{Re} u_k^* (u_k - v) = \operatorname{Re} w^* (u_k - v)$ and so

$$\lim \inf_{k \to \infty} \operatorname{Re} u_k^* (u_k - v) = \lim \inf_{k \to \infty} \operatorname{Re} w^* (u_k - v) = \operatorname{Re} w^* (u - v),$$

so from Formula 21,

$$\operatorname{Re} w^* (u - v) = \lim \inf_{k \to \infty} \operatorname{Re} u_k^* (u_k - v)$$

$$\geq \operatorname{Re} u^* (v) (u - v) > \operatorname{Re} w^* (u - v),$$

a contradiction. Thus, $w^* \in Tu$ and this proves the theorem.

6.5   Finite dimensional spaces

What can be said about finite dimensional locally convex topological vector spaces? In this section it will be shown that they are all like $\mathbb{R}^n$ or $\mathbb{C}^n$. So in finite dimensions, no extra generality is gained by introducing the notion of topological vector space.

Theorem 6.35 *Suppose X is a finite dimensional locally convex topological vector space. Let*

$$\{u_1, \cdots, u_m\}$$

be a basis for X and let

$$w_n = \sum_{k=1}^{m} c_k^n u_k$$

where $c_k^n \in \mathbb{F}$. Then

$$\lim_{n \to \infty} w_n = 0$$

if and only if

$$\lim_{n \to \infty} c_k^n = 0, \ k = 1, 2, \cdots, m. \tag{22}$$

Proof: First we show that if $c_k^n \to c_k$ in $\mathbb{F}$ then

$$\sum_{k=1}^{m} c_k^n u_k \to \sum_{k=1}^{m} c_k u_k. \tag{23}$$

To see this, let $A \subseteq \Psi$.

$$\rho_A \left(\sum_{k=1}^{m} c_k^n u_k - \sum_{k=1}^{m} c_k u_k \right) \le \sum_{k=1}^{m} |c_k^n - c_k| \rho_A (u_k) < r$$

whenever n is large enough. Since A and $r > 0$ are arbitrary, this establishes Formula 23. In particular, this proves the if part of the claimed equivalence.

Now suppose $\lim_{n \to \infty} w_n = 0$ and suppose Formula 22 does not hold. Then for some $\epsilon > 0$ there exists a subsequence, still denoted by n, such that

$$0 < \epsilon \le \sum_{k=1}^{m} |c_k^n|^2 \equiv (S(n))^2.$$

Then

$$\frac{w_n}{S(n)} = \sum_{k=1}^{m} (c_k^n / S(n)) u_k$$

and

$$(c_1^n, \cdots, c_m^n) S(n)^{-1} \equiv \mathbf{c}^n S(n)^{-1}$$

is a unit vector in $\mathbb{F}^m$. By the Heine Borel theorem there exists a subsequence, still denoted by n, and a unit vector $\mathbf{d}$ such that

$$\lim_{n \to \infty} \mathbf{c}^n S(n)^{-1} = \mathbf{d}$$

in $\mathbb{F}^m$. Then by Formula 23,

$$0 = \lim_{n \to \infty} w_n = \lim_{n \to \infty} \sum_{k=1}^{m} (c_k^n / S(n)) u_k = \sum_{k=1}^{m} d_k u_k,$$

contradicting the assumption that $\{u_1, \cdots, u_m\}$ is a basis. This proves the theorem.

It is possible to prove an improved version of this theorem if we assume also that every point of the topological vector space is a closed set.

Theorem 6.36 *Suppose X is a finite dimensional locally convex topological vector space of dimension m in which every point is a closed set and let*

$$\{u_1, \cdots, u_m\}$$

be a basis of X. Let $\theta : \mathbb{F}^m \rightarrow X$ be defined by

$$\theta(\mathbf{c}) = \sum_{i=1}^{m} c_i u_i.$$

Then θ is a homeomorphism.

Proof: First note that θ is $1-1$ and onto and is linear,

$$\theta(a\mathbf{c}_1 + b\mathbf{c}_2) = a\theta(\mathbf{c}_1) + b\theta(\mathbf{c}_2).$$

Also note that by Theorem 6.35, θ is continuous since that theorem says that

$$\mathbf{c}_n \rightarrow \mathbf{c} \text{ implies } \theta(\mathbf{c}_n) \rightarrow \theta(\mathbf{c}).$$

It remains to show that θ^{-1} is continuous.

Let S be the unit sphere in $\mathbb{F}^m$ and let B be the unit ball in $\mathbb{F}^m$.

$$S = \{\mathbf{c} : |\mathbf{c}| = 1\}, \ B = \{\mathbf{c} : |\mathbf{c}| < 1\}.$$

Then $\theta(S)$ is compact and does not contain 0 because θ is $1-1$. By Lemmas 6.12 and 6.13, we can say that $\theta(S)$ is closed because it is compact. Hence for some finite set $A \subseteq \Psi$,

$$0 \in B_A(0, r), \ B_A(0, r) \cap \theta(S) = \emptyset.$$

Since θ is $1-1$,

$$\theta^{-1}(B_A(0, r)) \cap S = \emptyset$$

and $\theta^{-1}(B_A(0, r))$ contains $\mathbf{0}$. If $p, q \in \theta^{-1}(B_A(0, r))$, then $\theta(p), \theta(q) \in B_A(0, r)$ and so

$$\theta(tp + (1-t)q) \in B_A(0, r)$$

for all $t \in [0, 1]$ since $B_A(0, r)$ is convex. Thus

$$tp + (1-t)q \in \theta^{-1}(B_A(0, r))$$

for each $t \in [0, 1]$ and so $\theta^{-1}(B_A(0, r))$ is convex. This implies it is connected. Since this set contains $\mathbf{0}$ and does not contain any points of

S, it must be a subset of B. Now we can apply Theorem 6.4 and Formula 2 to conclude that $\left(\theta^{-1}\right)_i$ is continuous. Thus θ^{-1} is continuous and this proves the theorem.

Theorem 6.37 *Let X be a locally convex topological vector space in which every point is a closed set. Then if*

$$Y = span\{u_1, \cdots, u_m\}$$

where $\{u_1, \cdots, u_m\}$ is a linearly independent set, then Y is closed.

Proof: Y, with the relative topology from X, is a locally convex topological vector space. The seminorms on Y are just those for X restricted to Y. If θ is as defined in Theorem 6.36, then θ is a homeomorphism of $\mathbb{F}^m$ and Y. Let $p \in \overline{Y}$ and let $B_A(0, r) \cap Y$ satisfy

$$\theta^{-1}\left(B_A(0, r) \cap Y\right) \subseteq B.$$

Then if $t = 2^{-1}\left(\rho_A(p)\right)^{-1} r$,

$$p \in tB_A(0, r) \cap \overline{Y}.$$

Now $\theta\left(t\overline{B}\right)$ is a closed set since θ is continuous and $\theta\left(t\overline{B}\right)$ is compact. Thus

$$tB_A(0, r) \cap Y \subseteq \theta(tB) \subseteq \theta\left(t\overline{B}\right).$$

If $p \notin \theta\left(t\overline{B}\right)$ then

$$p \in tB_A(0, r) \cap \left(\theta\left(t\overline{B}\right)\right)^C,$$

an open set. Also $p \in \overline{Y}$ so there exists

$$y \in Y \cap tB_A(0, r) \cap \left(\theta\left(t\overline{B}\right)\right)^C$$

contradicting

$$tB_A(0, r) \cap Y \subseteq \theta(tB).$$

Hence

$$y \in \theta\left(t\overline{B}\right) \subseteq Y$$

and this proves the theorem.

6.6 Exercises

1. Let X be a vector space with field of scalars $\mathbb{F} = \mathbb{R}$ or $\mathbb{C}$ and suppose X is also a topological space with topology τ, such that the operators

$$+ : X \times X \to X$$

and

$$\cdot : \mathbb{F} \times X \to X$$

are continuous. Such a space is called a topological vector space. Note that it is not assumed to be locally convex. Show that if

$$0 \in U \in \tau,$$

there exists W such that W is open, $0 \in W \subseteq U$ and W is balanced. By balanced we mean $\alpha W \subseteq W$ whenever $|\alpha| \le 1$. **Hint:** By continuity of $\cdot$, there exists $\delta > 0$ and an open set V containing 0 such that $\alpha V \subseteq U$ whenever $|\alpha| < \delta$. Consider

$$\cup \{\alpha V : |\alpha| < \delta\}.$$

2. Let X be a topological vector space. We say $\mathcal{B}_0$ is a local basis for τ at 0 if whenever $0 \in U \in \tau$, there exists $W \in \mathcal{B}_0$ such that $0 \in W \subseteq U \in \tau$. Let $\mathcal{B}_0$ be a basis for τ at 0 and let

$$\mathcal{B} \equiv \{W + x : x \in X \text{ and } W \in \mathcal{B}_0\}.$$

Show $\mathcal{B}$ is a basis for τ.

3. ↑ Let X be a topological vector space. Show it has a balanced local basis at 0.

4. ↑ We can define a topological vector space to be locally convex if it has a local basis at 0 of convex sets. Show that if X is locally convex, then it has a local basis at 0 of convex, balanced open sets. **Hint:** If $0 \in U$, and U is convex, Problem 1 shows there exists an open balanced set W such that $0 \in W \subseteq U$. Thus

$$0 \in W \subseteq \cap \{\omega U : |\omega| = 1\} \equiv C.$$

Show C is convex, $0 \in \text{interior}(C) \equiv C^o$, C^o is convex, and C^o is balanced. (Recall that the interior of C is the union of all open sets contained in C.)

5. ↑ Let X be a topological vector space which is locally convex as defined in Problem 4. Show there exists a collection of seminorms such that the topology determined by these seminorms in the manner described in this chapter coincides with the given topology on X. **Hint:** Let $\mathcal{B}_0$ be a local basis at 0 of balanced convex open sets as in Problem 4. For $U \in \mathcal{B}_0$, let ρ_U be the Minkowski functional

$$\rho_U(x) \equiv \inf\{t > 0 : x/t \in U\}$$

and consider the collection of seminorms

$$\Psi \equiv \{\rho_U : U \in \mathcal{B}_0\}.$$

Show for $U \in \mathcal{B}_0$, $B_{\rho_U}(0,1) = U$.

6. Let X be a topological vector space. Recall $f \in X'$ means f is linear and also $f^{-1}(U)$ is open whenever U is an open set in $\mathbb{F}$. It may be there are more open sets in the topology of X than are needed in order for this to happen. Define the weak topology to be the smallest topology on X with respect to which each $f \in X'$ is continuous. Show this definition yields the same topology as the weak topology described in the chapter. Note that this topology depends only on what is in X'. Compare Theorem 6.5.

7. Show that a closed subspace of a reflexive Banach space is reflexive. **Hint:** The proof of this is an exercise in the use of the Hahn Banach theorem. Let Y be the closed subspace of the reflexive space X and let $y^{**} \in Y''$. Then $i^{**}y^{**} \in X''$ and so $i^{**}y^{**} = Jx$ for some $x \in X$ because X is reflexive. Now argue that $x \in Y$ as follows. If $x \notin Y$, then there exists x^* such that $x^*(Y) = 0$ but $x^*(x) \neq 0$. Thus, $i^*x^* = 0$. Use this to get a contradiction. When you know that $x = y \in Y$, the Hahn Banach theorem implies i^* is onto Y' and for all $x^* \in X'$,

$$y^{**}(i^*x^*) = i^{**}y^{**}(x^*) = Jx(x^*) = x^*(x) = x^*(iy) = i^*x^*(y).$$

8. Suppose X is a Banach space and that $A : X \to X$ is a compact mapping. This means that A maps bounded subsets of X into totally bounded subsets of X. Suppose there exists a constant M independent of λ such that if

$$x = \lambda Ax$$

for $\lambda \in (0, 1)$, then $||x|| < M$. Show that then A has a fixed point. **Hint:** Define

$$\tilde{A}(x) \equiv \begin{cases} Ax & \text{if } ||Ax|| \leq M \\ \frac{MAx}{||Ax||} & \text{if } ||Ax|| > M \end{cases}$$

and argue that $\tilde{A}$ maps X into the bounded set, $\overline{B(0, M)}$, and that $\tilde{A}\left(\overline{B(0, M)}\right)$ is compact. Then apply the Schauder fixed point theorem and argue as in Theorem 6.28 that $\tilde{A}(x) = x$ for some x and that $||Ax|| \leq M$. This is known as the Schaefer fixed point theorem.

9. This problem considers a situation in which *a-priori* estimates described in the hypothesis of Theorem 6.28 are obtained. Suppose $\Psi(u) > 0$ a.e. and

$$||\mathbf{f}(t, \mathbf{x})|| \leq \Psi(|\mathbf{x}|).$$

Let

$$T_\infty \equiv \int_{|\mathbf{x}_0|}^\infty \frac{du}{\Psi(u)},$$

where T_∞ might equal ∞. Show that there exists a solution to Formula 12 if $T < T_\infty$. Try the theorem on the differential equations

$$y' = 1 + y^2, \quad y' = 1 + y^2 - \epsilon y^3.$$

Hint: Suppose $\mathbf{x}$ solves Formula 11. Then either $|\mathbf{x}(t)| \leq |\mathbf{x}_0|$ or $|\mathbf{x}(t)| > |\mathbf{x}_0|$. Suppose $|\mathbf{x}(t)| > |\mathbf{x}_0|$. Then this situation persists on an interval of the form $(\mu, t]$ where $|\mathbf{x}(\mu)| = |\mathbf{x}_0|$. For $s \in (\mu, t]$, show

$$|\mathbf{x}(s)|' \leq |\mathbf{x}'(s)| \leq \Psi(|\mathbf{x}(s)|)$$

and so

$$\Phi(|\mathbf{x}(t)|) \equiv \int_{|\mathbf{x}_0|}^{|\mathbf{x}(t)|} \frac{du}{\Psi(u)} \leq t - \mu \leq t.$$

Argue that

$$|\mathbf{x}(t)| \leq \Phi^{-1}(t) \leq \Phi^{-1}(T)$$

and therefore that for $t \in [0, T]$, $|\mathbf{x}(t)| \leq \max(\Phi^{-1}(T), |\mathbf{x}_0|)$.

Chapter 7
Measures and Measurable Functions

7.1 σ Algebras

This chapter is on the basics of measure theory. It deals with the abstract concept of a measure space. Let Ω be a set and let $\mathcal{F}$ be a collection of subsets of Ω satisfying

$$\emptyset \in \mathcal{F}, \ \Omega \in \mathcal{F}, \tag{1}$$

$$E \in \mathcal{F} \text{ implies } E^C \equiv \Omega \setminus E \in \mathcal{F},$$

$$\text{If } \{E_n\}_{n=1}^{\infty} \subseteq \mathcal{F}, \text{ then } \cup_{n=1}^{\infty} E_n \in \mathcal{F}. \tag{2}$$

Definition 7.1 A collection of subsets of a set, Ω, satisfying Formulas 1-2 is called a σ algebra.

As an example, let Ω be any set and let $\mathcal{F} = \mathcal{P}(\Omega)$, the set of all subsets of Ω (power set). This obviously satisfies Formulas 1-2.

Lemma 7.2 *Let $\mathcal{C}$ be a set whose elements are σ algebras of subsets of Ω. Then $\cap \mathcal{C}$ is a σ algebra also.*

Example 7.3 *Let (X, τ) be a topological space and let $\sigma(\tau) \equiv$ intersection of all σ algebras that contain τ. $\sigma(\tau)$ is called the σ algebra of Borel sets .*

This is a very important σ algebra and it will be referred to frequently as the Borel sets. Attempts to describe a typical Borel set are more trouble than they are worth and it is not easy to do so. Rather, one uses the definition just given in the example. Note, however, that all countable intersections of open sets and countable unions of closed sets are Borel sets. Such sets are called G_δ and F_σ respectively.

7.2 Monotone classes and algebras

Definition 7.4 $\mathcal{A}$ is said to be an algebra of subsets of Z if $Z \in \mathcal{A}, \phi \in \mathcal{A}$, and when $E, F \in \mathcal{A}$, $E \cup F$ and $E \setminus F$ are both in $\mathcal{A}$.

It is important to note that if $\mathcal{A}$ is an algebra, then it is also closed under finite intersections. Thus, $E \cap F = (E^C \cup F^C)^C \in \mathcal{A}$ because $E^C = Z \setminus E \in \mathcal{A}$ and similarly $F^C \in \mathcal{A}$.

Definition 7.5 $\mathcal{M} \subseteq \mathcal{P}(Z)$ is called a monotone class if
a.) ...$E_n \supseteq E_{n+1}...$, $E = \cap_{n=1}^{\infty} E_n$, and $E_n \in \mathcal{M}$, then $E \in \mathcal{M}$.
b.) ...$E_n \subseteq E_{n+1}...$, $E = \cup_{n=1}^{\infty} E_n$, and $E_n \in \mathcal{M}$, then $E \in \mathcal{M}$.
(In simpler notation, $E_n \downarrow E$ and $E_n \in \mathcal{M}$ implies $E \in \mathcal{M}$. $E_n \uparrow E$ and $E_n \in \mathcal{M}$ implies $E \in \mathcal{M}$.)

Lemma 7.6 *Suppose that $\mathcal{R}$ and $\mathcal{E}$ are subsets of $\mathcal{P}(Z)$ with the following properties.*

$$\emptyset, Z \in \mathcal{R}$$

$$\mathcal{E} = \textit{finite disjoint unions of sets of } \mathcal{R}$$

$$A \cap B \in \mathcal{R} \textit{ whenever } A, B \in \mathcal{R},$$

$$A \setminus B \in \mathcal{E} \textit{ whenever } A, B \in \mathcal{R}.$$

Then $\mathcal{E}$ is an algebra of sets of Z.

Proof: Note first that if $A \in \mathcal{R}$, then $A^C \in \mathcal{E}$ because $A^C = Z \setminus A$. Now suppose that E_1 and E_2 are in $\mathcal{E}$,

$$E_1 = \cup_{i=1}^{m} R_i, \quad E_2 = \cup_{j=1}^{n} R_j$$

where the R_i are disjoint sets in $\mathcal{R}$ and the R_j are disjoint sets in $\mathcal{R}$. Then

$$E_1 \cap E_2 = \cup_{i=1}^{m} \cup_{j=1}^{n} R_i \cap R_j$$

which is clearly an element of $\mathcal{E}$ because no two of the sets in the union can intersect and by assumption they are all in $\mathcal{R}$. Thus finite intersections of sets of $\mathcal{E}$ are in $\mathcal{E}$. If $E = \cup_{i=1}^{n} R_i$

$$E^C = \cap_{i=1}^{n} R_i^C = \textit{finite intersection of sets of } \mathcal{E}$$

which was just shown to be in $\mathcal{E}$. Thus if $E_1, E_2 \in \mathcal{E}$,

$$E_1 \setminus E_2 = E_1 \cap E_2^C \in \mathcal{E}$$

and

$$E_1 \cup E_2 = (E_1 \setminus E_2) \cup E_2 \in \mathcal{E}$$

from the definition of $\mathcal{E}$. This proves the lemma.

Corollary 7.7 *Let $(Z_1, \mathcal{R}_1, \mathcal{E}_1)$ and $(Z_2, \mathcal{R}_2, \mathcal{E}_2)$ be as just described in Lemma 7.6. Then $(Z_1 \times Z_2, \mathcal{R}, \mathcal{E})$ also satisfies the conditions of Lemma 7.6 if $\mathcal{R}$ is defined as*

$$\mathcal{R} \equiv \{R_1 \times R_2 : R_i \in \mathcal{R}_i\}$$

and

$$\mathcal{E} \equiv \{\text{ finite disjoint unions of sets of } \mathcal{R}\}.$$

Consequently, $\mathcal{E}$ is an algebra of sets.

Proof: It is clear $\emptyset, Z_1 \times Z_2 \in \mathcal{R}$. Let $R_1^1 \times R_2^1$ and $R_1^2 \times R_2^2$ be two elements of $\mathcal{R}$.

$$R_1^1 \times R_2^1 \cap R_1^2 \times R_2^2 = R_1^1 \cap R_1^2 \times R_2^1 \cap R_2^2 \in \mathcal{R}$$

by assumption.

$$R_1^1 \times R_2^1 \setminus \left(R_1^2 \times R_2^2\right) =$$

$$R_1^1 \times \left(R_2^1 \setminus R_2^2\right) \cup \left(R_1^1 \setminus R_1^2\right) \times \left(R_2^1 \cap R_2^2\right)$$

$$= R_1^1 \times A_2 \cup A_1 \times R_2$$

where $A_2 \in \mathcal{E}_2$, $A_1 \in \mathcal{E}_1$, and $R_2 \in \mathcal{R}_2$. Since the two sets in the above expression on the right do not intersect, and each A_i is a finite set of disjoint elements of $\mathcal{R}_i$, it follows the above expression is in $\mathcal{E}$. This proves the corollary. The following example will be referred to frequently.

Example 7.8 *Consider for $\mathcal{R}$, sets of the form $I = (a, b] \cap (-\infty, \infty)$ where $a \in [-\infty, \infty]$ and $b \in [-\infty, \infty]$. Then, clearly, $\emptyset, (-\infty, \infty) \in \mathcal{R}$ and it is not hard to see that all conditions for Corollary 7.7 are satisfied. Applying this corollary repeatedly, we find that for*

$$\mathcal{R} \equiv \left\{\prod_{i=1}^{n} I_i : I_i = (a_i, b_i] \cap (-\infty, \infty)\right\}$$

and $\mathcal{E}$ is defined as finite disjoint unions of sets of $\mathcal{R}$,

$$(\mathbb{R}^n, \mathcal{R}, \mathcal{E})$$

satisfies the conditions of Corollary 7.7 and in particular $\mathcal{E}$ is an algebra of sets of $\mathbb{R}^n$. It is clear that the same would hold if I were of the form $[a, b) \cap (-\infty, \infty)$.

Example 7.9 *It follows from Lemma 7.6 or more easily from Corollary 7.7 that the elementary sets form an algebra.*

Theorem 7.10 *(Monotone Class theorem) Let $\mathcal{A}$ be an algebra of subsets of Z and let $\mathcal{M}$ be a monotone class containing $\mathcal{A}$. Then $\mathcal{M} \supseteq \sigma(\mathcal{A})$, the smallest σ-algebra containing $\mathcal{A}$.*

Proof: We may assume $\mathcal{M}$ is the smallest monotone class containing $\mathcal{A}$. Such a smallest monotone class exists because the intersection of monotone classes containing $\mathcal{A}$ is a monotone class containing $\mathcal{A}$. We show that $\mathcal{M}$ is a σ-algebra. It will then follow $\mathcal{M} \supseteq \sigma(\mathcal{A})$. For $A \in \mathcal{A}$, define

$$\mathcal{M}_A \equiv \{B \in \mathcal{M} \text{ such that } A \cup B \in \mathcal{M}\}.$$

Clearly $\mathcal{M}_A$ is a monotone class containing $\mathcal{A}$. Hence $\mathcal{M}_A = \mathcal{M}$ because $\mathcal{M}$ is the smallest such monotone class. This shows that $A \cup B \in \mathcal{M}$ whenever $A \in \mathcal{A}$ and $B \in \mathcal{M}$. Now pick $B \in \mathcal{M}$ and define

$$\mathcal{M}_B \equiv \{D \in \mathcal{M} \text{ such that } D \cup B \in \mathcal{M}\}.$$

We just showed $\mathcal{A} \subseteq \mathcal{M}_B$. It is clear that $\mathcal{M}_B$ is a monotone class. Thus $\mathcal{M}_B = \mathcal{M}$ and it follows that $D \cup B \in \mathcal{M}$ whenever $D \in \mathcal{M}$ and $B \in \mathcal{M}$.

A similar argument shows that $D \setminus B \in \mathcal{M}$ whenever $D, B \in \mathcal{M}$. (For $A \in \mathcal{A}$, let

$$\mathcal{M}_A = \{B \in \mathcal{M} \text{ such that } B \setminus A \text{ and } A \setminus B \in \mathcal{M}\}.$$

Argue $\mathcal{M}_A$ is a monotone class containing $\mathcal{A}$, etc.)

Thus $\mathcal{M}$ is both a monotone class and an algebra. Hence, if $E \in \mathcal{M}$ then $Z \setminus E \in \mathcal{M}$. We want to show $\mathcal{M}$ is a σ-algebra. But if $E_i \in \mathcal{M}$ and $F_n = \cup_{i=1}^{n} E_i$, then $F_n \in \mathcal{M}$ and $F_n \uparrow \cup_{i=1}^{\infty} E_i$. Since $\mathcal{M}$ is a monotone class, $\cup_{i=1}^{\infty} E_i \in \mathcal{M}$ and so $\mathcal{M}$ is a σ-algebra. This proves the theorem.

Definition 7.11 Let $\mathcal{F}$ be a σ algebra of sets of Ω and let $\mu : \mathcal{F} \to [0, \infty]$. We call μ a measure if

$$\mu\left(\bigcup_{i=1}^{\infty} E_i\right) = \sum_{i=1}^{\infty} \mu(E_i) \tag{3}$$

whenever the E_i are disjoint sets of $\mathcal{F}$. $(\Omega, \mathcal{F}, \mu)$ is called a measure space. The elements of $\mathcal{F}$ are called the measurable sets.

Theorem 7.12 Let $\{E_m\}_{m=1}^{\infty}$ be a sequence of measurable sets in a measure space $(\Omega, \mathcal{F}, \mu)$. Then if $\cdots E_n \subseteq E_{n+1} \subseteq E_{n+2} \subseteq \cdots$,

$$\mu(\cup_{i=1}^{\infty} E_i) = \lim_{n \to \infty} \mu(E_n) \tag{4}$$

and if $\cdots E_n \supseteq E_{n+1} \supseteq E_{n+2} \supseteq \cdots$ and $\mu(E_1) < \infty$, then

$$\mu(\cap_{i=1}^{\infty} E_i) = \lim_{n \to \infty} \mu(E_n). \tag{5}$$

Proof: First note that $\cap_{i=1}^{\infty} E_i = (\cup_{i=1}^{\infty} E_i^C)^C \in \mathcal{F}$ so $\cap_{i=1}^{\infty} E_i$ is measurable. To show Formula 4, note that 4 is obviously true if $\mu(E_k) = \infty$ for any k. Therefore, assume $\mu(E_k) < \infty$ for all k. Thus

$$\mu(E_{k+1} \setminus E_k) = \mu(E_{k+1}) - \mu(E_k).$$

Hence by Formula 3,

$$\mu(\cup_{i=1}^{\infty} E_i) = \mu(E_1) + \sum_{k=1}^{\infty} \mu(E_{k+1} \setminus E_k) = \mu(E_1)$$

$$+ \sum_{k=1}^{\infty} \mu(E_{k+1}) - \mu(E_k)$$

$$= \mu(E_1) + \lim_{n \to \infty} \sum_{k=1}^{n} \mu(E_{k+1}) - \mu(E_k) = \lim_{n \to \infty} \mu(E_{n+1}).$$

This shows part 4. To verify Formula 5, since $\mu(E_1) < \infty$,

$$\mu(E_1) - \mu(\cap_{i=1}^{\infty} E_i) = \mu(E_1 \setminus \cap_{i=1}^{\infty} E_i) = \lim_{n \to \infty} \mu(E_1 \setminus \cap_{i=1}^{n} E_i)$$

$$= \mu(E_1) - \lim_{n \to \infty} \mu(\cap_{i=1}^{n} E_i) = \mu(E_1) - \lim_{n \to \infty} \mu(E_n),$$

where the second equality follows from part 4. Hence

$$\lim_{n \to \infty} \mu(E_n) = \mu(\cap_{i=1}^{\infty} E_i).$$

This proves the theorem.

Definition 7.13 Let $(\Omega, \mathcal{F}, \mu)$ be a measure space and let (X, τ) be a topological space. A function $f : \Omega \to X$ is said to be measurable if $f^{-1}(U) \in \mathcal{F}$ whenever $U \in \tau$. (Inverse images of open sets are measurable.)

Note the analogy with a continuous function mapping one topological space to another for which inverse images of open sets are open.

Definition 7.14 Let $\{a_n\}_{n=1}^{\infty} \subseteq X$, (X, τ) a topological space. Then

$$\lim_{n \to \infty} a_n = a$$

means that whenever $a \in U \in \tau$, there exists n_0 such that if $n > n_0$, then $a_n \in U$. (Every open set containing a also contains a_n for all but finitely many values of n.)

Usually X will be $\mathbb{R}$, $\mathbb{C}$, or $[0, \infty]$ with the usual topology. In the case of $[0, \infty]$, the sets $[0, b)$ and $(a, \infty]$ for all a and b positive numbers form a subbasis for the topology of this space. Other spaces which may be of interest include $\mathbb{R}^n$, $\mathbb{C}^n$, or even some infinite dimensional space. However, we will always assume (X, τ) satisfies the following properties.

$$\tau \text{ has a countable basis, } \mathcal{B}. \tag{6}$$

Whenever $U \in \mathcal{B}$, there exists a sequence of open sets, $\{V_m\}_{m=1}^{\infty}$, such that

$$\cdots V_m \subseteq \overline{V}_m \subseteq V_{m+1} \subseteq \cdots, \quad U = \bigcup_{m=1}^{\infty} V_m. \tag{7}$$

Note that $\mathbb{R}$, $\mathbb{C}$, $\mathbb{R}^n$, $\mathbb{C}^n$, and $[0, \infty]$ all have these properties.

Theorem 7.15 *Let f_n and f be functions mapping Ω to X where $\mathcal{F}$ is a σ algebra of measurable sets of Ω and (X, τ) is a topological space satisfying Formulas 6-7. Then if f_n is measurable, and $f(\omega) = \lim_{n \to \infty} f_n(\omega)$, it follows that f is also measurable. (Pointwise limits of measurable functions are measurable.)*

Proof: Let $\mathcal{B}$ be the countable basis of Formula 6 and let $U \in \mathcal{B}$. Let $\{V_m\}$ be the sequence of Formula 7. Since f is the pointwise limit of f_n,

$$f^{-1}(V_m) \subseteq \{\omega : f_k(\omega) \in V_m \text{ for all } k \text{ large enough}\} \subseteq f^{-1}(\overline{V}_m).$$

Therefore,

$$f^{-1}(U) = \cup_{m=1}^{\infty} f^{-1}(V_m) \subseteq \cup_{m=1}^{\infty} \cup_{n=1}^{\infty} \cap_{k=n}^{\infty} f_k^{-1}(V_m)$$

$$\subseteq \cup_{m=1}^{\infty} f^{-1}(\overline{V}_m) = f^{-1}(U).$$

It follows $f^{-1}(U) \in \mathcal{F}$ because it equals the expression in the middle which is measurable. Now let $W \in \tau$. Since $\mathcal{B}$ is countable, $W = \cup_{n=1}^{\infty} U_n$ for some sets $U_n \in \mathcal{B}$. Hence

$$f^{-1}(W) = \cup_{n=1}^{\infty} f^{-1}(U_n) \in \mathcal{F}.$$

This proves the theorem.

Example 7.16 *Let $X = [-\infty, \infty]$ and let a basis for a topology, τ, be sets of the form $[-\infty, a)$, (a, b), and $(a, \infty]$. Then it is clear that (X, τ) satisfies Formulas 6 - 7 with a countable basis, $\mathcal{B}$, given by sets of this form but with a and b rational.*

Definition 7.17 *Let $f_n : \Omega \to [-\infty, \infty]$.*

$$\limsup_{n \to \infty} f_n(\omega) = \lim_{n \to \infty} (\sup\{f_k(\omega) : k \geq n\}). \tag{8}$$

$$\lim_{n \to \infty} \inf f_n(\omega) = \lim_{n \to \infty} (\inf\{f_k(\omega) : k \geq n\}). \tag{9}$$

Note that in $[-\infty, \infty]$ with the topology just described, every increasing sequence converges and every decreasing sequence converges. This follows from Definition 7.14. Also, if

$$A_n(\omega) = \inf\{f_k(\omega) : k \geq n\}, \ B_n(\omega) = \sup\{f_k(\omega) : k \geq n\}.$$

It is clear that $B_n(\omega)$ is decreasing while $A_n(\omega)$ is increasing. Therefore, Formulas 8 and 9 always make sense unlike the limit.

Lemma 7.18 *Let $f : \Omega \to [-\infty, \infty]$ where $\mathcal{F}$ is a σ algebra of subsets of Ω. Then f is measurable if any of the following hold.*

$$f^{-1}((d, \infty]) \in \mathcal{F} \text{ for all finite } d,$$

$$f^{-1}([-\infty, c)) \in \mathcal{F} \text{ for all finite } c,$$

$$f^{-1}([d, \infty]) \in \mathcal{F} \text{ for all finite } d,$$

$$f^{-1}([-\infty, c]) \in \mathcal{F} \text{ for all finite } c.$$

Proof: First note that the first and the third are equivalent. To see this, note

$$f^{-1}([d, \infty]) = \cap_{n=1}^{\infty} f^{-1}((d - 1/n, \infty]),$$

$$f^{-1}((d, \infty]) = \cup_{n=1}^{\infty} f^{-1}([d + 1/n, \infty]).$$

Similarly, the second and fourth conditions are equivalent.

$$f^{-1}([-\infty, c]) = (f^{-1}((c, \infty]))^C$$

so the first and fourth conditions are equivalent. Thus all four conditions are equivalent and if any of them hold,

$$f^{-1}((a, b)) = f^{-1}([-\infty, b)) \cap f^{-1}((a, \infty]) \in \mathcal{F}.$$

Thus $f^{-1}(B) \in \mathcal{F}$ whenever B is a basic open set described in Example 7.16. Since every open set can be obtained as a countable union of these basic open sets, it follows that if any of the four conditions hold, then f is measurable. This proves the lemma.

Theorem 7.19 *Let $f_n : \Omega \to [-\infty, \infty]$ be measurable with respect to a σ algebra, $\mathcal{F}$, of subsets of Ω. Then $\limsup_{n \to \infty} f_n$ and $\liminf_{n \to \infty} f_n$ are measurable.*

Proof: Let $g_n(\omega) = \sup\{f_k(\omega) : k \geq n\}$. Then

$$g_n^{-1}((c, \infty]) = \cup_{k=n}^{\infty} f_k^{-1}((c, \infty]) \in \mathcal{F}.$$

Therefore g_n is measurable.

$$\limsup_{n\to\infty} f_n(\omega) = \lim_{n\to\infty} g_n(\omega)$$

and so by Theorem 7.15 $\limsup_{n\to\infty} f_n$ is measurable. Similar reasoning shows $\liminf_{n\to\infty} f_n$ is measurable.

Theorem 7.20 *Let $f_i, i = 1, \cdots, n$ be a measurable function mapping Ω to the topological space (X, τ) and suppose that τ has a countable basis, $\mathcal{B}$. Then $\mathbf{f} = (f_1 \cdots f_n)^T$ is a measurable function from Ω to $\prod_{i=1}^n X$. (Here it is understood that the topology of $\prod_{i=1}^n X$ is the standard product topology discussed in Chapter 1 and that $\mathcal{F}$ is the σ algebra of measurable subsets of Ω.)*

Proof: First we observe that sets of the form $\prod_{i=1}^n B_i$, $B_i \in \mathcal{B}$ form a countable basis for the product topology. Now

$$\mathbf{f}^{-1}(\prod_{i=1}^n B_i) = \cap_{i=1}^n f_i^{-1}(B_i) \in \mathcal{F}.$$

Since every open set is a countable union of these sets, it follows $\mathbf{f}^{-1}(U) \in \mathcal{F}$ for all open U.

Theorem 7.21 *Let $(\Omega, \mathcal{F})$ be a measure space and let $f_i, i = 1, \cdots, n$ be measurable functions mapping Ω to (X, τ), a topological space with a countable basis. Let $g : \prod_{i=1}^n X \to X$ be continuous and let $\mathbf{f} = (f_1 \cdots f_n)^T$. Then $g \circ \mathbf{f}$ is a measurable function.*

Proof: Let U be open.

$$(g \circ \mathbf{f})^{-1}(U) = \mathbf{f}^{-1}(g^{-1}(U)) = \mathbf{f}^{-1}(\text{open set}) \in \mathcal{F}$$

by Theorem 7.20.

Example 7.22 *Let $X = (-\infty, \infty]$ with a basis for the topology given by sets of the form (a, b) and $(c, \infty]$, a, b, c rational numbers. Let $+ : X \times X \to X$ be given by $+(x, y) = x + y$. Then $+$ is continuous; so if f, g are measurable functions mapping Ω to X, we may conclude by Theorem 7.21 that $f + g$ is also measurable. Also, if a, b are positive real numbers and $l(x, y) = ax + by$, then $l : X \times X \to X$ is continuous and so $l(f, g) = af + bg$ is measurable.*

Note that the basis given in this example provides the usual notions of convergence in $(-\infty, \infty]$. Theorems 7.20 and 7.21 imply that under appropriate conditions, sums, products, and, more generally, continuous functions of measurable functions are measurable. The following is also interesting.

Theorem 7.23 *Let* $f : \Omega \rightarrow X$ *be measurable. Then* $f^{-1}(B) \in \mathcal{F}$ *for every Borel set,* B, *of* (X, τ).

Proof: Let $\mathcal{S} \equiv \{B \subseteq X$ such that $f^{-1}(B) \in \mathcal{F}\}$. $\mathcal{S}$ contains all open sets. It is also clear that $\mathcal{S}$ is a σ algebra. Hence $\mathcal{S}$ contains the Borel sets because the Borel sets are defined as the intersection of all σ algebras containing the open sets.

The following theorem is often very useful when dealing with sequences of measurable functions.

Theorem 7.24 *(Egoroff) Let* $(\Omega, \mathcal{F}, \mu)$ *be a finite measure space*

$$(\mu(\Omega) < \infty)$$

and let f_n, f *be complex valued measurable functions such that*

$$\lim_{n \to \infty} f_n(\omega) = f(\omega)$$

for all $\omega \notin E$ *where* $\mu(E) = 0$. *Then for every* $\varepsilon > 0$, *there exists a set,*

$$F \supseteq E, \ \mu(F) < \varepsilon,$$

such that f_n *converges uniformly to* f *on* F^C.

Proof: Let $E_{km} = \{\omega : |f_n(\omega) - f(\omega)| \geq 1/m$ for some $n > k\}$. By Theorems 7.20 and 7.21,

$$\{\omega : |f_n(\omega) - f(\omega)| \geq 1/m\}$$

is measurable. Hence E_{km} is measurable because

$$E_{km} = \cup_{n=k+1}^{\infty}\{\omega : |f_n(\omega) - f(\omega)| \geq 1/m\}.$$

For fixed m, $\cap_{k=1}^{\infty} E_{km}$ is a measurable subset of E and so it has measure 0. Note also that

$$E_{km} \supseteq E_{(k+1)m}.$$

Since $\mu(E_{1m}) < \infty$,

$$0 = \mu(\cap_{k=1}^{\infty} E_{km}) = \lim_{k \to \infty} \mu(E_{km})$$

by Theorem 7.12. Let $k(m)$ be chosen such that $\mu(E_{k(m)m}) < \varepsilon 2^{-m}$. Let

$$F = E \cup \bigcup_{m=1}^{\infty} E_{k(m)m}.$$

Then $\mu(F) < \varepsilon$ because

$$\mu(F) \leq \mu(E) + \sum_{m=1}^{\infty} \mu\left(E_{k(m)m}\right).$$

Now let $\eta > 0$ be given and pick m_0 such that $m_0^{-1} < \eta$. If $\omega \in F^C$, then

$$\omega \in \bigcap_{m=1}^{\infty} E_{k(m)m}^C.$$

Hence $\omega \in E_{k(m_0)m_0}^C$ so

$$|f_n(\omega) - f(\omega)| < 1/m_0 < \eta$$

for all $n > k(m_0)$. This holds for all $\omega \in F^C$ and so f_n converges uniformly to f on F^C. This proves the theorem.

We conclude this chapter with a comment about notation. We say that something happens for μ a.e. ω and say μ almost everywhere if there exists a set E with $\mu(E) = 0$ and the thing takes place for all $\omega \notin E$. Thus $f(\omega) = g(\omega)$ a.e. if $f(\omega) = g(\omega)$ for all $\omega \notin E$ where $\mu(E) = 0$.

7.3   Exercises

1. Let $\Omega = \mathbb{N} = \{1, 2, \cdots\}$. Let $\mathcal{F} = \mathcal{P}(\mathbb{N})$ and let $\mu(S) =$ number of elements in S. Thus $\mu(\{1\}) = 1 = \mu(\{2\})$, $\mu(\{1, 2\}) = 2$, etc. Show $(\Omega, \mathcal{F}, \mu)$ is a measure space. It is called counting measure.

2. Let Ω be any uncountable set and let $\mathcal{F} = \{A \subseteq \Omega : \text{either } A \text{ or } A^C$ is countable$\}$. Let $\mu(A) = 1$ if A is uncountable and $\mu(A) = 0$ if A is countable. Show $(\Omega, \mathcal{F}, \mu)$ is a measure space.

3. Let $\mathcal{F}$ be a σ algebra of subsets of Ω and suppose $\mathcal{F}$ has infinitely many elements. Show that $\mathcal{F}$ is uncountable.

4. Prove Lemma 7.2.

5. We say g is Borel measurable if whenever U is open, $g^{-1}(U)$ is Borel. Let $f : \Omega \to X$ and let $g : X \to Y$ where X, Y are topological spaces and $\mathcal{F}$ is a σ algebra of sets of Ω. Suppose f is measurable and g is Borel measurable. Show $g \circ f$ is measurable.

6. Let $(\Omega, \mathcal{F})$ be a measure space and suppose $f : \Omega \to \mathbb{C}$. Show f is measurable if and only if $\operatorname{Re} f$ and $\operatorname{Im} f$ are measurable real-valued functions.

7. Let $(\Omega, \mathcal{F}, \mu)$ be a measure space. Define $\overline{\mu} : \mathcal{P}(\Omega) \to [0, \infty]$ by

$$\overline{\mu}(A) = \inf\{\mu(B) : B \supseteq A, \; B \in \mathcal{F}\}.$$

Show $\overline{\mu}$ satisfies

$$\overline{\mu}(\emptyset) = 0, \text{ if } A \subseteq B, \; \overline{\mu}(A) \leq \overline{\mu}(B), \; \overline{\mu}(\cup_{i=1}^{\infty} A_i) \leq \sum_{i=1}^{\infty} \overline{\mu}(A_i).$$

If $\overline{\mu}$ satisfies these conditions, it is called an outer measure. This shows every measure determines an outer measure on the power set.

8. Let $\{E_i\}$ be a sequence of measurable sets with the property that

$$\sum_{i=1}^{\infty} \mu(E_i) < \infty.$$

Let $S = \{\omega \in \Omega$ such that $\omega \in E_i$ for infinitely many values of $i\}$. Show $\mu(S) = 0$ and S is measurable. This is part of the Borel Cantelli lemma.

9. ↑ Let f_n, f be measurable functions with values in $\mathbb{C}$. We say that f_n converges in measure if

$$\lim_{n \to \infty} \mu(x \in \Omega : |f(x) - f_n(x)| \geq \varepsilon) = 0$$

for each fixed $\varepsilon > 0$. Prove the theorem of F. Riesz. If f_n converges to f in measure, then there exists a subsequence $\{f_{n_k}\}$ which converges to f a.e. **Hint:** Choose n_1 such that

$$\mu(x : |f(x) - f_{n_1}(x)| \geq 1) < 1/2.$$

Choose $n_2 > n_1$ such that

$$\mu(x : |f(x) - f_{n_2}(x)| \geq 1/2) < 1/2^2,$$

$n_3 > n_2$ such that

$$\mu(x : |f(x) - f_{n_3}(x)| \geq 1/3) < 1/2^3,$$

etc. Now consider what it means for $f_{n_k}(x)$ to fail to converge to $f(x)$. Then remember Problem 8.

10. Let $\mathcal{C} \equiv \{E_i\}_{i=1}^{\infty}$ be a countable collection of sets and let $\Omega_1 \equiv \cup_{i=1}^{\infty} E_i$. Show there exists an algebra of sets, $\mathcal{A}$, such that $\mathcal{A} \supseteq \mathcal{C}$ and $\mathcal{A}$ is countable. **Hint:** Let $\mathcal{C}_1$ denote all finite unions of sets of $\mathcal{C}$ and Ω_1. Thus $\mathcal{C}_1$ is countable. Now let $\mathcal{B}_1$ denote all complements with respect to Ω_1 of sets of $\mathcal{C}_1$. Let $\mathcal{C}_2$ denote all finite unions of sets of $\mathcal{B}_1 \cup \mathcal{C}_1$. Continue in this way, obtaining an increasing sequence $\mathcal{C}_n$, each of which is countable. Let

$$\mathcal{A} \equiv \cup_{i=1}^{\infty} \mathcal{C}_i.$$

Chapter 8
The Abstract Lebesgue Integral

In this chapter we develop the Lebesgue integral and present some of its most important properties. In all that follows μ will be a measure defined on a σ algebra $\mathcal{F}$ of subsets of Ω. We always define $0 \cdot \infty = 0$. This may seem somewhat arbitrary and this is so. However, a little thought will soon demonstrate that this is the right definition for this meaningless expression in the context of measure theory. To see this, consider the zero function defined on $\mathbb{R}$. What do we want the integral of this function to be? Obviously, by an analogy with the Riemann integral, we would want this to equal zero. Formally, it is zero times the length of the set or infinity. The following notation will be used.

For a set E,

$$\mathcal{X}_E(\omega) = \begin{cases} 1 \text{ if } \omega \in E, \\ 0 \text{ if } \omega \notin E. \end{cases}$$

This is called the characteristic function of E.

Definition 8.1 A function, s, is called simple if it is measurable and has only finitely many values. These values will never be $\pm\infty$.

Definition 8.2 If $s(x) \geq 0$ and s is simple,

$$\int s \equiv \sum_{i=1}^{m} a_i \mu(A_i)$$

where $A_i = \{\omega : s(x) = a_i\}$ and $a_1, \cdots, a_m$ are the distinct values of s.

Note that $\int s$ could equal $+\infty$ if $\mu(A_k) = \infty$ and $a_k > 0$ for some k, but $\int s$ is well defined because $s \geq 0$ and we use the convention that $0 \cdot \infty = 0$.

Lemma 8.3 *If $a, b \geq 0$ and if s and t are nonnegative simple functions, then*

$$\int as + bt \equiv a \int s + b \int t.$$

Proof: Let

$$s(\omega) = \sum_{i=1}^{n} \alpha_i \mathcal{X}_{A_i}(\omega), \; t(\omega) = \sum_{i=1}^{m} \beta_j \mathcal{X}_{B_j}(\omega)$$

where α_i are the distinct values of s and the β_j are the distinct values of t. Clearly $as + bt$ is a nonnegative simple function. Also,

$$(as + bt)(\omega) = \sum_{j=1}^{m} \sum_{i=1}^{n} (a\alpha_i + b\beta_j) \mathcal{X}_{A_i \cap B_j}(\omega)$$

The Abstract Lebesgue Integral

where the sets $A_i \cap B_j$ are disjoint. Now we don't know that all the values $a\alpha_i + b\beta_j$ are distinct, but we note that if $E_1, \cdots, E_r$ are disjoint measurable sets whose union is E, then $\alpha\mu(E) = \alpha \sum_{i=1}^r \mu(E_i)$. Thus

$$
\begin{aligned}
\int \alpha s + bt &= \sum_{j=1}^m \sum_{i=1}^n (a\alpha_i + b\beta_j)\mu(A_i \cap B_j) \\
&= a \sum_{i=1}^n \alpha_i \mu(A_i) + b \sum_{j=1}^m \beta_j \mu(B_j) \\
&= a \int s + b \int t.
\end{aligned}
$$

This proves the lemma.

Corollary 8.4 *Let* $s = \sum_{i=1}^n a_i \mathcal{X}_{E_i}$ *where* $a_i \geq 0$ *and the* E_i *are not necessarily disjoint. Then*

$$
\int s = \sum_{i=1}^n a_i \mu(E_i).
$$

Proof: $\int a\mathcal{X}_{E_i} = a\mu(E_i)$ so this follows from Lemma 8.3.

Now we are ready to define the Lebesgue integral of a nonnegative measurable function.

Definition 8.5 *Let* $f : \Omega \to [0, \infty]$ *be measurable. Then*

$$
\int f d\mu \equiv \sup\{\int s : 0 \leq s \leq f, \, s \text{ simple}\}.
$$

Lemma 8.6 *If* $s \geq 0$ *is a nonnegative simple function,* $\int s d\mu = \int s$. *Moreover, if* $f \geq 0$, *then* $\int f d\mu \geq 0$.

Proof: The second claim is obvious. To verify the first, suppose $0 \leq t \leq s$ and t is simple. Then clearly $\int t \leq \int s$ and so

$$
\int s d\mu = \sup\{\int t : 0 \leq t \leq s, \, t \text{ simple}\} \leq \int s.
$$

But $s \leq s$ and s is simple so $\int s d\mu \geq \int s$.

The next theorem is one of the big results that justifies the use of the Lebesgue integral.

Theorem 8.7 *(Monotone Convergence theorem) Let* $f \geq 0$ *and suppose* $\{f_n\}$ *is a sequence of nonnegative measurable functions satisfying*

$$
\lim_{n \to \infty} f_n(\omega) = f(\omega) \text{ for each } \omega.
$$

136

$$\cdots f_n(\omega) \le f_{n+1}(\omega) \cdots \tag{1}$$

Then f is measurable and

$$\int f d\mu = \lim_{n\to\infty} \int f_n d\mu.$$

Proof: First note that f is measurable by Theorem 7.15 since it is the limit of measurable functions. It is also clear from Formula 1 that $\lim_{n\to\infty} \int f_n d\mu$ exists because $\{\int f_n d\mu\}$ forms an increasing sequence. This limit may be $+\infty$ but in any case,

$$\lim_{n\to\infty} \int f_n d\mu \le \int f d\mu$$

because $\int f_n d\mu \le \int f d\mu$.

Let $\delta \in (0,1)$ and let s be a simple function with

$$0 \le s(\omega) \le f(\omega), \quad s(\omega) = \sum_{i=1}^{r} \alpha_i \mathcal{X}_{A_i}(\omega).$$

Then $(1-\delta)s(\omega) \le f(\omega)$ for all ω with strict inequality holding whenever $f(\omega) > 0$. Let

$$E_n = \{\omega : f_n(\omega) \ge (1-\delta)s(\omega)\} \tag{2}$$

Then

$$\cdots E_n \subseteq E_{n+1} \cdots, \quad \text{and} \quad \cup_{n=1}^{\infty} E_n = \Omega.$$

Therefore

$$\lim_{n\to\infty} \int s\mathcal{X}_{E_n} = \int s.$$

This follows from Theorem 7.12 which implies that $\alpha_i \mu(E_n \cap A_i) \to \alpha_i \mu(A_i)$. Thus, from Formula 2

$$\int f d\mu \ge \int f_n d\mu \ge \int f_n \mathcal{X}_{E_n} d\mu \ge (\int s\mathcal{X}_{E_n} d\mu)(1-\delta). \tag{3}$$

Letting $n \to \infty$ in Formula 3 we see that

$$\int f d\mu \ge \lim_{n\to\infty} \int f_n d\mu \ge (1-\delta) \int s. \tag{4}$$

Now let $\delta \downarrow 0$ in Formula 4 to obtain

$$\int f d\mu \ge \lim_{n\to\infty} \int f_n d\mu \ge \int s.$$

Now s was an arbitrary simple function less than or equal to f. Hence,

$$\int f d\mu \ge \lim_{n\to\infty} \int f_n d\mu \ge \sup\{\int s : 0 \le s \le f, \ s \text{ simple}\} \equiv \int f d\mu.$$

This proves the theorem.

The next theorem will be used frequently. It says roughly that measurable functions are pointwise limits of simple functions. This is similar to continuous functions being the limit of step functions.

Theorem 8.8 *Let $f \geq 0$ be measurable. Then there exists a sequence of simple functions $\{s_n\}$ satisfying*

$$0 \leq s_n(\omega) \tag{5}$$

$$\cdots s_n(\omega) \leq s_{n+1}(\omega) \cdots$$

$$f(\omega) = \lim_{n \to \infty} s_n(\omega) \text{ for all } \omega \in \Omega. \tag{6}$$

Before proving this, we give a definition.

Definition 8.9 If f, g are functions having values in $[0, \infty]$,

$$f \vee g = \max(f, g), \ f \wedge g = \min(f, g).$$

Note that if f, g have finite values,

$$f \vee g = 2^{-1}(f + g + |f - g|), \ f \wedge g = 2^{-1}(f + g - |f - g|).$$

From this observation, the following lemma is obvious.

Lemma 8.10 *If s, t are nonnegative simple functions, then*

$$s \vee t, \ s \wedge t$$

are also simple functions. (Recall $+\infty$ is not a value of either s or t.)

Proof of Theorem 8.8: Let

$$I = \{x : f(x) = +\infty\}.$$

Let $E_{nk} = f^{-1}([\frac{k}{n}, \frac{k+1}{n}))$. Let

$$t_n(\omega) = \sum_{k=0}^{2^n} \frac{k}{n} \mathcal{X}_{E_{nk}}(\omega) + n\mathcal{X}_I(\omega).$$

Then $t_n(\omega) \leq f(\omega)$ for all ω and $\lim_{n \to \infty} t_n(\omega) = f(\omega)$ for all ω. Let

$$s_1 = t_1, \ s_2 = t_1 \vee t_2, \ s_3 = t_1 \vee t_2 \vee t_3, \cdots.$$

Then the sequence $\{s_n\}$ satisfies Formulas 5-6 and this proves the theorem.

Next we show that the integral is linear on nonnegative functions. Roughly speaking, it shows the integral is trying to be linear and is only prevented from being linear at this point by not yet being defined on functions which could be negative or complex valued. We will define the integral for these functions soon and then this lemma will be the key to showing the integral is linear.

Lemma 8.11 *Let* $f, g \geq 0$ *be measurable. Let* $a, b \geq 0$ *be constants. Then*

$$\int (af + bg)d\mu = a \int f d\mu + b \int g d\mu.$$

Proof: Let $\{s_n\}$ and $\{\tilde{s}_n\}$ be increasing sequences of simple functions such that

$$\lim_{n \to \infty} s_n(\omega) = f(\omega), \quad \lim_{n \to \infty} \tilde{s}_n(\omega) = g(\omega).$$

Then by the monotone convergence theorem and Lemma 8.3,

$$
\begin{aligned}
\int (af + bg)d\mu &= \lim_{n \to \infty} \int (as_n + b\tilde{s}_n)d\mu \\
&= \lim_{n \to \infty} \int as_n + b\tilde{s}_n = \lim_{n \to \infty} a \int s_n + b \int \tilde{s}_n \\
&= \lim_{n \to \infty} a \int s_n d\mu + b \int \tilde{s}_n d\mu = a \int f d\mu + b \int g d\mu.
\end{aligned}
$$

This proves the lemma.

8.1   The space L^1

Now suppose f has complex values and is measurable. We need to define what is meant by the integral of such functions. First some theorems about measurability need to be shown.

Theorem 8.12 *Let* $f = u + iv$ *where* u, v *are real-valued functions. Then* f *is a measurable* $\mathbb{C}$ *valued function if and only if* u *and* v *are both measurable* $\mathbb{R}$ *valued functions.*

Proof: Suppose first that f is measurable. Let $V \subseteq \mathbb{R}$ be open.

$$u^{-1}(V) = \{\omega : u(\omega) \in V\} = \{\omega : f(\omega) \in V + i\mathbb{R}\} \in \mathcal{F},$$

$$v^{-1}(V) = \{\omega : v(\omega) \in V\} = \{\omega : f(\omega) \in \mathbb{R} + iV\} \in \mathcal{F}.$$

Now suppose u and v are real and measurable.

$$f^{-1}((a, b) + i(c, d)) = u^{-1}(a, b) \cap v^{-1}(c, d) \in \mathcal{F}.$$

Since every open set in $\mathbb{C}$ may be written as a countable union of open sets of the form $(a, b) + i(c, d)$, it follows that $f^{-1}(U) \in \mathcal{F}$ whenever U is open in $\mathbb{C}$. This proves the theorem.

Definition 8.13 $L^1(\Omega)$ is the space of complex valued measurable functions, f, satisfying

$$\int |f(\omega)| d\mu < \infty.$$

Note that if $f : \Omega \to \mathbb{C}$ is measurable, then by Theorem 7.21, $|f| : \Omega \to \mathbb{R}$ is also measurable.

Definition 8.14 If u is real-valued,

$$u^+ \equiv u \vee 0, \ u^- \equiv -(u \wedge 0).$$

Thus u^+ and u^- are both nonnegative and

$$u = u^+ - u^-, \ |u| = u^+ + u^-.$$

Definition 8.15 Let $f = u + iv$ where u, v are real-valued. Suppose $f \in L^1(\Omega)$. Then

$$\int f d\mu \equiv \int u^+ d\mu - \int u^- d\mu + i[\int v^+ d\mu - \int v^- d\mu].$$

Note that all this is well defined because $\int |f| d\mu < \infty$ and so

$$\int u^+ d\mu, \int u^- d\mu, \int v^+ d\mu, \int v^- d\mu$$

are all finite. The next theorem shows the integral is linear on $L^1(\Omega)$.

Theorem 8.16 $L^1(\Omega)$ *is a complex vector space and if* $a, b \in \mathbb{C}$ *and*

$$f, g \in L^1(\Omega),$$

then

$$\int af + bg d\mu = a \int f d\mu + b \int g d\mu. \tag{7}$$

Proof: First suppose f, g are real-valued and in $L^1(\Omega)$. We note that

$$h^+ = 2^{-1}(h + |h|), \ h^- = 2^{-1}(|h| - h)$$

whenever h is real-valued. Consequently,

$$f^+ + g^+ - (f^- + g^-) = (f + g)^+ - (f + g)^- = f + g.$$

140

Hence
$$f^+ + g^+ + (f+g)^- = (f+g)^+ + f^- + g^-. \tag{8}$$

From Lemma 8.11,
$$\int f^+ d\mu + \int g^+ d\mu + \int (f+g)^- d\mu = \int f^- d\mu + \int g^- d\mu + \int (f+g)^+ d\mu. \tag{9}$$

Since all integrals are finite,
$$\begin{aligned} \int (f+g) d\mu &\equiv \int (f+g)^+ d\mu - \int (f+g)^- d\mu \tag{10} \\ &= \int f^+ d\mu + \int g^+ d\mu - \left(\int f^- d\mu + \int g^- d\mu \right) \\ &\equiv \int f d\mu + \int g d\mu. \end{aligned}$$

Now suppose that c is a real constant and f is real-valued. Note
$$(cf)^- = -cf^+ \text{ if } c < 0, \quad (cf)^- = cf^- \text{ if } c \geq 0.$$
$$(cf)^+ = -cf^- \text{ if } c < 0, \quad (cf)^+ = cf^+ \text{ if } c \geq 0.$$

If $c < 0$, we use the above and Lemma 8.11 to write
$$\begin{aligned} \int cf d\mu &\equiv \int (cf)^+ d\mu - \int (cf)^- d\mu \\ &= -c \int f^- d\mu + c \int f^+ d\mu \equiv c \int f d\mu. \end{aligned}$$

Similarly, if $c \geq 0$,
$$\begin{aligned} \int cf d\mu &\equiv \int (cf)^+ d\mu - \int (cf)^- d\mu \\ &= c \int f^+ d\mu - c \int f^- d\mu \equiv c \int f d\mu. \end{aligned}$$

This shows Formula 7 holds if f, g, a, and b are all real-valued. To conclude, let $a = \alpha + i\beta$, $f = u + iv$ and use the preceding.
$$\begin{aligned} \int af d\mu &= \int (\alpha + i\beta)(u + iv) d\mu \\ &= \int (\alpha u - \beta v) + i(\beta u + \alpha v) d\mu \\ &= \alpha \int u d\mu - \beta \int v d\mu + i\beta \int u d\mu + i\alpha \int v d\mu \\ &= (\alpha + i\beta)\left(\int u d\mu + i \int v d\mu \right) = a \int f d\mu. \end{aligned}$$

Thus Formula 7 holds whenever f, g, a, and b are complex valued. It is obvious that $L^1(\Omega)$ is a vector space. This proves the theorem.

The next theorem, known as Fatou's lemma is another important theorem which justifies the use of the Lebesgue integral.

Theorem 8.17 *(Fatou's lemma) Let f_n be a nonnegative measurable function with values in $[0, \infty]$. Let $g(\omega) = \liminf_{n \to \infty} f_n(\omega)$. Then g is measurable and*

$$\int g \, d\mu \leq \lim_{n \to \infty} \inf \int f_n \, d\mu.$$

Proof: Let $g_n(\omega) = \inf\{f_k(\omega) : k \geq n\}$. Then

$$g_n^{-1}([a, \infty]) = \cap_{k=n}^{\infty} f_k^{-1}([a, \infty]) \in \mathcal{F}.$$

Thus g_n is measurable by Lemma 7.18. Also $g(\omega) = \lim_{n \to \infty} g_n(\omega)$ so g is measurable because it is the pointwise limit of measurable functions. Now the functions g_n form an increasing sequence of nonnegative measurable functions so the monotone convergence theorem applies. This yields

$$\int g \, d\mu = \lim_{n \to \infty} \int g_n \, d\mu \leq \lim_{n \to \infty} \inf \int f_n \, d\mu.$$

The last inequality holding because

$$\int g_n \, d\mu \leq \int f_n \, d\mu.$$

This proves the Theorem.

Theorem 8.18 *(Triangle inequality) Let $f \in L^1(\Omega)$. Then*

$$\left| \int f \, d\mu \right| \leq \int |f| \, d\mu.$$

Proof: $\int f \, d\mu \in \mathbb{C}$ so there exists $\alpha \in \mathbb{C}$, $|\alpha| = 1$ such that $|\int f \, d\mu| = \alpha \int f \, d\mu = \int \alpha f \, d\mu$. Hence

$$\begin{aligned}
\left| \int f \, d\mu \right| &= \int \alpha f \, d\mu = \int (\operatorname{Re}(\alpha f) + i \operatorname{Im}(\alpha f)) \, d\mu \\
&= \int \operatorname{Re}(\alpha f) \, d\mu = \int (\operatorname{Re}(\alpha f))^+ \, d\mu - \int (\operatorname{Re}(\alpha f))^- \, d\mu \\
&\leq \int (\operatorname{Re}(\alpha f))^+ + (\operatorname{Re}(\alpha f))^- \, d\mu \leq \int |\alpha f| \, d\mu = \int |f| \, d\mu
\end{aligned}$$

which proves the theorem.

Theorem 8.19 *(Dominated Convergence theorem) Let $f_n \in L^1(\Omega)$ and suppose*

$$f(\omega) = \lim_{n \to \infty} f_n(\omega),$$

and there exists a measurable function g, with values in $[0, \infty]$, such that

$$|f_n(\omega)| \le g(\omega) \text{ and } \int g(\omega) d\mu < \infty.$$

Then $f \in L^1(\Omega)$ and

$$\int f d\mu = \lim_{n \to \infty} \int f_n d\mu.$$

Proof: f is measurable by Theorem 7.15. Since $|f| \le g$, it follows that

$$f \in L^1(\Omega) \text{ and } |f - f_n| \le 2g.$$

By Fatou's lemma (Theorem 8.17),

$$
\begin{aligned}
\int 2g d\mu &\le \liminf_{n \to \infty} \int 2g - |f - f_n| d\mu \\
&= \int 2g d\mu - \limsup_{n \to \infty} \int |f - f_n| d\mu.
\end{aligned}
$$

Subtracting $\int 2g d\mu$,

$$0 \le -\limsup_{n \to \infty} \int |f - f_n| d\mu.$$

Hence

$$0 \ge \limsup_{n \to \infty} \left(\int |f - f_n| d\mu \right) \ge \limsup_{n \to \infty} \left| \int f d\mu - \int f_n d\mu \right|$$

which proves the theorem.

Definition 8.20 Let E be a measurable subset of Ω.

$$\int_E f d\mu \equiv \int f \mathcal{X}_E d\mu.$$

Also we may refer to $L^1(E)$. The σ algebra in this case is just

$$\{E \cap A : A \in \mathcal{F}\}$$

and the measure is μ restricted to this smaller σ algebra. Clearly, if $f \in L^1(\Omega)$, then

$$f \mathcal{X}_E \in L^1(E)$$

and if $f \in L^1(E)$, then letting $\tilde{f}$ be the 0 extension of f off of E, we see that $\tilde{f} \in L^1(\Omega)$.

8.2 Double sums of nonnegative terms

The definition of the Lebesgue integral and the monotone convergence theorem imply that the order of summation of a double sum of nonnegative terms can be interchanged and in fact the terms can be added in any order. To see this, let $\Omega = \mathbb{N} \times \mathbb{N}$ and let μ be counting measure defined on the set of all subsets of $\mathbb{N} \times \mathbb{N}$. Thus, $\mu(E) =$ the number of elements of E. Then $(\Omega, \mu, \mathcal{P}(\Omega))$ is a measure space and if $a : \Omega \to [0, \infty]$, then a is a measurable function. Following the usual notation, $a_{ij} \equiv a(i, j)$.

Theorem 8.21 *Let $a : \Omega \to [0, \infty]$. Then*

$$\sum_{i=1}^{\infty} \sum_{j=1}^{\infty} a_{ij} = \sum_{j=1}^{\infty} \sum_{i=1}^{\infty} a_{ij} = \int a \, d\mu = \sum_{k=1}^{\infty} a(\theta(k))$$

where θ is any one to one and onto map from $\mathbb{N}$ to Ω.

Proof: By the definition of the integral,

$$\sum_{j=1}^{n} \sum_{i=1}^{l} a_{ij} \leq \int a \, d\mu$$

for any n, l. Therefore, by the definition of what is meant by an infinite sum,

$$\sum_{j=1}^{\infty} \sum_{i=1}^{\infty} a_{ij} \leq \int a \, d\mu.$$

Now let $s \leq a$ and s is a nonnegative simple function. If $s(i, j) > 0$ for infinitely many values of $(i, j) \in \Omega$, then

$$\infty = \int a \, d\mu = \sum_{i=1}^{\infty} \sum_{j=1}^{\infty} a_{ij} = \sum_{j=1}^{\infty} \sum_{i=1}^{\infty} a_{ij} = \sum_{k=1}^{\infty} a(\theta(k))$$

and the conclusion of the theorem is established. Therefore, it suffices to assume $s(i, j) > 0$ for only finitely many values of $(i, j) \in \mathbb{N} \times \mathbb{N}$. Hence, for some $n > 1$,

$$\int s \leq \sum_{j=1}^{n} \sum_{i=1}^{n} a_{ij} \leq \sum_{j=1}^{\infty} \sum_{i=1}^{\infty} a_{ij}.$$

Since s is an arbitrary nonnegative simple function,

$$\int a \, d\mu \leq \sum_{j=1}^{\infty} \sum_{i=1}^{\infty} a_{ij}.$$

The same argument holds if i and j are interchanged. The last equation follows from the monotone convergence theorem.

8.3 Exercises

1. Let $\Omega = \mathbb{N} = \{1, 2, \cdots\}$ and $\mu(S) =$ number of elements in S. If

$$f : \Omega \to \mathbb{C}$$

what do we mean by $\int f d\mu$? Which functions are in $L^1(\Omega)$?

2. Give an example of a measure space, $(\Omega, \mu, \mathcal{F})$, and a sequence of nonnegative measurable functions $\{f_n\}$ converging pointwise to a function f, such that inequality is obtained in Fatou's lemma.

3. Let $(\Omega, \mathcal{F}, \mu)$ be a measure space and let $\mathfrak{S} \subseteq L^1(\Omega)$. We say that $\mathfrak{S}$ is uniformly integrable if for every $\varepsilon > 0$ there exists $\delta > 0$ such that for all $f \in \mathfrak{S}$

$$|\int_E f d\mu| < \varepsilon \text{ whenever } \mu(E) < \delta.$$

Show that $|\mathfrak{S}| \equiv \{|f| : f \in \mathfrak{S}\}$ is uniformly integrable if $\mathfrak{S}$ is. Also show that $\mathfrak{S}$ is uniformly integrable if $\mathfrak{S}$ is finite.

4. Suppose (Ω, μ) is a finite measure space and $\mathfrak{S} \subseteq L^1(\Omega)$. Show $\mathfrak{S}$ is uniformly integrable and bounded if and only if there exists an increasing step function h which satisfies

$$\lim_{t \to \infty} \frac{h(t)}{t} = \infty, \quad \sup\left\{\int_\Omega h(|f|) \, d\mu : f \in \mathfrak{S}\right\} < \infty.$$

Hint: Let $a_n(f) \equiv \int_{E_n(f)} |f| \, d\mu$ where

$$E_n(f) \equiv \{x \in \Omega : |f(x)| \in [n, n+1)\}.$$

Show first that there exists a constant C such that

$$\sum_{n=1}^\infty a_n(f) < C \text{ for all } f \in \mathfrak{S}.$$

Next use uniform integrability to show there exist an increasing sequence of integers m_k such that

$$\sup\left\{\sum_{n=m_k}^\infty a_n(f) : f \in \mathfrak{S}\right\} < \frac{1}{3^k}.$$

Now let $\lambda_l = 1$ if $l \le m_1$, $\lambda_k = 2^k$ if $k \in [m_k, m_{k+1})$. Now show there exists a constant K such that

$$\sup\left\{\sum_{n=1}^{\infty} \lambda_n a_n\,(f) : f \in \mathfrak{S}\right\} < K.$$

On $[n, n+1)$ let $h\,(x) \equiv n\lambda_n$. The converse is easier.

5. Let $(\Omega, \mathcal{F}, \mu)$ be a measure space and suppose $f \in L^1(\Omega)$ has the property that whenever $\mu(E) > 0$,

$$\frac{1}{\mu(E)}|\int_E f d\mu| \le C.$$

Show $|f(\omega)| \le C$ a.e.

6. Let $\{a_n\}, \{b_n\}$ be sequences in $[-\infty, \infty]$. Show

$$\lim_{n\to\infty} \sup\,(-a_n) = -\lim_{n\to\infty} \inf\,(a_n)$$

$$\lim_{n\to\infty} \sup\,(a_n + b_n) \le \lim_{n\to\infty} \sup\,a_n + \lim_{n\to\infty} \sup\,b_n$$

provided no sum is of the form $\infty - \infty$. Also show strict inequality can hold in the inequality. State and prove corresponding statements for $\lim\inf$.

7. Let $(\Omega, \mathcal{F}, \mu)$ be a measure space and suppose $f, g : \Omega \to [-\infty, \infty]$ are measurable. Prove the sets

$$\{\omega : f(\omega) < g(\omega)\} \text{ and } \{\omega : f(\omega) = g(\omega)\}$$

are measurable.

8. Let $\{f_n\}$ be a sequence of real or complex valued measurable functions. Let
$$S = \{\omega : \{f_n(\omega)\} \text{ converges}\}.$$

Show S is measurable.

9. In the monotone convergence theorem

$$0 \le \cdots \le f_n(\omega) \le f_{n+1}(\omega) \le \cdots.$$

The sequence of functions is increasing. In what way can "increasing" be replaced by "decreasing"?

10. Let $(\Omega, \mathcal{F}, \mu)$ be a measure space and suppose f_n converges uniformly to f and that f_n is in $L^1(\Omega)$. When can we conclude that

$$\lim_{n \to \infty} \int f_n \, d\mu = \int f \, d\mu?$$

11. Suppose $u_n(t)$ is a differentiable function for $t \in (a, b)$ and suppose that for $t \in (a, b)$,

$$|u_n(t)|, \ |u_n'(t)| < K_n$$

where $\sum_{n=1}^{\infty} K_n < \infty$. Show

$$\left(\sum_{n=1}^{\infty} u_n(t) \right)' = \sum_{n=1}^{\infty} u_n'(t).$$

12. Let $\{f_n\}$ be a sequence of functions in $L^1(\Omega, \mathcal{S}, \mu)$. Show, using Problem 10 of Chapter 7, there exists a σ-finite set of $\mathcal{S}$, Ω_1, and a σ algebra of subsets of $\Omega_1, \mathcal{S}_1$, such that $\mathcal{S}_1 \subseteq \mathcal{S}$, $f_n = 0$ off Ω_1, $f_n \in L^1(\Omega_1, \mathcal{S}_1, \mu)$, and $\mathcal{S}_1 = \sigma(\mathcal{A})$, the σ algebra generated by $\mathcal{A}$, for some $\mathcal{A}$ a countable algebra. **Hint:** Consider the set $\mathcal{E}_n$ defined as

$$\left\{ f_n^{-1}(B(z, r)) : z \in \mathbb{Q} + i\mathbb{Q}, r > 0, r \in \mathbb{Q}, \text{ and } 0 \notin \overline{B(z, r)} \right\}.$$

Then $\mathcal{E} \equiv \cup \mathcal{E}_n$ is a countable collection of sets. Use Problem 10 of Chapter 7 to get $\mathcal{A} \supseteq \mathcal{E}$ and then let $\mathcal{S} \equiv \sigma(\mathcal{A})$.

147

Chapter 9
The Construction Of Measures

9.1 Outer measures

We have impressive theorems about measure spaces and the abstract Lebesgue integral but a paucity of interesting examples. In this chapter, we discuss the method of outer measures due to Caratheodory (1918). This approach shows how to obtain measure spaces starting with an outer measure.

Definition 9.1 Let Ω be a nonempty set and let $\mu : \mathcal{P}(\Omega) \to [0, \infty]$ satisfy

$$\mu(\emptyset) = 0,$$

$$\text{If } A \subseteq B, \text{ then } \mu(A) \leq \mu(B),$$

$$\mu(\cup_{i=1}^{\infty} E_i) \leq \sum_{i=1}^{\infty} \mu(E_i).$$

Such a function is called an outer measure. For $E \subseteq \Omega$, we say E is μ measurable if for all $S \subseteq \Omega$,

$$\mu(S) = \mu(S \setminus E) + \mu(S \cap E). \tag{1}$$

To help in remembering 1, think of a measurable set, E, as a knife which is used to divide an arbitrary set, S, into the pieces, $S \setminus E$ and $S \cap E$. If E is a sharp knife, the amount of stuff after cutting is the same as the amount you started with. The measurable sets are like sharp knives. The idea is to show that the measurable sets form a σ algebra. First we give a definition and a lemma.

Definition 9.2 $(\mu \lfloor S)(A) \equiv \mu(S \cap A)$ for all $A \subseteq \Omega$. Thus $\mu \lfloor S$ is the name of a new outer measure.

Lemma 9.3 If A is μ measurable, then A is $\mu \lfloor S$ measurable.

Proof: Suppose A is μ measurable. We need to show that for all $T \subseteq \Omega$,
$$(\mu \lfloor S)(T) = (\mu \lfloor S)(T \cap A) + (\mu \lfloor S)(T \setminus A).$$
Thus we need to show

$$\mu(S \cap T) = \mu(T \cap A \cap S) + \mu(T \cap S \cap A^C). \tag{2}$$

But we know Formula 2 holds because A is measurable. Apply Definition 9.1 to $S \cap T$ instead of S.

The next theorem is the main result on outer measures. It is a very general result which applies whenever one has an outer measure on the power set of any set. This theorem will be referred to as Caratheodory's procedure in the rest of the book.

Theorem 9.4 *The collection of μ measurable sets, S, forms a σ algebra and*

$$\text{If } F_i \in S, \ F_i \cap F_j = \emptyset, \text{ then } \mu(\cup_{i=1}^{\infty} F_i) = \sum_{i=1}^{\infty} \mu(F_i). \tag{3}$$

If $\cdots F_n \subseteq F_{n+1} \subseteq \cdots$, then if $F = \cup_{n=1}^{\infty} F_n$ and $F_n \in S$, it follows that

$$\mu(F) = \lim_{n \to \infty} \mu(F_n). \tag{4}$$

If $\cdots F_n \supseteq F_{n+1} \supseteq \cdots$, and if $F = \cap_{n=1}^{\infty} F_n$ for $F_n \in S$ then if $\mu(F_1) < \infty$, we may conclude that

$$\mu(F) = \lim_{n \to \infty} \mu(F_n). \tag{5}$$

$$(S, \mu) \text{ is complete.} \tag{6}$$

By this we mean that if $F \in S$ and if $E \subseteq \Omega$ with $\mu(E \setminus F) + \mu(F \setminus E) = 0$, then $E \in S$.

Proof: First note that $\emptyset$ and Ω are obviously in S. Now suppose that $A, B \in S$. We show $A \setminus B = A \cap B^C$ is in S. Using the assumption that $B \in S$ in the second equation below, in which $S \cap A$ plays the role of S in the definition for B being μ measurable,

$$\mu(S \cap (A \cap B^C)) + \mu(S \setminus (A \cap B^C)) = \mu(S \cap A \cap B^C) + \mu(S \cap (A^C \cup B))$$

$$= \mu(S \cap (A^C \cup B)) + \mu(S \cap A) - \mu(S \cap A \cap B). \tag{7}$$

The following picture of $S \cap (A^C \cup B)$ may be of use.

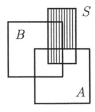

From Formula 7 and the picture, we see that

$$\mu(S \cap (A \cap B^C)) + \mu(S \setminus (A \cap B^C)) \le \mu(S \cap A \cap B) + \mu(S \setminus A)$$

$$+\mu(S \cap A) - \mu(S \cap A \cap B) = \mu(S \setminus A) + \mu(S \cap A) = \mu(S).$$

This has shown that if $A, B \in \mathcal{S}$, then $A \setminus B \in \mathcal{S}$. Since $\Omega \in \mathcal{S}$, this shows that $A \in \mathcal{S}$ if and only if $A^C \in \mathcal{S}$. Now if $A, B \in \mathcal{S}$,

$$A \cup B = (A^C \cap B^C)^C = (A^C \setminus B)^C \in \mathcal{S}.$$

By induction, if $A_1, \cdots, A_n \in \mathcal{S}$, then so is $\cup_{i=1}^n A_i$. If $A, B \in \mathcal{S}$, with $A \cap B = \emptyset$,

$$\mu(A \cup B) = \mu((A \cup B) \cap A) + \mu((A \cup B) \setminus A) = \mu(A) + \mu(B).$$

By induction, if $A_i \cap A_j = \emptyset$ and $A_i \in \mathcal{S}$,

$$\mu(\cup_{i=1}^n A_i) = \sum_{i=1}^n \mu(A_i).$$

Now let $A = \cup_{i=1}^\infty A_i$ where $A_i \cap A_j = \emptyset$ for $i \neq j$.

$$\sum_{i=1}^\infty \mu(A_i) \geq \mu(A) \geq \mu(\cup_{i=1}^n A_i) = \sum_{i=1}^n \mu(A_i).$$

Since this holds for all n, we conclude, since μ is assumed to be an outer measure, that

$$\mu(A) = \sum_{i=1}^\infty \mu(A_i)$$

which establishes Formula 3. Parts 4 and 5 follow from part 3 just as in the proof of Theorem 7.12. It remains to show $\mathcal{S}$ is closed under countable unions. We already know that if $A \in \mathcal{S}$, then $A^C \in \mathcal{S}$ and $\mathcal{S}$ is closed under finite unions. Let $A_i \in \mathcal{S}$, $A = \cup_{i=1}^\infty A_i$, $B_n = \cup_{i=1}^n A_i$. Then

$$
\begin{aligned}
\mu(S) &= \mu(S \cap B_n) + \mu(S \setminus B_n) \qquad (8) \\
&= (\mu \lfloor S)(B_n) + (\mu \lfloor S)(B_n^C).
\end{aligned}
$$

By Lemma 9.3 we know B_n is $(\mu \lfloor S)$ measurable and so is B_n^C. We want to show $\mu(S) \geq \mu(S \setminus A) + \mu(S \cap A)$. If $\mu(S) = \infty$, there is nothing to prove. Assume $\mu(S) < \infty$. Then we apply Parts 5 and 4 to Formula 8 and let $n \to \infty$. Thus

$$B_n \uparrow A, \quad B_n^C \downarrow A^C$$

and this yields

$$\mu(S) = (\mu \lfloor S)(A) + (\mu \lfloor S)(A^C) = \mu(S \cap A) + \mu(S \setminus A).$$

Thus $A \in \mathcal{S}$ and this proves Parts 3, 4, and 5.

Let $F \in \mathcal{S}$ and let $\mu(E \setminus F) + \mu(F \setminus E) = 0$. Then

$$
\begin{aligned}
\mu(S) &\le \mu(S \cap E) + \mu(S \setminus E) \\
&= \mu(S \cap E \cap F) + \mu(S \cap E \cap F^C) + \mu(S \cap E^C) \\
&\le \mu(S \cap F) + \mu(E \setminus F) + \mu(S \setminus F) + \mu(F \setminus E) \\
&= \mu(S \cap F) + \mu(S \setminus F) = \mu(S).
\end{aligned}
$$

Hence $\mu(S) = \mu(S \cap E) + \mu(S \setminus E)$ and so $E \in \mathcal{S}$. This shows Formula 6.

Where do outer measures come from? One way to obtain an outer measure is to start with a measure μ, defined on a σ algebra of sets, $\mathcal{S}$, and use the following definition of the outer measure induced by the measure.

Definition 9.5 Let μ be a measure defined on a σ algebra of sets, $\mathcal{S} \subseteq \mathcal{P}(\Omega)$. Then the outer measure induced by μ, denoted by $\overline{\mu}$ is defined on $\mathcal{P}(\Omega)$ as

$$
\overline{\mu}(E) = \inf\{\mu(V) : V \in \mathcal{S} \text{ and } V \supseteq E\}.
$$

The following lemma deals with the outer measure generated by a measure which is σ finite. It says that if the given measure is σ finite and complete then no new measurable sets are gained by going to the induced outer measure and then considering the measurable sets in the sense of Caratheodory.

Lemma 9.6 Let $(\Omega, \mathcal{S}, \mu)$ be any σ-finite measure space and let $\overline{\mu} : \mathcal{P}(\Omega) \to [0, \infty]$ be the outer measure induced by μ. Then $\overline{\mu}$ is an outer measure as claimed and if $\overline{\mathcal{S}}$ is the set of $\overline{\mu}$ measurable sets in the sense of Caratheodory, then $\overline{\mathcal{S}} \supseteq \mathcal{S}$ and $\overline{\mu} = \mu$ on $\mathcal{S}$. Furthermore, if $(\Omega, \mathcal{S}, \mu)$ is complete, then $\overline{\mathcal{S}} = \mathcal{S}$.

Proof: It is easy to see that $\overline{\mu}$ is an outer measure. Let $E \in \mathcal{S}$. We need to show $E \in \overline{\mathcal{S}}$ and $\overline{\mu}(E) = \mu(E)$. Let $S \subseteq \Omega$. We need to show

$$
\overline{\mu}(S) \ge \overline{\mu}(S \cap E) + \overline{\mu}(S \setminus E). \tag{9}
$$

If $\overline{\mu}(S) = \infty$, there is nothing to prove, so assume $\overline{\mu}(S) < \infty$. Thus there exists $T \in \mathcal{S}$, $T \supseteq S$, and

$$
\begin{aligned}
\overline{\mu}(S) &> \mu(T) - \varepsilon = \mu(T \cap E) + \mu(T \setminus E) - \varepsilon \\
&\ge \overline{\mu}(T \cap E) + \overline{\mu}(T \setminus E) - \varepsilon \\
&\ge \overline{\mu}(S \cap E) + \overline{\mu}(S \setminus E) - \varepsilon.
\end{aligned}
$$

Since ε is arbitrary, this proves Formula 9 and verifies $\mathcal{S} \subseteq \overline{\mathcal{S}}$. Now if $E \in \mathcal{S}$ and $V \supseteq E$,

$$
\mu(E) \le \mu(V).
$$

Hence, taking inf,
$$\mu(E) \le \overline{\mu}(E).$$
But also $\mu(E) \ge \overline{\mu}(E)$ since $E \in \mathcal{S}$ and $E \supseteq E$. Hence
$$\overline{\mu}(E) \le \mu(E) \le \overline{\mu}(E).$$

Now suppose $(\Omega, \mathcal{S}, \mu)$ is complete. Thus if $E, D \in \mathcal{S}$, and $\mu(E \setminus D) = 0$, then if
$$D \subseteq F \subseteq E,$$
it follows $F \in \mathcal{S}$. We know already that $\overline{\mathcal{S}} \supseteq \mathcal{S}$ so let $F \in \overline{\mathcal{S}}$. Using the assumption that the measure space is σ finite, let $\{B_n\} \subseteq \mathcal{S}$, $\cup B_n = \Omega$, $B_n \cap B_m = \emptyset$, $\mu(B_n) < \infty$. Let
$$E_n \supseteq F \cap B_n \; , \; \mu(E_n) = \overline{\mu}(F \cap B_n), \tag{10}$$
where $E_n \in \mathcal{S}$, and let
$$H_n \supseteq B_n \setminus F = B_n \cap F^C \; , \; \mu(H_n) = \overline{\mu}(B_n \setminus F), \tag{11}$$
where $H_n \in \mathcal{S}$. Now let $D_n = H_n^C \cap B_n$. Thus
$$H_n \supseteq B_n \cap F^C$$
and so
$$H_n^C \subseteq B_n^C \cup F$$
which implies
$$D_n = H_n^C \cap B_n \subseteq F \cap B_n.$$
We have
$$D_n \subseteq F \cap B_n \subseteq E_n, \; D_n, E_n \in \mathcal{S}.$$
Now from Formula 10 and Formula 11
$$\mu(E_n \setminus D_n) = \overline{\mu}((F \cap B_n) \setminus D_n) = \overline{\mu}\left(F \cap B_n \cap (B_n^C \cup H_n)\right)$$
$$= \overline{\mu}(F \cap H_n \cap B_n) = \overline{\mu}\left(F \cap (B_n \cap F^C) \cap B_n\right) = \overline{\mu}(\emptyset) = 0.$$
By completeness,
$$F \cap B_n \in \mathcal{S},$$
and it follows that
$$F = \cup_{n=1}^{\infty} F \cap B_n \in \mathcal{S}.$$
This proves the lemma.

Note that it was not necessary to assume μ was σ finite in order to consider $\overline{\mu}$ and conclude that $\overline{\mu} = \mu$ on $\mathcal{S}$. This is sometimes referred to as the process of completing a measure because $\overline{\mu}$ is a complete measure and $\overline{\mu}$ extends μ.

9.2 Regular measures

Usually Ω is not just a set. It is also a topological space. It is very important to consider how the measure is related to this topology.

Definition 9.7 Let μ be a measure on a σ algebra $\mathcal{S}$, of subsets of Ω, where (Ω, τ) is a topological space. We say μ is a Borel measure if $\mathcal{S}$ contains all Borel sets. We say μ is outer regular if μ is Borel and for all $E \in \mathcal{S}$,

$$\mu(E) = \inf\{\mu(V) : V \text{ is open and } V \supseteq E\}.$$

We say μ is inner regular if μ is Borel and

$$\mu(E) = \sup\{\mu(K) : K \subseteq E, \text{ and } K \text{ is compact}\}.$$

If the measure is both outer and inner regular, we say it is regular.

We will assume that (Ω, τ) is a topological space which satisfies the following two properties.

Property 1 There exist subsets of Ω, $\{\Omega_n\}$, with

$$\cdots \Omega_n \subseteq \text{interior}(\Omega_{n+1}) \subseteq \cdots$$

$$\Omega_n \text{ is compact}, \cup_{n=1}^{\infty}\Omega_n = \Omega.$$

Property 2 If K is a compact subset of Ω and $K \subseteq V$, an open set, then there exists a continuous function $f : \Omega \to [0, 1]$ such that $f(x) = 1$ if $x \in K$ and $f(x) = 0$ for all x outside of some compact subset of V containing K.

Property 2 is implied whenever the topological space (Ω, τ) is a locally compact Hausdorff space. This follows from the results on locally compact Hausdorff spaces and Urysohn's lemma which are presented in Chapter 1. In fact it can be shown that for a Hausdorff space, Property 2 is equivalent to the space being locally compact. We leave this as an exercise. Thus, we could state Property 2 more succinctly by saying (Ω, τ) is a locally compact Hausdorff space. However, it is the stated form of Property 2 which will actually be used in the arguments which follow.

Definition 9.8 We define $spt(f)$ (support of f) to be the closure of the set $\{x : f(x) \neq 0\}$. If V is an open set, $C_c(V)$ will be the set of continuous functions f, defined on Ω having $spt(f) \subseteq V$. Thus in Property 2, $f \in C_c(V)$.

Definition 9.9 A topological space (Ω, τ) is σ compact if $\Omega = \cup_{n=1}^{\infty} K_n$ where K_n is compact.

Theorem 9.10 *Any locally compact, σ compact Hausdorff space satisfies Property 1.*

Proof: Let $\mathcal{B}$ be the basis of τ consisting of open sets whose closures are compact. We may assume without loss of generality that for all n, $K_n \subseteq K_{n+1}$. If not, replace K_m with $\cup_{i=1}^{m} K_i$. Using the compactness of K_1, obtain finitely many sets of $\mathcal{B}$, $B_1, \cdots B_k$, which cover K_1. Let $\Omega_1 = \cup_{i=1}^{k} B_i$. Thus $\overline{\Omega}_1$ is compact and K_1 is contained in its interior. Now let $\overline{\Omega}_1 \cup K_2$ play the role of K_1 and obtain in the same way Ω_2 such that $\overline{\Omega}_2$ is compact and $\Omega_2 \supseteq \overline{\Omega}_1 \cup K_2$. Continuing in this way yields the desired sequence of sets.

Note that $\mathbb{R}^n$ with the usual topology satisfies both Properties.

If K is a compact subset of an open set, V, we say $K \prec \phi \prec V$ if

$$\phi \in C_c(V), \ \phi(K) = \{1\}, \ \phi(\Omega) \subseteq [0, 1].$$

Also for $\phi \in C_c(\Omega)$, we say $K \prec \phi$ if

$$\phi(\Omega) \subseteq [0, 1] \text{ and } \phi(K) = 1.$$

We say $\phi \prec V$ if

$$\phi(\Omega) \subseteq [0, 1] \text{ and } spt(\phi) \subseteq V.$$

Theorem 9.11 *(Partition of unity) Let K be a compact subset of a locally compact Hausdorff topological space satisfying Property 2 and suppose*

$$K \subseteq V = \cup_{i=1}^{n} V_i, \ V_i \ open.$$

Then there exist $\psi_i \prec V_i$ with

$$\sum_{i=1}^{n} \psi_i(x) = 1$$

for all $x \in K$.

Proof: Let $K_1 = K \setminus \cup_{i=2}^{n} V_i$. Thus K_1 is compact and $K_1 \subseteq V_1$. Let $K_1 \subseteq W_1 \subseteq \overline{W}_1 \subseteq V_1$ with $\overline{W}_1$ compact. To obtain W_1, use Property 2 to get f such that $K_1 \prec f \prec V_1$ and let $W_1 \equiv \{x : f(x) \neq 0\}$. Thus $W_1, V_2, \cdots V_n$ covers K and $\overline{W}_1 \subseteq V_1$. Let $K_2 = K \setminus (\cup_{i=3}^{n} V_i \cup W_1)$. Then K_2 is compact and $K_2 \subseteq V_2$. Let $K_2 \subseteq W_2 \subseteq \overline{W}_2 \subseteq V_2$ $\overline{W}_2$ compact. Continue this way finally obtaining $W_1, \cdots, W_n$, $K \subseteq W_1 \cup \cdots \cup W_n$, and $\overline{W}_i \subseteq V_i$ $\overline{W}_i$ compact. Now let $\overline{W}_i \subseteq U_i \subseteq \overline{U}_i \subseteq V_i$, $\overline{U}_i$ compact.

The Construction Of Measures

By Property 2, let $\overline{U}_i \prec \phi_i \prec V_i$, $\cup_{i=1}^n \overline{W}_i \prec \gamma \prec \cup_{i=1}^n U_i$. Define

$$\psi_i(x) = \begin{cases} \gamma(x)\phi_i(x)/\sum_{j=1}^n \phi_j(x) \text{ if } \sum_{j=1}^n \phi_j(x) \neq 0, \\ 0 \text{ if } \sum_{j=1}^n \phi_j(x) = 0. \end{cases}$$

If x is such that $\sum_{j=1}^n \phi_j(x) = 0$, then $x \notin \cup_{i=1}^n \overline{U}_i$. Consequently $\gamma(y) = 0$ for all y near x and so $\psi_i(y) = 0$ for all y near x. Hence ψ_i is continuous at such x. If $\sum_{j=1}^n \phi_j(x) \neq 0$, this situation persists near x and so ψ_i is continuous at such points. Therefore ψ_i is continuous. If $x \in K$, then $\gamma(x) = 1$ and so $\sum_{j=1}^n \psi_j(x) = 1$. Clearly $0 \leq \psi_i(x) \leq 1$ and $spt(\psi_j) \subseteq V_j$. This proves the theorem.

Definition 9.12 Let (Ω, τ) be a topological space. We say $L : C_c(\Omega) \to \mathbb{C}$ is a positive linear functional if L is linear,

$$L(af_1 + bf_2) = aLf_1 + bLf_2,$$

and if $Lf \geq 0$ whenever $f \geq 0$.

Theorem 9.13 *(Riesz representation theorem) Let (Ω, τ) be a locally compact Hausdorff space satisfying Properties 1 and 2 and let L be a positive linear functional on $C_c(\Omega)$. Then there exists a unique σ algebra $\mathcal{S}$ and a measure μ, such that*

$$\mu \text{ is complete, Borel, and regular,} \tag{12}$$

$$\mu(K) < \infty \text{ if } K \text{ is compact,} \tag{13}$$

$$Lf = \int f d\mu \text{ for all } f \in C_c(\Omega). \tag{14}$$

A measure satisfying the first and second conditions above is called a Radon measure. Thus the above theorem states that in the context of topological spaces satisfying Properties 1 and 2, every positive linear functional defined on the space $C_c(\Omega)$ is represented by a unique Radon measure. We will define an outer measure by the following definition, show that it, together with the σ algebra of sets measurable in the sense of Caratheodory, satisfies the conclusions of the theorem and then show this is the only σ algebra and measure which does so. Always, K will be a compact set and V will be an open set.

Definition 9.14 $\mu(V) \equiv \sup\{Lf : f \prec V\}$ for V open, $\mu(\emptyset) = 0$. $\mu(E) \equiv \inf\{\mu(V) : V \supseteq E\}$ for arbitrary sets E.

Lemma 9.15 μ *is a well-defined outer measure.*

Proof: First we need to check if $\mu(V) = \inf\{\mu(U) : U \supseteq V$ and U is open$\}$ whenever V is open. But this is clear because if $U \supseteq V$, U and V are open, then $\mu(U) \geq \mu(V)$ from the definition. Hence $\mu(V) \leq \mu(U)$ whenever $U \supseteq V$. Thus $\mu(V) \leq \inf\{\mu(U) : U \supseteq V, U$ open$\}$. But also $V \supseteq V$ so $\mu(V) \geq \inf\{\mu(U) : U \supseteq V, U$ open$\}$.

It remains to show that μ is an outer measure. Let $V = \cup_{i=1}^{\infty} V_i$ and let $f \prec V$. Then $spt(f) \subseteq \cup_{i=1}^{n} V_i$ for some n. Let $\psi_i \prec V_i$, $\sum_{i=1}^{n} \psi_i = 1$ on $spt(f)$.

$$Lf = \sum_{i=1}^{n} L(f\psi_i) \leq \sum_{i=1}^{n} \mu(V_i) \leq \sum_{i=1}^{\infty} \mu(V_i).$$

Hence

$$\mu(V) \leq \sum_{i=1}^{\infty} \mu(V_i)$$

since $f \prec V$ is arbitrary. Now let $E = \cup_{i=1}^{\infty} E_i$. We need show $\mu(E) \leq \sum_{i=1}^{\infty} \mu(E_i)$; so, without loss of generality, we may assume $\mu(E_i) < \infty$ for each i. Let $V_i \supseteq E_i$ with $\mu(E_i) + \varepsilon 2^{-i} > \mu(V_i)$.

$$\mu(E) \leq \mu(\cup_{i=1}^{\infty} V_i) \leq \sum_{i=1}^{\infty} \mu(V_i) \leq \varepsilon + \sum_{i=1}^{\infty} \mu(E_i).$$

Since ε was arbitrary, $\mu(E) \leq \sum_{i=1}^{\infty} \mu(E_i)$ which proves the lemma.

Lemma 9.16 *Let K be compact, $g \geq 0$, $g \in C_c(\Omega)$, and $g = 1$ on K. Then $\mu(K) \leq Lg$. Also $\mu(K) < \infty$ whenever K is compact.*

Proof: Let $\alpha \in (0,1)$ and $V_c = \{x : g(x) > \alpha\}$ and let $h \prec V_c$. Then $h \leq 1$ on V_c while $g\alpha^{-1} \geq 1$ on V_c and so $g\alpha^{-1} \geq h$ which implies

$$L(g\alpha^{-1}) \geq Lh,$$

which implies, since L is linear, that

$$Lg \geq \alpha Lh.$$

Since $h \prec V_c$ is arbitrary, and $K \subseteq V_c$,

$$Lg \geq \alpha\mu(V_c) \geq \alpha\mu(K).$$

Letting $\alpha \uparrow 1$ yields $Lg \geq \mu(K)$. This proves the first part of the lemma. The second assertion follows from this and Property 2. If K is given, let

$$K \prec g \prec \Omega$$

and so from what was just shown, $\mu(K) \leq Lg < \infty$. This proves the lemma.

Lemma 9.17 *If A and B are disjoint compact subsets of Ω, then $\mu(A \cup B) = \mu(A) + \mu(B)$.*

Proof: By Property 2, let $A \prec h \prec B^C$ and let $U_1 = h^{-1}((\frac{1}{2}, 1])$, $V_1 = h^{-1}([0, \frac{1}{2}))$. Then $A \subseteq U_1, B \subseteq V_1$ and $U_1 \cap V_1 = \emptyset$. Therefore we can obtain open sets, U and V, such that

$$U \supseteq A, \ V \supseteq B, \ U \cap V = \emptyset, \ \text{and} \ \mu(A \cup B) + \varepsilon \geq \mu(U \cup V).$$

(Recall from Lemma 9.16 $\mu(A \cup B) < \infty$.) Let $A \prec f \prec U$, $B \prec g \prec V$. Then by Lemma 9.16,

$$\mu(A \cup B) + \varepsilon \geq \mu(U \cup V) \geq L(f + g) = Lf + Lg \geq \mu(A) + \mu(B).$$

Since $\varepsilon > 0$ is arbitrary, this proves the lemma.

From Lemma 9.16 we make the following simple observation.

Lemma 9.18 *Let $f \in C_c(\Omega)$, $f(\Omega) \subseteq [0, 1]$. Then $\mu(spt(f)) \geq Lf$.*

Proof: Let $V \supseteq spt(f)$ and let $spt(f) \prec g \prec V$. Then $Lf \leq Lg \leq \mu(V)$ because $f \leq g$. Since this holds for all $V \supseteq spt(f)$, $Lf \leq \mu(spt(f))$ by definition of μ.

Lemma 9.19 *If K is compact there exists V open, $V \supseteq K$, such that $\mu(V \setminus K) \leq \varepsilon$. If V is open with $\mu(V) < \infty$, then there exists $K \subseteq V$ with $\mu(V \setminus K) \leq \varepsilon$.*

Proof: Let K be compact. Then from the definition of μ, there exists an open set U, with $\mu(U) < \infty$ and $U \supseteq K$. Suppose for every open set, V, containing $K, \mu(V \setminus K) > \varepsilon$. Then there exists $f \prec U \setminus K$ with $Lf > \varepsilon$. Consequently, $\mu(spt(f)) > Lf > \varepsilon$. Let $K_1 = spt(f)$ and repeat the construction with $U \setminus K_1$ in place of U. Continuing in this way we obtain a sequence of disjoint compact sets, $K, K_1, \cdots$ contained in U such that $\mu(K_i) > \varepsilon$. By Lemma 9.17

$$\mu(U) \geq \mu(K \cup \cup_{i=1}^r K_i) = \mu(K) + \sum_{i=1}^{r} \mu(K_i) \geq r\varepsilon$$

for all r, contradicting $\mu(U) < \infty$. This demonstrates the first part of the lemma.

To show the second part, employ a similar construction. Suppose $\mu(V \setminus K) > \varepsilon$ for all $K \subseteq V$. Then $\mu(V) > \varepsilon$ so there exists $f \prec V$ with $Lf > \varepsilon$. Let $K_1 = spt(f)$ so $\mu(spt(f)) > \varepsilon$. If $K_1 \cdots K_n$, disjoint, compact subsets of V have been chosen, let $g \prec (V \setminus \cup_{i=1}^{n} K_i)$ be such that $Lg > \varepsilon$. Hence $\mu(spt(g)) > \varepsilon$. Let $K_{n+1} = spt(g)$. In this way we generate a sequence of disjoint compact subsets of V, $\{K_i\}$ with $\mu(K_i) > \varepsilon$. Thus for any m, $K_1 \cdots K_m$ are all contained in V and are disjoint and compact. By Lemma 9.17

$$\mu(V) \geq \mu(\cup_{i=1}^{m} K_i) = \sum_{i=1}^{m} \mu(K_i) > m\varepsilon$$

for all m, a contradiction to $\mu(V) < \infty$. This proves the second part.

Lemma 9.20 *Let S be the σ algebra of μ measurable sets in the sense of Caratheodory. Then $S \supseteq$ Borel sets and μ is regular on S.*

Proof: Let C be a compact set and let V be an open set with $\mu(V) < \infty$. By the previous lemma, there exists an open set U containing C and a compact subset of V, K, such that $\mu(V \setminus K) < \varepsilon$ and $\mu(U \setminus C) < \varepsilon$. Then by Lemma 9.17,

$$
\begin{aligned}
\mu(V) &\geq \mu(K) \geq \mu((K \setminus U) \cup (K \cap C)) \\
&= \mu(K \setminus U) + \mu(K \cap C) \\
&\geq \mu(K \setminus C) + \mu(K \cap C) - \varepsilon \\
&\geq \mu(V \setminus C) + \mu(V \cap C) - 3\varepsilon
\end{aligned}
$$

because

$$
\begin{aligned}
\mu(V \cap C) &\leq \mu((V \setminus K) \cap C) + \mu(K \cap C) \\
&\leq \varepsilon + \mu(K \cap C)
\end{aligned}
$$

and

$$
\begin{aligned}
\mu(V \setminus C) &\leq \mu((V \setminus K) \setminus C) + \mu((V \cap K) \setminus C) \\
&\leq \varepsilon + \mu(K \setminus C).
\end{aligned}
$$

Since ε is arbitrary,

$$\mu(V) = \mu(V \setminus C) + \mu(V \cap C)$$

whenever C is compact and V is open. If $S \subseteq \Omega$, with $\mu(S) < \infty$, let $V \supseteq S$, $\mu(S) + \varepsilon > \mu(V)$. Then if C is compact,

$$
\begin{aligned}
\varepsilon + \mu(S) &> \mu(V) = \mu(V \setminus C) + \mu(V \cap C) \\
&\geq \mu(S \setminus C) + \mu(S \cap C).
\end{aligned}
$$

Since ε is arbitrary, this shows the compact sets are in S. (If $\mu(S) = \infty$ there is nothing to show.) But if C is closed, $C = \cup_{n=1}^{\infty} C \cap \Omega_n$ so S

contains the closed sets also. Hence $\mathcal{S}$ contains the Borel sets and this proves the first part of the lemma. It is clear that μ is outer regular. It only remains to show μ is inner regular.

Let $F \in \mathcal{S}$ and suppose $F \subseteq \Omega_n$. Then $F^C \cap \Omega_n$ has finite measure and is an element of $\mathcal{S}$. Consequently there exists $V \supseteq F^C \cap \Omega_n$ such that

$$\mu(V \setminus (\Omega_n \cap F^C)) < \varepsilon.$$

Thus

$$
\begin{aligned}
\mu(V \cap F) &\leq \mu((V \cap \Omega_n^C) \cup (V \cap F)) = \mu(V \cap (\Omega_n \cap F^C)^C) \\
&= \mu(V \setminus (\Omega_n \cap F^C)) < \varepsilon.
\end{aligned}
$$

Since $V \supseteq \Omega_n \cap F^C$, $V^C \subseteq \Omega_n^C \cup F$ so $\Omega_n \cap V^C \subseteq \Omega_n \cap F = F$. Hence $\Omega_n \cap V^C$ is a compact subset of F and

$$\mu(F) = \mu(V \cap F) + \mu(F \setminus V) < \varepsilon + \mu(F \setminus V) \leq \varepsilon + \mu(\Omega_n \cap V^C).$$

This shows μ is inner regular on elements of $\mathcal{S}$ which are contained in a compact set.

Now let $F \in \mathcal{S}$ be arbitrary. Let $K_n \subseteq F \cap \Omega_n$ and $\mu((F \cap \Omega_n) \setminus K_n) < \varepsilon 2^{-n}$. Let $H = \cup_{i=1}^{\infty} K_n$. Then

$$\mu(F \setminus H) \leq \sum_{n=1}^{\infty} \mu((F \cap \Omega_n) \setminus K_n) < \varepsilon.$$

If $\mu(F) = \infty$, then since $\mu(F) = \mu(F \setminus H) + \mu(H)$, it follows that $\mu(H) = \infty$ and so

$$\mu(F) = \mu(H) = \lim_{n \to \infty} \mu(\cup_{i=1}^{n} K_i).$$

If $\mu(F) < \infty$, then $\mu(F) < \varepsilon + \mu(H)$ and so

$$\mu(F) < \varepsilon + \mu(\cup_{i=1}^{n} K_i)$$

for some n. This proves Formula 12 of the Riesz representation theorem and completes the proof of this lemma. We have already established Formula 13 of this theorem.

It remains to show μ satisfies Formula 14.

Lemma 9.21 $\int f d\mu = Lf$ for all $f \in C_c(\Omega)$.

Proof: Let $f \in C_c(\Omega)$, f real-valued, and suppose $f(\Omega) \subseteq [a, b]$. Choose $t_0 < a$ and let $t_0 < t_1 < \cdots < t_n = b$, $t_i - t_{i-1} < \varepsilon$. Let

$$E_i = f^{-1}((t_{i-1}, t_i]) \cap spt(f). \tag{15}$$

Note that $\cup_{i=1}^n E_i$ is a closed set and in fact

$$\cup_{i=1}^n E_i = spt(f) \tag{16}$$

since $\Omega = \cup_{i=1}^n f^{-1}((t_{i-1}, t_i])$. Let $V_i \supseteq E_i, V_i$ is open and let V_i satisfy

$$f(x) < t_i + \varepsilon \text{ for all } x \in V_i, \tag{17}$$

$$\mu(V_i \setminus E_i) < \varepsilon/n.$$

By Theorem 9.11 there exists $h_i \in C_c(\Omega)$ such that

$$h_i \prec V_i, \ \sum_{i=1}^n h_i(x) = 1 \text{ on } spt(f).$$

Now note that for each i,

$$f(x)h_i(x) \le h_i(x)(t_i + \varepsilon).$$

(If $x \in V_i$, this follows from Formula 17. If $x \notin V_i$ both sides equal 0.) Therefore,

$$
\begin{aligned}
Lf \ &= \ L(\sum_{i=1}^n fh_i) \le L(\sum_{i=1}^n h_i(t_i + \varepsilon)) \\
&= \ \sum_{i=1}^n (t_i + \varepsilon)L(h_i) \\
&= \ \sum_{i=1}^n (|t_0| + t_i + \varepsilon)L(h_i) - |t_0|L\left(\sum_{i=1}^n h_i\right).
\end{aligned}
$$

Now note that $|t_0| + t_i + \varepsilon \ge 0$ and so from the definition of μ and Lemma 9.16, this is no larger than

$$\sum_{i=1}^n (|t_0| + t_i + \varepsilon)\mu(V_i) - |t_0|\mu(spt(f))$$

$$\le \sum_{i=1}^n (|t_0| + t_i + \varepsilon)(\mu(E_i) + \varepsilon/n) - |t_0|\mu(spt(f))$$

$$\le |t_0|\sum_{i=1}^n \mu(E_i) + |t_0|\varepsilon + \sum_{i=1}^n t_i\mu(E_i) + \varepsilon(|t_0| + |b|)$$

$$+\varepsilon\sum_{i=1}^n \mu(E_i) + \varepsilon^2 - |t_0|\mu(spt(f)).$$

161

From Formula 16 and Formula 15, the first and last terms cancel. Therefore this is no larger than

$$(2|t_0| + |b| + \mu(spt(f)) + \varepsilon)\varepsilon + \sum_{i=1}^{n} t_{i-1}\mu(E_i) + \varepsilon\mu(spt(f))$$

$$\leq \int f d\mu + (2|t_0| + |b| + 2\mu(spt(f)) + \varepsilon)\varepsilon.$$

Since $\varepsilon > 0$ is arbitrary,

$$Lf \leq \int f d\mu \tag{18}$$

for all $f \in C_c(\Omega)$, f real. Hence equality holds in Formula 18 because $L(-f) \leq -\int f d\mu$ so $L(f) \geq \int f d\mu$. Thus $Lf = \int f d\mu$ for all $f \in C_c(\Omega)$. Just apply the result for real functions to the real and imaginary parts of f. This proves the Lemma.

Now that we have shown that μ satisfies the conditions of the Riesz representation theorem, we show that μ is the only measure that does so.

Lemma 9.22 *The measure and σ algebra of Theorem 9.13 are unique.*

Proof: If $(\mu_1, \mathcal{S}_1)$ and $(\mu_2, \mathcal{S}_2)$ both work, let

$$K \subseteq V, \ K \prec f \prec V.$$

Then

$$\mu_1(K) \leq \int f d\mu_1 = Lf = \int f d\mu_2 \leq \mu_2(V).$$

Thus $\mu_1(K) \leq \mu_2(K)$ because of Formula 12. Similarly, $\mu_1(K) \geq \mu_2(K)$ and this shows that $\mu_1 = \mu_2$ on all compact sets. It follows from Formula 12 again that the two measures coincide on all open sets as well. Now let $E \in \mathcal{S}_1$, the σ algebra associated with μ_1, and let $E_n = E \cap \Omega_n$. By the regularity of the measures, there exist sets G and H such that G is a countable intersection of decreasing open sets and H is a countable union of increasing compact sets which satisfy

$$G \supseteq E_n \supseteq H, \ \mu_1(G \setminus H) = 0.$$

Since the two measures agree on all open and compact sets, it follows that $\mu_2(G) = \mu_1(G)$ and a similar equation holds for H in place of G. Therefore $\mu_2(G \setminus H) = \mu_1(G \setminus H) = 0$. By completeness of μ_2, $E_n \in \mathcal{S}_2$, the σ algebra associated with μ_2. Thus $E \in \mathcal{S}_2$ since $E = \cup_{n=1}^{\infty} E_n$. Similarly $\mathcal{S}_2 \subseteq \mathcal{S}_1$. Since the two σ algebras are equal and the two measures are equal on every open set, regularity of these measures shows they coincide on all measurable sets and this proves the theorem.

What can be said in the case where (Ω, τ) only satisfies Property 2? We did not use Property 1 until Lemma 9.20. In the case where Property 1 does not hold, we give the following alternative to Lemma 9.20.

Lemma 9.23 *Let $\mathcal{S}$ be the σ algebra of μ measurable sets in the sense of Caratheodory. Then $\mathcal{S} \supseteq$ Borel sets and μ is inner regular on every open set and for every $E \in \mathcal{S}$ with $\mu(E) < \infty$.*

Proof: Define

$$\mathcal{S}_1 = \{E \subseteq \Omega : E \cap K \in \mathcal{S}\}$$

for all compact K. The first part of the proof of Lemma 9.20 does not use Property 1 and demonstrates that the compact sets are in $\mathcal{S}$. Therefore, $\mathcal{S}_1$ contains the closed sets and $\mathcal{S}_1 \supseteq \mathcal{S}$. To see that $\mathcal{S}_1$ is closed with respect to taking complements, let $E \in \mathcal{S}_1$ and write

$$K = (E^C \cap K) \cup (E \cap K).$$

Then

$$E^C \cap K = K \setminus (E \cap K) \in \mathcal{S}.$$

Similarly $\mathcal{S}_1$ is closed under countable unions. Thus $\mathcal{S}_1$ is a σ algebra which contains the Borel sets since it contains the closed sets.

Now we show $\mathcal{S}_1 = \mathcal{S}$. Let $E \in \mathcal{S}_1$ and let V be an open set with $\mu(V) < \infty$ and choose $K \subseteq V$ such that $\mu(V \setminus K) < \varepsilon$. Then since $E \in \mathcal{S}_1$,

$$
\begin{aligned}
\mu(V) &= \mu(V \setminus (K \cap E)) + \mu(V \cap (K \cap E)) \\
&\geq \mu(V \setminus E) + \mu(V \cap E) - \varepsilon.
\end{aligned}
$$

Since ε is arbitrary,

$$\mu(V) = \mu(V \setminus E) + \mu(V \cap E).$$

Now let $S \subseteq \Omega$. If $\mu(S) = \infty$, then $\mu(S) = \mu(S \cap E) + \mu(S \setminus E)$. If $\mu(S) < \infty$, let

$$V \supseteq S, \ \mu(S) + \varepsilon \geq \mu(V).$$

Then

$$\mu(S) + \varepsilon \geq \mu(V) = \mu(V \setminus E) + \mu(V \cap E) \geq \mu(S \setminus E) + \mu(S \cap E).$$

Since ε is arbitrary, this shows that $E \in \mathcal{S}$ and so $\mathcal{S}_1 = \mathcal{S}$. Thus $\mathcal{S} \supseteq$ Borel sets as claimed.

We already know from Lemma 9.18 and the definition of μ that μ is inner regular on all open sets. We need to show that $\mu(F) = \sup\{\mu(K) : K \subseteq F\}$ for all $F \in \mathcal{S}$ with $\mu(F) < \infty$. This is done similarly to Lemma

9.20. Let U be an open set, $U \supseteq F$, $\mu(U) < \infty$. Let V be open, $V \supseteq F^C \cap U$, and $\mu(V \setminus (F^C \cap U)) < \varepsilon$. Then

$$
\begin{aligned}
\mu(V \cap F) &\leq \mu((V \cap U^C) \cup (V \cap F)) = \mu(V \cap (U \cap F^C)^C) \\
&= \mu(V \setminus (U \cap F^C)) < \varepsilon.
\end{aligned}
$$

Since $V \supseteq U \cap F^C$, $V^C \subseteq U^C \cup F$ so $U \cap V^C \subseteq U \cap F = F$. Hence $U \cap V^C$ is a subset of F. Now let $K \subseteq U$, $\mu(U \setminus K) < \varepsilon$. Thus $K \cap V^C$ is a compact subset of F and

$$
\begin{aligned}
\mu(F) &= \mu(V \cap F) + \mu(F \setminus V) \\
&< \varepsilon + \mu(F \setminus V) \leq \varepsilon + \mu(U \cap V^C) \leq 2\varepsilon + \mu(K \cap V^C).
\end{aligned}
$$

Since ε is arbitrary, this proves the second part of the lemma.

Lemma 9.21 holds with no change in its proof. This gives the existence part of the following theorem.

Theorem 9.24 *(Riesz representation theorem) Let (Ω, τ) be a locally compact Hausdorff space and let L be a positive linear functional on $C_c(\Omega)$. Then there exists a σ algebra $\mathcal{S}$ containing the Borel sets and a unique measure μ, defined on $\mathcal{S}$, such that*

$$\mu \text{ is complete, } \mu(K) < \infty \text{ for all } K \text{ compact,}$$

$$\mu(F) = \sup\{\mu(K) : K \subseteq F, \ K \text{ compact}\},$$

for all F open and for all $F \in \mathcal{S}$ with $\mu(F) < \infty$,

$$\mu(F) = \inf\{\mu(V) : V \supseteq F, \ V \text{ open}\}$$

for all $F \in \mathcal{S}$, and

$$\int f \, d\mu = Lf \quad \text{for all } f \in C_c(\Omega).$$

Proof: It only remains to prove uniqueness. Note that unlike Theorem 9.13, the assertion about uniqueness does not include the σ algebra. Suppose both μ_1 and μ_2 are measures on $\mathcal{S}$ satisfying the conclusions of the theorem. Then if K is compact and $V \supseteq K$, let $K \prec f \prec V$. Then

$$\mu_1(K) \leq \int f \, d\mu_1 = Lf = \int f \, d\mu_2 \leq \mu_2(V).$$

Thus $\mu_1(K) \leq \mu_2(K)$ for all K. Similarly, the inequality can be reversed and so we see that the two measures are equal on compact sets. By the assumption of inner regularity on open sets, the two measures are also equal on all open sets. By outer regularity, they are equal on all sets of $\mathcal{S}$. This proves the theorem.

The following theorem is an interesting application of the Riesz representation theorem for measures defined on subsets of $\mathbb{R}^n$.

Let M be a closed subset of $\mathbb{R}^n$. Then we may consider M as a locally compact metric space if we let the topology on M consist of intersections of open sets from the standard topology of $\mathbb{R}^n$ with M or equivalently, use the usual metric on $\mathbb{R}^n$ restricted to M.

Proposition 9.25 *Let τ be the relative topology of M consisting of intersections of open sets of $\mathbb{R}^n$ with M and let $\mathcal{B}$ be the Borel sets of the topological space (M, τ). Then*

$$\mathcal{B} = \mathcal{S} \equiv \{E \cap M : E \text{ is a Borel set of } \mathbb{R}^n\}.$$

Proof: It is clear that $\mathcal{S}$ defined above is a σ algebra containing τ and so $\mathcal{S} \supseteq \mathcal{B}$. Now let $\mathcal{R}$ and $\mathcal{E}$ be as described in Example 7.8. Let $R = \prod_{i=1}^{n} I_i \in \mathcal{R}$ where

$$I_i = (-\infty, \infty) \cap (a_i, b_i]$$

for $-\infty \leq a_i < b_i \leq \infty$. Letting $I_i^k = (-\infty, \infty) \cap (a_i, c_i^k)$ where $\lim_{k \to \infty} c_i^k = b_i$, and $c_i^k > b_i$ if $b_i < \infty$, and $c_i^k = \infty$ if $b_i = \infty$, it follows that for

$$R_k \equiv \prod_{i=1}^{n} I_i^k, \ R_k \cap M \in \tau \subseteq \mathcal{B}$$

and so $R \cap M = \cap_{k=1}^{\infty} R_k \cap M \in \mathcal{B}$. It follows that $A \cap M \in \mathcal{B}$ whenever $A \in \mathcal{E}$, finite disjoint unions of these sets. Now define

$$\mathcal{M} \equiv \{E \text{ Borel in } \mathbb{R}^n \text{ such that } E \cap M \in \mathcal{B}\}.$$

Then $\mathcal{M} \supseteq \mathcal{E}$ and is clearly a monotone class. Therefore, $\mathcal{M} \supseteq \sigma(\mathcal{E}) =$ Borel sets of $\mathbb{R}^n$ which shows $\mathcal{S} \subseteq \mathcal{B}$. This proves the proposition.

Theorem 9.26 *Suppose μ is a measure defined on the Borel sets of M where M is a closed subset of $\mathbb{R}^n$. Suppose also that μ is finite on compact sets. Then $\overline{\mu}$, the outer measure determined by μ, is a Radon measure on a σ algebra containing the Borel sets of (M, τ) where τ is the relative topology described above.*

Proof: Since μ is Borel and finite on compact sets, we may define a positive linear functional on $C_c(M)$ as

$$Lf \equiv \int_M f d\mu.$$

By the Riesz representation theorem, there exists a unique Radon measure and σ algebra, μ_1 and $\mathcal{S}(\mu_1)$ respectively, such that for all $f \in C_c(M)$,

$$\int_M f d\mu = \int_M f d\mu_1.$$

Let $\mathcal{R}$ and $\mathcal{E}$ be as described in Example 7.8 and let $C_r = (-r, r]^n$. Then if $R \in \mathcal{R}$, it follows that $R \cap C_r$ has the form,

$$R \cap C_r = \prod_{i=1}^{n} (a_i, b_i].$$

Let $f_k = \prod_{i=1}^{n} h_i^k$ where h_i^k is given by the following graph.

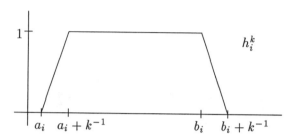

Then $spt\,(f_k) \subseteq [-r, r+1]^n \equiv K$. Thus

$$\int_{K \cap M} f_k \, d\mu = \int_M f_k \, d\mu = \int_M f_k \, d\mu_1 = \int_{K \cap M} f_k \, d\mu_1.$$

Since $f_k \to \mathcal{X}_{R \cap Q_r}$ pointwise and both measures are finite, on $K \cap M$, it follows from the dominated convergence theorem that

$$\mu\,(R \cap C_r \cap M) = \mu_1\,(R \cap C_r \cap M)$$

and so it follows that this holds for R replaced with any $A \in \mathcal{E}$. Now define

$$\mathcal{M} \equiv \{E \text{ Borel} : \mu\,(E \cap C_r \cap M) = \mu_1\,(E \cap C_r \cap M)\}.$$

Then $\mathcal{E} \subseteq \mathcal{M}$ and it is clear that $\mathcal{M}$ is a monotone class. By the theorem on monotone classes, it follows that $\mathcal{M} \supseteq \sigma\,(\mathcal{E})$, the smallest σ algebra containing $\mathcal{E}$. This σ algebra contains the open sets because

$$\prod_{i=1}^{n} (a_i, b_i) = \cup_{k=1}^{\infty} \prod_{i=1}^{n} (a_i, b_i - k^{-1}] \in \sigma\,(\mathcal{E})$$

and every open set can be written as a countable union of sets of the form $\prod_{i=1}^{n} (a_i, b_i)$. Therefore,

$$\mathcal{M} \supseteq \sigma\,(\mathcal{E}) \supseteq \text{Borel sets} \supseteq \mathcal{M}.$$

Thus,

$$\mu\,(E \cap C_r \cap M) = \mu_1\,(E \cap C_r \cap M)$$

for all E Borel. Letting $r \to \infty$, it follows that $\mu(E \cap M) = \mu_1(E \cap M)$ for all E a Borel set of $\mathbb{R}^n$. By Proposition 9.25 $\mu(F) = \mu_1(F)$ for all F Borel in (M, τ). Consequently,

$$
\begin{aligned}
\overline{\mu_1}(S) &\equiv \inf\{\mu_1(E) : E \supseteq S,\ E \in \mathcal{S}(\mu_1)\} \\
&= \inf\{\mu_1(F) : F \supseteq S,\ F \text{ Borel}\} \\
&= \inf\{\mu(F) : F \supseteq S,\ F \text{ Borel}\} \\
&= \overline{\mu}(S).
\end{aligned}
$$

Therefore, by Lemma 9.6, the $\overline{\mu}$ measurable sets consist of $\mathcal{S}(\mu_1)$ and $\overline{\mu} = \mu_1$ on $\mathcal{S}(\mu_1)$ and this shows $\overline{\mu}$ is regular as claimed. This proves the theorem.

The following theorem is quite useful. It states that under appropriate hypotheses every measurable function has a Borel measurable representative.

Theorem 9.27 *Let μ be the Radon measure of Theorem 9.13. Then if $f : \Omega \to [0, \infty]$, there exists $g : \Omega \to [0, \infty]$ such that g is Borel measurable and $g = f$ a.e.*

Proof: Let E be a measurable set and let $E_n \equiv \Omega_n \cap E$ so that E_n has finite measure. By outer regularity of μ, there exists $F_n \supseteq E_n$ such that F_n is a Borel set and $\mu(F_n \setminus E_n) = 0$. Let $F = \cup_{n=1}^{\infty} F_n$. Then F is a Borel set and $(F \setminus E) \subseteq \cup_{n=1}^{\infty}(F_n \setminus E_n)$, a set of measure zero. Therefore, for each E measurable, there exists F, a Borel set containing E such that $\mu(F \setminus E) = 0$. Now let s_n be an increasing sequence of nonnegative simple functions converging point wise to f. Say $s_n(\omega) = \sum_{i=1}^{N} c_i \mathcal{X}_{E_i}(\omega)$. Then pick $F_i \supseteq E_i$ such that $\mu(F_i \setminus E_i) = 0$ and F_i is a Borel set. Let $t_n(\omega) = \sum_{i=1}^{N} c_i \mathcal{X}_{F_i}(\omega)$ and define $\widetilde{s}_n(\omega) \equiv \max\{t_k(\omega) : k \leq n\}$. Then $\lim_{n \to \infty} \widetilde{s}_n(\omega) \equiv g(\omega)$ exists because $\widetilde{s}_n(\omega)$ is increasing. Also g is Borel measurable and

$$
\{\omega : g(\omega) \neq f(\omega)\} \subseteq \cup_{n=1}^{\infty}\{\omega : s_n(\omega) \neq \widetilde{s}_n(\omega)\},
$$

a set of measure zero. This proves the theorem.

For a complex valued measurable function, we can apply the above theorem to the positive and negative parts of the real and imaginary parts of the given function to obtain the following corollary.

Corollary 9.28 *In the situation of Theorem 9.27, let $f : \Omega \to \mathbb{C}$ be measurable. Then there exists a Borel measurable function, g, such that $f = g$ a.e.*

9.3 Lebesgue measure on $\mathbb{R}^1$

To construct Lebesgue measure on $\mathbb{R}^1$, we simply take for a positive linear functional the Riemann integral and denote by m the unique Radon measure representing this functional. From this it follows easily that $m(I)$ gives the length of an interval. Lebesgue measure for $\mathbb{R}^n$ will be discussed in Chapter 10.

9.4 Exercises

1. Obtain Lebesgue measure from the following outer measure.

$$\mu(A) = \inf\{\sum_{i=1}^{\infty} |I_i| \text{ such that } A \subseteq \cup_{i=1}^{\infty} I_i\}$$

where I_i is an interval of finite length and $|I_i| = $ length of I_i. Can you show that the measure obtained is a Radon measure without recourse to the Riesz representation theorem?

2. Recall that a bounded function defined on an interval $[a, b]$ is Riemann integrable, $f \in \mathcal{R}(a, b)$, if and only if there is a unique number, $\int_a^b f(x)dx$, between all upper and lower sums. Suppose $f \in \mathcal{R}(a, b)$. Show that $f\mathcal{X}_{[a,b]}$ is Lebesgue integrable and $\int_a^b f(x)dx = \int f\mathcal{X}_{[a,b]}dm$. **Hint:** You may want to begin by assuming $f \geq 0$. Next show f is measurable. Finally note that $\int_a^b f(x)dx$ and $\int f\mathcal{X}_{[a,b]}dm$ are both between every pair of upper and lower sums.

3. Let $F : \mathbb{R} \to [0, 1]$. F is increasing, continuous from the right and

$$\lim_{x \to -\infty} F(x) = 0, \ \lim_{x \to \infty} F(x) = 1.$$

(F is a probability distribution function.) For each $f \in C_c(\mathbb{R})$, let $Lf = \int f dF$, the Riemann Stieltjes integral. L provides a positive linear functional on $C_c(\mathbb{R})$. By Theorem 9.13 there is a unique Radon measure, μ, representing this functional. This is called Lebesgue Stieltjes measure. Find $\mu((a, b])$.

4. Consider the following nested sequence of compact sets, $\{P_n\}$. The set P_n consists of 2^n disjoint closed intervals contained in $[0, 1]$. The first interval, P_1, equals $[0, 1]$ and P_n is obtained from P_{n-1} by deleting the open interval which is the middle third of each closed interval

in P_n. Let $P = \cap_{n=1}^{\infty} P_n$. Show

$$P \neq \emptyset, \ m(P) = 0, \ P \sim [0,1].$$

(There is a 1-1 onto mapping of $[0,1]$ to P.) The set P is called the Cantor set.

5. ↑ Consider the sequence of functions defined in the following way. We let $f_1(x) = x$ on $[0,1]$. To get from f_n to f_{n+1}, let $f_{n+1} = f_n$ on all intervals where f_n is constant. If f_n is nonconstant on $[a,b]$, let $f_{n+1}(a) = f_n(a)$, $f_{n+1}(b) = f_n(b)$, f_{n+1} is piecewise linear and equal to $\frac{1}{2}(f_n(a) + f_n(b))$ on the middle third of $[a,b]$. Sketch a few of these and you will see the pattern. Show

$$\{f_n\} \text{ converges uniformly on } [0,1]. \tag{a.}$$

If $f(x) = \lim_{n \to \infty} f_n(x)$, show that

$$f(0) = 0, \ f(1) = 1, \ f \text{ is continuous,}$$

and

$$f'(x) = 0 \text{ for all } x \notin P \tag{b.}$$

where P is the Cantor set of Problem 4. This function is called the Cantor function.

6. Let $m(W) > 0$, W is measurable, $W \subseteq [a,b]$. Show there exists a nonmeasurable subset of W. **Hint:** Let $x \sim y$ if $x - y \in \mathbb{Q}$. Observe that $\sim$ is an equivalence relation on $\mathbb{R}$. Let $\mathcal{C}$ be the set of equivalence classes. Let $\mathcal{D} = \{C \cap W : C \in \mathcal{C} \text{ and } C \cap W \neq \emptyset\}$. If $C \cap W \in \mathcal{D}$, pick exactly one element of $C \cap W$. Denote by A the collection of these points. (Note the use of the axiom of choice.) Show

$$W \subseteq \cup_{r \in \mathbb{Q}} A + r \tag{a.}$$

$$A + r_1 \cap A + r_2 = \emptyset \text{ if } r_1 \neq r_2, \ r_i \in \mathbb{Q}. \tag{b.}$$

Observe that since $A \subseteq [a,b]$, if $|r| < 1$, then $A + r \subseteq [a-1, b+1]$. Use this to show that if $m(A) = 0$, or if $m(A) > 0$ a contradiction results.

7. ↑ Let $g(x) = x + f(x)$ where f is the strange function of Problem 5. Let P be the Cantor set of Problem 4. Let $[0,1] \setminus P = \cup_{j=1}^{\infty} I_j$ where I_j is open and $I_j \cap I_k = \emptyset$ if $j \neq k$. Show $m(g(I_j)) = m(I_j)$ so

$$m(g(\cup_{j=1}^{\infty} I_j)) = \sum_{j=1}^{\infty} m(g(I_j)) = \sum_{j=1}^{\infty} m(I_j) = 1.$$

Thus $m(g(P)) = 1$ because $g([0,1]) = [0,2]$. Let A be a non measurable subset of $g(P)$. Define $\phi(x) = \mathcal{X}_A(g(x))$. Thus $\phi(x) = 0$ unless $x \in P$. Tell why ϕ is measurable. (Recall $m(P) = 0$) Now observe that $\mathcal{X}_A(y) = \phi(g^{-1}(y))$ for $y \in [0,2]$. Tell why g^{-1} is continuous but $\phi \circ g^{-1}$ is not measurable. (This is an example of measurable $\circ$ continuous $\neq$ measurable.)

8. ↑ Do there exist Lebesgue measurable sets which are not Borel measurable? **Hint**: In Problem 7, ϕ is Lebesgue measurable. Now recall that Borel $\circ$ measurable = measurable.

9. Let E be countable subset of $\mathbb{R}$. Show $m(E) = 0$.

10. Give an example of sets $A_n \subseteq \mathbb{R}$ with $\cap_{n=1}^{\infty} A_n = \emptyset$, $A_n \supseteq A_{n+1}$, but $\lim_{n \to \infty} m(A_n) \neq 0$.

11. ↑ Given $1 > \varepsilon > 0$, show there exists an open set $E \subseteq [0,1]$ dense in $[0,1]$, and $m(E) = \varepsilon$. **Hint**: See the construction of the Cantor set in 4.

12. ↑ Construct a Borel set $E \subseteq \mathbb{R}$ such that $0 < m(E \cap I) < m(I)$ for every nontrivial interval I. Show that E can be constructed in such a way that $m(E) \leq \varepsilon$.

13. Let $f \in L^1(\Omega)$ for $(\Omega, \mu, \mathcal{S})$ a measure space. Show that $\{x : f(x) \neq 0\}$ has σ finite measure. **Hint**: ($\{x : f(x) \neq 0\} = \cup_{j=1}^{\infty} F_j$, $\mu(F_j) < \infty$).

14. Let $(\Omega, \mathcal{S}, \mu)$ be an arbitrary measure space and define $\bar{\mu} : \mathcal{P}(\Omega) \to [0, \infty]$ by

$$\bar{\mu}(S) = \inf\{\mu(E) : E \supseteq S \text{ and } E \in \mathcal{S}\}.$$

Show $\bar{\mu}$ is an outer measure. If $\overline{\mathcal{S}}$ is the set of $\bar{\mu}$ measurable sets in the sense of Caratheodory, show $\overline{\mathcal{S}} \supseteq \mathcal{S}$ and $\bar{\mu} = \mu$ on $\mathcal{S}$. This shows how any measure space can be completed and it also shows that the completion $\bar{\mu}$ comes from an outer measure.

15. Show that if a Hausdorff space satisfies Property 2, then it must be locally compact.

Chapter 10
Lebesgue Measure

10.1   Lebesgue measure

In this chapter, n dimensional Lebesgue measure and many of its properties are obtained from the Riesz representation theorem of Chapter 9. This is done by using a positive linear functional familiar to anyone who has had a course in calculus. The positive linear functional is

$$\Lambda f \equiv \int_{-\infty}^{\infty} \cdots \int_{-\infty}^{\infty} f(x_1, \cdots, x_n) dx_1 \cdots dx_n \tag{1}$$

for $f \in C_c(\mathbb{R}^n)$. This is the ordinary Riemann iterated integral and we need to observe that it makes sense.

Lemma 10.1 *Let $f \in C_c(\mathbb{R}^n)$ for $n \geq 2$. Then*

$$h(x_n) \equiv \int_{-\infty}^{\infty} \cdots \int_{-\infty}^{\infty} f(x_1 \cdots x_{n-1} x_n) dx_1 \cdots dx_{n-1}$$

is well defined and $h \in C_c(\mathbb{R})$.

Proof: Assume this is true for all $2 \leq k \leq n - 1$. Then fixing x_n,

$$x_{n-1} \to \int_{-\infty}^{\infty} \cdots \int_{-\infty}^{\infty} f(x_1 \cdots x_{n-2}, x_{n-1}, x_n) dx_1 \cdots dx_{n-2}$$

is a function in $C_c(\mathbb{R})$. Therefore, it makes sense to write

$$h(x_n) \equiv \int_{-\infty}^{\infty} \int_{-\infty}^{\infty} \cdots \int_{-\infty}^{\infty} f(x_1 \cdots x_{n-2}, x_{n-1}, x_n) dx_1 \cdots dx_{n-1}.$$

We need to verify $h \in C_c(\mathbb{R})$. Since f vanishes whenever $|\mathbf{x}|$ is large enough, it follows $h(x_n) = 0$ whenever $|x_n|$ is large enough. It only remains to show h is continuous. But f is uniformly continuous, so if $\varepsilon > 0$ is given there exists a δ such that

$$|f(\mathbf{x}_1) - f(\mathbf{x})| < \varepsilon$$

whenever $|\mathbf{x}_1 - \mathbf{x}| < \delta$. Thus, letting $|x_n - \bar{x}_n| < \delta$,

$$|h(x_n) - h(\bar{x}_n)| \leq$$

$$\int_{-\infty}^{\infty} \cdots \int_{-\infty}^{\infty} |f(x_1 \cdots x_{n-1}, x_n) - f(x_1 \cdots x_{n-1}, \bar{x}_n)| dx_1 \cdots dx_{n-1}$$

$$\leq \varepsilon(b-a)^{n-1}$$

where $spt(f) \subseteq [a,b]^n \equiv [a,b] \times \cdots \times [a,b]$. This argument also shows the lemma is true for $n = 2$. This proves the lemma.

From Lemma 10.1 it is clear that Formula 1 makes sense and also that Λ is a positive linear functional for $n = 1, 2, \cdots$.

Definition 10.2 m_n is the unique Radon measure representing Λ. Thus for all $f \in C_c(\mathbb{R}^n)$,

$$\Lambda f = \int f \, dm_n .$$

Let $R = \prod_{i=1}^n [a_i, b_i]$, $R_0 = \prod_{i=1}^n (a_i, b_i)$. What are $m_n(R)$ and $m_n(R_0)$? We show that both of these equal $\prod_{i=1}^n (b_i - a_i)$. To see this is the case, let k be large enough that

$$a_i + 1/k < b_i - 1/k$$

for $i = 1, \cdots, n$. Consider functions g_i^k and f_i^k having the following graphs.

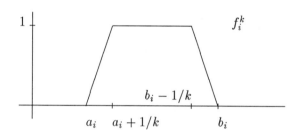

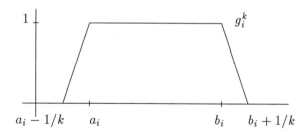

Let

$$g^k(\mathbf{x}) = \prod_{i=1}^n g_i^k(x_i), \quad f^k(\mathbf{x}) = \prod_{i=1}^n f_i^k(x_i).$$

Then

$$\prod_{i=1}^n (b_i - a_i + 2/k) \geq \Lambda g^k = \int g^k \, dm_n \geq m_n(R) \geq m_n(R_0)$$

$$\geq \int f^k \, dm_n = \Lambda f^k \geq \prod_{i=1}^{n} (b_i - a_i - 2/k).$$

Letting $k \to \infty$, it follows that

$$m_n(R) = m_n(R_0) = \prod_{i=1}^{n} (b_i - a_i)$$

as expected.

We say R is a half open box if

$$R = \prod_{i=1}^{n} [a_i, a_i + r).$$

Lemma 10.3 *Every open set is the countable disjoint union of half open boxes whose sides have length equal to 2^{-k} for some $k = 1, 2, \cdots$.*

Proof: Let

$$\mathcal{C}_k = \{\text{All half open boxes} \prod_{i=1}^{n} [a_i, a_i + 2^{-k}) \text{ where}$$

$$a_i = l2^{-k} \text{ for some integer } l.\}$$

Thus $\mathcal{C}_k$ consists of a countable disjoint collection of boxes whose union is $\mathbb{R}^n$. This is sometimes called a tiling of $\mathbb{R}^n$. Note that each box has diameter $2^{-k}\sqrt{n}$. Let U be open and let $\mathcal{B}_1 \equiv$ all sets of $\mathcal{C}_1$ which are contained in U. If $\mathcal{B}_1, \cdots, \mathcal{B}_k$ have been chosen, $\mathcal{B}_{k+1} \equiv$ all sets of $\mathcal{C}_{k+1}$ contained in

$$U \setminus \cup(\cup_{i=1}^{k} \mathcal{B}_i).$$

Let $\mathcal{B}_\infty = \cup_{i=1}^{\infty} \mathcal{B}_i$. We claim $\cup \mathcal{B}_\infty = U$. Clearly $\cup \mathcal{B}_\infty \subseteq U$. If $p \in U$, let k be the smallest integer such that p is contained in a box from $\mathcal{C}_k$ which is also a subset of U. Thus

$$p \in \cup \mathcal{B}_k \subseteq \cup \mathcal{B}_\infty.$$

Hence $\mathcal{B}_\infty$ is the desired countable disjoint collection of half open boxes whose union is U. This proves the lemma.

Lebesgue measure is translation invariant. This means roughly that if you take a Lebesgue measurable set, and slide it arround, it remains Lebesgue measurable and the measure does not change.

Theorem 10.4 *Lebesgue measure is translation invariant, i.e.,*

$$m_n(\mathbf{v}+E) = m_n(E),$$

for E Lebesgue measurable.

Proof: First note that if E is Borel, then so is $\mathbf{v} + E$. To show this, let
$$\mathcal{S} = \{E \in \text{Borel sets such that } \mathbf{v} + E \text{ is Borel}\}.$$
Then from Lemma 10.3, $\mathcal{S}$ contains the open sets and is easily seen to be a σ algebra, so $\mathcal{S} = $ Borel sets. Now let E be a Borel set. Choose V open such that
$$m_n(V) < m_n(E \cap B(0, k)) + \varepsilon, \quad V \supseteq E \cap B(0, k).$$
Then
$$m_n(\mathbf{v} + E \cap B(0, k)) \leq m_n(\mathbf{v} + V) = m_n(V)$$
$$\leq m_n(E \cap B(0, k)) + \varepsilon.$$
The equal sign is valid because the conclusion of Theorem 10.4 is clearly true for all open sets thanks to Lemma 10.3 and the simple observation that the theorem is true for boxes. Since ε is arbitrary,
$$m_n(\mathbf{v} + E \cap B(0, k)) \leq m_n(E \cap B(0, k)).$$
Letting $k \to \infty$,
$$m_n(\mathbf{v} + E) \leq m_n(E).$$
Since $\mathbf{v}$ is arbitrary,
$$m_n(-\mathbf{v} + (\mathbf{v} + E)) \leq m_n(E + \mathbf{v}).$$
Hence
$$m_n(\mathbf{v} + E) \leq m_n(E) \leq m_n(\mathbf{v} + E)$$
proving the theorem in the case where E is Borel. Now suppose that $m(S) = 0$. Then there exists $E \supseteq S$, E Borel, and $m(E) = 0$. Thus
$$m(S + \mathbf{v}) \leq m(E + \mathbf{v}) = m(E) = 0 = m(S).$$
Now let F be an arbitrary Lebesgue measurable set and let $F_r = F \cap B(0, r)$. Then there exists a Borel set E, $E \supseteq F_r$, and $m(E \setminus F_r) = 0$. Then since
$$(E + \mathbf{v}) \setminus (F_r + \mathbf{v}) = (E \setminus F_r) + \mathbf{v},$$
it follows
$$m((E + \mathbf{v}) \setminus (F_r + \mathbf{v})) = m((E \setminus F_r) + \mathbf{v}) = m(E \setminus F_r) = 0.$$
By completeness of m, $F_r + \mathbf{v}$ is Lebesgue measurable and $m(F_r + \mathbf{v}) = m(E + \mathbf{v})$. Hence
$$m(F_r) = m(E) = m(E + \mathbf{v}) = m(F_r + \mathbf{v}).$$
Letting $r \to \infty$, we obtain
$$m(F) = m(F + \mathbf{v})$$
and this proves the theorem.

10.2   Iterated integrals

The positive linear functional used to define Lebesgue measure was an iterated integral. Of course one could take the iterated integral in another order. What would happen to the resulting Radon measure if another order was used? This question will be considered in this section. First, here is a simple lemma.

Lemma 10.5 *If μ and ν are two Radon outer measures on $\mathbb{R}^n$ and $\mu = \nu$ on every half open box, then $\mu = \nu$.*

Proof: From Lemma 10.3, $\mu(U) = \nu(U)$ for all U open. By outer regularity, $\mu(E) = \nu(E)$ for all E measurable.

Corollary 10.6 *Let $\{1, 2, \cdots, n\} = \{k_1, k_2, \cdots, k_n\}$ and the k_i are distinct. For $f \in C_c(\mathbb{R}^n)$ let*

$$\tilde{\Lambda}f = \int_{-\infty}^{\infty} \cdots \int_{-\infty}^{\infty} f(x_1, \cdots, x_n) dx_{k_1} \cdots dx_{k_n},$$

an iterated integral in a different order. Then for all $f \in C_c(\mathbb{R}^n)$,

$$\Lambda f = \tilde{\Lambda}f,$$

and if $\tilde{m}_n$ is the Radon measure representing $\tilde{\Lambda}$ and m_n is the Radon measure representing m_n, then $m_n = \tilde{m}_n$.

Proof: Let $\tilde{m}_n$ be the Radon measure representing $\tilde{\Lambda}$. Then clearly $m_n = \tilde{m}_n$ on every half open box. By Lemma 10.5, $m_n = \tilde{m}_n$. Thus,

$$\Lambda f = \int f dm_n = \int f d\tilde{m}_n = \tilde{\Lambda}f.$$

Lemma 10.7 *Let E be any Borel set. Then*

$$\int_{\mathbb{R}^n} \mathcal{X}_E(\mathbf{x}) dm_n = \int \cdots \int \mathcal{X}_E(x_1, \cdots, x_n) dm(x_1) \cdots dm(x_n)$$

$$= \int \cdots \int \mathcal{X}_E(x_1, \cdots, x_n) dm(x_{k_1}) \cdots dm(x_{k_n}) \qquad (2)$$

where $\{k_1, \cdots, k_n\} = \{1, \cdots, n\}$ and everything which needs to be measurable is. Here the notation involving the iterated integrals refers to one-dimensional Lebesgue integrals.

Proof: Let $Q_k = (-k, k]^n$ and let

$\mathcal{M} \equiv \{$Borel sets, E, such that Formula 2 holds for $E \cap Q_k$

Lebesgue Measure

and there are no measurability problems}.

If $\mathcal{A}$ is the algebra of Example 7.8 of Chapter 7, we see easily that for all such sets, A of $\mathcal{A}$, Formula 2 holds for $A \cap Q_k$ and so $\mathcal{M} \supseteq \mathcal{A}$. Now the theorem on monotone classes implies $\mathcal{M} \supseteq \sigma(\mathcal{A})$ which, by Lemma 10.3, equals the Borel sets. Therefore, $\mathcal{M} =$ Borel sets. Letting $k \to \infty$, and using the monotone convergence theorem, yields the conclusion of the lemma.

The next theorem is referred to as Fubini's theorem. Although a more abstract version will be presented later, the version in the next theorem is particularly useful when dealing with Lebesgue measure.

Theorem 10.8 *Let $f \geq 0$ be Borel measurable. Then everything which needs to be measurable in order to write the following formulae is measurable, and*

$$\int_{\mathbb{R}^n} f(\mathbf{x}) \, dm_n = \int \cdots \int f(x_1, \cdots, x_n) \, dm(x_1) \cdots dm(x_n)$$

$$= \int \cdots \int f(x_1, \cdots, x_n) \, dm(x_{k_1}) \cdots dm(x_{k_n})$$

where $\{k_1, \cdots, k_n\} = \{1, \cdots, n\}$.

Proof: This follows from the previous lemma since the lemma holds for nonnegative simple functions in place of $\mathcal{X}_E$ and we may obtain f as the pointwise limit of an increasing sequence of nonnegative simple functions. The conclusion follows from the monotone convergence theorem.

Corollary 10.9 *Suppose f is complex valued and for some*

$$\{k_1, \cdots, k_n\} = \{1, \cdots, n\},$$

it follows that

$$\int \cdots \int |f(x_1, \cdots, x_n)| \, dm(x_{k_1}) \cdots dm(x_{k_n}) < \infty. \tag{3}$$

Then $f \in L^1(\mathbb{R}^n, m_n)$ and if $\{l_1, \cdots, l_n\} = \{1, \cdots, n\}$, then

$$\int_{\mathbb{R}^n} f \, dm_n = \int \cdots \int f(x_1, \cdots, x_n) \, dm(x_{l_1}) \cdots dm(x_{l_n}). \tag{4}$$

Proof: Applying Theorem 10.8 to the positive and negative parts of the real and imaginary parts of f, Formula 3 implies all these integrals are finite and all iterated integrals taken in any order are equal for these functions. Therefore, the definition of the integral implies Formula 4 holds.

176

10.3   Change of variables

In this section we show that if $F \in \mathcal{L}(\mathbb{R}^n, \mathbb{R}^n)$, then $m_n(F(E)) = \Delta F m_n(E)$ whenever E is Lebesgue measurable. The constant ΔF will also be shown to be $|\det(F)|$. In order to prove this theorem, we give a result from linear algebra which is useful in many different subjects. For example, a special case of this theorem is important in continuum mechanics and later in the book, we will use it in studying Hausdorff measures and the area and coarea formulas. This later application is why the theorem will be presented in a more general form than what is needed for the application of this section.

Theorem 10.10 *Let $F \in \mathcal{L}(\mathbb{R}^n, \mathbb{R}^m)$ where $m \geq n$. Then there exists $R \in \mathcal{L}(\mathbb{R}^n, \mathbb{R}^m)$ and $U \in \mathcal{L}(\mathbb{R}^n, \mathbb{R}^n)$ such that $U = U^*$, all eigenvalues of U are nonnegative,*

$$U^2 = F^*F, \ R^*R = I, \ F = RU,$$

and $|R\mathbf{x}| = |\mathbf{x}|$.

Proof: $(F^*F)^* = F^*F$ and so by linear algebra there is an orthonormal basis of eigenvectors, $\{\mathbf{v}_1, \cdots, \mathbf{v}_n\}$ such that

$$F^*F\mathbf{v}_i = \lambda_i \mathbf{v}_i.$$

It is also clear that $\lambda_i \geq 0$ because

$$\lambda_i(\mathbf{v}_i, \mathbf{v}_i) = (F^*F\mathbf{v}_i, \mathbf{v}_i) = (F\mathbf{v}_i, F\mathbf{v}_i) \geq 0.$$

Now if $\mathbf{u}, \mathbf{v} \in \mathbb{R}^n$, we define the tensor product $\mathbf{u} \otimes \mathbf{v} \in \mathcal{L}(\mathbb{R}^n, \mathbb{R}^n)$ by

$$\mathbf{u} \otimes \mathbf{v}(\mathbf{w}) \equiv (\mathbf{w}, \mathbf{v})\mathbf{u}.$$

Then $F^*F = \sum_{i=1}^n \lambda_i \mathbf{v}_i \otimes \mathbf{v}_i$ because both linear transformations agree on the basis $\{\mathbf{v}_1, \cdots, \mathbf{v}_n\}$. Let

$$U \equiv \sum_{i=1}^n \lambda_i^{1/2} \mathbf{v}_i \otimes \mathbf{v}_i.$$

Then $U^2 = F^*F$, $U = U^*$, and the eigenvalues of U, $\left\{\lambda_i^{1/2}\right\}_{i=1}^n$ are all nonnegative.

Now R is defined on $U(\mathbb{R}^n)$ by

$$RU\mathbf{x} \equiv F\mathbf{x}.$$

This is well defined because if $U\mathbf{x}_1 = U\mathbf{x}_2$, then $U^2(\mathbf{x}_1 - \mathbf{x}_2) = 0$ and so

$$0 = (F^*F(\mathbf{x}_1 - \mathbf{x}_2), \mathbf{x}_1 - \mathbf{x}_2) = |F(\mathbf{x}_1 - \mathbf{x}_2)|^2.$$

Now $|RU\mathbf{x}|^2 = |U\mathbf{x}|^2$ because

$$|RU\mathbf{x}|^2 = |F\mathbf{x}|^2 = (F\mathbf{x}, F\mathbf{x}) = (F^*F\mathbf{x}, \mathbf{x})$$
$$= (U^2\mathbf{x}, \mathbf{x}) = (U\mathbf{x}, U\mathbf{x}) = |U\mathbf{x}|^2.$$

Let $\{\mathbf{x}_1, \cdots, \mathbf{x}_r\}$ be an orthonormal basis for

$$U(\mathbb{R}^n)^\perp \equiv \{\mathbf{x} \in \mathbb{R}^n : (\mathbf{x}, \mathbf{z}) = 0 \text{ for all } \mathbf{z} \in U(\mathbb{R}^n)\}$$

and let $\{\mathbf{y}_1, \cdots, \mathbf{y}_p\}$ be an orthonormal basis for $F(\mathbb{R}^n)^\perp$. Then $p \geq r$ because if $\{F(\mathbf{z}_i)\}_{i=1}^s$ is an orthonormal basis for $F(\mathbb{R}^n)$, it follows that $\{U(\mathbf{z}_i)\}_{i=1}^s$ is orthonormal in $U(\mathbb{R}^n)$ because

$$(U\mathbf{z}_i, U\mathbf{z}_j) = (U^2\mathbf{z}_i, \mathbf{z}_j) = (F^*F\mathbf{z}_i, \mathbf{z}_j) = (F\mathbf{z}_i, F\mathbf{z}_j).$$

Therefore,

$$p + s = m \geq n = r + \dim U(\mathbb{R}^n) \geq r + s.$$

Now define $R \in \mathcal{L}(\mathbb{R}^n, \mathbb{R}^m)$ by $R\mathbf{x}_i \equiv \mathbf{y}_i$, $i = 1, \cdots, r$. Thus

$$\left| R\left(\sum_{i=1}^r c_i\mathbf{x}_i + U\mathbf{v} \right) \right|^2 = \left| \sum_{i=1}^r c_i\mathbf{y}_i + F\mathbf{v} \right|^2 = \sum_{i=1}^r |c_i|^2 + |F\mathbf{v}|^2$$

$$= \sum_{i=1}^r |c_i|^2 + |U\mathbf{v}|^2 = \left| \sum_{i=1}^r c_i\mathbf{x}_i + U\mathbf{v} \right|^2,$$

and so $|R\mathbf{z}| = |\mathbf{z}|$ which implies that for all $\mathbf{x}, \mathbf{y}$,

$$|\mathbf{x}|^2 + |\mathbf{y}|^2 + 2(\mathbf{x}, \mathbf{y}) = |\mathbf{x} + \mathbf{y}|^2 = |R(\mathbf{x} + \mathbf{y})|^2$$
$$= |\mathbf{x}|^2 + |\mathbf{y}|^2 + 2(R\mathbf{x}, R\mathbf{y}).$$

Therefore,

$$(\mathbf{x}, \mathbf{y}) = (R^*R\mathbf{x}, \mathbf{y})$$

for all $\mathbf{x}, \mathbf{y}$ and so $R^*R = I$ as claimed. This proves the theorem.

The following corollary follows as a simple consequence of this theorem.

Corollary 10.11 *Let $F \in \mathcal{L}(\mathbb{R}^n, \mathbb{R}^m)$ and suppose $n \geq m$. Then there exists a symmetric nonnegative element of $\mathcal{L}(\mathbb{R}^m, \mathbb{R}^m)$, U, and an element of $\mathcal{L}(\mathbb{R}^n, \mathbb{R}^m)$, R, such that*

$$F = UR, \quad RR^* = I.$$

Proof: We recall that if $M, L \in \mathcal{L}(\mathbb{R}^s, \mathbb{R}^p)$, then $L^{**} = L$ and $(ML)^* = L^*M^*$. Now apply Theorem 10.10 to $F^* \in \mathcal{L}(\mathbb{R}^m, \mathbb{R}^n)$. Thus,

$$F^* = R^*U$$

where R^* and U satisfy the conditions of that theorem. Then

$$F = UR$$

and $RR^* = R^{**}R^* = I$. This proves the corollary.

The next few lemmas involve the consideration of $F(E)$ where E is a measurable set. They show that $F(E)$ is Lebesgue measurable. We will have occasion to establish similar theorems in other contexts later in the book. In each case, the overall approach will be to show the mapping in question takes sets of measure zero to sets of measure zero and then to exploit the continuity of the mapping and the regularity and completeness of some measure to obtain the final result. The next lemma gives the first part of this procedure here.

Lemma 10.12 *Let $F \in \mathcal{L}(\mathbb{R}^n, \mathbb{R}^n)$, and suppose E is a measurable set having finite measure. Then*

$$\overline{m}_n\left(F\left(E\right)\right) \le \left(2\left\|F\right\|\sqrt{n}\right)^n m_n\left(E\right)$$

where $\overline{m}_n\left(\cdot\right)$ refers to the outer measure,

$$\overline{m}_n\left(S\right) \equiv \inf\left\{m_n\left(E\right) : E \supseteq S \text{ and } E \text{ is measurable}\right\}.$$

Proof: Let $\epsilon > 0$ be given and let $E \subseteq V$, an open set with $m\left(V\right) < m\left(E\right) + \epsilon$. Then let $V = \cup_{i=1}^{\infty} C_i$ where C_i is a half open box all of whose sides have length 2^{-l} for some $l \in \mathbb{N}$ and $C_i \cap C_j = \emptyset$ if $i \ne j$. Then $diam\left(C_i\right) = \sqrt{n}a_i$ where a_i is the length of the sides of C_i. Thus, if y_i is the center of C_i, then

$$B\left(Fy_i, \|F\| diam\left(C_i\right)\right) \supseteq FC_i.$$

Let C_i^* denote the cube with sides of length $2\|F\| diam\left(C_i\right)$ and center at Fy_i. Then

$$C_i^* \supseteq B\left(Fy_i, \|F\| diam\left(C_i\right)\right) \supseteq FC_i$$

and so

$$m_n\left(F\left(E\right)\right) \le m_n\left(F\left(V\right)\right) \le \sum_{i=1}^{\infty}\left(2\|F\| diam\left(C_i\right)\right)^n$$

$$\le \left(2\|F\|\sqrt{n}\right)^n \sum_{i=1}^{\infty}\left(m\left(C_i\right)\right) = \left(2\|F\|\sqrt{n}\right)^n m_n\left(V\right)$$

$$\le \left(2\|F\|\sqrt{n}\right)^n\left(m_n\left(E\right) + \epsilon\right).$$

Since $\epsilon > 0$ is arbitrary, this proves the lemma.

Lemma 10.13 *If E is Lebesgue measurable, then $F(E)$ is also Lebesgue measurable.*

Proof: First note that if K is compact, $F(K)$ is also compact and is therefore Borel measurable. Also, by Problem 3 of Chapter 2, if V is an open set, it can be written as a countable union of compact sets, $\{K_i\}$, and so

$$F(V) = \cup_{i=1}^{\infty} F(K_i)$$

which shows that $F(V)$ is Borel measurable also. Now take any Lebesgue measurable set, E, which is bounded and use regularity of Lebesgue measure to obtain open sets, V_i, and compact sets, K_i, such that

$$K_i \subseteq E \subseteq V_i,$$

$V_i \supseteq V_{i+1}$, $K_i \subseteq K_{i+1}$, $K_i \subseteq E \subseteq V_i$, and $m_n(V_i \setminus K_i) < 2^{-i}$. Let

$$Q \equiv \cap_{i=1}^{\infty} F(V_i), \quad P \equiv \cup_{i=1}^{\infty} F(K_i).$$

Thus both Q and P are Borel (hence Lebesgue) measurable. Observe that

$$Q \setminus P \subseteq \cap_{i,j} (F(V_i) \setminus F(K_j))$$

$$\subseteq \cap_{i=1}^{\infty} F(V_i) \setminus F(K_i) \subseteq \cap_{i=1}^{\infty} F(V_i \setminus K_i)$$

which means, by Lemma 10.12,

$$\overline{m}_n(Q \setminus P) \le \overline{m}_n(F(V_i \setminus K_i)) \le \left(2\|F\|\sqrt{n}\right)^n 2^{-i}$$

which implies $Q \setminus P$ is a set of Lebesgue measure zero since i is arbitrary. Also,

$$P \subseteq F(E) \subseteq Q.$$

By completeness of Lebesgue measure, this shows $F(E)$ is Lebesgue measurable.

If E is not bounded but is measurable, consider $E \cap B(0, k)$. Then

$$F(E) = \cup_{k=1}^{\infty} F(E \cap B(0, k))$$

and, thus, $F(E)$ is measurable. This proves the lemma.

Lemma 10.14 *Let $Q_0 \equiv [0, 1)^n$ and let $\Delta F \equiv m_n(F(Q_0))$. Then if Q is any half open box whose sides are of length 2^{-k}, $k \in \mathbb{N}$, and F is one to one, it follows*

$$m_n(FQ) = m_n(Q)\Delta F$$

and if F is not one to one we can say

$$m_n(FQ) \ge m_n(Q)\Delta F.$$

Proof: There are $\left(2^k\right)^n \equiv l$ nonintersecting half open boxes, Q_i, each having measure $\left(2^{-k}\right)^n$ whose union equals Q_0. If F is one to one, translation invariance of Lebesgue measure and the assumption F is linear imply

$$\left(2^k\right)^n m_n\left(F\left(Q\right)\right) = \sum_{i=1}^{l} m_n\left(F\left(Q_i\right)\right) =$$

$$m_n\left(\cup_{i=1}^{l} F\left(Q_i\right)\right) = m_n\left(F\left(Q_0\right)\right) = \Delta F. \qquad (*)$$

Therefore,

$$m_n\left(F\left(Q\right)\right) = \left(2^{-k}\right)^n \Delta F = m_n\left(Q\right) \Delta F.$$

If F is not one to one, the second equality sign in $(*)$ should be $\geq$. This proves the lemma.

Theorem 10.15 *If E is any Lebesgue measurable set, then*

$$m_n\left(F\left(E\right)\right) = \Delta F m_n\left(E\right).$$

If $R^* R = I$ and R preserves distances, then

$$\Delta R = 1.$$

Also, if $F, G \in \mathcal{L}\left(\mathbb{R}^n, \mathbb{R}^n\right),$ then

$$\Delta\left(FG\right) = \Delta F \Delta G. \qquad (5)$$

Proof: Let V be any open set and let $\{Q_i\}$ be half open disjoint boxes of the sort discussed earlier whose union is V and suppose first that F is one to one. Then

$$m_n\left(F\left(V\right)\right) = \sum_{i=1}^{\infty} m_n\left(F\left(Q_i\right)\right) = \Delta F \sum_{i=1}^{\infty} m_n\left(Q_i\right) = \Delta F m_n\left(V\right). \qquad (6)$$

Now let E be an arbitrary bounded measurable set and let V_i be a decreasing sequence of open sets containing E with

$$i^{-1} \geq m_n\left(V_i \setminus E\right).$$

Then let $S \equiv \cap_{i=1}^{\infty} F\left(V_i\right)$

$$S \setminus F\left(E\right) = \cap_{i=1}^{\infty}\left(F\left(V_i\right) \setminus F\left(E\right)\right) \subseteq \cap_{i=1}^{\infty} F\left(V_i \setminus E\right)$$

and so from Lemma 10.12,

$$m_n\left(S \setminus F\left(E\right)\right) \leq \lim_{i \to \infty} \sup\, m_n\left(F\left(V_i \setminus E\right)\right)$$

$$\leq \lim_{i \to \infty} \sup\, \left(2\left\|F\right\| \sqrt{n}\right)^n i^{-1} = 0.$$

Thus

$$m_n \left(F \left(E \right) \right) = m_n \left(S \right) = \lim_{i \to \infty} m_n \left(F \left(V_i \right) \right)$$

$$= \lim_{i \to \infty} \Delta F m_n \left(V_i \right) = \Delta F m_n \left(E \right). \tag{7}$$

If E is not bounded, apply the above to $E \cap B \left(0, k \right)$ and let $k \to \infty$.

To see the second claim of the theorem,

$$\Delta R m_n \left(B \left(0, 1 \right) \right) = m_n \left(R B \left(0, 1 \right) \right) = m_n \left(B \left(0, 1 \right) \right).$$

Now suppose F is not one to one. Then let $\{\mathbf{v}_1, \cdots, \mathbf{v}_n\}$ be an orthonormal basis of $\mathbb{R}^n$ such that for some $r < n, \{\mathbf{v}_1, \cdots, \mathbf{v}_r\}$ is an orthonormal basis for $F \left(\mathbb{R}^n \right)$. Let $R\mathbf{v}_i \equiv \mathbf{e}_i$ where the $\mathbf{e}_i$ are the standard unit basis vectors. Then $RF \left(\mathbb{R}^n \right) \subseteq span \left(\mathbf{e}_1, \cdots, \mathbf{e}_r \right)$ and so by what we know about the Lebesgue measure of boxes, whose sides are parallel to the coordinate axes,

$$m_n \left(RF \left(Q \right) \right) = 0$$

whenever Q is a box. Thus,

$$m_n \left(RF \left(Q \right) \right) = \Delta R m_n \left(F \left(Q \right) \right) = 0$$

and this shows that in the case where F is not one to one, $m_n \left(F \left(Q \right) \right) = 0$ which shows from Lemma 10.14 that $m_n \left(F \left(Q \right) \right) = \Delta F m_n \left(Q \right)$ even if F is not one to one. Therefore, Formula 6 continues to hold even if F is not one to one and this implies Formula 7. Formula 5 follows from this. This proves the theorem.

Lemma 10.16 *Suppose $U = U^*$ and U has all nonnegative eigenvalues, $\{\lambda_i\}$. Then*

$$\Delta U = \prod_{i=1}^{n} \lambda_i.$$

Proof: Suppose $U_0 \equiv \sum_{i=1}^{n} \lambda_i \mathbf{e}_i \otimes \mathbf{e}_i$. Note that

$$Q_0 = \left\{ \sum_{i=1}^{n} t_i \mathbf{e}_i : t_i \in [0, 1) \right\}.$$

Thus

$$U_0 \left(Q_0 \right) = \left\{ \sum_{i=1}^{n} \lambda_i t_i \mathbf{e}_i : t_i \in [0, 1) \right\}$$

and so

$$\Delta U_0 \equiv m_n \left(U_0 Q_0 \right) = \prod_{i=1}^{n} \lambda_i.$$

Now by linear algebra, since $U = U^*$,

$$U = \sum_{i=1}^{n} \lambda_i \mathbf{v}_i \otimes \mathbf{v}_i$$

where $\{\mathbf{v}_i\}$ is an orthonormal basis of eigenvectors. Define $R \in \mathcal{L}(\mathbb{R}^n, \mathbb{R}^n)$ such that

$$R\mathbf{v}_i = \mathbf{e}_i.$$

Then R preserves distances and $RUR^* = U_0$ where U_0 is given above. Therefore, if E is any measurable set,

$$\prod_{i=1}^{n} \lambda_i m_n(E) = \Delta U_0 m_n(E) = m_n(U_0(E)) = m_n(RUR^*(E))$$

$$= \Delta R \Delta U \Delta R^* m_n(E) = \Delta U m_n(E).$$

Hence $\prod_{i=1}^{n} \lambda_i = \Delta U$ as claimed. This proves the theorem.

Theorem 10.17 *Let $F \in \mathcal{L}(\mathbb{R}^n, \mathbb{R}^n)$. Then $\Delta F = |\det(F)|$. Thus*

$$m_n(F(E)) = |\det(F)| m_n(E)$$

for all E Lebesgue measurable.

Proof: By Theorem 10.10, $F = RU$ where R and U are described in that theorem. Then

$$\Delta F = \Delta R \Delta U = \Delta U = \det(U).$$

Now $F^*F = U^2$ and so $(\det(U))^2 = \det(U^2) = \det(F^*F) = (\det(F))^2$. Therefore,

$$\det(U) = |\det F|$$

and this proves the theorem.

10.4   Polar coordinates

One of the most useful of all techniques in establishing estimates which involve integrals taken with respect to Lebesgue measure on $\mathbb{R}^n$ is the technique of polar coordinates. This section presents the polar coordinate formula. To begin with we give a general lemma.

Lemma 10.18 *Let X and Y be topological spaces. Then if E is a Borel set in X and F is a Borel set in Y, then $E \times F$ is a Borel set in $X \times Y$.*

Proof: Let E be an open set in X and let

$$\mathcal{S}_E \equiv \{F \text{ Borel in } Y \text{ such that } E \times F \text{ is Borel in } X \times Y\}.$$

Then $\mathcal{S}_E$ contains the open sets and is clearly closed with respect to countable unions. Let $F \in \mathcal{S}_E$. Then

$$E \times F^C \cup E \times F = E \times Y = \text{ a Borel set.}$$

Therefore, since $E \times F$ is Borel, it follows $E \times F^C$ is Borel. Therefore, $\mathcal{S}_E$ is a σ algebra. It follows $\mathcal{S}_E = $ Borel sets, and so, we have shown-open $\times$ Borel $=$ Borel. Now let F be a fixed Borel set in Y and define

$$\mathcal{S}_F \equiv \{E \text{ Borel in } X \text{ such that } E \times F \text{ is Borel in } X \times Y\}.$$

The same argument which was just used shows $\mathcal{S}_F$ is a σ algebra containing the open sets. Therefore, $\mathcal{S}_F = $ the Borel sets, and this proves the lemma since F was an arbitrary Borel set.

Now we define the unit sphere in $\mathbb{R}^n$, S^{n-1}, by

$$S^{n-1} \equiv \{\mathbf{w} \in \mathbb{R}^n : |\mathbf{w}| = 1\}.$$

Then S^{n-1} is a compact metric space using the usual metric on $\mathbb{R}^n$. We define a map

$$\theta : S^{n-1} \times (0, \infty) \to \mathbb{R}^n \setminus \{\mathbf{0}\}$$

by

$$\theta(\mathbf{w}, \rho) \equiv \rho \mathbf{w}.$$

It is clear that θ is one to one and onto with a continuous inverse. Therefore, if $\mathcal{B}_1$ are the Borel sets in $S^{n-1} \times (0, \infty)$, and $\mathcal{B}$ are the Borel sets in $\mathbb{R}^n \setminus \{\mathbf{0}\}$, it follows

$$\mathcal{B} = \{\theta(F) : F \in \mathcal{B}_1\}. \tag{8}$$

Observe also that $\mathcal{B}_1 \equiv \mathcal{R}$ satisfies the conditions of Lemma 7.6 with Z defined as S^{n-1} and the same is true of the sets $(a, b] \cap (0, \infty)$ where $0 \leq a, b \leq \infty$ if Z is defined as $(0, \infty)$. By Corollary 7.7, finite disjoint unions of sets of the form

$$\{E \times I : E \text{ is Borel in } S^{n-1}$$

$$\text{and } I = (a, b] \cap (0, \infty) \text{ where } 0 \leq a, b \leq \infty\}$$

form an algebra of sets, $\mathcal{A}$. It is also clear that $\sigma(\mathcal{A})$ contains the open sets and so $\sigma(\mathcal{A}) = \mathcal{B}_1$ because every set in $\mathcal{A}$ is in $\mathcal{B}_1$ thanks to Lemma 10.18. Let $A_r \equiv S^{n-1} \times (0, r]$ and let

$$\mathcal{M} \equiv \left\{ F \in \mathcal{B}_1 : \int_{\mathbb{R}^n} \mathcal{X}_{\theta(F \cap A_r)} dm_n \right.$$

$$= \int_{(0,\infty)} \int_{S^{n-1}} \mathcal{X}_{\theta(F \cap A_r)} (\rho \mathbf{w}) \, \rho^{n-1} d\sigma dm \Bigg\},$$

where for E a Borel set in S^{n-1},

$$\sigma(E) \equiv n m_n \left(\theta \left(E \times (0, 1) \right) \right). \tag{9}$$

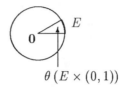

$$\theta \left(E \times (0, 1) \right)$$

Then if $F \in \mathcal{A}$, say $F = E \times (a, b]$, we can show $F \in \mathcal{M}$. This follows easily from the observation that

$$\int_{\mathbb{R}^n} \mathcal{X}_{\theta(F)} dm_n = \int_{\mathbb{R}^n} \mathcal{X}_{\theta(E \times (0,b])} (\mathbf{y}) \, dm_n - \int_{\mathbb{R}^n} \mathcal{X}_{\theta(E \times (0,a])} (\mathbf{y}) \, dm_n$$

$$= m_n \left(\theta(E \times (0,1)) b^n - m_n \left(\theta(E \times (0,1)) a^n = \sigma(E) (b^n - a^n), \right. \right.$$

a consequence of the change of variables theorem applied to $\mathbf{y} = a\mathbf{x}$, and

$$\int_{(0,\infty)} \int_{S^{n-1}} \mathcal{X}_{\theta(E \times (a,b])} (\rho \mathbf{w}) \, \rho^{n-1} d\sigma dm = \int_a^b \int_E \rho^{n-1} d\sigma d\rho$$

$$= \sigma(E) (b^n - a^n).$$

Since it is clear that $\mathcal{M}$ is a monotone class, it follows from the monotone class theorem that $\mathcal{M} = \mathcal{B}_1$. Letting $r \to \infty$, we may conclude that for all $F \in \mathcal{B}_1$,

$$\int_{\mathbb{R}^n} \mathcal{X}_{\theta(F)} dm_n = \int_{(0,\infty)} \int_{S^{n-1}} \mathcal{X}_{\theta(F)} (\rho \mathbf{w}) \, \rho^{n-1} d\sigma dm.$$

By Formula 8, if A is any Borel set in $\mathbb{R}^n$, then $A \setminus \{0\} = \theta(F)$ for some $F \in \mathcal{B}_1$. Thus

$$\int_{\mathbb{R}^n} \mathcal{X}_A dm_n = \int_{\mathbb{R}^n} \mathcal{X}_{\theta(F)} dm_n =$$

$$\int_{(0,\infty)} \int_{S^{n-1}} \mathcal{X}_{\theta(F)} (\rho \mathbf{w}) \, \rho^{n-1} d\sigma dm = \int_{(0,\infty)} \int_{S^{n-1}} \mathcal{X}_A (\rho \mathbf{w}) \, \rho^{n-1} d\sigma dm.$$

$$\tag{10}$$

With this preparation, it is easy to prove the main result which is the following theorem.

Theorem 10.19 *Let $f \geq 0$ and f is Borel measurable on $\mathbb{R}^n$. Then*

$$\int_{\mathbb{R}^n} f(\mathbf{y}) \, dm_n = \int_{(0,\infty)} \int_{S^{n-1}} f(\rho \mathbf{w}) \rho^{n-1} d\sigma dm \qquad (11)$$

where σ is defined by Formula 9 and $\mathbf{y} = \rho \mathbf{w}$, for $\mathbf{w} \in S^{n-1}$.

Proof: From Formula 10, Formula 11 holds for f replaced with a nonnegative simple function. Now the monotone convergence theorem applied to a sequence of simple functions increasing to f yields the desired conclusion.

10.5   Exercises

1. If $\mathcal{A}$ is the algebra of sets of Example 7.8, show $\sigma(\mathcal{A})$, the smallest σ algebra containing the algebra, is the Borel sets.

2. Suppose X, Y are two locally compact, σ compact, metric spaces. Let $\mathcal{A}$ be the collection of finite disjoint unions of sets of the form $E \times F$ where E and F are Borel sets. Show that $\mathcal{A}$ is an algebra and that the smallest σ algebra containing $\mathcal{A}$, $\sigma(\mathcal{A})$, contains the Borel sets of $X \times Y$. **Hint:** Show $X \times Y$, with the usual product topology, is a σ ompact metric space. Next show every open set can be written as a countable union of compact sets. Using this, show every open set can be written as a countable union of open sets of the form $U \times V$ where U is an open set in X and V is an open set in Y.

3. ↑ Suppose X, Y are two locally compact, σ compact, metric spaces and let μ and ν be Radon measures on X and Y respectively. Define for $f \in C_c(X \times Y)$,

$$\Lambda f \equiv \int_X \int_Y f(x,y) \, d\nu d\mu.$$

Show this is well defined and is a positive linear functional on

$$C_c(X \times Y).$$

Let $(\overline{\mu \times \nu})$ be the measure representing Λ. Show that for $f \geq 0$, and f Borel measurable,

$$\int_{X \times Y} f d(\overline{\mu \times \nu}) = \int_X \int_Y f(x,y) \, d\nu d\mu = \int_Y \int_X f(x,y) \, d\mu d\nu.$$

Hint: First show, using the dominated convergence theorem, that if $E \times F$ is the cartesian product of two Borel sets each of whom have finite measure, then

$$(\overline{\mu \times \nu})(E \times F) = \mu(E)\nu(F) = \int_X \int_Y \mathcal{X}_{E \times F}(x, y)\, d\mu d\nu.$$

4. Let $f{:}\mathbb{R}^n \to \mathbb{R}$ be defined by $f(\mathbf{x}) \equiv \left(1 + |\mathbf{x}|^2\right)^k$. Find the values of k for which f is in $L^1(\mathbb{R}^n)$. **Hint:** This is easy and reduces to a one-dimensional problem if you use the formula for integration using polar coordinates.

5. Let B be a Borel set in $\mathbb{R}^n$ and let $\mathbf{v}$ be a nonzero vector in $\mathbb{R}^n$. Suppose B has the following property. For each $\mathbf{x} \in \mathbb{R}^n$, $m(\{t : \mathbf{x} + t\mathbf{v} \in B\}) = 0$. Then show $m_n(B) = 0$. Note the condition on B says roughly that B is thin in one direction.

6. If $f : \mathbb{R}^n \to [0, \infty]$ is Lebesgue measurable, show there exists $g : \mathbb{R}^n \to [0, \infty]$ such that $g = f$ a.e. and g is Borel measurable.

7. ↑ Let $f \in L^1(\mathbb{R})$, $g \in L^1(\mathbb{R})$. Let

$$(f * g)(x) = \int_{\mathbb{R}} f(x - y)g(y)dy.$$

Show $|(f * g)(x)| < \infty$ a.e. and that

$$\|f * g\|_{L^1} \le \|f\|_{L^1}\|g\|_{L^1}, \quad \|f\|_{L^1} = \int |f|dx.$$

Hint: If f is Lebesgue measurable, there exists g Borel measurable with $g(x) = f(x)$ a.e.

8. ↑ Let $f : [0, \infty) \to \mathbb{R}$ be in $L^1(\mathbb{R}, m)$. The Laplace transform is given by $\widehat{f}(x) = \int_0^\infty e^{-xt} f(t)dt$. Let f, g be in $L^1(\mathbb{R}, m)$, and let $h(x) = \int_0^x f(x - t)g(t)dt$. Show $h \in L^1$, and $\widehat{h} = \widehat{f}\widehat{g}$.

9. Let $\mathcal{D}$ consist of functions, $g \in C_c(\mathbb{R}^n)$ which are of the form

$$g(\mathbf{x}) \equiv \prod_{i=1}^n g_i(x_i)$$

where each $g_i \in C_c(\mathbb{R})$. Show that if $f \in C_c(\mathbb{R}^n)$, then there exists a sequence of functions, $\{g_k\}$ in $\mathcal{D}$ which satisfies

$$\lim_{k \to \infty} \sup\{|f(\mathbf{x}) - g_k(\mathbf{x})| : \mathbf{x} \in \mathbb{R}^n\} = 0. \tag{*}$$

Now for $g \in \mathcal{D}$ given as above, let

$$\Lambda_0(g) \equiv \int \cdots \int \prod_{i=1}^{n} g_i(x_i) \, dm(x_1) \cdots dm(x_n),$$

and define, for arbitrary $f \in C_c(\mathbb{R}^n)$,

$$\Lambda f \equiv \lim_{k \to \infty} \Lambda_0 g_k$$

where * holds. Show this is a well-defined positive linear functional which yields Lebesgue measure. Establish all theorems in this chapter using this as a basis for the definition of Lebesgue measure. Note this approach is arguably less fussy than the presentation in the chapter. **Hint:** You might want to use the Stone Weierstrass theorem.

Chapter 11
Product Measure

In calculus, the actual evaluation of a multiple integral involves iterated integrals. An important manipulation is the change in order of integration. We are asked to believe that iterated integrals in different order are all equal and they all equal the value of the multiple integral. In fact, this is not always true. The following is an example.

Example 11.1 *Let $0 < \delta_1 < \delta_2 < \cdots < \delta_n \cdots < 1, \lim_{n \to 0} \delta_n = 1$. Let g_n be a real continuous function with $g_n = 0$ outside of (δ_n, δ_{n+1}) and $\int_0^1 g_n(x)dx = 1$ for all n. Define*

$$f(x, y) = \sum_{n=1}^{\infty} (g_n(x) - g_{n+1}(x))g_n(y).$$

Then you can show the following:
a.) f is continuous on $[0, 1) \times [0, 1)$
b.) $\int_0^1 \int_0^1 f(x, y)dydx = 1$, $\int_0^1 \int_0^1 f(x, y)dxdy = 0$.

Nevertheless, it is often the case that the iterated integrals are equal and give the value of an appropriate multiple integral. The best theorems of this sort are to be found in the theory of Lebesgue integration and this is what will be discussed in this chapter.

Definition 11.2 A measure space $(X, \mathcal{F}, \mu)$ is said to be σ finite if

$$X = \cup_{n=1}^{\infty} X_n, \quad X_n \in \mathcal{F}, \quad \mu(X_n) < \infty.$$

In the rest of this chapter, unless otherwise stated, $(X, \mathcal{S}, \mu)$ and $(Y, \mathcal{F}, \lambda)$ will be two σ finite measure spaces. Note that a Radon measure on a σ compact, locally compact space gives an example of a σ finite space. In particular, Lebesgue measure is σ finite.

Definition 11.3 A measurable rectangle is a set $A \times B \subseteq X \times Y$ where $A \in \mathcal{S}, B \in \mathcal{F}$. An elementary set will be any subset of $X \times Y$ which is a finite union of disjoint measurable rectangles. $\mathcal{S} \times \mathcal{F}$ will denote the smallest σ algebra of sets in $\mathcal{P}(X \times Y)$ containing all elementary sets.

Definition 11.4 Let $E \subseteq X \times Y$,

$$E_x = \{y \in Y : (x, y) \in E\},$$

$$E^y = \{x \in X : (x, y) \in E\}.$$

These are called the x and y sections.

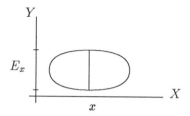

Theorem 11.5 *If $E \in \mathcal{S} \times \mathcal{F}$, then $E_x \in \mathcal{F}$ and $E^y \in \mathcal{S}$ for all $x \in X$ and $y \in Y$.*

Proof: Let

$$\mathcal{M} = \{E \subseteq \mathcal{S} \times \mathcal{F} \text{ such that } E_x \in \mathcal{F},$$

$$E^y \in \mathcal{S} \text{ for all } x \in X \text{ and } y \in Y.\}$$

Then $\mathcal{M}$ contains all measurable rectangles. If $E_i \in \mathcal{M}$,

$$(\cup_{i=1}^{\infty} E_i)_x = \cup_{i=1}^{\infty} (E_i)_x \in \mathcal{F}.$$

Similarly, $(\cup_{i=1}^{\infty} E_i)^y \in \mathcal{S}$. $\mathcal{M}$ is thus closed under countable unions. If $E \in \mathcal{M}$,

$$(E^C)_x = (E_x)^C \in \mathcal{F}.$$

Similarly, $(E^C)^y \in \mathcal{S}$. Thus $\mathcal{M}$ is closed under complementation. Therefore $\mathcal{M}$ is a σ-algebra containing the elementary sets. Hence, $\mathcal{M} \supseteq \mathcal{S} \times \mathcal{F}$. But $\mathcal{M} \subseteq \mathcal{S} \times \mathcal{F}$. Therefore $\mathcal{M} = \mathcal{S} \times \mathcal{F}$ and the theorem is proved.

It follows from Lemma 7.6 of Chapter 7 that the elementary sets form an algebra because clearly the intersection of two measurable rectangles is a measurable rectangle and

$$(A \times B) \setminus (A_0 \times B_0) = (A \setminus A_0) \times B \cup (A \cap A_0) \times (B \setminus B_0),$$

an elementary set. We use this in the next theorem.

Theorem 11.6 *If $(X, \mathcal{S}, \mu)$ and $(Y, \mathcal{F}, \lambda)$ are both finite measure spaces ($\mu(X)$, $\lambda(Y) < \infty$), then for every $E \in \mathcal{S} \times \mathcal{F}$,*
  a.) $x \to \lambda(E_x)$ is μ measurable, $y \to \mu(E^y)$ is λ measurable
  b.) $\int_X \lambda(E_x) d\mu = \int_Y \mu(E^y) d\lambda$.

Proof: Let $\mathcal{M} = \{E \in \mathcal{S} \times \mathcal{F}$ such that both a.) and b.) hold$\}$. Since μ and λ are both finite, the monotone convergence and dominated convergence theorems imply that $\mathcal{M}$ is a monotone class. Clearly $\mathcal{M}$ contains the algebra of elementary sets. By the monotone class theorem, $\mathcal{M} \supseteq \mathcal{S} \times \mathcal{F}$.

Theorem 11.7 *If $(X, \mathcal{S}, \mu)$ and $(Y, \mathcal{F}, \lambda)$ are both σ-finite measure spaces, then for every $E \in \mathcal{S} \times \mathcal{F}$,*
a.) $x \to \lambda(E_x)$ is μ measurable, $y \to \mu(E^y)$ is λ measurable.
b.) $\int_X \lambda(E_x)d\mu = \int_Y \mu(E^y)d\lambda$.

Proof: Let $X = \cup_{n=1}^{\infty} X_n, Y = \cup_{n=1}^{\infty} Y_n$ where,

$$X_n \subseteq X_{n+1}, Y_n \subseteq Y_{n+1}, \mu(X_n) < \infty, \lambda(Y_n) < \infty.$$

Let
$$\mathcal{S}_n = \{A \cap X_n : A \in \mathcal{S}\}, \quad \mathcal{F}_n = \{B \cap Y_n : B \in \mathcal{F}\}.$$

Thus $(X_n, \mathcal{S}_n, \mu)$ and $(Y_n, \mathcal{F}_n, \lambda)$ are both finite measure spaces.
Claim: If $E \in \mathcal{S} \times \mathcal{F}$, then $E \cap (X_n \times Y_n) \in \mathcal{S}_n \times \mathcal{F}_n$.
Proof: Let $\mathcal{M}_n = \{E \in \mathcal{S} \times \mathcal{F} : E \cap (X_n \times Y_n) \in \mathcal{S}_n \times \mathcal{F}_n\}$. Clearly $\mathcal{M}_n$ contains the algebra of elementary sets. It is also clear that $\mathcal{M}_n$ is a monotone class. Thus $\mathcal{M}_n = \mathcal{S} \times \mathcal{F}$.
Now let $E \in \mathcal{S} \times \mathcal{F}$. By Theorem 11.6,

$$\int_{X_n} \lambda((E \cap (X_n \times Y_n))_x)d\mu = \int_{Y_n} \mu((E \cap (X_n \times Y_n))^y)d\lambda \qquad (1)$$

where the integrands are measurable. Now

$$(E \cap (X_n \times Y_n))_x = \emptyset$$

if $x \notin X_n$ and a similar observation holds for the second integrand in Formula 1. Therefore,

$$\int_X \lambda((E \cap (X_n \times Y_n))_x)d\mu = \int_Y \mu((E \cap (X_n \times Y_n))^y)d\lambda.$$

Then letting $n \to \infty$, we use the monotone convergence theorem to get b.). The measurability assertions of a.) are valid because the limit of a sequence of measurable functions is measurable.

Definition 11.8 For $E \in \mathcal{S} \times \mathcal{F}$ and $(X, \mathcal{S}, \mu), (Y, \mathcal{F}, \lambda)$ σ-finite, $(\mu \times \lambda)(E) \equiv \int_X \lambda(E_x)d\mu = \int_Y \mu(E^y)d\lambda$.

This definition is well defined because of Theorem 11.7. We also have the following theorem.

Theorem 11.9 *If $A \in \mathcal{S}, B \in \mathcal{F}$, then $(\mu \times \lambda)(A \times B) = \mu(A)\lambda(B)$, and $\mu \times \lambda$ is a measure on $\mathcal{S} \times \mathcal{F}$ called product measure.*

The proof of Theorem 11.9 is obvious and is left to the reader. Use the Monotone Convergence theorem. The next theorem is one of several theorems due to Fubini and Tonelli. These theorems all have to do with interchanging the order of integration in a multiple integral. The main ideas are illustrated by the next theorem which is often referred to as Fubini's theorem.

Theorem 11.10 *Let $f : X \times Y \to [0, \infty]$ be measurable with respect to $\mathcal{S} \times \mathcal{F}$ and suppose μ and λ are σ-finite. Then*

$$\int_{X \times Y} f d(\mu \times \lambda) = \int_X \int_Y f(x, y) d\lambda d\mu = \int_Y \int_X f(x, y) d\mu d\lambda \quad (2)$$

and all integrals make sense.

Proof: For $E \in \mathcal{S} \times \mathcal{F}$, we note

$$\int_Y \mathcal{X}_E(x, y) d\lambda = \lambda(E_x), \quad \int_X \mathcal{X}_E(x, y) d\mu = \mu(E^y).$$

Thus from Definition 11.8, Formula 2 holds if $f = \mathcal{X}_E$. It follows that Formula 2 holds for every nonnegative simple function. By Theorem 8.8, there exists an increasing sequence, $\{f_n\}$, of simple functions converging pointwise to f. Then

$$\int_Y f(x, y) d\lambda = \lim_{n \to \infty} \int_Y f_n(x, y) d\lambda,$$

$$\int_X f(x, y) d\mu = \lim_{n \to \infty} \int_X f_n(x, y) d\mu.$$

This follows from the monotone convergence theorem. Since

$$x \to \int_Y f_n(x, y) d\lambda$$

is measurable with respect to $\mathcal{S}$, it follows that $x \to \int_Y f(x, y) d\lambda$ is also measurable with respect to $\mathcal{S}$. A similar conclusion can be drawn about $y \to \int_X f(x, y) d\mu$. Thus the two iterated integrals make sense. Since Formula 2 holds for f_n, another application of the Monotone Convergence theorem shows Formula 2 holds for f. This proves the theorem.

Corollary 11.11 *Let $f : X \times Y \to \mathbb{C}$ be $\mathcal{S} \times \mathcal{F}$ measurable. Suppose either $\int_X \int_Y |f| \, d\lambda d\mu$ or $\int_Y \int_X |f| \, d\mu d\lambda < \infty$. Then $f \in L^1(X \times Y, \mu \times \lambda)$ and*

$$\int_{X \times Y} f d(\mu \times \lambda) = \int_X \int_Y f d\lambda d\mu = \int_Y \int_X f d\mu d\lambda \quad (3)$$

with all integrals making sense.

Proof: Suppose first that f is real-valued. Apply Theorem 11.10 to f^+ and f^-. Formula 3 follows from observing that $f = f^+ - f^-$; and that all integrals are finite. If f is complex valued, consider real and imaginary parts. This proves the corollary.

How can we tell if f is $\mathcal{S} \times \mathcal{F}$ measurable? The following theorem gives a convenient way for many examples.

Theorem 11.12 *If X and Y are topological spaces having a countable basis of open sets and if $\mathcal{S}$ and $\mathcal{F}$ both contain the open sets, then $\mathcal{S} \times \mathcal{F}$ contains the Borel sets.*

Proof: We need to show $\mathcal{S} \times \mathcal{F}$ contains the open sets in $X \times Y$. If $\mathcal{B}$ is a countable basis for the topology of X and if $\mathcal{C}$ is a countable basis for the topology of Y, then

$$\{B \times C : B \in \mathcal{B}, C \in \mathcal{C}\}$$

is a countable basis for the topology of $X \times Y$. (Remember a basis for the topology of $X \times Y$ is the collection of sets of the form $U \times V$ where U is open in X and V is open in Y.) Thus every open set is a countable union of sets $B \times C$ where $B \in \mathcal{B}$ and $C \in \mathcal{C}$. Since $B \times C$ is a measurable rectangle, it follows that every open set in $X \times Y$ is in $\mathcal{S} \times \mathcal{F}$. This proves the theorem.

The importance of this theorem is that we can use it to assert that a function is product measurable if it is Borel measurable. For an example of how this can sometimes be done, see Problem 5 in this chapter.

Theorem 11.13 *Suppose $\mathcal{S}$ and $\mathcal{F}$ are Borel, μ and λ are regular on $\mathcal{S}$ and $\mathcal{F}$ respectively, and $\mathcal{S} \times \mathcal{F}$ is Borel. Then $\mu \times \lambda$ is regular on $\mathcal{S} \times \mathcal{F}$. (Recall Theorem 11.12 for a sufficient condition for $\mathcal{S} \times \mathcal{F}$ to be Borel.)*

Proof: Let $\mu(X_n) < \infty, \lambda(Y_n) < \infty$, and $X_n \uparrow X, Y_n \uparrow Y$. Let $R_n = X_n \times Y_n$ and define

$$\mathcal{G}_n = \{S \in \mathcal{S} \times \mathcal{F} : \mu \times \lambda \text{ is regular on } S \cap R_n\}.$$

By this we mean that for $S \in \mathcal{G}_n$

$$(\mu \times \lambda)(S \cap R_n) = \inf\{(\mu \times \lambda)(V) : V \text{ is open and } V \supseteq S \cap R_n\}$$

and

$$(\mu \times \lambda)(S \cap R_n) =$$

$$\sup\{(\mu \times \lambda)(K) : K \text{ is compact and } K \subseteq S \cap R_n\}.$$

If $P \times Q$ is a measurable rectangle, then

$$(P \times Q) \cap R_n = (P \cap X_n) \times (Q \cap Y_n).$$

193

Let $K_x \subseteq (P \cap X_n)$ and $K_y \subseteq (C \cap Y_n)$ be such that

$$\mu(K_x) + \varepsilon > \mu(P \cap X_n)$$

and

$$\lambda(K_y) + \varepsilon > \lambda(C \cap Y_n).$$

By Theorem 1.36 $K_x \times K_y$ is compact and from the definition of product measure,

$$(\mu \times \lambda)(K_x \times K_y) = \mu(K_x)\lambda(K_y)$$

$$\geq \mu(P \cap X_n)\lambda(C \cap Y_n) - \varepsilon(\lambda(C \cap Y_n) + \mu(P \cap X_n)) + \varepsilon^2.$$

Since ε is arbitrary, this verifies that $(\mu \times \lambda)$ is inner regular on $S \cap R_n$ whenever S is an elementary set. Similarly, $(\mu \times \lambda)$ is outer regular on $S \cap R_n$ whenever S is an elementary set. Thus $\mathcal{G}_n$ contains the elementary sets.

Next we show that $\mathcal{G}_n$ is a monotone class. If $S_k \downarrow S$ and $S_k \in \mathcal{G}_n$, let K_k be a compact subset of $S_k \cap R_n$ with

$$(\mu \times \lambda)(K_k) + \varepsilon 2^{-k} > (\mu \times \lambda)(S_k \cap R_n).$$

Let $K = \cap_{k=1}^\infty K_k$. Then

$$S \cap R_n \setminus K \subseteq \cup_{k=1}^\infty (S_k \cap R_n \setminus K_k).$$

Therefore

$$(\mu \times \lambda)(S \cap R_n \setminus K) \leq \sum_{k=1}^\infty (\mu \times \lambda)(S_k \cap R_n \setminus K_k)$$

$$\leq \sum_{k=1}^\infty \varepsilon 2^{-k} = \varepsilon.$$

Now let $V_k \supseteq S_k \cap R_n$, V_k is open and

$$(\mu \times \lambda)(S_k \cap R_n) + \varepsilon > (\mu \times \lambda)(V_k).$$

Let k be large enough that

$$(\mu \times \lambda)(S_k \cap R_n) - \varepsilon < (\mu \times \lambda)(S \cap R_n).$$

Then $(\mu \times \lambda)(S \cap R_n) + 2\varepsilon > (\mu \times \lambda)(V_k)$. This shows $\mathcal{G}_n$ is closed with respect to intersections of decreasing sequences of its elements. The consideration of increasing sequences is similar. By the monotone class theorem, $\mathcal{G}_n = \mathcal{S} \times \mathcal{F}$.

Now let $S \in \mathcal{S} \times \mathcal{F}$ and let $l < (\mu \times \lambda)(S)$. Then $l < (\mu \times \lambda)(S \cap R_n)$ for some n. It follows from the first part of this proof that there exists a compact subset of $S \cap R_n, K$, such that $(\mu \times \lambda)(K) > l$. It follows that $(\mu \times \lambda)$ is inner regular on $\mathcal{S} \times \mathcal{F}$. To verify that the product measure is outer regular on $\mathcal{S} \times \mathcal{F}$, let V_n be an open set such that

$$V_n \supseteq S \cap R_n, \ (\mu \times \lambda)(V_n \setminus (S \cap R_n)) < \varepsilon 2^{-n}$$

Let $V = \cup_{n=1}^{\infty} V_n$. Then $V \supseteq S$ and

$$V \setminus S \subseteq \cup_{n=1}^{\infty} V_n \setminus (S \cap R_n).$$

Thus,

$$(\mu \times \lambda)(V \setminus S) \leq \sum_{n=1}^{\infty} \varepsilon 2^{-n} = \varepsilon$$

and so

$$(\mu \times \lambda)(V) \leq \varepsilon + (\mu \times \lambda)(S).$$

This proves the theorem.

11.1   Completion of product measure

It can happen that even though $(X, \mathcal{S}, \mu)$ and $(Y, \mathcal{F}, \lambda)$ are complete measure spaces, $(X \times Y, \mathcal{S} \times \mathcal{F}, \mu \times \lambda)$ is not a complete measure space. That is, there can exist a set $E \in \mathcal{S} \times \mathcal{F}$ with $(\mu \times \lambda)(E) = 0$, but E has subsets which are not in $\mathcal{S} \times \mathcal{F}$. See Problem 6 in this section for an important example of this phenomenon. The next topic discusses an extension of product measure to give a complete measure space. We will always assume $(X, \mathcal{S}, \mu)$ and $(Y, \mathcal{F}, \lambda)$ are σ-finite and complete. For example, this is the case if $X = \mathbb{R}, \mu = m, \mathcal{S} =$ Lebesgue measurable sets.

Definition 11.14 Define an outer measure $\overline{\mu \times \lambda}$ by

$$(\overline{\mu \times \lambda})(A) = \inf \{(\mu \times \lambda)(B) : B \supseteq A, B \in \mathcal{S} \times \mathcal{F}\}.$$

It is clear that $(\overline{\mu \times \lambda})$ is an outer measure because if $A = \cup_{i=1}^{\infty} A_i$ and $(\overline{\mu \times \lambda})(A_i) = \infty$ for any i, then

$$(\overline{\mu \times \lambda})(A) \leq \sum_{i=1}^{\infty} (\overline{\mu \times \lambda})(A_i).$$

If $(\overline{\mu \times \lambda})(A_i) < \infty$ for all i, let $B_i \in \mathcal{S} \times \mathcal{F}, B_i \supseteq A_i$, and

$$(\mu \times \lambda)(B_i) < (\overline{\mu \times \lambda})(A_i) + \varepsilon 2^{-i}.$$

Then let $B = \cup_{i=1}^{\infty} B_i$. We have

$$(\overline{\mu \times \lambda})(A) \leq (\overline{\mu \times \lambda})(B) \leq (\mu \times \lambda)(B)$$

$$\leq \sum_{i=1}^{\infty} (\mu \times \lambda)(B_i) \leq \varepsilon + \sum_{i=1}^{\infty} (\overline{\mu \times \lambda})(A_i).$$

Since ε is arbitrary, this shows $\overline{\mu \times \lambda}$ is subadditive. The other requirements for an outer measure are obvious.

Definition 11.15 $\overline{\mathcal{S} \times \mathcal{F}}$ is the set of $\overline{\mu \times \lambda}$ measurable sets. Thus $E \in \overline{\mathcal{S} \times \mathcal{F}}$ if and only if for all $S \subseteq X \times Y$,

$$(\overline{\mu \times \lambda})(S) = (\overline{\mu \times \lambda})(S \cap E) + (\overline{\mu \times \lambda})(S \setminus E).$$

Theorem 11.16 *The following statements hold.*
   a.) $\mathcal{S} \times \mathcal{F} \subseteq \overline{\mathcal{S} \times \mathcal{F}}$
   b.) *If* $B \in \mathcal{S} \times \mathcal{F}$*, then* $(\overline{\mu \times \lambda})(B) = (\mu \times \lambda)(B)$
   c.) *If* $A \in \mathcal{P}(X \times Y)$*, then there exists* $B \in \overline{\mathcal{S} \times \mathcal{F}}$ *with*

$$B \supseteq A, \ (\overline{\mu \times \lambda})(A) = (\mu \times \lambda)(B).$$

Proof: Once a.) is shown, b.) and c.) are obvious. Let $S \subseteq X \times Y$ and let $B \in \mathcal{S} \times \mathcal{F}$. We need to show

$$(\overline{\mu \times \lambda})(S) = (\overline{\mu \times \lambda})(S \setminus B) + (\overline{\mu \times \lambda})(S \cap B). \qquad (4)$$

Since $\overline{\mu \times \lambda}$ is subadditive, this is obviously true if $(\overline{\mu \times \lambda})(S) = \infty$. So assume $(\overline{\mu \times \lambda})(S) < \infty$. Pick $D \in \mathcal{S} \times \mathcal{F}$ with

$$(\overline{\mu \times \lambda})(S) + \varepsilon \geq (\mu \times \lambda)(D), \ D \supseteq S.$$

Then

$$\begin{aligned}
(\overline{\mu \times \lambda})(S) \ &> \ (\mu \times \lambda)(D) - \varepsilon = (\mu \times \lambda)(D \cap B) + (\mu \times \lambda)(D \setminus B) - \varepsilon \\
&\geq \ (\overline{\mu \times \lambda})(S \cap B) + (\overline{\mu \times \lambda})(S \setminus B) - \varepsilon.
\end{aligned}$$

Since ε is arbitrary, this establishes Formula 4 and proves the theorem.

Lemma 11.17 *Let* $(\overline{\mu \times \lambda})(E) = 0$*. Then* $\lambda(E_x) = 0$ *for* μ *a.e.* x*. Furthermore, if* $A \in \overline{\mathcal{S} \times \mathcal{F}}$*, then there exists* $B \in \mathcal{S} \times \mathcal{F}$ *such that*

$$B \supseteq A, \ (\overline{\mu \times \lambda})(B \setminus A) = 0.$$

Proof: There exists $B \supseteq E$, $B \in \mathcal{S} \times \mathcal{F}$, and $(\mu \times \lambda)(B) = 0$. But

$$(\mu \times \lambda)(B) = \int_X \lambda(B_x) d\mu.$$

Therefore $\lambda(E_x) = 0$ for μ a.e. x. (Note that to write $\lambda(E_x)$ we needed the completeness of $(Y, \mathcal{F}, \lambda)$.) To complete the proof, let

$$R_n = X_n \times Y_n. \ \mu(X_n) < \infty, \lambda(Y_n) < \infty,$$

and $R_n \uparrow X \times Y$. Let $A_n = A \cap R_n$ and let $B_n \in \mathcal{S} \times \mathcal{F}$ with

$$B_n \supseteq A_n, \ B_n \subseteq B_{n+1},$$

and $(\overline{\mu \times \lambda})(B_n \setminus A_n) = 0$. Let $B = \cup B_n$. Then

$$(\overline{\mu \times \lambda})(B \setminus A) \le \sum_{i=1}^{\infty} (\overline{\mu \times \lambda})(B_i \setminus A_i) = 0.$$

Lemma 11.18 *Let $A \in \overline{\mathcal{S} \times \mathcal{F}}$. Then the following statements hold.*
a.) $y \to \mathcal{X}_A(x, y)$ is λ measurable for μ a.e. x
b.) $x \to \int_Y \mathcal{X}_A(x, y) d\lambda$ equals a μ measurable function. Thus the iterated integral $\int_X \int_Y \mathcal{X}_A d\lambda d\mu$ makes sense and furthermore,
c.) $(\overline{\mu \times \lambda})(A) = \int_{X \times Y} \mathcal{X}_A \, d(\overline{\mu \times \lambda}) = \int_X \int_Y \mathcal{X}_A d\lambda d\mu$.

Proof: By Lemma 11.17, let

$$B \supseteq A, B \in \mathcal{S} \times \mathcal{F}, (\overline{\mu \times \lambda})(B) = (\overline{\mu \times \lambda})(A),$$

and $(\overline{\mu \times \lambda})(B \setminus A) = 0$. Then $\lambda(B_x \setminus A_x) = 0 \ \mu$ a.e. x. It follows that for μ a.e. x,

$$\mathcal{X}_B(x, y) = \mathcal{X}_A(x, y)$$

for λ a.e. y. (If we fix x, $\mathcal{X}_B(x, y) \ne \mathcal{X}_A(x, y)$ exactly when $y \in B_x \setminus A_x$. Thus if x is such that

$$\lambda(B_x \setminus A_x) = 0,$$

it follows that $\mathcal{X}_B(x, y) = \mathcal{X}_A(x, y)$ for λ a.e. y.) Therefore, for μ a.e. x,

$$y \to \mathcal{X}_A(x, y) \text{ is measurable} \tag{5}$$

$$\int_Y \mathcal{X}_B(x, y) d\lambda = \int_Y \mathcal{X}_A(x, y) d\lambda \tag{6}$$

The reason for Formula 5 is this: The function $y \to \mathcal{X}_A(x, y)$ may be written as $\mathcal{X}_{A_x}(y)$. Also, $\mathcal{X}_{A_x}(y) = \mathcal{X}_{B_x}(y) - \mathcal{X}_{(B_x \setminus A_x)}(y)$ and B_x is λ measurable because $B \in S \times F$. Thus $\mathcal{X}_{A_x}(y)$ is the difference of two measurable functions, the second function being measurable because x was chosen so that $\lambda(B_x \setminus A_x) = 0$.
Of course Formula 6 follows from this whenever $\lambda(B_x \setminus A_x) = 0$. We know

$$x \to \int_Y \mathcal{X}_B(x, y) d\lambda$$

is μ measurable since $B \in \mathcal{S} \times \mathcal{F}$. Therefore, since Formula 6 holds for μ a.e. x, we have

$$x \to \int_Y \mathcal{X}_A(x, y) d\lambda$$

is μ measurable. So far we have established a.) and b.). To obtain c.) we note that by defining $\int_Y \mathcal{X}_A(x,y)d\lambda$ on the exceptional set of x to be anything we want (to write $\int_Y \mathcal{X}_A(x,y)d\lambda$ we need $\mathcal{X}_{A_x}(y)$ to be λ measurable which only happens for certain for a.e. x), we may regard $x \to \int_Y \mathcal{X}_A(x,y)d\lambda$ to be a μ measurable function so we can do $\int_X d\mu$ to it. This yields

$$\int_X \int_Y \mathcal{X}_A(x,y)d\lambda d\mu = \int_X \int_Y \mathcal{X}_B(x,y)d\lambda d\mu = \int_{X\times Y} \mathcal{X}_B(x,y)d(\mu\times\lambda)$$

$$= (\mu\times\lambda)(B) = (\overline{\mu\times\lambda})(A) = \int_{X\times Y} \mathcal{X}_A d(\overline{\mu\times\lambda}).$$

Corollary 11.19 *Let $s = \sum_{i=1}^n c_i \mathcal{X}_{A_i}$, $A_i \in \overline{\mathcal{S}\times\mathcal{F}}$, and $c_i \geq 0$. Then*
   a.) $\int_{X\times Y} s\, d(\overline{\mu\times\lambda}) = \int_X \int_Y s\, d\lambda d\mu$ where the iterated integral makes sense because there exists $B \in \mathcal{S}, \mu(B) = 0$, and
   b.) $y \to s(x,y)$ is λ measurable for $x \notin B$.
   c.) $x \to \int_Y s(x,y)d\lambda$ equals a μ measurable function on $X \setminus B$.

Here is another version of Fubini's theorem in this context.

Theorem 11.20 *Let $f \geq 0$ be $\overline{\mathcal{S}\times\mathcal{F}}$ measurable. Then*
   a.) $\int_{X\times Y} f d(\overline{\mu\times\lambda}) = \int_X \int_Y f d\lambda d\mu$ where the iterated integral makes sense because there exists $B \in \mathcal{S}$ with $\mu(B) = 0$ such that
   b.) $y \to f(x,y)$ is λ measurable for $x \notin B$.
   c.) $x \to \int_Y f(x,y)d\lambda$ equals a μ measurable function on $X \setminus B$.

Proof: Let $\{s_k\}$ be a sequence of $\overline{\mathcal{S}\times\mathcal{F}}$ measurable nonnegative simple functions increasing pointwise to f. By Corollary 11.19 there is a set $B_k \in \mathcal{S}$ with $\mu(B_k) = 0$ and for $x \notin B_k$ b.) and c.) of Corollary 11.19 hold. Let $B = \cup_{k=1}^\infty B_k$. Then $\mu(B) = 0$ and for $x \notin B$, b.) and c.) hold for $s = s_k$ for all k. Therefore using the monotone convergence theorem, b.) and c.) hold for f in place of s. Also, we obtain that for $x \notin B$,

$$\int_Y s_k(x,y)d\lambda \to \int_Y f(x,y)d\lambda.$$

Therefore another application of the Monotone Convergence theorem yields

$$\int_X \int_Y f d\lambda d\mu \equiv \int_{B^C} \int_Y f(x,y)d\lambda d\mu$$

$$= \lim_{k\to 0} \int_{B^C} \int_Y s_k(x,y)d\lambda d\mu$$

$$\equiv \lim_{k\to 0} \int_X \int_Y s_k(x,y)d\lambda d\mu$$

$$= \lim_{k\to 0} \int_{X\times Y} s_k d(\overline{\mu\times\lambda}) = \int_{X\times Y} f d(\overline{\mu\times\lambda}).$$

This proves the theorem.

As before, there is a corollary to this theorem.

Corollary 11.21 *Let $f : X \times Y \to \mathbb{C}$ be $\overline{\mathcal{S} \times \mathcal{F}}$ measurable, and suppose that either*

$$\int_X \int_Y |f| \, d\lambda d\mu < \infty$$

or

$$\int_Y \int_X |f| \, d\mu d\lambda < \infty.$$

Then $f \in L^1(X \times Y, \overline{\mu \times \lambda})$ and

$$\int_{X \times Y} f d(\overline{\mu \times \lambda}) = \int_X \int_Y f d\lambda d\mu = \int_Y \int_X f d\mu d\lambda$$

with the iterated integrals making sense.

From now on, we may write dx instead of dm to denote Lebesgue measure and we will occasionally use the traditional notation of calculus when referring to the Lebesgue integral with respect to Lebesgue measure.

11.2 Exercises

1. Let $\mathcal{A}$ be an algebra of sets in $\mathcal{P}(Z)$ and suppose μ and ν are two finite measures on $\sigma(\mathcal{A})$, the σ-algebra generated by $\mathcal{A}$. Show that if $\mu = \nu$ on $\mathcal{A}$, then $\mu = \nu$ on $\sigma(\mathcal{A})$.

2. ↑ Extend Problem 1 to the case where μ, ν are σ finite with

$$Z = \cup_{n=1}^{\infty} Z_n, \ Z_n \in \mathcal{A}$$

and $\mu(Z_n) < \infty$.

3. Show $\lim_{A \to \infty} \int_0^A \frac{\sin x}{x} dx = \frac{\pi}{2}$. **Hint:** Use $\frac{1}{x} = \int_0^\infty e^{-xt} dt$ and Fubini's theorem.

4. Suppose $g : \mathbb{R}^n \to \mathbb{R}$ has the property that g is continuous in each variable. Can we conclude that g is continuous? **Hint:** Consider

$$g(x, y) \equiv \begin{cases} \frac{xy}{x^2 + y^2} & \text{if } (x, y) \neq (0, 0), \\ 0 & \text{if } (x, y) = (0, 0). \end{cases}$$

5. Suppose $g : \mathbb{R}^n \to \mathbb{R}$ is continuous in every variable. Show that g is the pointwise limit of some sequence of continuous functions. Conclude that if g is continuous in each variable, then g is Borel measurable. Give an example of a Borel measurable function on $\mathbb{R}^n$ which is not continuous in each variable. **Hint:** In the case of $n = 2$ let

$$a_i \equiv \frac{i}{n}, i \in \mathbb{Z}$$

and for $(x, y) \in [a_{i-1}, a_i) \times \mathbb{R}$, we let

$$g_n (x, y) \equiv \frac{a_i - x}{a_i - a_{i-1}} g (a_{i-1}, y) + \frac{x - a_{i-1}}{a_i - a_{i-1}} g (a_i, y).$$

Show g_n converges to g and is continuous. Now use induction to verify the general result.

6. Show $(\mathbb{R}^2, m \times m, \mathcal{S} \times \mathcal{S})$ where $\mathcal{S}$ is the set of Lebesgue measurable sets is not a complete measure space. Show there exists $A \in \mathcal{S} \times \mathcal{S}$ and $E \subseteq A$ such that $(m \times m)(A) = 0$, but $E \notin \mathcal{S} \times \mathcal{S}$.

7. Recall that for

$$E \in \mathcal{S} \times \mathcal{F}, (\mu \times \lambda)(E) = \int_X \lambda(E_x) d\mu = \int_Y \mu(E^y) d\lambda.$$

Why is $\mu \times \lambda$ a measure on $\mathcal{S} \times \mathcal{F}$?

8. Suppose $G(x) = G(a) + \int_a^x g(t) \, dt$ where $g \in L^1$ and suppose $F(x) = F(a) + \int_a^x f(t) \, dt$ where $f \in L^1$. Show the usual formula for integration by parts holds,

$$\int_a^b fG \, dx = FG|_a^b - \int_a^b F g \, dx.$$

Hint: You might try replacing $G(x)$ with $G(a) + \int_a^x g(t) \, dt$ in the first integral on the left and then using Fubini's theorem.

9. Let $f : \Omega \to [0, \infty)$ be measurable where $(\Omega, \mathcal{S}, \mu)$ is a σ finite measure space. Let $\phi : [0, \infty) \to [0, \infty)$ satisfy: ϕ is increasing. Show

$$\int_X \phi(f(x)) d\mu = \int_0^\infty \phi'(t) \mu(x : f(x) > t) dt.$$

The function $t \to \mu(x : f(x) > t)$ is called the distribution function.
Hint:

$$\int_X \phi(f(x)) d\mu = \int_X \int_{\mathbb{R}} \mathcal{X}_{[0, f(x))} \phi'(t) dt dx.$$

Now try to use Fubini's theorem. Be sure to check that everything is appropriately measurable. In doing so, you may want to first consider $f(x)$ a nonnegative simple function. Is it necessary to assume $(\Omega, \mathcal{S}, \mu)$ is σ finite?

Chapter 12
The L^p Spaces

12.1 Basic inequalities and properties

The Lebesgue integral makes it possible to define and prove theorems about the space of functions described below. These L^p spaces are very useful in applications of real analysis and this chapter is about these spaces. In what follows $(\Omega, \mathcal{S}, \mu)$ will be a measure space.

Definition 12.1 Let $1 \leq p < \infty$. We define

$$L^p(\Omega) \equiv \{f \ : f \ \text{is measurable and} \ \int_\Omega |f(\omega)|^p d\mu < \infty\}$$

and

$$\|f\|_{L^p} \equiv \left(\int_\Omega |f|^p d\mu \right)^{\frac{1}{p}} \equiv \|f\|_p.$$

In fact $\| \ \|_p$ is a norm if things are interpreted correctly. First we need to obtain Holder's inequality. We will always use the following convention for each $p > 1$.

$$\frac{1}{p} + \frac{1}{q} = 1.$$

Often one uses p' instead of q in this context.

Theorem 12.2 *(Holder's inequality) If f and g are measurable functions, then if $p > 1$,*

$$\int |f| \, |g| \, d\mu \leq \left(\int |f|^p d\mu \right)^{\frac{1}{p}} \left(\int |g|^q d\mu \right)^{\frac{1}{q}}. \tag{1}$$

Proof: To begin with, we prove Young's inequality.

Lemma 12.3 *If $0 \leq a, b$ then $ab \leq \frac{a^p}{p} + \frac{b^q}{q}$.*

Proof: Consider the following picture:

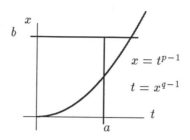

$$ab \le \int_0^a t^{p-1}\,dt + \int_0^b x^{q-1}\,dx = \frac{a^p}{p} + \frac{b^q}{q}.$$

Note equality occurs when $a^p = b^q$.

If either $\int |f|^p\,d\mu$ or $\int |g|^p\,d\mu$ equals ∞ or 0, the inequality 1 is obviously valid. Therefore assume both of these are less than ∞ and not equal to 0. By the lemma,

$$\int \frac{|f|}{||f||_p}\,\frac{|g|}{||g||_q}\,d\mu \le \frac{1}{p}\int \frac{|f|^p}{||f||_p^p}\,d\mu + \frac{1}{q}\int \frac{|g|^q}{||g||_q^q}\,d\mu = 1.$$

Hence,

$$\int |f|\,|g|\,d\mu \le ||f||_p\,||g||_q.$$

This proves Holder's inequality.

Corollary 12.4 *(Minkowski inequality) Let $1 \le p < \infty$. Then*

$$||f + g||_p \le ||f||_p + ||g||_p. \qquad (2)$$

Proof: If $p = 1$, this is obvious. Let $p > 1$. We can assume $||f||_p$ and $||g||_p < \infty$ and $||f + g||_p \ne 0$ or there is nothing to prove. Therefore,

$$\int |f + g|^p\,d\mu \le 2^{p-1}\left(\int |f|^p + |g|^p\,d\mu\right) < \infty.$$

Now

$$\int |f + g|^p\,d\mu \le$$

$$\int |f + g|^{p-1}|f|\,d\mu + \int |f + g|^{p-1}|g|\,d\mu$$

$$= \int |f + g|^{\frac{p}{q}}|f|\,d\mu + \int |f + g|^{\frac{p}{q}}|g|\,d\mu$$

$$\le \left(\int |f + g|^p\,d\mu\right)^{\frac{1}{q}}\left(\int |f|^p\,d\mu\right)^{\frac{1}{p}} + \left(\int |f + g|^p\,d\mu\right)^{\frac{1}{q}}\left(\int |g|^p\,d\mu\right)^{\frac{1}{p}}.$$

Dividing both sides by $(\int |f+g|^p d\mu)^{\frac{1}{q}}$ yields Formula 2. This proves the corollary.

This shows that if $f, g \in L^p$, then $f+g \in L^p$. Also, it is clear that if a is a constant and $f \in L^p$, then $af \in L^p$. Hence L^p is a vector space. Also we have the following from the Minkowski inequality and the definition of $\| \|_p$.

a.) $\|f\|_p \geq 0, \|f\|_p = 0$ if and only if $f = 0$ a.e.

b.) $\|af\|_p = |a| \|f\|_p$ if a is a scalar.

c.) $\|f + g\|_p \leq \|f\|_p + \|g\|_p$.

We see that $\| \|_p$ would be a norm if $\|f\|_p = 0$ implied $f = 0$. If we agree to identify all functions in L^p that differ only on a set of measure zero, then $\| \|_p$ is a norm and L^p is a normed vector space. We will do so from now on. Next we show that L^p is a complete normed vector space, a Banach space.

Theorem 12.5 *The following hold for $L^p(\Omega)$*

a.) $L^p(\Omega)$ is complete.

b.) If $\{f_n\}$ is a Cauchy sequence in $L^p(\Omega)$, then there exists $f \in L^p(\Omega)$ and a subsequence which converges a.e. to $f \in L^p(\Omega)$, and $\|f_n - f\|_p \to 0$.

Proof: Let $\{f_n\}$ be a Cauchy sequence in $L^p(\Omega)$. This means that for every $\varepsilon > 0$ there exists N such that if $n, m \geq N$, then $\|f_n - f_m\|_p < \varepsilon$. Now we will select a subsequence. Let n_1 be such that $\|f_n - f_m\|_p < 2^{-1}$ whenever $n, m \geq n_1$. Let n_2 be such that $n_2 > n_1$ and $\|f_n - f_m\|_p < 2^{-2}$ whenever $n, m \geq n_2$. If $n_1, \cdots, n_k$ have been chosen, let $n_{k+1} > n_k$ and whenever $n, m \geq n_{k+1}, \|f_n - f_m\|_p < 2^{-(k+1)}$. The subsequence will be $\{f_{n_k}\}$. Thus, $\|f_{n_k} - f_{n_{k+1}}\|_p < 2^{-k}$. Let

$$g_{k+1} = f_{n_{k+1}} - f_{n_k}.$$

Then by the Minkowski inequality,

$$\infty > \sum_{k=1}^{\infty} \|g_{k+1}\|_p \geq \sum_{k=1}^{m} \|g_{k+1}\|_p \geq \left\| \sum_{k=1}^{m} |g_{k+1}| \right\|_p$$

for all m. It follows that

$$\int \left(\sum_{k=1}^{m} |g_{k+1}| \right)^p d\mu \leq \left(\sum_{k=1}^{\infty} \|g_{k+1}\|_p \right)^p < \infty \tag{3}$$

for all m and so the monotone convergence theorem implies that the sum up to m in Formula 3 can be replaced by a sum up to ∞. Thus,

$$\sum_{k=1}^{\infty} |g_{k+1}(x)| < \infty \text{ a.e. } x.$$

Therefore, $\sum_{k=1}^{\infty} g_{k+1}(x)$ converges for a.e. x and for such x, let

$$f(x) = f_{n_1}(x) + \sum_{k=1}^{\infty} g_{k+1}(x).$$

Note that $\sum_{k=1}^{m} g_{k+1}(x) = f_{n_{m+1}}(x) - f_{n_1}(x)$. Therefore $\lim_{k\to\infty} f_{n_k}(x) = f(x)$ for all $x \notin E$ where $\mu(E) = 0$. If we redefine f_{n_k} to equal 0 on E and let $f(x) = 0$ for $x \in E$, it then follows that $\lim_{k\to\infty} f_{n_k}(x) = f(x)$ for all x. By Fatou's lemma,

$$||f - f_{n_k}||_p \leq \lim_{m\to\infty} \inf ||f_{n_m} - f_{n_k}||_p \leq \sum_{i=k}^{\infty} ||f_{n_{i+1}} - f_{n_i}||_p \leq 2^{-(k-1)}.$$

Therefore, $f \in L^p(\Omega)$ because

$$||f||_p \leq ||f - f_{n_k}||_p + ||f_{n_k}||_p < \infty,$$

and $\lim_{k\to\infty} ||f_{n_k} - f||_p = 0$. This proves b.). To see that the original Cauchy sequence converges to f in $L^p(\Omega)$, we write

$$||f - f_n||_p \leq ||f - f_{n_k}||_p + ||f_{n_k} - f_n||_p.$$

If $\varepsilon > 0$ is given, let $2^{-(k-1)} < \frac{\varepsilon}{2}$. Then if $n > n_k$,

$$||f - f_n|| < 2^{-(k-1)} + 2^{-k} < \frac{\varepsilon}{2} + \frac{\varepsilon}{2} = \varepsilon.$$

This proves part a.) and completes the proof of the theorem.

In working with the L^p spaces, the following inequality also known as Minkowski's inequality is very useful.

Theorem 12.6 *Let $(X, \mathcal{S}, \mu)$ and $(Y, \mathcal{F}, \lambda)$ be σ-finite measure spaces and let f be product measurable. Then the following inequality is valid for $p \geq 1$.*

$$\int_X \left(\int_Y |f(x,y)|^p \, d\lambda\right)^{\frac{1}{p}} d\mu \geq \left(\int_Y (\int_X |f(x,y)| \, d\mu)^p d\lambda\right)^{\frac{1}{p}}. \quad (4)$$

Proof: Let $X_n \uparrow X$, $Y_n \uparrow Y$, $\lambda(Y_n) < \infty$, $\mu(X_n) < \infty$, and let

$$f_m(x,y) = \begin{cases} f(x,y) & \text{if } |f(x,y)| \leq m, \\ m & \text{if } |f(x,y)| > m. \end{cases}$$

Thus

$$\left(\int_{Y_n} (\int_{X_k} |f_m(x,y)| d\mu)^p d\lambda\right)^{\frac{1}{p}} < \infty.$$

Let

$$J(y) = \int_{X_k} |f_m(x,y)|d\mu.$$

Then

$$\int_{Y_n} \left(\int_{X_k} |f_m(x,y)|d\mu \right)^p d\lambda = \int_{Y_n} J(y)^{p-1} \int_{X_k} |f_m(x,y)|d\mu \, d\lambda$$

$$= \int_{X_k} \int_{Y_n} J(y)^{p-1} |f_m(x,y)|d\lambda \, d\mu$$

by Fubini's theorem. Now we apply Holder's inequality and recall $p-1 = \frac{p}{q}$. This yields

$$\int_{Y_n} \left(\int_{X_k} |f_m(x,y)|d\mu \right)^p d\lambda$$

$$\leq \int_{X_k} \left(\int_{Y_n} J(y)^p \, d\lambda \right)^{\frac{1}{q}} \left(\int_{Y_n} |f_m(x,y)|^p \, d\lambda \right)^{\frac{1}{p}} d\mu$$

$$= \left(\int_{Y_n} J(y)^p \, d\lambda \right)^{\frac{1}{q}} \int_{X_k} \left(\int_{Y_n} |f_m(x,y)|^p \, d\lambda \right)^{\frac{1}{p}} d\mu$$

$$= \left(\int_{Y_n} (\int_{X_k} |f_m(x,y)|d\mu)^p d\lambda \right)^{\frac{1}{q}} \int_{X_k} \left(\int_{Y_n} |f_m(x,y)|^p \, d\lambda \right)^{\frac{1}{p}} d\mu.$$

Therefore,

$$\left(\int_{Y_n} \left(\int_{X_k} |f_m(x,y)|d\mu \right)^p d\lambda \right)^{\frac{1}{p}} \leq \int_{X_k} \left(\int_{Y_n} |f_m(x,y)|^p \, d\lambda \right)^{\frac{1}{p}} d\mu. \quad (5)$$

To obtain Formula 4 let $m \to \infty$ and use the Monotone Convergence theorem to replace f_m by f in Formula 5. Next let $k \to \infty$ and use the same theorem to replace X_k with X. Finally let $n \to \infty$ and use the Monotone Convergence theorem again to replace Y_n with Y. This yields Formula 4.

Next, we develop some properties of the L^p spaces.

12.2   Density of simple functions

Theorem 12.7 Let $p \geq 1$ and let $(\Omega, \mathcal{S}, \mu)$ be a measure space. Then the simple functions are dense in $L^p(\Omega)$.

Proof: By breaking an arbitrary function into real and imaginary parts and then considering the positive and negative parts of these, we see that there is no loss of generality in assuming f has values in $[0, \infty]$. By Theorem 8.8, there is an increasing sequence of simple functions, $\{s_n\}$, converging pointwise to $f(x)^p$. Let $t_n(x) = s_n(x)^{\frac{1}{p}}$. Thus, $t_n(x) \uparrow f(x)$. Now

$$|f(x) - t_n(x)| \leq |f(x)|.$$

By the Dominated Convergence theorem, we may conclude

$$0 = \lim_{n \to \infty} \int |f(x) - t_n(x)|^p d\mu.$$

Thus simple functions are dense in L^p.

Recall that for Ω a topological space, $C_c(\Omega)$ is the space of continuous functions with compact support in Ω. Also recall the following definition.

Definition 12.8 Let $(\Omega, \mathcal{S}, \mu)$ be a measure space and suppose (Ω, τ) is also a topological space. Then $(\Omega, \mathcal{S}, \mu)$ is called a regular measure space if the σ-algebra of Borel sets is contained in $\mathcal{S}$ and for all $E \in \mathcal{S}$,

$$\mu(E) = \inf\{\mu(V) : V \supseteq E \text{ and } V \text{ open}\}$$

and

$$\mu(E) = \sup\{\mu(K) : K \subseteq E \text{ and } K \text{ is compact }\}.$$

For example any Radon measure on a σ-compact, locally compact metric space yields an example of a regular measure space by Chapter 9. Recall also the notation that $spt\,(f)$ denotes the closure of the set of nonzero values of f.

Lemma 12.9 *Let Ω be a locally compact metric space and let K be a compact subset of V, an open set. Then there exists a continuous function $f : \Omega \to [0, 1]$ such that $f(x) = 1$ for all $x \in K$ and $spt(f)$ is a compact subset of V.*

Proof: Let $K \subseteq W \subseteq \overline{W} \subseteq V$ and $\overline{W}$ is compact. Define f by

$$f(x) = \frac{dist(x, W^C)}{dist(x, K) + dist(x, W^C)}.$$

It is not necessary to be in a metric space to do this. See Urysohn's lemma for example (Problem 10 of Chapter 1). It is just less trouble and since we are mainly interested in $\mathbb{R}^n$ and subsets of $\mathbb{R}^n$, this is the case presented here.

Theorem 12.10 *Let $(\Omega, \mathcal{S}, \mu)$ be a regular measure space as in Definition 12.8 where the conclusion of Lemma 12.9 holds. Then $C_c(\Omega)$ is dense in $L^p(\Omega)$.*

Proof: Let $f \in L^p(\Omega)$ and pick a simple function, s, such that $||s - f||_p < \frac{\varepsilon}{2}$ where $\varepsilon > 0$ is arbitrary. Let

$$s(x) = \sum_{i=1}^{m} c_i \mathcal{X}_{E_i}(x)$$

where $c_1, \cdots, c_m$ are the distinct nonzero values of s. Thus the E_i are disjoint and $\mu(E_i) < \infty$ for each i. Therefore there exist compact sets, K_i and open sets, V_i, such that $K_i \subseteq E_i \subseteq V_i$ and

$$\sum_{i=1}^{m} |c_i| \mu(V_i \setminus K_i)^{\frac{1}{p}} < \frac{\varepsilon}{2}.$$

Let $h_i \in C_c(\Omega)$ satisfy

$$
\begin{aligned}
h_i(x) &= 1 \text{ for } x \in K_i \\
spt(h_i) &\subseteq V_i.
\end{aligned}
$$

Let

$$g = \sum_{i=1}^{m} c_i h_i.$$

Then by Minkowski's inequality,

$$
\begin{aligned}
||g - s||_p &\leq \left(\int_\Omega (\sum_{i=1}^{m} |c_i| \, |h_i(x) - \mathcal{X}_{E_i}(x)|)^p d\mu \right)^{\frac{1}{p}} \\
&\leq \sum_{i=1}^{m} \left(\int_\Omega |c_i|^p |h_i(x) - \mathcal{X}_{E_i}(x)|^p d\mu \right)^{\frac{1}{p}} \\
&\leq \sum_{i=1}^{m} |c_i| \mu(V_i \setminus K_i)^{\frac{1}{p}} < \frac{\varepsilon}{2}.
\end{aligned}
$$

Therefore,

$$||f - g||_p \leq ||f - s||_p + ||s - g||_p < \frac{\varepsilon}{2} + \frac{\varepsilon}{2} = \varepsilon.$$

This proves the theorem.

12.3 Continuity of translation

Definition 12.11 Let f be a function defined on $U \subseteq \mathbb{R}^n$ and let $\mathbf{w} \in \mathbb{R}^n$. Then $f_{\mathbf{w}}$ will be the function defined on $\mathbf{w} + U$ by

$$f_{\mathbf{w}}(\mathbf{x}) = f(\mathbf{x} - \mathbf{w}).$$

Theorem 12.12 *(Continuity of translation in L^p) Let $f \in L^p(\mathbb{R}^n)$ with $\mu = m$, Lebesgue measure. Then*

$$\lim_{||\mathbf{w}|| \to 0} ||f_{\mathbf{w}} - f||_p = 0.$$

Proof: Let $\varepsilon > 0$ be given and let $g \in C_c(\mathbb{R}^n)$ with $||g - f||_p < \frac{\varepsilon}{3}$. Since Lebesgue measure is translation invariant $(m(\mathbf{w}+E) = m(E))$, $||g_{\mathbf{w}} - f_{\mathbf{w}}||_p = ||g - f||_p < \frac{\varepsilon}{3}$. Therefore

$$
\begin{aligned}
||f - f_{\mathbf{w}}||_p &\leq ||f - g||_p + ||g - g_{\mathbf{w}}||_p + ||g_{\mathbf{w}} - f_{\mathbf{w}}|| \\
&< \frac{2\varepsilon}{3} + ||g - g_{\mathbf{w}}||_p.
\end{aligned}
$$

But $\lim_{||\mathbf{w}|| \to 0} g_{\mathbf{w}}(\mathbf{x}) = g(\mathbf{x})$ uniformly in x because g is uniformly continuous. Therefore, whenever $||\mathbf{w}||$ is small enough, $||g - g_{\mathbf{w}}||_p < \frac{\varepsilon}{3}$. Thus $||f - f_{\mathbf{w}}||_p < \varepsilon$ whenever $||\mathbf{w}||$ is sufficiently small. This proves the theorem.

12.4 Separability

Theorem 12.13 *For $p \geq 1$, $L^p(\mathbb{R}^n, m)$ is separable. This means there exists a countable set, $\mathcal{D}$, such that if $f \in L^p(\mathbb{R}^n)$ and $\varepsilon > 0$, there exists $g \in \mathcal{D}$ such that $||f - g||_p < \varepsilon$.*

Proof: Let Q be all functions of the form $c\mathcal{X}_{[\mathbf{a},\mathbf{b})}$ where

$$[\mathbf{a}, \mathbf{b}) \equiv [a_1, b_1) \times [a_2, b_2) \times \cdots \times [a_n, b_n),$$

and both a_i, b_i are rational, while c has rational real and imaginary parts. Let $\mathcal{D}$ be the set of all finite sums of functions in Q. Thus, $\mathcal{D}$ is countable. We now show that $\mathcal{D}$ is dense in $L^p(\mathbb{R}^n, m)$. To see this, we need to show that for every $f \in L^p(\mathbb{R}^n)$, there exists an element of $\mathcal{D}$, s such that $||s - f||_p < \varepsilon$. By Theorem 12.10 we can assume without loss of generality that $f \in C_c(\mathbb{R}^n)$. Let $\mathcal{P}_m$ consist of all sets of the form $[\mathbf{a}, \mathbf{b})$ where $a_i = j2^{-m}$ and $b_i = (j + 1)2^{-m}$ for j an integer. Thus $\mathcal{P}_m$ consists of a tiling of $\mathbb{R}^n$ into half open rectangles having diameters $2^{-m}n^{\frac{1}{2}}$. There are countably many of these rectangles; so, let $\mathcal{P}_m = \{[\mathbf{a}_i, \mathbf{b}_i)\}$ and $\mathbb{R}^n = \cup_{i=1}^{\infty}[\mathbf{a}_i, \mathbf{b}_i)$. Let c_i^m be complex numbers with rational real and complex parts satisfying

$$|f(\mathbf{a}_i) - c_i^m| < 5^{-m},$$

$$|c_i^m| \leq |f(\mathbf{a}_i)|. \tag{6}$$

Let $s_m(x) = \sum_{i=1}^{\infty} c_i^m \mathcal{X}_{[a_i, b_i)}$. Since $f(a_i) = 0$ except for finitely many values of i, Formula 6 implies $s_m \in \mathcal{D}$. It is also clear that, since f is uniformly continuous, $\lim_{m \to \infty} s_m(x) = f(x)$ uniformly in x. Hence $\lim_{x \to 0} ||s_m - f||_p = 0$.

Corollary 12.14 *Let Ω be any Lebesgue measurable subset of $\mathbb{R}^n$. Then $L^p(\Omega)$ is separable. Here $L^p(\Omega) \equiv L^p(\Omega, m L\Omega)$.*

Proof: Let $\widetilde{\mathcal{D}}$ be the restrictions of $\mathcal{D}$ to Ω. If $f \in L^p(\Omega)$, let F be the zero extension of f to all of $\mathbb{R}^n$. Let $\varepsilon > 0$ be given. By Theorem 12.13 there exists $s \in \mathcal{D}$ such that $||F - s||_p < \varepsilon$. Thus

$$||s - f||_{L^p(\Omega)} \leq ||s - F||_{L^p(\mathbb{R}^n)} < \varepsilon$$

and so the countable set $\widetilde{\mathcal{D}}$ is dense in $L^p(\Omega)$.

12.5   Mollifiers and density of smooth functions

Definition 12.15 Let U be an open subset of $\mathbb{R}^n$. $C_c^{\infty}(U)$ is the vector space of all infinitely differentiable functions which equal zero for all **x** outside of some compact set contained in U.

Example 12.16 *Let $U = \{\mathbf{x} \in \mathbb{R}^n : ||\mathbf{x}|| < 2\}$*

$$\psi(x) = \begin{cases} \exp\left[\left(||\mathbf{x}||^2 - 1\right)^{-1}\right] & \text{if } ||\mathbf{x}|| < 1, \\ 0 \text{ if } ||x|| \geq 1. \end{cases}$$

Then a little work shows $\psi \in C_c^{\infty}(U)$. The following also is easily obtained.

Lemma 12.17 *Let U be any open set. Then $C_c^{\infty}(U) \neq \emptyset$.*

Definition 12.18 Let $U = \{\mathbf{x} \in \mathbb{R}^n : ||\mathbf{x}|| < 1\}$. A sequence $\{\psi_m\} \subseteq C_c^{\infty}(U)$ is called a mollifier (sometimes an approximate identity) if

$$\psi_m(\mathbf{x}) \geq 0, \ \psi_m(\mathbf{x}) = 0, \text{ if } ||\mathbf{x}|| \geq \frac{1}{m},$$

and $\int \psi_m(\mathbf{x}) = 1$.

As before, $\int f(\mathbf{x}, \mathbf{y}) d\mu(\mathbf{y})$ will mean **x** is fixed and the function $\mathbf{y} \to f(\mathbf{x}, \mathbf{y})$ is being integrated. We may also write dx for $dm(x)$ in the case of Lebesgue measure.

Example 12.19 *Let*

$$\psi \in C_c^\infty(B(0,1)) \ \ (B(0,1) = \{\mathbf{x} : ||\mathbf{x}|| < 1\})$$

with $\psi(\mathbf{x}) \geq 0$ and $\int \psi dm = 1$. Let $\psi_m(\mathbf{x}) = c_m \psi(m\mathbf{x})$ where c_m is chosen in such a way that $\int \psi_m dm = 1$. By the change of variables theorem of Chapter 10 we see that $c_m = m^n$.

Definition 12.20 A function, f, is said to be in $L^1_{loc}(\mathbb{R}^n, \mu)$ if f is μ measurable and if $|f|\mathcal{X}_K \in L^1(\mathbb{R}^n, \mu)$ for every compact set, K. Here μ is a Radon measure on $\mathbb{R}^n$. Usually $\mu = m$, Lebesgue measure. When this is so, we write $L^1_{loc}(\mathbb{R}^n)$ or $L^p(\mathbb{R}^n)$, etc. If $f \in L^1_{loc}(\mathbb{R}^n)$, and $g \in C_c(\mathbb{R}^n)$,

$$f * g(\mathbf{x}) \equiv \int f(\mathbf{x} - \mathbf{y})g(\mathbf{y})dm = \int f(\mathbf{y})g(\mathbf{x} - \mathbf{y})dm.$$

Theorem 12.21 *Let K be a compact subset of an open set, U. Then there exists a function, $h \in C_c^\infty(U)$, such that $h(\mathbf{x}) = 1$ for all $\mathbf{x} \in K$ and $h(\mathbf{x}) \in [0,1]$ for all $\mathbf{x}$.*

Proof: Let $r > 0$ be small enough that $K + B(0, 3r) \subseteq U$. Let $K_r = K + B(0, r)$.

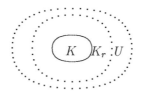

Consider $\mathcal{X}_{K_r} * \psi_m$ where ψ_m is a mollifier. Let m be so large that $\frac{1}{m} < r$. Then it is straightforward to verify that $h = \mathcal{X}_{K_r} * \psi_m$ satisfies the desired conclusions.

Theorem 12.22 *For each $p \geq 1$, $C_c^\infty(\mathbb{R}^n)$ is dense in $L^p(\mathbb{R}^n)$.*

Proof: Let $f \in L^p(\mathbb{R}^n)$ and let $\varepsilon > 0$ be given. Choose $g \in C_c(\mathbb{R}^n)$ such that $||f - g||_p < \frac{\varepsilon}{2}$. Now let $g_m = g * \psi_m$ where $\{\psi_m\}$ is a mollifier.

$$[g_m(\mathbf{x} + h\mathbf{e}_i) - g_m(\mathbf{x})]h^{-1}$$

$$= h^{-1} \int g(\mathbf{y})[\psi_m(\mathbf{x} + h\mathbf{e}_i - \mathbf{y}) - \psi_m(\mathbf{x} - \mathbf{y})]dm.$$

The integrand is dominated by $C|g(\mathbf{y})|h$ for some constant C depending on

$$\max\{|\partial\psi_{mi}(\mathbf{x})/\partial x_j| : \mathbf{x} \in \mathbb{R}^n, i, j \in \{1, 2, \cdots, n\}\}.$$

By the Dominated Convergence theorem, the limit as $h \to 0$ exists and yields

$$\frac{\partial g_m(\mathbf{x})}{\partial x_i} = \int g(\mathbf{y}) \frac{\partial \psi_m(\mathbf{x} - \mathbf{y})}{\partial x_i} dy.$$

Similarly, all other partial derivatives exist and are continuous as are all higher order derivatives. Consequently, $g_m \in C_c^\infty(\mathbb{R}^n)$. It vanishes if $\mathbf{x} \notin spt(g) + B(0, \frac{1}{m})$.

$$
\begin{aligned}
\|g - g_m\|_p &= \left(\int |g(\mathbf{x}) - \int g(\mathbf{x} - \mathbf{y}) \psi_m(\mathbf{y}) dm(\mathbf{y})|^p \, dm(\mathbf{x}) \right)^{\frac{1}{p}} \\
&\leq \left(\int (\int |g(\mathbf{x}) - g(\mathbf{x} - \mathbf{y})| \psi_m(\mathbf{y}) dm(\mathbf{y}))^p \, dm(\mathbf{x}) \right)^{\frac{1}{p}} \\
&\leq \int \left(\int |g(\mathbf{x}) - g(\mathbf{x} - \mathbf{y})|^p \, dm(\mathbf{x}) \right)^{\frac{1}{p}} \psi_m(\mathbf{y}) dm(\mathbf{y}) \\
&= \int_{B(0,\frac{1}{m})} \|g - g_\mathbf{y}\|_p \psi_m(\mathbf{y}) dm(\mathbf{y}) \\
&< \frac{\varepsilon}{2}
\end{aligned}
$$

whenever m is large enough. This follows from Theorem 12.12. Theorem 12.6 was used to obtain the third inequality. There is no measurability problem because the function

$$(\mathbf{x}, \mathbf{y}) \to |g(\mathbf{x}) - g(\mathbf{x} - \mathbf{y})| \psi_m(\mathbf{y})$$

is continuous so it is surely Borel measurable, hence product measurable. Thus when m is large enough,

$$\|f - g_m\|_p \leq \|f - g\|_p + \|g - g_m\|_p < \frac{\varepsilon}{2} + \frac{\varepsilon}{2} = \varepsilon.$$

This proves the theorem.

12.6 Exercises

1. Let E be a Lebesgue measurable set in $\mathbb{R}$. Suppose $m(E) > 0$. Consider the set

$$E - E = \{x - y : x \in E, y \in E\}.$$

Show that $E - E$ contains an interval. **Hint:** Let

$$f(x) = \int \mathcal{X}_E(t) \mathcal{X}_E(x + t) dt.$$

Note f is continuous at 0 and $f(0) > 0$. Remember continuity of translation in L^p.

2. Let $\phi \in C_c^\infty(\mathbb{R}^n)$ and let $\phi_m(x) = \phi(mx)m^n$. Suppose $\phi \geq 0$ and $\int_{\mathbb{R}^n} \phi(y)dy = 1$. Show

   a.) $\int_{\mathbb{R}^n} \phi_m(y)dy = 1$.

   b.) If $f \in L^p(\mathbb{R}^n)$, $\lim_{m\to\infty} f * \phi_m = f$ in $L^p(\mathbb{R}^n)$.

   c.) $f * \phi_m$ is a function in $C_c^\infty(\mathbb{R}^n)$.

3. Let $\phi : \mathbb{R} \to \mathbb{R}$ be convex. This means

$$\phi(\lambda x + (1 - \lambda)y) \leq \lambda\phi(x) + (1 - \lambda)\phi(y)$$

   whenever $\lambda \in [0, 1]$. Show that if ϕ is convex, then ϕ is continuous.

4. ↑ Prove Jensen's inequality. If $\phi : \mathbb{R} \to \mathbb{R}$ is convex, $\mu(\Omega) = 1$, and $f : \Omega \to \mathbb{R}$ is in $L^1(\Omega)$, then $\phi(\int_\Omega f \, d\mu) \leq \int_\Omega \phi(f)d\mu$. **Hint:** Let $s = \int_\Omega f \, d\mu$ and show there exists λ such that $\phi(s) \leq \phi(t) + \lambda(s - t)$ for all t.

5. Let $\frac{1}{p} + \frac{1}{p'} = 1$, $p > 1$, let $f \in L^p(\mathbb{R})$, $g \in L^{p'}(\mathbb{R})$. Show $f * g$ is uniformly continuous on $\mathbb{R}$ and $|(f * g)(x)| \leq ||f||_{L^p}||g||_{L^{p'}}$.

6. $B(p, q) = \int_0^1 x^{p-1}(1 - x)^{q-1}dx, \Gamma(p) = \int_0^\infty e^{-t}t^{p-1}dt$ for $p, q > 0$. The first of these is called the beta function, while the second is the gamma function. Show a.) $\Gamma(p + 1) = p\Gamma(p)$;   b.) $\Gamma(p)\Gamma(q) = B(p, q)\Gamma(p+q)$. We will use the result of this problem in Chapter 19 in the discussion of Hausdorff measures.

7. Let $f \in C_c(0, \infty)$ and define $F(x) = \frac{1}{x} \int_0^x f(t)dt$. Show

$$||F||_{L^p(0,\infty)} \leq \frac{p}{p - 1}||f||_{L^p(0,\infty)} \text{ whenever } p > 1.$$

   Hint: Use $xF' = f - F$ and integrate $\int_0^\infty |F(x)|^p dx$ by parts.

8. ↑ Now suppose $f \in L^p(0, \infty)$, $p > 1$, and f not necessarily in $C_c(0, \infty)$. Note that $F(x) = \frac{1}{x} \int_0^x f(t)dt$ still makes sense for each $x > 0$. Is the inequality of Problem 7 still valid? Why? This inequality is called Hardy's inequality.

9. When does equality hold in Holder's inequality? **Hint:** First suppose $f, g \geq 0$. This isolates the most interesting aspect of the question.

10. ↑ Consider Hardy's inequality of Problems 7 and 8. Show equality holds only if $f = 0$ a.e. **Hint:** If equality holds, we can assume $f \geq 0$.

Why? You might then establish $(p - 1) \int_0^\infty F^p dx = p \int_0^\infty F^{\frac{p}{q}} f \, dx$ and use Problem 9.

11. ↑ In Hardy's inequality, show the constant $p(p - 1)^{-1}$ cannot be improved. Also show that if $f > 0$ and $f \in L^1$, then $F \notin L^1$ so it is important that $p > 1$. **Hint:** Try $f(x) = x^{-\frac{1}{p}} \mathcal{X}_{[A^{-1}, A]}$.

12. A set of functions, $\Phi \subseteq L^1$, is uniformly integrable if for all $\varepsilon > 0$ there exists a $\sigma > 0$ such that $\left| \int_E f \, du \right| < \varepsilon$ whenever $\mu(E) < \sigma$. Prove Vitali's Convergence theorem: Let $\{f_n\}$ be uniformly integrable, $\mu(\Omega) < \infty$, $f_n(x) \to f(x)$ a.e., and $|f(x)| < \infty$ a.e. Then $f \in L^1$ and $\lim_{n \to \infty} \int_\Omega |f_n - f| d\mu = 0$. **Hint:** See the problem on uniform integrability in Chapter 7.

13. ↑ Show the Vitali Convergence theorem implies the Dominated Convergence theorem for finite measure spaces.

14. ↑ Suppose $\mu(\Omega) < \infty$, $\{f_n\} \subseteq L^1(\Omega)$, and

$$\int_\Omega h(|f_n|) \, d\mu < C$$

for all n where h is a continuous, nonnegative function satisfying

$$\lim_{t \to \infty} \frac{h(t)}{t} = \infty.$$

Show $\{f_n\}$ is uniformly integrable.

15. ↑ Give an example of a situation in which the Vitali Convergence theorem applies, but the Dominated Convergence theorem does not.

16. We say $f \in L^\infty(\Omega, \mu)$ if there exists a set of measure zero, E, and a constant $C < \infty$ such that $|f(x)| \leq C$ for all $x \notin E$.

$$\|f\|_\infty \equiv \inf\{C : |f(x)| \leq C \text{ a.e.}\}.$$

Show $\| \ \|_\infty$ is a norm on $L^\infty(\Omega, \mu)$ provided we identify f and g if $f(x) = g(x)$ a.e. Show $L^\infty(\Omega, \mu)$ is complete.

17. Suppose $f \in L^\infty \cap L^1$. Show $\lim_{p \to \infty} \|f\|_{L^p} = \|f\|_\infty$.

18. Suppose $\phi : \mathbb{R} \to \mathbb{R}$ and $\phi(\int_0^1 f(x) dx) \leq \int_0^1 \phi(f(x)) dx$ for every real bounded measurable f. Can it be concluded that ϕ is convex?

19. Suppose $\mu(\Omega) < \infty$. Show that if $1 \leq p < q$, then $L^q(\Omega) \subseteq L^p(\Omega)$.

20. Show $L^1(\mathbb{R}) \not\subseteq L^2(\mathbb{R})$ and $L^2(\mathbb{R}) \not\subseteq L^1(\mathbb{R})$ if Lebesgue measure is used.

21. Show that if $x \in [0, 1]$ and $p \geq 2$, then

$$(\frac{1+x}{2})^p + (\frac{1-x}{2})^p \leq \frac{1}{2}(1 + x^p).$$

Note this is obvious if $p = 2$. Use this to conclude the following inequality valid for all $z, w \in \mathbb{C}$ and $p \geq 2$.

$$\left|\frac{z+w}{2}\right|^p + \left|\frac{z-w}{2}\right|^p \leq \frac{|z|^p}{2} + \frac{|w|^p}{2}.$$

Hint: For the first part, divide both sides by x^p, let $y = \frac{1}{x}$ and show the resulting inequality is valid for all $y \geq 1$. If $|z| \geq |w| > 0$, this takes the form

$$|\frac{1}{2}(1 + re^{i\theta})|^p + |\frac{1}{2}(1 - re^{i\theta})|^p \leq \frac{1}{2}(1 + r^p)$$

whenever $0 \leq \theta < 2\pi$ and $r \in [0, 1]$. Show the expression on the left is maximized when $\theta = 0$ and use the first part.

22. ↑ If $p \geq 2$, establish Clarkson's inequality. Whenever $f, g \in L^p$,

$$\left\|\frac{1}{2}(f + g)\right\|_p^p + \left\|\frac{1}{2}(f - g)\right\|_p^p \leq \frac{1}{2}\|f\|^p + \frac{1}{2}\|g\|_p^p.$$

For more on Clarkson inequalities (there are others), see Hewitt and Stromberg [27] or Ray [41].

23. ↑ Show that for $p \geq 2$, L^p is uniformly convex. This means that if $\{f_n\}, \{g_n\} \subseteq L^p$, $\|f_n\|_p$, $\|g_n\|_p \leq 1$, and $\|f_n + g_n\|_p \to 2$, then $\|f_n - g_n\|_p \to 0$.

24. Suppose that $\theta \in [0, 1]$ and $r, s, q > 0$ with

$$\frac{1}{q} = \frac{\theta}{r} + \frac{1-\theta}{s}.$$

show that

$$(\int |f|^q d\mu)^{1/q} \leq ((\int |f|^r d\mu)^{1/r})^\theta ((\int |f|^s d\mu)^{1/s})^{1-\theta}.$$

If $q, r, s \geq 1$ this says that

$$\|f\|_q \leq \|f\|_r^\theta \|f\|_s^{1-\theta}.$$

Hint:

$$\int |f|^q \, d\mu = \int |f|^{q\theta} |f|^{q(1-\theta)} d\mu.$$

Now note that $1 = \frac{\theta q}{r} + \frac{q(1-\theta)}{s}$ and use Holder's inequality.

25. Generalize Theorem 12.6 as follows. Let $0 \leq p_1 \leq p_2 < \infty$. Then

$$\left(\int_Y \left(\int_X |f(x,y)|^{p_1} \, d\mu \right)^{p_2/p_1} d\lambda \right)^{1/p_2}$$

$$\leq \left(\int_X \left(\int_Y |f(x,y)|^{p_2} \, d\lambda \right)^{p_1/p_2} d\mu \right)^{1/p_1}.$$

Chapter 13
Representation Theorems

13.1 Radon Nikodym Theorem

This chapter is on various representation theorems. The first theorem, the Radon Nikodym Theorem, is a representation theorem for one measure in terms of another. The approach given here is due to Von Neumann and depends on the Riesz representation theorem for Hilbert space.

Definition 13.1 Let μ and λ be two measures defined on a σ-algebra, $\mathcal{S}$, of subsets of a set, Ω. We say that λ is absolutely continuous with respect to μ and write $\lambda << \mu$ if $\lambda(E) = 0$ whenever $\mu(E) = 0$.

Theorem 13.2 *(Radon Nikodym) Let λ and μ be finite measures defined on a σ-algebra, $\mathcal{S}$, of subsets of Ω. Suppose $\lambda << \mu$. Then there exists $f \in L^1(\Omega, \mu)$ such that $f(x) \geq 0$ and*

$$\lambda(E) = \int_E f \, d\mu.$$

Proof: Let $\Lambda : L^2(\Omega, \mu + \lambda) \to \mathbb{C}$ be defined by

$$\Lambda g = \int_\Omega g \, d\lambda.$$

By Holder's inequality,

$$|\Lambda g| \leq \left(\int_\Omega 1^2 d\lambda \right)^{1/2} \left(\int_\Omega |g|^2 \, d(\lambda + \mu) \right)^{1/2} = \lambda(\Omega)^{1/2} \|g\|_2$$

and so $\Lambda \in (L^2(\Omega, \mu + \lambda))'$. By the Riesz representation theorem in Hilbert space, there exists $h \in L^2(\Omega, \mu + \lambda)$ with

$$\Lambda g = \int_\Omega g \, d\lambda = \int_\Omega hg d(\mu + \lambda). \qquad (1)$$

Letting $E = \{x \in \Omega : \operatorname{Im} h(x) > 0\}$, and letting $g = \mathcal{X}_E$, 1 implies

$$\lambda(E) = \int_E (\operatorname{Re} h + i \, \operatorname{Im} h) d(\mu + \lambda). \qquad (2)$$

Since the left side of Formula 2 is real, this shows $(\mu + \lambda)(E) = 0$. Similarly, if

$$E = \{x \in \Omega : \operatorname{Im} h(x) < 0\},$$

then $(\mu + \lambda)(E) = 0$. Thus we may assume h is real-valued. Now let $E = \{x : h(x) < 0\}$ and let $g = \mathcal{X}_E$. Then from Formula 2

$$\lambda(E) = \int_E h \, d(\mu + \lambda).$$

Since $h(x) < 0$ on E, it follows $(\mu + \lambda)(E) = 0$ or else the right side of this equation would be negative. Thus we can take $h \geq 0$. Now let $E = \{x : h(x) \geq 1\}$ and let $g = \mathcal{X}_E$. Then

$$\lambda(E) = \int_E h \, d(\mu + \lambda) \geq \mu(E) + \lambda(E).$$

Therefore $\mu(E) = 0$. Since $\lambda << \mu$, it follows that $\lambda(E) = 0$ also. Thus we can assume

$$0 \leq h(x) < 1$$

for all x. From Formula 1, whenever $g \in L^2(\Omega, \mu + \lambda)$,

$$\int_\Omega g(1 - h) d\lambda = \int_\Omega hg d\mu. \tag{3}$$

Let $g(x) = \sum_{i=0}^n h^i(x) \mathcal{X}_E(x)$ in Formula 3. This yields

$$\int_E (1 - h^{n+1}(x)) d\lambda = \int_E \sum_{i=1}^{n+1} h^i(x) d\mu. \tag{4}$$

Let $f(x) = \sum_{i=1}^\infty h^i(x)$ and use the Monotone Convergence theorem in Formula 4 to let $n \to \infty$ and conclude

$$\lambda(E) = \int_E f \, d\mu.$$

We know $f \in L^1(\Omega, \mu)$ because λ is finite. This proves the theorem.
The f in the theorem is sometimes denoted by

$$\frac{d\lambda}{d\mu}.$$

The next corollary is a generalization to σ finite measure spaces.

Corollary 13.3 *Suppose* $\lambda << \mu$ *and there exist sets* $S_n \in \mathcal{S}$ *with*

$$S_n \cap S_m = \emptyset, \ \cup_{n=1}^\infty S_n = \Omega,$$

and $\lambda(S_n), \mu(S_n) < \infty$. *Then there exists* $f \geq 0$, f *measurable, and*

$$\lambda(E) = \int_E f \, d\mu$$

for all $E \in \mathcal{S}$. The function f is $\mu + \lambda$ a.e. unique.

Proof: Let $\mathcal{S}_n = \{E \cap S_n : E \in \mathcal{S}\}$. Clearly $\mathcal{S}_n$ is a σ algebra of subsets of S_n, λ, μ are both finite measures on $\mathcal{S}_n$, and $\lambda << \mu$. Thus, by Theorem 13.2, there exists an $\mathcal{S}_n$ measurable function f_n, $f_n(x) \geq 0$, with

$$\lambda(E) = \int_E f_n \, d\mu$$

for all $E \in \mathcal{S}_n$. Define $f(x) = f_n(x)$ for $x \in S_n$. Then f is measurable because

$$f^{-1}((a, \infty]) = \cup_{n=1}^{\infty} f_n^{-1}((a, \infty]) \in \mathcal{S}.$$

Also, for $E \in \mathcal{S}$,

$$\begin{aligned}
\lambda(E) &= \sum_{n=1}^{\infty} \lambda(E \cap S_n) = \sum_{n=1}^{\infty} \int \mathcal{X}_{E \cap S_n}(x) f_n(x) d\mu \\
&= \sum_{n=1}^{\infty} \int \mathcal{X}_{E \cap S_n}(x) f(x) d\mu \\
&= \int_E f \, d\mu.
\end{aligned}$$

To see f is unique, suppose f_1 and f_2 both work and consider

$$E \equiv \{x : f_1(x) - f_2(x) > 0\}.$$

Then

$$0 = \lambda(E \cap S_n) - \lambda(E \cap S_n) = \int_{E \cap S_n} f_1(x) - f_2(x) d\mu.$$

Hence $\mu(E \cap S_n) = 0$ so $\mu(E) = 0$. Hence $\lambda(E) = 0$ also. Similarly

$$(\mu + \lambda)(\{x : f_2(x) - f_1(x) > 0\}) = 0.$$

This version of the Radon Nikodym theorem will suffice for most applications, but more general versions are available. To see one of these, one can read the treatment in Hewitt and Stromberg. This involves the notion of decomposable measure spaces, a generalization of $\sigma-$ finite.

13.2   Vector measures

The next topic will use the Radon Nikodym theorem. It is the topic of vector and complex measures. Here we are mainly concerned with complex measures although a vector measure can have values in any topological vector space. Later we will return to this topic in the context of Banach space valued measures.

Definition 13.4 Let $(V, || \cdot ||)$ be a normed linear space and let $(\Omega, \mathcal{S})$ be a measure space. We call a function $\mu : \mathcal{S} \to V$ a vector measure if μ is countably additive. That is, if $\{E_i\}_{i=1}^{\infty}$ is a sequence of disjoint sets of $\mathcal{S}$,

$$\mu(\cup_{i=1}^{\infty} E_i) = \sum_{i=1}^{\infty} \mu(E_i).$$

Definition 13.5 Let $(\Omega, \mathcal{S})$ be a measure space and let μ be a vector measure defined on $\mathcal{S}$. A subset, $\pi(E)$, of $\mathcal{S}$ is called a partition of E if $\pi(E)$ consists of finitely many disjoint sets of $\mathcal{S}$ and $\cup \pi(E) = E$. Let

$$|\mu|(E) = \sup\{ \sum_{F \in \pi(E)} ||\mu(F)|| : \pi(E) \text{ is a partition of } E\}.$$

$|\mu|$ is called the total variation of μ.

The next theorem may seem a little surprising. It states that, if finite, the total variation is a nonnegative measure.

Theorem 13.6 *If $|\mu|(\Omega) < \infty$, then $|\mu|$ is a measure on $\mathcal{S}$.*

Proof: Let $E_1 \cap E_2 = \emptyset$ and let $\{A_1^i \cdots A_{n_i}^i\} = \pi(E_i)$ with

$$|\mu|(E_i) - \varepsilon < \sum_{j=1}^{n_i} ||\mu(A_j^i)|| \quad i = 1, 2.$$

Let $\pi(E_1 \cup E_2) = \pi(E_1) \cup \pi(E_2)$. Then

$$|\mu|(E_1 \cup E_2) \geq \sum_{F \in \pi(E_1 \cup E_2)} ||\mu(F)|| > |\mu|(E_1) + |\mu|(E_2) - 2\varepsilon.$$

Since $\varepsilon > 0$ was arbitrary, it follows that

$$|\mu|(E_1 \cup E_2) \geq |\mu|(E_1) + |\mu|(E_2). \tag{5}$$

Let $\{E_j\}_{j=1}^{\infty}$ be a sequence of disjoint sets of $\mathcal{S}$. Let $E_{\infty} = \cup_{j=1}^{\infty} E_j$ and let

$$\{A_1, \cdots, A_n\} = \pi(E_{\infty})$$

be such that

$$|\mu|(E_{\infty}) - \varepsilon < \sum_{i=1}^{n} ||\mu(A_i)||.$$

But $||\mu(A_i)|| \le \sum_{j=1}^{\infty} ||\mu(A_i \cap E_j)||$. Therefore,

$$
\begin{aligned}
|\mu|(E_\infty) - \varepsilon \;\; &< \;\; \sum_{i=1}^{n} \sum_{j=1}^{\infty} ||\mu(A_i \cap E_j)|| \\
&= \;\; \sum_{j=1}^{\infty} \sum_{i=1}^{n} ||\mu(A_i \cap E_j)|| \\
&\le \;\; \sum_{j=1}^{\infty} |\mu|(E_j).
\end{aligned}
$$

The interchange in order of integration follows from Fubini's theorem, and the last inequality follows because $A_1 \cap E_j, \cdots, A_n \cap E_j$ is a partition of E_j.

Since $\varepsilon > 0$ is arbitrary, this shows

$$
|\mu|(\cup_{j=1}^{\infty} E_j) \le \sum_{j=1}^{\infty} |\mu|(E_j).
$$

By induction, Formula 5 implies that whenever the E_i are distinct,

$$
|\mu|(\cup_{j=1}^{n} E_j) \ge \sum_{j=1}^{n} |\mu|(E_j).
$$

Therefore,

$$
\sum_{j=1}^{\infty} |\mu|(E_j) \ge |\mu|(\cup_{j=1}^{\infty} E_j) \ge |\mu|(\cup_{j=1}^{n} E_j) \ge \sum_{j=1}^{n} |\mu|(E_j).
$$

Since n is arbitrary, this implies

$$
|\mu|(\cup_{j=1}^{\infty} E_j) = \sum_{j=1}^{\infty} |\mu|(E_j)
$$

which proves the theorem.

In the case where $V = \mathbb{C}$, it is automatically the case that $|\mu|(\Omega) < \infty$. This is proved in Rudin [46]. We will not need to use this fact, so it is left for the interested reader to look up.

The next theorem follows from the Radon Nikodym theorem and, in the case of complex vector measures considered in this chapter, can be proved from this theorem. It will serve as the basis for the Riesz representation theorems for L^p on σ finite measure spaces.

Theorem 13.7 *Let* $(\Omega, \mathcal{S})$ *be a measure space and let* $\lambda : \mathcal{S} \to \mathbb{C}$ *be a complex vector measure with* $|\lambda|(\Omega) < \infty$. *Let* $\mu : \mathcal{S} \to [0, \mu(\Omega)]$ *be a* *finite measure such that $\lambda << \mu$. Then there exists a unique $f \in L^1(\Omega)$ such that for all $E \in \mathcal{S}$,*

$$\int_E f d\mu = \lambda(E).$$

Proof: It is clear that $\operatorname{Re}\lambda$ and $\operatorname{Im}\lambda$ are real-valued vector measures on $\mathcal{S}$. Since $|\lambda|(\Omega) < \infty$, it follows easily that $|\operatorname{Re}\lambda|(\Omega)$ and $|\operatorname{Im}\lambda|(\Omega) < \infty$. Therefore, each of

$$\frac{|\operatorname{Re}\lambda| + \operatorname{Re}\lambda}{2}, \quad \frac{|\operatorname{Re}\lambda| - \operatorname{Re}(\lambda)}{2}, \quad \frac{|\operatorname{Im}\lambda| + \operatorname{Im}\lambda}{2}, \quad \text{and} \quad \frac{|\operatorname{Im}\lambda| - \operatorname{Im}(\lambda)}{2}$$

are finite measures on $\mathcal{S}$. It is also clear that each of these finite measures are absolutely continuous with respect to μ. Thus there exist unique nonnegative functions in $L^1(\Omega)$, f_1, f_2, g_1, g_2 such that for all $E \in \mathcal{S}$,

$$
\begin{aligned}
\frac{1}{2}(|\operatorname{Re}\lambda| + \operatorname{Re}\lambda)(E) &= \int_E f_1 d\mu, \\
\frac{1}{2}(|\operatorname{Re}\lambda| - \operatorname{Re}\lambda)(E) &= \int_E f_2 d\mu, \\
\frac{1}{2}(|\operatorname{Im}\lambda| + \operatorname{Im}\lambda)(E) &= \int_E g_1 d\mu, \\
\frac{1}{2}(|\operatorname{Im}\lambda| - \operatorname{Im}\lambda)(E) &= \int_E g_2 d\mu.
\end{aligned}
$$

Now let $f = f_1 - f_2 + i(g_1 - g_2)$.

The following corollary is about representing a vector measure in terms of its total variation.

Corollary 13.8 *Let λ be a complex vector measure with $|\lambda|(\Omega) < \infty$. Then there exists a unique $f \in L^1(\Omega)$ such that $\lambda(E) = \int_E f d|\lambda|$. Furthermore, $|f| = 1$ a.e. This is called the polar decomposition of λ.*

Proof: First we note that $\lambda << |\lambda|$ and so such an L^1 function exists and is unique. We have to show $|f| = 1$ a.e.

Lemma 13.9 *Suppose $(\Omega, \mathcal{S}, \mu)$ is a measure space and f is a function in $L^1(\Omega, \mu)$ with the property that*

$$\left| \int_E f \, d\mu \right| \leq \mu(E)$$

for all $E \in \mathcal{S}$. Then $|f| \leq 1$ a.e.

Proof of the lemma: Consider the following picture.

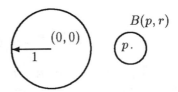

where $B(p, r) \cap B(0, 1) = \emptyset$. Let $E = f^{-1}(B(p, r))$. If $\mu(E) \neq 0$ then

$$\left| \frac{1}{\mu(E)} \int_E f \, d\mu - p \right| = \left| \frac{1}{\mu(E)} \int_E (f - p) d\mu \right|$$

$$\leq \frac{1}{\mu(E)} \int_E |f - p| d\mu < r.$$

Hence

$$\left| \frac{1}{\mu(E)} \int_E f d\mu \right| > 1,$$

contradicting the assumption of the lemma. It follows $\mu(E) = 0$. Since $\{z \in \mathbb{C} : |z| > 1\}$ can be covered by countably many such balls, it follows that

$$\mu(f^{-1}(\{z \in \mathbb{C} : |z| > 1\})) = 0.$$

Thus $|f(x)| \leq 1$ a.e. as claimed. This proves the lemma.

To finish the proof of Corollary 13.8, if $|\lambda|(E) \neq 0$,

$$\left| \frac{\lambda(E)}{|\lambda|(E)} \right| = \left| \frac{1}{|\lambda|(E)} \int_E f \, d|\lambda| \right| \leq 1.$$

Therefore $|f| \leq 1$, $|\lambda|$ a.e. Now let

$$E_n = \{x \in \Omega : |f(x)| \leq 1 - \frac{1}{n}\}.$$

If $|\lambda|(E_n) > 0$, let

$$0 < \varepsilon < \frac{|\lambda|(E_n)}{n}$$

and pick a partition, $\{F_1, \cdots, F_m\}$, of E_n such that

$$\sum_{i=1}^{m} |\lambda(F_i)| \geq |\lambda|(E_n) - \varepsilon.$$

Then since $\lambda(F_i) = \int_{F_i} f d|\lambda|$ and $|f| < 1 - \frac{1}{n}$ on E_n,

$$|\lambda|(E_n) - \varepsilon \leq \sum_{i=1}^{m} |\lambda(F_i)| \leq \sum_{i=1}^{m} (1 - \frac{1}{n})|\lambda|(F_i)$$

$$= |\lambda|(E_n)(1 - \frac{1}{n}).$$

Hence, from the choice of ε,

$$\varepsilon \geq \frac{|\lambda|(E_n)}{n} > \varepsilon,$$

a contradiction. Thus $|\lambda|(E_n) = 0$. But $\{x \in \Omega : |f(x)| < 1\} = \cup_{n=1}^{\infty} E_n$. So $|\lambda|(\{x \in \Omega : |f(x)| < 1\}) = 0$. This proves Corollary 13.8.

13.3   Representation theorems for the dual space of L^p the σ finite case

In Chapter 3 we discussed the definition of a Banach space and the dual space of a Banach space. We also saw in Chapter 12 that the L^p spaces are Banach spaces. The next topic deals with the dual space of L^p for $p \geq 1$ in the case where the measure space is σ finite or finite.

Theorem 13.10 *(Riesz representation theorem) Let $p > 1$ and let $(\Omega, \mathcal{S}, \mu)$ be a finite measure space. If $\Lambda \in (L^p(\Omega))'$, then there exists a unique $h \in L^q(\Omega)$ $(\frac{1}{p} + \frac{1}{q} = 1)$ such that*

$$\Lambda f = \int_\Omega h f d\mu.$$

Proof: (Uniqueness) If h_1 and h_2 both represent Λ, consider

$$f = |h_1 - h_2|^{q-2}(\overline{h_1} - \overline{h_2}).$$

By Holder's inequality, it is easy to see that $f \in L^p(\Omega)$. Thus

$$0 = \Lambda f - \Lambda f =$$

$$\int h_1 |h_1 - h_2|^{q-2}(\overline{h_1} - \overline{h_2}) - h_2 |h_1 - h_2|^{q-2}(\overline{h_1} - \overline{h_2}) d\mu$$

$$= \int |h_1 - h_2|^q d\mu.$$

Therefore $h_1 = h_2$ and this proves uniqueness.

Now let $\lambda(E) = \Lambda(\mathcal{X}_E)$. Let $A_1, \cdots, A_n$ be a partition of Ω.

$$|\Lambda \mathcal{X}_{A_i}| = w_i(\Lambda \mathcal{X}_{A_i}) = \Lambda(w_i \mathcal{X}_{A_i})$$

for some $w_i \in \mathbb{C}$, $|w_i| = 1$. Thus

$$\sum_{i=1}^n |\lambda(A_i)| = \sum_{i=1}^n |\Lambda(\mathcal{X}_{A_i})| = \Lambda(\sum_{i=1}^n w_i \mathcal{X}_{A_i})$$

$$\leq ||\Lambda||(\int |\sum_{i=1}^{n} w_i \mathcal{X}_{A_i}|^p d\mu)^{\frac{1}{p}} = ||\Lambda||(\int_{\Omega} d\mu)^{\frac{1}{p}} = ||\Lambda||\mu(\Omega)^{\frac{1}{p}}.$$

Therefore $|\lambda|(\Omega) < \infty$. Also, if $\{E_i\}_{i=1}^{\infty}$ is a sequence of disjoint sets of $\mathcal{S}$, let

$$F_n = \cup_{i=1}^{n} E_i, \ F = \cup_{i=1}^{\infty} E_i.$$

Then by the Dominated Convergence theorem,

$$||\mathcal{X}_{F_n} - \mathcal{X}_F||_p \to 0.$$

Therefore,

$$\lambda(F) = \Lambda(\mathcal{X}_F) = \lim_{n \to \infty} \Lambda(\mathcal{X}_{F_n}) = \lim_{n \to \infty} \sum_{k=1}^{n} \Lambda(\mathcal{X}_{E_k}) = \sum_{k=1}^{\infty} \lambda(E_k).$$

This shows λ is a complex measure with $|\lambda|$ finite. It is also clear that $\lambda << \mu$. Therefore, by the Radon Nikodym theorem, there exists $h \in L^1(\Omega)$ with

$$\lambda(E) = \int_E h d\mu = \Lambda(\mathcal{X}_E).$$

Now let $s = \sum_{i=1}^{m} c_i \mathcal{X}_{E_i}$ be a simple function. We have

$$\Lambda(s) = \sum_{i=1}^{m} c_i \Lambda(\mathcal{X}_{E_i}) = \sum_{i=1}^{m} c_i \int_{E_i} h d\mu = \int h s d\mu. \qquad (6)$$

Let $E_n = \{x : |h(x)| \leq n\}$. Thus $|h\mathcal{X}_{E_n}| \leq n$. Then

$$|h\mathcal{X}_{E_n}|^{q-2}(\overline{h}\mathcal{X}_{E_n}) \in L^p(\Omega).$$

By continuity of Λ and density of simple functions in $L^p(\Omega)$ (Theorem 12.7), it follows that

$$||h\mathcal{X}_{E_n}||_q^q = \int h|h\mathcal{X}_{E_n}|^{q-2}(\overline{h}\mathcal{X}_{E_n})d\mu = \Lambda(|h\mathcal{X}_{E_n}|^{q-2}(\overline{h}\mathcal{X}_{E_n}))$$

$$\leq ||\Lambda|| \ || \ |h\mathcal{X}_{E_n}|^{q-2}\overline{h}\mathcal{X}_{E_n}||_p = ||\Lambda|| \ ||h\mathcal{X}_{E_n}||_q^{\frac{q}{p}}.$$

Therefore, since $q - \frac{q}{p} = 1$, it follows that

$$||h\mathcal{X}_{E_n}||_q \leq ||\Lambda||.$$

Letting $n \to \infty$, the Monotone Convergence theorem implies

$$||h||_q \leq ||\Lambda||. \qquad (7)$$

Now that h has been shown to be in $L^q(\Omega)$, it follows from Formula 6 and the density of the simple functions, Theorem 12.7, that

$$\Lambda f = \int hf$$

for all $f \in L^p(\Omega)$. This proves Theorem 13.10.

Corollary 13.11 *If h is the function of Theorem 13.10 representing Λ, then $||h||_q = ||\Lambda||$.*

Proof: $||\Lambda|| = \sup\{\int hf : ||f||_p \leq 1\} \leq ||h||_q \leq ||\Lambda||$ by Formula 7, and Holder's inequality.

To represent elements of the dual space of $L^1(\Omega)$, we need another Banach space.

Definition 13.12 Let $(\Omega, \mathcal{S}, \mu)$ be a measure space. $L^\infty(\Omega)$ is the vector space of measurable functions such that for some $M > 0$, $|f(x)| \leq M$ for all x outside of some set of measure zero ($|f(x)| \leq M$ a.e.). We define $f = g$ when $f(x) = g(x)$ a.e. and $||f||_\infty \equiv \inf\{M : |f(x)| \leq M \text{ a.e.}\}$.

Theorem 13.13 $L^\infty(\Omega)$ *is a Banach space.*

Proof: It is clear that $L^\infty(\Omega)$ is a vector space and it is routine to verify that $|| \ ||_\infty$ is a norm.

To verify completeness, let $\{f_n\}$ be a Cauchy sequence in $L^\infty(\Omega)$. Let

$$|f_n(x) - f_m(x)| \leq ||f_n - f_m||_\infty$$

for all $x \notin E_{nm}$, a set of measure 0. Let $E = \cup_{n,m} E_{nm}$. Thus $\mu(E) = 0$ and for each $x \notin E$, $\{f_n(x)\}_{n=1}^\infty$ is a Cauchy sequence in $\mathbb{C}$. Let

$$f(x) = \begin{cases} 0 \text{ if } x \in E \\ \lim_{n \to \infty} f_n(x) \text{ if } x \notin E \end{cases} = \lim_{n \to \infty} \mathcal{X}_{E^c}(x) f_n(x).$$

Then f is clearly measurable because it is the limit of measurable functions. If

$$F_n = \{x : |f_n(x)| > ||f_n||_\infty\}$$

and $F = \cup_{n=1}^\infty F_n$, it follows $\mu(F) = 0$ and that for $x \notin F \cup E$,

$$|f(x)| \leq \lim_{n \to \infty} \inf |f_n(x)| \leq \lim_{n \to \infty} \inf ||f_n||_\infty < \infty$$

because $\{f_n\}$ is a Cauchy sequence. Thus $f \in L^\infty(\Omega)$. Let n be large enough that whenever $m > n$,

$$||f_m - f_n|| < \varepsilon.$$

Thus, if $x \notin E$,

$$\begin{aligned}
|f(x) - f_n(x)| &= \lim_{m \to \infty} |f_m(x) - f_n(x)| \\
&\leq \lim_{m \to \infty} \inf ||f_m - f_n||_\infty < \varepsilon.
\end{aligned}$$

Hence $||f - f_n||_\infty < \varepsilon$ for all n large enough. This proves the theorem. The next theorem is the Riesz representation theorem for $\left(L^1(\Omega)\right)'$.

Theorem 13.14 *(Riesz representation theorem) Let $(\Omega, \mathcal{S}, \mu)$ be a finite measure space. If $\Lambda \in (L^1(\Omega))'$, then there exists a unique $h \in L^\infty(\Omega)$ such that*

$$\Lambda(f) = \int_\Omega hf \, d\mu$$

for all $f \in L^1(\Omega)$.

Proof: Just as in the proof of Theorem 13.10, there exists a unique $h \in L^1(\Omega)$ such that for all simple functions, s,

$$\Lambda(s) = \int hs \, d\mu. \tag{8}$$

To show $h \in L^\infty(\Omega)$, let $\varepsilon > 0$ be given and let

$$E = \{x : |h(x)| \geq ||\Lambda|| + \varepsilon\}.$$

Let $|k| = 1$ and $hk = |h|$. Since the measure space is finite, $k \in L^1(\Omega)$. Let $\{s_n\}$ be a sequence of simple functions converging to k in $L^1(\Omega)$, and pointwise. Also let $|s_n| \leq 1$. Therefore

$$\Lambda(k\mathcal{X}_E) = \lim_{n \to \infty} \Lambda(s_n \mathcal{X}_E) = \lim_{n \to \infty} \int_E hs_n = \int_E hk$$

where the last equality holds by the Dominated Convergence theorem. Therefore,

$$\begin{aligned}
||\Lambda||\mu(E) &\geq |\Lambda(k\mathcal{X}_E)| = |\int_\Omega hk\mathcal{X}_E d\mu| = \int_E |h| d\mu \\
&\geq (||\Lambda|| + \varepsilon)\mu(E).
\end{aligned}$$

It follows that $\mu(E) = 0$. Since $\varepsilon > 0$ was arbitrary, $||\Lambda|| \geq ||h||_\infty$. Now we have seen that $h \in L^\infty(\Omega)$, the density of the simple functions and Formula 8 imply

$$\Lambda f = \int_\Omega hf d\mu \, , \ ||\Lambda|| \geq ||h||_\infty. \tag{9}$$

This proves the existence part of the theorem. To verify uniqueness, suppose h_1 and h_2 both represent Λ and let $f \in L^1(\Omega)$ be such that $|f| \leq 1$ and $f(h_1 - h_2) = |h_1 - h_2|$. Then

$$0 = \Lambda f - \Lambda f = \int (h_1 - h_2)f = \int |h_1 - h_2|.$$

Thus $h_1 = h_2$.

Corollary 13.15 *If h is the function in $L^\infty(\Omega)$ representing $\Lambda \in (L^1(\Omega))'$, then $||h||_\infty = ||\Lambda||$.*

Proof: $||\Lambda|| = \sup\{|\int hf d\mu| : ||f||_1 \leq 1\} \leq ||h||_\infty \leq ||\Lambda||$ by Formula 9.

Next we extend these results to the σ finite case. This involves a clever trick.

Theorem 13.16 *(Riesz representation theorem) Let (Ω, S, μ) be σ finite and let*

$$\Lambda \in (L^p(\Omega, \mu))', \; p \geq 1.$$

Then there exists a unique $h \in L^q(\Omega, \mu)$, $L^\infty(\Omega, \mu)$ if $p = 1$ such that

$$\Lambda f = \int hf d\mu.$$

Also $||h|| = ||\Lambda||$. ($||h|| = ||h||_q$ if $p > 1$, $||h||_\infty$ if $p = 1$). Here

$$\frac{1}{p} + \frac{1}{q} = 1.$$

Proof: Let $\{\Omega_n\}$ be a sequence of disjoint elements of S having the property that

$$0 < \mu(\Omega_n) < \infty, \; \cup_{n=1}^\infty \Omega_n = \Omega.$$

Define

$$r(x) = \sum_{n=1}^\infty \frac{1}{n^2} \mathcal{X}_{\Omega_n}(x) \mu(\Omega_n)^{-1}, \quad \tilde{\mu}(E) = \int_E r d\mu.$$

Thus $\tilde{\mu}(\Omega) = \sum_{n=1}^\infty \frac{1}{n^2} < \infty$ so $\tilde{\mu}$ is a finite measure. Also, if $f \geq 0$ and measurable with respect to S,

$$\int f d\tilde{\mu} = \int f \, r d\mu.$$

It is also clear that $\tilde{\mu} << \mu$ and $\tilde{\mu} >> \mu$. Let $\eta : L^p(\Omega, \tilde{\mu}) \to L^p(\Omega, \mu)$ be given by

$$\eta f = r^{\frac{1}{p}} f.$$

Then η is 1-1 onto and

$$||\eta f||_{L^p(\mu)} = ||f||_{L^p(\widetilde{\mu})}.$$

Let $\Lambda \in (L^p(\mu))'$. Applying the Riesz representation theorem for finite measure spaces, there exists a unique $h \in L^q(\widetilde{\mu})$ $(L^\infty(\widetilde{\mu})$ if $p = 1)$ such that

$$\Lambda(\eta f) = (\eta^* \Lambda) f = \int h f d\widetilde{\mu} \tag{10}$$

for all $f \in L^p(\widetilde{\mu})$. Also

$$||h|| = ||\eta^* \Lambda|| = ||\Lambda||$$

where $||h|| = ||h||_{L^q(\widetilde{\mu})}$ if $p > 1$ and $||h|| = ||h||_{L^\infty(\widetilde{\mu})}$ if $p = 1$. If $p > 1$, this yields

$$\Lambda(\eta f) = \int h f d\widetilde{\mu} = \int h r^{\frac{1}{q}} (\eta f) d\mu.$$

Thus $h r^{\frac{1}{q}}$ represents Λ. Also

$$||h r^{\frac{1}{q}}||_{L^q(\mu)} = ||h||_{L^q(\widetilde{\mu})} = ||\Lambda||.$$

If $p = 1$, equation 10 implies

$$\Lambda(\eta f) = \int h f d\widetilde{\mu} = \int h (\eta f) d\mu.$$

Also $||h||_{L^\infty(\mu)} = ||h||_{L^\infty(\widetilde{\mu})} = ||\Lambda||$. Uniqueness is done as before and this proves the theorem.

With the Riesz representation theorem, it is easy to show that

$$L^p(\Omega), \ p > 1$$

is a reflexive Banach space.

Theorem 13.17 *For $(\Omega, \mathcal{S}, \mu)$ a σ finite measure space and $p > 1$, $L^p(\Omega)$ is reflexive.*

Proof: Let $\delta_r : (L^r(\Omega))' \to L^{r'}(\Omega)$ be defined for $\frac{1}{r} + \frac{1}{r'} = 1$ by

$$\int (\delta_r \Lambda) g \, d\mu = \Lambda g$$

for all $g \in L^r(\Omega)$. From Theorem 13.16 δ_r is 1-1, onto, continuous and linear. By the Open Map theorem, δ_r^{-1} is also 1-1, onto, and continuous

($\delta_r \Lambda$ equals the representor of Λ). Thus δ_r^* is also 1-1, onto, and continuous by Corollary 3.28. Now observe that $J = \delta_p^* \circ \delta_q^{-1}$. To see this, let $z^* \in (L^q)'$, $y^* \in (L^p)'$,

$$\begin{aligned} \delta_p^* \circ \delta_q^{-1}(\delta_q z^*)(y^*) &= (\delta_p^* z^*)(y^*) \\ &= z^*(\delta_p y^*) \\ &= \int (\delta_q z^*)(\delta_p y^*) d\mu, \end{aligned}$$

$$\begin{aligned} J(\delta_q z^*)(y^*) &= y^*(\delta_q z^*) \\ &= \int (\delta_p y^*)(\delta_p z^*) d\mu. \end{aligned}$$

Therefore $\delta_p^* \circ \delta_q^{-1} = J$ on $\delta_q(L^q)' = L^p$. But the two δ maps are onto and so J is also onto.

13.4   Riesz Representation theorem for non σ finite measure spaces

It is not necessary to assume μ is either finite or σ finite to establish the Riesz representation theorem for $1 < p < \infty$. This involves the notion of uniform convexity. First we recall Clarkson's inequality for $p \geq 2$. This was Problem 22 in Chapter 12.

Lemma 13.18 *(Clarkson inequality $p \geq 2$) For $p \geq 2$,*

$$||\frac{f+g}{2}||_p^p + ||\frac{f-g}{2}||_p^p \leq \frac{1}{2}||f||_p^p + \frac{1}{2}||g||_p^p.$$

Definition 13.19 A Banach space, X, is said to be uniformly convex if whenever $||x_n|| \leq 1$ and $||\frac{x_n+x_m}{2}|| \to 1$ as $n, m \to \infty$, then $\{x_n\}$ is a Cauchy sequence and $x_n \to x$ where $||x|| = 1$.

Observe that Clarkson's inequality implies L^p is uniformly convex for all $p \geq 2$. Uniformly convex spaces have a very nice property which is described in the following lemma. Roughly, this property is that any element of the dual space achieves its norm at some point of the closed unit ball.

Lemma 13.20 *Let X be uniformly convex and let $L \in X'$. Then there exists $x \in X$ such that*

$$||x|| = 1, \ Lx = ||L||.$$

Proof: Let $||\tilde{x}_n|| \leq 1$ and $|L\tilde{x}_n| \to ||L||$. Let $x_n = w_n\tilde{x}_n$ where $|w_n| = 1$ and

$$w_n L\tilde{x}_n = |L\tilde{x}_n|.$$

Thus $Lx_n = |Lx_n| = |L\tilde{x}_n| \to ||L||$.

$$Lx_n \to ||L||, \; ||x_n|| \leq 1.$$

We can assume, without loss of generality, that

$$Lx_n = |Lx_n| \geq \frac{||L||}{2}$$

and $L \neq 0$.

Claim $||\frac{x_n+x_m}{2}|| \to 1$ as $n, m \to \infty$.

Proof of Claim: Let n, m be large enough that $Lx_n, Lx_m \geq ||L|| - \frac{\varepsilon}{2}$ where $0 < \varepsilon$. Then $||x_n + x_m|| \neq 0$ because if it equals 0, then $x_n = -x_m$ so $-Lx_n = Lx_m$ but both Lx_n and Lx_m are positive. Therefore we can consider $\frac{x_n+x_m}{||x_n+x_m||}$ a vector of norm 1. Thus,

$$||L|| \geq |L\frac{(x_n + x_m)}{||x_n + x_m||}| \geq \frac{2||L|| - \varepsilon}{||x_n + x_m||}.$$

Hence

$$||x_n + x_m|| \, ||L|| \geq 2||L|| - \varepsilon.$$

Since $\varepsilon > 0$ is arbitrary, $\lim_{n,m\to\infty} ||x_n+x_m|| = 2$. This proves the claim.

By uniform convexity, $\{x_n\}$ is Cauchy and $x_n \to x$, $||x|| = 1$. Thus $Lx = \lim_{n\to\infty} Lx_n = ||L||$. This proves Lemma 13.20.

The proof of the Riesz representation theorem will be based on the following lemma which says that if you can show certain things, then you can represent a linear functional.

Lemma 13.21 *(McShane) Let X be a complex normed linear space and let $L \in X'$. Suppose there exists $x \in X$, $||x|| = 1$ with $Lx = ||L|| \neq 0$. Let $y \in X$ and let $\psi_y(t) = ||x + ty||$ for $t \in \mathbb{R}$. Suppose $\psi'_y(0)$ exists for each $y \in X$. Then for all $y \in X$,*

$$\psi'_y(0) + i\psi'_{-iy}(0) = ||L||^{-1}Ly.$$

Proof: Suppose first that $||L|| = 1$. Then

$$L(x + t(y - L(y)x)) = Lx = 1 = ||L||.$$

Therefore, $||x + t(y - L(y)x)|| \geq 1$. Also for small t, $|L(y)t| < 1$, and so

$$1 \leq ||x + t(y - L(y)x)|| = ||(1 - L(y)t)x + ty||$$

$$\leq |1 - L(y) t| \left\| x + \frac{t}{1 - L(y) t} y \right\|.$$

This implies

$$\frac{1}{|1 - tL(y)|} \leq \left\| x + \frac{t}{1 - L(y)t} y \right\|. \tag{11}$$

Using the formula for the sum of a geometric series,

$$\frac{1}{1 - tLy} = 1 + tLy + o(t)$$

where $\lim_{t \to 0} o(t)(t^{-1}) = 0$. Using this in Formula 11, we obtain

$$|1 + tL(y) + o(t)| \leq \|x + ty + o(t)\|$$

where $o(t)t^{-1} \to 0$ as $t \to 0$. Now if $t > 0$, since $\|x\| = 1$, we have

$$
\begin{aligned}
(\psi_y(t) - \psi_y(0))t^{-1} &= (\|x + ty\| - \|x\|)t^{-1} \\
&\geq (|1 + t\, L(y)| - 1)t^{-1} + \frac{o(t)}{t} \\
&\geq \operatorname{Re} L(y) + \frac{o(t)}{t}.
\end{aligned}
$$

If $t < 0$,

$$(\psi_y(t) - \psi_y(0))t^{-1} \leq \operatorname{Re} L(y) + \frac{o(t)}{t}.$$

Since $\psi'_y(0)$ is assumed to exist, this shows

$$\psi'_y(0) = \operatorname{Re} L(y). \tag{12}$$

Now

$$Ly = \operatorname{Re} L(y) + i \,\operatorname{Im} L(y)$$

so

$$L(-iy) = -i(Ly) = -i \,\operatorname{Re} L(y) + \operatorname{Im} L(y)$$

and

$$L(-iy) = \operatorname{Re} L(-iy) + i \operatorname{Im} L(-iy).$$

Hence

$$\operatorname{Re} L(-iy) = \operatorname{Im} L(y).$$

Consequently, by Formula 12

$$Ly = \operatorname{Re} L(y) + i \,\operatorname{Im} L(y) = \operatorname{Re} L(y) + i \operatorname{Re} L(-iy)$$

$$= \psi'_y(0) + i \,\psi'_{-iy}(0).$$

This proves the lemma when $||L|| = 1$. For arbitrary $L \neq 0$, let $Lx = ||L||, ||x|| = 1$. Then from above, if $L_1 y \equiv ||L||^{-1} L(y)$, $||L_1|| = 1$ and so from what was just shown,

$$L_1(y) = \frac{L(y)}{||L||} = \psi'_y(0) + i\psi_{-iy}(0)$$

and this proves McShane's lemma.

We will use the uniform convexity and this lemma to prove the Riesz representation theorem next. Let $p \geq 2$ and let $\eta : L^q \to (L^p)'$ be defined by

$$\eta(g)(f) = \int_\Omega gf \, d\mu. \tag{13}$$

Theorem 13.22 *(Riesz representation theorem $p \geq 2$) The map η is 1-1, onto, continuous, and*

$$||\eta g|| = ||g||, \; ||\eta|| = 1.$$

Proof: Obviously η is linear. Suppose $\eta g = 0$. Then $0 = \int gf \, d\mu$ for all $f \in L^p$. Let

$$f = |g|^{q-2}\overline{g}.$$

Then $f \in L^p$ and so $0 = \int |g|^q d\mu$. Hence $g = 0$ and η is 1-1. That $\eta g \in (L^p)'$ is obvious from the Holder inequality. In fact,

$$|\eta(g)(f)| \leq ||g||_q ||f||_p,$$

and so $||\eta(g)|| \leq ||g||_q$. To see that equality holds, let

$$f = |g|^{q-2}\overline{g} \, ||g||_q^{1-q}.$$

Then $||f||_p = 1$ and

$$\eta(g)(f) = \int_\Omega |g|^q d\mu ||g||_q^{1-q} = ||g||_q.$$

Thus $||\eta|| = 1$. It remains to show η is onto. Let $L \in (L^p)'$. We need show $L = \eta g$ for some $g \in L^q$. Without loss of generality, we may assume $L \neq 0$. Let

$$Lg = ||L||, \; g \in L^p, \; ||g|| = 1.$$

We can assert the existence of such a g by Lemma 13.20. For $f \in L^p$,

$$\psi_f(t) \equiv ||g + tf||_p \equiv \phi_f(t)^{\frac{1}{p}}.$$

We show $\phi'_f(0)$ exists. Let $[g = 0]$ denote the set $\{x : g(x) = 0\}$.

$$\frac{\phi_f(t) - \phi_f(0)}{t} =$$

235

$$\frac{1}{t}\int(|g+tf|^p - |g|^p)d\mu = \frac{1}{t}\int_{[g=0]}|t|^p\,|f|^p d\mu$$

$$+\int_{[g\neq 0]} p|g(x) + s(x)f(x)|^{p-2}\,\mathrm{Re}[(g(x) + s(x)f(x))\bar{f}(x)]d\mu \qquad (14)$$

where the Mean Value theorem is being used and $s(x)$ is between 0 and t, the integrand in the second integral of Formula 14 equaling

$$\frac{1}{t}(|g(x) + tf(x)|^p - |g(x)|^p).$$

Now if $|t| < 1$, the integrand in the last integral of Formula 14 is bounded by

$$p\left[\frac{(|g(x)| + |f(x)|)^p}{q} + \frac{|f(x)|^p}{p}\right]$$

which is a function in L^1 since f, g are in L^p(we used the inequality $ab \leq \frac{a^q}{q} + \frac{b^p}{p}$). Because of this, we can apply the Dominated Convergence theorem and obtain

$$\phi'_f(0) = p\int |g(x)|^{p-2}\,\mathrm{Re}(g(x)\bar{f}(x))d\mu.$$

Hence

$$\psi'_f(0) = ||g||^{\frac{-p}{q}}\int |g(x)|^{p-2}\,\mathrm{Re}(g(x)\bar{f}(x))d\mu.$$

Note $\frac{1}{p} - 1 = -\frac{1}{q}$. Therefore,

$$\psi'_{-if}(0) = ||g||^{\frac{-p}{q}}\int |g(x)|^{p-2}\,\mathrm{Re}(ig(x)\bar{f}(x))d\mu.$$

But $\mathrm{Re}(ig\bar{f}) = \mathrm{Im}(-g\bar{f})$ and so by the McShane lemma,

$$\begin{aligned}
Lf &= ||L||\,||g||^{\frac{-p}{q}}\int |g(x)|^{p-2}[\mathrm{Re}(g(x)\bar{f}(x)) + i\,\mathrm{Re}(ig(x)\bar{f}(x))]d\mu \\
&= ||L||\,||g||^{\frac{-p}{q}}\int |g(x)|^{p-2}[\mathrm{Re}(g(x)\bar{f}(x)) + i\,\mathrm{Im}(-g(x)\bar{f}(x))]d\mu \\
&= ||L||\,||g||^{\frac{-p}{q}}\int |g(x)|^{p-2}\bar{g}(x)f(x)d\mu.
\end{aligned}$$

This shows that

$$L = \eta(||L||\,||g||^{\frac{-p}{q}}|g|^{p-2}\bar{g})$$

and verifies η is onto. This proves the theorem.

To prove the Riesz representation theorem for $1 < p < 2$, one can verify that L^p is uniformly convex and then repeat the above argument. Note that no reference to $p > 1$ was used in the proof. Unfortunately, this requires Clarkson's Inequalities for $p \in (1, 2)$ which are more technical than the case where $p \geq 2$. To see this done see Hewitt & Stromberg [27] or Ray [41]. Here we take a different approach using the Milman theorem which states that uniform convexity implies the space is Reflexive.

Theorem 13.23 *(Riesz representation theorem) Let $1 < p < \infty$ and let $\eta : L^q \to (L^p)'$ be given by Formula 13. Then η is 1-1, onto, and $||\eta g|| = ||g||$.*

Proof: Everything is the same as the proof of Theorem 13.22 except for the assertion that η is onto. Suppose $1 < p < 2$. (The case $p \geq 2$ was done in Theorem 13.22.) Then $q > 2$ and so we know from Theorem 13.22 that $\overline{\eta} : L^p \to (L^q)'$ defined as

$$\overline{\eta}f(g) \equiv \int_\Omega fg d\mu$$

is onto and $||\overline{\eta}f|| = ||f||$. Then $\overline{\eta}^* : (L^q)'' \to (L^p)'$ is also 1-1, onto, and $||\overline{\eta}^*L|| = ||L||$. By Milman's theorem, J is onto from $L^q \to (L^q)''$. This occurs because of the uniform convexity of L^q which follows from Clarkson's inequality. Thus both maps in the following diagram are 1-1 and onto.

$$L^q \xrightarrow{J} (L^q)'' \xrightarrow{\overline{\eta}^*} (L^p)'.$$

Now if $g \in L^q$, $f \in L^p$, then

$$\overline{\eta}^* J(g)(f) = Jg(\overline{\eta}f) = (\overline{\eta}f)(g) = \int_\Omega fg \, d\mu.$$

Thus if $\eta : L^q \to (L^p)'$ is the mapping of Formula 13, this shows $\eta = \overline{\eta}^* J$. Also

$$||\eta g|| = ||\eta^* Jg|| = ||Jg|| = ||g||.$$

This proves the theorem.

In the case where $p = 1$, it is also possible to give the Riesz representation theorem in a more general context than σ finite spaces. To see this done, see Hewitt and Stromberg [27]. The dual space of L^∞ has also been studied. See Dunford and Schwartz [16].

13.5   The dual space of $C(X)$

Next we represent the dual space of $C(X)$ where X is a compact Hausdorff space. It will turn out to be a space of measures.

Let $L \in C(X)'$. Also denote by $C^+(X)$ the set of nonnegative continuous functions defined on X. Define for $f \in C^+(X)$

$$\lambda(f) = \sup\{|Lg| : |g| \leq f\}.$$

Note that $\lambda(f) < \infty$ because $|Lg| \leq ||L|| \, ||g|| \leq ||L|| \, ||f||$ for $|g| \leq f$.

Lemma 13.24 *If* $c \geq 0$, $\lambda(cf) = c\lambda(f)$, $f_1 \leq f_2$ *implies* $\lambda f_1 \leq \lambda f_2$, *and*

$$\lambda(f_1 + f_2) = \lambda(f_1) + \lambda(f_2).$$

Proof: The first two assertions are easy to see so we consider the third. Let $|g_j| \leq f_j$ and let $\widetilde{g}_j = e^{i\theta_j} g_j$ where θ_j is chosen such that $e^{i\theta_j} Lg_j = |Lg_j|$. Thus $L\widetilde{g}_j = |Lg_j|$. Then

$$|\widetilde{g}_1 + \widetilde{g}_2| \leq f_1 + f_2.$$

Hence

$$|Lg_1| + |Lg_2| = L\widetilde{g}_1 + L\widetilde{g}_2 =$$

$$L(\widetilde{g}_1 + \widetilde{g}_2) = |L(\widetilde{g}_1 + \widetilde{g}_2)| \leq \lambda(f_1 + f_2). \tag{15}$$

Choose g_1 and g_2 such that $|Lg_i| + \varepsilon > \lambda(f_i)$. Then Formula 15 shows

$$\lambda(f_1) + \lambda(f_2) - 2\varepsilon \leq \lambda(f_1 + f_2).$$

Since $\varepsilon > 0$ is arbitrary, it follows that

$$\lambda(f_1) + \lambda(f_2) \leq \lambda(f_1 + f_2). \tag{16}$$

Now let $|g| \leq f_1 + f_2$, $|Lg| \geq \lambda(f_1 + f_2) - \varepsilon$. Let

$$h_i(x) = \begin{cases} \frac{f_i(x)g(x)}{f_1(x)+f_2(x)} & \text{if } f_1(x) + f_2(x) > 0, \\ 0 & \text{if } f_1(x) + f_2(x) = 0. \end{cases}$$

Then h_i is continuous and $h_1(x) + h_2(x) = g(x)$, $|h_i| \leq f_i$. Therefore,

$$\begin{aligned} -\varepsilon + \lambda(f_1 + f_2) & \leq & |Lg| \leq |Lh_1 + Lh_2| \leq |Lh_1| + |Lh_2| \\ & \leq & \lambda(f_1) + \lambda(f_2). \end{aligned}$$

Since $\varepsilon > 0$ is arbitrary, this shows with Formula 16 that

$$\lambda(f_1 + f_2) \leq \lambda(f_1) + \lambda(f_2) \leq \lambda(f_1 + f_2)$$

which proves the lemma.

Let $C(X; \mathbb{R})$ be the real-valued functions in $C(X)$ and define

$$\Lambda_R(f) = \lambda f^+ - \lambda f^-$$

for $f \in C(X; \mathbb{R})$. Using Lemma 13.24 and the identity

$$(f_1 + f_2)^+ + f_1^- + f_2^- = f_1^+ + f_2^+ + (f_1 + f_2)^-$$

to write

$$\lambda(f_1 + f_2)^+ - \lambda(f_1 + f_2)^- = \lambda f_1^+ - \lambda f_1^- + \lambda f_2^+ - \lambda f_2^-,$$

we see that $\Lambda_R(f_1 + f_2) = \Lambda_R(f_1) + \Lambda_R(f_2)$. To show that Λ_R is linear, we need to verify that $\Lambda_R(cf) = c\Lambda_R(f)$ for all $c \in \mathbb{R}$. But

$$(cf)^\pm = cf^\pm,$$

if $c \geq 0$ while

$$(cf)^+ = -c(f)^-,$$

if $c < 0$ and

$$(cf)^- = (-c)f^+,$$

if $c < 0$. Thus, if $c < 0$,

$$\Lambda_R(cf) = \lambda(cf)^+ - \lambda(cf)^- = \lambda(-cf)^- - \lambda(-c)f^+$$
$$= -c\lambda(f^-) + c\lambda(f^+) = c(\lambda(f^+) - \lambda(f^-)).$$

(If this looks familiar it may be because we used this approach earlier in defining the integral of a real-valued function.) Now let

$$\Lambda f = \Lambda_R(\operatorname{Re} f) + i\Lambda_R(\operatorname{Im} f)$$

for arbitrary $f \in C(X)$. It is easy to see that Λ is a positive linear functional on $C(X)$ $(= C_c(X)$ since X is compact). By Theorem 9.13, there exists a unique Radon outer measure μ such that

$$\Lambda f = \int_X f \, d\mu$$

for all $f \in C(X)$. Thus $\Lambda(1) = \mu(X)$. Now we present the Riesz representation theorem for $C(X)'$.

Theorem 13.25 *Let $L \in (C(X))'$ then there exists a Radon measure μ and a function $\sigma \in L^\infty(X, \mu)$ such that*

$$L(f) = \int_X f \, \sigma \, d\mu.$$

Proof: Let $f \in C(X)$. Then there exists a unique Radon measure μ such that

$$|Lf| \leq \Lambda(|f|) = \int_X |f| d\mu = \|f\|_1.$$

Since μ is a Radon measure, we know $C(X)$ is dense in $L^1(X, \mu)$. Therefore L extends uniquely to an element of $(L^1(X, \mu))'$. By the Riesz representation theorem for L^1, there exists a unique $\sigma \in L^\infty(X, \mu)$ such that

$$Lf = \int_X f \, \sigma \, d\mu$$

for all $f \in C(X)$.

It is possible to give a simple generalization of the above theorem. For X a locally compact Hausdorff space, we will denote by $\tilde{X}$ the one point compactification of X. Thus, $\tilde{X} = X \cup \{\infty\}$ and the topology of $\tilde{X}$ consists of the usual topology of X along with all complements of compact sets which are defined as the open sets containing ∞. We also define by $C_0(X)$ the space of continuous functions, f, defined on X such that in the topology of $\tilde{X}$, $\lim_{x \to \infty} f(x) = 0$. For this space of functions, $\|f\|_0 \equiv \sup \{|f(x)| : x \in X\}$ is a norm which makes this into a Banach space. Then the generalization is the following corollary.

Corollary 13.26 *Let $L \in (C_0(X))'$ where X is a locally compact Hausdorff space. Then there exists $\sigma \in L^\infty(X, \mu)$ for μ a Radon measure such that for all $f \in C_0(X)$,*

$$L(f) = \int_X f\sigma d\mu.$$

Proof: Let

$$\tilde{D} \equiv \left\{ f \in C\left(\tilde{X}\right) : f(\infty) = 0 \right\}.$$

Thus $\tilde{D}$ is a closed subspace of the Banach space $C\left(\tilde{X}\right)$. Let $\theta : C_0(X) \to \tilde{D}$ be defined by

$$\theta f(x) = \begin{cases} f(x) & \text{if } x \in X, \\ 0 & \text{if } x = \infty. \end{cases}$$

Then θ is an isometry of $C_0(X)$ and $\tilde{D}$. It follows we have the following diagram.

$$C_0(X)' \xleftarrow{\theta^*} \left(\tilde{D}\right)' \xleftarrow{i^*} C\left(\tilde{X}\right)'$$

$$C_0(X) \xrightarrow{\theta} \tilde{D} \xrightarrow{i} C\left(\tilde{X}\right)$$

By the Hahn Banach theorem, there exists $L_1 \in C\left(\tilde{X}\right)'$ such that $\theta^* i^* L_1 = L$. Now we apply Theorem 13.25 to get the existence of a Radon measure, μ_1, on $\tilde{X}$ and a function $\sigma \in L^\infty\left(\tilde{X}, \mu_1\right)$, such that

$$L_1 g = \int_{\tilde{X}} g\sigma d\mu_1.$$

Letting the σ algebra of μ_1 measurable sets be denoted by $\mathcal{S}_1$, we define

$$\mathcal{S} \equiv \{E \setminus \{\infty\} : E \in \mathcal{S}_1\}$$

and let μ be the restriction of μ_1 to $\mathcal{S}$. If $f \in C_0(X)$,

$$Lf = \theta^* i^* L_1 f \equiv L_1 i \theta f = L_1 \theta f = \int_{\widetilde{X}} \theta f \sigma d\mu_1 = \int_X f \sigma d\mu.$$

This proves the corollary.

13.6   Exercises

1. Suppose $\lambda(E) = \int_E f d\mu$ where λ and μ are two measures and $f \in L^1(\mu)$. Show $\lambda << \mu$.

2. Suppose λ, μ are two finite measures defined on a σ algebra $\mathcal{S}$. Show $\lambda = \lambda_S + \lambda_A$ where λ_S and λ_A are finite measures satisfying

    $$\lambda_S(E) = \lambda_S(E \cap S) , \ \mu(S) = 0 \text{ for some } S \subseteq \mathcal{S},$$

    $$\lambda_A << \mu.$$

    This is called the Lebesgue decomposition. **Hint**: This is just a generalization of the Radon Nikodym theorem. In the proof of this theorem, let

    $$S = \{x : h(x) = 1\}, \ \lambda_S(E) = \lambda(E \cap S),$$

    $$\lambda_A(E) = \lambda(E \cap S^c).$$

    We write $\mu \perp \lambda_S$ and $\lambda_A << \mu$ in this situation.

3. ↑ Generalize the result of Problem 2 to the case where μ is σ finite and λ is finite.

4. Let F be a nondecreasing right continuous, bounded function,

    $$\lim_{x \to -\infty} F(x) = 0.$$

    Let $Lf = \int f dF$ where $f \in C_c(\mathbb{R})$ and this is just the Riemann Stieltjes integral. Let λ be the Radon measure representing L. Show

    $$\lambda((a, b]) = F(b) - F(a) , \ \lambda((a, b)) = F(b-) - F(a).$$

5. ↑ Using Problems 2, 3, and 4, show there is a bounded nondecreasing function $G(x)$ such that $G(x) \leq F(x)$ and $G(x) = \int_{-\infty}^x \ell(t) dt$ for some $\ell \geq 0$, $\ell \in L^1(m)$. Also, if $F(x) - G(x) = S(x)$, then $S(x)$

241

Representation Theorems

is non decreasing and if λ_S is the measure representing $\int f dS$, then $\lambda_S \perp m$. **Hint:** Consider $G(x) = \lambda_A((-\infty, x])$.

6. Let λ and μ be two measures defined on $\mathcal{S}$, a σ algebra of subsets of Ω. Suppose μ is σ finite and $g \geq 0$ with g measurable. Show that

$$g = \frac{d\lambda}{d\mu}, \ (\lambda(E) = \int_E g d\mu)$$

if and only if for all $A \in \mathcal{S}$,and $\alpha, \beta, \geq 0$,

$$\lambda(A \cap \{x : g(x) \geq \alpha\}) \geq \alpha\mu(A \cap \{x : g(x) \geq \alpha\}),$$

$$\lambda(A \cap \{x : g(x) < \beta\}) \leq \beta\mu(A \cap \{x : g(x) < \beta\}).$$

Hint: To show $g = \frac{d\lambda}{d\mu}$ from the two conditions, use the conditions to argue that for $\mu(A) < \infty$,

$$\beta\mu(A \cap \{x : g(x) \in [\alpha, \beta)\}) \geq \lambda(A \cap \{x : g(x) \in [\alpha, \beta)\})$$

$$\geq \alpha\mu(A \cap \{x : g(x) \in [\alpha, \beta)\}).$$

7. Let $r, p, q \in (1, \infty)$ satisfy

$$\frac{1}{r} = \frac{1}{p} + \frac{1}{q} - 1$$

and let $f \in L^p(\mathbb{R}^n)$, $g \in L^q(\mathbb{R}^n)$, $f, g \geq 0$. Young's inequality says that

$$\|f * g\|_r \leq \|g\|_q \|f\|_p.$$

Prove Young's inequality by justifying or modifying the steps in the following argument using Problem 25 of Chapter 12. Let

$$h \in L^{r'}(\mathbb{R}^n) \cdot \left(\frac{1}{r} + \frac{1}{r'} = 1\right)$$

$$\int f * g(\mathbf{x}) |h(\mathbf{x})| dx = \int \int f(\mathbf{y}) g(\mathbf{x} - \mathbf{y}) |h(\mathbf{x})| dx dy.$$

Let $r\theta = p$ so $\theta \in (0, 1)$, $p'(1 - \theta) = q$, $p' \geq r'$. Then the above

$$= \int \int f(\mathbf{y}) g(\mathbf{x} - \mathbf{y})^\theta g(\mathbf{x} - \mathbf{y})^{1-\theta} |h(\mathbf{x})| dy dx$$

$$\leq \int \left(\int \left(g(\mathbf{x} - \mathbf{y})^{1-\theta} |h(\mathbf{x})|\right)^{r'} dx\right)^{1/r'} \cdot$$

$$\left(\int \left(f(\mathbf{y}) g(\mathbf{x} - \mathbf{y})^\theta\right)^r dx\right)^{1/r} dy$$

242

$$\leq \left[\int \left(\int \left(g\left(\mathbf{x} - \mathbf{y} \right)^{1-\theta} \left| h\left(\mathbf{x} \right) \right| \right)^{r'} dx \right)^{p'/r'} dy \right]^{1/p'} \cdot$$

$$\left[\int \left(\int \left(f\left(\mathbf{y} \right) g\left(\mathbf{x} - \mathbf{y} \right)^{\theta} \right)^{r} dx \right)^{p/r} dy \right]^{1/p}$$

$$\leq \left[\int \left(\int \left(g\left(\mathbf{x} - \mathbf{y} \right)^{1-\theta} \left| h\left(\mathbf{x} \right) \right| \right)^{p'} dy \right)^{r'/p'} dx \right]^{1/r'} \cdot$$

$$\left[\int f\left(\mathbf{y} \right)^{p} \left(\int g\left(\mathbf{x} - \mathbf{y} \right)^{\theta r} dx \right)^{p/r} dy \right]^{1/p}$$

$$= \left[\int \left| h\left(\mathbf{x} \right) \right|^{r'} \left(\int g\left(\mathbf{x} - \mathbf{y} \right)^{(1-\theta)p'} dy \right)^{r'/p'} dx \right]^{1/r'} \left\| g \right\|_q^{q/r} \left\| f \right\|_p$$

$$= \left\| g \right\|_q^{q/r} \left\| g \right\|_q^{q/p'} \left\| f \right\|_p \left\| h \right\|_{r'} = \left\| g \right\|_q \left\| f \right\|_p \left\| h \right\|_{r'} .$$

Therefore $\left\| f * g \right\|_r \leq \left\| g \right\|_q \left\| f \right\|_p$. Does this inequality continue to hold if r, p, q are only assumed to be in $[1, \infty]$? Explain.

8. Let $X = [0, \infty]$ with a subbasis for the topology given by sets of the form $[0, b)$ and $(a, \infty]$. Show that X is a compact set. Consider all functions of the form

$$\sum_{k=0}^{n} a_k e^{-xkt}$$

where $t > 0$. Show that this collection of functions is an algebra of functions in $C\left(X \right)$ if we define

$$e^{-t\infty} \equiv 0,$$

and that it separates the points and annihilates no point. Conclude that this algebra is dense in $C\left(X \right)$.

9. ↑ Suppose $f \in C\left(X \right)$ and for all t large enough, say $t \geq \delta$,

$$\int_0^{\infty} e^{-tx} f\left(x \right) dx = 0.$$

Show that $f\left(x \right) = 0$ for all $x \in (0, \infty)$. **Hint:** Show the measure given by $d\mu = e^{-\delta x} dx$ is a regular measure and so $C\left(X \right)$ is dense in $L^2\left(X, \mu \right)$. Now use Problem 8 to conclude that $f\left(x \right) = 0$ a.e.

10. ↑ Suppose f is continuous on $(0, \infty)$, and for some $\gamma > 0$,

$$\left| f\left(x \right) e^{-\gamma x} \right| \leq C$$

for all $x \in (0, \infty)$, and suppose also that for all t large enough,

$$\int_0^\infty f(x) e^{-tx} dx = 0.$$

Show $f(x) = 0$ for all $x \in (0, \infty)$. A common procedure in elementary differential equations classes is to obtain the Laplace transform of an unknown function and then, using a table of Laplace transforms, find what the unknown function is. Can the result of this problem be used to justify this procedure?

11. It was shown that L^p for $p > 1$ is reflexive. It follows from the Eberlein Smulian theorem that L^p is weakly sequentially compact. This problem considers the case of weak sequential compactness in

$$L^1(\Omega, \mathcal{S}, \mu).$$

Let $K \subseteq L^1(\Omega, \mathcal{S}, \mu)$ be such that for all $f \in K$,

$$\|f\|_1 \leq C, \tag{17}$$

and if $E_n \downarrow \emptyset$ then for all $\epsilon > 0$ there exists n_ϵ such that if $n \geq n_\epsilon$ then

$$\left| \int_{E_n} f d\mu \right| < \epsilon \tag{18}$$

for all $f \in K$. Show that then K is weakly sequentially compact. This is actually an equivalence. For more on this theorem see [16]. **Hint:** Let $\{f_n\} \subseteq K$ and use the Cantor diagonal process to obtain $\{g_n\}$, a subsequence such that $\int_E g_n d\mu$ converges for all $E \in \mathcal{A}$, where $\mathcal{E}, \mathcal{A}, \mathcal{S}_1, \Omega_1$ are as in Problem 12 of Chapter 8. Use the Monotone Class theorem and Formula 18 to conclude this convergence takes place for all $E \in \mathcal{S}_1$. Picking a uniformly bounded representative of $h \in L^\infty(\Omega_1, \mathcal{S}_1, \mu)$ let $\{s_k\}$ be a sequence of simple functions converging uniformly to h on Ω_1. Then use Formula 17 to show

$$\int g_n h d\mu \text{ converges.}$$

Chapter 14
Fundamental Theorem of Calculus

One of the most remarkable theorems in Lebesgue integration is the Lebesgue fundamental theorem of calculus which says that if f is a function in L^1, then the indefinite integral,

$$x \rightarrow \int_a^x f(t)\, dt,$$

can be differentiated for a.e. x and gives $f(x)$ a.e. This is a very significant generalization of the usual fundamental theorem of calculus found in calculus. To prove this theorem, we use a covering theorem due to Vitali and theorems on maximal functions. This approach leads to very general results without very many painful technicalities. We will be in the context of $(\mathbb{R}^n, \mathcal{S}, m)$ where m is n-dimensional Lebesgue measure. When this important theorem is established, it will be used to prove the very useful theorem about change of variables in multiple integrals.

By Lemma 9.6 of Chapter 9 and the completeness of m, we know that the Lebesgue measurable sets are exactly those measurable in the sense of Caratheodory. Also, we can regard m as an outer measure defined on all of $\mathcal{P}(\mathbb{R}^n)$. We will use the following notation.

$$B(\mathbf{p}, r) = \{\mathbf{x} : |\mathbf{x} - \mathbf{p}| < r\}. \tag{1}$$

If

$$B = B(\mathbf{p}, r), \text{ then } \hat{B} = B(\mathbf{p}, 5r).$$

14.1 The Vitali covering theorem

Lemma 14.1 *Let $\mathcal{F}$ be a collection of balls as in Formula 1. Suppose*

$$\infty > M \equiv \sup\{r : B(\mathbf{p}, r) \in \mathcal{F}\} > 0.$$

Then there exists $\mathcal{G} \subseteq \mathcal{F}$ such that

$$\text{if } B(\mathbf{p}, r) \in \mathcal{G} \text{ then } r > \frac{M}{2}, \tag{2}$$

$$\text{if } B_1, B_2 \in \mathcal{G} \text{ then } B_1 \cap B_2 = \emptyset, \tag{3}$$

$\mathcal{G}$ is maximal with respect to Formulas 2 and 3.

Proof: Let $\mathcal{H} = \{\mathcal{B} \subseteq \mathcal{F}$ such that Formula 2 and Formula 3 hold$\}$. Obviously $\mathcal{H} \neq \emptyset$ because there exists $B(\mathbf{p}, r) \in \mathcal{F}$ with $r > \frac{M}{2}$. Partially order $\mathcal{H}$ by set inclusion and use the Hausdorff maximal theorem (see the appendix on set theory) to let $\mathcal{C}$ be a maximal chain in $\mathcal{H}$. Clearly $\cup\mathcal{C}$ satisfies Formula 2 and Formula 3. If $\cup\mathcal{C}$ is not maximal with respect to these two properties, then $\mathcal{C}$ was not a maximal chain. Let $\mathcal{G} = \cup\mathcal{C}$.

Theorem 14.2 *(Vitali) Let $\mathcal{F}$ be a collection of balls and let*

$$A \equiv \cup\{B : B \in \mathcal{F}\}.$$

Suppose

$$\infty > M \equiv \sup\{r : B(\mathbf{p}, r) \in \mathcal{F}\} > 0.$$

Then there exists $\mathcal{G} \subseteq \mathcal{F}$ such that $\mathcal{G}$ consists of disjoint balls and

$$A \subseteq \cup\{\widehat{B} : B \in \mathcal{G}\}.$$

Proof: Let $\mathcal{G}_1 \subseteq \mathcal{F}$ satisfy

$$B(\mathbf{p}, r) \in \mathcal{G}_1 \text{ implies } r > \frac{M}{2}, \tag{4}$$

$$B_1, B_2 \in \mathcal{G}_1 \text{ implies } B_1 \cap B_2 = \emptyset, \tag{5}$$

$\mathcal{G}_1$ is maximal with respect to Formulas 4, and 5.

Suppose $\mathcal{G}_1, \cdots, \mathcal{G}_{m-1}$ have been chosen, $m \geq 2$. Let

$$\mathcal{F}_m = \{B \in \mathcal{F} : B \subseteq \mathbb{R}^n \setminus \cup\{\mathcal{G}_1 \cup \cdots \cup \mathcal{G}_{m-1}\}\}.$$

Let $\mathcal{G}_m \subseteq \mathcal{F}_m$ satisfy the following.

$$B(\mathbf{p}, r) \in \mathcal{G}_m \text{ implies } r > \frac{M}{2^m}, \tag{6}$$

$$B_1, B_2 \in \mathcal{G}_m \text{ implies } B_1 \cap B_2 = \emptyset, \tag{7}$$

$\mathcal{G}_m$ is a maximal subset of $\mathcal{F}_m$ with respect to Formulas 6 and 7. If $\mathcal{F}_m = \emptyset$, $\mathcal{G}_m = \emptyset$. Define

$$\mathcal{G} \equiv \cup_{k=1}^{\infty} \mathcal{G}_k.$$

Thus $\mathcal{G}$ is a collection of disjoint balls in $\mathcal{F}$. We need to show $\{\widehat{B} : B \in \mathcal{G}\}$ covers A. Let $\mathbf{x} \in A$. Then $\mathbf{x} \in B(\mathbf{p}, r) \in \mathcal{F}$. Pick m such that

$$\frac{M}{2^m} < r \leq \frac{M}{2^{m-1}}.$$

We claim $\mathbf{x} \in \widehat{B}$ for some $B \in \mathcal{G}_1 \cup \cdots \cup \mathcal{G}_m$. To see this, note that $B(\mathbf{p}, r)$ must intersect some set of $\mathcal{G}_1 \cup \cdots \cup \mathcal{G}_m$ because if it didn't, then

$$\mathcal{G}_m \cup \{B(\mathbf{p}, r)\} = \mathcal{G}'_m$$

would satisfy Formulas 6 and 7, and $\mathcal{G}'_m \supsetneq \mathcal{G}_m$ contradicting the maximality of $\mathcal{G}_m$. Let the set intersected be $B(\mathbf{p_0}, r_0)$. Thus $r_0 > M2^{-m}$.

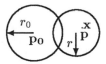

Then if $\mathbf{x} \in B(\mathbf{p}, r)$,

$$
\begin{aligned}
|\mathbf{x} - \mathbf{p_0}| &\leq |\mathbf{x} - \mathbf{p}| + |\mathbf{p} - \mathbf{p_0}| < r + r_0 + r \\
&\leq \frac{2M}{2^{m-1}} + r_0 < 4r_0 + r_0 = 5r_0
\end{aligned}
$$

since $r_0 > M/2^{-m}$. Hence $B(\mathbf{p}, r) \subseteq B(\mathbf{p_0}, 5r_0)$ and this proves the theorem.

14.2   Differentiation with respect to Lebesgue measure

The covering theorem just presented will now be used to establish the fundamental theorem of calculus. In discussing this, we introduce the space of functions which is locally integrable in the following definition. This space of functions is the most general one for which the maximal function defined below makes sense.

Definition 14.3 $f \in L^1_{loc}(\mathbb{R}^n)$ means $f\mathcal{X}_{B(0,R)} \in L^1(\mathbb{R}^n)$ for all $R > 0$. For $f \in L^1_{loc}(\mathbb{R}^n)$, the Hardy Littlewood Maximal Function, Mf, is defined by

$$
Mf(\mathbf{x}) \equiv \sup_{r>0} \frac{1}{m(B(\mathbf{x}, r))} \int_{B(\mathbf{x},r)} |f(\mathbf{y})| dy.
$$

Theorem 14.4   If $f \in L^1(\mathbb{R}^n)$, then for $\alpha > 0$,

$$
\overline{m}([Mf > \alpha]) \leq \frac{5^n}{\alpha} ||f||_1.
$$

(Here and elsewhere, $[Mf > \alpha] \equiv \{\mathbf{x} \in \mathbb{R}^n : Mf(\mathbf{x}) > \alpha\}$ with other occurrences of $[\ ]$ being defined similarly.)

Proof: Let $S \equiv [Mf > \alpha]$. For $\mathbf{x} \in S$, choose $r_\mathbf{x} > 0$ with

$$
\frac{1}{m(B(\mathbf{x}, r_\mathbf{x}))} \int_{B(\mathbf{x}, r_\mathbf{x})} |f| \, dm > \alpha.
$$

The $r_\mathbf{x}$ are all bounded because

$$
m(B(\mathbf{x}, r_\mathbf{x})) < \frac{1}{\alpha} \int_{B(\mathbf{x}, r_\mathbf{x})} |f| \, dm < \frac{1}{\alpha} ||f||_1.
$$

By the Vitali covering theorem, there are disjoint balls $B(\mathbf{x}_i, r_i)$ such that

$$S \subseteq \cup_{\mathbf{x} \in S} B(\mathbf{x}, r_{\mathbf{x}}) \subseteq \cup_{i=1}^{\infty} B(\mathbf{x}_i, 5r_i)$$

and

$$\frac{1}{m(B(\mathbf{x}_i, r_i))} \int_{B(\mathbf{x}_i, r_i)} |f| \, dm > \alpha.$$

Therefore

$$
\begin{aligned}
\overline{m}(S) &\leq \sum_{i=1}^{\infty} m(B(\mathbf{x}_i, 5r_i)) = 5^n \sum_{i=1}^{\infty} m(B(\mathbf{x}_i, r_i)) \\
&\leq \frac{5^n}{\alpha} \sum_{i=1}^{\infty} \int_{B(\mathbf{x}_i, r_i)} |f| \, dm \\
&\leq \frac{5^n}{\alpha} \int_{\mathbb{R}^n} |f| \, dm,
\end{aligned}
$$

the last inequality being valid because the balls $B(\mathbf{x}_i, r_i)$ are disjoint. This proves the theorem.

Lemma 14.5 *Let $f \geq 0$, and $f \in L^1$, then*

$$\lim_{r \to 0} \frac{1}{m(B(\mathbf{x}, r))} \int_{B(\mathbf{x}, r)} f(\mathbf{y}) dy = f(\mathbf{x}) \quad a.e. \ \mathbf{x}.$$

Proof: Let $\alpha > 0$ and let

$$B_\alpha = \left[\limsup_{r \to 0} \left| \frac{1}{m(B(\mathbf{x}, r))} \int_{B(\mathbf{x}, r)} f(\mathbf{y}) dy - f(\mathbf{x}) \right| > \alpha \right].$$

Then for any $g \in C_c(\mathbb{R}^n)$, B_α equals

$$\left[\limsup_{r \to 0} \left| \frac{1}{m(B(\mathbf{x}, r))} \int_{B(\mathbf{x}, r)} f(\mathbf{y}) - g(\mathbf{y}) dy - (f(\mathbf{x}) - g(\mathbf{x})) \right| > \alpha \right]$$

because for any $g \in C_c(\mathbb{R}^n)$,

$$\lim_{r \to 0} \frac{1}{m(B(\mathbf{x}, r))} \int_{B(\mathbf{x}, r)} g(\mathbf{y}) dy = g(\mathbf{x}).$$

Thus

$$B_\alpha \subseteq [M(|f - g|) + |f - g| > \alpha],$$

and so

$$B_\alpha \subseteq \left[M(|f - g|) > \frac{\alpha}{2} \right] \cup \left[|f - g| > \frac{\alpha}{2} \right].$$

Now

$$\frac{\alpha}{2} m \left(\left[|f - g| > \frac{\alpha}{2} \right] \right) = \frac{\alpha}{2} \int_{[|f-g|>\frac{\alpha}{2}]} dx$$

$$\leq \int_{[|f-g|>\frac{\alpha}{2}]} |f - g| dx \leq ||f - g||_1.$$

Therefore by Theorem 14.4,

$$\overline{m}(B_\alpha) \leq \left(\frac{2(5^n)}{\alpha} + \frac{2}{\alpha}\right) ||f - g||_1.$$

Since $C_c(\mathbb{R}^n)$ is dense in $L^1(\mathbb{R}^n)$ and g is arbitrary, this estimate shows $\overline{m}(B_\alpha) = 0$. It follows by Lemma 9.6, since B_α is measurable in the sense of Caratheodory, that B_α is Lebesgue measurable and $m(B_\alpha) = 0$.

$$\left[\limsup_{r\to 0} \left|\frac{1}{m(B(\mathbf{x}, r))} \int_{B(\mathbf{x}, r)} f(\mathbf{y})dy - f(\mathbf{x})\right| > 0\right] \subseteq \cup_{m=1}^\infty B_{\frac{1}{m}} \quad (8)$$

and each set $B_{1/m}$ has measure 0 so the set on the left in Formula 8 is also Lebesgue measurable and has measure 0. Thus, if $\mathbf{x}$ is not in this set,

$$0 = \limsup_{r\to 0} \left|\frac{1}{m(B(\mathbf{x}, r))} \int_{B(\mathbf{x}, r)} f(\mathbf{y})dy - f(\mathbf{x})\right| \geq$$

$$\geq \liminf_{r\to 0} \left|\frac{1}{m(B(\mathbf{x}, r))} \int_{B(\mathbf{x}, r)} f(\mathbf{y})dy - f(\mathbf{x})\right| \geq 0.$$

This proves the lemma.

Corollary 14.6 *If $f \geq 0$ and $f \in L_{loc}^1(\mathbb{R}^n)$, then*

$$\lim_{r\to 0} \frac{1}{m(B(\mathbf{x}, r))} \int_{B(\mathbf{x}, r)} f(\mathbf{y})dy = f(\mathbf{x}) \quad a.e.\ \mathbf{x}. \quad (9)$$

Proof: Apply Lemma 14.5 to $f\mathcal{X}_{B(0,R)}$ for $R = 1, 2, 3, \cdots$. Thus Formula 9 holds for a.e. $\mathbf{x} \in B(0, R)$ for each $R = 1, 2, \cdots$.

Theorem 14.7 *(Fundamental Theorem of Calculus) Let $f \in L_{loc}^1(\mathbb{R}^n)$. Then there exists a set of measure $0, B$, such that if $\mathbf{x} \notin B$, then*

$$\lim_{r\to 0} \frac{1}{m(B(\mathbf{x}, r))} \int_{B(\mathbf{x}, r)} |f(\mathbf{y}) - f(\mathbf{x})| dy = 0.$$

Proof: Let $\{d_i\}_{i=1}^\infty$ be a countable dense subset of $\mathbb{C}$. By Corollary 14.6, there exists a set of measure $0, B_i$, such that if $\mathbf{x} \notin B_i$

$$\lim_{r\to 0} \frac{1}{m(B(\mathbf{x}, r))} \int_{B(\mathbf{x}, r)} |f(\mathbf{y}) - d_i| dy = |f(\mathbf{x}) - d_i|. \quad (10)$$

Let $B = \cup_{i=1}^{\infty} B_i$ and let $\mathbf{x} \notin B$. Pick d_i such that $|f(\mathbf{x}) - d_i| < \frac{\varepsilon}{2}$. Then

$$\frac{1}{m(B(\mathbf{x},r))} \int_{B(\mathbf{x},r)} |f(\mathbf{y}) - f(\mathbf{x})| dy \leq \frac{1}{m(B(\mathbf{x},r))} \int_{B(\mathbf{x},r)} |f(\mathbf{y}) - d_i| dy$$

$$+ \frac{1}{m(B(\mathbf{x},r))} \int_{B(\mathbf{x},r)} |f(\mathbf{x}) - d_i| dy$$

$$\leq \frac{\varepsilon}{2} + \frac{1}{m(B(\mathbf{x},r))} \int_{B(\mathbf{x},r)} |f(\mathbf{y}) - d_i| dy.$$

By Formula 10

$$\frac{1}{m(B(\mathbf{x},r))} \int_{B(\mathbf{x},r)} |f(\mathbf{y}) - f(\mathbf{x})| dy \leq \varepsilon$$

whenever r is small enough. This proves the theorem.

Definition 14.8 For B the set of Theorem 14.7, B^C is called the Lebesgue set or the set of Lebesgue points.

Corollary 14.9 *Let $f \in L^1_{loc}(\mathbb{R}^n)$. Then*

$$\lim_{r \to 0} \frac{1}{m(B(\mathbf{x},r))} \int_{B(\mathbf{x},r)} f(\mathbf{y}) dy = f(\mathbf{x}) \quad a.e. \ \mathbf{x}.$$

The next corollary is a one dimensional version of what was just presented.

Corollary 14.10 *Let $f \in L^1(\mathbb{R})$ and let*

$$F(x) = \int_{-\infty}^{x} f(t) dt.$$

Then for a.e. x, $F'(x) = f(x)$.

Proof: For $h > 0$

$$\frac{1}{h} \int_{x}^{x+h} |f(y) - f(x)| dy \leq 2\left(\frac{1}{2h}\right) \int_{x-h}^{x+h} |f(y) - f(x)| dy$$

By Theorem 14.7, this converges to 0 a.e. Similarly

$$\frac{1}{h} \int_{x-h}^{x} |f(y) - f(x)| dy$$

converges to 0 a.e. x.

$$\left| \frac{F(x+h) - F(x)}{h} - f(x) \right| \leq \frac{1}{h} \int_{x}^{x+h} |f(y) - f(x)| dy$$

and

$$\left| \frac{F(x) - F(x-h)}{h} - f(x) \right| \leq \frac{1}{h} \int_{x-h}^{x} |f(y) - f(x)| dy.$$

Therefore,

$$\lim_{h \to 0} \frac{F(x+h) - F(x)}{h} = f(x) \quad \text{a.e. } x$$

This proves the corollary.

14.3   The change of variables theorem for multiple integrals

As an application, we discuss a change of variables theorem for multiple integrals. In the discussion of Lebesgue measure, a simple change of variables theorem has been presented in the case where the function was a linear transformation. All change of variables theorems depend not so much on the global behaviour of the transformation as on its local behaviour which is controled by the derivative of the function, a linear transformation. We will use the Radon Nikodym Theorem to get the existence of a certain function related to the change of variables and then we will use the fundamental theorem of calculus to identify this function in terms of the derivative of the given function.

In this section U is an open set and $V = \mathbf{h}(U) \subseteq \mathbb{R}^n$ for $\mathbf{h} \in C^1(U)$. The first lemma is a nice result sometimes discussed in advanced calculus.

Lemma 14.11   *Let K be compact and let $\mathcal{C}$ be an open cover of K. Then there exists $r > 0$ such that whenever $\mathbf{x} \in K$, $B(\mathbf{x}, r)$ is contained in some set of $\mathcal{C}$ (this r is called a Lesbesgue number).*

Proof: For each $\mathbf{x} \in K$, let $r_{\mathbf{x}}$ be such that $B(\mathbf{x}, 2r_{\mathbf{x}})$ is contained in some set of $\mathcal{C}$. Let $B(\mathbf{x}_1, r_{\mathbf{x}_1}), \cdots, B(\mathbf{x}_m, r_{\mathbf{x}_m})$ cover K. Let $0 < r < \min(r_{\mathbf{x}_1}, \cdots, r_{\mathbf{x}_m})$. If $\mathbf{x} \in K$, then $\mathbf{x} \in B(\mathbf{x}_i, r_{\mathbf{x}_i})$ for some i. Thus $B(\mathbf{x}, r) \subseteq B(\mathbf{x}_i, r_{\mathbf{x}_i} + r) \subseteq B(\mathbf{x}_i, 2r_{\mathbf{x}_i})$ which is a subset of some set of $\mathcal{C}$.

Lemma 14.12   *Let $K \subseteq U$, for K compact. Then there exist constants $r, M > 0$ such that if $\mathbf{x} \in K$ and $|\mathbf{x} - \mathbf{y}| < r$, then $|\mathbf{h}(\mathbf{x}) - \mathbf{h}(\mathbf{y})| < M|\mathbf{x} - \mathbf{y}|$.*

Proof: For each $\mathbf{x} \in K$, let $B(\mathbf{x}, \delta_{\mathbf{x}})$ satisfy $\overline{B(\mathbf{x}, \delta_{\mathbf{x}})} \subseteq U$. Let

$$B(\mathbf{x}_1, \delta_{\mathbf{x}_1}), \cdots, B(\mathbf{x}_m, \delta_{\mathbf{x}_m})$$

cover K and let r be a Lebesgue number for $\{B(\mathbf{x}_i, \delta_{\mathbf{x}_i})\}_{i=1}^{m}$. Let

$$\infty > M > \max\{\|D\mathbf{h}(\mathbf{x})\| : \mathbf{x} \in \bigcup_{i=1}^{m} \overline{B(\mathbf{x}_i, \delta_{\mathbf{x}_i})}\}.$$

Thus, for $\mathbf{x} \in K$, $B(\mathbf{x}, r) \subseteq B(\mathbf{x}_i, \delta_{\mathbf{x}_i})$ for some i. Therefore, for $|\mathbf{y} - \mathbf{x}| < r$,

$$
\begin{aligned}
|\mathbf{h}(\mathbf{y}) - \mathbf{h}(\mathbf{x})| &\leq \int_0^1 |D\mathbf{h}(\mathbf{x} + t(\mathbf{y} - \mathbf{x}))(\mathbf{y} - \mathbf{x})| dt \\
&\leq \int_0^1 \|D\mathbf{h}(\mathbf{x} + t(\mathbf{y} - \mathbf{x}))\| dt |\mathbf{y} - \mathbf{x}| \\
&< M|\mathbf{y} - \mathbf{x}|.
\end{aligned}
$$

This proves the lemma.

We want to show that $\mathbf{h}$ maps Lebesgue measurable sets to Lebesgue measurable sets. This is done in two steps. First we show $\mathbf{h}$ maps sets of measure zero to sets of measure zero. Next we use continuity of $\mathbf{h}$ and regularity of Lebesgue measure to finish the argument.

Lemma 14.13 *If $A \subseteq U$ and $m(A) = 0$, then $\overline{m}(\mathbf{h}(A)) = 0$ and so $\mathbf{h}(A)$ is Lebesgue measurable.*

Proof: Let $W_k \equiv \cup \left\{ B\left(\mathbf{x}, \frac{1}{k}\right) : \mathbf{x} \notin U \right\}$ and let $K_k \equiv W_k^C \cap \overline{B(\mathbf{0}, k)}$. Then $K_k \uparrow U$, and K_k is compact. Let $r_k, M_k > 0$ be the constants of Lemma 14.12 associated with K_k, let $A_k = K_k \cap A$, and let $O_k \subseteq U$ be open, bounded, $O_k \supseteq A_k$, and $m(O_k) < \varepsilon$. For each $\mathbf{x} \in A_k$, let $B(\mathbf{x}, r_{\mathbf{x}}) \equiv B_{\mathbf{x}}$ be such that

$$
0 < 5r_{\mathbf{x}} < r_k \ , \ B_{\mathbf{x}} \subseteq O_k.
$$

By Vitali's covering theorem, there exists

$$
\{B_i\}_{i=1}^\infty \subseteq \{B_{\mathbf{x}} : \mathbf{x} \in A_k\}
$$

with $B_i \cap B_j = \emptyset$, and

$$
A_k \subseteq \bigcup_{i=1}^\infty \hat{B}_i.
$$

Then

$$
\overline{m}(\mathbf{h}(A_k)) \leq \sum_{i=1}^\infty \overline{m}(\mathbf{h}(\hat{B}_i)) \leq \sum_{i=1}^\infty M_k^n 5^n m(B_i)
$$

$$
\leq M_k^n 5^n m(O_k) \leq M_k^n 5^n \varepsilon.
$$

Since ε is arbitrary, $\overline{m}(\mathbf{h}(A_k)) = 0$. Now $\mathbf{h}(A) = \cup_{k=1}^\infty \mathbf{h}(A_k)$ and so $\overline{m}(\mathbf{h}(A)) = 0$ also. This shows $\mathbf{h}(A)$ is Lebesgue measurable.

In the next corollary, the regularity of Lebesgue measure and Lemma 14.13 are used to show $\mathbf{h}$ maps Lebesgue measurable sets to Lebesgue measurable sets.

The change of variables theorem for multiple integrals

Corollary 14.14 *If A is Lebesgue measurable, then so is $\mathbf{h}(A)$.*

Proof: Assume first that $m(A) < \infty$. Let $K_k \subseteq A$, K_k compact,

$$K_k \subseteq K_{k+1}, \ m(A) = \lim_{k \to \infty} m(K_k).$$

Let $F = \cup_{k=1}^{\infty} K_k$. Then $F \subseteq A$, $m(A \setminus F) = 0$, $\mathbf{h}(F) = \cup_{k=1}^{\infty} \mathbf{h}(K_k)$. Thus $\mathbf{h}(F)$ is measurable because it is a countable union of compact sets. Now $\mathbf{h}(A) = \mathbf{h}(F) \cup \mathbf{h}(A \setminus F)$. By Lemma 14.13, $\overline{m} \ (\mathbf{h}(A \setminus F)) = 0$ and so $\mathbf{h}(A \setminus F)$ is Lebesgue measurable. Since $\mathbf{h}(A)$ is the union of two Lebesgue measurable sets, it follows $\mathbf{h}(A)$ is also Lebesgue measurable. This proves the corollary if $m(A) < \infty$.

For arbitrary $A \subseteq U$, let $K_k \uparrow U$, K_k compact, and let $A_k = A \cap K_k$. Then $\mathbf{h}(A) = \cup_{k=1}^{\infty} \mathbf{h}(A_k)$ and $\mathbf{h}(A_k)$ is measurable so $\mathbf{h}(A)$ is also measurable. This proves the corollary.

Next we will use this information to present a measure which is absolutely continuous with respect to Lebesgue measure.

Corollary 14.15 *Let $\mathbf{h}$ be 1 - 1 and let $\mu(A) = m(\mathbf{h}(A))$ for all $A \in \mathcal{S}$. Then μ is a measure and $\mu << m$.*

Corollary 14.16 *There exists $f \geq 0$, f measurable, $f\mathcal{X}_K \in L^1$ whenever K is a compact subset of U, and for all $E \subseteq U$, $E \in \mathcal{S}$,*

$$m(\mathbf{h}(E)) = \mu(E) = \int_E f dx. \tag{11}$$

Proof: By the corollary to the Radon Nikodym Theorem in Chapter 13, there exists $f \geq 0$, f measurable such that Formula 11 holds. This is because we can let $K_k \uparrow U$, K_k compact. Then μ and m are both finite on K_k. The condition that $f\mathcal{X}_K \in L^1$ follows from Formula 11 and this proves the corollary.

Next we identify f. Let $\mathbf{x}$ be a Lebesgue point of f. Then by the fundamental theorem of calculus,

$$\frac{m(\mathbf{h}(B(\mathbf{x}, r)))}{m(B(\mathbf{x}, r))} = \frac{1}{m(B(\mathbf{x}, r))} \int_{B(\mathbf{x}, r)} f(\mathbf{y}) dy \to f(\mathbf{x}) \tag{12}$$

as $r \to 0$. For $\mathbf{y} \in B(\mathbf{x}, r)$,

$$\mathbf{h}(\mathbf{y}) = \mathbf{h}(\mathbf{x}) + D\mathbf{h}(\mathbf{x})(\mathbf{y} - \mathbf{x}) + o(|\mathbf{y} - \mathbf{x}|)$$

where $|o(|\mathbf{y} - \mathbf{x}|)| < \varepsilon r$ whenever r is small enough. We always let r be small enough. Hence

$$\mathbf{h}(B(\mathbf{x}, r)) \subseteq \mathbf{h}(\mathbf{x}) + D\mathbf{h}(\mathbf{x})B(0, r) + B(0, \varepsilon r).$$

Case 1: $\det(Dh(x)) = 0$.

Then $Dh(x)B(0, r)$ lies in a set, D, of diameter no larger than

$$2\|Dh(x)\|r$$

which is a subset of an $n - 1$ dimensional subspace of $\mathbb{R}^n$. Hence

$$
\begin{aligned}
m(h(B(x, r))) &\leq m(D + B(0, \varepsilon r)) \\
&\leq (2\|Dh(x)\|r + 2\varepsilon r)^{n-1} 2\varepsilon r \\
&= C_n r^n \varepsilon.
\end{aligned}
$$

From Formula 12,

$$0 \leq f(x) = \lim_{r \to 0} \frac{m(h(B(x, r)))}{m(B(x, r))} \leq \frac{C_n r^n \varepsilon}{\alpha(n) r^n} = C_n \varepsilon.$$

Since ε is arbitrary,

$$f(x) = 0 = |\det Dh(x)|.$$

Case 2: $\det Dh(x) \neq 0$.

Then for $y \in B(x, r)$,

$$h(y) - h(x) = Dh(x)[y - x + (Dh(x))^{-1} o(|y - x|)].$$

Now for r small enough,

$$|(Dh(x))^{-1} o(|y - x|)| < \varepsilon r,$$

whenever $y \in B(x, r)$. Therefore, for small r,

$$
\begin{aligned}
h(B(x, r)) - h(x) &\subseteq Dh(x)[B(0, r) + B(0, \varepsilon r)] \\
&\subseteq Dh(x)[B(0, r(1 + \varepsilon))].
\end{aligned}
$$

It follows from this and the theorem on change of variables for a constant linear transformation in Chapter 11 that

$$\frac{m(h(B(x, r)))}{m(B(x, r))} \leq |\det Dh(x)|(1 + \varepsilon)^n.$$

From Formula 12, and since $\varepsilon > 0$ is arbitrary,

$$f(x) \leq |\det Dh(x)|.$$

This has proved the following lemma.

The change of variables theorem for multiple integrals

Lemma 14.17 *If E is Lebesgue measurable and* h *is 1 - 1 and* C^1,

$$\int_V \mathcal{X}_{\mathbf{h}(E)}(\mathbf{y})dy = \int_E f(\mathbf{x})dx \le \int_U \mathcal{X}_E(\mathbf{x})|\det D\mathbf{h}(\mathbf{x})|dx.$$

Corollary 14.18 *Let* $g : U \to [0, \infty]$ *be measurable and let* h *be* $1 - 1$ *and* C^1, $\mathbf{h} : U \to V$. *Then*

$$\int_U g(\mathbf{x})|\det D\mathbf{h}(\mathbf{x})|dx \ge \int_V g(\mathbf{h}^{-1}(\mathbf{y}))dy. \tag{13}$$

Proof: First we note that $g \circ \mathbf{h}^{-1}$ is measurable because if O is open in $[0, \infty]$,

$$
\begin{aligned}
(g \circ \mathbf{h}^{-1})^{-1}(O) &= \mathbf{h}(g^{-1}(O)) = \mathbf{h}(\text{measurable set}) \\
&= \text{measurable set}
\end{aligned}
$$

by Corollary 14.14. If $g = \mathcal{X}_E$, then

$$\mathcal{X}_E(\mathbf{h}^{-1}(\mathbf{y})) = \mathcal{X}_{\mathbf{h}(E)}(\mathbf{y})$$

and so by Lemma 14.17,

$$\int_V \mathcal{X}_E(\mathbf{h}^{-1}(\mathbf{y}))dy = \int_V \mathcal{X}_{\mathbf{h}(E)}(\mathbf{y})dy \le \int_U \mathcal{X}_E(\mathbf{x})|\det D\mathbf{h}(\mathbf{x})|dx.$$

Thus Formula 13 holds for all simple functions. Approximating g with an increasing sequence of simple functions, the Monotone Convergence Theorem implies Formula 13.

In fact, we can easily get more than this. This is done in the next corollary.

Corollary 14.19 *Let* V *be open and let* h *and* $\mathbf{h}^{-1}$ *both be* C^1. *Then*

$$\int_V g(\mathbf{y})dy = \int_U g(\mathbf{h}(\mathbf{x}))|\det D\mathbf{h}(\mathbf{x})|dx$$

whenever $g \ge 0$ *is measurable,* $g : V \to [0, \infty]$.

Proof: $g \circ \mathbf{h}$ is measurable because

$$
\begin{aligned}
(g \circ \mathbf{h})^{-1}(\text{open}) &= \mathbf{h}^{-1}(g^{-1}(\text{open})) \\
&= \mathbf{h}^{-1}(\text{measurable}) \\
&= \text{measurable}
\end{aligned}
$$

by Corollary 14.14. Therefore Formula 13 applied to $g \circ \mathbf{h}$ in place of g yields

$$\int_V g(\mathbf{y})dy \le \int_U g(\mathbf{h}(\mathbf{x}))|\det D\mathbf{h}(\mathbf{x})|dx. \tag{14}$$

255

We can reverse the roles of U and V and $\mathbf{h}$ and $\mathbf{h}^{-1}$. Thus

$$
\begin{aligned}
\int_V g(\mathbf{y})dy &\leq \int_U g(\mathbf{h}(\mathbf{x}))|\det\ D\mathbf{h}(\mathbf{x})|dx \\
&\leq \int g(\mathbf{h}(\mathbf{h}^{-1}(\mathbf{y}))|\det\ D\mathbf{h}(\mathbf{h}^{-1}(\mathbf{y})||\det(D\mathbf{h}^{-1})(\mathbf{y})|dy \\
&= \int_V g(\mathbf{y})|\det(\mathbf{h}\circ\mathbf{h}^{-1})(\mathbf{y})|dy \\
&= \int_V g(\mathbf{y})dy.
\end{aligned}
$$

Therefore we actually obtain equality in Formula 14. This proves Corollary 14.19.

We want to remove some assumptions. We only want to assume $\mathbf{h}$ is 1-1 and C^1 and drop the assumption that V is open and $\mathbf{h}^{-1}$ is C^1. To do so we need another lemma.

Lemma 14.20 *(Sard) Let $\mathbf{h} : U \to V$ be C^1. Let $S = \{\mathbf{x} \in U : \det(\ D\mathbf{h}(\mathbf{x})) = 0\}$. Then $m(\mathbf{h}(S)) = 0$.*

Proof: First observe that S is closed because $D\mathbf{h}$ is continuous. Thus there are no measurability questions about $\mathbf{h}(S)$. Let K be a compact subset of U and let O be a bounded open set,

$$
U \supseteq O \supseteq K \cap S.
$$

For each $\mathbf{x} \in K \cap S$, let $B_\mathbf{x}$ be a ball centered at $\mathbf{x}$ with radius $r_\mathbf{x}$ such that $\mathbf{x} \in \widehat{B}_\mathbf{x} \subseteq O$. Since $D\mathbf{h}(\mathbf{x})$ maps $\mathbb{R}^n$ into an $n-1$ dimensional subspace, we can also assume that $B_\mathbf{x}$ is small enough that

$$
m(\mathbf{h}(\widehat{B}_\mathbf{x})) \leq M^{n-1}r_\mathbf{x}^n 5^n \varepsilon
$$

where $M = 2\max\{\|D\mathbf{h}(\mathbf{x})\| : \mathbf{x} \in K\} + 1$ and ε is a small positive number. By the Vitali Covering Theorem, there are disjoint balls $\{B_i\} \subseteq \{B_\mathbf{x}\}_{\mathbf{x}\in K\cap S}$ such that the $\widehat{B}$ cover $S \cap K$.

$$
\begin{aligned}
m(\mathbf{h}(S\cap K)) &\leq \sum_{i=1}^\infty m(\mathbf{h}(\widehat{B}_i)) \\
&\leq (\sum_{i=1}^\infty r_i^n)M^{n-1}5^n\varepsilon \\
&\leq C(\sum_{i=1}^\infty m(B_i))M^{n-1}5^n\varepsilon \\
&\leq CM^{n-1}5^n m(O)\varepsilon
\end{aligned}
$$

where C depends only on n. Since $m(O)$ is finite and $\varepsilon > 0$ is arbitrary, this shows

$$m(\mathbf{h}(S \cap K)) = 0.$$

Since this holds for all compact subsets K, it follows $m(\mathbf{h}(S)) = 0$ and this proves the lemma.

Theorem 14.21 *(Change of Variables) Let $\mathbf{h}$ be $C^1(U)$ and 1-1. Then if $g \geq 0$ and g is Lebesgue measurable,*

$$\int_{\mathbf{h}(U)} g(\mathbf{y})dy = \int_U g(\mathbf{h}(\mathbf{x}))|\det\ D\mathbf{h}(\mathbf{x})|dx.$$

Proof: Let

$$U_+ = \{\mathbf{x} : |\det D\mathbf{h}(\mathbf{x})| > 0\}, \quad S = \{\mathbf{x} : |\det D\mathbf{h}(\mathbf{x})| = 0\}. \tag{15}$$

By the Inverse Function Theorem, $\mathbf{h}$ is 1-1 on U_+, $\mathbf{h}(U_+)$ is open, and $\mathbf{h}^{-1}$ is C^1 on $\mathbf{h}(U_+)$. By Lemma 14.20, $m(\mathbf{h}(S)) = 0$ and so

$$\int_{\mathbf{h}(U)} g(\mathbf{y})dy = \int_{\mathbf{h}(U_+)} g(\mathbf{y})dy.$$

Thus,

$$\int_{\mathbf{h}(U)} g(\mathbf{y})dy = \int_{\mathbf{h}(U_+)} g(\mathbf{y})dy = \int_{U_+} g(\mathbf{h}(\mathbf{x}))|\det D\mathbf{h}(\mathbf{x})|dx$$
$$= \int_U g(\mathbf{h}(\mathbf{x}))|\det D\mathbf{h}(\mathbf{x})|dx.$$

This proves the theorem.

Next we give a version of this theorem which considers the case where $\mathbf{h}$ is only C^1, not necessarily 1-1. For U_+ and S given in Formula 15, Lemma 14.20 implies $m(\mathbf{h}(S)) = 0$. For $\mathbf{x} \in U_+$, the inverse function theorem implies there exists an open set $B_{\mathbf{x}}$ such that

$$\mathbf{x} \in B_{\mathbf{x}} \subseteq U_+, \ \mathbf{h} \text{ is } 1-1 \text{ on } B_{\mathbf{x}}.$$

Let $\{B_i\}$ be a countable subset of $\{B_{\mathbf{x}}\}_{\mathbf{x} \in U_+}$ such that

$$U_+ = \cup_{i=1}^{\infty} B_i.$$

Let $E_1 = B_1$. If $E_1, \cdots, E_k$ have been chosen, $E_{k+1} = B_{k+1} \setminus \cup_{i=1}^k E_i$. Thus

$$\cup_{i=1}^{\infty} E_i = U_+, \ \mathbf{h} \text{ is } 1-1 \text{ on } E_i, \ E_i \cap E_j = \emptyset,$$

and each E_i is a Borel set contained in the open set B_i. Now we define

$$n(\mathbf{y}) = \sum_{i=1}^{\infty} \mathcal{X}_{\mathbf{h}(E_i)}(\mathbf{y}) + \mathcal{X}_{\mathbf{h}(S)}(\mathbf{y}).$$

Thus $n(\mathbf{y}) \geq 0$ and is Borel measurable.

Lemma 14.22 *Let $F \subseteq V = \mathbf{h}(U)$ be measurable. Then*

$$\int_{\mathbf{h}(U)} n(\mathbf{y}) \mathcal{X}_F(\mathbf{y}) dy = \int_U \mathcal{X}_F(\mathbf{h}(\mathbf{x})) |\det D\mathbf{h}(\mathbf{x})| dx.$$

Proof: Using Lemma 14.20 and the Monotone Convergence Theorem or Fubini's Theorem,

$$
\begin{aligned}
\int_{\mathbf{h}(U)} n(\mathbf{y}) \mathcal{X}_F(\mathbf{y}) dy &= \int_{\mathbf{h}(U)} \left(\sum_{i=1}^{\infty} \mathcal{X}_{\mathbf{h}(E_i)}(\mathbf{y}) + \mathcal{X}_{\mathbf{h}(S)}(\mathbf{y}) \right) \mathcal{X}_F(\mathbf{y}) dy \\
&= \sum_{i=1}^{\infty} \int_{\mathbf{h}(U)} \mathcal{X}_{\mathbf{h}(E_i)}(\mathbf{y}) \mathcal{X}_F(\mathbf{y}) dy \\
&= \sum_{i=1}^{\infty} \int_{\mathbf{h}(B_i)} \mathcal{X}_{\mathbf{h}(E_i)}(\mathbf{y}) \mathcal{X}_F(\mathbf{y}) dy \\
&= \sum_{i=1}^{\infty} \int_{B_i} \mathcal{X}_{E_i}(\mathbf{x}) \mathcal{X}_F(\mathbf{h}(\mathbf{x})) |\det D\mathbf{h}(\mathbf{x})| dx \\
&= \sum_{i=1}^{\infty} \int_U \mathcal{X}_{E_i}(\mathbf{x}) \mathcal{X}_F(\mathbf{h}(\mathbf{x})) |\det D\mathbf{h}(\mathbf{x})| dx \\
&= \int_U \sum_{i=1}^{\infty} \mathcal{X}_{E_i}(\mathbf{x}) \mathcal{X}_F(\mathbf{h}(\mathbf{x})) |\det D\mathbf{h}(\mathbf{x})| dx
\end{aligned}
$$

$$= \int_{U_+} \mathcal{X}_F(\mathbf{h}(\mathbf{x})) |\det D\mathbf{h}(\mathbf{x})| dx = \int_U \mathcal{X}_F(\mathbf{h}(\mathbf{x})) |\det D\mathbf{h}(\mathbf{x})| dx.$$

This proves the lemma.

Definition 14.23 For $\mathbf{y} \in V = \mathbf{h}(U)$,

$$\#(\mathbf{y}) = |\mathbf{h}^{-1}(\mathbf{y})|.$$

Thus $\#(\mathbf{y}) \equiv$ number of elements in $\mathbf{h}^{-1}(\mathbf{y})$.

We observe that

$$\#(\mathbf{y}) = n(\mathbf{y}) \quad \text{a.e.} \tag{16}$$

And thus $\#$ is a measurable function. This follows because $n(\mathbf{y}) = \#(\mathbf{y})$ if $\mathbf{y} \notin \mathbf{h}(S)$, a set of measure 0.

Theorem 14.24 *Let $g \geq 0$, g measurable, and let $\mathbf{h}$ be $C^1(U)$. Then*

$$\int_{\mathbf{h}(U)} \#(\mathbf{y}) g(\mathbf{y}) dy = \int_U g(\mathbf{h}(\mathbf{x})) |\det D\mathbf{h}(\mathbf{x})| dx. \tag{17}$$

Proof: From Formula 16 and Lemma 14.22, Formula 17 holds for all g, a nonnegative simple function. Approximating an arbitrary $g \geq 0$ with an increasing pointwise convergent sequence of simple functions yields Formula 17 for $g \geq 0$, g measurable. This proves the theorem.

14.4   Exercises

1. Let E be a Lebesgue measurable set. We say $\mathbf{x} \in E$ is a point of density if
    $$\lim_{r \to 0} \frac{m(E \cap B(\mathbf{x}, r))}{m(B(\mathbf{x}, r))} = 1.$$
    Show that a.e. point of E is a point of density.

2. Let $(\Omega, \mathcal{S}, \mu)$ be any σ finite measure space, $f \geq 0$, f real-valued, and measurable. Let ϕ be an increasing C^1 function with $\phi(0) = 0$. Show
    $$\int_{\Omega} \phi \circ f d\mu = \int_0^\infty \phi'(t) \mu([f(x) > t]) dt.$$
    Hint:
    $$\int_{\Omega} \phi(f(x)) d\mu = \int_{\Omega} \int_0^{f(x)} \phi'(t) dt d\mu = \int_{\Omega} \int_0^\infty \mathcal{X}_{[0,f(x))}(t) \phi'(t) dt d\mu.$$
    Argue $\phi'(t) \mathcal{X}_{[0,f(x))}(t)$ is product measurable and use Fubini's theorem. The function $t \to \mu([f(x) > t])$ is called the distribution function.

3. Let f be in $L^1_{loc}(\mathbb{R}^n)$. Show Mf is Borel measurable.

4. If $f \in L^p, 1 < p < \infty$, show $Mf \in L^p$ and
    $$\|Mf\|_p \leq A(p, n)\|f\|_p.$$
    Hint: Let
    $$f_1(\mathbf{x}) \equiv \begin{cases} f(\mathbf{x}) & \text{if } |f(\mathbf{x})| > \alpha/2, \\ 0 & \text{if } |f(\mathbf{x})| \leq \alpha/2. \end{cases}$$
    Argue $[Mf(\mathbf{x}) > \alpha] \subseteq [Mf_1(\mathbf{x}) > \alpha/2]$. Then by Problem 2,
    $$\int (Mf)^p dx = \int_0^\infty p\alpha^{p-1} m([Mf > \alpha]) d\alpha$$
    $$\leq \int_0^\infty p\alpha^{p-1} m([Mf_1 > \alpha/2]) d\alpha.$$

Now use Theorem 14.4 and Fubini's Theorem as needed.

5. Show $|f(\mathbf{x})| \leq Mf(\mathbf{x})$ at every Lebesgue point of f whenever $f \in L^1(\mathbb{R}^n)$.

6. The result of this Problem is sometimes called the Vitali Covering Theorem. It will be used in Chapter 20. Let $E \subseteq \mathbb{R}^n$ be Lebesgue measurable, $m(E) < \infty$, and let $\mathcal{F}$ be a collection of balls that cover E in the sense of Vitali. This means that if $\mathbf{x} \in E$ and $\varepsilon > 0$, then there exists $B \in \mathcal{F}$, diameter of $B < \varepsilon$ and $\mathbf{x} \in B$. Show there exists a countable sequence of disjoint balls of $\mathcal{F}$, $\{B_j\}$, such that $m(E \setminus \cup_{j=1}^\infty B_j) = 0$. **Hint:** Let $E \subseteq U$, U open and

$$m(E) > (1 - 10^{-n})m(U).$$

Let $\{B_j\}$ be disjoint,

$$E \subseteq \cup_{j=1}^\infty \hat{B}_j, \ B_j \subseteq U.$$

Thus

$$m(E) \leq 5^n m(\cup_{j=1}^\infty B_j).$$

Then

$$m(E) > (1 - 10^{-n})m(U)$$
$$\geq (1 - 10^{-n})[m(E \setminus \cup_{j=1}^\infty B_j) + m(\cup_{j=1}^\infty B_j)]$$
$$\geq (1 - 10^{-n})[m(E \setminus \cup_{j=1}^\infty B_j) + 5^{-n}m(E)].$$

Hence

$$m(E \setminus \cup_{j=1}^\infty B_j) \leq (1 - 10^{-n})^{-1}(1 - (1 - 10^{-n})5^{-n})m(E).$$

Let $(1 - 10^{-n})^{-1}(1 - (1 - 10^{-n})5^{-n}) < \theta < 1$ and pick N_1 large enough that

$$\theta m(E) \geq m(E \setminus \cup_{j=1}^{N_1} \bar{B}_j).$$

Let $\mathcal{F}_1 = \{B \in \mathcal{F} : B_j \cap B = \emptyset, \ j = 1, \cdots, N_1\}$. If $E \setminus \cup_{j=1}^{N_1} \bar{B}_j \neq \emptyset$, then $\mathcal{F}_1 \neq \emptyset$ and covers $E \setminus \cup_{j=1}^{N_1} \bar{B}_j$ in the sense of Vitali. Repeat the same argument, letting $E \setminus \cup_{j=1}^{N_1} \bar{B}_j$ play the role of E.

7. Suppose λ is a Radon measure on $\mathbb{R}^n$, and $\lambda(S) < \infty$ where $m_n(S) = 0$ and $\lambda(E) = \lambda(E \cap S)$. (If $\lambda(E) = \lambda(E \cap S)$ where $m_n(S) = 0$ we say $\lambda \perp m_n$.) Show that for m_n a.e. $\mathbf{x}$ the following holds. If $B_i \downarrow \{\mathbf{x}\}$, then $\lim_{i \to \infty} \frac{\lambda(B_i)}{m_n(B_i)} = 0$. **Hint:** You might try this. Set $\varepsilon, r > 0$, and let

$$E_\varepsilon = \{\mathbf{x} \in S^C : \text{ there exists } \{B_i^{\mathbf{x}}\}, B_i^{\mathbf{x}} \downarrow \{\mathbf{x}\} \text{ with } \frac{\lambda(B_i^{\mathbf{x}})}{m_n(B_i^{\mathbf{x}})} \geq \varepsilon\}.$$

Let K be compact, $\lambda(S \setminus K) < r\varepsilon$. Let $\mathcal{F}$ consist of those balls just described that do not intersect K and which have radius < 1. This is a Vitali cover of E_ε. Let $B_1, \cdots, B_k$ be disjoint balls from $\mathcal{F}$ and

$$\overline{m}_n(E_\varepsilon \setminus \cup_{i=1}^k B_i) < r.$$

Then

$$\overline{m}_n(E_\varepsilon) < r + \sum_{i=1}^k m_n(B_i) < r + \varepsilon^{-1} \sum_{i=1}^k \lambda(B_i) =$$

$$r + \varepsilon^{-1} \sum_{i=1}^k \lambda(B_i \cap S) \le r + \varepsilon^{-1}\lambda(S \setminus K) < 2r.$$

Since r is arbitrary, $m_n(E_\varepsilon) = 0$. Consider $E = \cup_{k=1}^\infty E_{k-1}$ and let $\mathbf{x} \notin S \cup E$.

8. ↑ Is it necessary to assume $\lambda(S) < \infty$ in Problem 7? Explain.

9. ↑ Let S be an increasing function on $\mathbb{R}$ which is right continuous,

$$\lim_{x \to -\infty} S(x) = 0,$$

and S is bounded. Let λ be the measure representing $\int f \, dS$. Thus $\lambda((-\infty, x]) = S(x)$. Suppose $\lambda \perp m$. Show $S'(x) = 0$ m a.e. **Hint:**

$$0 \le h^{-1}(S(x+h) - S(x))$$

$$= \frac{\lambda((x, x+h])}{m((x, x+h])} \le 3\frac{\lambda((x-h, x+2h))}{m((x-h, x+2h))}.$$

Now apply Problem 7. Similarly $h^{-1}(S(x) - S(x-h)) \to 0$.

10. ↑ Let f be increasing, bounded above and below, and right continuous. Show $f'(x)$ exists a.e. **Hint:** See Problem 5 of Chapter 13.

11. ↑ Suppose $|f(x) - f(y)| \le K|x - y|$. Show there exists $g \in L^\infty(\mathbb{R})$, $\|g\|_\infty \le K$, and

$$f(y) - f(x) = \int_x^y g(t)dt.$$

Hint: Let $F(x) = Kx + f(x)$ and let λ be the measure representing $\int f \, dF$. Show $\lambda << m$. What does this imply about the differentiability of a Lipschitz continuous function?

12. ↑ Let f be increasing. Show $f'(x)$ exists a.e.

13. Let $f(x) = x^2$. Thus $\int_{-1}^{1} f(x)dx = 2/3$. Let's change variables. $u = x^2$, $du = 2xdx = 2u^{1/2}dx$. Thus

$$2/3 = \int_{-1}^{1} x^2 dx = \int_{1}^{1} u/2u^{1/2}du = 0.$$

Can this be done correctly using Theorem 14.24? Explain.

14. Consider the construction employed to obtain the Cantor set, but instead of removing the middle third interval, remove only enough that the sum of the lengths of all the open intervals which are removed is less than one. That which remains is called a fat Cantor set. Show it is a compact set which has measure greater than one which contains no interval and has the property that every point is a limit point of the set. Let P be such a fat Cantor set and consider

$$f(x) = \int_{0}^{x} \mathcal{X}_{PC}(t)\, dt.$$

Show that f is a strictly increasing function which has the property that its derivative equals zero on a set of positive measure.

Chapter 15
General Radon Measures

15.1 Besicovitch covering theorem

The fundamental theorem of calculus presented above for Lebesgue measures can be generalized to arbitrary Radon measures. It turns out that the same approach works if a different covering theorem is employed instead of the Vitali theorem. This covering theorem is the Besicovitch covering theorem which we present in this section. It is necessary because for a general Radon measure, μ, it is no longer the case that the measure is translation invariant. This implies that there is no way to estimate $\mu\left(\widehat{B}\right)$ in terms of $\mu(B)$ and thus the Vitali covering theorem is of no use. The following theorem is the Besicovitch covering theorem. Note that the balls in the covering are not enlarged as they are in the Vitali theorem.

Theorem 15.1 *There exists a constant N_n, depending only on n with the following property. If $\mathcal{F}$ is any collection of nonempty balls in $\mathbb{R}^n$ with*

$$\sup\{diam(B) : B \in \mathcal{F}\} < \infty$$

and if A is the set of centers of the balls in $\mathcal{F}$, then there exist subsets of $\mathcal{F}$, $\mathcal{G}_1, \cdots, \mathcal{G}_n$, such that each $\mathcal{G}_i$ is a countable collection of disjoint balls from $\mathcal{F}$ and

$$A \subseteq \cup_{i=1}^{N_n}\{B : B \in \mathcal{G}_i\}.$$

Proof: To begin with assume the set of centers A is bounded. Let

$$D \equiv \sup\{diam(B) : B \in \mathcal{F}\}.$$

Choose any $B_1 \equiv B(\mathbf{a}_1, r_1) \in \mathcal{F}$ such that $r_1 \geq \frac{3}{4}\left(\frac{D}{2}\right)$. Assuming that for $j \geq 2, B_1, \cdots, B_{j-1}$ have been chosen, choose B_j as follows. If $A_j \equiv A \setminus \cup_{i=1}^{j-1} B_j = \emptyset$, then stop and set $J \equiv j-1$. If $A_j \neq \emptyset$, then choose $B_j = B(\mathbf{a}_j, r_j) \in \mathcal{F}$ such that $\mathbf{a}_j \in A_j$ and

$$r_j \geq \frac{3}{4}\sup\{r : B(\mathbf{a}, r) \in \mathcal{F}, \mathbf{a} \in A_j\}.$$

If $A_j \neq \emptyset$ for all j, then set $J = \infty$. This process defines a sequence of balls from $\mathcal{F}$,

$$B_j = B(\mathbf{a}_j, r_j),$$

which is either finite or countably infinite.

Claim 1: If $i < j$, then $r_j \leq \frac{4}{3} r_i$.

Proof: Assume $i < j$. Then $A_j \subseteq A_i$ and therefore, $\mathbf{a}_j \in A_i$. Therefore

$$r_i \geq \frac{3}{4} \sup \{r : B(\mathbf{a}, r) \in \mathcal{F}, \mathbf{a} \in A_i\} \geq \frac{3}{4} r_j.$$

Claim 2: $\left\{ B\left(\mathbf{a}_j, \frac{r_j}{3}\right) \right\}_{j=1}^{J}$ are disjoint.

Proof: Assume $i < j$. Then $\mathbf{a}_j \notin B_i$ and so $|\mathbf{a}_i - \mathbf{a}_j| > r_i$. Now if there exists

$$\mathbf{x} \in B\left(\mathbf{a}_i, \frac{r_i}{3}\right) \cap B\left(\mathbf{a}_j, \frac{r_j}{3}\right),$$

then

$$|\mathbf{a}_i - \mathbf{a}_j| \leq |\mathbf{a}_i - \mathbf{x}| + |\mathbf{x} - \mathbf{a}_j| \leq \frac{1}{3}(r_i + r_j)$$

$$\leq \frac{1}{3}\left(r_i + \frac{4}{3} r_i\right) < r_i,$$

a contradiction.

Claim 3: If $J = \infty$, then $\lim_{j \to \infty} r_j = 0$.

Proof: Since $A + B(\mathbf{0}, D)$ is bounded, $m_n(A + B(\mathbf{0}, D)) < \infty$. Therefore,

$$\sum_j \left(\frac{1}{3}\right)^n \alpha(n) r_j^n = \sum_j m_n\left(B\left(\mathbf{a}_j, \frac{r_j}{3}\right)\right) \leq m_n(A + B(\mathbf{0}, D)) < \infty$$

and so $r_j \to 0$ as $j \to \infty$.

Claim 4: $A \subseteq \cup_{j=1}^{J} B_j$.

Proof: There are two cases depending on whether $J = \infty$ or is finite. Suppose first that $J = \infty$ and let $\mathbf{a} \in A$. There exists r such that $B(\mathbf{a}, r) \in \mathcal{F}$. Now, by Claim 3, there exists r_j such that $r_j < \frac{3}{4} r$. If $\mathbf{a} \notin \cup_{i=1}^{j-1} B_i$, then $\mathbf{a} \in A_j$ and

$$r_j \geq \frac{3}{4} \sup \{r : B(\mathbf{a}, r) \in \mathcal{F}, \mathbf{a} \in A_j\} \geq \frac{3}{4} r,$$

a contradiction. Thus

$$\mathbf{a} \in \cup_{i=1}^{j-1} B_i \subseteq \cup_{i=1}^{\infty} B_i.$$

In the other case where $J < \infty$, there exists j such that $A_j = A \setminus \cup_{j=1}^{J} B_j = \emptyset$. But then $A \subseteq \cup_{j=1}^{J} B_j$.

Lemma 15.2 *There exists a constant M_n depending only on n such that for each $1 \leq k < J$, M_n exceeds the number of sets B_j for $j < k$ which have nonempty intersection with B_k.*

Proof: These sets B_j which intersect B_k are of two types. Either they have large radius, $r_j > 10 r_k$, or they have small radius, $r_j \leq 10 r_k$.

In this argument we will denote by $card\,(S)$ the number of elements in the set S. Define for fixed k,

$$I \equiv \{j : 1 \leq j < k,\ B_j \cap B_k \neq \emptyset,\ r_j \leq 10r_k\},$$

$$K \equiv \{j : 1 \leq j < k,\ B_j \cap B_k \neq \emptyset,\ r_j > 10r_k\}.$$

Claim 5: $B\left(\mathbf{a}_j, \frac{r_j}{3}\right) \subseteq B\left(\mathbf{a}_k, 15r_k\right)$ for $j \in I$.

Proof: Let $j \in I$. Then $B_j \cap B_k \neq \emptyset$ and $r_j \leq 10r_k$. Now if

$$\mathbf{x} \in B\left(\mathbf{a}_j, \frac{r_j}{3}\right),$$

then

$$|\mathbf{x} - \mathbf{a}_k| \ \leq \ |\mathbf{x} - \mathbf{a}_j| + |\mathbf{a}_j - \mathbf{a}_k| \leq \frac{r_j}{3} + r_j + r_k =$$

$$\frac{4}{3}r_j + r_k \ \leq \ \frac{43}{3}r_k < 15r_k.$$

Therefore, $B\left(\mathbf{a}_j, \frac{r_j}{3}\right) \subseteq B\left(\mathbf{a}_k, 15r_k\right)$.

Claim 6: $Card\,(I) \leq 60^n$.

Proof:

$$\alpha\,(n)\,15^n r_k^n \ = \ m_n\left(B\left(\mathbf{a}_k, 15r_k\right)\right) \geq \sum_{j \in I} m_n\left(B\left(\mathbf{a}_j, \frac{r_j}{3}\right)\right)$$

$$= \ \sum_{j \in I} \alpha\,(n)\left(\frac{r_j}{3}\right)^n \geq \sum_{j \in I} \alpha\,(n)\left(\frac{r_k}{4}\right)^n \left[\text{since } r_k \leq \frac{4}{3}r_j\right]$$

$$= \ card\,(I)\,\alpha\,(n)\left(\frac{r_k}{4}\right)^n$$

and so it follows that $card\,(I) \leq 60^n$ as claimed.

Claim 7: $Card\,(K) \leq L_n$ where L_n is a constant depending only on n.

Proof: Let B_i and B_j be balls from $\mathcal{F}$ such that i and j are in K with $i < j$. Consider the following diagram in which we will establish a lower bound on the angle θ.

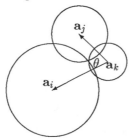

From the picture, and the assumption that $i, j \in K$,

$$\begin{cases} \mathbf{a}_k \notin B_j,\ r_i > 10r_k, \\ \mathbf{a}_j \notin B_i,\ r_j > 10r_k, \\ B_j \cap B_k \neq \emptyset,\ B_i \cap B_k \neq \emptyset. \end{cases}$$

Then

$$10r_k < r_i < |\mathbf{a}_i - \mathbf{a}_k| \le r_i + r_k, |\mathbf{a}_i - \mathbf{a}_j| \ge r_i,$$

and

$$10r_k < r_j < |\mathbf{a}_j - \mathbf{a}_k| \le r_j + r_k < \frac{4}{3}r_i + r_k.$$

By the law of cosines and the above,

$$
\begin{aligned}
\cos(\theta) &= \frac{|\mathbf{a}_i - \mathbf{a}_k|^2 + |\mathbf{a}_j - \mathbf{a}_k|^2 - |\mathbf{a}_i - \mathbf{a}_j|^2}{2|\mathbf{a}_i - \mathbf{a}_k||\mathbf{a}_j - \mathbf{a}_k|} \\
&\le \frac{(r_i + r_k)^2 + |\mathbf{a}_j - \mathbf{a}_k|^2 - r_i^2}{2r_i|\mathbf{a}_j - \mathbf{a}_k|} \\
&= \frac{2r_i r_k + r_k^2 + |\mathbf{a}_j - \mathbf{a}_k|^2}{2r_i|\mathbf{a}_j - \mathbf{a}_k|} \\
&= \frac{r_k}{|\mathbf{a}_j - \mathbf{a}_k|} + \frac{r_k^2}{2r_i|\mathbf{a}_j - \mathbf{a}_k|} + \frac{|\mathbf{a}_j - \mathbf{a}_k|}{2r_i}.
\end{aligned}
$$

By Claim 1 and the assumption that $r_i \ge 10r_k$ and $r_j \ge 10r_k$, the above is no larger than

$$
\begin{aligned}
&\frac{r_k}{r_j} + \frac{r_k^2}{2r_i r_j} + \frac{r_j + r_k}{2r_i} \\
&\le \frac{1}{10} + \frac{1}{200} + \left(\frac{4}{3}\right)\left(\frac{1}{2}\right) + \left(\frac{1}{20}\right) < .822.
\end{aligned}
$$

Therefore, $\theta \ge \cos^{-1}(.822) \equiv \theta_0$ whenever $i, j \in K$.

Now consider the ball $B(0,1)$. Choose $r_0 > 0$ such that if $\mathbf{x} \in \partial B(0,1)$ and $\mathbf{y}, \mathbf{z} \in B(\mathbf{x}, r_0)$, then the angle between the vectors $\mathbf{y}$ and $\mathbf{z}$ will be less than θ_0. Since $\partial B(0,1)$ is compact, there exists a finite subcover of $\partial B(0,1)$ consisting of such balls $B(\mathbf{x}, r_0)$ for $\mathbf{x} \in \partial B(0,1)$. Let L_n be the number of sets in this cover of $\partial B(0,1)$. Next define $\phi(\mathbf{t}) \equiv r_k \mathbf{t} + \mathbf{a}_k$. Clearly ϕ is a homeomorphism. Also ϕ maps $B(0,1)$ onto $B(\mathbf{a}_k, r_k)$ and maps $B(\mathbf{x}, r_0)$, $\mathbf{x} \in \partial B(0,1)$ onto $B(r_k \mathbf{x} + \mathbf{a}_k, r_0 r_k)$. Therefore,

$$\partial B(\mathbf{a}_k, r_k)$$

can be covered by L_n balls of the form

$$B(r_k \mathbf{x} + \mathbf{a}_k, r_0 r_k), \quad r_k \mathbf{x} + \mathbf{a}_k \in \partial B(\mathbf{a}_k, r_k).$$

Now if

$$\mathbf{p}, \mathbf{q} \in B(r_k \mathbf{x} + \mathbf{a}_k, r_0 r_k),$$

then $\mathbf{p} = r_k \mathbf{u} + \mathbf{a}_k$ and $\mathbf{q} = r_k \mathbf{v} + \mathbf{a}_k$ where

$$\mathbf{u}, \mathbf{v} \in B(\mathbf{x}_i, r_0)$$

where $B(\mathbf{x}_i, r_0)$ is one of the L_n balls that cover $B(0,1)$. Also $\mathbf{p} - \mathbf{a}_k = r_k\mathbf{u}$ and $\mathbf{q} - \mathbf{a}_k = r_k\mathbf{v}$. Hence, if α is the angle between the vectors $\mathbf{p} - \mathbf{a}_k$ and $\mathbf{q} - \mathbf{a}_k$, the angle between the vectors $\mathbf{u}$ and $\mathbf{v}$ is also equal to α. Thus $\alpha < \theta_0$.

We showed above that if $i, j \in K$ then the angle between the vectors $\mathbf{a}_j - \mathbf{a}_k$ and $\mathbf{a}_i - \mathbf{a}_k$ is greater than θ_0. Therefore, the rays from $\mathbf{a}_k$ having direction vectors $\mathbf{a}_j - \mathbf{a}_k$ and $\mathbf{a}_i - \mathbf{a}_k$ cannot both pass through the same ball $B(r_k\mathbf{x} + \mathbf{a}_k, r_0 r_k)$ which shows there are at most L_n intersections of $B(\mathbf{a}_k, r_k)$ with $B(\mathbf{a}_i, r_i)$ for $i \in K$. Therefore, $card(K) \leq L_n$ and we may take $M_n = L_n + 60^n + 1$. This completes the proof of Lemma 15.2.

Next we will subdivide the balls $\{B_i\}_{i=1}^{J}$ into M_n subsets $\mathcal{G}_1, \cdots, \mathcal{G}_{M_n}$ each of which consists of disjoint balls. We accomplish this in the following way. Let $B_1 \in \mathcal{G}_1$. If $B_1, \cdots, B_k$ have each been assigned to one of the sets $\mathcal{G}_1, \cdots, \mathcal{G}_{M_n}$, let $B_{k+1} \in \mathcal{G}_r$ where r is the smallest index having the property that B_{k+1} does not intersect any of the balls already in $\mathcal{G}_r$. There must exist such an index $r \in \{1, \cdots, M_n\}$ because otherwise $B_{k+1} \cap B_j \neq \emptyset$ for at least M_n values of $j < k+1$ contradicting Lemma 15.2. By Claim 4

$$A \subseteq \cup_{i=1}^{M_n} \{B : B \in \mathcal{G}_i\} = \cup_{j=1}^{J} B_j.$$

This proves Theorem 15.1 in the case where A is bounded. To complete the proof of this theorem, we remove this restriction.

Define

$$A_l \equiv A \cap \{\mathbf{x} \in \mathbb{R}^n : 10(l-1)D \leq |\mathbf{x}| < 10lD\}, \ l = 1, 2, \cdots$$

and

$$\mathcal{F}_l = \{B(\mathbf{a}, r) : B(\mathbf{a}, r) \in \mathcal{F} \text{ and } \mathbf{a} \in A_l\}.$$

Then since D is an upper bound for all the diameters of these balls,

$$(\cup\mathcal{F}_l) \cap (\cup\mathcal{F}_m) = \emptyset \tag{1}$$

whenever $m \geq l + 2$. Therefore, applying what was just shown to the pair $(A_l, \mathcal{F}_l)$, there exist subsets of $\mathcal{F}_l$, $\mathcal{G}_1^l \cdots \mathcal{G}_{M_n}^l$ such that each $\mathcal{G}_i^l$ is a countable collection of disjoint balls of $\mathcal{F}_l \subseteq \mathcal{F}$ and

$$A_l \subseteq \cup_{i=1}^{M_n} \{B : B \in \mathcal{G}_i^l\}.$$

Now let $\mathcal{G}_j \equiv \cup_{l=1}^{\infty} \mathcal{G}_j^{2l-1}$ for $1 \leq j \leq M_n$ and for $1 \leq j \leq M_n$, let $\mathcal{G}_{j+M_n} \equiv \cup_{l=1}^{\infty} \mathcal{G}_j^{2l}$. Thus, letting $N_n \equiv 2M_n$,

$$A = \cup_{l=1}^{\infty} A_{2l} \cup \cup_{l=1}^{\infty} A_{2l-1} \subseteq \cup_{j=1}^{N_n} \{B : B \in \mathcal{G}_j\}$$

and by Formula 1, each $\mathcal{G}_j$ is a countable set of disjoint balls of $\mathcal{F}$. This proves the Besicovitch covering theorem.

267

15.2 Differentiation with respect to Radon measures

In this section the Besicovitch covering theorem will be used to give a generalization of the Lebesgue differentiation theorem to general Radon measures. In what follows, μ will be a Radon measure,

$$Z \equiv \{\mathbf{x} \in \mathbb{R}^n : \mu(B(\mathbf{x},r)) = 0 \text{ for some } r > 0\},$$

$$\fint_{B(\mathbf{x},r)} f\,d\mu \equiv \begin{cases} 0 \text{ if } \mathbf{x} \in Z, \\ \frac{1}{\mu(B(\mathbf{x},r))} \int_{B(\mathbf{x},r)} f\,d\mu \text{ if } \mathbf{x} \notin Z, \end{cases}$$

and we define the maximal function $Mf : \mathbb{R}^n \to [0, \infty]$ by

$$Mf(\mathbf{x}) \equiv \sup_{r \le 1} \fint_{B(\mathbf{x},r)} |f|\,d\mu.$$

Lemma 15.3 *Z is measurable and $\mu(Z) = 0$.*

Proof: For each $\mathbf{x} \in Z$, there exists a ball $B(\mathbf{x},r)$ with $\mu(B(\mathbf{x},r)) = 0$. Let $\mathcal{C}$ be the collection of these balls. Since $\mathbb{R}^n$ has a countable basis, a countable subset, $\widetilde{\mathcal{C}}$, of $\mathcal{C}$ also covers Z. Let

$$\widetilde{\mathcal{C}} = \{B_i\}_{i=1}^\infty.$$

Then letting $\overline{\mu}$ denote the outer measure determined by μ,

$$\overline{\mu}(Z) \le \sum_{i=1}^\infty \overline{\mu}(B_i) = \sum_{i=1}^\infty \mu(B_i) = 0$$

by Lemma 9.6 of Chapter 9. Therefore, Z is measurable and has measure zero as claimed.

Theorem 15.4 *Let μ be a Radon measure and let $f \in L^1(\mathbb{R}^n, \mu)$. Then*

$$\lim_{r \to 0} \fint_{B(\mathbf{x},r)} |f(\mathbf{y}) - f(\mathbf{x})|\,d\mu(\mathbf{y}) = 0$$

for μ a.e. $\mathbf{x} \in \mathbb{R}^n$.

Proof: We begin the proof with the following claim.
Claim 1: Let $A = \{\mathbf{x} \in \mathbb{R}^n : Mf(\mathbf{x}) > \epsilon\}$. Then $\overline{\mu}(A) \le N_n \epsilon^{-1} \|f\|_1$.
Proof: First note that $A \cap Z = \emptyset$. For each $\mathbf{x} \in A$ there exists a ball $B_{\mathbf{x}} = B(\mathbf{x},r_{\mathbf{x}})$ with $r_{\mathbf{x}} \le 1$ and

$$\mu(B_{\mathbf{x}})^{-1} \int_{B(\mathbf{x},r_{\mathbf{x}})} |f|\,d\mu > \epsilon.$$

Differentiation with respect to Radon measures

Let $\mathcal{F}$ be this collection of balls so that A is the set of centers of balls of $\mathcal{F}$. By the Besicovitch covering theorem,

$$A \subseteq \cup_{i=1}^{N_n} \{B : B \in \mathcal{G}_i\}$$

where $\mathcal{G}_i$ is a collection of disjoint balls of $\mathcal{F}$. Now for some i,

$$\overline{\mu}(A)/N_n \leq \mu(\cup\{B : B \in \mathcal{G}_i\})$$

because if this is not so, then

$$\overline{\mu}(A) \leq \sum_{i=1}^{N_n} \mu(\cup\{B : B \in \mathcal{G}_i\}) < \sum_{i=1}^{N_n} \frac{\overline{\mu}(A)}{N_n} = \overline{\mu}(A),$$

a contradiction. Therefore for this i,

$$\frac{\overline{\mu}(A)}{N_n} \leq \mu(\cup\{B : B \in \mathcal{G}_i\}) = \sum_{B \in \mathcal{G}_i} \mu(B) \leq \sum_{B \in \mathcal{G}_i} \epsilon^{-1} \int_B |f|\, d\mu$$

$$\leq \epsilon^{-1} \int_{\mathbb{R}^n} |f|\, d\mu = \epsilon^{-1} \|f\|_1 \, .$$

This shows Claim 1.

Claim 2: If g is any continuous function defined on $\mathbb{R}^n$, then

$$\lim_{r \to 0} \fint_{B(\mathbf{x},r)} |g(\mathbf{y}) - g(\mathbf{x})|\, d\mu(\mathbf{y}) = 0$$

and if $\mathbf{x} \notin Z$,

$$\lim_{r \to 0} \frac{1}{\mu(B(\mathbf{x},r))} \int_{B(\mathbf{x},r)} g(\mathbf{y})\, d\mu(\mathbf{y}) = g(\mathbf{x}). \tag{2}$$

Proof: If $\mathbf{x} \in Z$ there is nothing to prove. If $\mathbf{x} \notin Z$, then since g is continuous at $\mathbf{x}$, whenever r is small enough,

$$\fint_{B(\mathbf{x},r)} |g(\mathbf{y}) - g(\mathbf{x})|\, d\mu(\mathbf{y})$$

$$= \frac{1}{\mu(B(\mathbf{x},r))} \int_{B(\mathbf{x},r)} |g(\mathbf{y}) - g(\mathbf{x})|\, d\mu(\mathbf{y})$$

$$\leq \frac{1}{\mu(B(\mathbf{x},r))} \int_{B(\mathbf{x},r)} \epsilon\, d\mu(\mathbf{y}) = \epsilon.$$

Formula 2 follows from the above and the triangle inequality. This proves the claim.

Now let $g \in C_c(\mathbb{R}^n)$ and $\mathbf{x} \notin Z$. Then

$$\fint_{B(\mathbf{x},r)} |f(\mathbf{y}) - f(\mathbf{x})| \, d\mu(\mathbf{y})$$

$$\leq \fint_{B(\mathbf{x},r)} |f(\mathbf{y}) - g(\mathbf{y})| \, d\mu(\mathbf{y})$$

$$+ \fint_{B(\mathbf{x},r)} |g(\mathbf{y}) - f(\mathbf{x})| \, d\mu(\mathbf{y}).$$

It follows

$$\limsup_{r \to 0} \fint_{B(\mathbf{x},r)} |f(\mathbf{y}) - f(\mathbf{x})| \, d\mu(\mathbf{y})$$

$$\leq \limsup_{r \to 0} \fint_{B(\mathbf{x},r)} |f(\mathbf{y}) - g(\mathbf{y})| \, d\mu(\mathbf{y})$$

$$+ \limsup_{r \to 0} \fint_{B(\mathbf{x},r)} |g(\mathbf{y}) - f(\mathbf{x})| \, d\mu(\mathbf{y}).$$

Now from Claim 2, (Formula 2), and the observation that

$$\mathbf{y} \to |g(\mathbf{y}) - f(\mathbf{x})|$$

is continuous,

$$= \limsup_{r \to 0} \fint_{B(\mathbf{x},r)} |f(\mathbf{y}) - g(\mathbf{y})| \, d\mu(\mathbf{y}) + |g(\mathbf{x}) - f(\mathbf{x})|$$

$$\leq M(|f - g|)(\mathbf{x}) + |g(\mathbf{x}) - f(\mathbf{x})|.$$

Letting

$$U(\mathbf{x}) = \limsup_{r \to 0} \fint_{B(\mathbf{x},r)} |f(\mathbf{y}) - f(\mathbf{x})| \, d\mu(\mathbf{y}),$$

it follows that

$$\{\mathbf{x} \in \mathbb{R}^n : U(\mathbf{x}) > \alpha\} \subseteq \left\{\mathbf{x} \in \mathbb{R}^n : M(|f - g|)(\mathbf{x}) > \frac{\alpha}{2}\right\}$$
$$\cup \left\{\mathbf{x} \in \mathbb{R}^n : |g(\mathbf{x}) - f(\mathbf{x})| > \frac{\alpha}{2}\right\}.$$

Now it is clear that

$$\overline{\mu}\left(\left\{\mathbf{x} \in \mathbb{R}^n : |g(\mathbf{x}) - f(\mathbf{x})| > \frac{\alpha}{2}\right\}\right) \leq \frac{2}{\alpha} \|f - g\|_1,$$

and so by Claim 1

$$\overline{\mu}(\{\mathbf{x} \in \mathbb{R}^n : U(\mathbf{x}) > \alpha\}) \leq N_n \frac{2}{\alpha} \|f - g\|_1 + \frac{2}{\alpha} \|f - g\|_1.$$

Since $g \in C_c(\mathbb{R}^n)$ is arbitrary and $C_c(\mathbb{R}^n)$ is dense in $L^1(\mathbb{R}^n, \mu)$ (because μ is a Radon measure), it follows that

$$\bar{\mu}(\{\mathbf{x} \in \mathbb{R}^n : U(\mathbf{x}) > \alpha\}) = 0$$

which implies $\{\mathbf{x} \in \mathbb{R}^n : U(\mathbf{x}) > \alpha\}$ is a measurable set of measure zero. Since $\alpha > 0$ is arbitrary, this also shows $\{\mathbf{x} \in \mathbb{R}^n : U(\mathbf{x}) > 0\} \equiv E$ is a measurable set of measure zero. By Lemma 15.3 the set $E \cup Z$ is a set of measure zero and if $\mathbf{x} \notin E \cup Z$, the above has shown

$$0 \le \liminf_{r \to 0} \fint_{B(\mathbf{x},r)} |f(\mathbf{y}) - f(\mathbf{x})| \, d\mu(y)$$

$$\le \limsup_{r \to 0} \fint_{B(\mathbf{x},r)} |f(\mathbf{y}) - f(\mathbf{x})| \, d\mu(y) = 0$$

which proves the theorem.

The following corollary is the main result. We refer to this as the Lebesgue Besicovitch Differentiation theorem.

Corollary 15.5 *If $f \in L^1_{loc}(\mathbb{R}^n, \mu)$,*

$$\lim_{r \to 0} \fint_{B(\mathbf{x},r)} |f(\mathbf{y}) - f(\mathbf{x})| \, d\mu(y) \quad \mu \ a.e. \ \mathbf{x}. \tag{3}$$

Proof: If f is replaced by $f\mathcal{X}_{B(\mathbf{0},k)}$ then the conclusion 3 holds for all $\mathbf{x} \notin F_k$ where F_k is a set of μ measure 0. Letting $k = 1, 2, \cdots$, and $F \equiv \cup_{k=1}^{\infty} F_k$, it follows that F is a set of measure zero and for any $\mathbf{x} \notin F$, and $k \in \{1, 2, \cdots\}$, 3 holds if f is replaced by $f\mathcal{X}_{B(\mathbf{0},k)}$. Picking any such $\mathbf{x}$, and letting $k > |\mathbf{x}| + 1$, this shows

$$\lim_{r \to 0} \fint_{B(\mathbf{x},r)} |f(\mathbf{y}) - f(\mathbf{x})| \, d\mu(y)$$

$$= \lim_{r \to 0} \fint_{B(\mathbf{x},r)} |f\mathcal{X}_{B(\mathbf{0},k)}(\mathbf{y}) - f\mathcal{X}_{B(\mathbf{0},k)}(\mathbf{x})| \, d\mu(y) = 0.$$

This proves the corollary.

15.3 Slicing measures

Let μ be a finite Radon measure. In this section we will show that we can write a formula of the following form.

$$\mu(F) = \int_F d\mu = \int_{\mathbb{R}^n} \int_{\mathbb{R}^m} \mathcal{X}_F(\mathbf{x}, \mathbf{y}) \, d\nu_{\mathbf{x}}(y) \, d\alpha(x)$$

where $\alpha(E) = \mu(E \times \mathbb{R}^m)$. When this is done, the measures, $\nu_\mathbf{x}$, are called slicing measures and we are showing that an integral with respect to μ can be written as an iterated integral in terms of the measure α and the slicing measures, $\nu_\mathbf{x}$. This is like going backwards in the construction of product measure. We are starting with a measure, μ, defined on the Cartesian product and producing α and an infinite family of slicing measures from it whereas in the construction of product measure, we start with two measures and produce a new measure on a σ algebra of subsets of the Cartesian product of two spaces. First we give two technical lemmas.

Lemma 15.6 *The space* $C_c(\mathbb{R}^m)$ *with the norm*

$$\|f\| \equiv \sup\{|f(\mathbf{y})| : \mathbf{y} \in \mathbb{R}^m\}$$

is separable.

Proof: Let $\mathcal{D}_l$ consist of all functions which are of the form

$$\sum_{|\alpha| \leq N} a_\alpha \mathbf{y}^\alpha \left(dist\left(\mathbf{y}, B(0, l+1)^C\right)\right)^{n_\alpha}$$

where $a_\alpha \in \mathbb{Q}$, α is a multi-index, and n_α is a positive integer. Then $\mathcal{D}_l$ is countable, separates the points of $\overline{B(0,l)}$ and annihilates no point of $\overline{B(0,l)}$. By the Stone Weierstrass theorem $\mathcal{D}_l$ is dense in the space $C\left(\overline{B(0,l)}\right)$ and so $\cup\{\mathcal{D}_l : l \in \mathbb{N}\}$ is a countable dense subset of $C_c(\mathbb{R}^m)$.

Lemma 15.7 *If* μ *and* ν *are two Radon measures defined on* σ *algebras,* $\mathcal{S}_\mu$ *and* $\mathcal{S}_\nu$, *of subsets of* $\mathbb{R}^n$ *and if* $\mu(V) = \nu(V)$ *for all* V *open, then* $\mu = \nu$ *and* $\mathcal{S}_\mu = \mathcal{S}_\nu$.

Proof: Let $\overline{\mu}$ and $\overline{\nu}$ be the outer measures determined by μ and ν. Then if E is a Borel set,

$$\begin{aligned}
\mu(E) &= \inf\{\mu(V) : V \supseteq E \text{ and } V \text{ is open}\} \\
&= \inf\{\nu(V) : V \supseteq E \text{ and } V \text{ is open}\} = \nu(E).
\end{aligned}$$

Now if S is any subset of $\mathbb{R}^n$,

$$\overline{\mu}(S) \equiv \inf\{\mu(E) : E \supseteq S, E \in \mathcal{S}_\mu\}$$

$$= \inf\{\mu(F) : F \supseteq S, F \text{ is Borel}\}$$

$$= \inf\{\nu(F) : F \supseteq S, F \text{ is Borel}\}$$

$$= \inf\{\nu(E) : E \supseteq S, E \in \mathcal{S}_\nu\} \equiv \overline{\nu}(S)$$

where the second and fourth equalities follow from the outer regularity which is assumed to hold for both measures. Therefore, the two outer measures are identical and so the measurable sets determined in the sense of Caratheodory are also identical. By Lemma 9.6 of Chapter 9 this implies both of the given σ algebras are equal and $\mu = \nu$.

The main result in the section is the following theorem.

Theorem 15.8 *Let μ be a finite Radon measure on $\mathbb{R}^{n+m}$ defined on a σ algebra, $\mathcal{F}$. Then there exists a unique finite Radon measure, α, defined on a σ algebra, $\mathcal{S}$, of sets of $\mathbb{R}^n$ which satisfies*

$$\alpha(E) = \mu(E \times \mathbb{R}^m) \tag{4}$$

for all E Borel. There also exists a Borel set of α measure zero, N, such that for each $\mathbf{x} \notin N$, there exists a Radon probability measure $\nu_{\mathbf{x}}$ such that if f is a nonnegative Borel measurable function or a Borel measurable function in $L^1(\mu)$,

$$\mathbf{x} \to \int_{\mathbb{R}^m} f(\mathbf{x}, \mathbf{y})\, d\nu_{\mathbf{x}}(\mathbf{y}) \text{ is } \alpha \text{ measurable} \tag{5}$$

and

$$\int_{\mathbb{R}^{n+m}} f(\mathbf{x}, \mathbf{y})\, d\mu = \int_{\mathbb{R}^n} \left(\int_{\mathbb{R}^m} f(\mathbf{x}, \mathbf{y})\, d\nu_{\mathbf{x}}(\mathbf{y}) \right) d\alpha(x). \tag{6}$$

If $\widehat{\nu}_{\mathbf{x}}$ is any other collection of Radon measures satisfying Formulas 5 and 6, then $\widehat{\nu}_{\mathbf{x}} = \nu_{\mathbf{x}}$ for α a.e. $\mathbf{x}$.

Proof: First we deal with the uniqueness of α. Suppose α_1 is another Radon measure satisfying Formula 4. Then in particular, α_1 and α agree on open sets and so the two measures are the same by Lemma 15.7.

To establish the existence of α, define α_0 on Borel sets by

$$\alpha_0(E) = \mu(E \times \mathbb{R}^m).$$

Thus α_0 is a finite Borel measure and so it is finite on compact sets. Therefore Theorem 9.26 implies the existence of α.

Next we consider the uniqueness of $\nu_{\mathbf{x}}$. Suppose $\nu_{\mathbf{x}}$ and $\widehat{\nu}_{\mathbf{x}}$ satisfy all conclusions of the theorem with exceptional sets denoted by N and $\widehat{N}$ respectively. Then, enlarging N and $\widehat{N}$, we may also assume, using Lemma 15.3, that for $\mathbf{x} \notin N \cup \widehat{N}$, $\alpha(B(\mathbf{x}, r)) > 0$ whenever $r > 0$. Now let

$$A = \prod_{i=1}^{n} (a_i, b_i]$$

where a_i and b_i are rational. Thus there are countably many such sets. Then from the conclusion of the theorem, if $\mathbf{x}_0 \notin N \cup \widehat{N}$,

$$\frac{1}{\alpha(B(\mathbf{x}_0, r))} \int_{B(\mathbf{x}_0, r)} \int_{\mathbb{R}^m} \mathcal{X}_A(\mathbf{y})\, d\nu_{\mathbf{x}}(\mathbf{y})\, d\alpha$$

$$= \frac{1}{\alpha\left(B\left(\mathbf{x}_0, r\right)\right)} \int_{B(\mathbf{x}_0, r)} \int_{\mathbb{R}^m} \mathcal{X}_A\left(\mathbf{y}\right) d\widehat{\nu}_{\mathbf{x}}\left(y\right) d\alpha,$$

and by the Lebesgue Besicovitch Differentiation theorem, there exists a set of α measure zero, E_A, such that if $\mathbf{x}_0 \notin E_A \cup N \cup \widehat{N}$, then we may take the limit in the above as $r \to 0$ and conclude that

$$\nu_{\mathbf{x}_0}\left(A\right) = \widehat{\nu}_{\mathbf{x}_0}\left(A\right).$$

Letting E denote the union of all the sets E_A for A as described above, it follows that E is a set of measure zero and if $\mathbf{x}_0 \notin E \cup N \cup \widehat{N}$ then $\nu_{\mathbf{x}_0}\left(A\right) = \widehat{\nu}_{\mathbf{x}_0}\left(A\right)$ for all such sets A. But every open set can be written as a disjoint union of sets of this form and so for all such $\mathbf{x}_0$, $\nu_{\mathbf{x}_0}\left(V\right) = \widehat{\nu}_{\mathbf{x}_0}\left(V\right)$ for all V open. By Lemma 15.7 this shows the two measures are equal and proves the uniqueness assertion for $\nu_{\mathbf{x}}$. It remains to show the existence of the measures $\nu_{\mathbf{x}}$. This will be done with the aid of the following lemma.

Lemma 15.9 *There exists a set of α measure 0, N, independent of $f \in C_c\left(\mathbb{R}^{n+m}\right)$ such that if $\mathbf{x} \notin N$, $\alpha\left(B\left(\mathbf{x}, r\right)\right) > 0$ for all $r > 0$ and*

$$\lim_{r \to 0} \frac{1}{\alpha\left(B\left(\mathbf{x}, r\right)\right)} \int_{B(\mathbf{x}, r) \times \mathbb{R}^m} f d\mu = g_f\left(\mathbf{x}\right)$$

where g_f is a α measurable function with the property that

$$\int_{\mathbb{R}^n} g_f\left(\mathbf{x}\right) d\alpha = \int_{\mathbb{R}^n \times \mathbb{R}^m} f d\mu.$$

Proof: Let $f \in C_c\left(\mathbb{R}^{n+m}\right)$ and let

$$\eta_f\left(E\right) \equiv \int_{E \times \mathbb{R}^m} f d\mu.$$

Then η_f is a finite measure with $\eta_f \ll \alpha$. By the Radon Nikodym theorem, there exists a Borel measurable function $\widetilde{g}_f$ such that for all Borel E,

$$\int_{E \times \mathbb{R}^m} f d\mu = \int_E \widetilde{g}_f d\alpha.$$

By the theory of differentiation for Radon measures, there exists a set of α measure zero, N_f such that if $\mathbf{x} \notin N_f$, then $\alpha\left(B\left(\mathbf{x}, r\right)\right) > 0$ for all $r > 0$ and

$$\lim_{r \to 0} \frac{1}{\alpha\left(B\left(\mathbf{x}, r\right)\right)} \int_{B(\mathbf{x}, r) \times \mathbb{R}^m} f d\mu = \lim_{r \to 0} \fint_{B(\mathbf{x}, r)} \widetilde{g}_f d\alpha = \widetilde{g}_f\left(\mathbf{x}\right).$$

Let $\mathcal{D}$ be a countable dense subset of $C_c\left(\mathbb{R}^{n+m}\right)$ and let

$$N \equiv \cup\left\{N_f : f \in \mathcal{D}\right\}.$$

Then if $f \in C_c(\mathbb{R}^{n+m})$ is arbitrary, and $\mathbf{x} \notin N$,

$$\left| \frac{1}{\alpha\left(B\left(\mathbf{x}, r_1\right)\right)} \int_{B\left(\mathbf{x}, r_1\right) \times \mathbb{R}^m} f d\mu - \frac{1}{\alpha\left(B\left(\mathbf{x}, r_2\right)\right)} \int_{B\left(\mathbf{x}, r_2\right) \times \mathbb{R}^m} f d\mu \right| <$$

$$\frac{\epsilon}{2} + \left| \frac{1}{\alpha\left(B\left(\mathbf{x}, r_1\right)\right)} \int_{B\left(\mathbf{x}, r_1\right) \times \mathbb{R}^m} g d\mu - \frac{1}{\alpha\left(B\left(\mathbf{x}, r_2\right)\right)} \int_{B\left(\mathbf{x}, r_2\right) \times \mathbb{R}^m} g d\mu \right|$$

where $g \in \mathcal{D}$ is chosen close enough to f. Therefore, taking r_i small enough, we find the right side is less than ϵ. Since ϵ is arbitrary, this shows the limit exists. We define this limit which exists for all $f \in C_c(\mathbb{R}^{n+m})$ and $\mathbf{x} \notin N$ as $g_f(\mathbf{x})$. By the first part of the argument, $g_f(\mathbf{x}) = \tilde{g}_f(\mathbf{x})$ a.e. Thus, $g_f(\mathbf{x})$ is α measurable because it equals a Borel measurable function α a.e. The final formula follows from

$$\int_{\mathbb{R}^n} g_f(\mathbf{x}) \, d\alpha = \int_{\mathbb{R}^n} \tilde{g}_f(\mathbf{x}) \, d\alpha = \int_{\mathbb{R}^n \times \mathbb{R}^m} f d\mu.$$

Continuing with the proof of the theorem, let $\mathbf{x} \notin N$ and let $f(\mathbf{z}, \mathbf{y}) \equiv \psi(\mathbf{z}) \phi(\mathbf{y})$ where ψ and ϕ are continuous functions with compact support in $\mathbb{R}^n$ and $\mathbb{R}^m$ respectively. Suppose first that $\psi(\mathbf{x}) = 1$. Then we may define a positive linear functional

$$L_{\mathbf{x}} \phi \equiv \lim_{r \to 0} \frac{1}{\alpha\left(B\left(\mathbf{x}, r\right)\right)} \int_{B\left(\mathbf{x}, r\right) \times \mathbb{R}^m} \psi(\mathbf{z}) \phi(\mathbf{y}) \, d\mu(z, y).$$

This functional may appear to depend on the choice of ψ satisfying $\psi(\mathbf{x}) = 1$ but this is not the case because ψ is continuous. Let $\nu_{\mathbf{x}}$ be the Radon measure representing $L_{\mathbf{x}}$. Thus replacing an arbitrary $\psi \in C_c(\mathbb{R}^n)$ with $\frac{\psi}{\psi(\mathbf{x})}$, in the case when $\psi(\mathbf{x}) \neq 0$, we always obtain

$$\psi(\mathbf{x}) \int_{\mathbb{R}^m} \phi(\mathbf{y}) \, d\nu_{\mathbf{x}}(\mathbf{y}) = \lim_{r \to 0} \frac{1}{\alpha\left(B\left(\mathbf{x}, r\right)\right)} \int_{B\left(\mathbf{x}, r\right) \times \mathbb{R}^m} \psi \phi d\mu.$$

By Lemma 15.9,

$$\int_{\mathbb{R}^n} \int_{\mathbb{R}^m} \psi \phi d\nu_{\mathbf{x}} d\alpha = \int_{\mathbb{R}^n} g_{\psi\phi} d\alpha = \int_{\mathbb{R}^n \times \mathbb{R}^m} \psi \phi d\mu.$$

Letting ψ_k, ϕ_k increase to 1 pointwise, we may apply the monotone convergence theorem to find

$$\int_{\mathbb{R}^n} \nu_{\mathbf{x}}(\mathbb{R}^m) \, d\alpha = \int_{\mathbb{R}^n \times \mathbb{R}^m} d\mu = \mu(\mathbb{R}^n \times \mathbb{R}^m) < \infty \tag{7}$$

showing that $\mathbf{x} \to \nu_{\mathbf{x}}(\mathbb{R}^m)$ is a function in $L^1(\alpha)$. In particular, this function is finite α a.e.

Now we show that $\psi\phi$ can be replaced by $\mathcal{X}_B$ where B is an arbitrary Borel set in $\mathbb{R}^n \times \mathbb{R}^m$. To do so, let

$$R_1 \equiv \prod_{i=1}^{n}(a_i, b_i], \quad R_2 \equiv \prod_{i=1}^{m}(c_i, d_i]$$

and let ψ_k be a sequence of functions in $C_c(\mathbb{R}^n)$ which is bounded, piecewise linear in each variable, and converging pointwise to $\mathcal{X}_{R_1}$. Also let ϕ_k be a similar sequence converging pointwise to $\mathcal{X}_{R_2}$. Then by the dominated convergence theorem,

$$\int_{\mathbb{R}^n \times \mathbb{R}^m} \mathcal{X}_{R_1}(\mathbf{x}) \mathcal{X}_{R_2}(\mathbf{y}) \, d\mu = \lim_{k \to \infty} \int_{\mathbb{R}^n \times \mathbb{R}^m} \psi_k(\mathbf{x}) \phi_k(\mathbf{y}) \, d\mu$$

$$= \lim_{k \to \infty} \int_{\mathbb{R}^n} \psi_k(\mathbf{x}) \left(\int_{\mathbb{R}^m} \phi_k(\mathbf{y}) \, d\nu_{\mathbf{x}} \right) d\alpha. \tag{8}$$

Since $\nu_{\mathbf{x}}$ is finite α a.e., it follows that for α a.e. $\mathbf{x}$,

$$\lim_{k \to \infty} \int_{\mathbb{R}^m} \phi_k(\mathbf{y}) \, d\nu_{\mathbf{x}} = \int_{\mathbb{R}^m} \mathcal{X}_{R_1}(\mathbf{y}) \, d\nu_{\mathbf{x}}.$$

Since the ϕ_k are uniformly bounded, Formula 7 implies the existence of a dominating function for the integrand in Formula 8. Therefore, we may take the limit inside the integrals and obtain

$$\int_{\mathbb{R}^n \times \mathbb{R}^m} \mathcal{X}_{R_1 \times R_2}(\mathbf{x}, \mathbf{y}) \, d\mu = \int_{\mathbb{R}^n} \int_{\mathbb{R}^m} \mathcal{X}_{R_1 \times R_2}(\mathbf{x}, \mathbf{y}) \, d\nu_{\mathbf{x}} d\alpha.$$

Now let $Q_k \equiv (-k, k]^n \times (-k, k]^m$ and define

$$\mathcal{M} \equiv \{B \text{ a Borel set in } \mathbb{R}^{n+m} \text{ such that}$$

$$\int_{\mathbb{R}^n \times \mathbb{R}^m} \mathcal{X}_{B \cap Q_k}(\mathbf{x}, \mathbf{y}) \, d\mu = \int_{\mathbb{R}^n} \int_{\mathbb{R}^m} \mathcal{X}_{B \cap Q_k}(\mathbf{x}, \mathbf{y}) \, d\nu_{\mathbf{x}} d\alpha\}.$$

We just showed that $\mathcal{M}$ contains the algebra of sets $\mathcal{E}$, described in Example 7.8 of Chapter 7. It follows from the dominated convergence theorem that $\mathcal{M}$ is a monotone class. Therefore, $\mathcal{M}$ must equal the Borel sets. Letting $k \to \infty$ and applying the monotone convergence theorem, we see

$$\int_{\mathbb{R}^n \times \mathbb{R}^m} \mathcal{X}_B(\mathbf{x}, \mathbf{y}) \, d\mu = \int_{\mathbb{R}^n} \int_{\mathbb{R}^m} \mathcal{X}_B(\mathbf{x}, \mathbf{y}) \, d\nu_{\mathbf{x}} d\alpha$$

for all B Borel. It follows this equation holds for all nonnegative Borel simple functions in place of $\mathcal{X}_B$ and, therefore, for all nonnegative Borel measurable functions, f.

To see $\nu_{\mathbf{x}}$ is a probability measure, let $B = E \times \mathbb{R}^m$. Then from the above equation,

$$\mu\left(E \times \mathbb{R}^m\right) = \alpha\left(E\right) = \int_E \nu_{\mathbf{x}}\left(\mathbb{R}^m\right) d\alpha$$

so for a.e. $\mathbf{x}$,

$$1 = \lim_{r \to 0} \frac{1}{\alpha\left(B\left(\mathbf{x}, r\right)\right)} \int_{B(\mathbf{x}, r)} \nu_z\left(\mathbb{R}^m\right) d\alpha\left(z\right) = \nu_{\mathbf{x}}\left(\mathbb{R}^m\right).$$

This proves the theorem.

15.4   Young measures

If F is a bounded continuous function and $\{f_k\}$ is a sequence of functions in $L^\infty\left(U\right)$ such that their L^∞ norms are uniformly bounded, then we can use Corollary 6.21, the Riesz representation theorem, and the Banach Alouglu theorem to obtain a weak* convergent subsequence for $F \circ f_k$. It would be nice if we could say something like $F \circ f_{k_i}$ converges weak * to $F \circ f$ where f_{k_i} converges weak* to f but this will not happen in general. We can say something of this sort however. It turns out there exists a collection of probability measures called Young's measures, $\nu_{\mathbf{x}}$ determined by an appropriate subsequence of $\{f_k\}$, just as f in the above wished for situation is determined by a subsequence of $\{f_k\}$, such that for any bounded continuous F,

$$F \circ f_{k_i} \to \overline{F}$$

in the weak * topology for $\overline{F}$ defined by

$$\overline{F}\left(\mathbf{x}\right) \equiv \int F\left(\mathbf{y}\right) d\nu_{\mathbf{x}}\left(y\right).$$

More precisely, we will prove the following theorem.

Theorem 15.10 *Let U be a bounded open set in $\mathbb{R}^n$ and let $\{f_k\}$ be contained in a bounded subset of $L^\infty\left(U; \mathbb{R}^m\right)$. Then there exists a subsequence k_i and probability measures $\nu_{\mathbf{x}}$ for m_n a.e. $\mathbf{x} \in U$ such that if F is any bounded continuous function defined on $\mathbb{R}^m$, then*

$$F\left(f_{k_i}\right) \to \overline{F} \ weak * \ in \ L^\infty\left(U\right)$$

where

$$\overline{F}\left(\mathbf{x}\right) \equiv \int_{\mathbb{R}^m} F\left(\mathbf{y}\right) d\nu_{\mathbf{x}}\left(y\right).$$

Before proving the theorem, we note that $L^\infty (U; \mathbb{R}^m)$ is the space of functions having values in $\mathbb{R}^m$ such that each component is in $L^\infty (U)$.

Proof of the theorem: Let A be a Borel set in $\mathbb{R}^n \times \mathbb{R}^m$. Then define

$$\mu_k (A) \equiv \int_U \mathcal{X}_A (\mathbf{x}, \mathbf{f}_k (\mathbf{x})) \, dx.$$

There are no measurability problems because $\mathcal{X}_A$ is Borel and the integrand is $\mathcal{X}_A \circ \mathbf{h}$ where

$$\mathbf{h} (\mathbf{x}) \equiv (\mathbf{x}, \mathbf{f}_k (\mathbf{x}))^T$$

and is Lebesgue measurable. Now

$$\mu_k (E \times \mathbb{R}^m) = \int_U \mathcal{X}_{E \times \mathbb{R}^m} (\mathbf{x}, \mathbf{f}_k (\mathbf{x})) \, dx$$

$$= \int_U \mathcal{X}_E (\mathbf{x}) \, dx = m_n (E \cap U).$$

In particular, this shows $\mu_k (\mathbb{R}^n \times \mathbb{R}^m) = m_n (U)$ and so μ_k is a finite Borel measure. It follows from Theorem 9.26 that μ_k can be extended to a Radon measure, still denoted by μ_k. Now μ_k is an element of $C_0 (\mathbb{R}^n \times \mathbb{R}^m)'$ if we define

$$\mu_k (g) \equiv \int_{\mathbb{R}^n \times \mathbb{R}^m} g \, d\mu_k$$

because if $||g|| \equiv \sup \{ |g (\mathbf{x})| : \mathbf{x} \in \mathbb{R}^n \times \mathbb{R}^m \} \leq 1$, then

$$|\mu_k (g)| \leq \int_{\mathbb{R}^n \times \mathbb{R}^m} |g| \, d\mu_k = \mu_k (\mathbb{R}^n \times \mathbb{R}^m) = m_n (U) < \infty.$$

Since $C_0 (\mathbb{R}^n \times \mathbb{R}^m)$ is separable, Corollary 6.21 implies there is a subsequence k_i such that

$$\mu_{k_i} \to \mu \text{ weak } * \text{ in } C_0 (\mathbb{R}^n \times \mathbb{R}^m)'.$$

where μ is a Radon measure. Letting $g (\mathbf{x}) = 1$ on $B (0, r)$,

$$g \in C_0 (\mathbb{R}^n \times \mathbb{R}^m),$$

and $0 \leq g (\mathbf{x}) \leq 1$,

$$\mu (B (0, r)) \leq \int_{\mathbb{R}^n \times \mathbb{R}^m} g \, d\mu \leq \epsilon + \int_{\mathbb{R}^n \times \mathbb{R}^m} g \, d\mu_{k_i} \leq m_n (U) + \epsilon$$

whenever k_i is large enough. Hence μ is a finite Radon measure.

Let E and G be Borel sets.

$$
\int_{\mathbb{R}^n \times \mathbb{R}^m} \mathcal{X}_E(\mathbf{x}) \mathcal{X}_G(\mathbf{y}) \, d\mu_k = \mu_k(E \times G) \equiv \int_U \mathcal{X}_{E \times G}(\mathbf{x}, \mathbf{f}_k(\mathbf{x})) \, dx
$$

$$
= \int_U \mathcal{X}_E(\mathbf{x}) \mathcal{X}_G(\mathbf{f}_k(\mathbf{x})) \, dx.
$$

It follows that if s_l and t_l are simple functions,

$$
\int_{\mathbb{R}^n \times \mathbb{R}^m} s_l(\mathbf{x}) t_l(\mathbf{y}) \, d\mu_k = \int_U s_l(\mathbf{x}) t_l(\mathbf{f}_k(\mathbf{x})) \, dx.
$$

Letting $\xi \in C_c(U)$ and $F \in C_c(\mathbb{R}^m)$, we can have $\lim_{l \to \infty} s_l(\mathbf{x}) = \xi(\mathbf{x})$ and $\lim_{l \to \infty} t_l(\mathbf{y}) = F(\mathbf{y})$ pointwise where the two sequences of simple functions are also bounded independent of $l, \mathbf{x}$, and $\mathbf{y}$. Then by the dominated convergence theorem,

$$
\int_{\mathbb{R}^n \times \mathbb{R}^m} \xi(\mathbf{x}) F(\mathbf{y}) \, d\mu_{k_i} = \int_U \xi(\mathbf{x}) F(\mathbf{f}_{k_i}(\mathbf{x})) \, dx.
$$

Letting $i \to \infty$,

$$
\int_{\mathbb{R}^n \times \mathbb{R}^m} \xi(\mathbf{x}) F(\mathbf{y}) \, d\mu = \lim_{i \to \infty} \int_U \xi(\mathbf{x}) F(\mathbf{f}_{k_i}(\mathbf{x})) \, dx.
$$

Defining $\alpha(E) \equiv \mu(E \times \mathbb{R}^m)$ for all E Borel, Theorem 15.8 implies there are probability measures $\nu_{\mathbf{x}}$ for α a.e. $\mathbf{x}$ such that

$$
\int_{\mathbb{R}^n} \xi(\mathbf{x}) \int_{\mathbb{R}^m} F(\mathbf{y}) \, d\nu_{\mathbf{x}}(x) \, d\alpha = \lim_{i \to \infty} \int_U \xi(\mathbf{x}) F(\mathbf{f}_{k_i}(\mathbf{x})) \, dx. \tag{9}
$$

Claim: $\alpha(E) = m_n(U \cap E)$.

Proof of claim: In what follows, K will be compact and V will be open. Let V be an open set in $\mathbb{R}^n$ and let $g \prec V \times \mathbb{R}^m$ be such that

$$
\int_{\mathbb{R}^{n+m}} g \, d\mu + \epsilon \geq \mu(V \times \mathbb{R}^m).
$$

Then

$$
\mu(V \times \mathbb{R}^m) \leq \int_{\mathbb{R}^{n+m}} g \, d\mu + \epsilon = \lim_{i \to \infty} \int_{\mathbb{R}^{n+m}} g \, d\mu_{k_i} + \epsilon
$$

$$
\leq \lim_{i \to \infty} \mu_{k_i}(V \times \mathbb{R}^m) + \epsilon = m_n(U \cap V) + \epsilon.
$$

Thus $\mu(V \times \mathbb{R}^m) \leq m_n(V \cap U)$. Now let $r > \sup\{\|\mathbf{f}_k\|_\infty : k = 1, 2, \cdots\}$. Then

$$
\mu_k(E \times \mathbb{R}^m) \equiv \int_U \mathcal{X}_{E \times \mathbb{R}^m}(\mathbf{x}, \mathbf{f}_k(\mathbf{x})) \, dx
$$

$$
= \int_U \mathcal{X}_{E \times \overline{B(0,r)}}(\mathbf{x}, \mathbf{f}_k(\mathbf{x})) \, dx = \mu_k\left(E \times \overline{B(0,r)}\right).
$$

Now let K be compact in $\mathbb{R}^n$ and choose g such that

$$K \times \overline{B(0,r)} \prec g, \int_{\mathbb{R}^{n+m}} g d\mu - \epsilon \le \mu\left(K \times \overline{B(0,r)}\right).$$

Then

$$
\begin{aligned}
\mu\left(K \times \mathbb{R}^m\right) &\ge \mu\left(K \times \overline{B(0,r)}\right) \ge \int_{\mathbb{R}^{n+m}} g d\mu - \epsilon \\
&= \lim_{i \to \infty} \int_{\mathbb{R}^{n+m}} g d\mu_{k_i} - \epsilon \ge \lim_{i \to \infty} \mu_{k_i}\left(K \times \overline{B(0,r)}\right) - \epsilon \\
&= \lim_{i \to \infty} \mu_{k_i}\left(K \times \mathbb{R}^m\right) - \epsilon = m_n\left(K \cap U\right) - \epsilon.
\end{aligned}
$$

and so

$$\mu\left(K \times \mathbb{R}^m\right) \ge m_n\left(K \cap U\right).$$

Now let E be an arbitrary Borel set in $\mathbb{R}^n$ and let $E_p \equiv E \cap B(0,p)$. Choose V and K such that

$$K \subseteq E_p \subseteq V, \ m_n\left(V \setminus K\right) < \epsilon.$$

Then

$$m_n\left(K \cap U\right) \le \mu\left(K \times \mathbb{R}^m\right)$$

$$\le \mu\left(E_p \times \mathbb{R}^m\right) \le \mu\left(V \times \mathbb{R}^m\right) \le m_n\left(V \cap U\right),$$

and also

$$
\begin{aligned}
m_n\left(K \cap U\right) &= \mu_{k_i}\left(K \times \mathbb{R}^m\right) \le \mu_{k_i}\left(E_p \times \mathbb{R}^m\right) \\
&= m_n\left(E_p \cap U\right) \le \mu_{k_i}\left(V \times \mathbb{R}^m\right) = m_n\left(V \cap U\right).
\end{aligned}
$$

Thus both $\mu\left(E_p \times \mathbb{R}^m\right)$ and $m_n\left(E_p \cap U\right)$ are contained in the interval of length no more than 2ϵ,

$$\left[m_n\left(K \cap U\right), m_n\left(V \cap U\right)\right].$$

Since ϵ is arbitrary, this shows $\mu\left(E_p \times \mathbb{R}^m\right) = m_n\left(E_p \cap U\right)$. Letting $p \to \infty$ yields the conclusion of the claim.

With the above claim, Formula 9 implies that if $\xi \in C_c\left(\mathbb{R}^n\right)$ and $F \in C_c\left(\mathbb{R}^m\right)$, then

$$\int_U \xi\left(\mathbf{x}\right) \int_{\mathbb{R}^m} F\left(\mathbf{y}\right) d\nu_{\mathbf{x}}\left(x\right) dx = \lim_{i \to \infty} \int_U \xi\left(\mathbf{x}\right) F\left(\mathbf{f}_{k_i}\left(\mathbf{x}\right)\right) dx.$$

In fact this formula holds for an arbitrary bounded and continuous F. As earlier in the proof, let

$$r > \sup\left\{\|f_k\|_\infty, \ k = 1, 2 \cdots\right\}.$$

Now let F be an arbitrary bounded and continuous function and define a sequence of functions with compact support, $\{F_l\}$, such that

$$\lim_{l \to \infty} F_l(\mathbf{y}) \to F(\mathbf{y}), \ F_l(\mathbf{y}) = F(\mathbf{y})$$

if $|\mathbf{y}| \le r$. Then,

$$\lim_{i \to \infty} \int_U \xi(\mathbf{x}) F(\mathbf{f}_{k_i}(\mathbf{x})) \, dx \ = \ \lim_{i \to \infty} \int_U \xi(\mathbf{x}) F_l(\mathbf{f}_{k_i}(\mathbf{x})) \, dx$$

$$= \ \int_U \xi(\mathbf{x}) \int_{\mathbb{R}^m} F_l(\mathbf{y}) \, d\nu_{\mathbf{x}} dx.$$

Now let $l \to \infty$ to obtain

$$\lim_{i \to \infty} \int_U \xi(\mathbf{x}) F(\mathbf{f}_{k_i}(\mathbf{x})) \, dx = \int_U \xi(\mathbf{x}) \int_{\mathbb{R}^m} F(\mathbf{y}) \, d\nu_{\mathbf{x}} dx.$$

Since $C_c(U)$ is dense in $L^1(U)$, this shows

$$F(\mathbf{f}_{k_i}) \to \overline{F} \text{ weak } * \text{ in } L^\infty(U)$$

where

$$\overline{F}(\mathbf{x}) \equiv \int_{\mathbb{R}^m} F(\mathbf{y}) \, d\nu_{\mathbf{x}}$$

and this proves the theorem.

15.5   Exercises

1. Let a positive linear functional, L, be defined on $C_c(\mathbb{R})$ by

    $$Lf \equiv \sum_{k \in \mathbb{N} \cap spt(f)} f(k)$$

    and let μ be the Radon measure representing this functional. What is $\int f d\mu$? What is

    $$\lim_{r \to \infty} \frac{1}{\mu(B(x, r))} \int_{B(x,r)} f(x) \, d\mu?$$

    More generally, let F be an increasing function and let $L(f) \equiv \int f dF$ where the integral is the Riemann Steiltjes integral. What does the Lebesgue Besicovitch differentiation theorem give in this example?

2. Let $E \subseteq \mathbb{R}^n$ be μ measurable for μ a Radon measure, $\mu(E) < \infty$, and let $\mathcal{F}$ be a collection of closed balls that cover E in the following

way. If $\mathbf{x} \in E$ and $\varepsilon > 0$, then there exists $B \in \mathcal{F}$, centered at $\mathbf{x}$, diameter of $B < \varepsilon$. Show there exists a countable sequence of disjoint balls of $\mathcal{F}$, $\{B_j\}$, such that $\mu(E \setminus \cup_{j=1}^{\infty} B_j) = 0$. **Hint:** In the Besicovitch theorem, there is a constant N_n. Show that for some finite list of open sets of $\mathcal{F}$ contained in U, $B_1, \cdots, B_{m_1}$,

$$\sum_{i=1}^{m_1} \mu(B_i) > \frac{1}{2N_n} \mu(U)$$

where $U \supseteq E$ and satisfies $\mu(E) > (1 - r)\mu(U)$ for r a small positive number small enough that

$$\lambda \equiv (1 - r)^{-1} \left(1 - \frac{1}{2N_n}\right) < 1.$$

Show from the above that

$$\lambda \mu(E) \geq \left(1 - \frac{1}{2N_n}\right) \mu(U)$$

$$\geq \mu(U \setminus \cup_{i=1}^{m_1} B_i) \geq \mu(E \setminus \cup_{i=1}^{m_1} B_i).$$

Now iterate this estimate.

3. In the section on slicing measure, Formula 6 could have been written as

$$\int_{\mathbb{R}^{n+m}} f d\mu = \int_{\mathbb{R}^n} \left(\int_{\mathbf{h}^{-1}(\mathbf{x})} f d\nu_{\mathbf{x}}\right) d\alpha$$

where $\mathbf{h} : \mathbb{R}^{n+m} \to \mathbb{R}^n$ is given as $\mathbf{h}(\mathbf{x}, \mathbf{y}) \equiv \mathbf{x}$. Generalize this theorem on slicing measures to include an arbitrary continuous function, $\mathbf{h} : \mathbb{R}^{n+m} \to \mathbb{R}^n$. **Hint:** In defining $L_{\mathbf{x}}$ for $\mathbf{x} \notin N$, let

$$L_{\mathbf{x}} \psi \equiv \lim_{r \to 0} \frac{1}{\alpha(B(\mathbf{x},r))} \int_{\mathbf{h}^{-1}(B(\mathbf{x},r))} \widetilde{\psi} d\mu$$

where $\widetilde{\psi}$ is in $C_c(\mathbb{R}^{n+m})$ and equals ψ on $\mathbf{h}^{-1}(\mathbf{x})$. You may want to use the Tietze extension theorem to accomplish this extension of ψ.

Chapter 16
Fourier Transforms

16.1 The Schwartz class

The Fourier transform of a function in $L^1(\mathbb{R}^n)$ is given by

$$Ff(\mathbf{t}) \equiv (2\pi)^{-n/2} \int_{\mathbb{R}^n} e^{-i\mathbf{t}\cdot\mathbf{x}} f(\mathbf{x}) dx.$$

However, we want to take the Fourier transform of many other kinds of functions. In particular we want to take the Fourier transform of functions in $L^2(\mathbb{R}^n)$ which is not a subset of $L^1(\mathbb{R}^n)$. Thus the above integral may not make sense. In defining what is meant by the Fourier Transform of more general functions, it is convenient to use a special class of functions known as the Schwartz class which is a subset of $L^p(\mathbb{R}^n)$ for all $p \geq 1$. The procedure is to define the Fourier transform of functions in the Schwartz class and then use this to define the Fourier transform of a more general function in terms of what happens when it is multiplied by the Fourier transform of functions in the Schwartz class.

The functions in the Schwartz class are infinitely differentiable and they vanish very rapidly as $|x| \to \infty$ along with all their partial derivatives. To describe precisely what we mean by this, we need to present some notation.

Definition 16.1 $\alpha = (\alpha_1, \cdots, \alpha_n)$ for $\alpha_1 \cdots \alpha_n$ positive integers is called a multi-index. For α a multi-index, $|\alpha| \equiv \alpha_1 + \cdots + \alpha_n$ and if $\mathbf{x} \in \mathbb{R}^n$,

$$\mathbf{x} = (x_1, \cdots, x_n),$$

and f a function, we define

$$\mathbf{x}^\alpha \equiv x_1^{\alpha_1} x_2^{\alpha_2} \cdots x_n^{\alpha_n}, \ D^\alpha f(\mathbf{x}) \equiv \frac{\partial^{|\alpha|} f(\mathbf{x})}{\partial x_1^{\alpha_1} \partial x_2^{\alpha_2} \cdots \partial x_n^{\alpha_n}}.$$

Definition 16.2 We say $f \in \mathfrak{S}$, the Schwartz class, if $f \in C^\infty(\mathbb{R}^n)$ and for all positive integers N,

$$\rho_N(f) < \infty$$

where

$$\rho_N(f) = \sup\{(1 + |\mathbf{x}|^2)^N |D^\alpha f(\mathbf{x})| : \mathbf{x} \in \mathbb{R}^n, \ |\alpha| \leq N\}.$$

Thus $f \in \mathfrak{S}$ if and only if $f \in C^{\infty}(\mathbb{R}^n)$ and

$$\sup\{|\mathbf{x}^{\beta} D^{\alpha} f(\mathbf{x})| : \mathbf{x} \in \mathbb{R}^n\} < \infty \tag{1}$$

for all multi indices α and β.

Also note that if $f \in \mathfrak{S}$, then $p(f) \in \mathfrak{S}$ for any polynomial, p with $p(0) = 0$ and that

$$\mathfrak{S} \subseteq L^p(\mathbb{R}^n) \cap L^{\infty}(\mathbb{R}^n)$$

for any $p \geq 1$.

Definition 16.3 (Fourier transform on $\mathfrak{S}$) For $f \in \mathfrak{S}$,

$$Ff(\mathbf{t}) \equiv (2\pi)^{-n/2} \int_{\mathbb{R}^n} e^{-i\mathbf{t}\cdot\mathbf{x}} f(\mathbf{x}) dx,$$

$$F^{-1}f(\mathbf{t}) \equiv (2\pi)^{-n/2} \int_{\mathbb{R}^n} e^{i\mathbf{t}\cdot\mathbf{x}} f(\mathbf{x}) dx.$$

Here $\mathbf{x} \cdot \mathbf{t} = \sum_{i=1}^{n} x_i t_i$.

It will follow from the development given below that $(F \circ F^{-1})(f) = f$ and $(F^{-1} \circ F)(f) = f$ whenever $f \in \mathfrak{S}$, thus justifying the above notation.

Theorem 16.4 If $f \in \mathfrak{S}$, then Ff and $F^{-1}f$ are also in $\mathfrak{S}$.

Proof: To begin with, let $\alpha = e_j = (0, 0, \cdots, 1, 0, \cdots, 0)$, the 1 in the jth slot.

$$\frac{F^{-1}f(\mathbf{t} + h e_j) - F^{-1}f(\mathbf{t})}{h} = (2\pi)^{-n/2} \int_{\mathbb{R}^n} e^{i\mathbf{t}\cdot\mathbf{x}} f(\mathbf{x})\left(\frac{e^{i h x_j} - 1}{h}\right) dx. \tag{2}$$

Consider the integrand in 2.

$$\left| e^{i\mathbf{t}\cdot\mathbf{x}} f(\mathbf{x})\left(\frac{e^{i h x_j} - 1}{h}\right) \right| = |f(\mathbf{x})| \left| \left(\frac{e^{i(h/2)x_j} - e^{-i(h/2)x_j}}{h}\right) \right|$$

$$= |f(\mathbf{x})| \left| \frac{i \sin\left((h/2) x_j\right)}{(h/2)} \right|$$

$$\leq |f(\mathbf{x})| |x_j|$$

and this is a function in $L^1(\mathbb{R}^n)$ because $f \in \mathfrak{S}$. Therefore by the Dominated Convergence Theorem,

$$\frac{\partial F^{-1}f(\mathbf{t})}{\partial t_j} = (2\pi)^{-n/2} \int_{\mathbb{R}^n} e^{i\mathbf{t}\cdot\mathbf{x}} i x_j f(\mathbf{x}) dx$$

$$= i(2\pi)^{-n/2} \int_{\mathbb{R}^n} e^{i\mathbf{t}\cdot\mathbf{x}} \mathbf{x}^{e_j} f(\mathbf{x}) dx.$$

Now $\mathbf{x}^{e_j} f(\mathbf{x}) \in \mathfrak{S}$ and so we may continue in this way and take derivatives indefinitely. Thus $F^{-1}f \in C^{\infty}(\mathbb{R}^n)$ and from the above argument,

$$D^{\alpha} F^{-1}f(\mathbf{t}) = (2\pi)^{-n/2} \int_{\mathbb{R}^n} e^{i\mathbf{t}\cdot\mathbf{x}}(i\mathbf{x})^{\alpha} f(\mathbf{x})d\mathbf{x}.$$

To complete showing $F^{-1}f \in \mathfrak{S}$,

$$\mathbf{t}^{\beta} D^{\alpha} F^{-1}f(\mathbf{t}) = (2\pi)^{-n/2} \int_{\mathbb{R}^n} e^{i\mathbf{t}\cdot\mathbf{x}}\mathbf{t}^{\beta}(i\mathbf{x})^{\alpha} f(\mathbf{x})d\mathbf{x}.$$

Integrate this integral by parts to get

$$\mathbf{t}^{\beta} D^{\alpha} F^{-1}f(\mathbf{t}) = (2\pi)^{-n/2} \int_{\mathbb{R}^n} i^{|\beta|} e^{i\mathbf{t}\cdot\mathbf{x}} D^{\beta}((i\mathbf{x})^{\alpha} f(\mathbf{x}))d\mathbf{x}. \qquad (3)$$

Here is how this is done.

$$\int_{\mathbb{R}} e^{it_j x_j} t_j^{\beta_j}(i\mathbf{x})^{\alpha} f(\mathbf{x})dx_j = \frac{e^{it_j x_j}}{it_j} t_j^{\beta_j}(i\mathbf{x})^{\alpha} f(\mathbf{x}) |_{-\infty}^{\infty} +$$
$$i\int_{\mathbb{R}} e^{it_j x_j} t_j^{\beta_j - 1} D^{e_j}((i\mathbf{x})^{\alpha} f(\mathbf{x}))dx_j$$

where the boundary term vanishes because $f \in \mathfrak{S}$. Returning to Formula 3, we use Formula 1, and the fact that $|e^{ia}| = 1$ to conclude

$$|\mathbf{t}^{\beta} D^{\alpha} F^{-1}f(\mathbf{t})| \leq C \int_{\mathbb{R}^n} |D^{\beta}((i\mathbf{x})^{a} f(\mathbf{x}))|d\mathbf{x} < \infty.$$

It follows $F^{-1}f \in \mathfrak{S}$. Similarly $Ff \in \mathfrak{S}$ whenever $f \in \mathfrak{S}$.

Theorem 16.5 $F \circ F^{-1}(f) = f$ and $F^{-1} \circ F(f) = f$ whenever $f \in \mathfrak{S}$.

Before proving this theorem, we need a lemma.

Lemma 16.6

$$(2\pi)^{-n/2} \int_{\mathbb{R}^n} e^{(-1/2)\mathbf{u}\cdot\mathbf{u}}du = 1, \qquad (4)$$

$$(2\pi)^{-n/2} \int_{\mathbb{R}^n} e^{(-1/2)(\mathbf{u}-i\mathbf{a})\cdot(\mathbf{u}-i\mathbf{a})}du = 1. \qquad (5)$$

Proof:

$$\left(\int_{\mathbb{R}} e^{-x^2/2}dx\right)^2 = \int_{\mathbb{R}}\int_{\mathbb{R}} e^{-(x^2+y^2)/2}dxdy$$
$$= \int_0^{\infty}\int_0^{2\pi} re^{-r^2/2}d\theta dr = 2\pi.$$

Therefore

$$\int_{\mathbb{R}^n} e^{(-1/2)\mathbf{u}\cdot\mathbf{u}}du = \prod_{i=1}^{n}\int_{\mathbb{R}} e^{-x_j^2/2}dx_j = (2\pi)^{n/2}.$$

This proves Formula 4. To prove Formula 5 it suffices to show that

$$\int_{\mathbb{R}} e^{(-1/2)(x-ia)^2}dx = (2\pi)^{1/2}. \tag{6}$$

Define $h(a)$ to be the left side of Formula 6. Thus

$$h(a) = (\int_{\mathbb{R}} e^{(-1/2)x^2}\cos(ax)dx)e^{a^2/2}.$$

We know $h(0) = (2\pi)^{1/2}$.

$$h'(a) = ah(a) + e^{a^2/2}\frac{d}{da}(\int_{\mathbb{R}} e^{-x^2/2}\cos(ax)dx). \tag{7}$$

Forming difference quotients and using the Dominated Convergence Theorem, we can take $\frac{d}{da}$ inside the integral in Formula 7 to obtain

$$-\int_{\mathbb{R}} xe^{(-1/2)x^2}\sin(ax)dx.$$

Integrating this by parts yields

$$\frac{d}{da}(\int_{\mathbb{R}} e^{-x^2/2}\cos(ax)dx) = -a(\int_{\mathbb{R}} e^{-x^2/2}\cos(ax)dx).$$

Therefore

$$\begin{aligned} h'(a) &= ah(a) - ae^{a^2/2}\int_{\mathbb{R}} e^{-x^2/2}\cos(ax)dx \\ &= ah(a) - ah(a) = 0. \end{aligned}$$

This proves the lemma since $h(0) = (2\pi)^{1/2}$.

Proof of Theorem 16.5 Let

$$g_\varepsilon(\mathbf{x}) = e^{(-\varepsilon^2/2)\mathbf{x}\cdot\mathbf{x}}.$$

Thus $0 \le g_\varepsilon(\mathbf{x}) \le 1$ and $\lim_{\varepsilon\to 0+} g_\varepsilon(\mathbf{x}) = 1$. By the Dominated Convergence Theorem,

$$(F \circ F^{-1})f(\mathbf{x}) = \lim_{\varepsilon\to 0}(2\pi)^{\frac{-n}{2}}\int_{\mathbb{R}^n} F^{-1}f(\mathbf{t})g_\varepsilon(\mathbf{t})e^{-i\mathbf{t}\cdot\mathbf{x}}dt.$$

Therefore,

$$
\begin{aligned}
(F \circ F^{-1}) f(\mathbf{x}) &= \lim_{\varepsilon \to 0} (2\pi)^{-n} \int_{\mathbb{R}^n} \int_{\mathbb{R}^n} e^{i\mathbf{y} \cdot \mathbf{t}} f(\mathbf{y}) g_\varepsilon(\mathbf{t}) e^{-i\mathbf{x} \cdot \mathbf{t}} dy \, dt \qquad (8) \\
&= \lim_{\varepsilon \to 0} (2\pi)^{-n} \int_{\mathbb{R}^n} \int_{\mathbb{R}^n} e^{i(\mathbf{y} - \mathbf{x}) \cdot \mathbf{t}} f(\mathbf{y}) g_\varepsilon(\mathbf{t}) dt \, dy \\
&= \lim_{\varepsilon \to 0} (2\pi)^{-\frac{n}{2}} \int_{\mathbb{R}^n} f(\mathbf{y}) [(2\pi)^{-\frac{n}{2}} \int_{\mathbb{R}^n} e^{i(\mathbf{y} - \mathbf{x}) \cdot \mathbf{t}} g_\varepsilon(\mathbf{t}) dt] dy.
\end{aligned}
$$

Consider [ ] in Formula 8. This equals

$$
(2\pi)^{-\frac{n}{2}} \varepsilon^{-n} \int_{\mathbb{R}^n} e^{-\frac{1}{2}(\mathbf{u} - i\mathbf{a}) \cdot (\mathbf{u} - i\mathbf{a})} e^{-\frac{1}{2} |\frac{\mathbf{x} - \mathbf{y}}{\varepsilon}|^2} du,
$$

where $\mathbf{a} = \varepsilon^{-1}(\mathbf{y} - \mathbf{x})$, and $|\mathbf{z}| = (\mathbf{z} \cdot \mathbf{z})^{\frac{1}{2}}$. Applying Lemma 16.6

$$
\begin{aligned}
(2\pi)^{-\frac{n}{2}} [\ ] &= (2\pi)^{-\frac{n}{2}} \varepsilon^{-n} e^{-\frac{1}{2} |\frac{\mathbf{x} - \mathbf{y}}{\varepsilon}|^2} \\
&\equiv m_\varepsilon(\mathbf{y} - \mathbf{x}) = m_\varepsilon(\mathbf{x} - \mathbf{y})
\end{aligned}
$$

and by Lemma 16.6,

$$
\int_{\mathbb{R}^n} m_\varepsilon(\mathbf{y}) dy = 1. \qquad (9)
$$

Thus from Formula 8,

$$
\begin{aligned}
(F \circ F^{-1}) f(\mathbf{x}) &= \lim_{\varepsilon \to 0} \int_{\mathbb{R}^n} f(\mathbf{y}) m_\varepsilon(\mathbf{x} - \mathbf{y}) dy \qquad (10) \\
&= \lim_{\varepsilon \to 0} f * m_\varepsilon(\mathbf{x}).
\end{aligned}
$$

$$
\int_{|\mathbf{y}| \geq \delta} m_\varepsilon(\mathbf{y}) dy = (2\pi)^{-\frac{n}{2}} \Big(\int_{|\mathbf{y}| \geq \delta} e^{-\frac{1}{2} |\frac{\mathbf{y}}{\varepsilon}|^2} dy \Big) \varepsilon^{-n}.
$$

Using polar coordinates,

$$
= (2\pi)^{-n/2} \int_\delta^\infty \int_{S^{n-1}} e^{(-1/(2\varepsilon^2)) \rho^2} \rho^{n-1} d\omega \, d\rho \, \varepsilon^{-n}
$$

$$
= (2\pi)^{-n/2} \Big(\int_{\delta/\varepsilon}^\infty e^{(-1/2) \rho^2} \rho^{n-1} d\rho \Big) C_n.
$$

This clearly converges to 0 as $\varepsilon \to 0+$ because of the Dominated Convergence Theorem and the fact that $\rho^{n-1} e^{-\rho^2/2}$ is in $L^1(\mathbb{R})$. Hence

$$
\lim_{\varepsilon \to 0} \int_{|\mathbf{y}| \geq \delta} m_\varepsilon(\mathbf{y}) dy = 0.
$$

Let δ be small enough that $|f(\mathbf{x}) - f(\mathbf{x} - \mathbf{y})| < \eta$ whenever $|\mathbf{y}| \leq \delta$. Therefore, from Formulas 9 and 10,

$$
|f(\mathbf{x}) - (F \circ F^{-1}) f(\mathbf{x})| = \lim_{\varepsilon \to 0} |f(\mathbf{x}) - f * m_\varepsilon(\mathbf{x})|
$$

$$\leq \limsup_{\varepsilon \to 0} \int_{\mathbb{R}^n} |f(\mathbf{x}) - f(\mathbf{x} - \mathbf{y})| m_\varepsilon(\mathbf{y}) dy$$

$$\leq \limsup_{\varepsilon \to 0} \left(\int_{|\mathbf{y}| > \delta} |f(\mathbf{x}) - f(\mathbf{x} - \mathbf{y})| m_\varepsilon(\mathbf{y}) dy + \right.$$

$$\left. \int_{|\mathbf{y}| \leq \delta} |f(\mathbf{x}) - f(\mathbf{x} - \mathbf{y})| m_\varepsilon(\mathbf{y}) dy \right)$$

$$\leq \limsup_{\varepsilon \to 0} ((\int_{|\mathbf{y}| > \delta} m_\varepsilon(\mathbf{y}) dy) 2 ||f||_\infty + \eta) = \eta.$$

Since $\eta > 0$ is arbitrary, $f(\mathbf{x}) = (F \circ F^{-1}) f(\mathbf{x})$ whenever $f \in \mathfrak{S}$. This proves Theorem 16.5 and justifies the notation in Definition 16.3.

16.2 Fourier transforms of functions in $L^2(\mathbb{R}^n)$

With this preparation, we are ready to begin the consideration of Ff and $F^{-1} f$ for $f \in L^2(\mathbb{R}^n)$. First note that the formula given for Ff and $F^{-1} f$ when $f \in \mathfrak{S}$ will not work for $f \in L^2(\mathbb{R}^n)$ unless f is also in $L^1(\mathbb{R}^n)$. The following theorem will make possible the definition of Ff and $F^{-1} f$ for arbitrary $f \in L^2(\mathbb{R}^n)$.

Theorem 16.7 For $\phi \in \mathfrak{S}$, $||F\phi||_2 = ||F^{-1}\phi||_2 = ||\phi||_2$.

Proof: First note that for $\psi \in \mathfrak{S}$,

$$F(\bar{\psi}) = \overline{F^{-1}(\psi)}, \; F^{-1}(\bar{\psi}) = \overline{F(\psi)}. \qquad (11)$$

This follows from the definition. Let $\phi, \psi \in \mathfrak{S}$.

$$\int_{\mathbb{R}^n} (F\phi(\mathbf{t})) \psi(\mathbf{t}) dt = (2\pi)^{-n/2} \int_{\mathbb{R}^n} \int_{\mathbb{R}^n} \psi(\mathbf{t}) \phi(\mathbf{x}) e^{-i\mathbf{t} \cdot \mathbf{x}} dx dt \quad (12)$$

$$= \int_{\mathbb{R}^n} \phi(\mathbf{x}) (F\psi(\mathbf{x})) dx.$$

Similarly,

$$\int_{\mathbb{R}^n} \phi(\mathbf{x}) (F^{-1}\psi(\mathbf{x})) dx = \int_{\mathbb{R}^n} (F^{-1}\phi(\mathbf{t})) \psi(\mathbf{t}) dt. \qquad (13)$$

Now, Formula 11 - 13 imply

$$\int_{\mathbb{R}^n} |\phi(\mathbf{x})|^2 dx = \int_{\mathbb{R}^n} \phi(\mathbf{x}) \overline{F^{-1}(F\phi(\mathbf{x}))} dx$$

$$= \int_{\mathbb{R}^n} \phi(\mathbf{x}) F(\overline{F\phi(\mathbf{x})}) dx$$

$$= \int_{\mathbb{R}^n} F\phi(\mathbf{x})(\overline{F\phi(\mathbf{x})})dx$$

$$= \int_{\mathbb{R}^n} |F\phi|^2 dx.$$

Similarly

$$||\phi||_2 = ||F^{-1}\phi||_2.$$

This proves the theorem.

Theorem 16.8 *If $f \in L^2(\mathbb{R}^n)$, there exists a unique $\hat{f} \in L^2(\mathbb{R}^n)$ and $\tilde{f} \in L^2(\mathbb{R}^n)$ such that for all $\phi \in \mathfrak{S}$,*

$$\int_{\mathbb{R}^n} \hat{f}(\mathbf{x})\phi(\mathbf{x})dx = \int_{\mathbb{R}^n} f(\mathbf{x})F\phi(\mathbf{x})dx, \qquad (14)$$

$$\int_{\mathbb{R}^n} \tilde{f}(\mathbf{x})\phi(\mathbf{x})dx = \int_{\mathbb{R}^n} f(\mathbf{x})F^{-1}\phi(\mathbf{x})dx. \qquad (15)$$

Proof: Let

$$L\phi = \int_{\mathbb{R}^n} f(\mathbf{x})F\phi(\mathbf{x})dx \qquad (16)$$

for $\phi \in \mathfrak{S}$. Then L is linear on $\mathfrak{S}$ and

$$|L\phi| \leq ||f||_2||F\phi||_2 = ||f||_2||\phi||_2.$$

Since $\mathfrak{S}$ is dense in $L^2(\mathbb{R}^n)$, L extends uniquely to an element of $(L^2(\mathbb{R}^n))'$ still denoted by L. Let $\hat{f}$ be the representor of L. Thus $\hat{f} \in L^2(\mathbb{R}^n)$ and

$$\int_{\mathbb{R}^n} \phi(\mathbf{x})\hat{f}(\mathbf{x})dx = L\phi = \int_{\mathbb{R}^n} f(\mathbf{x})F\phi(\mathbf{x})dx.$$

This proves Formula 14. Formula 15 is similar.

Definition 16.9 For $f \in L^2(\mathbb{R}^n)$, Ff and $F^{-1}f$ are the unique elements of $L^2(\mathbb{R}^n)$ satisfying

$$\int_{\mathbb{R}^n} (Ff)\phi dx = \int_{\mathbb{R}^n} f(F\phi)dx,$$

$$\int_{\mathbb{R}^n} (F^{-1}f)\phi dx = \int_{\mathbb{R}^n} f(F^{-1}\phi)dx,$$

respectively for all $\phi \in \mathfrak{S}$.

By Formula 12 and Formula 13, this is well defined.

Theorem 16.10 *(Plancherel) For $f \in L^2(\mathbb{R}^n)$.*

$$(F^{-1} \circ F)(f) = f = (F \circ F^{-1})(f), \qquad (17)$$

$$||f||_2 = ||Ff||_2 = ||F^{-1}f||_2. \tag{18}$$

Proof: Let $\phi \in \mathfrak{S}$.

$$\int_{\mathbb{R}^n} (F^{-1} \circ F)(f)\phi dx \quad = \quad \int_{\mathbb{R}^n} Ff \, F^{-1}\phi dx = \int_{\mathbb{R}^n} f \, F(F^{-1}(\phi))dx$$

$$= \quad \int f\phi dx.$$

Thus $(F^{-1} \circ F)(f) = f$ because $\mathfrak{S}$ is dense in $L^2(\mathbb{R}^n)$. Similarly $(F \circ F^{-1})(f) = f$. This proves Formula 17. To show Formula 18, we use the density of $\mathfrak{S}$ to write

$$\begin{aligned}
||f||_2 \quad &= \quad \sup\{\int_{\mathbb{R}^n} f\phi dx : ||\phi||_2 \le 1, \ \phi \in \mathfrak{S}\} \\
&= \quad \sup\{|\int_{\mathbb{R}^n} f \, F\phi dx : ||\phi||_2 \le 1, \phi \in \mathfrak{S}\} \\
&= \quad \sup\{|\int_{\mathbb{R}^n} (Ff)\, \phi dx : ||\phi||_2 \le 1, \phi \in \mathfrak{S}\} \\
&= \quad ||Ff||_2.
\end{aligned}$$

Similarly,

$$||f||_2 = ||F^{-1}f||_2.$$

This proves the theorem.

The following corollary is a simple generalization of this.

Corollary 16.11 *For $f, g \in L^2(\mathbb{R}^n)$,*

$$(f, g) = (Ff, Fg) = (F^{-1}f, F^{-1}g).$$

Proof: First note

$$\text{Re}(f, ig) = \text{Im}(f, g).$$

By Theorem 16.10,

$$(f + g, f + g) = (Ff + Fg, Ff + Fg).$$

Hence, using the axioms of the inner product and Plancherel's theorem,

$$\text{Re}(f, g) = \text{Re}(Ff, Fg).$$

Thus

$$\text{Im}(f, g) = \text{Re}(f, ig) = \text{Re}(Ff, iFg) = \text{Im}(Ff, Fg).$$

Since $\text{Re}(f, g) = \text{Re}(Ff, Fg)$, and $\text{Im}(f, g) = \text{Im}(Ff, Fg)$, it follows

$$(f, g) = (Ff, Fg).$$

Similarly,
$$(f, g) = (F^{-1}f, F^{-1}g).$$

This proves the corollary.

How do we compute Ff and $F^{-1}f$?

Theorem 16.12 *For $f \in L^2(\mathbb{R}^n)$, and let E_r be a bounded measurable set with $E_r \uparrow \mathbb{R}^n$. Let $f_r = f\mathcal{X}_{E_r}$. Then*
$$Ff = \lim_{r \to \infty} Ff_r, \quad F^{-1}f = \lim_{r \to \infty} F^{-1}f_r.$$

Proof: $\|f - f_r\|_2 \to 0$ and so $\|Ff - Ff_r\|_2 \to 0$ and $\|F^{-1}f - F^{-1}f_r\|_2 \to 0$ by Plancherel's Theorem. This proves the theorem.

What are Ff_r and $F^{-1}f_r$? Let $\phi \in \mathfrak{S}$

$$\int_{\mathbb{R}^n} Ff_r \phi \, dx = \int_{\mathbb{R}^n} f_r F\phi \, dx$$
$$= (2\pi)^{-\frac{n}{2}} \int_{\mathbb{R}^n} \int_{\mathbb{R}^n} f_r(\mathbf{x}) e^{-i\mathbf{x}\cdot\mathbf{y}} \phi(\mathbf{y}) dy dx$$
$$= \int_{\mathbb{R}^n} [(2\pi)^{-\frac{n}{2}} \int_{\mathbb{R}^n} f_r(\mathbf{x}) e^{-i\mathbf{x}\cdot\mathbf{y}} dx] \phi(\mathbf{y}) dy.$$

Since this holds for all $\phi \in \mathfrak{S}$, a dense subset of $L^2(\mathbb{R}^n)$, it follows that

$$Ff_r(\mathbf{y}) = (2\pi)^{-\frac{n}{2}} \int_{\mathbb{R}^n} f_r(\mathbf{x}) e^{-i\mathbf{x}\cdot\mathbf{y}} dx.$$

Similarly
$$F^{-1}f_r(\mathbf{y}) = (2\pi)^{-\frac{n}{2}} \int_{\mathbb{R}^n} f_r(\mathbf{x}) e^{i\mathbf{x}\cdot\mathbf{y}} dx.$$

This shows that to take the Fourier transform of a function in $L^2(\mathbb{R}^n)$, it suffices to take the limit as $r \to \infty$ in $L^2(\mathbb{R}^n)$ of $(2\pi)^{-\frac{n}{2}} \int_{\mathbb{R}^n} f_r(\mathbf{x}) e^{-i\mathbf{x}\cdot\mathbf{y}} dx$. A similar procedure works for the inverse Fourier transform.

Definition 16.13 For $f \in L^1(\mathbb{R}^n)$, define

$$Ff(\mathbf{x}) \equiv (2\pi)^{-n/2} \int_{\mathbb{R}^n} e^{-i\mathbf{x}\cdot\mathbf{y}} f(\mathbf{y}) \, dy,$$

$$F^{-1}f(\mathbf{x}) \equiv (2\pi)^{-n/2} \int_{\mathbb{R}^n} e^{i\mathbf{x}\cdot\mathbf{y}} f(\mathbf{y}) \, dy.$$

Thus, for $f \in L^1(\mathbb{R}^n)$, Ff and $F^{-1}f$ are in $L^\infty(\mathbb{R}^n)$.

Corollary 16.14 Let $h \in L^2(\mathbb{R}^n)$ and let $f \in L^1(\mathbb{R}^n)$. Then $h * f \in L^2(\mathbb{R}^n)$,
$$F^{-1}(h * f) = (2\pi)^{n/2} F^{-1}hF^{-1}f,$$

$$F\left(h * f\right) = \left(2\pi\right)^{n/2} FhFf,$$

and

$$||h * f||_2 \leq ||h||_2 \, ||f||_1 \, . \tag{19}$$

Proof: Without loss of generality, we may assume h and f are both Borel measurable. Then an application of Minkowski's inequality yields

$$\left(\int_{\mathbb{R}^n} \left(\int_{\mathbb{R}^n} |h\left(\mathbf{x} - \mathbf{y}\right)| \, |f\left(\mathbf{y}\right)| \, dy\right)^2 dx\right)^{1/2} \leq ||f||_1 \, ||h||_2 \, . \tag{20}$$

Hence $\int |h\left(\mathbf{x} - \mathbf{y}\right)| \, |f\left(\mathbf{y}\right)| \, dy < \infty$ a.e. $\mathbf{x}$ and

$$\mathbf{x} \rightarrow \int h\left(\mathbf{x} - \mathbf{y}\right) f\left(\mathbf{y}\right) dy$$

is in $L^2\left(\mathbb{R}^n\right)$. Let $E_r \uparrow \mathbb{R}^n$, $m\left(E_r\right) < \infty$. Thus,

$$h_r \equiv \mathcal{X}_{E_r} h \in L^2\left(\mathbb{R}^n\right) \cap L^1\left(\mathbb{R}^n\right),$$

and letting $\phi \in \mathfrak{S}$,

$$\int F\left(h_r * f\right)\left(\phi\right) dx$$

$$\equiv \int \left(h_r * f\right)\left(F\phi\right) dx$$

$$= \left(2\pi\right)^{-n/2} \int \int \int h_r\left(\mathbf{x} - \mathbf{y}\right) f\left(\mathbf{y}\right) e^{-i\mathbf{x} \cdot \mathbf{t}} \phi\left(\mathbf{t}\right) dt \, dy \, dx$$

$$= \left(2\pi\right)^{-n/2} \int \int \left(\int h_r\left(\mathbf{x} - \mathbf{y}\right) e^{-i\left(\mathbf{x} - \mathbf{y}\right) \cdot \mathbf{t}} dx\right) f\left(\mathbf{y}\right) e^{-i\mathbf{y} \cdot \mathbf{t}} dy \phi\left(\mathbf{t}\right) dt$$

$$= \int \left(2\pi\right)^{n/2} Fh_r\left(\mathbf{t}\right) Ff\left(\mathbf{t}\right) \phi\left(\mathbf{t}\right) dt.$$

Since ϕ is arbitrary and $\mathfrak{S}$ is dense in $L^2\left(\mathbb{R}^n\right)$,

$$F\left(h_r * f\right) = \left(2\pi\right)^{n/2} Fh_r Ff.$$

Now by Minkowski's Inequality, $h_r * f \rightarrow h * f$ in $L^2\left(\mathbb{R}^n\right)$ and also it is clear that $h_r \rightarrow h$ in $L^2\left(\mathbb{R}^n\right)$; so, by Plancherel's theorem, we may take the limit in the above and conclude

$$F\left(h * f\right) = \left(2\pi\right)^{n/2} FhFf.$$

The assertion for F^{-1} is similar and Formula 19 follows from Formula 20.

16.3   Tempered distributions

In this section we give an introduction to the general theory of Fourier transforms. Recall that $\mathfrak{S}$ is the set of all $\phi \in C^\infty (\mathbb{R}^n)$ such that for $N = 1, 2, \cdots$,

$$\rho_N (\phi) \equiv \sup_{|\alpha| \leq N, \mathbf{x} \in \mathbb{R}^n} \left(1 + |\mathbf{x}|^2\right)^N |D^\alpha \phi (\mathbf{x})| < \infty.$$

Thus ρ_N is a seminorm and $\mathfrak{S}$ together with this collection of semi norms is a topological vector space with a basis for the topology given by sets of the form

$$B^N (\phi, r) \equiv \{\psi \in \mathfrak{S} \text{ such that } \rho_N (\psi - \phi) < r\}.$$

Then $\mathfrak{S}'$, the space of continuous linear functions defined on $\mathfrak{S}$ mapping $\mathfrak{S}$ to $\mathbb{C}$, are called tempered distributions. Thus,

Definition 16.15 $f \in \mathfrak{S}'$ means $f : \mathfrak{S} \to \mathbb{C}$, f is linear, and f is continuous.

How can we verify $f \in \mathfrak{S}'$? The following lemma is about this question along with the similar question of when a linear map from $\mathfrak{S}$ to $\mathfrak{S}$ is continuous.

Lemma 16.16 *Let* $f : \mathfrak{S} \to \mathbb{C}$ *be linear and let* $L : \mathfrak{S} \to \mathfrak{S}$ *be linear. Then* f *is continuous if and only if*

$$|f (\phi)| \leq C \rho_N (\phi) \qquad\qquad (a.)$$

for some N. *Also,* L *is continuous if and only if for each* N, *there exists* M *such that*

$$\rho_N (L\phi) \leq C \rho_M (\phi) \qquad\qquad (b.)$$

for some C *independent of* ϕ.

Proof: It suffices to verify continuity at 0 because f and L are linear. We verify $(b.)$. Let $0 \in U$ where U is an open set. Then $0 \in B^N (0, r) \subseteq U$ for some $r > 0$ and N. Then if M and C are as described in $(b.)$, and $\psi \in B^M (0, C^{-1}r)$, we have

$$\rho_N (L\psi) \leq C \rho_M (\psi) < r;$$

so, this shows

$$B^M (0, C^{-1}r) \subseteq L^{-1} (B^N (0, r)) \subseteq L^{-1} (U),$$

which shows that L is continuous at 0. The argument for f and the only if part of the proof is left for the reader.

The key to extending the Fourier transform to $\mathfrak{S}'$ is the following theorem which states that F and F^{-1} are continuous. This is analogous to the procedure in defining the Fourier transform for functions in $L^2(\mathbb{R}^n)$. Recall we proved these mappings preserved the L^2 norms of functions in $\mathfrak{S}$.

Theorem 16.17 *For F and F^{-1} the Fourier and inverse Fourier transforms,*

$$F\phi(\mathbf{x}) \equiv (2\pi)^{-n/2} \int_{\mathbb{R}^n} e^{-i\mathbf{x}\cdot\mathbf{y}} \phi(\mathbf{y})\, dy,$$

$$F^{-1}\phi(\mathbf{x}) \equiv (2\pi)^{-n/2} \int_{\mathbb{R}^n} e^{i\mathbf{x}\cdot\mathbf{y}} \phi(\mathbf{y})\, dy,$$

F and F^{-1} are continuous linear maps from $\mathfrak{S}$ to $\mathfrak{S}$.

Proof: Let $|\alpha| \leq N$ where $N > 0$, and let $\mathbf{x} \neq \mathbf{0}$. Then

$$\left(1 + |\mathbf{x}|^2\right)^N \left|D^\alpha F^{-1}\phi(\mathbf{x})\right| \equiv C_n \left(1 + |\mathbf{x}|^2\right)^N \left|\int_{\mathbb{R}^n} e^{i\mathbf{x}\cdot\mathbf{y}} \mathbf{y}^\alpha \phi(\mathbf{y})\, dy\right|. \tag{21}$$

Suppose $x_j^2 \geq 1$ for $j \in \{i_1, \cdots, i_r\}$ and $x_j^2 < 1$ if $j \notin \{i_1, \cdots, i_r\}$. Then after integrating by parts in the integral of Formula 21, we obtain the following for the right side of Formula 21:

$$C_n \left(1 + |\mathbf{x}|^2\right)^N \left|\int_{\mathbb{R}^n} e^{i\mathbf{x}\cdot\mathbf{y}} D^\beta \left(\mathbf{y}^\alpha \phi(\mathbf{y})\right)\, dy\right| \prod_{j \in \{i_1, \cdots, i_r\}} x_j^{-2N}, \tag{22}$$

where

$$\beta \equiv 2N \sum_{k=1}^r \mathbf{e}_{i_k},$$

the vector with 0 in the *jth* slot if $j \notin \{i_1, \cdots, i_r\}$ and a $2N$ in the *jth* slot for $j \in \{i_1, \cdots, i_r\}$. Now letting $C(n, N)$ denote a generic constant depending only on n and N, the product rule and a little estimating yields

$$\left|D^\beta \left(\mathbf{y}^\alpha \phi(\mathbf{y})\right)\right| \left(1 + |\mathbf{y}|^2\right)^{nN} \leq \rho_{2Nn}(\phi)\, C(n, N).$$

Also the function $\mathbf{y} \to \left(1 + |\mathbf{y}|^2\right)^{-nN}$ is integrable. Therefore, the expression in Formula 22 is no larger than

$$C(n, N) \left(1 + |\mathbf{x}|^2\right)^N \prod_{j \in \{i_1, \cdots, i_r\}} x_j^{-2N} \rho_{2Nn}(\phi)$$

$$\leq C\left(n, N\right) \left(1 + n + \sum_{j \in \{i_1, \cdots, i_r\}} |x_j|^2\right)^N \prod_{j \in \{i_1, \cdots, i_r\}} x_j^{-2N} \rho_{2Nn}\left(\phi\right)$$

$$\leq C\left(n, N\right) \sum_{j \in \{i_1, \cdots, i_r\}} |x_j|^{2N} \prod_{j \in \{i_1, \cdots, i_r\}} |x_j|^{-2N} \rho_{2Nn}\left(\phi\right)$$

$$\leq C\left(n, N\right) \rho_{2Nn}\left(\phi\right).$$

Therefore, if $x_j^2 \geq 1$ for some j,

$$\left(1 + |\mathbf{x}|^2\right)^N \left|D^\alpha F^{-1}\phi\left(\mathbf{x}\right)\right| \leq C\left(n, N\right) \rho^{2Nn}\left(\phi\right).$$

If $x_j^2 < 1$ for all j, we can use Formula 21 to obtain

$$\left(1 + |\mathbf{x}|^2\right)^N \left|D^\alpha F^{-1}\phi\left(\mathbf{x}\right)\right| \leq C_n \left(1 + n\right)^N \left|\int_{\mathbb{R}^n} e^{i\mathbf{x}\cdot\mathbf{y}} \mathbf{y}^\alpha \phi\left(\mathbf{y}\right) dy\right|$$

$$\leq C\left(n, N\right) \rho_R\left(\phi\right)$$

for some R depending on N and n. Let $M \geq \max\left(R, 2nN\right)$ and we see

$$\rho_N\left(F^{-1}\phi\right) \leq C\rho_M\left(\phi\right)$$

where M and C depend only on N and n. By Lemma 16.16, F^{-1} is continuous. Similarly, F is continuous. This proves the theorem.

Definition 16.18 For $f \in \mathfrak{S}'$, we define Ff and $F^{-1}f$ in $\mathfrak{S}'$ by

$$Ff\left(\phi\right) \equiv f\left(F\phi\right), \ F^{-1}f\left(\phi\right) \equiv f\left(F^{-1}\phi\right).$$

To see this is a good definition, consider the following.

$$\left|Ff\left(\phi\right)\right| \equiv \left|f\left(F\phi\right)\right| \leq C\rho_N\left(F\phi\right) \leq C\rho_M\left(\phi\right).$$

Also note that F and F^{-1} are both one to one and onto. This follows from the fact that these mappings map $\mathfrak{S}$ one to one onto $\mathfrak{S}$.

What are some examples of things in $\mathfrak{S}'$? In answering this question, we will use the following lemma.

Lemma 16.19 If $f \in L^1_{loc}\left(\mathbb{R}^n\right)$ and $\int_{\mathbb{R}^n} f\phi dx = 0$ for all $\phi \in C_c^\infty\left(\mathbb{R}^n\right)$, then $f = 0$ a.e.

Proof: It is enough to verify this for $f \geq 0$. Let

$$E \equiv \{\mathbf{x} : f\left(\mathbf{x}\right) \geq r\}, \ E_R \equiv E \cap B\left(0, R\right).$$

Let K_n be an increasing sequence of compact sets and let V_n be a decreasing sequence of open sets satisfying

$$K_n \subseteq E_R \subseteq V_n, \ m\left(V_n \setminus K_n\right) \leq 2^{-n}, V_1 \text{ is bounded.}$$

Let

$$\phi_n \in C_c^\infty\left(V_n\right), \ K_n \prec \phi_n \prec V_n.$$

Then $\phi_n\left(\mathbf{x}\right) \to \mathcal{X}_{E_R}\left(\mathbf{x}\right)$ a.e. because the set where $\phi_n\left(\mathbf{x}\right)$ fails to converge is contained in the set of all $\mathbf{x}$ which are in infinitely many of the sets $V_n \setminus K_n$. This set has measure zero and ; so, by the dominated convergence theorem,

$$0 = \lim_{n \to \infty} \int_{\mathbb{R}^n} f\phi_n \, dx = \lim_{n \to \infty} \int_{V_1} f\phi_n \, dx = \int_{E_R} f dx \geq rm\left(E_R\right).$$

Thus, $m\left(E_R\right) = 0$ and therefore $m\left(E\right) = 0$. Since $r > 0$ is arbitrary, it follows

$$m\left(\left[\mathbf{x} : f\left(\mathbf{x}\right) > 0\right]\right) = 0.$$

This proves the lemma.

Theorem 16.20 *Let f be a measurable function with polynomial growth,*

$$|f\left(\mathbf{x}\right)| \leq C\left(1 + |\mathbf{x}|^2\right)^N \ \textit{ for some } N,$$

or let $f \in L^p\left(\mathbb{R}^n\right)$ for some $p \in [1, \infty]$. Then $f \in \mathfrak{S}'$ if we define

$$f\left(\phi\right) \equiv \int f\phi dx.$$

Proof: Let f have polynomial growth first. Then

$$
\begin{aligned}
\int |f| \, |\phi| \, dx \ &\leq \ C \int \left(1 + |\mathbf{x}|^2\right)^{nN} |\phi| \, dx \\
&\leq \ C \int \left(1 + |\mathbf{x}|^2\right)^{nN} \left(1 + |\mathbf{x}|^2\right)^{-2nN} dx \, \rho_{2nN}\left(\phi\right) \\
&\leq \ C\left(N, n\right) \rho_{2nN}\left(\phi\right) < \infty.
\end{aligned}
$$

Therefore we can define

$$f\left(\phi\right) \equiv \int f\phi dx$$

and it follows that

$$|f(\phi)| \leq C\left(N, n\right) \rho_{2nN}\left(\phi\right).$$

By Lemma 16.16, $f \in \mathfrak{S}'$. Next suppose $f \in L^p(\mathbb{R}^n)$.

$$\int |f| |\phi| \, dx \leq \int |f| \left(1 + |\mathbf{x}|^2\right)^{-M} dx \, \rho_M(\phi)$$

where we choose M large enough that $\left(1 + |\mathbf{x}|^2\right)^{-M} \in L^{p'}$. Then by Holder's Inequality,

$$|f(\phi)| \leq \|f\|_p \, C_n \rho_M(\phi).$$

By Lemma 16.16, $f \in \mathfrak{S}'$. This proves the theorem.

Definition 16.21 If $f \in \mathfrak{S}'$ and $\phi \in \mathfrak{S}$, then $\phi f \in \mathfrak{S}'$ if we define

$$\phi f(\psi) \equiv f(\phi \psi).$$

We need to verify that with this definition, $\phi f \in \mathfrak{S}'$. It is clearly linear. There exist constants C and N such that

$$
\begin{aligned}
|\phi f(\psi)| &\equiv |f(\phi \psi)| \leq C \rho_N(\phi \psi) \\
&= C \sup_{\mathbf{x} \in \mathbb{R}^n, |\alpha| \leq N} \left(1 + |\mathbf{x}|^2\right)^N |D^\alpha(\phi \psi)| \\
&\leq C(\phi, n, N) \rho_N(\psi).
\end{aligned}
$$

Thus by Lemma 16.16, $\phi f \in \mathfrak{S}'$.

The next topic is that of convolution. This was discussed in Corollary 16.14 but we review it here. This corollary implies that if $f \in L^2(\mathbb{R}^n) \subseteq \mathfrak{S}'$ and $\phi \in \mathfrak{S}$, then

$$f * \phi \in L^2(\mathbb{R}^n), \quad \|f * \phi\|_2 \leq \|f\|_2 \|\phi\|_1,$$

and also

$$F(f * \phi)(\mathbf{x}) = F\phi(\mathbf{x}) Ff(\mathbf{x}) (2\pi)^{n/2}, \tag{23}$$

$$F^{-1}(f * \phi)(\mathbf{x}) = F^{-1}\phi(\mathbf{x}) F^{-1}f(\mathbf{x}) (2\pi)^{n/2}.$$

By Definition 16.21,

$$F(f * \phi) = (2\pi)^{n/2} F\phi Ff$$

whenever $f \in L^2(\mathbb{R}^n)$ and $\phi \in \mathfrak{S}$.

Now it is easy to see the proper way to define $f * \phi$ when f is only in $\mathfrak{S}'$ and $\phi \in \mathfrak{S}$.

Definition 16.22 Let $f \in \mathfrak{S}'$ and let $\phi \in \mathfrak{S}$. Then we define

$$f * \phi \equiv (2\pi)^{n/2} F^{-1}(F\phi Ff).$$

Theorem 16.23 *Let $f \in \mathfrak{S}'$ and let $\phi \in \mathfrak{S}$.*

$$F(f * \phi) = (2\pi)^{n/2} F\phi F f, \tag{24}$$

$$F^{-1}(f * \phi) = (2\pi)^{n/2} F^{-1}\phi F^{-1}f.$$

Proof: Note that Formula 24 follows from Definition 16.22 and both assertions hold for $f \in \mathfrak{S}$. Next we write for $\psi \in \mathfrak{S}$,

$$\left(\psi * F^{-1}F^{-1}\phi\right)(\mathbf{x})$$

$$= \left(\int\int\int \psi(\mathbf{x} - \mathbf{y})\, e^{i\mathbf{y}\cdot\mathbf{y}_1} e^{i\mathbf{y}_1\cdot\mathbf{z}}\phi(\mathbf{z})\, dz dy_1 dy\right)(2\pi)^n$$

$$= \left(\int\int\int \psi(\mathbf{x} - \mathbf{y})\, e^{-i\mathbf{y}\cdot\tilde{\mathbf{y}}_1} e^{-i\tilde{\mathbf{y}}_1\cdot\mathbf{z}}\phi(\mathbf{z})\, dz d\tilde{y}_1 dy\right)(2\pi)^n$$

$$= (\psi * FF\phi)(\mathbf{x}).$$

Now for $\psi \in \mathfrak{S}$,

$$(2\pi)^{n/2} F\left(F^{-1}\phi F^{-1}f\right)(\psi) \equiv (2\pi)^{n/2}\left(F^{-1}\phi F^{-1}f\right)(F\psi) \equiv$$

$$(2\pi)^{n/2} F^{-1}f\left(F^{-1}\phi F\psi\right) \equiv (2\pi)^{n/2} f\left(F^{-1}\left(F^{-1}\phi F\psi\right)\right) =$$

$$f\left((2\pi)^{n/2} F^{-1}\left(\left(FF^{-1}F^{-1}\phi\right)(F\psi)\right)\right) \equiv$$

$$f\left(\psi * F^{-1}F^{-1}\phi\right) = f\left(\psi * FF\phi\right)$$

$$(2\pi)^{n/2} F^{-1}\left(F\phi Ff\right)(\psi) \equiv (2\pi)^{n/2}\left(F\phi Ff\right)\left(F^{-1}\psi\right) \equiv$$

$$(2\pi)^{n/2} Ff\left(F\phi F^{-1}\psi\right) \equiv (2\pi)^{n/2} f\left(F\left(F\phi F^{-1}\psi\right)\right) =$$

by Formula 23,

$$= f\left(F\left((2\pi)^{n/2}\left(F\phi F^{-1}\psi\right)\right)\right)$$

$$= f\left(F\left(F^{-1}(FF\phi * \psi)\right)\right) = f(\psi * FF\phi).$$

Comparing the above shows

$$(2\pi)^{n/2} F\left(F^{-1}\phi F^{-1}f\right) = (2\pi)^{n/2} F^{-1}(F\phi Ff) \equiv f * \phi$$

and ; so,

$$(2\pi)^{n/2}\left(F^{-1}\phi F^{-1}f\right) = F^{-1}(f * \phi)$$

which proves the theorem.

16.4   Exercises

1. Let $f \in L^1(\mathbb{R}^n)$,

$$Ff(t) \equiv (2\pi)^{-n/2} \int_{\mathbb{R}^n} e^{-it\cdot x} f(\mathbf{x}) dx,$$

$$F^{-1}f(t) \equiv (2\pi)^{-n/2} \int_{\mathbb{R}^n} e^{it\cdot x} f(\mathbf{x}) dx.$$

Show that $F^{-1}f$ and Ff are both continuous and bounded. Show also that

$$\lim_{|\mathbf{x}| \to \infty} F^{-1}f(\mathbf{x}) = \lim_{|\mathbf{x}| \to \infty} Ff(\mathbf{x}) = 0.$$

2. Suppose $f \in L^1(\mathbb{R}^n)$ and $||g_k - f||_1 \to 0$. Show Fg_k and $F^{-1}g_k$ converge uniformly to Ff and $F^{-1}f$ respectively.

3. Suppose $F^{-1}f \in L^1(\mathbb{R}^n)$. Observe that just as in Theorem 16.5,

$$(F \circ F^{-1})f(\mathbf{x}) = \lim_{\varepsilon \to 0} f * m_\varepsilon(\mathbf{x}).$$

Use this to argue that if f and $F^{-1}f \in L^1(\mathbb{R}^n)$, then

$$(F \circ F^{-1}) f(\mathbf{x}) = f(\mathbf{x}) \text{ a.e. } x.$$

Similarly

$$(F^{-1} \circ F)f(\mathbf{x}) = f(\mathbf{x}) \text{ a.e.}$$

if f and $Ff \in L^1(\mathbb{R}^n)$. **Hint**: Show $f * m_\varepsilon \to f$ in $L^1(\mathbb{R}^n)$. Thus there is a subsequence $\varepsilon_k \to 0$ such that $f * m_{\varepsilon_k}(\mathbf{x}) \to f(\mathbf{x})$ a.e.

4. ↑ Show that if $F^{-1}f \in L^1$ or $Ff \in L^1$, then f equals a continuous bounded function a.e.

5. Let $f, g \in L^1(\mathbb{R}^n)$. Show $f * g \in L^1$ and $F(f * g) = (2\pi)^{n/2} FfFg$.

6. ↑ Suppose $f, g \in L^1(\mathbb{R})$ and $Ff = Fg$. Show $f = g$ a.e.

7. ↑ Suppose $f * f = f$ or $f * f = 0$ and $f \in L^1(\mathbb{R})$. Show $f = 0$.

8. Suppose $f \in \mathfrak{S}$, $f_{x_j} \in L^1(\mathbb{R}^n)$. Show $F(f_{x_j})(t) = it_j Ff(t)$.

9. Let $f \in \mathfrak{S}$ and let k be a positive integer.

$$||f||_{k,2} \equiv (||f||_2^2 + \sum_{|\alpha| \le k} ||D^\alpha f||_2^2)^{1/2}.$$

One could also define

$$|||f|||_{k,2} \equiv \left(\int_{R^n} |Ff(\mathbf{x})|^2 (1+|\mathbf{x}|^2)^k \, dx \right)^{1/2}.$$

Show both $|| \; ||_{k,2}$ and $||| \; |||_{k,2}$ are norms on $\mathfrak{S}$ and that they are equivalent. These are Sobolev space norms. What are some possible advantages of the second norm over the first? **Hint:** For which values of k do these make sense?

10. ↑ Define $H^k(\mathbb{R}^n)$ by $f \in L^2(\mathbb{R}^n)$ such that

$$\left(\int |Ff(x)|^2 (1+|x|^2)^k \, dx \right)^{\frac{1}{2}} < \infty,$$

$$|||f|||_{k,2} \equiv \left(\int |Ff(x)|^2 (1+|x|^2)^k \, dx \right)^{\frac{1}{2}}.$$

Show $H^k(\mathbb{R}^n)$ is a Banach space, and that if k is a positive integer, $H^k(\mathbb{R}^n) = \{ f \in L^2(\mathbb{R}^n) :$ there exists $\{u_j\} \subseteq \mathfrak{S}$ with $||u_j - f||_2 \to 0$ and $\{u_j\}$ is a Cauchy sequence in $|| \; ||_{k,2}$ of Problem 9}. This is one way to define Sobolev Spaces. **Hint:** One way to do the second part of this is to let $g_s \to Ff$ in $L^2((1+|x|^2)^k \, dx)$ where $g_s \in C_c(\mathbb{R}^n)$. We can do this because $(1+|x|^2)^k \, dx$ is a Radon measure. By convolving with a mollifier, we can, without loss of generality, assume $g_s \in C_c^\infty(\mathbb{R}^n)$. Thus $g_s = Ff_s$, $f_s \in \mathfrak{S}$. Then by Problem 9, f_s is Cauchy in the norm $|| \; ||_{k,2}$.

11. If Y is a closed subspace of a reflexive Banach space X, show Y is reflexive.

12. ↑ Show $H^k(\mathbb{R}^n)$ is a Hilbert space.

13. ↑ If $2k > n$, show that if $f \in H^k(\mathbb{R}^n)$, then f equals a bounded continuous function a.e. **Hint:** Show $Ff \in H^k(\mathbb{R}^n)$, and then use Problem 4. To do this, write

$$|Ff(x)| = |Ff(x)|(1+|x|^2)^{\frac{k}{2}}(1+|x|^2)^{\frac{-k}{2}}.$$

So

$$\int |Ff(x)| \, dx = \int |Ff(x)|(1+|x|^2)^{\frac{k}{2}}(1+|x|^2)^{\frac{-k}{2}} \, dx.$$

Use Holder's Inequality. This is an example of a Sobolev Embedding Theorem.

14. Show $\left\{ (2m\pi)^{-(1/2)} e^{i\frac{n}{m}x} \right\}_{n\in\mathbb{Z}}$ is a complete orthonormal set in

$$L^2(-m\pi, m\pi)$$

and suppose f is a function in $L^2(\mathbb{R})$ satisfying $Ff(t) = 0$ if $|t| > m\pi$. Show that if this is so, then

$$f(x) = \frac{1}{\pi} \sum_{n\in\mathbb{Z}} f\left(\frac{-n}{m}\right) \frac{\sin(\pi(mx+n))}{mx+\pi}.$$

This is sometimes called the Shannon sampling theorem.

Chapter 17
Probability

17.1 Random vectors

The purpose of this chapter is to consider some of the main theorems in probability which are closely related to measure theory and Lebesgue integration. This section is on random vectors and random variables, random vectors with values in $\mathbb{R}$ or $\mathbb{C}$. We will denote by $\mathbb{F}$ either $\mathbb{C}$ or $\mathbb{R}$ with the understanding that $\mathbb{F} = \mathbb{C}$ unless specified otherwise. In dealing with random vectors, we will need to consider the integral of a vector valued function. This is easy to do by considering it componentwise. Thus, if ν is a measure on a σ algebra of sets of Ω and

$$\mathbf{f} = (f_1, \cdots, f_p)^T$$

is a vector valued measurable function defined on Ω having values in $\mathbb{F}^p$, we define

$$\int \mathbf{f} \, d\nu \equiv \left(\int f_1 \, d\nu, \cdots, \int f_p \, d\nu \right)^T$$

The notation $L^1(\Omega; \mathbb{F}^p)$ will denote those vector valued measurable functions for which $\int |\mathbf{f}| \, d\nu < \infty$. Note that $\mathbf{f} \in L^1(\Omega; \mathbb{F}^p)$ if and only if $f_i \in L^1(\Omega)$ for each component function.

Definition 17.1 We say $\mathbf{X}$ is a random vector if $\mathbf{X}$ is a measurable function from some probability space, $(\Omega, \mathcal{S}, P)$ to $\mathbb{F}^n$ where here P is called a probability measure because $P(\Omega) = 1$. We will also follow the convention of denoting the random vector by capital letters and its values by lower case letters. Thus, if $\mathbf{X}$ is a random vector, $\mathbf{x}$ will refer to its values.

In this section the underlying probability space is not so much the thing of interest as the probability that the random vector has values in some Borel subset of $\mathbb{F}^n$. We may define a Radon probability measure, $\lambda_{\mathbf{X}}$ as follows. For E a Borel set in $\mathbb{F}^n$,

$$\widehat{\lambda}_{\mathbf{X}}(E) \equiv P(\{\omega : \mathbf{X}(\omega) \in E\}).$$

Now since $\widehat{\lambda}_{\mathbf{X}}$ is a probability measure, it is certainly finite on the compact sets. Therefore, by theorem 9.26 there is a unique Radon measure extending $\widehat{\lambda}_{\mathbf{X}}$ which will be denoted as $\lambda_{\mathbf{X}}$. This proves the following lemma.

Lemma 17.2 *Let* **X** *be a random vector. Then there exists a unique σ algebra containing the Borel sets of $\mathbb{F}^n$ and Radon measure, $\lambda_{\mathbf{X}}$ defined on this σ algebra which satisfies*

$$\lambda_{\mathbf{X}}(E) \equiv P(\{\omega : \mathbf{X}(\omega) \in E\})$$

for every E a Borel set in $\mathbb{F}^n$.

In probability and statistics, the notion of the distribution function of a random variable is also very important.

Definition 17.3 For **X** a random vector with values in $\mathbb{R}^p$, we define the distribution function, $F_{\mathbf{X}}$ as $F_{\mathbf{X}}(\mathbf{x}) = P(\mathbf{X} \in R_{\mathbf{x}})$ where

$$R_{\mathbf{x}} \equiv \prod_{i=1}^{p} (-\infty, x_i].$$

Lemma 17.4 *Let* **X** *and* **Y** *be random vectors with values in $\mathbb{R}^p$. Then $F_{\mathbf{X}}(\mathbf{x}) = F_{\mathbf{Y}}(\mathbf{y})$ for all $\mathbf{x} \in \mathbb{R}^p$ if and only if $\lambda_{\mathbf{X}} = \lambda_{\mathbf{Y}}$.*

Proof: The if part is obvious. Suppose the distribution functions coincide. Then $\lambda_{\mathbf{X}} = \lambda_{\mathbf{Y}}$ on the algebra of half open rectangles described in example 7.8. Therefore, an application of the monotone class theorem shows $\lambda_{\mathbf{X}} = \lambda_{\mathbf{Y}}$ on the Borel sets. Therefore, since the two measures are Radon measures which are equal on open sets, they are equal. This proves the lemma.

Definition 17.5 For **X** a random vector with values in $\mathbb{F}^n$ we define

$$E(\mathbf{X}) \equiv \int_{\mathbb{F}^n} \mathbf{x} d\lambda_{\mathbf{X}}.$$

The symbol, $E(\mathbf{X})$ is called the expected value of **X**. Of course, for it to make sense we must be assuming that $\int_{\mathbb{F}^n} |\mathbf{x}| d\lambda_{\mathbf{X}} < \infty$. We can also obtain the expected value of a random vector as an integral over the underlying probability space.

Lemma 17.6 *For all* $\mathbf{X} \in L^1(\Omega; \mathbb{F}^p)$,

$$\int_{\Omega} \mathbf{X}(\omega) dP(\omega) = E(\mathbf{X}) \equiv \int_{\mathbb{R}^p} \mathbf{x} d\lambda_{\mathbf{X}}(x). \tag{1}$$

Proof: Suppose first $X(\omega) = \mathcal{X}_A(\omega)$ where $A \in \mathcal{S}$. Then $\int_{\Omega} \mathcal{X}_A dP(\omega) = P(A)$. Also $\int_{\mathbb{R}} x d\lambda_X(x) = \int_{\{1\}} d\lambda_X(x) = P(X = 1) = P(A)$. It follows Formula 1 holds whenever X is a nonnegative simple function. Consequently, the monotone convergence theorem implies Formula 1 holds for all nonnegative random variables and hence for all $\mathbf{X} \in L^1\left(\Omega; \mathbb{F}^p\right)$ after considering the component functions of $\mathbf{X}$.

For a random vector, $\mathbf{X}$, certain functions of $\mathbf{X}$ have expected values which are so important in applications that they are given special symbols.

Definition 17.7 For $\mathbf{X}$ a random vector,

$$\mu \equiv E\left(\mathbf{X}\right), \ \Sigma \equiv E\left(\left(\mathbf{X} - \mu\right)\left(\mathbf{X} - \mu\right)^*\right).$$

In this definition, μ is called the mean and Σ is called the covariance matrix. Thus

$$\Sigma_{ij} = E\left(\left(X_i - \mu_i\right)\left(X_j - \mu_j\right)\right)$$

where $\mu = \left(\mu_1, \cdots, \mu_n\right)$ and $\mathbf{X} = \left(X_1, \cdots, X_n\right)$.

Definition 17.8 Let $\mathbf{X}$ be a random vector. The function

$$\mathbf{t} \rightarrow E\left(e^{i\mathbf{t}\cdot\mathbf{X}}\right) \equiv \phi_{\mathbf{X}}\left(\mathbf{t}\right)$$

is called the characteristic function of $\mathbf{X}$.

Definition 17.9 If $\mathbf{X}$ is a random vector with values in $\mathbb{R}^p$ and $E\left(e^{\mathbf{t}\cdot\mathbf{X}}\right)$ is finite, we denote this expected value by $M_{\mathbf{X}}\left(\mathbf{t}\right)$ and call the function $\mathbf{t} \rightarrow M_{\mathbf{X}}\left(\mathbf{t}\right)$ the moment generating function of $\mathbf{X}$.

Now we give some important theorems about some of these definitions. The following theorem gives a convenient way to verify that $\lambda_{\mathbf{X}} = \lambda_{\mathbf{Y}}$.

Theorem 17.10 *Let $\mathbf{X}$ and $\mathbf{Y}$ be random vectors with values in $\mathbb{R}^p$ and suppose $E\left(e^{i\mathbf{t}\cdot\mathbf{X}}\right) = E\left(e^{i\mathbf{t}\cdot\mathbf{Y}}\right)$ for all $\mathbf{t} \in \mathbb{R}^p$. Then $\lambda_{\mathbf{X}} = \lambda_{\mathbf{Y}}$.*

Proof: Both $\lambda_{\mathbf{X}}$ and $\lambda_{\mathbf{Y}}$ are in $\mathfrak{S}'$ if we define for $\psi \in \mathfrak{S}$, $\lambda_{\mathbf{X}}\left(\psi\right) \equiv \int_{\mathbb{R}^p} \psi d\lambda_{\mathbf{X}}$. Then letting $\psi \in \mathfrak{S}$ and using Fubini's theorem,

$$
\begin{aligned}
\int_{\mathbb{R}^p}\int_{\mathbb{R}^p} e^{i\mathbf{t}\cdot\mathbf{y}}\psi\left(\mathbf{t}\right)dtd\lambda_{\mathbf{Y}} &= \int_{\mathbb{R}^p}\int_{\mathbb{R}^p} e^{i\mathbf{t}\cdot\mathbf{y}}d\lambda_{\mathbf{Y}}\psi\left(\mathbf{t}\right)dt \\
&= \int_{\mathbb{R}^p} E\left(e^{i\mathbf{t}\cdot\mathbf{Y}}\right)\psi\left(\mathbf{t}\right)dt \\
&= \int_{\mathbb{R}^p} E\left(e^{i\mathbf{t}\cdot\mathbf{X}}\right)\psi\left(\mathbf{t}\right)dt \\
&= \int_{\mathbb{R}^p}\int_{\mathbb{R}^p} e^{i\mathbf{t}\cdot\mathbf{x}}d\lambda_{\mathbf{X}}\psi\left(\mathbf{t}\right)dt \\
&= \int_{\mathbb{R}^p}\int_{\mathbb{R}^p} e^{i\mathbf{t}\cdot\mathbf{x}}\psi\left(\mathbf{t}\right)dtd\lambda_{\mathbf{X}}.
\end{aligned}
$$

Thus $\lambda_{\mathbf{Y}} \left(F^{-1}\psi \right) = \lambda_{\mathbf{X}} \left(F^{-1}\psi \right)$ and since $\psi \in \mathfrak{S}$ is arbitrary, $\lambda_{\mathbf{X}} = \lambda_{\mathbf{Y}}$ in $\mathfrak{S}'$. But $\mathfrak{S}$ is dense in $C_0 \left(\mathbb{R}^p \right)$ and so $\lambda_{\mathbf{X}} = \lambda_{\mathbf{Y}}$ as measures. This proves the theorem.

Not all random vectors have moment generating functions but when they do, the moment generating function is often much easier to use than the characteristic function.

Lemma 17.11 *Suppose* $\mathbf{X}$ *is a random vector with values in* $\mathbb{R}^p$ *which has a moment generating function for* $|\mathbf{t}| \leq r$. *Then* $E \left(|\mathbf{X}|^m \right) < \infty$ *for all* $m \in \mathbb{N}$. *Also, if* $|\mathbf{v}| < \frac{r}{2}$ *then*

$$\sum_{n=0}^{\infty} \int_{\mathbb{R}^p} \frac{|(\mathbf{v} \cdot \mathbf{x})^n|}{n!} d\lambda_{\mathbf{X}} < \infty. \tag{2}$$

Proof: Let $\{\mathbf{u}_1, \cdots, \mathbf{u}_l\}$ be a set of unit vectors with the property that for all $\mathbf{x} \in \mathbb{R}^p$, $(\mathbf{x} \cdot \mathbf{u}_i) \geq \frac{1}{2} |\mathbf{x}|$ for some $\mathbf{u}_i$. Set

$$S_i \equiv \left\{ \mathbf{x} \in \mathbb{R}^p : \mathbf{x} \cdot \mathbf{u}_i \geq \frac{1}{2} |\mathbf{x}| \right\}.$$

For $|\mathbf{v}| < \frac{r}{2}$,

$$\sum_{n=0}^{\infty} \int_{\mathbb{R}^p} \frac{|(\mathbf{v} \cdot \mathbf{x})^n|}{n!} d\lambda_{\mathbf{X}} \leq \sum_{i=1}^{l} \int_{S_i} \sum_{n=0}^{\infty} \frac{r^n 2^{-n} |\mathbf{x}|^n}{n!} d\lambda_{\mathbf{X}}$$

$$= \sum_{i=1}^{l} \int_{S_i} e^{(r/2)|\mathbf{x}|} d\lambda_{\mathbf{X}}.$$

Now for $\mathbf{x} \in S_i$, $\mathbf{x} \cdot \mathbf{u}_i \geq \frac{1}{2} |\mathbf{x}|$ and so $\mathbf{x} \cdot r\mathbf{u}_i \geq \frac{r}{2} |\mathbf{x}|$ which implies the above is no smaller than

$$\leq \sum_{i=1}^{l} \int_{S_i} e^{\mathbf{x} \cdot r\mathbf{u}_i} d\lambda_{\mathbf{X}} \leq \sum_{i=1}^{l} \int_{\mathbb{R}^p} e^{\mathbf{x} \cdot r\mathbf{u}_i} d\lambda_{\mathbf{X}} < \infty.$$

This shows Formula 2.

To verify $E \left(|\mathbf{X}|^m \right) < \infty$ for all m, suppose not. Then for some m, $E \left(|\mathbf{X}|^m \right) = \infty$ and consequently, for some j, $\int_{S_j} |\mathbf{x}|^m \, d\lambda_{\mathbf{X}} = \infty$. There exists a constant, C, such that for all $\mathbf{x} \in S_j$,

$$|\mathbf{x}|^m \leq C e^{(r/2)|\mathbf{x}|} \leq C e^{\mathbf{x} \cdot r\mathbf{u}_j}.$$

Consequently,

$$\infty = \int_{S_j} |\mathbf{x}|^m \, d\lambda_{\mathbf{X}} \leq C \int_{S_j} e^{(r/2)|\mathbf{x}|} d\lambda_{\mathbf{X}} \leq C \int_{S_j} e^{\mathbf{x} \cdot r\mathbf{u}_j} d\lambda_{\mathbf{X}} < \infty,$$

a contradiction. This proves the lemma.

Corollary 17.12 *If the moment generating function, $M_{\mathbf{X}}(\mathbf{t})$, exists for all $|\mathbf{t}| \leq r, r > 0$, then*

$$E\left(\prod_{i=1}^{p} |X_i|^{r_i}\right) < \infty \tag{3}$$

and

$$E\left(\prod_{i=1}^{p} (X_i)^{r_i}\right) = \frac{\partial^r M_{\mathbf{X}}(0)}{\partial t_1^{r_1} \cdots \partial t_p^{r_p}}. \tag{4}$$

Proof: The first assertion is obvious after replacing $|X_i|$ with $|\mathbf{X}|$ and using Lemma 17.11. From the first assertion, the second is a routine exercise in the dominated convergence theorem to take derivatives inside the integral.

Next we show that if $M_{\mathbf{X}}(\mathbf{t}) = M_{\mathbf{Y}}(\mathbf{t})$, then $\lambda_{\mathbf{X}} = \lambda_{\mathbf{Y}}$ by verifying that $\phi_{\mathbf{X}} = \phi_{\mathbf{Y}}$.

Lemma 17.13 *Suppose $M_{\mathbf{X}}(\mathbf{t})$ exists for all $|\mathbf{t}| \leq r, r > 0$. Then for all $\mathbf{t} \in \mathbb{R}^p$,*

$$\phi_{\mathbf{X}}(\mathbf{t}+\mathbf{v}) = \phi_{\mathbf{X}}(\mathbf{t}) + \sum_{k=1}^{\infty} \frac{D^k \phi_{\mathbf{X}}(\mathbf{t}) \mathbf{v}^k}{k!} \tag{5}$$

for all $|\mathbf{v}| < \frac{r}{2}$.

Proof: From Taylor's formula, Theorem 24.52 found in the appendix on Taylor series,

$$\phi_{\mathbf{X}}(\mathbf{t}+\mathbf{v}) = \phi_{\mathbf{X}}(\mathbf{t}) + \sum_{k=1}^{m} \frac{D^k \phi_{\mathbf{X}}(\mathbf{t}) \mathbf{v}^k}{k!} + \frac{D^{m+1} \phi_{\mathbf{X}}(\mathbf{t}+\lambda\mathbf{v})}{(m+1)!}\mathbf{v}^{m+1}$$

for some $\lambda \in [0,1]$. We show the last term converges to 0 as $m \to \infty$ whenever $|\mathbf{v}| < \frac{r}{2}$.

A routine computation shows

$$D^{m+1}\phi_{\mathbf{X}}(\mathbf{t}) \mathbf{v}^{m+1} = \int_{\mathbb{R}^p} e^{i\mathbf{t}\cdot\mathbf{x}} i^m (\mathbf{x}\cdot\mathbf{v})^m \, d\lambda_{\mathbf{X}} \tag{6}$$

and so

$$\frac{\left|D^{m+1}\phi_{\mathbf{X}}(\mathbf{t}) \mathbf{v}^{m+1}\right|}{(m+1)!} = \frac{1}{(m+1)!} \left|\int_{\mathbb{R}^p} e^{i\mathbf{t}\cdot\mathbf{x}} i^m (\mathbf{x}\cdot\mathbf{v})^m \, d\lambda_{\mathbf{X}}\right|$$

$$\leq \frac{1}{(m+1)!} \int_{\mathbb{R}^p} |(\mathbf{x}\cdot\mathbf{v})|^m \, d\lambda_{\mathbf{X}}$$

which converges to 0 by Lemma 17.11. This verifies Formula 5.

We could also note that $\phi_{\mathbf{X}}(\mathbf{t})$ is an analytic function on $\mathbb{C}^n$ due to the existence of the moments, and then apply the main theorem in the appendix on Taylor series to get Formula 5 directly. This theorem is long and hard however and so it seems better to avoid it by using the arguments given here.

Theorem 17.14 *Suppose $M_{\mathbf{X}}(\mathbf{t}) = M_{\mathbf{Y}}(\mathbf{t})$ for all $|\mathbf{t}| \leq r, r > 0$. Then $\phi_{\mathbf{X}}(\mathbf{t}) = \phi_{\mathbf{Y}}(\mathbf{t})$ for all $\mathbf{t} \in \mathbb{R}^p$ and, consequently, $\lambda_{\mathbf{X}} = \lambda_{\mathbf{Y}}$.*

Proof: By Lemma 17.13 Formula 5 holds for both $\phi_{\mathbf{X}}$ and $\phi_{\mathbf{Y}}$ for all $\mathbf{t}$ and for all $|\mathbf{v}| < \frac{r}{2}$. From Formula 6 and Corollary 17.12, $D^k \phi_{\mathbf{X}}(0) = D^k \phi_{\mathbf{Y}}(0)$ for all $k = 0, 1, \cdots$. Letting $\mathbf{t} = 0$ in Formula 5, this shows $\phi_{\mathbf{X}}(\mathbf{v}) = \phi_{\mathbf{Y}}(\mathbf{v})$ whenever $|\mathbf{v}| < \frac{r}{2}$. Now suppose $\phi_{\mathbf{X}}(\mathbf{v}) = \phi_{\mathbf{Y}}(\mathbf{v})$ for all $|\mathbf{v}| < m\left(\frac{r}{2}\right)$. Let

$$m\left(\frac{r}{2}\right) \leq |\mathbf{w}| < (m+1)\frac{r}{2}.$$

Then $|\mathbf{w} - \mathbf{v}| < m\left(\frac{r}{2}\right)$ where $\mathbf{v} \equiv \frac{\mathbf{w}}{|\mathbf{w}|}\frac{r}{2}$. Hence

$$\phi_{\mathbf{X}}(\mathbf{w}) = \phi_{\mathbf{X}}(\mathbf{w} - \mathbf{v} + \mathbf{v}) = \phi_{\mathbf{X}}(\mathbf{w} - \mathbf{v}) + \sum_{k=1}^{\infty} \frac{D^k \phi_{\mathbf{X}}(\mathbf{w} - \mathbf{v})\mathbf{v}^k}{k!}$$

$$= \phi_{\mathbf{Y}}(\mathbf{w} - \mathbf{v}) + \sum_{k=1}^{\infty} \frac{D^k \phi_{\mathbf{Y}}(\mathbf{w} - \mathbf{v})\mathbf{v}^k}{k!} = \phi_{\mathbf{Y}}(\mathbf{w}).$$

Therefore, $\phi_{\mathbf{X}} = \phi_{\mathbf{Y}}$ on $\mathbb{R}^p$ and this proves the theorem.

17.2 Conditional probability and independence

The notion of conditional probability is about the probability that one random vector, $\mathbf{X}$, has values in some set given knowledge of the value of another random vector, $\mathbf{Y}$. This is referred to as conditioning on $\mathbf{Y}$. There are significant problems which result in trying to assign a meaning to conditional probability given values of a random vector, $\mathbf{Y}$, which may have zero probability. Typically, many cases are considered and finally a rather abstract notion of conditional probability emerges. In this section, the theory of conditional probability will be based on slicing measures, defined in Chapter 15. This approach is more direct and may also be more consistent with the way most people actually think about conditional probability.

If $\mathbf{X}, \mathbf{Y}$ are two random vectors defined on a probability space having values in $\mathbb{F}^{p_1}$ and $\mathbb{F}^{p_2}$ respectively, and if E is a Borel set in the appropriate space, then $(\mathbf{X}, \mathbf{Y})$ is a random vector with values in $\mathbb{F}^{p_1} \times \mathbb{F}^{p_2}$ and $\lambda_{(\mathbf{X},\mathbf{Y})}(E \times \mathbb{F}^{p_2}) = \lambda_{\mathbf{X}}(E)$, $\lambda_{(\mathbf{X},\mathbf{Y})}(\mathbb{F}^{p_1} \times E) = \lambda_{\mathbf{Y}}(E)$. Thus, by the section on slicing measures in Chapter 15, there exist probability measures, denoted here by $\lambda_{\mathbf{X}|\mathbf{y}}$ and $\lambda_{\mathbf{Y}|\mathbf{x}}$, such that whenever E is a Borel set in $\mathbb{F}^{p_1} \times \mathbb{F}^{p_2}$,

$$\int_{\mathbb{F}^{p_1} \times \mathbb{F}^{p_2}} \mathcal{X}_E d\lambda_{(\mathbf{X},\mathbf{Y})} = \int_{\mathbb{F}^{p_1}} \int_{\mathbb{F}^{p_2}} \mathcal{X}_E d\lambda_{\mathbf{Y}|\mathbf{x}} d\lambda_{\mathbf{X}},$$

and

$$\int_{\mathbb{F}^{p_1} \times \mathbb{F}^{p_2}} \mathcal{X}_E d\lambda_{(\mathbf{X},\mathbf{Y})} = \int_{\mathbb{F}^{p_2}} \int_{\mathbb{F}^{p_1}} \mathcal{X}_E d\lambda_{\mathbf{X}|\mathbf{y}} d\lambda_{\mathbf{Y}}.$$

Definition 17.15 Let $\mathbf{X}$ and $\mathbf{Y}$ be two random vectors defined on a probability space. The conditional probability measure of $\mathbf{Y}$ given $\mathbf{X}$ is the measure $\lambda_{\mathbf{Y}|\mathbf{x}}$ in the above. Similarly the conditional probability measure of $\mathbf{X}$ given $\mathbf{Y}$ is the measure $\lambda_{\mathbf{X}|\mathbf{y}}$.

More generally, we can use the theory of slicing measures to consider any finite list of random vectors, $\{\mathbf{X}_i\}$, defined on a probability space with $\mathbf{X}_i \in \mathbb{F}^{p_i}$, and write the following for E a Borel set in $\prod_{i=1}^n \mathbb{F}^{p_i}$.

$$\int_{\mathbb{F}^{p_1} \times \cdots \times \mathbb{F}^{p_n}} \mathcal{X}_E d\lambda_{(\mathbf{X}_1, \cdots, \mathbf{X}_n)} = \int_{\mathbb{F}^{p_1}} \int_{\mathbb{F}^{p_2} \times \cdots \times \mathbb{F}^{p_n}} \mathcal{X}_E d\lambda_{(\mathbf{X}_2, \cdots, \mathbf{X}_n)|\mathbf{x}_1} d\lambda_{\mathbf{X}_1}$$

$$= \int_{\mathbb{F}^{p_1}} \int_{\mathbb{F}^{p_2}} \int_{\mathbb{F}^{p_3} \times \cdots \times \mathbb{F}^{p_n}} \mathcal{X}_E d\lambda_{(\mathbf{X}_3, \cdots, \mathbf{X}_n)|\mathbf{x}_1 \mathbf{x}_2} d\lambda_{\mathbf{X}_2|\mathbf{x}_1} d\lambda_{\mathbf{X}_1}$$

$$\vdots$$

$$= \int_{\mathbb{F}^{p_1}} \cdots \int_{\mathbb{F}^{p_n}} \mathcal{X}_E d\lambda_{\mathbf{X}_n|\mathbf{x}_1 \mathbf{x}_2 \cdots \mathbf{x}_{n-1}} d\lambda_{\mathbf{X}_{n-1}|\mathbf{x}_1 \cdots \mathbf{x}_{n-2}} \cdots d\lambda_{\mathbf{X}_2|\mathbf{x}_1} d\lambda_{\mathbf{X}_1}. \quad (7)$$

Obviously, we could have obtained any order in the iterated integrals by simply modifying the "given" variables, those occurring after the symbol |, to be those which have been integrated in an outer level of the iterated integral.

Definition 17.16 Let $\{\mathbf{X}_1, \cdots, \mathbf{X}_n\}$ be random vectors defined on a probability space having values in $\mathbb{F}^{p_1}, \cdots, \mathbb{F}^{p_n}$ respectively. We say the random vectors are independent if for every E a Borel set in $\mathbb{F}^{p_1} \times \cdots \times \mathbb{F}^{p_n}$,

$$\int_{\mathbb{F}^{p_1} \times \cdots \times \mathbb{F}^{p_n}} \mathcal{X}_E d\lambda_{(\mathbf{X}_1, \cdots, \mathbf{X}_n)}$$

$$= \int_{\mathbb{F}^{p_1}} \cdots \int_{\mathbb{F}^{p_n}} \mathcal{X}_E d\lambda_{\mathbf{X}_n} d\lambda_{\mathbf{X}_{n-1}} \cdots d\lambda_{\mathbf{X}_2} d\lambda_{\mathbf{X}_1} \quad (8)$$

and the iterated integration may be taken in any order. If $\mathcal{A}$ is any set of random vectors defined on a probability space, we say $\mathcal{A}$ is independent if any finite set of random vectors from $\mathcal{A}$ is independent.

Thus, we say the random vectors are independent exactly when the dependence on the givens in Formula 7 can be dropped.

Proposition 17.17 *Equations 8 and 7 hold with $\mathcal{X}_E$ replaced by any nonnegative Borel measurable function and for any bounded continuous function.*

Proof: The two equations hold for simple functions in place of $\mathcal{X}_E$ and so an application of the monotone convergence theorem applied to an increasing sequence of simple functions converging pointwise to a given nonnegative Borel measurable function yields the conclusion of the proposition in the case of the nonnegative Borel function. For a bounded continuous function, we can apply the result just established to the positive and negative parts of the real and imaginary parts of the function.

Lemma 17.18 *Let $\mathbf{X}_1, \cdots, \mathbf{X}_n$ be random vectors with values in $\mathbb{F}^{p_1}, \cdots, \mathbb{F}^{p_n}$ respectively and let $\mathbf{g} : \mathbb{F}^{p_1} \times \cdots \times \mathbb{F}^{p_n} \to \mathbb{F}^k$ be Borel measurable. Then $\mathbf{g}(\mathbf{X}_1, \cdots, \mathbf{X}_n)$ is a random vector with values in $\mathbb{F}^k$ and if $h : \mathbb{F}^k \to [0, \infty)$, then*

$$\int_{\mathbb{F}^k} h(\mathbf{y}) \, d\lambda_{\mathbf{g}(\mathbf{X}_1, \cdots, \mathbf{X}_n)}(y) =$$

$$\int_{\mathbb{F}^{p_1} \times \cdots \times \mathbb{F}^{p_n}} h(\mathbf{g}(\mathbf{x}_1, \cdots, \mathbf{x}_n)) \, d\lambda_{(\mathbf{X}_1, \cdots, \mathbf{X}_n)}. \tag{9}$$

If $\mathbf{X}_i$ is a random vector with values in $\mathbb{F}^{p_i}, i = 1, 2, \cdots$ and if $\mathbf{g}_i : \mathbb{F}^{p_i} \to \mathbb{F}^{k_i}$, where $\mathbf{g}_i$ is Borel measurable, then the random vectors $\mathbf{g}_i(\mathbf{X}_i)$ are also independent.

Proof: First let E be a Borel set in $\mathbb{F}^k$. From the definition,

$$\int_{\mathbb{F}^k} \mathcal{X}_E \, d\lambda_{\mathbf{g}(\mathbf{X}_1, \cdots, \mathbf{X}_n)} = \int_{\mathbb{F}^{p_1} \times \cdots \times \mathbb{F}^{p_n}} \mathcal{X}_{\mathbf{g}^{-1}(E)} \, d\lambda_{(\mathbf{X}_1, \cdots, \mathbf{X}_n)}$$

$$= \int_{\mathbb{F}^{p_1} \times \cdots \times \mathbb{F}^{p_n}} \mathcal{X}_E(\mathbf{g}(\mathbf{x}_1, \cdots, \mathbf{x}_n)) \, d\lambda_{(\mathbf{X}_1, \cdots, \mathbf{X}_n)}.$$

This proves Formula 9 in the case when h is $\mathcal{X}_E$. To prove it in the general case, approximate the nonnegative Borel measurable function with simple functions for which the formula is true, and use the monotone convergence theorem.

It remains to prove the last assertion that functions of independent random vectors are also independent random vectors. Let E be a Borel set in $\mathbb{F}^{k_1} \times \cdots \times \mathbb{F}^{k_n}$. Then for

$$\pi_i(\mathbf{x}_1, \cdots, \mathbf{x}_n) \equiv \mathbf{x}_i,$$

$$\int_{\mathbb{F}^{k_1} \times \cdots \times \mathbb{F}^{k_n}} \mathcal{X}_E \, d\lambda_{(\mathbf{g}_1(\mathbf{X}_1), \cdots, \mathbf{g}_n(\mathbf{X}_n))}$$

$$\equiv \int_{\mathbb{F}^{p_1} \times \cdots \times \mathbb{F}^{p_n}} \mathcal{X}_E \circ (g_1 \circ \pi_1, \cdots, g_n \circ \pi_n) \, d\lambda_{(\mathbf{X}_1, \cdots, \mathbf{X}_n)}$$

$$= \int_{\mathbb{F}^{p_1}} \cdots \int_{\mathbb{F}^{p_n}} \mathcal{X}_E \circ (g_1 \circ \pi_1, \cdots, g_n \circ \pi_n) \, d\lambda_{\mathbf{X}_n} \cdots d\lambda_{\mathbf{X}_1}$$

$$= \int_{\mathbb{F}^{k_1}} \cdots \int_{\mathbb{F}^{k_n}} \mathcal{X}_E \, d\lambda_{g_n(\mathbf{X}_n)} \cdots d\lambda_{g_1(\mathbf{X}_1)}$$

and this proves the last assertion.

How do independent random variables occur in statistics? Roughly, it is like this. An observation is made, the outcome of which will be a number (vector) and the value of the observation is taken to be the value of an underlying random variable (vector). If a sequence of observations is made in such a way that the outcomes of each observation do not depend on the outcomes of the other observations, then the alleged random variables whose values are being observed are considered independent. If the distribution function controlling the probability of outcomes is assumed to be the same for each observation, (for example $N_1\left(\mu, \sigma^2\right)$), the random variables are called identically distributed and the phrase "random sample from $N_1\left(\mu, \sigma^2\right)$" is used. Of course another description would be substituted for $N_1\left(\mu, \sigma^2\right)$ if another distribution function were assumed to be in control.

If the above discussion seems a little mystical, there is still a mathematical question about existence of independent random variables with given distribution functions which needs to be resolved. We have the following existence theorem.

Proposition 17.19 *Let $\nu_1, \cdots, \nu_n$ be Radon probability measures defined on $\mathbb{R}^p$. Then there exists a probability space and independent random vectors $\{\mathbf{X}_1, \cdots, \mathbf{X}_n\}$ defined on this probability space such that $\lambda_{\mathbf{X}_i} = \nu_i$.*

Proof: Let $(\Omega, \mathcal{S}, P) \equiv \left((\mathbb{R}^p)^n, \mathcal{S}_1 \times \cdots \times \mathcal{S}_n, \nu_1 \times \cdots \times \nu_n\right)$ where this is just the product σ algebra and product measure which satisfies the following for measurable rectangles.

$$(\nu_1 \times \cdots \times \nu_n) \left(\prod_{i=1}^n E_i \right) = \prod_{i=1}^n \nu_i\left(E_i\right).$$

Now let $\mathbf{X}_i\left(\mathbf{x}_1, \cdots, \mathbf{x}_i, \cdots, \mathbf{x}_n\right) = \mathbf{x}_i$. Then from the definition, if E is a Borel set in $\mathbb{R}^p$,

$$\lambda_{\mathbf{X}_i}(E) \equiv P\left\{\mathbf{X}_i \in E\right\}$$

$$= (\nu_1 \times \cdots \times \nu_n) \left(\mathbb{R}^p \times \cdots \times E \times \cdots \times \mathbb{R}^p\right) = \nu_i\left(E\right).$$

Let $\mathcal{M}$ consist of all Borel sets of $(\mathbb{R}^p)^n$ such that

$$\int_{\mathbb{R}^p} \cdots \int_{\mathbb{R}^p} \mathcal{X}_E\left(\mathbf{x}_1, \cdots, \mathbf{x}_n\right) d\lambda_{\mathbf{X}_1} \cdots d\lambda_{\mathbf{X}_n} = \int_{(\mathbb{R}^p)^n} \mathcal{X}_E \, d\lambda_{(\mathbf{X}_1, \cdots, \mathbf{X}_n)}.$$

We see from what was just shown and the definition of $(\nu_1 \times \cdots \times \nu_n)$ that $\mathcal{M}$ contains all sets of the form $\prod_{i=1}^n E_i$ where each $E_i \in$ Borel sets of $\mathbb{R}^p$. Therefore, $\mathcal{M}$ contains the algebra of all finite disjoint unions of such sets. It is also clear that $\mathcal{M}$ is a monotone class and so by the theorem on monotone classes, $\mathcal{M}$ equals the Borel sets. Therefore, the given random vectors are independent and this proves the proposition.

Lemma 17.20 *If $\{X_i\}_{i=1}^n$ are independent random variables having values in $\mathbb{F}$,*

$$E\left(\prod_{i=1}^n X_i\right) = \prod_{i=1}^n E(X_i).$$

Proof: By Lemma 17.18 and denoting by P the product, $\prod_{i=1}^n X_i$,

$$E\left(\prod_{i=1}^n X_i\right) = \int_{\mathbb{F}} z d\lambda_P(z) = \int_{\mathbb{F}\times\mathbb{F}} \prod_{i=1}^n x_i d\lambda_{(X_1,\cdots,X_n)}$$

$$= \int_{\mathbb{F}} \cdots \int_{\mathbb{F}} \prod_{i=1}^n x_i d\lambda_{X_1} \cdots d\lambda_{X_n} = \prod_{i=1}^n E(X_i).$$

There is a way to tell if random vectors are independent by using their characteristic functions.

Proposition 17.21 *If $\mathbf{X}_1$ and $\mathbf{X}_2$ are random vectors having values in $\mathbb{R}^{p_1}$ and $\mathbb{R}^{p_2}$ respectively, then the random vectors are independent if and only if*

$$E\left(e^{iP}\right) = \prod_{j=1}^2 E\left(e^{i\mathbf{t}_j \cdot \mathbf{X}_j}\right)$$

where $P \equiv \sum_{j=1}^2 \mathbf{t}_j \cdot \mathbf{X}_j$ for $\mathbf{t}_j \in \mathbb{R}^{p_j}$.

The proof of this proposition will depend on the following lemma.

Lemma 17.22 *Let $\mathbf{Y}$ be a random vector with values in $\mathbb{R}^p$ and let f be bounded and measurable with respect to the Radon measure, $\lambda_{\mathbf{Y}}$, and satisfy*

$$\int f(\mathbf{y}) e^{i\mathbf{t}\cdot\mathbf{y}} d\lambda_{\mathbf{Y}} = 0$$

for all $\mathbf{t} \in \mathbb{R}^p$. Then $f(\mathbf{y}) = 0$ for $\lambda_{\mathbf{Y}}$ a.e. $\mathbf{y}$.

Proof: The proof is just like the proof of Theorem 17.10 applied to the measure, $f(\mathbf{y}) d\lambda_{\mathbf{Y}}$. Thus we may conclude $\int_E f(\mathbf{y}) d\lambda_{\mathbf{Y}} = 0$ for all E Borel. Hence $f(\mathbf{y}) = 0$ a.e.

Proof of the proposition: If the $\mathbf{X}_j$ are independent, the formula follows from Lemma 17.20 and Lemma 17.18.

Now suppose the formula holds. Then

$$\int_{\mathbb{R}^{p_2}} \int_{\mathbb{R}^{p_1}} e^{i t_1 \cdot \mathbf{x}_1} e^{i t_2 \cdot \mathbf{x}_2} d\lambda_{\mathbf{X}_1} d\lambda_{\mathbf{X}_2} = E\left(e^{iP}\right)$$

$$= \int_{\mathbb{R}^{p_2}} \int_{\mathbb{R}^{p_1}} e^{i t_1 \cdot \mathbf{x}_1} e^{i t_2 \cdot \mathbf{x}_2} d\lambda_{\mathbf{X}_1 | \mathbf{x}_2} d\lambda_{\mathbf{X}_2}.$$

Now we apply Lemma 17.22 to conclude that

$$\int_{\mathbb{R}^{p_1}} e^{i t_1 \cdot \mathbf{x}_1} d\lambda_{\mathbf{X}_1} = \int_{\mathbb{R}^{p_1}} e^{i t_1 \cdot \mathbf{x}_1} d\lambda_{\mathbf{X}_1 | \mathbf{x}_2} \tag{10}$$

for $\lambda_{\mathbf{X}_2}$ a.e. $\mathbf{x}_2$, the exceptional set depending on t_1. Therefore, taking the union of all exceptional sets corresponding to $t_1 \in \mathbb{Q}^{p_1}$, it follows by continuity and the dominated convergence theorem that Formula 10 holds for all t_1 whenever $\mathbf{x}_2$ is not an element of this exceptional set of measure zero. Therefore, for such $\mathbf{x}_2$, Theorem 17.10 applies and we may conclude $\lambda_{\mathbf{X}_1 | \mathbf{x}_2} = \lambda_{\mathbf{X}_1}$ for $\lambda_{\mathbf{X}_2}$ a.e. $\mathbf{x}_2$. Hence, if E is a Borel set in $\mathbb{R}^{p_1} \times \mathbb{R}^{p_2}$, $\int_{\mathbb{R}^{p_1+p_2}} \mathcal{X}_E d\lambda_{(\mathbf{X}_1, \mathbf{X}_2)} = \int_{\mathbb{R}^{p_2}} \int_{\mathbb{R}^{p_1}} \mathcal{X}_E d\lambda_{\mathbf{X}_1 | \mathbf{x}_2} d\lambda_{\mathbf{X}_2}$ $= \int_{\mathbb{R}^{p_2}} \int_{\mathbb{R}^{p_1}} \mathcal{X}_E d\lambda_{\mathbf{X}_1} d\lambda_{\mathbf{X}_2}$. A repeat of the above argument will give the iterated integral in the reverse order or else one could apply Fubini's theorem to obtain this. This proves the proposition.

The proposition also holds if 2 is replaced with n and the argument is a longer version of what was just presented. We also have the following simple corollary.

Corollary 17.23 *Suppose $M_{\mathbf{X}_i}(t_i)$ exists for $i = 1, 2$ and $\mathbf{X}_j$ has values in $\mathbb{R}^{p_j}$. Then $\mathbf{X}_1$ and $\mathbf{X}_2$ are independent if and only if*

$$E\left(e^{P}\right) = \prod_{i=1}^{2} E\left(e^{t_j \cdot \mathbf{X}_j}\right) \tag{11}$$

where $P = t_1 \cdot \mathbf{X}_1 + t_2 \cdot \mathbf{X}_2$.

Proof: The only if part of the assertion is obvious from Lemma 17.18 and Lemma 17.20.

The if part follows from Theorem 17.14 and Proposition 17.19. Let $\widetilde{\mathbf{X}}_1 \equiv (\mathbf{X}_1, 0)$ and $\widetilde{\mathbf{X}}_2 \equiv (0, \mathbf{X}_2)$. Thus $\widetilde{\mathbf{X}}_1$ and $\widetilde{\mathbf{X}}_2$ map into $\mathbb{R}^{p_1+p_2}$ and letting $t = (t_1, t_2)$, Formula 11 is equivalent to $E\left(e^{t \cdot \left(\widetilde{\mathbf{X}}_1 + \widetilde{\mathbf{X}}_2\right)}\right) = E\left(e^{t \cdot \widetilde{\mathbf{X}}_1}\right) E\left(e^{t \cdot \widetilde{\mathbf{X}}_2}\right)$. By Proposition 17.19, we may take $\mathbf{Y}_1$ and $\mathbf{Y}_2$ to be such that $\mathbf{Y}_1$ and $\mathbf{Y}_2$ are independent, map into $\mathbb{R}^{p_1+p_2}$, and $\lambda_{\mathbf{Y}_j} = \lambda_{\widetilde{\mathbf{X}}_j}$. Then since $\mathbf{Y}_1$ and $\mathbf{Y}_2$ are independent,

$$E\left(e^{t \cdot (\mathbf{Y}_1 + \mathbf{Y}_2)}\right) = E\left(e^{t \cdot \mathbf{Y}_1}\right) E\left(e^{t \cdot \mathbf{Y}_2}\right)$$

$$= E\left(e^{t\cdot\widetilde{\mathbf{X}}_1}\right) E\left(e^{t\cdot\widetilde{\mathbf{X}}_2}\right) = E\left(e^{t\cdot\left(\widetilde{\mathbf{X}}_1+\widetilde{\mathbf{X}}_2\right)}\right).$$

By Theorem 17.14 $E\left(e^{it\cdot(\mathbf{Y}_1+\mathbf{Y}_2)}\right) = E\left(e^{it\cdot\left(\widetilde{\mathbf{X}}_1+\widetilde{\mathbf{X}}_2\right)}\right)$ and $E\left(e^{it\cdot\widetilde{\mathbf{X}}_j}\right) =$ $E\left(e^{it\cdot\mathbf{Y}_j}\right)$. Therefore, since the Y_j are independent,

$$E\left(e^{it\cdot\left(\widetilde{\mathbf{X}}_1+\widetilde{\mathbf{X}}_2\right)}\right) = E\left(e^{it\cdot(\mathbf{Y}_1+\mathbf{Y}_2)}\right)$$

$$= E\left(e^{it\cdot\mathbf{Y}_1}\right) E\left(e^{it\cdot\mathbf{Y}_2}\right) = E\left(e^{it\cdot\widetilde{\mathbf{X}}_1}\right) E\left(e^{it\cdot\widetilde{\mathbf{X}}_2}\right)$$

and now it follows from Proposition 17.21 that the two vectors $\widetilde{\mathbf{X}}_1$ and $\widetilde{\mathbf{X}}_2$ are independent. From this it follows from Lemma 17.18 that $\mathbf{X}_1$ and $\mathbf{X}_2$ are independent.

This corollary shows that one can resolve questions of independence by using moment generating functions when they exist. As with the corresponding theorem about characteristic functions there is a generalization to m random vectors.

An important idea in statistics is the idea of sufficient statistics. This is described next.

Definition 17.24 Let $\mathbf{X}$ be a random vector with values in $\mathbb{F}^p$ and suppose $\lambda_{\mathbf{X}} = \mu(\theta)$ where θ is a parameter which determines different elements of the family of Radon measures, $\mu(\theta)$. Let $\mathbf{T}(\mathbf{X}) = \mathbf{T} \in \mathbb{F}^r$. We say the random vector, $\mathbf{T}$, is a sufficient statistic for θ if for all Borel $E \in \mathbb{F}^p$,

$$\int_{\mathbb{F}^p} \mathcal{X}_E(\mathbf{x}) \, d\lambda_{\mathbf{X}|t}$$

is independent of θ.

The reason $\mathbf{T}$ is "sufficient" for θ is that the lack of dependence of the above expression on θ implies that when the value of $\mathbf{T}$ is known, nothing more can be found out about θ from the knowledge of probabilities of $\mathbf{X}$. It is important to have a reasonably simple way to identify sufficient statistics and the next theorem gives such a way.

Theorem 17.25 *Suppose there exists a continuous function, g_θ, for each parameter θ and a Radon measure, η, such that for all E a Borel set in $\mathbb{F}^p$,*

$$\int_{\mathbb{F}^p} \mathcal{X}_E(x) \, d\lambda_{\mathbf{X}}(x) = \int_{\mathbb{F}^p} \mathcal{X}_E(x) \, g_\theta(\mathbf{T}(x)) \, d\eta(x).$$

Then $\mathbf{T}$ is a sufficient statistic for θ.

The proof depends on the following lemma.

Conditional probability and independence

Lemma 17.26 *Let F be a Borel set in $\mathbb{F}^r$. Then $\lambda_{\mathbf{T}|\mathbf{x}}(F) = \mathcal{X}_F(\mathbf{T}(\mathbf{x}))$ for $\lambda_{\mathbf{X}}$ a.e. $\mathbf{x}$.*

Proof: Let E be a Borel set in $\mathbb{F}^p$. Then $\int_E \lambda_{\mathbf{T}|\mathbf{x}}(F) \, d\lambda_{\mathbf{X}}$

$$\equiv \int_{E \times F} d\lambda_{(\mathbf{X},\mathbf{T})} \equiv P(\mathbf{X} \in E \text{ and } \mathbf{T} \in F)$$

$$= P(\mathbf{X} \in E \text{ and } \mathbf{X} \in \mathbf{T}^{-1}(F)) = P(\mathbf{X} \in E \cap \mathbf{T}^{-1}(F)).$$

Also

$$\int_E \mathcal{X}_F(\mathbf{T}(\mathbf{x})) \, d\lambda_{\mathbf{X}} \equiv P(\mathbf{X} \in E \cap \mathbf{T}^{-1}(F)).$$

Since this holds for all E Borel, the lemma is proved.

Proof of theorem: By the Lebesgue Besicovitch differentiation theorem, for $\lambda_{\mathbf{T}}$ a.e. $\mathbf{t}_1$,

$$\int_{\mathbb{F}^p} \mathcal{X}_E(\mathbf{x}) \, d\lambda_{\mathbf{X}|\mathbf{t}_1}$$

$$= \lim_{r \to 0} \frac{1}{\lambda_{\mathbf{T}}(B(\mathbf{t}_1, r))} \int_{\mathbb{F}^r} \mathcal{X}_{B(\mathbf{t}_1, r)}(\mathbf{t}) \int_{\mathbb{F}^p} \mathcal{X}_E(\mathbf{x}) \, d\lambda_{\mathbf{X}|\mathbf{t}} d\lambda_{\mathbf{T}}$$

$$= \lim_{r \to 0} \frac{1}{\lambda_{\mathbf{T}}(B(\mathbf{t}_1, r))} \int_{\mathbb{F}^p} \int_{\mathbb{F}^r} \mathcal{X}_{B(\mathbf{t}_1, r)}(\mathbf{t}) \, d\lambda_{\mathbf{T}|\mathbf{x}} \mathcal{X}_E(\mathbf{x}) \, d\lambda_{\mathbf{X}}$$

which, by the lemma, equals

$$= \lim_{r \to 0} \frac{1}{\lambda_{\mathbf{T}}(B(\mathbf{t}_1, r))} \int_{\mathbb{F}^p} \mathcal{X}_{B(\mathbf{t}_1, r)}(\mathbf{T}(\mathbf{x})) \mathcal{X}_E(\mathbf{x}) \, d\lambda_{\mathbf{X}}. \tag{12}$$

Now $\lambda_{\mathbf{T}}(B(\mathbf{t}_1, r)) \equiv P(\mathbf{T} \in B(\mathbf{t}_1, r)) = P(\mathbf{X} \in \mathbf{T}^{-1}(B(\mathbf{t}_1, r)))$ and so by the assumption of the theorem,

$$\lambda_{\mathbf{T}}(B(\mathbf{t}_1, r)) = \int_{\mathbb{F}^p} \mathcal{X}_{B(\mathbf{t}_1, r)}(\mathbf{T}(\mathbf{x})) \, d\lambda_{\mathbf{X}}$$

$$= \int_{\mathbb{F}^p} \mathcal{X}_{B(\mathbf{t}_1, r)}(\mathbf{T}(\mathbf{x})) g_\theta(\mathbf{T}(\mathbf{x})) \, d\eta$$

$$= \int_{\mathbb{F}^p} \mathcal{X}_{B(\mathbf{t}_1, r)}(\mathbf{T}(\mathbf{x})) (g_\theta(\mathbf{T}(\mathbf{x})) - g_\theta(\mathbf{t}_1)) \, d\eta$$

$$+ \int_{\mathbb{F}^p} \mathcal{X}_{B(\mathbf{t}_1, r)}(\mathbf{T}(\mathbf{x})) \, d\eta g_\theta(\mathbf{t}_1).$$

Therefore using the assumption of the theorem on the integral in Formula 12, $\int_{\mathbb{F}^p} \mathcal{X}_E(\mathbf{x}) \, d\lambda_{\mathbf{X}|\mathbf{t}_1} = \lim_{r \to 0} \frac{A_r}{B_r}$ where

$$A_r = \int_{\mathbb{F}^p} \mathcal{X}_{B(\mathbf{t}_1, r)}(\mathbf{T}(\mathbf{x})) \mathcal{X}_E(\mathbf{x}) (g_\theta(\mathbf{T}(\mathbf{x})) - g_\theta(\mathbf{t}_1)) \, d\eta$$

$$+ \int_{\mathbb{F}^p} \mathcal{X}_{B(\mathbf{t}_1, r)} \left(\mathbf{T} \left(\mathbf{x} \right) \right) \mathcal{X}_E \left(\mathbf{x} \right) d\eta g_\theta \left(\mathbf{t}_1 \right)$$

and

$$B_r = \int_{\mathbb{F}^p} \mathcal{X}_{B(\mathbf{t}_1, r)} \left(\mathbf{T} \left(\mathbf{x} \right) \right) \left(g_\theta \left(\mathbf{T} \left(\mathbf{x} \right) \right) - g_\theta \left(\mathbf{t}_1 \right) \right) d\eta$$

$$+ \int_{\mathbb{F}^p} \mathcal{X}_{B(\mathbf{t}_1, r)} \left(\mathbf{T} \left(\mathbf{x} \right) \right) d\eta g_\theta \left(\mathbf{t}_1 \right).$$

Then this limit equals

$$= \lim_{r \to 0} \frac{\int_{\mathbb{F}^p} \mathcal{X}_{B(\mathbf{t}_1, r)} \left(\mathbf{T} \left(\mathbf{x} \right) \right) \mathcal{X}_E \left(\mathbf{x} \right) d\eta}{\int_{\mathbb{F}^p} \mathcal{X}_{B(\mathbf{t}_1, r)} \left(\mathbf{T} \left(\mathbf{x} \right) \right) d\eta}$$

because of the continuity of g_θ. This proves the theorem.

It is not necessary to assume g_θ is continuous. A more general result is outlined in Problem 21.

17.3   Conditional expectation

Definition 17.27 Let $\mathbf{X}$ and $\mathbf{Y}$ be random vectors having values in $\mathbb{F}^{p_1}$ and $\mathbb{F}^{p_2}$ respectively. Then if

$$\int |\mathbf{x}| \, d\lambda_{\mathbf{X}|\mathbf{y}} \left(x \right) < \infty,$$

we define

$$E \left(\mathbf{X}|\mathbf{y} \right) \equiv \int \mathbf{x} d\lambda_{\mathbf{X}|\mathbf{y}} \left(x \right).$$

Proposition 17.28 Suppose $\int_{\mathbb{F}^{p_1} \times \mathbb{F}^{p_2}} |\mathbf{x}| \, d\lambda_{(\mathbf{X}, \mathbf{Y})} \left(x \right) < \infty$. Then $E \left(\mathbf{X}|\mathbf{y} \right)$ exists for $\lambda_{\mathbf{Y}}$ a.e. $\mathbf{y}$ and

$$\int_{\mathbb{F}^{p_2}} E \left(\mathbf{X}|\mathbf{y} \right) d\lambda_{\mathbf{Y}} = \int_{\mathbb{F}^{p_1}} \mathbf{x} d\lambda_{\mathbf{X}} \left(x \right) = E \left(\mathbf{X} \right).$$

Proof: $\infty > \int_{\mathbb{F}^{p_1} \times \mathbb{F}^{p_2}} |\mathbf{x}| \, d\lambda_{(\mathbf{X}, \mathbf{Y})} = \int_{\mathbb{F}^{p_2}} \int_{\mathbb{F}^{p_1}} |\mathbf{x}| \, d\lambda_{\mathbf{X}|\mathbf{y}} \left(x \right) d\lambda_{\mathbf{Y}} \left(y \right)$ and so $\int_{\mathbb{F}^{p_1}} |\mathbf{x}| \, d\lambda_{\mathbf{X}|\mathbf{y}} \left(x \right) < \infty$, $\lambda_{\mathbf{Y}}$ a.e. Now

$$\int_{\mathbb{F}^{p_2}} E \left(\mathbf{X}|\mathbf{y} \right) d\lambda_{\mathbf{Y}}$$

$$= \int_{\mathbb{F}^{p_2}} \int_{\mathbb{F}^{p_1}} \mathbf{x} d\lambda_{\mathbf{X}|\mathbf{y}} \left(x \right) d\lambda_{\mathbf{Y}} \left(y \right) = \int_{\mathbb{F}^{p_1} \times \mathbb{F}^{p_2}} \mathbf{x} d\lambda_{(\mathbf{X}, \mathbf{Y})}$$

$$= \int_{\mathbb{F}^{p_1}} \int_{\mathbb{F}^{p_2}} \mathbf{x} d\lambda_{\mathbf{Y}|\mathbf{x}} \left(y \right) d\lambda_{\mathbf{X}} \left(x \right) = \int_{\mathbb{F}^{p_2}} \mathbf{x} d\lambda_{\mathbf{X}} \left(x \right) = E \left(\mathbf{X} \right).$$

Definition 17.29 Let $\{X_n\}$ be any sequence, finite or infinite, of random variables with values in $\mathbb{R}$ which are defined on some probability space, $(\Omega, \mathcal{S}, P)$. We say $\{X_n\}$ is a Martingale if

$$E(X_n | x_{n-1}, \cdots, x_1) = x_{n-1}$$

and we say $\{X_n\}$ is a submartingale if

$$E(X_n | x_{n-1}, \cdots, x_1) \geq x_{n-1}.$$

Next we define what is meant by an upcrossing.

Definition 17.30 Let $\{x_i\}_{i=1}^{I}$ be any sequence of real numbers, $I \leq \infty$. Define an increasing sequence of integers $\{m_k\}$ as follows. m_1 is the first integer ≥ 1 such that $x_{m_1} \leq a$, m_2 is the first integer larger than m_1 such that $x_{m_2} \geq b$, m_3 is the first integer larger than m_2 such that $x_{m_3} \leq a$, etc. Then each sequence, $\{x_{m_{2k-1}}, \cdots, x_{m_{2k}}\}$, is called an upcrossing of $[a, b]$.

Proposition 17.31 *Let $\{X_i\}_{i=1}^{n}$ be a finite sequence of real random variables defined on Ω where $(\Omega, \mathcal{S}, P)$ is a probability space. Let $U_{[a,b]}(\omega)$ denote the number of upcrossings of $X_i(\omega)$ of the interval $[a, b]$. Then $U_{[a,b]}$ is a random variable.*

Proof: Let $X_0(\omega) \equiv a + 1$, let $Y_0(\omega) \equiv 0$, and let $Y_k(\omega)$ remain 0 for $k = 0, \cdots, l$ until $X_l(\omega) \leq a$. When this happens (if ever), $Y_{l+1}(\omega) \equiv 1$. Then let $Y_i(\omega)$ remain 1 for $i = l + 1, \cdots, r$ until $X_r(\omega) \geq b$ when $Y_{r+1}(\omega) \equiv 0$. Let $Y_k(\omega)$ remain 0 for $k \geq r + 1$ until $X_k(\omega) \leq a$ when $Y_k(\omega) \equiv 1$ and continue in this way. Thus the upcrossings of $X_i(\omega)$ are identified as unbroken strings of ones with a zero at each end, with the possible exception of the last string of ones which may be missing the zero at the upper end and may or may not be an upcrossing.

Note also that Y_0 is measurable because it is identically equal to 0 and that if Y_k is measurable, then Y_{k+1} is measurable because the only change in going from k to $k + 1$ is a change from 0 to 1 or from 1 to 0 on a measurable set determined by X_k. Now let

$$Z_k(\omega) = \begin{cases} 1 & \text{if } Y_k(\omega) = 1 \text{ and } Y_{k+1}(\omega) = 0, \\ 0 & \text{otherwise,} \end{cases}$$

if $k < n$ and

$$Z_n(\omega) = \begin{cases} 1 & \text{if } Y_n(\omega) = 1 \text{ and } X_n(\omega) \geq b, \\ 0 & \text{otherwise.} \end{cases}$$

Thus $Z_k(\omega) = 1$ exactly when an upcrossing has been completed and each Z_i is a random variable.

$$U_{[a,b]}(\omega) = \sum_{k=1}^{n} Z_k(\omega)$$

so $U_{[a,b]}$ is a random variable as claimed.

The following corollary collects some key observations found in the above construction.

Corollary 17.32 $U_{[a,b]}(\omega) \leq$ *the number of unbroken strings of ones in the sequence,* $\{Y_k(\omega)\}$ *there being at most one unbroken string of ones which produces no upcrossing. Also*

$$Y_i(\omega) = \psi_i\left(\{X_j(\omega)\}_{j=1}^{i-1}\right), \tag{13}$$

where ψ_i is some function of the past values of $X_j(\omega)$.

Lemma 17.33 *(upcrossing lemma) Let $\{X_i\}_{i=1}^n$ be a submartingale and suppose*

$$E(|X_n|) < \infty.$$

Then

$$E\left(U_{[a,b]}\right) \leq \frac{E(|X_n|) + |a|}{b - a}.$$

Proof: Let $\phi(x) \equiv a + (x - a)^+$. Thus ϕ is a convex and increasing function.

$$\phi(X_{k+r}) - \phi(X_k) = \sum_{i=k+1}^{k+r} \phi(X_i) - \phi(X_{i-1})$$

$$= \sum_{i=k+1}^{k+r} (\phi(X_i) - \phi(X_{i-1})) Y_i + \sum_{i=k+1}^{k+r} (\phi(X_i) - \phi(X_{i-1}))(1 - Y_i).$$

The upcrossings of $\phi(X_i)$ are exactly the same as the upcrossings of X_i and from Formula 13,

$$E\left(\sum_{i=k+1}^{k+r} (\phi(X_i) - \phi(X_{i-1}))(1 - Y_i)\right)$$

$$= \sum_{i=k+1}^{k+r} \int_{\mathbb{R}^i} (\phi(x_i) - \phi(x_{i-1}))\left(1 - \psi_i\left(\{x_j\}_{j=1}^{i-1}\right)\right) d\lambda_{(X_1,\cdots,X_i)}$$

$$= \sum_{i=k+1}^{k+r} \int_{\mathbb{R}^{i-1}} \int_{\mathbb{R}} (\phi(x_i) - \phi(x_{i-1})) \cdot$$

$$\left(1 - \psi_i\left(\{x_j\}_{j=1}^{i-1}\right)\right) d\lambda_{X_i|x_1\cdots x_{i-1}} d\lambda_{(X_1,\cdots,X_{i-1})}$$

$$= \sum_{i=k+1}^{k+r} \int_{\mathbb{R}^{i-1}} \left(1 - \psi_i\left(\{x_j\}_{j=1}^{i-1}\right)\right).$$

318

$$\int_{\mathbb{R}} \left(\phi \left(x_i \right) - \phi \left(x_{i-1} \right) \right) d\lambda_{X_i \mid x_1 \cdots x_{i-1}} d\lambda_{(X_1, \cdots, X_{i-1})}$$

By Jensen's inequality, Problem 4 of Chapter 12,

$$\geq \sum_{i=k+1}^{k+r} \int_{\mathbb{R}^{i-1}} \left(1 - \psi_i \left(\{ x_j \}_{j=1}^{i-1} \right) \right) \cdot$$

$$\left[\phi \left(E \left(X_i \mid x_1, \cdots, x_{i-1} \right) \right) - \phi \left(x_{i-1} \right) \right] d\lambda_{(X_1, \cdots, X_{i-1})}$$

$$\geq \sum_{i=k+1}^{k+r} \int_{\mathbb{R}^{i-1}} \left(1 - \psi_i \left(\{ x_j \}_{j=1}^{i-1} \right) \right) \left[\phi \left(x_{i-1} \right) - \phi \left(x_{i-1} \right) \right] d\lambda_{(X_1, \cdots, X_{i-1})} = 0$$

because of the assumption that our sequence of random variables is a submartingale and the observation that ϕ is both convex and increasing.
Now let the unbroken strings of ones for $\{ Y_i \left(\omega \right) \}$ be

$$\{ k_1, \cdots, k_1 + r_1 \}, \{ k_2, \cdots, k_2 + r_2 \}, \cdots, \{ k_m, \cdots, k_m + r_m \} \qquad (14)$$

where $m = V \left(\omega \right) \equiv$ the number of unbroken strings of ones in the sequence $\{ Y_i \left(\omega \right) \}$. By Corollary 17.32 $V \left(\omega \right) \geq U_{[a,b]} \left(\omega \right)$.

$$\phi \left(X_n \left(\omega \right) \right) - \phi \left(X_1 \left(\omega \right) \right)$$

$$= \sum_{k=1}^{n} \left(\phi \left(X_k \left(\omega \right) \right) - \phi \left(X_{k-1} \left(\omega \right) \right) \right) Y_k \left(\omega \right)$$

$$+ \sum_{k=1}^{n} \left(\phi \left(X_k \left(\omega \right) \right) - \phi \left(X_{k-1} \left(\omega \right) \right) \right) \left(1 - Y_k \left(\omega \right) \right).$$

Summing the first sum over the unbroken strings of ones (the terms in which $Y_i \left(\omega \right) = 0$ contribute nothing), implies

$$\phi \left(X_n \left(\omega \right) \right) - \phi \left(X_1 \left(\omega \right) \right)$$

$$\geq U_{[a,b]} \left(\omega \right) \left(b - a \right) + 0 +$$

$$\sum_{k=1}^{n} \left(\phi \left(X_k \left(\omega \right) \right) - \phi \left(X_{k-1} \left(\omega \right) \right) \right) \left(1 - Y_k \left(\omega \right) \right) \qquad (15)$$

where the zero on the right side results from a string of ones which does not produce an upcrossing. It is here that we use $\phi \left(x \right) \geq a$. Such a string begins with $\phi \left(X_k \left(\omega \right) \right) = a$ and results in an expression of the form $\phi \left(X_{k+m} \left(\omega \right) \right) - \phi \left(X_k \left(\omega \right) \right) \geq 0$ since $\phi \left(X_{k+m} \left(\omega \right) \right) \geq a$. If we had not replaced X_k with $\phi \left(X_k \right)$, it would have been possible for $\phi \left(X_{k+m} \left(\omega \right) \right)$ to be less than a and the zero in the above could have been a negative number.

Therefore from Formula 15,

$$(b - a) E\left(U_{[a,b]}\right) \leq E\left(\phi\left(X_n\right) - \phi\left(X_1\right)\right) \leq E\left(\phi\left(X_n\right) - a\right)$$
$$= E\left(\left(X_n - a\right)^+\right) \leq |a| + E\left(|X_n|\right)$$

and this proves the lemma.

Theorem 17.34 *(submartingale convergence theorem) Let* $\{X_i\}_{i=1}^{\infty}$ *be a submartingale with* $K \equiv \sup\{E(|X_n|) : n \geq 1\} < \infty$. *Then there exists a random variable,* X_{∞}, *such that* $E\left(|X_{\infty}|\right) \leq K$ *and* $\lim_{n \to \infty} X_n(\omega) = X_{\infty}(\omega)$ *a.e.*

Proof: Let $a, b \in \mathbb{Q}$ and let $a < b$. Let $U_{[a,b]}^n(\omega)$ be the number of upcrossings of $\{X_i(\omega)\}_{i=1}^n$. Then let

$$U_{[a,b]}(\omega) \equiv \lim_{n \to \infty} U_{[a,b]}^n(\omega) = \text{ number of upcrossings of } \{X_i\}.$$

By the upcrossing lemma,

$$E\left(U_{[a,b]}^n\right) \leq \frac{E\left(|X_n|\right) + |a|}{b - a} \leq \frac{K + |a|}{b - a}$$

and so by the monotone convergence theorem,

$$E\left(U_{[a,b]}\right) \leq \frac{K + |a|}{b - a} < \infty$$

which shows $U_{[a,b]}(\omega)$ is finite a.e., for all $\omega \notin S_{[a,b]}$ where $P\left(S_{[a,b]}\right) = 0$. Define

$$S \equiv \cup \left\{S_{[a,b]} : a, b \in \mathbb{Q}, a < b\right\}.$$

Then $P(S) = 0$ and if $\omega \notin S$, $\{X_k\}_{k=1}^{\infty}$ has only finitely many upcrossings of every interval having rational endpoints. Thus, for $\omega \notin S$, $\limsup_{k \to \infty} X_k(\omega) = \liminf_{k \to \infty} X_k(\omega) = \lim_{k \to \infty} X_k(\omega) \equiv X_{\infty}(\omega)$. Letting $X_{\infty}(\omega) = 0$ for $\omega \in S$, Fatou's lemma implies

$$\int_{\Omega} |X_{\infty}| \, dP = \int_{\Omega} \lim_{n \to \infty} \inf |X_n| \, dP \leq \lim_{n \to \infty} \inf \int_{\Omega} |X_n| \, dP \leq K$$

and so this proves the theorem.

17.4   Conditional expectation given a σ algebra

There is a more general notion of conditional expectation known as conditional expectation given a σ algebra. It is also a more natural way to do the mathematics although it seems somewhat more abstract at first.

Conditional expectation given a σ algebra

In this section it will be shown how this form of conditional expectation is related to the version just described.

Definition 17.35 For $\mathbf{Y}$ a random vector with values in $\mathbb{F}^{p_2}$,

$$\sigma(\mathbf{Y}) \equiv \{\mathbf{Y}^{-1}(F) : F \text{ is a Borel set in } \mathbb{F}^{p_2}\}.$$

It follows that $\sigma(\mathbf{Y})$ is the smallest σ algebra with respect to which, $\mathbf{Y}$ is a measurable $\mathbb{F}^{p_2}$ valued function.

Definition 17.36 Let $\mathbf{X}$ and $\mathbf{Y}$ be random vectors with $E(|\mathbf{X}|) < \infty$. Then we define a new random vector, $E(\mathbf{X}|\mathbf{Y})$, by

$$E(\mathbf{X}|\mathbf{Y})(\omega) = E(\mathbf{X}|\mathbf{Y}(\omega)).$$

Lemma 17.37 *For $E(|\mathbf{X}|) < \infty$, there is a representative of the function $E(\mathbf{X}|\cdot)$ such that the function $E(\mathbf{X}|\mathbf{Y})$ is measurable with respect to $\sigma(\mathbf{Y})$.*

Proof: From Theorem 15.8,

$$\mathbf{y} \to E(\mathbf{X}|\mathbf{y}) \equiv \int x \, d\lambda_{\mathbf{X}|\mathbf{y}}(x)$$

is a $\lambda_{\mathbf{Y}}$ measurable function of $\mathbf{y}$. Since $\lambda_{\mathbf{Y}}$ is a Radon measure, we may pick a Borel measurable representative and assume $E(\mathbf{X}|\cdot)$ is Borel measurable.

$$E(\mathbf{X}|\mathbf{Y}) \equiv E(\mathbf{X}|\cdot) \circ \mathbf{Y},$$

and so

$$E(\mathbf{X}|\mathbf{Y})^{-1}(\text{open set}) = \mathbf{Y}^{-1}\left(E(\mathbf{X}|\cdot)^{-1}(\text{open set})\right) \in \sigma(\mathbf{Y}).$$

This proves the lemma.

Next let $A \in \sigma(\mathbf{Y})$ be arbitrary. Thus $A = \mathbf{Y}^{-1}(F)$ for F a Borel set in $\mathbb{F}^{p_2}$. By Lemma 17.37,

$$\int_A E(\mathbf{X}|\mathbf{Y})(\omega) \, dP(\omega) = \int_\Omega \mathcal{X}_{\mathbf{Y}^{-1}(F)}(\omega) E(\mathbf{X}|\mathbf{Y})(\omega) \, dP(\omega)$$

$$= \int_{\mathbb{F}^{p_2}} \mathcal{X}_F(\mathbf{y}) E(\mathbf{X}|\mathbf{y}) \, d\lambda_{\mathbf{Y}}(\mathbf{y})$$

$$= \int_{\mathbb{F}^{p_2}} \int_{\mathbb{F}^{p_1}} \mathcal{X}_F(\mathbf{y}) x \, d\lambda_{\mathbf{X}|\mathbf{y}}(x) \, d\lambda_{\mathbf{Y}}(\mathbf{y})$$

$$\equiv \int_{\mathbb{F}^{p_1} \times \mathbb{F}^{p_2}} \mathcal{X}_F(\mathbf{y}) x \, d\lambda_{(\mathbf{X},\mathbf{Y})}$$

$$= \int_{\mathbb{F}^{p_1}} \mathbf{z} d\lambda_{\mathcal{X}_F(\mathbf{Y})\mathbf{X}}(z)$$

$$= \int_{\Omega} \mathcal{X}_F(\mathbf{Y}(\omega)) \mathbf{X}(\omega) dP$$

$$= \int_{\Omega} \mathcal{X}_{\mathbf{Y}^{-1}(F)}(\omega) \mathbf{X}(\omega) dP = \int_A \mathbf{X}(\omega) dP. \qquad (16)$$

Now we are ready to define conditional expectation with respect to a σ algebra and show it is a generalization of that which was presented in the last section.

Definition 17.38 If $\mathcal{S}_1 \subseteq \mathcal{S}$, and $\mathbf{X}, \mathbf{Y}$ are random vectors with

$$\mathbf{X} \in L^1(\Omega; \mathbb{F}^{p_1}, \mathcal{S}),$$

$E(\mathbf{X}|\mathcal{S}_1)$ is defined to be a $\mathcal{S}_1$ measurable function defined on Ω such that for all $A \in \mathcal{S}_1$,

$$\int_A E(\mathbf{X}|\mathcal{S}_1) dP = \int_A \mathbf{X} dP. \qquad (17)$$

By 16 $E(\mathbf{X}|\mathbf{Y})$ satisfies the conditions of the definition in the case where $\mathcal{S}_1 = \sigma(\mathbf{Y})$. Next we show $E(\mathbf{X}|\mathcal{S}_1)$ exists for arbitrary σ algebras, $\mathcal{S}_1$.

Proposition 17.39 *Whenever $\mathcal{S}_1 \subseteq \mathcal{S}$ is a σ algebra and*

$$\mathbf{X} \in L^1(\Omega; \mathbb{F}^{p_1}, \mathcal{S})$$

is a random vector measurable with respect to $\mathcal{S}$, there exists a unique $\mathcal{S}_1$ measurable function, $E(\mathbf{X}|\mathcal{S}_1)$, in $L^1(\Omega; \mathbb{F}^{p_1}, \mathcal{S}_1)$ such that for all $A \in \mathcal{S}_1$, Formula 17 holds.

Proof: Let

$$\nu(A) \equiv \int_A \mathbf{X}(\omega) dP$$

for all $A \in \mathcal{S}_1$. Then $\nu << P$ and $|\nu|$ is clearly finite. Thus, by the Radon Nikodym theorem, there exists a unique $\mathcal{S}_1$ measurable function, $E(\mathbf{X}|\mathcal{S}_1)$, such that

$$\int_A E(\mathbf{X}|\mathcal{S}_1) dP = \int_A \mathbf{X}(\omega) dP.$$

This proves the proposition.

In particular this shows that for $\mathbf{X} \in L^1(\Omega; \mathbb{F}^{p_1})$, and $\mathbf{Y}$ measurable,

$$E(\mathbf{X}|\mathbf{Y}) = E(\mathbf{X}|\sigma(\mathbf{Y})).$$

The foregoing discussion implies a very significant observation which is stated as the next proposition.

Proposition 17.40 *Let* $\mathbf{X}$ *and* $\mathbf{Y}$ *be random vectors with* $\mathbf{X} \in L^1(\Omega)$. *Then if* $\mathbf{X} \in \sigma(\mathbf{Y})$, *there exists a Borel function* $\mathbf{g}$ *such that*

$$\mathbf{X}(\omega) = \mathbf{g}(\mathbf{Y}(\omega)) \quad a.e.$$

Proof: Suppose $\mathbf{X} \in \sigma(\mathbf{Y})$. By Theorem 15.8, on slicing measure, $E(\mathbf{X}|\cdot)$ can be taken as Borel measurable. Since $\mathbf{X}$ is $\sigma(\mathbf{Y})$ measurable, $E(\mathbf{X}|\mathbf{Y}) \equiv E(\mathbf{X}|\sigma(\mathbf{Y})) = \mathbf{X}$ and so we let $\mathbf{g} = E(\mathbf{X}|\cdot)$.

Finally we note that if $\{Y_i\}$ is a sequence of random variables and $\mathbf{Y}^n = (Y_1, \cdots, Y_n)$, then the σ algebras $\sigma(\mathbf{Y}^n) \equiv \sigma(Y_1, \cdots, Y_n)$ are increasing. To see this, let $A \in \sigma(Y_1, \cdots, Y_n)$ so $A = (\mathbf{Y}^n)^{-1}(F)$ where F is Borel in $\mathbb{F}^n$. Then $A = (\mathbf{Y}^{n+1})^{-1}(F \times \mathbb{F}) \in \sigma(Y_1, \cdots, Y_{n+1})$.

There are several properties of conditional expectation which are important. The first of these is a generalization of Jensen's inequality but before this we need a lemma about convex functions.

Lemma 17.41 *Let* I *be an open interval on* $\mathbb{R}$ *and let* ϕ *be a convex function defined on* I. *Then there exists a sequence* $\{(a_n, b_n)\}$ *such that*

$$\phi(x) = \sup\{a_n x + b_n, n = 1, \cdots\}.$$

Proof: Let $x \in I$ and let $t > x$. Then by convexity of ϕ,

$$\frac{\phi(x + \lambda(t - x)) - \phi(x)}{\lambda(t - x)} \leq \frac{\phi(x)(1 - \lambda) + \lambda\phi(t) - \phi(x)}{\lambda(t - x)}$$

$$= \frac{\phi(t) - \phi(x)}{t - x}.$$

Therefore $t \to \frac{\phi(t) - \phi(x)}{t - x}$ is increasing if $t > x$. If $t < x$

$$\frac{\phi(x + \lambda(t - x)) - \phi(x)}{\lambda(t - x)} \geq \frac{\phi(x)(1 - \lambda) + \lambda\phi(t) - \phi(x)}{\lambda(t - x)}$$

$$= \frac{\phi(t) - \phi(x)}{t - x}$$

and so $t \to \frac{\phi(t) - \phi(x)}{t - x}$ is increasing for $t \neq x$. Let

$$a_x \equiv \inf\left\{\frac{\phi(t) - \phi(x)}{t - x} : t > x\right\}.$$

Then if $t_1 < x$, and $t > x$,

$$\frac{\phi(t_1) - \phi(x)}{t_1 - x} \leq a_x \leq \frac{\phi(t) - \phi(x)}{t - x}.$$

Thus for all $t \in I$,

$$\phi(t) \geq a_x(t - x) + \phi(x). \tag{18}$$

Pick $t_2 > x$. Then for all $t \in (x, t_2)$

$$a_x \leq \frac{\phi(t) - \phi(x)}{t - x} \leq \frac{\phi(t_2) - \phi(x)}{t_2 - x}$$

and so

$$a_x(t - x) + \phi(x) \leq \phi(t) \leq \left(\frac{\phi(t_2) - \phi(x)}{t_2 - x}\right)(t - x) + \phi(x). \tag{19}$$

Pick $t_3 < x$. Then for $t_3 < t < x$

$$a_x \geq \frac{\phi(t) - \phi(x)}{t - x} \geq \frac{\phi(t_3) - \phi(x)}{t_3 - x}$$

and so

$$a_x(t - x) + \phi(x) \leq \phi(t) \leq \left(\frac{\phi(t_3) - \phi(x)}{t_3 - x}\right)(t - x) + \phi(x). \tag{20}$$

Formulas 19 and 20 show that ϕ is continuous.

$$\psi(x) \equiv \sup\{a_r(x - r) + \phi(r) : r \in \mathbb{Q} \cap I\}.$$

Then ψ is convex on I so ψ is continuous. Also $\psi(r) \geq \phi(r)$ so by Formula 18,

$$\psi(x) \geq \phi(x) \geq \sup\{a_r(x - r) + \phi(r)\} \equiv \psi(x).$$

Thus $\psi(x) = \phi(x)$ and we let $\mathbb{Q} \cap I = \{r_n\}$, and pick $a_n = a_{r_n}$ and $b_n = a_{r_n} r_n + \phi(r_n)$. This proves the lemma.

Lemma 17.42 *If $X \leq Y$, then $E(X|\mathcal{S}_1) \leq E(Y|\mathcal{S}_1)$ a.e. Also*

$$X \to E(X|\mathcal{S}_1)$$

is linear.

Proof: Let

$$A = \{\omega : E(X|\mathcal{S}_1)(\omega) > E(Y|\mathcal{S}_1)(\omega)\}.$$

Then if A has positive measure,

$$\int_A E(X|\mathcal{S}_1)\, dP \equiv \int_A X\, dP$$

$$\leq \int_A Y \, dP \equiv \int_A E\left(Y|\mathcal{S}_1\right) dP < \int_A E\left(X|\mathcal{S}_1\right) dP,$$

a contradiction. Hence $E\left(X|\mathcal{S}_1\right) \leq E\left(Y|\mathcal{S}_1\right)$ a.e. as claimed. It is obvious $X \to E\left(X|\mathcal{S}_1\right)$ is linear.

Theorem 17.43 *(Jensen's inequality)Let $X\left(\omega\right) \in I$ and let $\phi : I \to \mathbb{R}$ be convex. Suppose*

$$E\left(|X|\right), E\left(|\phi\left(X\right)|\right) < \infty.$$

Then

$$\phi\left(E\left(X|\mathcal{S}_1\right)\right) \leq E\left(\phi\left(X\right)|\mathcal{S}_1\right).$$

Proof: Let $\phi\left(x\right) = \sup\left\{a_n x + b_n\right\}$.

$$\frac{1}{P\left(A\right)} \int_A E\left(X|\mathcal{S}_1\right) dP = \frac{1}{P\left(A\right)} \int_A X \, dP \in I \text{ a.e.}$$

whenever $P\left(A\right) \neq 0$. Hence $E\left(X|\mathcal{S}_1\right)\left(\omega\right) \in I$ a.e. and so it makes sense to consider $\phi\left(E\left(X|\mathcal{S}_1\right)\right)$. Now

$$a_n E\left(X|\mathcal{S}_1\right)\left(\omega\right) + b_n = E\left(a_n X + b_n|\mathcal{S}_1\right) \leq E\left(\phi\left(X\right)|\mathcal{S}_1\right)\left(\omega\right).$$

Thus

$$\sup\left\{a_n E\left(X|\mathcal{S}_1\right)\left(\omega\right) + b_n\right\}$$
$$= \phi\left(E\left(X|\mathcal{S}_1\right)\left(\omega\right)\right) \leq E\left(\phi\left(X\right)|\mathcal{S}_1\right)\left(\omega\right) \text{ a.e.}$$

which proves the theorem.

Definition 17.44 Let $\mathcal{S}_k$ be an increasing sequence of σ algebras which are subsets of $\mathcal{S}$ and X_k be a sequence of real-valued random variables with $E\left(|X_k|\right) < \infty$ such that X_k is $\mathcal{S}_k$ measurable. Then this sequence is called a martingale if

$$E\left(X_{k+1}|\mathcal{S}_k\right) = X_k,$$

a submartingale if

$$E\left(X_{k+1}|\mathcal{S}_k\right) \geq X_k,$$

and a supermartingale if

$$E\left(X_{k+1}|\mathcal{S}_k\right) \leq X_k.$$

Lemma 17.45 *Let ϕ be a convex and increasing function and suppose $\left\{\left(X_n, \mathcal{S}_n\right)\right\}$ is a submartingale. Then if $E\left(|\phi\left(X_n\right)|\right) < \infty$, it follows $\left\{\left(\phi\left(X_n\right), \mathcal{S}_n\right)\right\}$ is also a submartingale.*

Proof: It is given that $E\left(X_{n+1}, \mathcal{S}_n\right) \geq X_n$ and so

$$\phi\left(X_n\right) \leq \phi\left(E\left(X_{n+1}, \mathcal{S}_n\right)\right) \leq E\left(\phi\left(X_{n+1}\right) | \mathcal{S}_n\right)$$

by Jensen's inequality.

As before, we have the upcrossing lemma.

Lemma 17.46 *(upcrossing lemma) Let $\{(X_i, \mathcal{S}_i)\}_{i=1}^{n}$ be a submartingale and let $U_{[a,b]}\left(\omega\right)$ be the number of upcrossings of $[a, b]$. Then*

$$E\left(U_{[a,b]}\right) \leq \frac{E\left(|X_n|\right) + |a|}{b - a}.$$

Proof: As before, we let $\phi\left(x\right) \equiv a + \left(x - a\right)^{+}$ so that ϕ is an increasing convex function. By Lemma 17.45 it follows that $\{(\phi\left(X_k\right), \mathcal{S}_k)\}$ is also a submartingale. As before,

$$\phi\left(X_{k+r}\right) - \phi\left(X_k\right) = \sum_{i=k+1}^{k+r} \phi\left(X_i\right) - \phi\left(X_{i-1}\right)$$

$$= \sum_{i=k+1}^{k+r} \left(\phi\left(X_i\right) - \phi\left(X_{i-1}\right)\right) Y_i + \sum_{i=k+1}^{k+r} \left(\phi\left(X_i\right) - \phi\left(X_{i-1}\right)\right)\left(1 - Y_i\right).$$

Observe that Y_i is $\mathcal{S}_{i-1}$ measurable from its construction in Proposition 17.31. Therefore, letting

$$A_i \equiv \{\omega : Y_i\left(\omega\right) = 0\},$$

$$E\left(\sum_{i=k+1}^{k+r} \left(\phi\left(X_i\right) - \phi\left(X_{i-1}\right)\right)\left(1 - Y_i\right)\right)$$

$$= \sum_{i=k+1}^{k+r} \int_{\Omega} \left(\phi\left(X_i\right) - \phi\left(X_{i-1}\right)\right)\left(1 - Y_i\right) dP$$

$$= \sum_{i=k+1}^{k+r} \int_{A_i} \left(\phi\left(X_i\right) - \phi\left(X_{i-1}\right)\right) dP$$

$$= \sum_{i=k+1}^{k+r} \int_{A_i} E\left(\phi\left(X_i\right), \mathcal{S}_{i-1}\right) dP - \int_{A_i} \phi\left(X_{i-1}\right) dP$$

$$\geq \sum_{i=k+1}^{k+r} \int_{A_i} \phi\left(X_{i-1}\right) dP - \int_{A_i} \phi\left(X_{i-1}\right) dP = 0.$$

The rest of the argument is identical with the proof of Lemma 17.33 starting with Formula 14.

Theorem 17.47 *(submartingale convergence theorem) Let*

$$\{(X_i, \mathcal{S}_i)\}_{i=1}^{\infty}$$

be a submartingale with $K \equiv \sup E(|X_n|) < \infty$. *Then there exists a random variable,* X, *such that* $E(|X|) \leq K$ *and*

$$\lim_{n \to \infty} X_n(\omega) = X(\omega) \quad a.e.$$

Proof: The proof of this theorem in this context is identical with the proof of Theorem 17.34.

It is also important to consider the case when the σ algebras are decreasing.

Definition 17.48 Let $(\Omega, \mathcal{S}, P)$ be a probability space and let $\{\mathcal{S}_k\}$ be either a finite or infinite sequence of σ algebras which are subsets of $\mathcal{S}$. Suppose also that $\mathcal{S}_k \supseteq \mathcal{S}_{k+1}$. Then the sequence $\{(X_k, \mathcal{S}_k)\}$ is called a reverse submartingale if X_k is a real random variable in $L^1(\Omega)$ which is $\mathcal{S}_k$ measurable, and $E(X_k | \mathcal{S}_{k+1}) \geq X_{k+1}$. If the inequality is replaced with equality, we call the sequence a reverse martingale.

Thus if $\{(X_k, \mathcal{S}_k)\}$ is a reverse submartingale, for each n,

$$(X_n, \mathcal{S}_n), \cdots, (X_1, \mathcal{S}_1)$$

is a submartingale.

Theorem 17.49 *(reverse submartingale convergence theorem) Let*

$$\{(X_k, \mathcal{S}_k)\}_{k=1}^{\infty}$$

be a reverse submartingale. Then there exists a random variable X_{∞} *such that* $X_{\infty}(\omega) = \lim_{n \to \infty} X_n(\omega)$ *a.e. If* $\sup\{E(|X_k|)\} = K < \infty$, *then* $X_{\infty} \in L^1(\Omega)$.

Proof: Let $[a, b]$ be an interval with rational endpoints and let $U_{[a,b]}^n$ denote the upcrossings of the submartingale $(X_n, \mathcal{S}_n), \cdots, (X_1, \mathcal{S}_1)$. Then by the upcrossing lemma,

$$E\left(U_{[a,b]}^n\right) \leq \frac{E(|X_1|) + |a|}{b - a} \tag{21}$$

which is a constant independent of n. Therefore, letting $n \to \infty$ and denoting $U_{[a,b]}$ as the pointwise limit of $U_{[a,b]}^n$, it follows from the monotone convergence theorem that Formula 21 holds with $U_{[a,b]}$ in place of $U_{[a,b]}^n$. Therefore, there exists a set of measure zero, $S_{[a,b]}$, such that if $\omega \notin S_{[a,b]}$, $U_{[a,b]}(\omega)$ is finite. Let

$$S \equiv \cup \left\{ S_{[a,b]} : a, b \in \mathbb{Q}, \text{ and } a < b \right\}.$$

Then $P(S) = 0$ and if $\omega \notin S$, then $U_{[a,b]}(\omega)$ is finite for all intervals with rational endpoints. This implies

$$\lim_{n \to \infty} \sup X_n(\omega) - \lim_{n \to \infty} \inf X_n(\omega) = 0.$$

Let $X_\infty(\omega) = \lim_{n \to \infty} X_n(\omega)$ for $\omega \in S$, and $X_\infty(\omega) = 0$ for $\omega \in S$. If

$$\sup \left\{ \int_\Omega |X_k| \, dP \right\} = K < \infty$$

then Fatou's lemma implies $X_\infty \in L^1(\Omega)$. This proves the theorem.

Let $Y \in L^1(\Omega)$ and let $\{S_n\}$ be a sequence of σ algebras which are subsets of S. Suppose the sequence of σ algebras is increasing. Then it is clear that $\{(E(Y|S_n), S_n)\}$ is a martingale. To see this note

$$E(E(Y|S_{n+1})|S_n) = E(Y|S_n).$$

This follows directly from Definition 17.38. Indeed, if $A \in S_n$,

$$\int_A E(E(Y|S_{n+1})|S_n) \, dP \equiv \int_A E(Y|S_{n+1}) \, dP \equiv \int_A Y \, dP$$

and so from the definition of what we mean by conditional expectation,

$$E(E(Y|S_{n+1})|S_n) = E(Y|S_n).$$

It follows that if the σ algebras are increasing, then $\{(X_n, S_n)\}$ is a martingale if $X_n \equiv E(Y|S_n)$. Similarly, if the σ algebras are decreasing, $\{(X_n, S_n)\}$ is a reverse martingale.

The next two theorems deal with the case just described, the first for increasing σ algebras and the second for decreasing σ algebras.

Theorem 17.50 *Suppose $Y \in L^1(\Omega)$ and let $X_n \equiv E(Y|S_n)$ where S_n is an increasing sequence of σ algebras contained in S. Define S_∞ to be the smallest σ algebra which contains $\cup_{n=1}^\infty S_n$. Then $X_n \to E(Y|S_\infty)$ a.e. and in $L^1(\Omega)$.*

Proof: First note that

$$E(|X_n|) = E(|E(Y|S_n)|) \leq E(E(|Y||S_n)) = E(|Y|) < \infty$$

and so by the submartingale convergence theorem, $X_n(\omega) \to X_\infty(\omega)$ a.e. where $E(|X_\infty|) \leq E(|Y|)$ also. Let $A_c^n \equiv \{\omega \in \Omega : |X_n| \geq c\}$. Then

$$\int_{A_c^n} |X_n| \, dP = \int_{A_c^n} |E(Y|S_n)| \, dP \leq \int_{A_c^n} |Y| \, dP.$$

Also

$$P(A_o^n) \leq c^{-1} E\left(|X_n|\right) \leq c^{-1} E\left(E\left(|Y| \, | \mathcal{S}_n\right)\right) = c^{-1} E\left(|Y|\right) \qquad (22)$$

which converges to 0 independent of n as $c \to \infty$. Choosing c large enough, we may assume

$$\int_{A_c^n} |X_\infty| \, dP, \text{ and } \int_{A_c^n} |X_n| \, dP \leq \epsilon$$

for all n.

$$\int_\Omega |X_n(\omega) - X_\infty(\omega)| \, dP \;=\; \int_{A_c^n} |X_n(\omega) - X_\infty(\omega)| \, dP$$

$$+ \int_{\{\omega : |X_n(\omega)| < c\}} |X_\infty - X_n| \, dP$$

$$\leq 2\epsilon + \int_{\{\omega : |X_n(\omega)| < c\}} |X_\infty - X_n| \, dP. \qquad (23)$$

The integrand in this last integral is dominated by $c + |X_\infty|$, a function in $L^1(\Omega)$, and so by the dominated convergence theorem, this integral converges to 0. Since ϵ is arbitrary, this shows $X_n \to X_\infty$ in $L^1(\Omega)$ as well as pointwise. It remains to show that $X_\infty = E\left(Y| \mathcal{S}_\infty\right)$. It is clearly $\mathcal{S}_\infty$ measurable.

Let $A \in \mathcal{S}_n$. Then

$$\int_A Y \, dP \equiv \int_A E\left(Y| \mathcal{S}_n\right) dP = \int_A X_n \, dP \to \int_A X_\infty \, dP.$$

Therefore,

$$\int_A Y \, dP = \int_A X_\infty \, dP \qquad (24)$$

for all $A \in \cup_{n=1}^\infty \mathcal{S}_n$, an algebra of sets. If

$$\mathcal{M} \equiv \{A \in \mathcal{S}_\infty : \text{ Formula 24 holds}\}$$

It follows easily that $\mathcal{M}$ is a monotone class containing the algebra, $\cup_{n=1}^\infty \mathcal{S}_n$. Therefore $\mathcal{M} = \mathcal{S}_\infty$ by the monotone class lemma and this shows $X_\infty = E\left(Y| \mathcal{S}_\infty\right)$.

Theorem 17.51 *Suppose $\{\mathcal{S}_n\}$ is a decreasing sequence of σ algebras, $Y \in L^1(\Omega)$, and $\mathcal{S}_\infty \equiv \cap_{n=1}^\infty \mathcal{S}_n$. Letting $X_n \equiv E\left(Y| \mathcal{S}_n\right)$, it follows $X_n \to E\left(Y| \mathcal{S}_\infty\right)$ a.e. and in L^1.*

Proof:

$$E\left(|X_n|\right) = E\left(|E\left(Y| \mathcal{S}_n\right)|\right) \leq E\left(E\left(|Y| \, | \mathcal{S}_n\right)\right) \equiv E\left(|Y|\right) < \infty$$

and so by the reverse submartingale theorem, there exists a random variable $X_\infty \in L^1(\Omega)$ such that $X_n(\omega) \to X_\infty(\omega)$ a.e. Recall that the set on which convergence to X_∞ does not take place is the union of sets $S_{[a,b]}$ where $S_{[a,b]}$ is the set where the upcrossing function, $U_{[a,b]}$, equals infinity. Now if we define by $U_{[a,b],n}$ the upcrossings of $\{X_n, X_{n+1}, \cdots\}$ we see that the set where $U_{[a,b]}$ equals infinity is the same as the set where $U_{[a,b],n}$ equals infinity and this set is in $\mathcal{S}_n$. Therefore, $S_{[a,b]} \in \mathcal{S}_\infty = \cap_{n=1}^\infty \mathcal{S}_n$. Thus $S \equiv \cup_{n=1}^\infty \{S_{[a,b]} : a, b \in \mathbb{Q}\}$ is in $\mathcal{S}_\infty$ also. Redefining X_∞ to equal 0 on S, it follows

$$\lim_{n \to \infty} X_n(\omega) \mathcal{X}_{S^C}(\omega) = X_\infty(\omega)$$

for all ω and $X_n \mathcal{X}_{S^C}$ is measurable with respect to $\mathcal{S}_n$. Then since $X_n \mathcal{X}_{S^C}$ is $\mathcal{S}_k$ measurable whenever $n > k$, it follows that X_∞ is $\mathcal{S}_\infty$ measurable.

Just as in Theorem 17.50, the Estimates 22 and 23 hold and so the sequence converges in $L^1(\Omega)$ also. It remains to verify that $X_\infty = E(Y|\mathcal{S}_\infty)$. To do this, let $A \in \mathcal{S}_\infty$. As in the proof of Theorem 17.50,

$$\int_A Y \, dP \equiv \int_A E(Y|\mathcal{S}_n) \, dP = \int_A X_n \, dP \to \int_A X_\infty \, dP$$

which shows $X_\infty = E(Y|\mathcal{S}_\infty)$.

17.5 Strong law of large numbers

Definition 17.52 Let $\{X_i\}_{i=1}^\infty$ be a sequence of random variables.

$$\sigma(X_1, X_2, \cdots)$$

is defined to be the smallest σ algebra in $\mathcal{S}$ such that each of the random variables is measurable. We also define $E(X|X_1, \cdots) \equiv E(X|\sigma(X_1, \cdots))$.

Thus $\sigma(X_1, \cdots)$ is the smallest σ algebra containing the σ algebras

$$\{\sigma(X_1, \cdots, X_n)\}.$$

The idea of independent random variables is that the value of one does not depend on the values of the others. This next lemma, although it is technical, should make good sense heuristically. It states roughly that if X depends only on the variables $X_n, X_{n+1}, \cdots$ and $\{X_i\}$ is a collection of independent random variables, then $E(X) = E(X|X_1, \cdots, X_{n-1})$.

Lemma 17.53 Let $\{X_i\}_{i=1}^\infty$ be a sequence of independent random variables and suppose $X \in \sigma(X_n, \cdots)$ with $E(|X|) < \infty$. Then

$$E(X|X_1, \cdots, X_{n-1}) = E(X) \quad a.e.$$

Proof: Let $m > 0$ and define $Y_m \equiv E(X|X_n, \cdots, X_{m+n})$. By Proposition 17.40,

$$Y_m(\omega) = g(X_n(\omega), \cdots, X_{m+n}(\omega)) \text{ a.e.}$$

where g is a Borel function. (In fact, $g(\cdot) = E(X|\cdot)$.)

$$E(g(X_n, \cdots, X_{n+m})|x_1, \cdots, x_{n-1})$$

$$\equiv \int_{\mathbb{F}^{m+1}} g(x_n, \cdots, x_{n+m}) \, d\lambda_{(X_n, \cdots, X_{n+m})|x_1 \cdots x_{n-1}}.$$

Letting $B \in \sigma(X_1, \cdots, X_{n-1})$ and $B = (X_1, \cdots, X_{n-1})^{-1}(A)$ where A is a Borel set in $\mathbb{F}^{n-1}$, thus

$$\int_B E(g(X_n, \cdots, X_{n+m})|X_1, \cdots, X_{n-1}) \, dP$$

$$= \int_A E(g(X_n, \cdots, X_{n+m})|x_1, \cdots, x_{n-1}) \, d\lambda_{(X_1, \cdots, X_{n-1})}$$

$$= \int_{\mathbb{F}^{n-1}} \int_{\mathbb{F}^{m+1}} \mathcal{X}_A(x_1, \cdots, x_{n-1}) \cdot$$

$$g(x_n, \cdots, x_{n+m}) \, d\lambda_{(X_n, \cdots, X_{n+m})|x_1 \cdots x_{n-1}} d\lambda_{(X_1, \cdots, X_{n-1})}$$

$$= \int_{\mathbb{F}^{m+1}} \int_{\mathbb{F}^{n-1}} \mathcal{X}_A(x_1, \cdots, x_{n-1}) \cdot$$

$$g(x_n, \cdots, x_{n+m}) \, d\lambda_{(X_1, \cdots, X_{n-1})|x_n \cdots x_{n+m}} d\lambda_{(X_n, \cdots, X_{n+m})}$$

$$= \int_{\mathbb{F}^{m+1}} g(x_n, \cdots, x_{n+m}) \, d\lambda_{(X_n, \cdots, X_{n+m})} P(B)$$

$$= E(g(X_n, \cdots, X_{n+m})) P(B).$$

It follows that for all $B \in \sigma(X_1, \cdots, X_{n-1})$,

$$\int_B E(g(X_n, \cdots, X_{n+m})|X_1, \cdots, X_{n-1}) \, dP$$

$$= \int_B E(g(X_n, \cdots, X_{n+m})) \, dP.$$

Therefore,

$$
\begin{aligned}
E(Y_m|X_1, \cdots, X_{n-1}) &= E(g(X_n, \cdots, X_{n+m})|X_1, \cdots, X_{n-1}) \\
&= E(g(X_n, \cdots, X_{n+m})) = E(Y_m) \\
&= E(E(X|X_n, \cdots, X_{m+n})) = E(X). \quad (25)
\end{aligned}
$$

By Theorem 17.50

$$Y_m \to E(X|X_n, \cdots)$$

in L^1 and a.e. But $E(X|X_n, \cdots) = X$ because X is measurable with respect to $\sigma(X_n, \cdots)$. Let $A \in \sigma(X_1, \cdots, X_{n-1})$. Then

$$\left| \int_A E(Y_m|X_1, \cdots, X_{n-1}) \, dP - \int_A E(X|X_1, \cdots, X_{n-1}) \, dP \right|$$

$$= \left| \int_A (Y_m - X) \, dP \right|$$

$$\leq \int_A |Y_m - X| \, dP$$

which converges to 0. Therefore,

$$\int_A X \, dP \equiv \int_A E(X|X_1, \cdots, X_{n-1}) \, dP =$$

$$\lim_{m \to \infty} \int_A E(Y_m|X_1, \cdots, X_{n-1}) \, dP = \int_A E(X) \, dP$$

from Formula 25. Thus, $E(X) = E(X|X_1, \cdots, X_{n-1})$ as claimed.

The next theorem gives an interesting application of the limit theorems just presented. It deals with a tail function, so called because it is measurable with respect to $\sigma(X_n, \cdots)$ for all n.

Theorem 17.54 *Let $\{X_i\}_{i=1}^{\infty}$ be a sequence of independent random variables and suppose $X \in L^1(\Omega)$ is such that $X \in \sigma(X_n, \cdots)$ for all n. Then $X(\omega) = constant$ a.e.*

Proof: From Lemma 17.53,

$$E(X) = E(X|X_1, \cdots, X_{n-1}).$$

By Theorem 17.50,

$$E(X|X_1, \cdots, X_{n-1}) \to E(X|X_1, \cdots) = X$$

a.e. Thus, $X = E(X)$ a.e. and this proves the theorem.

It is easy to extend this theorem by removing the assumption that $X \in L^1$. See Problem 19 for this generalization.

Corollary 17.55 *(Kolmogorov 0-1 law) Let $A \in \sigma(X_n, X_{n+1}, \cdots)$ for all n, where $\{X_i\}_{i=1}^{\infty}$ is a sequence of independent random variables. Then $P(A)$ equals either 0 or 1.*

Proof: $\mathcal{X}_A$ is measurable in $\sigma(X_n, \cdots)$ and so by Theorem 17.54 $\mathcal{X}_A(\omega) = constant$ a.e. Thus $P(A) = 0$ or 1.

We say the random variables, $\{X_i\}_{i=1}^{\infty}$ are identically distributed if $\lambda_{X_i} = \lambda_{X_j}$ for all i, j.

The following technical lemma is used to prove the famous strong law of large numbers. It says roughly that if you know the sum of n independent identically distributed random variables, then the expected value of any of the random variables in the sum, given the sum, should be the same for each of the random variables in the sum.

Lemma 17.56 *Let $\{X_i\}_{i=1}^{\infty}$ be a sequence of independent identically distributed random variables with $E\left(|X_i|\right) < \infty$. Also let*

$$S_n \equiv \sum_{i=1}^{n} X_i.$$

Then for $i = 1, \cdots, n$,

$$E\left(X_i | S_n X_{n+1}, \cdots, X_{n+r}\right) = \frac{S_n}{n} \ a.e.; \tag{26}$$

Also

$$\sigma\left(S_n, X_{n+1}, \cdots, X_{n+r}\right) = \sigma\left(S_n, S_{n+1}, \cdots, S_{n+r}\right). \tag{27}$$

Proof: Let B be a Borel set in $\mathbb{R}^{r+1}$

$$\int_B E\left(X_i | s_n, x_{n+1}, \cdots, x_{n+r}\right) d\lambda_{(S_n X_{n+1}, \cdots, X_{n+r})} =$$

$$\int_{\mathbb{R}^{r+1}} \int_{\mathbb{R}} x_i \mathcal{X}_B\left(s_n, x_{n+1}, \cdots, x_{n+r}\right) d\lambda_{X_i | s_n x_{n+1} \cdots x_{n+r}} d\lambda_{(S_n, X_{n+1}, \cdots, X_{n+r})}$$

$$= \int_{\mathbb{R}^{r+2}} x_i \mathcal{X}_B\left(s_n, x_{n+1}, \cdots, x_{n+r}\right) d\lambda_{(X_i, S_n, X_{n+1}, \cdots, X_{n+r})}$$

$$= \int_{\mathbb{R}^{n+r}} x_i \mathcal{X}_B\left(\sum_{i=1}^{n} x_i, x_{n+1}, \cdots, x_{n+r}\right) d\lambda_{(X_1, \cdots, X_n X_{n+1}, \cdots, X_{n+r})}$$

$$= \int_{\mathbb{R}} \cdots \int_{\mathbb{R}} x_i \mathcal{X}_B\left(\sum_{i=1}^{n} x_i, x_{n+1}, \cdots, x_{n+r}\right) d\lambda_{X_1} \cdots d\lambda_{X_{n+r}} \equiv l_i\left(B\right). \tag{28}$$

Since the order of integration is immaterial, it follows $l_i\left(B\right) = l_j\left(B\right) \equiv l\left(B\right)$. Also,

$$E\left(S_n | S_n X_{n+1}, \cdots, X_{n+r}\right) = S_n$$

and so, letting $A = \left(S_n, X_{n+1}, \cdots, X_{n+r}\right)^{-1}\left(B\right)$,

$$\int_A S_n \, dP = \int_A E\left(S_n | S_n, X_{n+1}, \cdots, X_{n+r}\right) dP$$

$$= \sum_{i=1}^{n} \int_{A} E\left(X_i | S_n, X_{n+1}, \cdots, X_{n+r}\right) dP$$

$$= \sum_{i=1}^{n} \int_{B} E\left(X_i | s_n x_{n+1}, \cdots, x_{n+r}\right) d\lambda_{(S_n X_{n+1}, \cdots, X_{n+r})} = nl\left(B\right).$$

Hence, from Formula 28,

$$l\left(B\right) = \int_{B} E\left(X_i | s_n x_{n+1}, \cdots, x_{n+r}\right) d\lambda_{(S_n X_{n+1}, \cdots, X_{n+r})} =$$

$$\int_{A} E\left(X_i | S_n, X_{n+1}, \cdots, X_{n+r}\right) dP = \frac{1}{n} \int_{A} S_n dP$$

and this happens for all $A \in \sigma\left(S_n, X_{n+1}, \cdots, X_{n+r}\right)$. Hence

$$E\left(X_i | S_n X_{n+1}, \cdots, X_{n+r}\right) = \frac{S_n}{n} \text{ a.e.}$$

and this verifies Formula 26.

It remains to verify Formula 27. This is easy to see if we note that

$$\left(S_n, X_{n+1}, \cdots, X_{n+r}\right)$$

is a Borel function of $\left(S_n, S_{n+1}, \cdots, S_{n+r}\right)$ and that $\left(S_n, S_{n+1}, \cdots, S_{n+r}\right)$ is a Borel function of

$$\left(S_n, X_{n+1}, \cdots, X_{n+r}\right).$$

Thus, if B is a Borel set,

$$\left(S_n, S_{n+1}, \cdots, S_{n+r}\right)^{-1}\left(B\right)$$

$$= \left(S_n, X_{n+1}, \cdots, X_{n+r}\right)^{-1}\left(g^{-1}\left(B\right)\right) \in \sigma\left(S_n, X_{n+1}, \cdots, X_{n+r}\right)$$

for some Borel measurable function $\mathbf{g}$. It follows that

$$\sigma\left(S_n, S_{n+1}, \cdots, S_{n+r}\right) \subseteq \sigma\left(S_n, X_{n+1}, \cdots, X_{n+r}\right).$$

The other inclusion is established the same way. This proves the lemma.

With this lemma, it is easy to give a proof of the Strong law of large numbers.

Theorem 17.57 *(Strong law of large numbers) Let $\{X_i\}_{i=1}^{\infty}$ be a sequence of independent and identically distributed real random variables with*

$$E\left(|X_i|\right) < \infty.$$

Then

$$\lim_{n \to \infty} \frac{S_n\left(\omega\right)}{n} = E\left(X_1\right)$$

a.e. and in $L^1(\Omega)$.

Proof: From Lemma 17.56, $E(X_i|S_n S_{n+1}, \cdots, S_{n+r}) = \frac{S_n}{n}$. By Theorem 17.50

$$\lim_{r \to \infty} E(X_i|S_n, S_{n+1}, \cdots, S_{n+r}) = E(X_i|S_n, \cdots)$$

a.e. and in $L^1(\Omega)$. Hence

$$\frac{S_n}{n} = E(X_i|S_n, \cdots) \text{ a.e.}$$

for all $i \leq n$. In particular this holds for $i = 1$. Now by Theorem 17.51

$$E(X_1|S_n \cdots) = \frac{S_n}{n} \to E(X_1|S_\infty)$$

a.e. and in $L^1(\Omega)$ where $S_\infty = \cap_{n=1}^\infty \sigma(S_n \cdots)$. But $\lim_{n \to \infty} \frac{S_n}{n} = \lim_{n \to \infty} \frac{1}{n} \sum_{k=m}^n X_k$ and so $\lim_{n \to \infty} \frac{S_n}{n}$ is a tail function. By Theorem 17.54, this function equals a constant, C, a.e. Therefore, the L^1 convergence implies $\lim_{n \to \infty} \int \frac{S_n}{n} dP = \lim_{n \to \infty} E(X_1) = E(X_1) = \int C dP = C$, and so $C = E(X_1)$. This proves the theorem.

17.6   The normal distribution

Definition 17.58 We say a random vector, $\mathbf{X}$, with values in $\mathbb{R}^p$ has a multivariate normal distribution and write $\mathbf{X} \sim N_p(\mu, \Sigma)$ if for all Borel $E \subseteq \mathbb{R}^p$,

$$\lambda_{\mathbf{X}}(E) = \int_{\mathbb{R}^p} \mathcal{X}_E(\mathbf{x}) \frac{1}{(2\pi)^{p/2} \det(\Sigma)^{1/2}} e^{\frac{-1}{2}(\mathbf{x}-\mu)^* \Sigma^{-1}(\mathbf{x}-\mu)} d\mathbf{x}$$

for μ a given vector and Σ a given positive definite symmetric matrix.

Theorem 17.59 For $\mathbf{X} \sim N_p(\mu, \Sigma)$, $\mu = E(\mathbf{X})$ and

$$\Sigma = E((\mathbf{X} - \mu)(\mathbf{X} - \mu)^*).$$

This shows the two parameters in the definition of the multivariate normal are not mislabelled.

Proof: Let R be an orthogonal transformation such that

$$R\Sigma R^* = D = diag(\sigma_1^2, \cdots, \sigma_p^2).$$

Changing the variable by $\mathbf{x} - \mu = R^*\mathbf{y}$,

$$
\begin{aligned}
E(\mathbf{X}) &\equiv \int_{\mathbb{R}^p} \mathbf{x} e^{\frac{-1}{2}(\mathbf{x}-\mu)^*\Sigma^{-1}(\mathbf{x}-\mu)} dx \left(\frac{1}{(2\pi)^{p/2} \det(\Sigma)^{1/2}} \right) \\
&= \int_{\mathbb{R}^p} (R^*\mathbf{y} + \mu) e^{-\frac{1}{2}\mathbf{y}^*D^{-1}\mathbf{y}} dy \left(\frac{1}{(2\pi)^{p/2} \prod_{i=1}^p \sigma_i} \right) \\
&= \mu \int_{\mathbb{R}^p} e^{-\frac{1}{2}\mathbf{y}^*D^{-1}\mathbf{y}} dy \left(\frac{1}{(2\pi)^{p/2} \prod_{i=1}^p \sigma_i} \right) = \mu
\end{aligned}
$$

by Fubini's theorem and the easy to establish formula (Lemma 16.6)

$$
\frac{1}{\sqrt{2\pi}\sigma} \int_{\mathbb{R}} e^{-\frac{y^2}{2\sigma^2}} dy = 1.
$$

Next let $M \equiv E\left((\mathbf{X} - \mu)(\mathbf{X} - \mu)^*\right)$. Thus, changing the variable as above,

$$
R M R^* = \int_{\mathbb{R}^p} \mathbf{y}\mathbf{y}^* e^{-\frac{1}{2}\mathbf{y}^*D^{-1}\mathbf{y}} dy \left(\frac{1}{(2\pi)^{p/2} \prod_{i=1}^p \sigma_i} \right).
$$

Now

$$
(R M R^*)_{ij} = \int_{\mathbb{R}^p} y_i y_j e^{-\frac{1}{2}\mathbf{y}^*D^{-1}\mathbf{y}} dy \left(\frac{1}{(2\pi)^{p/2} \prod_{i=1}^p \sigma_i} \right) = 0,
$$

so; $R M R^*$ is a diagonal matrix.

$$
(R M R^*)_{ii} = \int_{\mathbb{R}^p} y_i^2 e^{-\frac{1}{2}\mathbf{y}^*D^{-1}\mathbf{y}} dy \left(\frac{1}{(2\pi)^{p/2} \prod_{i=1}^p \sigma_i} \right).
$$

Using Fubini's theorem and the easy to establish equations, (Lemma 16.6),

$$
\frac{1}{\sqrt{2\pi}\sigma} \int_{\mathbb{R}} e^{-\frac{y^2}{2\sigma^2}} dy = 1, \quad \frac{1}{\sqrt{2\pi}\sigma} \int_{\mathbb{R}} y^2 e^{-\frac{y^2}{2\sigma^2}} dy = \sigma^2,
$$

we see $(R M R^*)_{ii} = \sigma_i^2$. Hence $R M R^* = D$ and so $M = R^* D R = \Sigma$. This proves the theorem.

Theorem 17.60 *Suppose $\mathbf{X}_1 \sim N_p(\mu_1, \Sigma_1)$, $\mathbf{X}_2 \sim N_p(\mu_2, \Sigma_2)$ and the two random vectors are independent. Then*

$$
\mathbf{X}_1 + \mathbf{X}_2 \sim N_p(\mu_1 + \mu_2, \Sigma_1 + \Sigma_2).
$$

Proof: Since $\mathbf{X}_1$ and $\mathbf{X}_2$ are independent, $e^{it\cdot\mathbf{X}_1}$ and $e^{it\cdot\mathbf{X}_2}$ are also independent. Hence

$$
E\left(e^{it\cdot\mathbf{X}_1+\mathbf{X}_2}\right) = E\left(e^{it\cdot\mathbf{X}_1}\right) E\left(e^{it\cdot\mathbf{X}_2}\right).
$$

The normal distribution

Now we find $E\left(e^{it\cdot\mathbf{X}}\right)$ for $\mathbf{X} \sim N_p\left(\mu, \Sigma\right)$.

$$E\left(e^{it\cdot\mathbf{X}}\right) \equiv \frac{1}{(2\pi)^{p/2}\left(\det\Sigma\right)^{1/2}} \int_{\mathbb{R}^p} e^{it\cdot\mathbf{x}} e^{-\frac{1}{2}(\mathbf{x}-\mu)^*\Sigma^{-1}(\mathbf{x}-\mu)} dx.$$

Let R be an orthogonal transformation such that

$$R\Sigma R^* = D = diag\left(\sigma_1^2, \cdots, \sigma_p^2\right).$$

Then we let $R\left(\mathbf{x}-\mu\right) = \mathbf{y}$. Then

$$E\left(e^{it\cdot\mathbf{X}}\right) = \frac{1}{(2\pi)^{p/2}\prod_{i=1}^p \sigma_i} \int_{\mathbb{R}^p} e^{it\cdot(R^*\mathbf{y}+\mu)} e^{-\frac{1}{2}\mathbf{y}^*D^{-1}\mathbf{y}} dx.$$

Therefore

$$E\left(e^{it\cdot\mathbf{X}}\right) = \frac{1}{(2\pi)^{p/2}\prod_{i=1}^p \sigma_i} \int_{\mathbb{R}^p} e^{is\cdot(\mathbf{y}+R\mu)} e^{-\frac{1}{2}\mathbf{y}^*D^{-1}\mathbf{y}} dx$$

where $\mathbf{s} = R\mathbf{t}$. This equals

$$e^{it\cdot\mu} \prod_{i=1}^p \left(\int_{\mathbb{R}} e^{is_i y_i} e^{-\frac{1}{2\sigma_i^2}y_i^2} dy_i\right) \frac{1}{\sqrt{2\pi}\sigma_i}$$

$$= e^{it\cdot\mu} \prod_{i=1}^p \left(\int_{\mathbb{R}} e^{is_i\sigma_i u} e^{-\frac{1}{2}u^2} du\right) \frac{1}{\sqrt{2\pi}}$$

$$= e^{it\cdot\mu} \prod_{i=1}^p e^{-\frac{1}{2}s_i^2\sigma_i^2} \frac{1}{\sqrt{2\pi}} \int_{\mathbb{R}} e^{-\frac{1}{2}(u-is_i\sigma_i)^2} du$$

$$= e^{it\cdot\mu} e^{-\frac{1}{2}\sum_{i=1}^p s_i^2\sigma_i^2} = e^{it\cdot\mu} e^{-\frac{1}{2}\mathbf{t}^*\Sigma\mathbf{t}}$$

by Lemma 16.6. Thus,

$$\begin{aligned} E\left(e^{it\cdot\mathbf{X}_1+\mathbf{X}_2}\right) &= E\left(e^{it\cdot\mathbf{X}_1}\right) E\left(e^{it\cdot\mathbf{X}_2}\right) \\ &= e^{it\cdot\mu_1} e^{-\frac{1}{2}\mathbf{t}^*\Sigma_1\mathbf{t}} e^{it\cdot\mu_2} e^{-\frac{1}{2}\mathbf{t}^*\Sigma_2\mathbf{t}} \\ &= e^{it\cdot(\mu_1+\mu_2)} e^{-\frac{1}{2}\mathbf{t}^*(\Sigma_1+\Sigma_2)\mathbf{t}} \end{aligned}$$

which is the characteristic function of a random vector distributed as

$$N_p\left(\mu_1 + \mu_2, \Sigma_1 + \Sigma_2\right).$$

17.7   The central limit theorem

The central limit theorem is important in probability and statistics because it is a limit theorem for a certain sequence of random variables which are obtained from a given sequence of random variables about which very little need be known.

Theorem 17.61 *(Central limit theorem) Let $X_1, X_2, \cdots$ be independent real-valued random variables with finite positive variances, $\sigma_1^2, \sigma_2^2, \cdots$ and means, $\mu_1, \mu_2, \cdots$. Also assume that for all $\epsilon > 0$,*

$$\lim_{n\to\infty} \frac{1}{s_n^2} \sum_{k=1}^{n} \int_{|x-\mu_k|>\epsilon s_n} (x - \mu_k)^2 \, d\lambda_{X_k} = 0 \tag{29}$$

where

$$s_n^2 \equiv \sum_{i=1}^{n} \sigma_i^2. \tag{30}$$

Then if

$$Z_n \equiv \sum_{k=1}^{n} \frac{X_k - \mu_k}{s_n}, \tag{31}$$

it follows that for all $x \in \mathbb{R}$,

$$\lim_{n\to\infty} F_{Z_n}(x) \equiv \lim_{n\to\infty} P(Z_n \leq x) = F_W(x) \tag{32}$$

where $W \sim N_1(0, 1)$. Thus

$$F_W(x) \equiv \frac{1}{\sqrt{2\pi}} \int_{-\infty}^{x} e^{-\frac{t^2}{2}} \, dt. \tag{33}$$

Proof: Let $Y_k^n = \frac{X_k - \mu_k}{s_n}$. Then

$$E(Y_k^n) = 0, \; E\left((Y_k^n)^2\right) = \frac{\sigma_k^2}{s_n^2}$$

and Formula 29 reduces to

$$\lim_{n\to\infty} \sum_{k=1}^{n} \int_{|y|>\epsilon} y^2 d\lambda_{Y_k^n} = 0.$$

To see this, let $E \equiv \left\{ x : \frac{|x-\mu_k|}{s_n} > \epsilon \right\}$ and let $g(x) = \mathcal{X}_E(x) \frac{(x-\mu_k)^2}{s_n^2}$.

$$\begin{aligned}
\frac{1}{s_n^2} \int_{|x-\mu_k|>\epsilon s_n} (x - \mu_k)^2 \, d\lambda_{X_k} &= \int_{\mathbb{R}} g(x) \, d\lambda_{X_k} \\
&= \int_{\mathbb{R}} z \, d\lambda_{g(X_k)} \\
&= \int_{\mathbb{R}} z \, d\lambda_{h(Y_k^n)}
\end{aligned}$$

where $h(y) = \mathcal{X}_F(y) y^2$ for $F \equiv \{y : |y| > \epsilon\}$. This equals

$$\int_{\mathbb{R}} h(y) \, d\lambda_{Y_k^n} = \int_{|y| > \epsilon} y^2 d\lambda_{Y_k^n}.$$

Now let $W_k^n \sim N_1\left(0, \frac{\sigma_k^2}{s_n^2}\right)$ and $\{W_1^n, \cdots, W_n^n\}$ are independent. By Theorem 17.60,

$$\sum_{k=1}^{n} W_k^n \equiv W \sim N_1(0, 1).$$

Define $BC^3(\mathbb{R})$ to be the space of functions, f, defined on $\mathbb{R}$ such that f, f', f'', and f''' all exist and are continuous and bounded on $\mathbb{R}$. We make this space of functions into a Banach space by using the norm

$$\|f\|_3 \equiv \sum_{i=0}^{3} \left\|f^{(i)}\right\|_\infty, \quad \|g\|_\infty \equiv \sup\{|g(x)| : x \in \mathbb{R}\}.$$

Let $BC(\mathbb{R})$ be the space of functions which are bounded and continuous with the $\|\cdot\|_\infty$ norm.

Now define operators A_k and $B_k \in \mathcal{L}(BC^3(\mathbb{R}), BC(\mathbb{R}))$ by

$$A_k f(x) \equiv \int_{\mathbb{R}} f(x + y) \, d\lambda_{Y_k^n}, \quad B_k f(x) \equiv \int_{\mathbb{R}} f(x + y) \, d\lambda_{W_k^n}.$$

Lemma 17.62 A_k, B_k *map* $BC^3(\mathbb{R})$ *to* $BC^3(\mathbb{R})$ *and when considered as elements of* $\mathcal{L}(BC^3(\mathbb{R}), BC(\mathbb{R}))$,

$$\|A_k\|, \|B_k\| \leq 1. \tag{34}$$

Whenever $\{k_1, \cdots, k_r\} \subseteq \{1, \cdots, n\}$,

$$\|A_{k_1} A_{k_2} \cdots A_{k_r} - B_{k_1} B_{k_2} \cdots B_{k_r}\| \leq \sum_{i=1}^{r} \|A_{k_i} - B_{k_i}\|. \tag{35}$$

Proof: All the above assertions are obvious except the last. The last is true if $r = 1$. Suppose true for $r < n$. Then

$$\|A_{k_1} A_{k_2} \cdots A_{k_{r+1}} - B_{k_1} B_{k_2} \cdots B_{k_{r+1}}\|$$

$$\leq \|A_{k_1} A_{k_2} \cdots A_{k_{r+1}} - B_{k_1} A_{k_2} \cdots A_{k_{r+1}}\|$$

$$+ \|B_{k_1} A_{k_2} \cdots A_{k_{r+1}} - B_{k_1} B_{k_2} \cdots B_{k_{r+1}}\|$$

$$\leq \|A_{k_1} - B_{k_1}\| + \|A_{k_2} \cdots A_{k_{r+1}} - B_{k_2} \cdots B_{k_{r+1}}\|$$

$$\leq \sum_{i=1}^{r+1} \|A_{k_i} - B_{k_i}\|$$

by induction.

Note that by Lemma 17.18,

$$\int_{\mathbb{R}} f\left(x+z\right) d\lambda_{Z_n}\left(z\right) = \int_{\mathbb{R}} \cdots \int_{\mathbb{R}} f\left(x+y_1+\cdots+y_n\right) d\lambda_{Y_1^n} \cdots d\lambda_{Y_n^n}$$
$$= A_1 A_2, \cdots, A_n f\left(x\right)$$

and

$$\int_{\mathbb{R}} f\left(x+z\right) d\lambda_W\left(z\right) = \int_{\mathbb{R}} \cdots \int_{\mathbb{R}} f\left(x+y_1+\cdots+y_n\right) d\lambda_{W_1^n} \cdots d\lambda_{W_n^n}$$
$$= B_1 B_2, \cdots, B_n f\left(x\right).$$

Let $f \in BC^3\left(\mathbb{R}\right)$. Then

$$f\left(x+y\right) = f\left(x\right) + f'\left(x\right)y + \frac{1}{2}f''\left(x\right)y^2 + \frac{1}{3!}f'''\left(\xi_1\right)y^3$$
$$= f\left(x\right) + f'\left(x\right)y + \frac{1}{2}f''\left(\xi_2\right)y^2. \tag{36}$$

Therefore, using Formula 36,

$$A_k f\left(x\right) \equiv \int_{\mathbb{R}} f\left(x\right) + f'\left(x\right)y + \frac{1}{2}f''\left(x\right)y^2 + \frac{1}{3!}f'''\left(\xi_1\right)y^3 d\lambda_{Y_k^n}$$

$$= f\left(x\right) + \frac{1}{2}f''\left(x\right)\frac{\sigma_k^2}{s_n^2} + \int_{|y|\le\epsilon}\frac{1}{3!}f'''\left(\xi_1\right)y^3 d\lambda_{Y_k^n} +$$
$$\int_{|y|>\epsilon}\left(\frac{1}{2}f''\left(\xi_2\right)y^2 - \frac{1}{2}f''\left(x\right)y^2\right) d\lambda_{Y_k^n}.$$

The sum of the last two terms is dominated by

$$\frac{\epsilon\,||f||_3}{6}\frac{\sigma_k^2}{s_n^2} + ||f||_3 \int_{|y|>\epsilon} y^2 d\lambda_{Y_k^n}.$$

Next we consider $B_k f$

$$B_k f\left(x\right) \equiv \int_{\mathbb{R}} f\left(x\right) + f'\left(x\right)y + \frac{1}{2}f''\left(x\right)y^2 + \frac{1}{3!}f'''\left(\xi_1\right)y^3 d\lambda_{W_k^n}$$

$$= f\left(x\right) + \frac{1}{2}f''\left(x\right)\frac{\sigma_k^2}{s_n^2} + \int_{\mathbb{R}}\frac{1}{3!}f'''\left(\xi_1\right)y^3 \frac{s_n}{\sqrt{2\pi}\sigma_k}e^{-\frac{s_n^2 y^2}{2\sigma_k^2}}dy. \tag{37}$$

The last term in Formula 37 is dominated by

$$\frac{2\,||f||_3}{3!}\int_0^\infty y^3 \frac{s_n}{\sqrt{2\pi}\sigma_k}e^{-\frac{s_n^2 y^2}{2\sigma_k^2}}dy = \frac{2\,||f||_3}{3\sqrt{2\pi}}\left(\frac{\sigma_k}{s_n}\right)^3$$

Therefore,

$$|A_k f(x) - B_k f(x)|$$

$$\leq \frac{2\,\|f\|_3}{3\sqrt{2\pi}}\left(\frac{\sigma_k}{s_n}\right)^3 + \frac{\epsilon\,\|f\|_3}{6}\frac{\sigma_k^2}{s_n^2} + \|f\|_3 \int_{|y|>\epsilon} y^2 d\lambda_{Y_k^n}$$

and it follows

$$\|A_k - B_k\| \leq \frac{2}{3\sqrt{2\pi}}\left(\frac{\sigma_k}{s_n}\right)^3 + \frac{\epsilon}{6}\frac{\sigma_k^2}{s_n^2} + \int_{|y|>\epsilon} y^2 d\lambda_{Y_k^n}.$$

Therefore, by Lemma 17.62,

$$\|A_1 \cdots A_n - B_1 \cdots B_n\|$$

$$\leq \sum_{k=1}^{n} \frac{2}{3\sqrt{2\pi}}\left(\frac{\sigma_k}{s_n}\right)^3 + \frac{\epsilon}{6}\frac{\sigma_k^2}{s_n^2} + \int_{|y|>\epsilon} y^2 d\lambda_{Y_k^n}$$

$$= \frac{\epsilon}{6} + \sum_{k=1}^{n} \int_{|y|>\epsilon} y^2 d\lambda_{Y_k^n} + \frac{2}{3\sqrt{2\pi}} \max\left\{\left(\frac{\sigma_k}{s_n}\right), k = 1, \cdots, n\right\}.$$

Suppose the maximum occurs when $k = l$.

$$\frac{\sigma_l^2}{s_n^2} = \int_{|y|>\epsilon} y^2 d\lambda_{Y_l^n} + \int_{|y|\leq\epsilon} y^2 d\lambda_{Y_l^n}$$

$$\leq \epsilon^2 + \int_{|y|>\epsilon} y^2 d\lambda_{Y_l^n} \leq \epsilon^2 + \sum_{k=1}^{n} \int_{|y|>\epsilon} y^2 d\lambda_{Y_k^n}.$$

Therefore,

$$\|A_1 \cdots A_n - B_1 \cdots B_n\|$$

$$\leq \frac{\epsilon}{6} + \left(\epsilon^2 + \sum_{k=1}^{n} \int_{|y|>\epsilon} y^2 d\lambda_{Y_k^n}\right)^{1/2} \frac{2}{3\sqrt{2\pi}} +$$

$$+ \sum_{k=1}^{n} \int_{|y|>\epsilon} y^2 d\lambda_{Y_k^n} = e_n.$$

Now let $x \in \mathbb{R}$ and let $f \in BC^3(\mathbb{R})$ be such that f is decreasing, $f(y) = 1$ if $y \leq 0$, $f(y) = 0$ if $y \geq \epsilon$.

$$\left|\int_{\mathbb{R}} f(x+y) d\lambda_{Z_n} - \int_{\mathbb{R}} f(x+y) d\lambda_W\right|$$

$$\leq \|A_1 \cdots A_n - B_1 \cdots B_n\| \|f\|_3$$
$$\leq e_n \|f\|_3.$$

Since $\limsup_{n\to\infty} e_n \le 2\epsilon$ and ϵ is arbitrary, it follows that

$$\lim_{n\to\infty} \int_{\mathbb{R}} f(x+y)\, d\lambda_{Z_n} = \int_{\mathbb{R}} f(x+y)\, d\lambda_W.$$

$$F_W(-x) \le \int_{\mathbb{R}} f(x+y)\, d\lambda_W \le F_W(-x+\epsilon),$$

$$F_{Z_n}(-x+\epsilon) \ge \int_{\mathbb{R}} f(x+y)\, d\lambda_{Z_n} \ge F_{Z_n}(-x).$$

Then since x is arbitrary,

$$\liminf_{n\to\infty} F_{Z_n}(-x) \ge \liminf_{n\to\infty} \int_{\mathbb{R}} f(x+\epsilon+y)\, d\lambda_{Z_n}$$

$$= \int_{\mathbb{R}} f(x+\epsilon+y)\, d\lambda_W \ge F_W(-x-\epsilon)$$

and so

$$F_W(-x+\epsilon) \ge \limsup_{n\to\infty} \int_{\mathbb{R}} f(x+y)\, d\lambda_W =$$

$$\limsup_{n\to\infty} \int_{\mathbb{R}} f(x+y)\, d\lambda_{Z_n} \ge \limsup_{n\to\infty} F_{Z_n}(-x)$$

$$\ge \liminf_{n\to\infty} F_{Z_n}(-x) \ge F_W(-x-\epsilon).$$

Therefore

$$F_W(-x+\epsilon) \ge \limsup_{n\to\infty} F_{Z_n}(-x) \ge \liminf_{n\to\infty} F_{Z_n}(-x) \ge F_W(-x-\epsilon).$$

Letting $\epsilon \to 0$ yields

$$F_W(-x) = \lim_{n\to\infty} F_{Z_n}(-x)$$

which proves the theorem.

17.8 The continuity theorem

There is another approach to the study of limit theorems which is based on characteristic functions. The fundamental result which makes this approach possible is a theorem which states that if the characteristic functions converge pointwise, then one can assert something about the convergence of the distribution functions. Before proving and stating this theorem, we need some lemmas and definitions.

Definition 17.63 Let $\{\mathbf{X}_n\}$ be random vectors with values in $\mathbb{R}^p$. We say $\{\lambda_{\mathbf{X}_n}\}_{n=1}^{\infty}$ is tight if for all $\epsilon > 0$ there exists $r > 0$ such that if $u \leq r$, there exists $N(u)$ such that

$$\sup\left\{\lambda_{\mathbf{X}_n}\left[\mathbf{x} : |x_j| \geq \frac{2}{u}\right], \, j = 1, \cdots, p\right\} < \epsilon$$

whenever $n > N(u)$.

In words, this says that whenever l is large enough the measure of the set where $|\mathbf{x}| \geq l$ is ultimately small.

Lemma 17.64 *If $\mathbf{X}_n, \mathbf{X}$ are random vectors with values in $\mathbb{R}^p$ such that*

$$\lim_{n \to \infty} \phi_{\mathbf{X}_n}(\mathbf{t}) = \phi_{\mathbf{X}}(\mathbf{t})$$

for all $\mathbf{t}$, then $\{\lambda_{\mathbf{X}_n}\}_{n=1}^{\infty}$ is tight.

Proof: Let $\mathbf{e}_j$ be the *jth* standard unit basis vector.

$$
\begin{aligned}
\frac{1}{u}\int_{-u}^{u}(1 - \phi_{\mathbf{X}_n}(t\mathbf{e}_j))\,dt &= \int_{\mathbb{R}^p}\frac{1}{u}\int_{-u}^{u}1 - e^{itx_j}\,dt\,d\lambda_{\mathbf{X}_n}(x) \\
&= 2\int_{\mathbb{R}^p}\left(1 - \frac{\sin(ux_j)}{ux_j}\right)d\lambda_{\mathbf{X}_n}(x) \\
&\geq 2\int_{[|x_j| \geq \frac{2}{u}]}\left(1 - \frac{1}{|ux_j|}\right)d\lambda_{\mathbf{X}_n}(x) \\
&\geq \int_{[|x_j| \geq \frac{2}{u}]}1\,d\lambda_{\mathbf{X}_n}(x) \\
&= \lambda_{\mathbf{X}_n}\left[\mathbf{x} : |x_j| \geq \frac{2}{u}\right].
\end{aligned}
$$

If $\epsilon > 0$ is given, there exists $r > 0$ such that if $u \leq r$,

$$\frac{1}{u}\int_{-u}^{u}(1 - \phi_{\mathbf{X}}(t\mathbf{e}_j))\,dt < \epsilon$$

for all $j = 1, \cdots, p$ and so, by the dominated convergence theorem, the same is true with $\phi_{\mathbf{X}_n}$ in place of $\phi_{\mathbf{X}}$ provided n is large enough, say $n \geq N(u)$. Thus, if $u \leq r$, and $n \geq N(u)$,

$$\lambda_{\mathbf{X}_n}\left[\mathbf{x} : |x_j| \geq \frac{2}{u}\right] < \epsilon$$

for all $j \in \{1, \cdots, p\}$.

Lemma 17.65 *If $\phi_{\mathbf{X}_n}(\mathbf{t}) \to \phi_{\mathbf{X}}(\mathbf{t})$ for all $\mathbf{t}$, then $\lambda_{\mathbf{X}_n}$ converges weak $*$ in $\mathfrak{S}'$ to $\lambda_{\mathbf{X}}$.*

Proof: First note that if $\mathbf{X}$ is any random vector,

$$F^{-1}(\lambda_{\mathbf{X}}) = \phi_{\mathbf{X}}(\cdot)(2\pi)^{-(n/2)} \in L^{\infty}.$$

By the dominated convergence theorem

$$\int_{\mathbb{R}^p} \phi_{\mathbf{X}_n}(\mathbf{t})\,\psi(\mathbf{t})\,dt \to \int_{\mathbb{R}^p} \phi_{\mathbf{X}}(\mathbf{t})\,\psi(\mathbf{t})\,dt$$

whenever $\psi \in \mathfrak{S}$. Thus $F^{-1}\lambda_{\mathbf{X}_n}$ converges weak$*$ in $\mathfrak{S}'$ to $F^{-1}\lambda_{\mathbf{X}}$. Therefore, $FF^{-1}\lambda_{\mathbf{X}_n}$ converges weak$*$ in $\mathfrak{S}'$ to $FF^{-1}\lambda_{\mathbf{X}}$ because if $\psi \in \mathfrak{S}$,

$$\begin{aligned}
\lambda_{\mathbf{X}_n}(\psi) &= FF^{-1}\lambda_{\mathbf{X}_n}(\psi) \equiv F^{-1}\lambda_{\mathbf{X}_n}(F\psi) \to F^{-1}\lambda_{\mathbf{X}}(F\psi) \\
&\equiv FF^{-1}\lambda_{\mathbf{X}}(\psi) = \lambda_{\mathbf{X}}(\psi).
\end{aligned}$$

Lemma 17.66 *If $\phi_{\mathbf{X}_n}(\mathbf{t}) \to \phi_{\mathbf{X}}(\mathbf{t})$, then if ψ is any bounded uniformly continuous function,*

$$\lim_{n \to \infty} \int_{\mathbb{R}^p} \psi\,d\lambda_{\mathbf{X}_n} = \int_{\mathbb{R}^p} \psi\,d\lambda_{\mathbf{X}}.$$

Proof: Let $\epsilon > 0$ be given, let ψ be a bounded function in $C^{\infty}(\mathbb{R}^p)$. Now let $\eta \in C_c^{\infty}(Q_r)$ where $Q_r \equiv (-r, r)^p$ satisfy the additional requirement that $\eta = 1$ on $Q_{r/2}$ and $\eta(\mathbf{x}) \in [0, 1]$ for all $\mathbf{x}$. By Lemma 17.64 the set, $\{\lambda_{\mathbf{X}_n}\}_{n=1}^{\infty}$, is tight and so if $\epsilon > 0$ is given, there exists r sufficiently large and $n_0(r)$ such that whenever $n \geq n_0$,

$$\int_{[\mathbf{x} \notin Q_{r/2}]} |1 - \eta|\,|\psi|\,d\lambda_{\mathbf{X}_n} < \frac{\epsilon}{3},$$

and

$$\int_{[\mathbf{x} \notin Q_{r/2}]} |1 - \eta|\,|\psi|\,d\lambda_{\mathbf{X}} < \frac{\epsilon}{3}.$$

For such $n \geq n_0$,

$$\left| \int_{\mathbb{R}^p} \psi\,d\lambda_{\mathbf{X}_n} - \int_{\mathbb{R}^p} \psi\,d\lambda_{\mathbf{X}} \right| \leq \left| \int_{\mathbb{R}^p} \psi\,d\lambda_{\mathbf{X}_n} - \int_{\mathbb{R}^p} \psi\eta\,d\lambda_{\mathbf{X}_n} \right| +$$

$$\left| \int_{\mathbb{R}^p} \psi\eta\,d\lambda_{\mathbf{X}_n} - \int_{\mathbb{R}^p} \psi\eta\,d\lambda_{\mathbf{X}} \right| + \left| \int_{\mathbb{R}^p} \psi\eta\,d\lambda_{\mathbf{X}} - \int_{\mathbb{R}^p} \psi\,d\lambda_{\mathbf{X}} \right|$$

$$\leq \frac{2\epsilon}{3} + \left| \int_{\mathbb{R}^p} \psi\eta\,d\lambda_{\mathbf{X}_n} - \int_{\mathbb{R}^p} \psi\eta\,d\lambda_{\mathbf{X}} \right| < \epsilon$$

whenever n is large enough by Lemma 17.65 because $\psi\eta \in \mathfrak{S}$. This establishes the conclusion of the lemma in the case where ψ is also infinitely differentiable. To consider the general case, let ψ only be uniformly continuous and let $\psi_k = \psi * \phi_k$ where ϕ_k is a mollifier whose support is in $(-(1/k), (1/k))^p$. Then ψ_k converges uniformly to ψ and so the desired conclusion follows for ψ after a routine estimate.

Definition 17.67 Let μ be a Radon measure on $\mathbb{R}^p$. We say a Borel set, A, is a μ continuity set if $\mu(\partial A) = 0$ where $\partial A \equiv \overline{A} \setminus int(A)$.

The main result is the following continuity theorem. More can be said about the equivalence of various criteria [6].

Theorem 17.68 *If $\phi_{\mathbf{X}_n}(t) \to \phi_{\mathbf{X}}(t)$ then $\lambda_{\mathbf{X}_n}(A) \to \lambda_{\mathbf{X}}(A)$ whenever A is a $\lambda_{\mathbf{X}}$ continuity set.*

Proof: First suppose K is a closed set and let

$$\psi_k(\mathbf{x}) \equiv (1 - kdist(\mathbf{x}, K))^+.$$

Thus, since K is closed $\lim_{k \to \infty} \psi_k(\mathbf{x}) = \mathcal{X}_K(\mathbf{x})$. Choose k large enough that

$$\int_{\mathbb{R}^p} \psi_k d\lambda_{\mathbf{X}} \le \lambda_{\mathbf{X}}(K) + \epsilon.$$

Then by Lemma 17.66, applied to the bounded uniformly continuous function ψ_k,

$$\limsup_{n \to \infty} \lambda_{\mathbf{X}_n}(K) \le \limsup_{n \to \infty} \int \psi_k d\lambda_{\mathbf{X}_n} = \int \psi_k d\lambda_{\mathbf{X}} \le \lambda_{\mathbf{X}}(K) + \epsilon.$$

Since ϵ is arbitrary, this shows

$$\limsup_{n \to \infty} \lambda_{\mathbf{X}_n}(K) \le \lambda_{\mathbf{X}}(K)$$

for all K closed.

Next suppose V is open and let

$$\psi_k(\mathbf{x}) = 1 - \left(1 - kdist\left(\mathbf{x}, V^C\right)\right)^+.$$

Thus $\psi_k(\mathbf{x}) \in [0, 1]$, $\psi_k = 1$ if $dist(\mathbf{x}, V^C) \ge 1/k$, and $\psi_k = 0$ on V^C. Since V is open, it follows

$$\lim_{k \to \infty} \psi_k(\mathbf{x}) = \mathcal{X}_V(\mathbf{x}).$$

Choose k large enough that

$$\int \psi_k d\lambda_{\mathbf{X}} \ge \lambda_{\mathbf{X}}(V) - \epsilon.$$

Then by Lemma 17.66,

$$\lim_{n \to \infty} \inf \lambda_{\mathbf{X}_n}(V) \geq \lim_{n \to \infty} \inf \int \psi_k(\mathbf{x}) \, d\lambda_{\mathbf{X}_n} = \int \psi_k(\mathbf{x}) \, d\lambda_{\mathbf{X}} \geq \lambda_{\mathbf{X}}(V) - \epsilon$$

and since ϵ is arbitrary,

$$\lim_{n \to \infty} \inf \lambda_{\mathbf{X}_n}(V) \geq \lambda_{\mathbf{X}}(V).$$

Now let $\lambda_{\mathbf{X}}(\partial A) = 0$ for A a Borel set.

$$\lambda_{\mathbf{X}}(int(A)) \leq \lim_{n \to \infty} \inf \lambda_{\mathbf{X}_n}(int(A)) \leq \lim_{n \to \infty} \inf \lambda_{\mathbf{X}_n}(A) \leq$$
$$\lim_{n \to \infty} \sup \lambda_{\mathbf{X}_n}(A) \leq \lim_{n \to \infty} \sup \lambda_{\mathbf{X}_n}(\overline{A}) \leq \lambda_{\mathbf{X}}(\overline{A}).$$

But $\lambda_{\mathbf{X}}(int(A)) = \lambda_{\mathbf{X}}(\overline{A})$ by assumption and so $\lim_{n \to \infty} \lambda_{\mathbf{X}_n}(A) = \lambda_{\mathbf{X}}(A)$ as claimed.

As an application of this theorem the following is a version of the central limit theorem in the situation in which the limit distribution is multivariate normal. It concerns a sequence of random vectors, $\{\mathbf{X}_k\}_{k=1}^{\infty}$, which are identically distributed, have finite mean μ, and satisfy

$$E\left(|\mathbf{X}_k|^2\right) < \infty. \tag{38}$$

Theorem 17.69 *Let $\{\mathbf{X}_k\}_{k=1}^{\infty}$ be random vectors satisfying Formula 38, which are independent and identically distributed with mean μ and positive definite covariance $\mathbf{\Sigma}$. Let*

$$\mathbf{Z}_n \equiv \sum_{j=1}^{n} \frac{\mathbf{X}_j - \mu}{\sqrt{n}}. \tag{39}$$

Then for $\mathbf{Z} \sim N_p(\mathbf{0}, \mathbf{\Sigma})$,

$$\lim_{n \to \infty} F_{\mathbf{Z}_n}(\mathbf{x}) = F_{\mathbf{Z}}(\mathbf{x}) \tag{40}$$

for all $\mathbf{x}$.

Proof: The characteristic function of $\mathbf{Z}_n$ is given by

$$\phi_{\mathbf{Z}_n}(\mathbf{t}) = E\left(e^{i\mathbf{t} \cdot \sum_{j=1}^{n} \frac{\mathbf{X}_j - \mu}{\sqrt{n}}}\right) = \prod_{j=1}^{n} E\left(e^{i\mathbf{t} \cdot \left(\frac{\mathbf{X}_j - \mu}{\sqrt{n}}\right)}\right).$$

By Taylor's theorem,

$$e^{ix} = 1 + ix - \frac{e^{i\theta x} x^2}{2}$$

for some $\theta \in [0,1]$ which depends on x. Denoting $\mathbf{X}_j$ as $\mathbf{X}$, this implies

$$e^{it\cdot\left(\frac{\mathbf{X}-\mu}{\sqrt{n}}\right)} = 1 + it\cdot\frac{\mathbf{X}-\mu}{\sqrt{n}} - e^{i\theta t\cdot\frac{\mathbf{X}-\mu}{\sqrt{n}}}\frac{(\mathbf{t}\cdot(\mathbf{X}-\mu))^2}{2n}$$

where θ depends on $\mathbf{X}$ and $\mathbf{t}$ and is in $[0,1]$. This equals

$$1 + it\cdot\frac{\mathbf{X}-\mu}{\sqrt{n}} - \frac{(\mathbf{t}\cdot(\mathbf{X}-\mu))^2}{2n}$$

$$+ \left(1 - e^{i\theta t\cdot\frac{\mathbf{X}-\mu}{\sqrt{n}}}\right)\frac{(\mathbf{t}\cdot(\mathbf{X}-\mu))^2}{2n}.$$

Thus

$$\phi_{\mathbf{Z}_n}(\mathbf{t}) = \prod_{j=1}^{n}\left[1 - E\left(\frac{(\mathbf{t}\cdot(\mathbf{X}-\mu))^2}{2n}\right)\right.$$

$$\left. + E\left(\left(1 - e^{i\theta t\cdot\frac{\mathbf{X}-\mu}{\sqrt{n}}}\right)\frac{(\mathbf{t}\cdot(\mathbf{X}-\mu))^2}{2n}\right)\right]$$

$$= \prod_{j=1}^{n}\left[1 - \frac{1}{2n}\mathbf{t}^*\Sigma\mathbf{t} + \frac{1}{2n}E\left(\left(1 - e^{i\theta t\cdot\frac{\mathbf{X}-\mu}{\sqrt{n}}}\right)(\mathbf{t}\cdot(\mathbf{X}-\mu))^2\right)\right]. \qquad (41)$$

Now we have the simple identity for complex numbers whose moduli are no larger than one,

$$|z_1\cdots z_n - w_1\cdots w_n| \le \sum_{k=1}^{n}|z_k - w_k|. \qquad (42)$$

Also for each $\mathbf{t}$, and all n large enough,

$$\left|\frac{1}{2n}E\left(\left(1 - e^{i\theta t\cdot\frac{\mathbf{X}-\mu}{\sqrt{n}}}\right)(\mathbf{t}\cdot(\mathbf{X}-\mu))^2\right)\right| < 1.$$

Applying Formula 42 to Formula 41,

$$\phi_{\mathbf{Z}_n}(\mathbf{t}) = \prod_{j=1}^{n}\left(1 - \frac{1}{2n}\mathbf{t}^*\Sigma\mathbf{t}\right) + e_n$$

where

$$|e_n| \le \sum_{j=1}^{n}\left|\frac{1}{2n}E\left(\left(1 - e^{i\theta t\cdot\frac{\mathbf{X}-\mu}{\sqrt{n}}}\right)(\mathbf{t}\cdot(\mathbf{X}-\mu))^2\right)\right|$$

$$= \frac{1}{2}\left|E\left(\left(1 - e^{i\theta t\cdot\frac{\mathbf{X}-\mu}{\sqrt{n}}}\right)(\mathbf{t}\cdot(\mathbf{X}-\mu))^2\right)\right|$$

which converges to 0 by the Dominated Convergence theorem. Therefore,

$$\lim_{n \to \infty} \left| \phi_{\mathbf{Z}_n} (\mathbf{t}) - \left(1 - \frac{\mathbf{t}^* \Sigma \mathbf{t}}{2n} \right)^n \right| = 0$$

and so

$$\lim_{n \to \infty} \phi_{\mathbf{Z}_n} (\mathbf{t}) = e^{-\frac{1}{2} \mathbf{t}^* \Sigma \mathbf{t}} = \phi_{\mathbf{Z}} (\mathbf{t})$$

where $\mathbf{Z} \sim N_p (\mathbf{0}, \Sigma)$. Therefore, $F_{\mathbf{Z}_n} (\mathbf{x}) \to F_{\mathbf{Z}} (\mathbf{x})$ for all $\mathbf{x}$ because $R_{\mathbf{x}} \equiv \prod_{k=1}^{p} (-\infty, x_k]$ is a set of $\lambda_{\mathbf{Z}}$ continuity due to the assumption that $\lambda_{\mathbf{Z}} << m_p$ which is implied by $\mathbf{Z} \sim N_p (\mathbf{0}, \Sigma)$. This proves the theorem.

Suppose $\mathbf{X}$ is a random vector with covariance Σ and mean μ, and suppose also that Σ^{-1} exists. Consider $\Sigma^{-(1/2)} (\mathbf{X} - \mu) \equiv \mathbf{Y}$. Then $E(\mathbf{Y}) = 0$ and

$$\begin{aligned} E(\mathbf{Y}\mathbf{Y}^*) &= E\left(\Sigma^{-(1/2)} (\mathbf{X} - \mu)(\mathbf{X}^* - \mu) \Sigma^{-(1/2)} \right) \\ &= \Sigma^{-(1/2)} E\left((\mathbf{X} - \mu)(\mathbf{X}^* - \mu) \right) \Sigma^{-(1/2)} = I. \end{aligned}$$

Thus $\mathbf{Y}$ has zero mean and covariance I. Therefore, we have the following corollary to Theorem 17.69.

Corollary 17.70 *Let independent identically distributed random variables* $\{\mathbf{X}_j\}_{j=1}^{\infty}$ *have mean* μ *and positive definite covariance* Σ *where* Σ^{-1} *exists. Then if*

$$\mathbf{Z}_n \equiv \sum_{j=1}^{n} \Sigma^{-(1/2)} \frac{(\mathbf{X}_j - \mu)}{\sqrt{n}},$$

it follows that for $\mathbf{Z} \sim N_p (\mathbf{0}, I)$,

$$F_{\mathbf{Z}_n} (\mathbf{x}) \to F_{\mathbf{Z}} (\mathbf{x})$$

for all $\mathbf{x}$.

17.9   Exercises

1. A random variable, X, has the chi square distribution (written as $X \sim \mathcal{X}^2 (r)$) with r degrees of freedom, if it has a probability density function given as

$$f(x) = \begin{cases} \frac{1}{\Gamma((r/2))2^{(r/2)}} x^{(r/2)-1} e^{-(x/2)} & \text{if } 0 < x < \infty, \\ 0 & \text{if } x \leq 0. \end{cases}$$

This means that

$$P(X \in E) = \int_E f(x)\, dx.$$

For $X \sim \mathcal{X}^2(r)$ show $M_X(t) = (1-2t)^{-(r/2)}$ for $|t| < 1/2$.

2. ↑ Let $X_i \sim \mathcal{X}^2(r_i)$ for $i = 1, 2, \cdots, n$ and let the X_i be independent. Show

$$\sum_{i=1}^{n} X_i \sim \mathcal{X}^2(r)$$

where $r = \sum_{i=1}^{n} r_i$. **Hint:** Use moment generating functions.

3. ↑ Suppose $X \sim N_1(\mu, \sigma^2)$. Show $V \equiv \frac{(X-\mu)^2}{\sigma^2} \sim \mathcal{X}^2(1)$.

4. ↑ Let $X_i \sim N_1(\mu, \sigma^2)$, let $(X_1 \cdots X_n)$ be independent, and let

$$Y \equiv \sum_{i=1}^{n} \left(\frac{X_i - \mu}{\sigma}\right)^2.$$

Show $Y \sim \mathcal{X}^2(n)$. **Hint:** Use Problems 2 and 3.

5. ↑ Let $X_1, \cdots, X_n$ be independent random variables distributed as $N_1(\mu, \sigma^2)$. Let $\overline{X} \equiv \frac{1}{n}\sum_{i=1}^{n} X_i$. Show $\overline{X} \sim N_1(\mu, \sigma^2/n)$.

6. ↑ Let $X_1, \cdots, X_n$ be as in Problem 5 and consider the random vector

$$(X_1 - \overline{X}, \cdots, X_n - \overline{X}) \equiv \mathbf{Z}.$$

Show

$$E\left(e^{t\overline{X} + \mathbf{t} \cdot \mathbf{Z}}\right) = E\left(e^{t\overline{X}}\right) E\left(e^{\mathbf{t} \cdot \mathbf{Z}}\right).$$

Conclude from this and Corollary 17.23 that $\overline{X}$ and $\mathbf{Z}$ are independent. Now use Lemma 17.18 to conclude $\overline{X}$ and

$$S^2 \equiv \frac{1}{n}\sum_{i=1}^{n} (X_i - \overline{X})^2$$

are independent.

7. ↑ Show $\frac{nS^2}{\sigma^2} \sim \mathcal{X}^2(n-1)$. In statistics this may be used to obtain a confidence interval for the variance of the assumed underlying normal distribution. **Hint:** See Problems 5, 4, and 2. Use moment generating functions.

349

8. In general, we say a random vector, $\mathbf{Z}$, having values in $\mathbb{R}^p$, has a probability density function if there exists a Borel measurable function, f, defined on $\mathbb{R}^p$ such that $f(\mathbf{x}) \geq 0$ and $P(\mathbf{Z} \in E) = \int_E f(\mathbf{z})\, dz$. Thus in this case the measure $\lambda_{\mathbf{Z}}$ is absolutely continuous with respect to Lebesgue measure and the density function is the Radon Nikodym derivative. Consider the case where $\mathbf{Z} = (\mathbf{X}, \mathbf{Y})$ and $\mathbf{X}$ has values in $\mathbb{R}^{p_1}$ while $\mathbf{Y}$ has values in $\mathbb{R}^{p_2}$ and we write the density function of $\mathbf{Z}$ as $g(\mathbf{x}, \mathbf{y})$. Recall $\lambda_{\mathbf{X}}(E) = \lambda_{(\mathbf{X}, \mathbf{Y})}(E \times \mathbb{R}^{p_2})$ and show that the random vector, $\mathbf{X}$, has a density function given as $g_1(\mathbf{x}) \equiv \int_{\mathbb{R}^{p_2}} g(\mathbf{x}, \mathbf{y})\, dy$ while $\mathbf{Y}$ has a density function given as $g_2(\mathbf{y}) = \int_{\mathbb{R}^{p_1}} g(\mathbf{x}, \mathbf{y})\, dx$. Assuming that $g_1(\mathbf{x}) > 0$ for all $\mathbf{x}$, show the measure, $\lambda_{\mathbf{X}|\mathbf{y}}$, is obtained as

$$\lambda_{\mathbf{X}|\mathbf{y}}(E) = \int_E \frac{g(\mathbf{x}, \mathbf{y})}{g_2(\mathbf{y})}\, dx.$$

What is the description of $\lambda_{\mathbf{Y}|\mathbf{x}}$? What does it mean in terms of density functions to say the two random vectors, $\mathbf{X}$ and $\mathbf{Y}$ are independent?

9. ↑ Let $g(\mathbf{x})$ be a density function for a random vector $\mathbf{X}$ having values in $\mathbb{R}^p$. Assume also that an open set, U, is defined as

$$P(\mathbf{X} \in F) = P(\mathbf{X} \in F \cap U)$$

(thus $g(\mathbf{x}) = 0$ is $\mathbf{x} \notin U$) and let $\mathbf{h}: U \to \mathbb{R}^p$ be one to one satisfying

$$P(\mathbf{X} \in \mathbf{h}^{-1}(\mathbf{h}(S))) = 0$$

where S is the set of points where $\mathbf{h}'(\mathbf{x})$ is singular. Let $\mathbf{Y} = \mathbf{h}(\mathbf{X})$ and show using the change of variables formula that $\mathbf{Y}$ has a density function given as

$$g \circ \mathbf{h}^{-1}(\mathbf{y})(J\mathbf{h})^{-1}(\mathbf{h}^{-1}(\mathbf{y})) \mathcal{X}_{\mathbf{h}(S)^C}(\mathbf{y}).$$

Now generalize this formula to the case where $\mathbf{h}$ is only assumed to be $C^1(U)$. **Hint:** Use the construction found in the proof of Theorem 14.24 about the change of variables for C^1 mappings which may not be one to one.

10. ↑ This problem introduces the t distribution. Let $W \sim N_1(0, 1)$ and let $V \sim \mathcal{X}^2(r)$ be two independent random variables. An example of this is the case when $X_1, \cdots, X_n$ is a random sample from $N_1(\mu, \sigma^2)$ and we let $V = \frac{nS^2}{\sigma^2}$ and $W = \frac{n\overline{X}}{\sigma} - \mu$. We know these random variables are independent by Problem 6 and the second is $N_1(0, 1)$

by a simple exercise in moment generating functions. Define the random variable

$$T = \frac{W}{\sqrt{V/r}}$$

and notice how the parameter σ disappears if W and V are as just described. Explain why the density of the random vector (W, U) is

$$\phi(w, v) \equiv$$

$$\begin{cases} \frac{1}{\sqrt{2\pi}} e^{-\frac{w^2}{2}} \frac{1}{\Gamma(r/2)2^{r/2}} v^{r/2-1} e^{-v/2} & \text{if } (w, v) \in ((-\infty, \infty) \times (0, \infty)) \\ 0 \text{ elsewhere} \end{cases}.$$

Then consider the change of variables

$$t = \frac{w}{\sqrt{v/r}} \text{ and } u = v.$$

Show the density function of the random vector (t, u) is

$$g(t, u) = \phi\left(\frac{t\sqrt{u}}{\sqrt{r}}, u\right) \frac{\sqrt{u}}{\sqrt{r}}.$$

Now use Problem 8 to show the distribution function of T is

$$g_1(t) = \frac{\Gamma[(r+1)/2]}{\sqrt{\pi r}\Gamma(r/2)} \frac{1}{(1+t^2/r)^{(r+1)/2}}, \text{ for } t \in (-\infty, \infty).$$

11. Assume $\{F_n\}$ is a sequence of monotonically increasing functions on $\mathbb{R}$ having values in $[0,1]$. Prove there exists a function F and a subsequence $\{n_k\}$ such that $F(x) = \lim_{k \to \infty} F_{n_k}(x)$ for every $x \in \mathbb{R}$. This is Helley's selection theorem. **Hint**: Use the Cantor Diagonal process to get a subsequence, $\{F_k\}$, converging at every rational number. Define $F(r) = \lim_{k \to \infty} F_k(r)$ all $r \in \mathbb{Q}$. Let $F(x) = \sup\{F(r) : r \le x\}$. Show F is monotone and well defined. From this, argue that $F_k(x)$ converges to $F(x)$ at every point of continuity of F in addition to every rational number. Next argue that F is discontinuous at only countably many points. Take a further subsequence which converges at these points also. Then modify F at these points so that the result yields the limit of the new subsequence there.

12. Let $\mathbf{X}$ be a random vector with values in $\mathbb{F}^p$ and let $\mathbf{f} : \mathbb{F}^p \to \mathbb{F}^r$ be Borel measurable and in L^1. Show the usual triangle inequality holds.

$$\left| \int_{\mathbb{F}^p} \mathbf{f}(\mathbf{x}) \, d\lambda_{\mathbf{X}} \right| \le \int_{\mathbb{R}} |\mathbf{f}(\mathbf{x})| \, d\lambda_{\mathbf{X}}.$$

13. ↑ Prove the Rao Blackwell theorem which states: Let $\mathbf{X}$ and $\mathbf{Y}$ be random vectors with $E(\mathbf{Y}) = \mu$. Then

$$E(E(\mathbf{Y}|\mathbf{X})) = \mu, \text{ and } E\left(|\mathbf{Y} - \mu|^2\right) \geq E\left(|E(\mathbf{Y}|\mathbf{X}) - \mu|^2\right)$$

Thus, in one dimension, $E(Y|X)$ has the same mean but smaller variance than Y. **Hint:** This is a simple exercise in Jensen's inequality and Problem 12.

14. Suppose $\mathbf{X}$ is a random vector with density function f_θ where θ is a parameter. Suppose

$$f_\theta(\mathbf{x}) = g_\theta(\mathbf{T}(\mathbf{x})) h(\mathbf{x})$$

where g_θ is a continuous function. Show, using Theorem 17.25, that $\mathbf{T}$ is sufficient for θ. This is called the Fisher Neyman criterion.

15. ↑ Let $\mathbf{X} = (\mathbf{X}_1, \cdots, \mathbf{X}_n)$ where $\mathbf{X}_i \sim N_p(\mu, \Sigma)$ and assume Σ is known. Show that $\overline{\mathbf{X}} \equiv \frac{1}{n} \sum_{i=1}^n \mathbf{X}_i$ is sufficient for μ.

16. ↑ Let $\mathbf{X}$ be a random vector such that $\lambda_{\mathbf{X}} = \mu(\theta)$ for θ a parameter. We say a random vector $\mathbf{U}$ is unbiased for θ if $E(\mathbf{U}) = \theta$ and we say a random vector, $\mathbf{V}$, is a "best" statistic for θ if it is unbiased and if $\mathbf{U}$ is any unbiased statistic, then

$$E\left(|\mathbf{V} - \theta|^2\right) \leq E\left(|\mathbf{U} - \theta|^2\right).$$

Suppose $\mathbf{T} = \mathbf{T}(\mathbf{X})$ is sufficient for θ. Show $E(\mathbf{U}|\mathbf{T})$ does not depend on θ and that $E(\mathbf{U}|\mathbf{T})$ is unbiased for θ and is a "better" statistic for θ than $\mathbf{U}$. Conclude that if there exists a best statistic for θ, it must be a Borel measurable function of $\mathbf{T}$.

17. ↑ Let $\{\mu(\theta)\}$ be a family of Radon probability measures on $\mathbb{F}^p$ indexed by θ. This family of measures is called complete if whenever $g \in \cap_\theta L^1(\mu(\theta))$,

$$\int_{\mathbb{F}^p} g(\mathbf{x}) \, d\mu(\theta)(x) = 0$$

for all θ implies

$$g(\mathbf{x}) = 0$$

for $\mu(\theta)$ a.e. $\mathbf{x}$ for all θ. Show that if $\mathbf{U}(\mathbf{X})$ and $\mathbf{V}(\mathbf{X})$ are both unbiased statistics for θ, then

$$\mathbf{U}(\mathbf{x}) = \mathbf{V}(\mathbf{x})$$

for $\mu(\theta)$ a.e. $\mathbf{x}$ for all θ. Conclude that if there exists a sufficient statistic for θ, $\mathbf{T}(\mathbf{X})$, then there exists a unique best statistic for θ which is a function of $\mathbf{T}$.

18. Let $A_n \in \mathcal{S}$ where $(\Omega, \mathcal{S}, P)$ is a probability space. Such a set is called an event. A set of events $\{A_n\}$ is said to be independent if the random variables $\{\mathcal{X}_{A_n}\}$ are independent. Show $P\left(\cap_{n=1}^N A_n\right) = \prod_{n=1}^N P(A_n)$ if the events are independent. Suppose $\{A_n\}_{n=1}^\infty$ are independent events and $\sum_{n=1}^\infty P(A_n) = \infty$. Show that then

$$P\left(\cap_{n=1}^\infty \cup_{k=n}^\infty A_k\right) = 1.$$

This is called the second Borel Cantelli lemma. **Hint:** It is enough to show $P\left(\cap_{k=n}^N A_k^C\right) \to 0$ as $N \to \infty$. Note

$$P\left(\cap_{k=n}^N A_k^C\right) = \prod_{n=1}^N P\left(A_k^C\right) = \prod_{n=1}^N \left(1 - P(A_k)\right) \le \prod_{n=1}^N e^{-P(A_k)}.$$

19. The conclusion of Theorem 17.54 holds without the assumption that $X \in L^1(\Omega)$. **Hint:** Let

$$f_n(x) = \begin{cases} x & \text{if } |x| < n, \\ n & \text{if } x \ge n, \\ -n & \text{if } x \le -n. \end{cases}$$

Then $f_n \circ X \in L^1$ so it is a.e. equal to a constant, c_n. Now $f_n \circ X \in \sigma(X_n, \cdots)$ for all n and so $f_n(X(\omega)) = c_n$ for all $\omega \notin E_n$, with $P(E_n) = 0$. Now consider $\omega \notin \cup E_n$ and take a limit as $n \to \infty$.

20. Let $\mathbf{T} : \mathbb{R}^p \to \mathbb{R}^r$ be a Borel measurable random vector and let η be a Radon measure on a σ algebra of sets of $\mathbb{R}^p$ satisfying

$$\eta\left(\mathbf{T}^{-1}(B(\mathbf{0}, r))\right) < \infty$$

for all $r > 0$. Using Theorem 9.26, show there exists a Radon measure, μ on a σ algebra of sets of $\mathbb{R}^r$, such that whenever $g \ge 0$ is a Borel measurable function,

$$\int_{\mathbb{R}^p} g(\mathbf{T}(\mathbf{x})) \, d\eta(x) = \int_{\mathbb{R}^r} g(\mathbf{t}) \, d\mu(t).$$

Hint: $\mu(E) \equiv \eta\left(\mathbf{T}^{-1}(E)\right)$ for E a Borel set.

21. ↑ It is not necessary to assume g_θ is continuous in Theorem 17.25. Suppose η and μ are as described in Problem 20 and suppose $g_\theta \in$

$L^1_{loc}(\mathbb{R}^r, \mu)$. Show that the conclusion of Theorem 17.25 still holds. **Hint:** Change variables as in Problem 20 and consider

$$\int_{\mathbb{F}^r} \mathcal{X}_{B(\mathbf{t}_1, r)}(\mathbf{t})\left(g_\theta(\mathbf{t}) - g_\theta(\mathbf{t}_1)\right) d\mu$$

$$= 0 \cdot \mu\left(\{\mathbf{t}_1\}\right) + \int_{B(\mathbf{t}_1, r)\setminus\{\mathbf{t}_1\}}\left(g_\theta(\mathbf{t}) - g_\theta(\mathbf{t}_1)\right) d\mu,$$

along with the assumption that $g_\theta \in L^1_{loc}$ and the observation that

$$\cap\{B(\mathbf{t}_1, r)\setminus\{\mathbf{t}_1\} : r > 0\} = \emptyset.$$

Now apply this to the last part of the proof of Theorem 17.25.

Chapter 18
Weak Derivatives

18.1   Test functions and weak derivatives

In elementary courses in mathematics, functions are often thought of as things which have a formula associated with them and it is the formula which receives the most attention. For example, in beginning calculus courses the derivative of a function is defined as the limit of a difference quotient. We start with one function which we tend to identify with a formula and, by taking a limit, we get another formula for the derivative. A jump in abstraction occurs as soon as we encounter the derivative of a function of n variables where the derivative is defined as a certain linear transformation which is determined not by a formula but by what it does to vectors. When this is understood, we see that it reduces to the usual idea in one dimension. The idea of weak partial derivatives goes further in the direction of defining something in terms of what it does rather than by a formula, and extra generality is obtained when it is used. In particular, it is possible to differentiate almost anything if we use a weak enough notion of what we mean by the derivative. This has the advantage of letting us talk about a weak partial derivative of a function without having to agonize over the important question of existence but it has the disadvantage of not allowing us to say very much about this weak partial derivative. Nevertheless, it is the idea of weak partial derivatives which makes it possible to use functional analytic techniques in the study of partial differential equations and we will show in this chapter that the concept of weak derivative is useful for unifying the discussion of some very important theorems. We will also show that certain things we wish were true, such as the equality of mixed partial derivatives, are true within the context of weak derivatives.

Let $\Omega \subseteq \mathbb{R}^n$. A distribution on Ω is defined to be a linear functional on $C_c^\infty(\Omega)$, called the space of test functions. The space of all such linear functionals will be denoted by $\mathcal{D}^*(\Omega)$. Actually, more is sometimes done here. One imposes a topology on $C_c^\infty(\Omega)$ making it into a topological vector space, and when this has been done, $\mathcal{D}'(\Omega)$ is defined as the dual space of this topological vector space. To see this, consult the book by Yosida [56] or the book by Rudin [47].

Example: The space $L_{loc}^1(\Omega)$ may be considered as a subset of $\mathcal{D}^*(\Omega)$ as follows.

$$f(\phi) \equiv \int_\Omega f(\mathbf{x})\,\phi(\mathbf{x})\,dx$$

for all $\phi \in C_c^\infty(\Omega)$. Recall that $f \in L_{loc}^1(\Omega)$ if $f \mathcal{X}_K \in L^1(\Omega)$ whenever K is compact.

Example: $\delta_x \in \mathcal{D}^*(\Omega)$ where $\delta_{\mathbf{x}}(\phi) \equiv \phi(\mathbf{x})$.

It will be observed from the above two examples and a little thought that $\mathcal{D}^*(\Omega)$ is truly enormous. We shall define the derivative of a distribution in such a way that it agrees with the usual notion of a derivative on those distributions which are also continuously differentiable functions. With this in mind, let f be a smooth function defined on Ω. Then $D_{x_i} f$ makes sense and for $\phi \in C_c^\infty(\Omega)$

$$D_{x_i} f(\phi) \equiv \int_\Omega D_{x_i} f(\mathbf{x}) \phi(\mathbf{x}) \, dx = -\int_\Omega f D_{x_i} \phi \, dx = -f(D_{x_i}\phi).$$

Motivated by this we make the following definition.

Definition 18.1 For $T \in \mathcal{D}^*(\Omega)$

$$D_{x_i} T(\phi) \equiv -T(D_{x_i}\phi).$$

Of course one can continue taking derivatives indefinitely. Thus,

$$D_{x_i x_j} T \equiv D_{x_i}\left(D_{x_j} T\right)$$

and it is clear that all mixed partial derivatives are equal because this holds for the functions in $C_c^\infty(\Omega)$. Thus we can differentiate virtually anything, even functions that may be discontinuous everywhere. However the notion of "derivative" is very weak, hence the name, "weak derivatives".

Example: Let $\Omega = \mathbb{R}$ and let

$$H(x) \equiv \begin{cases} 1 \text{ if } x \geq 0, \\ 0 \text{ if } x < 0. \end{cases}$$

Then

$$DH(\phi) = -\int H(x) \phi'(x) \, dx = \phi(0) = \delta_0(\phi).$$

Note that in this example, DH is not a function.

What happens when Df is a function?

Theorem 18.2 *Let $\Omega = (a, b)$ and suppose that f and Df are both in $L^1(a, b)$. Then f is equal to a continuous function a.e., still denoted by f and*

$$f(x) = f(a) + \int_a^x Df(t) \, dt.$$

In proving Theorem 18.2 we shall use the following lemma.

Lemma 18.3 *Let $T \in \mathcal{D}^* (a, b)$ and suppose $DT = 0$. Then there exists a constant C such that*

$$T(\phi) = \int_a^b C\phi dx.$$

Proof: $T(D\phi) = 0$ for all $\phi \in C_c^\infty (a, b)$ from the definition of $DT = 0$. Let

$$\phi_0 \in C_c^\infty (a, b), \quad \int_a^b \phi_0(x)\, dx = 1,$$

and let

$$\psi_\phi(x) = \int_a^x [\phi(t) - \left(\int_a^b \phi(y)\, dy \right) \phi_0(t)] dt$$

for $\phi \in C_c^\infty (a, b)$. Thus $\psi_\phi \in C_c^\infty (a, b)$ and

$$D\psi_\phi = \phi - \left(\int_a^b \phi(y)\, dy \right) \phi_0.$$

Therefore,

$$\phi = D\psi_\phi + \left(\int_a^b \phi(y)\, dy \right) \phi_0$$

and so

$$T(\phi) = T(D\psi_\phi) + \left(\int_a^b \phi(y)\, dy \right) T(\phi_0) = \int_a^b T(\phi_0) \phi(y)\, dy.$$

Let $C = T\phi_0$. This proves the lemma.

Proof of Theorem 18.2 Since f and Df are both in $L^1 (a, b)$,

$$Df(\phi) - \int_a^b Df(x) \phi(x)\, dx = 0.$$

Consider

$$f(\cdot) - \int_a^{(\cdot)} Df(t)\, dt$$

and let $\phi \in C_c^\infty (a, b)$.

$$D \left(f(\cdot) - \int_a^{(\cdot)} Df(t)\, dt \right)(\phi)$$

$$\equiv - \int_a^b f(x) \phi'(x)\, dx + \int_a^b \left(\int_a^x Df(t)\, dt \right) \phi'(x)\, dx$$

$$= Df(\phi) + \int_a^b \int_t^b Df(t)\, \phi'(x)\, dx\, dt$$

$$= Df(\phi) - \int_a^b Df(t)\, \phi(t)\, dt = 0.$$

By Lemma 18.3, there exists a constant, C, such that

$$\left(f(\cdot) - \int_a^{(\cdot)} Df(t)\, dt \right)(\phi) = \int_a^b C\phi(x)\, dx$$

for all $\phi \in C_c^\infty(a, b)$. Thus

$$\int_a^b \left\{ \left(f(x) - \int_a^x Df(t)\, dt \right) - C \right\} \phi(x)\, dx = 0$$

for all $\phi \in C_c^\infty(a, b)$. It follows from Lemma 18.6 in the next section that

$$f(x) - \int_a^x Df(t)\, dt - C = 0 \text{ a.e. } x.$$

Thus we let $f(a) = C$ and write

$$f(x) = f(a) + \int_a^x Df(t)\, dt.$$

This proves Theorem 18.2.

Theorem 18.2 says that

$$f(x) = f(a) + \int_a^x Df(t)\, dt$$

whenever it makes sense to write $\int_a^x Df(t)\, dt$, if Df is interpreted as a weak derivative. Somehow, this is the way it ought to be. It follows from the fundamental theorem of calculus in Chapter 14 that $f'(x)$ exists for a.e. x where the derivative is taken in the sense of a limit of difference quotients and $f'(x) = Df(x)$. This raises an interesting question. Suppose f is continuous on $[a, b]$ and $f'(x)$ exists in the classical sense for a.e. x. Does it follow that

$$f(x) = f(a) + \int_a^x f'(t)\, dt?$$

The answer is no. To see an example, consider Problem 5 of Chapter 9 which gives an example of a function which is continuous on $[0, 1]$, has a zero derivative for a.e. x but climbs from 0 to 1 on $[0, 1]$. Thus this function is not recovered from integrating its classical derivative.

In summary, if the notion of weak derivative is used, one can at least give meaning to the derivative of almost anything, the mixed partial derivatives are always equal, and, in one dimension, one can recover the function from integrating its derivative. None of these claims are true for the classical derivative. Thus weak derivatives are convenient and rule out pathologies.

18.2 Weak derivatives in L_{loc}^p

Definition 18.4 We say $f \in L_{loc}^p(\mathbb{R}^n)$ if $f\mathcal{X}_K \in L^p$ whenever K is compact.

Definition 18.5 For $\alpha = (k_1, \cdots, k_n)$ where the k_i are nonnegative integers, we define

$$|\alpha| \equiv \sum_{i=1}^{n} |k_{x_i}|, \; D^\alpha f(\mathbf{x}) \equiv \frac{\partial^{|\alpha|} f(\mathbf{x})}{\partial x_{k_1} \partial x_{k_2} \cdots \partial x_{k_n}}.$$

We want to consider the case where u and $D^\alpha u$ for $|\alpha| = 1$ are each in $L_{loc}^p(\mathbb{R}^n)$. The next lemma is the one alluded to in the proof of Theorem 18.2.

Lemma 18.6 *Suppose $f \in L_{loc}^1(\mathbb{R}^n)$ and suppose*

$$\int f\phi dx = 0$$

for all $\phi \in C_c^\infty(\mathbb{R}^n)$. Then $f(\mathbf{x}) = 0$ a.e. $\mathbf{x}$.

Proof: Without loss of generality f is real-valued. Let

$$E \equiv \{\mathbf{x} : f(\mathbf{x}) > \epsilon\}$$

and let

$$E_m \equiv E \cap B(0, m).$$

We show that $m(E_m) = 0$. If not, there exists an open set, V, and a compact set K satisfying

$$K \subseteq E_m \subseteq V \subseteq B(0, m), \; m(V \setminus K) < 4^{-1} m(E_m),$$

$$\int_{V \setminus K} |f| \, dx < \epsilon 4^{-1} m(E_m).$$

Let H and W be open sets satisfying

$$K \subseteq H \subseteq \overline{H} \subseteq W \subseteq \overline{W} \subseteq V$$

and let

$$\overline{H} \prec g \prec W$$

where the symbol, $\prec$, has the same meaning as it does in Chapter 9. Then let ϕ_δ be a mollifier and let $h \equiv g * \phi_\delta$ for δ small enough that

$$K \prec h \prec V.$$

Thus

$$0 = \int fh dx = \int_K f dx + \int_{V \setminus K} fh dx$$
$$\geq \epsilon m(K) - \epsilon 4^{-1} m(E_m)$$
$$\geq \epsilon(m(E_m) - 4^{-1} m(E_m)) - \epsilon 4^{-1} m(E_m)$$
$$\geq 2^{-1} \epsilon m(E_m).$$

Therefore, $m(E_m) = 0$, a contradiction. Thus

$$m(E) \leq \sum_{m=1}^{\infty} m(E_m) = 0$$

and so, since $\epsilon > 0$ is arbitrary,

$$m(\{ \mathbf{x} : f(\mathbf{x}) > 0 \}) = 0.$$

Similarly $m(\{ \mathbf{x} : f(\mathbf{x}) < 0 \}) = 0$. This proves the lemma.

This lemma allows the following definition.

Definition 18.7 We say for $u \in L^1_{loc}(\mathbb{R}^n)$ that $D^\alpha u \in L^1_{loc}(\mathbb{R}^n)$ if there exists a function $g \in L^1_{loc}(\mathbb{R}^n)$, necessarily unique by Lemma 18.6, such that for all $\phi \in C^\infty_c(\mathbb{R}^n)$,

$$\int g\phi dx = D^\alpha u(\phi) = \int (-1)^{|\alpha|} u(D^\alpha \phi) dx.$$

We call g $D^\alpha u$ when this occurs.

Lemma 18.8 Let $u \in L^1_{loc}$ and suppose $u_{,i} \in L^1_{loc}$, where the subscript on the u following the comma denotes the ith weak partial derivative. Then if ϕ_ϵ is a mollifier and $u_\epsilon \equiv u * \phi_\epsilon$, we can conclude that $u_{\epsilon,i} \equiv u_{,i} * \phi_\epsilon$.

Proof: If $\psi \in C^\infty_c(\mathbb{R}^n)$, then

$$\int u(\mathbf{x} - \mathbf{y}) \psi_{,i}(\mathbf{x}) dx = \int u(\mathbf{z}) \psi_{,i}(\mathbf{z} + \mathbf{y}) dz$$
$$= -\int u_{,i}(\mathbf{z}) \psi(\mathbf{z} + \mathbf{y}) dz$$
$$= -\int u_{,i}(\mathbf{x} - \mathbf{y}) \psi(\mathbf{x}) dx.$$

Therefore,

$$u_{\epsilon,i}(\psi) = -\int u_\epsilon \psi_{,i} = -\int \int u(\mathbf{x} - \mathbf{y}) \phi_\epsilon(\mathbf{y}) \psi_{,i}(\mathbf{x}) d y dx$$

360

$$= -\int\int u(\mathbf{x} - \mathbf{y})\,\psi_{,i}(\mathbf{x})\,\phi_\epsilon(\mathbf{y})\,dxdy$$

$$= \int\int u_{,i}(\mathbf{x} - \mathbf{y})\,\psi(\mathbf{x})\,\phi_\epsilon(\mathbf{y})\,dxdy$$

$$= \int u_{,i} * \phi_\epsilon(\mathbf{x})\,\psi(\mathbf{x})\,dx.$$

The technical questions about product measurability in the use of Fubini's theorem may be resolved by picking a Borel measurable representative for u. This proves the lemma.

Next we discuss a form of the product rule.

Lemma 18.9 *Let* $\psi \in C^\infty(\mathbb{R}^n)$ *and suppose* $u, u_{,i} \in L^p_{loc}(\mathbb{R}^n)$. *Then* $(u\psi)_{,i}$ *and* $u\psi$ *are in* L^p_{loc} *and*

$$(u\psi)_{,i} = u_{,i}\psi + u\psi_{,i}.$$

Proof: Let $\psi \in C^\infty_c(\mathbb{R}^n)$ then

$$(u\psi)_{,i}(\phi) = -\int u\psi\phi_{,i}dx$$

$$= -\int u[(\psi\phi)_{,i} - \phi\psi_{,i}]dx$$

$$= \int(u_{,i}\psi\phi + u\psi_{,i}\phi)\,dx.$$

This proves the lemma. We recall the notation for the gradient of a function.

$$\nabla u(\mathbf{x}) \equiv (u_{,1}(\mathbf{x}) \cdots u_{,n}(\mathbf{x}))^T$$

thus

$$Du(\mathbf{x})\,\mathbf{v} = \nabla u(\mathbf{x}) \cdot \mathbf{v}.$$

18.3 Morrey's inequality

The following inequality will be called Morrey's inequality. It relates an expression which is given pointwise to an integral of the *pth* power of the derivative.

Lemma 18.10 *Let* $u \in C^1(\mathbb{R}^n)$ *and* $p > n$. *Then there exists a constant,* C, *depending only on* n *such that for any* $\mathbf{x}, \mathbf{y} \in \mathbb{R}^n$,

$$|u(\mathbf{x}) - u(\mathbf{y})|$$

$$\leq C\left(\int_{B(\mathbf{x},2|\mathbf{x}-\mathbf{y}|)} |\nabla u(\mathbf{z})|^p dz\right)^{1/p}\left(|\mathbf{x} - \mathbf{y}|^{(1-n/p)}\right).$$

Proof: In the argument C will be a generic constant which depends on n.

$$\int_{B(\mathbf{x},r)} |u(\mathbf{x}) - u(\mathbf{y})|\, dy = \int_0^r \int_{S^{n-1}} |u(\mathbf{x} + \rho\omega) - u(\mathbf{x})|\, \rho^{n-1} d\sigma d\rho$$

$$\leq \int_0^r \int_{S^{n-1}} \int_0^\rho |\nabla u(\mathbf{x} + t\omega)|\, \rho^{n-1} dt d\sigma d\rho$$

$$\leq \int_{S^{n-1}} \int_0^r \int_0^r |\nabla u(\mathbf{x} + t\omega)|\, \rho^{n-1} d\rho dt d\sigma$$

$$\leq Cr^n \int_{S^{n-1}} \int_0^r |\nabla u(\mathbf{x} + t\omega)|\, dt d\sigma$$

$$= Cr^n \int_{S^{n-1}} \int_0^r \frac{|\nabla u(\mathbf{x} + t\omega)|}{t^{n-1}} t^{n-1} dt d\sigma$$

$$= Cr^n \int_{B(\mathbf{x},r)} \frac{\|\nabla u(\mathbf{z})\|}{\|\mathbf{z} - \mathbf{x}\|^{n-1}} dz.$$

Thus if we define

$$\fint_E f\, dx = \frac{1}{m(E)} \int_E f\, dx,$$

then

$$\fint_{B(\mathbf{x},r)} |u(\mathbf{y}) - u(\mathbf{x})|\, dy \leq C \int_{B(\mathbf{x},r)} |\nabla u(\mathbf{z})|\, \|\mathbf{z} - \mathbf{x}\|^{1-n} dz. \qquad (1)$$

Now let $r = |\mathbf{x} - \mathbf{y}|$ and

$$U = B(\mathbf{x}, r),\ V = B(\mathbf{y}, r),\ W = U \cap V.$$

Thus W equals the intersection of two balls of radius r with the center of one on the boundary of the other. It is clear there exists a constant, C, depending only on n such that

$$\frac{m(W)}{m(U)} = \frac{m(W)}{m(V)} = C.$$

Then from Formula 1,

$$|u(\mathbf{x}) - u(\mathbf{y})| = \fint_W |u(\mathbf{x}) - u(\mathbf{y})|\, dz$$

$$\leq \fint_W |u(\mathbf{x}) - u(\mathbf{z})|\, dz + \fint_W |u(\mathbf{z}) - u(\mathbf{y})|\, dz$$

$$= \frac{C}{m(U)} \left[\int_W |u(\mathbf{x}) - u(\mathbf{z})|\, dz + \int_W |u(\mathbf{z}) - u(\mathbf{y})|\, dz \right]$$

362

$$\leq C \left[\fint_U |u(\mathbf{x}) - u(\mathbf{z})| \, dz + \fint_V |u(\mathbf{y}) - u(\mathbf{z})| \, dz \right]$$

$$\leq C \left[\int_U |\nabla u(\mathbf{z})| \, |\mathbf{z} - \mathbf{x}|^{1-n} \, dz + \int_V |\nabla u(\mathbf{z})| \, |\mathbf{z} - \mathbf{y}|^{1-n} \, dz \right]. \qquad (2)$$

Consider the first of these two integrals. This is no smaller than

$$\leq C \left(\int_{B(\mathbf{x},r)} |\nabla u(\mathbf{z})|^p \, dz \right)^{1/p} \left(\int_{B(\mathbf{x},r)} (|\mathbf{z} - \mathbf{x}|^{1-n})^{p/(p-1)} \, dz \right)^{(p-1)/p}$$

$$= C \left(\int_{B(\mathbf{x},r)} |\nabla u(\mathbf{z})|^p \, dz \right)^{1/p} \left(\int_0^r \int_{S^{n-1}} \rho^{p(1-n)/(p-1)} \rho^{n-1} \, d\sigma d\rho \right)^{(p-1)/p}$$

$$= C \left(\int_{B(\mathbf{x},r)} |\nabla u(\mathbf{z})|^p \, dz \right)^{1/p} \left(\int_0^r \rho^{(1-n)/(p-1)} \, d\rho \right)^{(p-1)/p}$$

$$= C \left(\int_{B(\mathbf{x},r)} |\nabla u(\mathbf{z})|^p \, dz \right)^{1/p} r^{(1-n/p)} \qquad (3)$$

$$\leq C \left(\int_{B(\mathbf{x},2|\mathbf{x}-\mathbf{y}|)} |\nabla u(\mathbf{z})|^p \, dz \right)^{1/p} r^{(1-n/p)}. \qquad (4)$$

The second integral in Formula 2 is dominated by the same expression found in Formula 3 except the ball over which the integral is taken is centered at $\mathbf{y}$ not $\mathbf{x}$. Thus this integral is also dominated by the expression in Formula 4 and so,

$$|u(\mathbf{x}) - u(\mathbf{y})| \leq C \left(\int_{B(\mathbf{x},2|\mathbf{x}-\mathbf{y}|)} |\nabla u(\mathbf{z})|^p \, dz \right)^{1/p} |\mathbf{x} - \mathbf{y}|^{(1-n/p)} \qquad (5)$$

which proves the lemma.

18.4   Rademacher's theorem

Next we extend this inequality to the case where we only have u and $u_{,i}$ in L^p_{loc} for $p > n$. This leads to an elegant proof of the differentiability a.e. of a Lipschitz continuous function. Let $\psi_k \in C^\infty_c(\mathbb{R}^n)$, $\psi_k \geq 0$, and $\psi_k(\mathbf{z}) = 1$ for all $\mathbf{z} \in B(\mathbf{0}, k)$. Then

$$u\psi_k, \ (u\psi_k)_{,i} \in L^p(\mathbb{R}^n).$$

Let ϕ_ϵ be a mollifier and consider

$$(u\psi_k)_\epsilon \equiv u\psi_k * \phi_\epsilon.$$

By Lemma 18.8,

$$(u\psi_k)_{\epsilon,i} = (u\psi_k)_{,i} * \phi_\epsilon.$$

Therefore

$$(u\psi_k)_{\epsilon,i} \to (u\psi_k)_{,i} \text{ in } L^p(\mathbb{R}^n) \tag{6}$$

and

$$(u\psi_k)_\epsilon \to u\psi_k \text{ in } L^p(\mathbb{R}^n) \tag{7}$$

as $\epsilon \to 0$. By Formula 7, there exists a subsequence $\epsilon \to 0$ such that

$$(u\psi_k)_\epsilon(\mathbf{z}) \to u\psi_k(\mathbf{z}) \text{ a.e.} \tag{8}$$

Since $\psi_k(\mathbf{z}) = 1$ for $|\mathbf{z}| < k$, this shows

$$(u\psi_k)_\epsilon(\mathbf{z}) \to u(\mathbf{z}) \tag{9}$$

and for a.e. $\mathbf{z}$ with $|\mathbf{z}| < k$. Denoting the exceptional set of Formula 9 by E_k, let

$$\mathbf{x}, \mathbf{y} \notin \cup_{k=1}^\infty E_k \equiv E.$$

Also let k be so large that

$$B(\mathbf{0}, k) \supseteq B(\mathbf{x}, 2|\mathbf{x} - \mathbf{y}|).$$

Then by Formula 5,

$$|(u\psi_k)_\epsilon(\mathbf{x}) - (u\psi_k)_\epsilon(\mathbf{y})|$$

$$\leq C \left(\int_{B(\mathbf{x}, 2|\mathbf{y} - \mathbf{x}|)} |\nabla(u\psi_k)_\epsilon|^p \, dz \right)^{1/p} |\mathbf{x} - \mathbf{y}|^{(1-n/p)}$$

where C depends only on n. Now by Formula 8, there exists a subsequence, $\epsilon \to 0$, such that Formula 9, holds for $\mathbf{z} = \mathbf{x}, \mathbf{y}$. Thus, from Formula 6,

$$|u(\mathbf{x}) - u(\mathbf{y})| \leq C \left(\int_{B(\mathbf{x}, 2|\mathbf{y} - \mathbf{x}|)} |\nabla u|^p \, dz \right)^{1/p} |\mathbf{x} - \mathbf{y}|^{(1-n/p)}. \tag{10}$$

Redefining u on E, in the case where $p > n$, we can obtain Formula 10 for all $\mathbf{x}, \mathbf{y}$. This has proved the following theorem.

Theorem 18.11 *Suppose $u, u_{,i} \in L^p_{loc}(\mathbb{R}^n)$ for $i = 1, \cdots, n$ and $p > n$. Then u has a representative, still denoted by u, such that for all $\mathbf{x}, \mathbf{y} \in \mathbb{R}^n$,*

$$|u(\mathbf{x}) - u(\mathbf{y})| \leq C \left(\int_{B(\mathbf{x}, 2|\mathbf{y} - \mathbf{x}|)} |\nabla u|^p \, dz \right)^{1/p} |\mathbf{x} - \mathbf{y}|^{(1-n/p)}.$$

The next corollary is a very remarkable result. It says that not only is u continuous by virtue of having weak partial derivatives in L^p for large p, but also it is differentiable a.e.

Corollary 18.12 *Let* $u, u_{,i} \in L_{loc}^p (\mathbb{R}^n)$ *for* $i = 1, \cdots, n$ *and* $p > n$. *Then the representative of* u *described in Theorem 18.11 is differentiable a.e.*

Proof: Consider

$$|u(\mathbf{y}) - u(\mathbf{x}) - \nabla u(\mathbf{x}) \cdot (\mathbf{y} - \mathbf{x})|$$

where $u_{,i}$ is a representative of $u_{,i}$, an element of L^p. Define

$$g(\mathbf{z}) \equiv u(\mathbf{z}) + \nabla u(\mathbf{x}) \cdot (\mathbf{y} - \mathbf{z}).$$

Then the above expression is of the form

$$|g(\mathbf{y}) - g(\mathbf{x})|$$

and

$$\nabla g(\mathbf{z}) = \nabla u(\mathbf{z}) - \nabla u(\mathbf{x}).$$

Therefore $g \in L_{loc}^p (\mathbb{R}^n)$ and $g_{,i} \in L_{loc}^p (\mathbb{R}^n)$. It follows from Theorem 18.11 that

$$|g(\mathbf{y}) - g(\mathbf{x})| = |u(\mathbf{y}) - u(\mathbf{x}) - \nabla u(\mathbf{x}) \cdot (\mathbf{y} - \mathbf{x})|$$

$$\leq C \left(\int_{B(\mathbf{x}, 2|\mathbf{y} - \mathbf{x}|)} |\nabla g(\mathbf{z})|^p dz \right)^{1/p} |\mathbf{x} - \mathbf{y}|^{(1 - n/p)}$$

$$= C \left(\int_{B(\mathbf{x}, 2|\mathbf{y} - \mathbf{x}|)} |\nabla u(\mathbf{z}) - \nabla u(\mathbf{x})|^p dz \right)^{1/p} |\mathbf{x} - \mathbf{y}|^{(1 - n/p)}$$

$$= C \left(\fint_{B(\mathbf{x}, 2|\mathbf{y} - \mathbf{x}|)} |\nabla u(\mathbf{z}) - \nabla u(\mathbf{x})|^p dz \right)^{1/p} |\mathbf{x} - \mathbf{y}|.$$

This last expression is $o(|\mathbf{y} - \mathbf{x}|)$ at every Lebesgue point, $\mathbf{x}$, of ∇u. This proves the corollary and shows ∇u is the gradient a.e.

Now suppose u is Lipschitz on $\mathbb{R}^n$,

$$|u(\mathbf{x}) - u(\mathbf{y})| \leq K |\mathbf{x} - \mathbf{y}|$$

for some constant K. We define $Lip(u)$ as the smallest value of K that works in this inequality. The following corollary is known as Rademacher's theorem. It states that every Lipschitz function is differentiable a.e.

Weak Derivatives

Corollary 18.13 *If u is Lipschitz continuous then u is differentiable a.e. and $||Du||_\infty \leq Lip(u)$.*

Proof: Let

$$D_{e_i}^h u(x) \equiv h^{-1}[u(x + he_i) - u(x)].$$

Then $D_{e_i}^h u$ is bounded in $L^\infty(\mathbb{R}^n)$ and

$$||D_{e_i}^h u||_\infty \leq Lip(u).$$

It follows that $D_{e_i}^h u$ is contained in a ball in $L^\infty(\mathbb{R}^n)$, the dual space of $L^1(\mathbb{R}^n)$. By Theorem 6.19 and Corollary 6.21 in Chapter 6 on weak* topologies, there is a subsequence $h \to 0$ such that

$$D_{e_i}^h u \rightharpoonup w, \ ||w||_\infty \leq Lip(u)$$

where the convergence takes place in the weak * topology of $L^\infty(\mathbb{R}^n)$. Let $\phi \in C_c^\infty(\mathbb{R}^n)$. Then

$$\int w\phi dx = \lim_{h\to 0} \int D_{e_i}^h u\phi dx$$

$$= \lim_{h\to 0} \int u(\mathbf{x}) \frac{(\phi(\mathbf{x} - he_i) - \phi(\mathbf{x}))}{h} dx$$

$$= -\int u(\mathbf{x})\phi_{,i}(\mathbf{x}) dx.$$

Thus $w = u_{,i}$ and we see that $u_{,i} \in L^\infty(\mathbb{R}^n)$ for each i. Hence $u, u_{,i} \in L_{loc}^p(\mathbb{R}^n)$ for all $p > n$ and so u is differentiable a.e. and ∇u is given in terms of the weak derivatives of u by Corollary 18.12. This proves the corollary.

18.5 Exercises

1. Let K be a bounded subset of $L^p(\mathbb{R}^n)$ and suppose that for all $\epsilon > 0$, there exist a $\delta > 0$ and G such that $\overline{G}$ is compact such that if $|\mathbf{h}| < \delta$, then

$$\int |u(\mathbf{x} + \mathbf{h}) - u(\mathbf{x})|^p dx < \epsilon^p$$

for all $u \in K$ and

$$\int_{\mathbb{R}^n \setminus \overline{G}} |u(\mathbf{x})|^p dx < \epsilon^p$$

for all $u \in K$. Show that K is precompact in $L^p(\mathbb{R}^n)$. **Hint:** Let ϕ_k be a mollifier and consider

$$K_k \equiv \{u * \phi_k : u \in K\}.$$

Verify the conditions of the Ascoli Arzela theorem for these functions defined on $\overline{G}$ and show there is an ϵ net for each $\epsilon > 0$. Can you modify this to let an arbitrary open set take the place of $\mathbb{R}^n$?

2. For an arbitrary open set $U \subseteq \mathbb{R}^n$, we define $X^{1p}(U)$ as the set of all functions in $L^p(U)$ whose weak partial derivatives are also in $L^p(U)$. Here we say a function in $L^p(U), g$ equals $u_{,i}$ if and only if

$$\int_U g\phi dx = -\int_U u\phi_{,i}dx$$

for all $\phi \in C_c^\infty(U)$. The norm in this space is given by

$$\|u\|_{1p} \equiv \left(\int_U |u|^p + |\nabla u|^p \, dx\right)^{1/p}.$$

Then we define the Sobolev space $W^{1p}(U)$ to be the closure of $C^\infty(\overline{U})$ in $X^{1p}(U)$ where $C^\infty(\overline{U})$ is defined to be restrictions of all functions in $C_c^\infty(\mathbb{R}^n)$ to U. Show that this definition of weak derivative is well defined and that $X^{1p}(U)$ is a reflexive Banach space. **Hint:** To do this, show the operator $u \to u_{,i}$ is a closed operator and that $X^{1p}(U)$ can be considered as a closed subspace of $L^p(U)^{n+1}$. Show that in general the product of reflexive spaces is reflexive and then recall the Problem 7 of Chapter 6 which states that a closed subspace of a reflexive Banach space is reflexive. Thus, conclude that $W^{1p}(U)$ is also a reflexive Banach space.

3. Theorem 18.11 shows that if the weak derivatives of a function $u \in L^p(\mathbb{R}^n)$ are in $L^p(U)$, for $p > n$, then the function has a continuous representative. (In fact, one can conclude more than continuity from this theorem.) It is also important to consider the case when $p < n$. To aid in the study of this case which will be carried out in the next few problems, show the following inequality for $n \geq 2$.

$$\int_{\mathbb{R}^n} \prod_{j=1}^n |w_j(\mathbf{x})| \, dm_n \leq \prod_{i=1}^n \left(\int_{\mathbb{R}^{n-1}} |w_j(\mathbf{x})|^{n-1} \, dm_{n-1}\right)^{1/n-1}$$

where w_j does not depend on the jth component of $\mathbf{x}$, x_j. **Hint:** First show it is true for $n = 2$ and then use Holder's inequality and induction. You might benefit from first trying the case $n = 3$ to get the idea.

4. ↑ Show that if $\phi \in C_c^\infty(\mathbb{R}^n)$, then

$$\|\phi\|_{n/(n-1)} \leq \frac{1}{\sqrt[n]{n}} \sum_{j=1}^{n} \left\| \frac{\partial \phi}{\partial x_j} \right\|_1.$$

Hint: First show that if $a_i \geq 0$, then

$$\prod_{i=1}^{n} a_i^{1/n} \leq \frac{1}{\sqrt[n]{n}} \sum_{j=1}^{n} a_j.$$

Then observe that

$$|\phi(\mathbf{x})| \leq \int_{-\infty}^{\infty} |\phi_{,j}(\mathbf{x})| \, dx_j$$

so

$$\|\phi\|_{n/(n-1)}^{n/(n-1)} = \int |\phi|^{n/(n-1)} \, dm_n$$

$$\leq \int \prod_{j=1}^{n} \left(\int_{-\infty}^{\infty} |\phi_{,j}(\mathbf{x})| \, dx_j \right)^{1/(n-1)} \, dm_n$$

$$\leq \prod_{j=1}^{n} \left(\int |\phi_{,j}(\mathbf{x})| \, dm_n \right)^{1/(n-1)}.$$

Hence

$$\|\phi\|_{n/(n-1)} \leq \prod_{j=1}^{n} \left(\int |\phi_{,j}(\mathbf{x})| \, dm_n \right)^{1/n}.$$

5. ↑ Show that if $\phi \in C_c^\infty(\mathbb{R}^n)$, then if $\frac{1}{q} = \frac{1}{p} - \frac{1}{n}$, where $p < n$, then

$$\|\phi\|_q \leq \frac{1}{\sqrt[n]{n}} \frac{(n-1)p}{n-p} \sum_{j=1}^{n} \|\phi_{,i}\|_p.$$

Also show that if $u \in W^{1p}(\mathbb{R}^n)$, then $u \in L^q(\mathbb{R}^n)$ and the inclusion map is continuous. This is part of the Sobolev embedding theorem. For more on Sobolev spaces see Adams [2]. **Hint:** Let $r > 1$. Then $|\phi|^r \in C_c^\infty(\mathbb{R}^n)$ and

$$\left| |\phi|_{,i}^r \right| = r |\phi|^{r-1} |\phi_{,i}|.$$

Now apply the result of Problem 4 to write

$$\left(\int |\phi|^{\frac{rn}{n-1}} \, dm_n \right)^{(n-1)/n} \leq \frac{r}{\sqrt[n]{n}} \sum_{i=1}^{n} \int |\phi|^{r-1} |\phi_{,i}| \, dm_n$$

$$\leq \frac{r}{\sqrt[n]{n}} \sum_{i=1}^{n} \left(\int |\phi_{,i}|^p \right)^{1/p} \left(\int \left(|\phi|^{r-1} \right)^{p/(p-1)} dm_n \right)^{(p-1)/p}.$$

Now choose r such that

$$\frac{(r-1)\,p}{p-1} = \frac{rn}{n-1}$$

so that the last term on the right can be cancelled with the first term on the left and simplify.

Chapter 19
Hausdorff Measures

In this chapter we discuss Hausdorff measures, a generalization of Lebesgue measure. Hausdorff measure is defined in terms of coverings of sets which have small diameters and the quantity which is important is the diameters of the sets. The concept will be used to present a simple and unified treatment of surface measure in later chapters. To study Hausdorff measures, we will use many ideas from earlier chapters, in particular the following lemma found at the beginning of Chapter 14.

Lemma 19.1 *Let $(\Omega, \mathcal{S}, \mu)$ be any σ-finite measure space and define*

$$\overline{\mu} : \mathcal{P}(\Omega) \to [0, \infty]$$

by

$$\overline{\mu}(E) = \inf\{\mu(V) : V \in \mathcal{S} \text{ and } V \supseteq E\}.$$

Then $\overline{\mu}$ is an outer measure and if $\overline{\mathcal{S}}$ is the set of $\overline{\mu}$ measurable sets in the sense of Caratheodory, then $\overline{\mathcal{S}} \supseteq \mathcal{S}$ and $\overline{\mu} = \mu$ on $\mathcal{S}$. Furthermore, if $(\Omega, \mathcal{S}, \mu)$ is complete, then $\overline{\mathcal{S}} = \mathcal{S}$.

Let $\alpha(n)$ be the volume of the unit ball in $\mathbb{R}^n$. Thus the volume of $B(0, r)$ in $\mathbb{R}^n$ is $\alpha(n)r^n$. There is a very important and interesting inequality known as the isodiametric inequality which says that if A is any set in $\mathbb{R}^n$, then

$$\overline{m}(A) \leq \alpha(n)(2^{-1}\text{diam}(A))^n.$$

This inequality may seem obvious at first but it is not really. The reason it is not is that there are sets which are not subsets of any sphere having the same diameter as the set. For example, consider an equilateral triangle.

Lemma 19.2 *Let $f : \mathbb{R}^{n-1} \to [0, \infty)$ be Borel measurable and let*

$$S = \{(\mathbf{x}, y) : |y| < f(\mathbf{x})\}.$$

Then S is a Borel set in $\mathbb{R}^n$.

Proof: Set s_k be an increasing sequence of Borel measurable functions converging pointwise to f.

$$s_k(\mathbf{x}) = \sum_{m=1}^{N_k} c_m^k \mathcal{X}_{E_m^k}(\mathbf{x}).$$

Let

$$S_k = \cup_{m=1}^{N_k} E_m^k \times (-c_m^k, c_m^k).$$

Then $(\mathbf{x},y) \in S_k$ if and only if $f(\mathbf{x}) > 0$ and $|y| < s_k(\mathbf{x}) \leq f(\mathbf{x})$. It follows that $S_k \subseteq S_{k+1}$ and

$$S = \cup_{k=1}^{\infty} S_k.$$

But each S_k is a Borel set and so S is also a Borel set. This proves the lemma.

Let P_i be the projection onto

$$span\{\mathbf{e}_1, \cdots, \mathbf{e}_{i-1}, \mathbf{e}_{i+1}, \cdots, \mathbf{e}_n\}.$$

Recall from Chapter 11 the notation

$$A_{P_i\mathbf{x}} \equiv \{x_i : (x_1, \cdots, x_i, \cdots, x_n) \in A\}$$

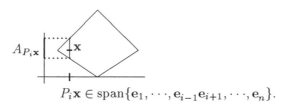

$$P_i\mathbf{x} \in span\{\mathbf{e}_1, \cdots, \mathbf{e}_{i-1}\mathbf{e}_{i+1}, \cdots, \mathbf{e}_n\}.$$

Lemma 19.3 *Let $A \subseteq \mathbb{R}^n$ be a Borel set. Then $P_i\mathbf{x} \to m(A_{P_i\mathbf{x}})$ is a Borel measurable function defined on $P_i(\mathbb{R}^n)$.*

Proof: Let $R_k = (-k, k)^n$ and let

$$\mathcal{M} = \{A \in \text{Borel sets such that } P_i\mathbf{x} \to m((A \cap R_k)_{P_i\mathbf{x}})$$

is Borel measurable$\}$

Let $\mathcal{A}$ denote the algebra of disjoint unions of sets of the form $\prod_{i=1}^{n} A_i$ where A_i is Borel. Since the Cartesian product of Borel sets is also Borel, each of these sets is a Borel set. Furthermore it is clear that $\mathcal{M} \supseteq \mathcal{A}$ and that $\mathcal{M}$ is a monotone class. Hence, by the monotone class theorem, $\mathcal{M} \supseteq \sigma(\mathcal{A})$, the σ−algebra generated by $\mathcal{A}$. But $\sigma(\mathcal{A})$ contains the open sets and so $\sigma(\mathcal{A})$ contains the Borel sets. Therefore $\mathcal{M} = \sigma(\mathcal{A}) =$ Borel sets. Now

$$m(A_{P_i\mathbf{x}}) = \lim_{k \to \infty} m((A \cap R_k)_{P_i\mathbf{x}})$$

so this proves the lemma.

Now let $A \subseteq \mathbb{R}^n$ be Borel. Let P_i be the projection onto

$$span\{\mathbf{e}_1, \cdots, \mathbf{e}_{i-1}, \mathbf{e}_{i+1}, \cdots, \mathbf{e}_n\}$$

as just described.

$$A_{P_i\mathbf{x}} = \{y \in \mathbb{R} : P_i\mathbf{x} + y\mathbf{e}_i \in A\}$$

Thus for $\mathbf{x} = (x_1, \cdots, x_n)$,

$$A_{P_i\mathbf{x}} = \{y \in \mathbb{R} : (x_1, \cdots, x_{i-1}, y, x_{i+1}, \cdots, x_n) \in A\}.$$

Since A is Borel, it follows from Lemma 19.2 that

$$P_i\mathbf{x} \to m(A_{P_i\mathbf{x}})$$

is a Borel measurable function on $P_i\mathbb{R}^n = \mathbb{R}^{n-1}$.

19.1 Steiner symmetrization

We define

$$S(A, \mathbf{e}_i) \equiv \{\mathbf{x} = P_i\mathbf{x} + y\mathbf{e}_i : |y| < 2^{-1}m(A_{P_i\mathbf{x}})\}$$

Lemma 19.4 *Let A be a Borel subset of $\mathbb{R}^n$. Then $S(A, \mathbf{e}_i)$ satisfies*

$$P_i\mathbf{x} + y\mathbf{e}_i \in S(A, \mathbf{e}_i) \text{ if and only if } P_i\mathbf{x} - y\mathbf{e}_i \in S(A, \mathbf{e}_i),$$

$$S(A, \mathbf{e}_i) \text{ is a Borel set in } \mathbb{R}^n,$$

$$m_n(S(A, \mathbf{e}_i)) = m_n(A), \tag{1}$$

$$diam(S(A, \mathbf{e}_i)) \leq diam(A). \tag{2}$$

Proof: The first assertion is obvious from the definition. The Borel measurability of $S(A, \mathbf{e}_i)$ follows from the definition and Lemmas 19.2 and 19.1. To show Formula 1,

$$
\begin{aligned}
m_n(S(A, \mathbf{e}_i)) &= \int_{P_i\mathbb{R}^n} \int_{-2^{-1}m(A_{P_i\mathbf{x}})}^{2^{-1}m(A_{P_i\mathbf{x}})} dx_i dx_1 \cdots dx_{i-1} dx_{i+1} \cdots dx_n \\
&= \int_{P_i\mathbb{R}^n} m(A_{P_i\mathbf{x}}) dx_1 \cdots dx_{i-1} dx_{i+1} \cdots dx_n \\
&= m(A).
\end{aligned}
$$

Now suppose $\mathbf{x}_1$ and $\mathbf{x}_2 \in S(A, \mathbf{e}_i)$

$$\mathbf{x}_1 = P_i\mathbf{x}_1 + y_1\mathbf{e}_i, \quad \mathbf{x}_2 = P_i\mathbf{x}_2 + y_2\mathbf{e}_i.$$

For $\mathbf{x} \in A$ define

$$l(\mathbf{x}) = \sup\{y : P_i\mathbf{x} + y\mathbf{e}_i \in A\}.$$

$$g(\mathbf{x}) = \inf\{y : P_i\mathbf{x} + y\mathbf{e}_i \in A\}.$$

Then it is clear that

$$l(\mathbf{x}_1) - g(\mathbf{x}_1) \geq m(A_{P_i\mathbf{x}_1}) \geq 2|y_1|,$$

$$l(\mathbf{x}_2) - g(\mathbf{x}_2) \geq m(A_{P_i\mathbf{x}_2}) \geq 2|y_2|.$$

Therefore if $|y_1| \geq |y_2|$, then

$$|y_1 - y_2| \leq 2|y_1| \leq l(\mathbf{x}_1) - g(\mathbf{x}_1) \tag{3}$$

and if $|y_1| \leq |y_2|$, then

$$|y_1 - y_2| \leq 2|y_2| \leq l(\mathbf{x}_2) - g(\mathbf{x}_2).$$

Claim: $|y_1 - y_2| \leq |l(\mathbf{x}_1) - g(\mathbf{x}_2)|$ or $|y_1 - y_2| \leq |l(\mathbf{x}_2) - g(\mathbf{x}_1)|$.
   Proof of Claim: If not,

$$2|y_1 - y_2| \geq |l(\mathbf{x}_1) - g(\mathbf{x}_2)| + |l(\mathbf{x}_2) - g(\mathbf{x}_1)|$$

$$\geq |l(\mathbf{x}_1) - g(\mathbf{x}_1) + g(\mathbf{x}_1) - g(\mathbf{x}_2) + l(\mathbf{x}_2) - g(\mathbf{x}_2) + g(\mathbf{x}_2) - g(\mathbf{x}_1)|$$

$$\geq l(\mathbf{x}_1) - g(\mathbf{x}_1) + l(\mathbf{x}_2) - g(\mathbf{x}_2).$$

If $|y_1| \geq |y_2|$ then this and the second half of Formula 3 implies

$$\begin{aligned} 2|y_1| &\geq l(\mathbf{x}_1) - g(\mathbf{x}_1) + l(\mathbf{x}_2) - g(\mathbf{x}_2) \tag{4} \\ &\geq 2|y_1| + l(\mathbf{x}_2) - g(\mathbf{x}_2). \end{aligned}$$

But since $\mathbf{x}_2 \in S(A, \mathbf{e}_i)$, $m(A_{P_i\mathbf{x}_2}) > 0$ and so $l(\mathbf{x}_2) - g(\mathbf{x}_2) > 0$. This contradicts Formula 4. A similar contradiction results if $|y_1| \leq |y_2|$.
   From the claim

$$\begin{aligned} |\mathbf{x}_1 - \mathbf{x}_2| &= (|P_i\mathbf{x}_1 - P_i\mathbf{x}_2|^2 + |y_1 - y_2|^2)^{1/2} \\ &\leq (|P_i\mathbf{x}_1 - P_i\mathbf{x}_2|^2 + (z_1 - z_2 + 2\varepsilon)^2)^{1/2} \\ &\leq \mathrm{diam}(A) + O(\sqrt{\varepsilon}) \end{aligned}$$

where z_1 and z_2 are such that $P_i\mathbf{x}_1 + z_1\mathbf{e}_i \in A$, $P_i\mathbf{x}_2 + z_2\mathbf{e}_i \in A$, and either

$$|z_1 - l(\mathbf{x}_1)| < \varepsilon \text{ and } |z_2 - g(\mathbf{x}_2)| < \varepsilon$$

or

$$|z_1 - g(\mathbf{x}_1)| < \varepsilon \text{ and } |z_2 - l(\mathbf{x}_2)| < \varepsilon,$$

depending on which inequality holds in the claim. Since $\mathbf{x}_1, \mathbf{x}_2$ are arbitrary elements of $S(A, \mathbf{e}_i)$ and ε is arbitrary, this proves Formula 2.
   The next lemma says that if A is already symmetric with respect to the *j*th direction, then this symmetry is not destroyed by taking $S(A, \mathbf{e}_i)$.

Lemma 19.5 *Suppose A is a Borel set in $\mathbb{R}^n$ such that $P_j\mathbf{x} + \mathbf{e}_jx_j \in A$ if and only if $P_j\mathbf{x} + (-x_j)\mathbf{e}_j \in A$. Then if $i \neq j$, $P_j\mathbf{x} + \mathbf{e}_jx_j \in S(A, \mathbf{e}_i)$ if and only if $P_j\mathbf{x} + (-x_j)\mathbf{e}_j \in S(A, \mathbf{e}_i)$.*

Proof: By definition,

$$P_j \mathbf{x} + \mathbf{e}_j x_j \in S(A, \mathbf{e}_i)$$

if and only if

$$|x_i| < 2^{-1} m(A_{P_i(P_j \mathbf{x} + \mathbf{e}_j x_j)}).$$

Now

$$x_i \in A_{P_i(P_j \mathbf{x} + \mathbf{e}_j x_j)}$$

if and only if

$$x_i \in A_{P_i(P_j \mathbf{x} + (-x_j) \mathbf{e}_j)}$$

by the assumption on A which says that A is symmetric in the $\mathbf{e}_j$ direction. Hence

$$P_j \mathbf{x} + \mathbf{e}_j x_j \in S(A, \mathbf{e}_i)$$

if and only if

$$|x_i| < 2^{-1} m(A_{P_i(P_j \mathbf{x} + (-x_j) \mathbf{e}_j)})$$

if and only if

$$P_j \mathbf{x} + (-x_j) \mathbf{e}_j \in S(A, \mathbf{e}_i).$$

This proves the lemma.

19.2 The isodiametric inequality

The next theorem is called the isodiametric inequality. It is the key result used to compare Lebesgue and Hausdorff measures.

Theorem 19.6 *Let A be any Lebesgue measurable set in $\mathbb{R}^n$. Then*

$$m_n(A) \leq \alpha(n)(2^{-1} diam(A))^n.$$

Proof: Suppose first that A is Borel. Let $A_1 = S(A, \mathbf{e}_1)$ and let $A_k = S(A_{k-1}, \mathbf{e}_k)$. Then by the preceding lemmas, A_n is a Borel set, $\text{diam}(A_n) \leq \text{diam}(A)$, $m_n(A_n) = m_n(A)$, and A_n is symmetric. Thus $x \in A_n$ if and only if $-x \in A_n$. It follows that

$$A_n \subseteq \overline{B(0, 2^{-1} \text{diam}(A_n))}.$$

(If $\mathbf{x} \in \mathbf{A}_n \setminus \overline{B(0, 2^{-1} \text{diam}(A_n))}$, then $-\mathbf{x} \in \mathbf{A}_n \setminus \overline{B(0, 2^{-1} \text{diam}(A_n))}$ and so $\text{diam}(A_n) \geq 2|\mathbf{x}| > \text{diam}(A_n)$.) Therefore,

$$m_n(A_n) \leq \alpha(n)(2^{-1} \text{diam}(A_n))^n.$$

It remains to establish this inequality for arbitrary measurable sets. Letting A be such a set, let $\{K_n\}$ be an increasing sequence of compact subsets of A such that

$$m(A) = \lim_{k \to \infty} m(K_k).$$

Then

$$
\begin{aligned}
m(A) &= \lim_{k \to \infty} m(K_k) \leq \lim_{k \to \infty} \sup \, \alpha(n)(2^{-1}\mathrm{diam}(K_k))^n \\
&\leq \alpha(n)(2^{-1}\mathrm{diam}(A))^n.
\end{aligned}
$$

This proves the theorem.

Corollary 19.7 *Let $\overline{m_n}(E) = \inf\{m_n(F) : F$ is Lebesgue measurable and $F \supseteq E\}$. Then $\overline{m_n}$ is an outer measure, the σ algebra of measurable sets is just the collection of Lebesgue measurable sets and $\overline{m_n} = m_n$ on this σ algebra. Furthermore, for any $E \subseteq \mathbb{R}^n$,*

$$\overline{m_n}(E) \leq \alpha(n)(2^{-1}diam(E))^n.$$

Proof: The first part of this corollary is implied by Lemma 19.1. For the second part, let $E \subseteq \mathbb{R}^n$.

$$\overline{m_n}(E) \leq \overline{m_n}(\overline{E}) = m_n(\overline{E})$$

$$\leq \alpha(n)(2^{-1}\mathrm{diam}(\overline{E}))^n = \alpha(n)(2^{-1}\mathrm{diam}(E))^n.$$

This proves the corollary.

What is $\alpha(n)$?

Theorem 19.8 $\alpha(n) = \pi^{n/2}(\Gamma(n/2 + 1))^{-1}$ *where $\Gamma(s)$ is the gamma function*

$$\Gamma(s) = \int_0^\infty e^{-t}t^{s-1}dt.$$

Proof: First let $n = 1$.

$$
\begin{aligned}
\Gamma(3/2) &= \int_0^\infty e^{-t}t^{1/2}dt = \int_0^\infty 2ue^{-u^2}u\,du \\
&= \int_0^\infty e^{-u^2}du = 2^{-1}\pi^{1/2}.
\end{aligned}
$$

Thus $2 = \pi^{1/2}(\Gamma(1/2 + 1))^{-1} = \alpha(1)$ and this shows the theorem is true if $n = 1$. Assume the theorem is true for n and let B_k be the unit ball in $\mathbb{R}^k$. Then by the result in $\mathbb{R}^n$,

$$m_{n+1}(B_{n+1}) = \int_{-1}^1 \alpha(n)(1 - x_{n+1}^2)^{n/2}dx_{n+1}$$

$$= 2\alpha(n) \int_0^1 (1-t^2)^{n/2} dt.$$

Doing an integration by parts, this equals

$$= 2\alpha(n)n \int_0^1 t^2 (1-t^2)^{(n-2)/2} dt$$

$$= 2\alpha(n)n \frac{1}{2} \int_0^1 u^{1/2}(1-u)^{n/2-1} du$$

$$= n\alpha(n) \int_0^1 u^{3/2-1}(1-u)^{n/2-1} du.$$

By Problem 6 of Chapter 12,

$$= n\alpha(n)\Gamma(3/2)\Gamma(n/2)(\Gamma((n+3)/2))^{-1}$$

$$= n\pi^{n/2}(\Gamma(n/2+1))^{-1}(\Gamma((n+3)/2))^{-1}\Gamma(3/2)\Gamma(n/2)$$
$$= n\pi^{n/2}(\Gamma(n/2)(n/2))^{-1}(\Gamma((n+1)/2+1))^{-1}\Gamma(3/2)\Gamma(n/2)$$
$$= 2\pi^{n/2}\Gamma(3/2)(\Gamma((n+1)/2+1))^{-1}$$
$$= \pi^{(n+1)/2}(\Gamma((n+1)/2+1))^{-1}$$

which proves the theorem. The introduction of the gamma function and the subsequent simplification was accomplished using the following identities discussed in Problem 6 of Chapter 12.

$$\Gamma(p)\Gamma(q) = \int_0^1 x^{p-1}(1-x)^{q-1} dx \, \Gamma(p+q).$$

$$s\Gamma(s) = \Gamma(s+1).$$

19.3   Hausdorff measures

With this preparation we are ready to define the Hausdorff measures. Let

$$0 \le s < \infty, \; \alpha(s) = \pi^{s/2}(\Gamma(s/2+1))^{-1}$$

and let $E \subseteq \mathbb{R}^n$.

$$\mathcal{H}_\delta^s(E) \equiv \inf\{\sum_{j=1}^\infty \alpha(s)(2^{-1}\mathrm{diam}(C_j))^s : E \subseteq \cup_{j=1}^\infty C_j, \mathrm{diam}(C_j) \le \delta\}$$

$$\mathcal{H}^s(E) \equiv \lim_{\delta \to 0} \mathcal{H}_\delta^s(E).$$

Lemma 19.9 $\mathcal{H}^s$ and $\mathcal{H}^s_\delta$ are outer measures.

Proof: It is clear that $\mathcal{H}^s(\emptyset) = 0$ and if $A \subseteq B$, then $\mathcal{H}^s(A) \leq \mathcal{H}^s(B)$ with similar assertions valid for $\mathcal{H}^s_\delta$. Suppose $E = \cup_{i=1}^\infty E_i$ and $\mathcal{H}^s_\delta(E_i) < \infty$ for each i. Let $\{C^i_j\}_{j=1}^\infty$ be a covering of E_i with

$$\sum_{j=1}^\infty \alpha(s)(2^{-1}\mathrm{diam}(C^i_j))^s - \varepsilon/2^i < \mathcal{H}^s_\delta(E_i)$$

and $\mathrm{diam}(C^i_j) \leq \delta$. Then

$$
\begin{aligned}
\mathcal{H}^s_\delta(E) &\leq \sum_{i=1}^\infty \sum_{j=1}^\infty \alpha(s)(2^{-1}\mathrm{diam}(C^i_j))^s \\
&\leq \sum_{i=1}^\infty \mathcal{H}^s_\delta(E_i) + \varepsilon/2^i \\
&\leq \varepsilon + \sum_{i=1}^\infty \mathcal{H}^s_\delta(E_i).
\end{aligned}
$$

It follows that since $\varepsilon > 0$ is arbitrary,

$$\mathcal{H}^s_\delta(E) \leq \sum_{i=1}^\infty \mathcal{H}^s_\delta(E_i)$$

which shows $\mathcal{H}^s_\delta$ is an outer measure. Now notice that $\mathcal{H}^s_\delta(E)$ is increasing as $\delta \to 0$. Picking a sequence δ_k decreasing to 0, we can use the monotone convergence theorem to conclude

$$\mathcal{H}^s(E) \leq \sum_{i=1}^\infty \mathcal{H}^s(E_i).$$

This proves the lemma.

Next we wish to show that the σ algebra of $\mathcal{H}^s$ measurable sets includes the Borel sets. To this end, we give a very interesting condition known as Caratheodory's criterion which is a sufficient condition for this to occur.

19.4   Properties of Hausdorff measure

Theorem 19.10 Let $\overline{\mu}$ be an outer measure on the subsets of (X, d), a metric space. If
$$\overline{\mu}(A \cup B) = \overline{\mu}(A) + \overline{\mu}(B)$$

whenever dist(A, B) > 0, then the σ algebra of measurable sets contains the Borel sets.

Proof: We only need show that closed sets are in $\mathcal{S}$, the σ-algebra of measurable sets, because then the open sets are also in $\mathcal{S}$ and so $\mathcal{S} \supseteq$ Borel sets. Let K be closed and let S be a subset of Ω. We need to show $\mu(S) \geq \mu(S \cap K) + \mu(S \setminus K)$. Therefore, we may assume without loss of generality that $\mu(S) < \infty$. Let

$$K_n = \{x : dist(x, K) \leq \frac{1}{n}\} = \text{closed set}$$

$(x \to dist(x, K)$ is continuous.)

$$\mu(S) \geq \mu((S \cap K) \cup (S \setminus K_n)) = \mu(S \cap K) + \mu(S \setminus K_n) \qquad (5)$$

by assumption, since $S \cap K$ and $S \setminus K_n$ are a positive distance apart. Now

$$\mu(S \setminus K_n) \leq \mu(S \setminus K) \leq \mu(S \setminus K_n) + \mu((K_n \setminus K) \cap S). \qquad (6)$$

We look at $\mu((K_n \setminus K) \cap S)$. Note that

$$K_n \setminus K = \cup_{k=n}^{\infty} K_k \setminus K_{k+1}$$

because K is closed. Therefore

$$\mu(S \cap (K_n \setminus K)) \leq \sum_{k=n}^{\infty} \mu(S \cap (K_k \setminus K_{k+1})). \qquad (7)$$

Now

$$\sum_{k=1}^{\infty} \mu(S \cap (K_k \setminus K_{k+1})) = \sum_{k \text{ even}} \mu(S \cap (K_k \setminus K_{k+1})) +$$
$$+ \sum_{k \text{ odd}} \mu(S \cap (K_k \setminus K_{k+1})). \qquad (8)$$

Note that if $A = \cup_{i=1}^{\infty} A_i$ and the distance between any pair of sets is positive, then

$$\mu(A) = \sum_{i=1}^{\infty} \mu(A_i),$$

because

$$\sum_{i=1}^{\infty} \mu(A_i) \geq \mu(A) \geq \mu(\cup_{i=1}^{n} A_i) = \sum_{i=1}^{n} \mu(A_i).$$

Therefore, from Formula 8,

$$\sum_{k=1}^{\infty} \mu(S \cap (K_k \setminus K_{k+1}))$$

$$= \mu(\bigcup_{k \text{ even}} S \cap (K_k \setminus K_{k+1})) + \mu(\bigcup_{k \text{ odd}} S \cap (K_k \setminus K_{k+1}))$$

$$< 2\mu(S) < \infty.$$

Therefore from Formula 7

$$\lim_{n \to \infty} \mu(S \cap (K_n \setminus K)) = 0.$$

From Formula 6

$$0 \leq \mu(S \setminus K) - \mu(S \setminus K_n) \leq \mu(S \cap (K_n \setminus K))$$

and so

$$\lim_{n \to \infty} \mu(S \setminus K_n) = \mu(S \setminus K).$$

From Formula 5

$$\mu(S) \geq \mu(S \cap K) + \mu(S \setminus K).$$

This shows $K \in \mathcal{S}$ and proves the theorem.

Theorem 19.11 *The σ algebra of $\mathcal{H}^s$ measurable sets contains the Borel sets and $\mathcal{H}^s$ has the property that for all $E \subseteq \mathbb{R}^n$, there exists a Borel set $F \supseteq E$ such that $\mathcal{H}^s(F) = \mathcal{H}^s(E)$.*

Proof: Let $\text{dist}(A, B) = 2\delta_0 > 0$. We need to show

$$\mathcal{H}^s(A) + \mathcal{H}^s(B) = \mathcal{H}^s(A \cup B).$$

Let $\{C_j\}_{j=1}^{\infty}$ be a covering of $A \cup B$ such that $\text{diam}(C_j) \leq \delta < \delta_0$ for each j and

$$\mathcal{H}_\delta^s(A \cup B) + \varepsilon > \sum_{j=1}^{\infty} \alpha(s)(2^{-1}\text{diam}(C_j))^s.$$

Thus

$$\mathcal{H}_\delta^s(A \cup B) + \varepsilon > \sum_{j \in J_1} \alpha(s)(2^{-1}\text{diam}(C_j))^s + \sum_{j \in J_2} \alpha(s)(2^{-1}\text{diam}(C_j))^s$$

where

$$J_1 = \{j : C_j \cap A \neq \emptyset\}, \ J_2 = \{j : C_j \cap B \neq \emptyset\}.$$

Since $\text{dist}(A, B) = 2\delta_0$, $J_1 \cap J_2 = \emptyset$ and so this inequality is valid. It follows that

$$\mathcal{H}_\delta^s(A \cup B) + \varepsilon > \mathcal{H}_\delta^s(A) + \mathcal{H}_\delta^s(B).$$

Letting $\delta \to 0$, and noting $\varepsilon > 0$ was arbitrary, yields

$$\mathcal{H}^s(A \cup B) = \mathcal{H}^s(A) + \mathcal{H}^s(B)$$

since $\mathcal{H}^s$ is an outer measure. By Caratheodory's criterion, $\mathcal{H}^s$ is a Borel measure.

To verify the second assertion, note first that there is no loss of generality in letting $\mathcal{H}^s(E) < \infty$. Let

$$E \subseteq \cup_{j=1}^{\infty} C_j, \ \mathrm{diam}(C_j) < \delta,$$

and

$$\mathcal{H}^s_\delta(E) + \delta > \sum_{j=1}^{\infty} \alpha(s)(2^{-1}\mathrm{diam}(C_j))^s.$$

Let

$$F_\delta = \cup_{j=1}^{\infty} \overline{C_j}.$$

Thus $F_\delta \supseteq E$ and

$$\mathcal{H}^s_\delta(E) \ \leq \ \mathcal{H}^s_\delta(F_\delta) \leq \sum_{j=1}^{\infty} \alpha(s)(2^{-1}\mathrm{diam}(\overline{C_j}))^s$$

$$= \ \sum_{j=1}^{\infty} \alpha(s)(2^{-1}(\mathrm{diam}(C_j))^s < \delta + \mathcal{H}^s_\delta(E).$$

Let $\delta_k \to 0$ and let $F = \cap_{k=1}^{\infty} F_{\delta_k}$. Then $F \supseteq E$ and

$$\mathcal{H}^s_{\delta_k}(E) \leq \mathcal{H}^s_{\delta_k}(F) \leq \mathcal{H}^s_{\delta_k}(F_\delta) \leq \delta_k + \mathcal{H}^s_{\delta_k}(E).$$

Letting $k \to \infty$,

$$\mathcal{H}^s(E) \leq \mathcal{H}^s(F) \leq \mathcal{H}^s(E)$$

and this proves the theorem.

A measure satisfying the conclusion of 19.11 is sometimes called a Borel regular measure.

Lemma 19.12 *If n is a positive integer, and if $m_n(A) = 0$, then*

$$\mathcal{H}^n(A) = 0.$$

Proof: Suppose A is a bounded set with $m_n(A) = 0$. Let $V \supseteq A$, V is open, bounded and $m_n(V) < \varepsilon 5^{-n}$. For $\delta > 0$, let $\mathcal{B}$ be a collection of balls of diameter $\leq 5^{-1}\delta$ with $V = \cup \mathcal{B}$. By Vitali's covering theorem in Chapter 14, there is a sequence $\{B_i\}$ of these balls with $B_i \cap B_j = \emptyset$ and

$$V \subseteq \sum_{i=1}^{\infty} m_n(\widehat{B_i})$$

where $\widehat{B_i}$ is the ball with the same center having five times the radius. Then

$$\mathcal{H}_\delta^n(A) \leq \sum_{i=1}^\infty \alpha(n)(2^{-1}\text{diam}(\widehat{B_i}))^n$$

$$\leq \sum_{i=1}^\infty m(\widehat{B_i}) = 5^n \sum_{i=1}^\infty m_n(B_i)$$

$$\leq 5^n m_n(V) < \varepsilon.$$

Since ε is arbitrary, $\mathcal{H}_\delta^n(A) = 0$. Since δ is arbitrary, $\mathcal{H}(A) = 0$. If A is not bounded, apply the above argument to $A_r = A \cap B(0, r)$ to conclude $\mathcal{H}(A_r) = 0$ and so $\mathcal{H}(A) = 0$. This proves the lemma.

Lemma 19.13 *Let U be an open set with $m_n(U) < \infty$ and let $\delta > 0$. Then there exists a sequence of disjoint balls $\{B_i\}$, $B_i \subseteq U$, and $diam(B_i) < \delta$ such that*

$$m_n(U \setminus (\cup_{i=1}^\infty B_i)) = 0.$$

Proof: Using Vitali's theorem, we can get a sequence of disjoint balls $\{B_i\}$ with $\text{diam}(B_i) < \delta$ and $\overline{B_i} \subseteq U$ such that $m_n(U) \leq 5^n \sum_{i=1}^\infty m_n(B_i)$. Then for some number m_1,

$$m_n(U) \leq 10^n \sum_{i=1}^{m_1} m_n(\overline{B_i}).$$

Then

$$m_n(U \setminus (\cup_{i=1}^{m_1} \overline{B_i})) = m_n(U) - \sum_{i=1}^{m_1} m_n(\overline{B_i}) \leq (1 - 10^{-n})m_n(U).$$

Let $1 - 10^{-n} < r < 1$. Then

$$m_n(U \setminus (\cup_{i=1}^{m_1} \overline{B_i})) \leq r m_n(u).$$

Now apply the above argument to $U \setminus (\cup_{i=1}^{m_1} \overline{B_i})$ obtaining

$$m_n(U \setminus (\cup_{i=1}^{m_2} \overline{B_i})) \leq r^2 m_n(U).$$

Continue in this way getting

$$m_n(U \setminus (\cup_{i=1}^{m_k} \overline{B_i})) \leq r^k m_n(U).$$

Then the promised sequence of balls is $\{B_i\}$. We have

$$m_n(U \setminus (\cup_{i=1}^\infty \overline{B_i})) = m_n(U \setminus (\cup_{i=1}^\infty B_i)) = 0.$$

This proves the lemma.

Lemma 19.14 *Let U be a bounded open set. Then $\mathcal{H}^n(U) = m_n(U)$.*

Proof: Let $m_n(U \setminus (\cup_{i=1}^\infty B_i)) = 0$ where $\text{diam}(B_i) \leq \delta$ and

$$B_i \cap B_j = \emptyset, B_i \subseteq U.$$

Then by Lemma 19.12

$$\mathcal{H}_\delta^n(U) = \mathcal{H}_\delta^n(\cup_{i=1}^\infty B_i) \leq \sum_{i=1}^\infty m_n(B_i) = m_n(\cup_{i=1}^\infty B_i) = m_n(U).$$

Since this holds for every $\delta > 0$,

$$\mathcal{H}^n(U) \leq m_n(U).$$

Now to establish the other inequality, let

$$U = \cup_{i=1}^\infty C_i, \ \text{diam}(C_j) \leq \delta,$$

and

$$\mathcal{H}_\delta^n(U) + \varepsilon > \sum_{i=1}^\infty \alpha(n)(2^{-1}\text{diam}(C_i))^n.$$

By the isodiametric inequality, Theorem 19.6

$$\begin{aligned}
\varepsilon + \mathcal{H}_\delta^n(U) &> \sum_{i=1}^\infty \alpha(n)(2^{-1}\text{diam}(C_i))^n \\
&\geq \sum_{i=1}^\infty \alpha(n)(2^{-1}\text{diam}(\overline{C_i}))^n \\
&\geq \sum_{i=1}^\infty m_n(\overline{C_i}) \geq m_n(\cup_{i=1}^\infty \overline{C_i}) \\
&\geq m_n(\cup_{i=1}^\infty C_i) = m_n(U).
\end{aligned}$$

Letting $\delta \to 0$,

$$\varepsilon + \mathcal{H}^n(U) \geq m_n(U).$$

Since ε is arbitrary, this proves the lemma.

Theorem 19.15 $\mathcal{H}^n = \overline{m_n}$ *and so the σ algebra of $\mathcal{H}^n$ measurable sets coincides with the σ algebra of Lebesgue measurable sets and the two measures are equal on this σ algebra.*

Proof: Let $\prod_{i=1}^n [a_i, b_i)$ be a half open rectangle. Then we just showed that

$$\mathcal{H}^n(\prod_{i=1}^n (a_i - k^{-1}, b_i)) = m_n(\prod_{i=1}^n (a_i - k^{-1}, b_i)).$$

Letting $k \to \infty$, we see that $\mathcal{H}^n(R) = m_n(R)$ for every half open rectangle R of the form $\prod_{i=1}^n [a_i, b_i)$. By Lemma 10.3, this shows $m_n = \mathcal{H}^n$ on every open set, bounded or not. Let $\mathcal{A}$ be the algebra of sets consisting of $\emptyset, \mathbb{R}^n$, and all finite disjoint unions of half open rectangles. Then we just showed that $m_n = \mathcal{H}^n$ on this algebra. Let $R_k = \prod_{k=1}^n [-k, k)$ and define

$$\mathcal{M} = \{E \text{ Borel such that } m_n(R_k \cap E) = \mathcal{H}^n(R_k \cap E)\}.$$

Then $\mathcal{M}$ contains $\mathcal{A}$ and $\mathcal{M}$ is a monotone class. By the monotone class theorem $\mathcal{M} \supseteq \sigma(\mathcal{A})$, the smallest σ algebra containing $\mathcal{A}$. But $\sigma(\mathcal{A})$ contains the open sets by Lemma 10.3 and so $\sigma(\mathcal{A}) = $ Borel sets. Thus,

$$m_n(R_k \cap E) = \mathcal{H}^n(R_k \cap E)$$

for each E Borel. Letting $k \to \infty$, we see that $m_n = \mathcal{H}^n$ on the Borel sets.

Now let $E \subseteq \mathbb{R}^n$ be arbitrary. By Theorem 19.11, there exists a Borel set $F_1 \supseteq E$ with

$$\mathcal{H}^n(F_1) = \mathcal{H}^n(E).$$

Also there exists a Borel set $F_2 \supseteq E$ such that

$$\overline{m_n}(F_2) = m_n(F_2) = \overline{m_n}(E).$$

Let $F = F_1 \cap F_2$. Then

$$\mathcal{H}^n(E) = \mathcal{H}^n(F) = m_n(F) = \overline{m_n}(E).$$

Thus $\overline{m_n} = \mathcal{H}^n$ on $\mathcal{P}(\mathbb{R}^n)$. By Lemma 19.1 $m_n = \mathcal{H}^n$ and the σ algebra of $\overline{m_n}$ measurable sets equals that of $\mathcal{H}^n$ measurable sets which equals the σ algebra of Lebesgue measurable sets. This proves the theorem.

This gives another way to think of Lebesgue measure which is a particularly nice way because it is coordinate free, depending only on the notion of distance.

For $s < n$, note that $\mathcal{H}^n$ is not a Radon measure because it will not generally be finite on compact sets. For example, let $n = 2$ and consider $\mathcal{H}^1(L)$ where L is a line segment joining $(0,0)$ to $(1,0)$. Then $\mathcal{H}^1(L)$ is no smaller than $\mathcal{H}^1(L)$ when L is considered a subset of $\mathbb{R}^1, n = 1$. Thus by what was just shown, $\mathcal{H}^1(L) \geq 1$. Hence $\mathcal{H}^1([0,1] \times [0,1]) = \infty$. The situation is this: L is a one-dimensional object inside $\mathbb{R}^2$ and $\mathcal{H}^1$ is giving a one-dimensional measure of this object. In fact, we can use these Hausdorff measures to make such heuristic remarks as these precise. We define the Hausdorff dimension of a set, A, as

$$\dim(A) = \inf\{s : \mathcal{H}^s(A) = 0\}$$

We can also use Hausdorff measure to give a unified discussion of the notion of change of variables theorems for Lipschitz maps and of surface measure. To do so, we give some introductory results here. Recall that for $L \in \mathcal{L}\left(\mathbb{R}^k, \mathbb{R}^l\right), L^*$ is defined by

$$(L\mathbf{u}, \mathbf{v}) = (\mathbf{u}, L^*\mathbf{v}).$$

Also recall Theorem 10.10 of Chapter 10. We state it here for convenience. We say $U \in \mathcal{L}\left(\mathbb{R}^n, \mathbb{R}^n\right)$ is symmetric if $U = U^*$ and for such U, we say it is nonnegative if all its eigenvalues are nonnegative or equivalently, if for all $\mathbf{x} \in \mathbb{R}^n$,

$$(U\mathbf{x}, \mathbf{x}) \geq 0.$$

Lemma 19.16 *Let $F \in \mathcal{L}(\mathbb{R}^n, \mathbb{R}^m)$ where $n \leq m$. Then there exists a symmetric nonnegative element of $\mathcal{L}(\mathbb{R}^n, \mathbb{R}^n), U$, and an element of $\mathcal{L}(\mathbb{R}^n, \mathbb{R}^m), R$, such that*

$$F = RU, \ (U\mathbf{x}, \mathbf{x}) \geq 0, \ U = U^*, \ R^*R = I, U^2 = F^*F,$$

and R preserves distances,

$$|R\mathbf{u}| = |\mathbf{u}|.$$

Lemma 19.17 *Let $R \in \mathcal{L}(\mathbb{R}^n, \mathbb{R}^m), \ n \leq m$, and $R^*R = I$. Then if $A \subseteq \mathbb{R}^n$,*

$$\mathcal{H}^n(RA) = \mathcal{H}^n(A).$$

Proof: Note that $|R(\mathbf{x} - \mathbf{y})| = |\mathbf{x} - \mathbf{y}|$. Now suppose A is bounded, $R(A) \subseteq \cup_{j=1}^\infty C_j$, $\operatorname{diam}(C_j) \leq \delta$, and

$$\mathcal{H}_\delta^n(RA) + \varepsilon > \sum_{j=1}^\infty \alpha(n)(2^{-1}(\operatorname{diam}(C_j)))^n.$$

Replacing each C_j with $C_j \cap (RA)$,

$$\mathcal{H}_\delta^n(RA) + \varepsilon \ > \ \sum_{j=1}^\infty \alpha(n)(2^{-1}(\operatorname{diam}(C_j \cap (RA))))^n$$

$$\geq \ \sum_{j=1}^\infty \alpha(n) \left(2^{-1}\left(\operatorname{diam}\left(R^{-1}\left(C_j \cap (RA)\right)\right)\right)\right)^n$$

$$\geq \ \mathcal{H}_\delta^n(A).$$

Thus $\mathcal{H}_\delta^n(RA) \geq \mathcal{H}_\delta^n(A)$. Now let $A \subseteq \cup_{j=1}^\infty C_j$, $\operatorname{diam}(C_j) \leq \delta$, and

$$\mathcal{H}_\delta^n(A) + \varepsilon \geq \sum_{j=1}^n \alpha(n) \left(2^{-1}\left(\operatorname{diam}\left(C_j\right)\right)\right)^n$$

Then

$$\mathcal{H}_\delta^n(A) + \varepsilon \ \geq \ \sum_{j=1}^n \alpha(n) \left(2^{-1}\left(\mathrm{diam}\left(C_j\right)\right)\right)^n$$

$$= \ \sum_{j=1}^n \alpha(n) \left(2^{-1}\left(\mathrm{diam}\left(RC_j\right)\right)\right)^n$$

$$\geq \ \mathcal{H}_\delta^n(RA).$$

Hence $\mathcal{H}_\delta^n(RA) = \mathcal{H}_\delta^n(A)$. Letting $\delta \to 0$ yields the desired conclusion in the case where A is bounded. For the general case, let $A_r = A \cap B\left(0, r\right)$. Then $\mathcal{H}^n(RA_r) = \mathcal{H}^n(A_r)$. Now let $r \to \infty$. This proves the lemma.

Lemma 19.18 *Let* $F \in \mathcal{L}(\mathbb{R}^n, \mathbb{R}^m), n \leq m,$ *and let* $F = RU$ *where* R *and* U *are described in Lemma 19.16. Then if* $A \subseteq \mathbb{R}^n$ *is Lebesgue measurable,*

$$\mathcal{H}^n(FA) = \det(U)m_n(A).$$

Proof:

$$\mathcal{H}^n(FA) = \mathcal{H}^n(RUA)$$

$$= \mathcal{H}^n(UA) = m_n(UA) = \det(U)m_n(A).$$

Definition 19.19 We define J to equal $\det(U)$. Thus

$$J = \det((F^*F)^{1/2}) = (\det(F^*F))^{1/2}.$$

Chapter 20
The Area Formula

20.1 Lipschitz mappings

This chapter is on a generalization of the change of variables formula found in Chapter 14. In this section, Ω will be a Lebesgue measurable set in $\mathbb{R}^n$ and $\mathbf{h} : \Omega \to \mathbb{R}^m$ will be Lipschitz. In the case of the earlier change of variables formula, the mapping was from a subset of $\mathbb{R}^n$, U, to a subset of $\mathbb{R}^n$ and Lebesgue measure was involved in both U and $\mathbf{h}(U)$. Here m can be larger than n and instead of Lebesgue measure on $\mathbf{h}(U)$, we use Hausdorff measure. Also, $\mathbf{h}$ is only assumed to be Lipschitz, not C^1. It turns out all the earlier results have a pleasing generalization in this setting. This generalization also makes possible the consideration of surface measure and the divergence theorem. First, we give a simple extension property of Lipschitz maps.

Theorem 20.1 *If* $\mathbf{h} : \Omega \to \mathbb{R}^m$ *is Lipschitz, then there exists* $\bar{\mathbf{h}} : \mathbb{R}^n \to \mathbb{R}^m$ *which extends* $\mathbf{h}$ *and is also Lipschitz.*

Proof: It suffices to assume $m = 1$ because if this is shown, it may be applied to the components of $\mathbf{h}$ to get the desired result. Suppose

$$|h(\mathbf{x}) - h(\mathbf{y})| \leq K |\mathbf{x} - \mathbf{y}|. \tag{1}$$

Define

$$\bar{h}(\mathbf{x}) \equiv \inf\{h(\mathbf{w}) + K |\mathbf{x} - \mathbf{w}| : \mathbf{w} \in \Omega\}. \tag{2}$$

If $\mathbf{x} \in \Omega$, then for all $\mathbf{w} \in \Omega$,

$$h(\mathbf{w}) + K |\mathbf{x} - \mathbf{w}| \geq h(\mathbf{x})$$

by Formula 1. This shows $h(\mathbf{x}) \leq \bar{h}(\mathbf{x})$. But also we can take $\mathbf{w} = \mathbf{x}$ in Formula 2 which yields $\bar{h}(\mathbf{x}) \leq h(\mathbf{x})$. Therefore $\bar{h}(\mathbf{x}) = h(\mathbf{x})$ if $\mathbf{x} \in \Omega$.

Now suppose $\mathbf{x}, \mathbf{y} \in \mathbb{R}^n$ and consider $|\bar{h}(\mathbf{x}) - \bar{h}(\mathbf{y})|$. Without loss of generality we may assume $\bar{h}(\mathbf{x}) \geq \bar{h}(\mathbf{y})$. (If not, repeat the following argument with $\mathbf{x}$ and $\mathbf{y}$ interchanged.) Pick $\mathbf{w} \in \Omega$ such that

$$h(\mathbf{w}) + K |\mathbf{y} - \mathbf{w}| - \epsilon < \bar{h}(\mathbf{y}).$$

Then

$$|\bar{h}(\mathbf{x}) - \bar{h}(\mathbf{y})| = \bar{h}(\mathbf{x}) - \bar{h}(\mathbf{y}) \leq h(\mathbf{w}) + K |\mathbf{x} - \mathbf{w}| -$$

$$[h(\mathbf{w}) + K|\mathbf{y} - \mathbf{w}| - \epsilon] \leq K|\mathbf{x} - \mathbf{y}| + \epsilon.$$

Since ϵ is arbitrary,

$$|\bar{h}(\mathbf{x}) - \bar{h}(\mathbf{y})| \leq K|\mathbf{x} - \mathbf{y}|$$

and this proves the theorem.

We will use $\bar{\mathbf{h}}$ to denote a Lipschitz extension of the Lipschitz function $\mathbf{h}$. The next lemma is an application of the Vitali covering theorem. It states that every open set can be filled with disjoint balls except for a set of measure zero.

Lemma 20.2 *Let V be an open set in $\mathbb{R}^r, m_r(V) < \infty$. Then there exists a sequence of disjoint open balls $\{B_i\}$ having radii less than δ and a set of measure 0, T, such that*

$$V = (\cup_{i=1}^{\infty} B_i) \cup T.$$

Proof: See Lemma 19.13 of Chapter 19. This also is Problem 6 in Chapter 14.

We wish to show that $\mathbf{h}$ maps Lebesgue measurable sets to $\mathcal{H}^n$ measurable sets. In showing this the key result is the next lemma which states that $\mathbf{h}$ maps sets of measure zero to sets of measure zero.

Lemma 20.3 *If $m_n(T) = 0$ then $\mathcal{H}^n(\bar{\mathbf{h}}(T)) = 0$.*

Proof: Let V be an open set containing T whose measure is less than ϵ. Now using the Vitali covering theorem, there exists a sequence of disjoint balls $\{B_i\}$, $B_i = B(\mathbf{x}_i, r_i)$, whose radii are less than δ which are contained in V such that the sequence of enlarged balls, $\{\widehat{B}_i\}$, having the same center but 5 times the radius, covers T. Then

$$\mathcal{H}^n_{5Lip(\bar{\mathbf{h}})\delta}(\bar{\mathbf{h}}(T)) \leq \mathcal{H}^n_{5Lip(\bar{\mathbf{h}})\delta}\left(\bar{\mathbf{h}}\left(\cup_{i=1}^{\infty}\widehat{B}_i\right)\right)$$

$$\leq \sum_{i=1}^{\infty} \mathcal{H}^n_{5Lip(\bar{\mathbf{h}})\delta}\left(\bar{\mathbf{h}}\left(\widehat{B}_i\right)\right)$$

$$\leq \sum_{i=1}^{\infty} \alpha(n)(Lip(\bar{\mathbf{h}}))^n 5^n r_i^n = 5^n(Lip(\bar{\mathbf{h}}))^n \sum_{i=1}^{\infty} m_n(B_i)$$

$$\leq (Lip(\bar{\mathbf{h}}))^n 5^n m_n(V) \leq \epsilon(Lip(\bar{\mathbf{h}}))^n 5^n.$$

Since we can do this for each δ, let $\delta \to 0$ to conclude

$$\mathcal{H}^n(\bar{\mathbf{h}}(T)) \leq \epsilon(Lip(\bar{\mathbf{h}}))^n 5^n.$$

Since ϵ is arbitrary, this proves the lemma.

Actually, the argument in this lemma holds in other contexts which do not imply $\mathbf{h}$ is Lipschitz continuous. For one such example, see Problem 13.

With the conclusion of this lemma, the next lemma is fairly easy to obtain.

Lemma 20.4 *If A is Lebesgue measurable, then $\bar{\mathbf{h}}(A)$ is $\mathcal{H}^n$ measurable. Furthermore,*

$$\mathcal{H}^n\left(\bar{\mathbf{h}}(A)\right) \leq \left(Lip\left(\bar{\mathbf{h}}\right)\right)^n m_n(A). \tag{3}$$

Proof: Let $A_k = A \cap B(0,k)$, $k \in \mathbb{N}$. We establish Formula 3 for A_k in place of A and then let $k \to \infty$ to obtain Formula 3. Let $V \supseteq A_k$ and let $m_n(V) < \infty$. By Lemma 20.2, there is a sequence of disjoint balls $\{B_i\}$, of radius less than δ and a set of measure 0, T, such that

$$V = \cup_{i=1}^\infty B_i \cup T, \ B_i = B(x_i, r_i).$$

Let $b(\delta) \equiv Lip\left(\bar{\mathbf{h}}\right)\delta$. Then by Lemma 20.3,

$$\mathcal{H}^n_{b(\delta)}\left(\bar{\mathbf{h}}(A_k)\right) \leq \mathcal{H}^n_{b(\delta)}\left(\bar{\mathbf{h}}(V)\right)$$

$$\leq \mathcal{H}^n_{b(\delta)}\left(\bar{\mathbf{h}}(\cup_{i=1}^\infty B_i)\right) + \mathcal{H}^n_{b(\delta)}\left(\bar{\mathbf{h}}(T)\right) = \mathcal{H}^n_{b(\delta)}\left(\bar{\mathbf{h}}(\cup_{i=1}^\infty B_i)\right)$$

$$\leq \sum_{i=1}^\infty \mathcal{H}^n_{b(\delta)}\left(\bar{\mathbf{h}}(B_i)\right) \leq \sum_{i=1}^\infty \mathcal{H}^n_{b(\delta)}\left(B\left(\bar{\mathbf{h}}(x_i), Lip\left(\bar{\mathbf{h}}\right)r_i\right)\right)$$

$$\leq \sum_{i=1}^\infty \alpha(n)\left(Lip\left(\bar{\mathbf{h}}\right)r_i\right)^n = Lip\left(\bar{\mathbf{h}}\right)^n \sum_{i=1}^\infty m_n(B_i) = Lip\left(\bar{\mathbf{h}}\right)^n m_n(V).$$

Letting $\delta \to 0$,

$$\mathcal{H}^n\left(\bar{\mathbf{h}}(A_k)\right) \leq Lip\left(\bar{\mathbf{h}}\right)^n m_n(V).$$

Since V is an arbitrary open set containing A_k, it follows from regularity of Lebesgue measure that

$$\mathcal{H}^n\left(\bar{\mathbf{h}}(A_k)\right) \leq Lip\left(\bar{\mathbf{h}}\right)^n m_n(A_k). \tag{4}$$

Now let $k \to \infty$ to obtain Formula 3. This proves the formula. It remains to show $\bar{\mathbf{h}}(A)$ is $\mathcal{H}^n$ measurable.

By inner regularity of Lebesgue measure, there exists a set, F, which is the countable union of compact sets and a set T with $m_n(T) = 0$ such that

$$F \cup T = A_k.$$

Then $\bar{\mathbf{h}}(F) \subseteq \bar{\mathbf{h}}(A_k) \subseteq \bar{\mathbf{h}}(F) \cup \bar{\mathbf{h}}(T)$. By continuity of $\bar{\mathbf{h}}$, $\bar{\mathbf{h}}(F)$ is a countable union of compact sets and so it is Borel. By Formula 4 with T in place of A_k,

$$\mathcal{H}^n\left(\bar{\mathbf{h}}(T)\right) = 0$$

and so $\bar{\mathbf{h}}(T)$ is $\mathcal{H}^n$ measurable. Therefore, $\bar{\mathbf{h}}(A_k)$ is $\mathcal{H}^n$ measurable because $\mathcal{H}^n$ is a complete measure and we have exhibited $\bar{\mathbf{h}}(A_k)$ between two $\mathcal{H}^n$ measurable sets whose difference has measure 0. Now

$$\bar{\mathbf{h}}(A) = \cup_{k=1}^{\infty} \bar{\mathbf{h}}(A_k)$$

so $\bar{\mathbf{h}}(A)$ is also $\mathcal{H}^n$ measurable and this proves the lemma.

The following lemma, found in Rudin [47], is interesting for its own sake and will serve as the basis for many of the theorems and lemmas which follow. Its proof is based on the Brouwer fixed point theorem, a short proof of which is given in the on the appendix devoted to this theorem. The idea is that if a continuous function mapping a ball in $\mathbb{R}^k$ to $\mathbb{R}^k$ doesn't move any point very much, then the image of the ball must contain a slightly smaller ball.

Lemma 20.5 Let $B = B(0, r)$, a ball in $\mathbb{R}^k$ and let $\mathbf{F} : \overline{B} \to \mathbb{R}^k$ be continuous and suppose for some $\epsilon < 1$,

$$|\mathbf{F}(\mathbf{v}) - \mathbf{v}| < \epsilon r$$

for all $\mathbf{v} \in \overline{B}$. Then

$$\mathbf{F}\left(\overline{B}\right) \supseteq B(0, r(1-\epsilon)).$$

Proof: Suppose $\mathbf{a} \in B(0, r(1-\epsilon)) \setminus \mathbf{F}\left(\overline{B}\right)$ and let

$$\mathbf{G}(\mathbf{v}) \equiv \frac{r(\mathbf{a} - \mathbf{F}(\mathbf{v}))}{|\mathbf{a} - \mathbf{F}(\mathbf{v})|}.$$

If $|\mathbf{v}| = r$,

$$\mathbf{v} \cdot (\mathbf{a} - \mathbf{F}(\mathbf{v})) = \mathbf{v} \cdot \mathbf{a} - \mathbf{v} \cdot \mathbf{F}(\mathbf{v})$$

$$= \mathbf{v} \cdot \mathbf{a} - \mathbf{v} \cdot (\mathbf{F}(\mathbf{v}) - \mathbf{v}) - r^2$$

$$< r^2(1-\epsilon) + \epsilon r^2 - r^2 = 0.$$

Then for $|\mathbf{v}| = r$, $\mathbf{G}(\mathbf{v}) \neq \mathbf{v}$ because we just showed that $\mathbf{v} \cdot \mathbf{G}(\mathbf{v}) < 0$ but $\mathbf{v} \cdot \mathbf{v} = r^2 > 0$. If $|\mathbf{v}| < r$, it follows that $\mathbf{G}(\mathbf{v}) \neq \mathbf{v}$ because $|\mathbf{G}(\mathbf{v})| = r$ but $|\mathbf{v}| < r$. This lack of a fixed point contradicts the Brouwer fixed point theorem and this proves the lemma.

We are interested in generalizing the change of variables formula. Since $\mathbf{h}$ is only Lipschitz, $D\mathbf{h}(\mathbf{x})$ may not exist for all $\mathbf{x}$ but from the theorem of Rademacher in Chapter 18, $D\bar{\mathbf{h}}(\mathbf{x})$ exists a.e. $\mathbf{x}$. Also, thanks to Lemma 19.16 of Chapter 19, when $D\bar{\mathbf{h}}(\mathbf{x})$ exists,

$$D\bar{\mathbf{h}}(\mathbf{x}) = R(\mathbf{x})U(\mathbf{x})$$

where $(U(\mathbf{x})\mathbf{u}, \mathbf{v}) = (U(\mathbf{x})\mathbf{v}, \mathbf{u})$, $(U(\mathbf{x})\mathbf{u}, \mathbf{u}) \geq 0$ and $R^*R = I$.

Lemma 20.6 *In this situation, $|R^*\mathbf{u}| \leq |\mathbf{u}|$.*

Proof: First note that

$$
\begin{aligned}
(\mathbf{u} - RR^*\mathbf{u}, RR^*\mathbf{u}) &= (\mathbf{u}, RR^*\mathbf{u}) - |RR^*\mathbf{u}|^2 - \\
&= |R^*\mathbf{u}|^2 - |R^*\mathbf{u}|^2 = 0,
\end{aligned}
$$

and so

$$
\begin{aligned}
|\mathbf{u}|^2 &= |\mathbf{u} - RR^*\mathbf{u} + RR^*\mathbf{u}|^2 \\
&= |\mathbf{u} - RR^*\mathbf{u}|^2 + |RR^*\mathbf{u}|^2 \\
&= |\mathbf{u} - RR^*\mathbf{u}|^2 + |R^*\mathbf{u}|^2.
\end{aligned}
$$

This proves the lemma.

Lemma 20.7 *Let $T \subseteq \mathbb{R}^m$. Then*

$$\mathcal{H}^n(T) \geq \mathcal{H}^n(RR^*T) = \mathcal{H}^n(R^*T).$$

Proof: The equality holds because R preserves distances. Suppose without loss of generality, that $\mathcal{H}^n(T) < \infty$ and let $\{C_i\}$ cover T, with $diam(C_i) \leq \delta$, and

$$\mathcal{H}^n_\delta(T) > \sum_{i=1}^n \alpha(n)\left(\frac{diam(C_i)}{2}\right)^n - \epsilon.$$

Then since R^* decreases lengths, and $\{RR^*C_i\}$ covers RR^*T,

$$\mathcal{H}^n_\delta(T) > \sum_{i=1}^n \alpha(n)\left(\frac{diam(C_i)}{2}\right)^n - \epsilon$$

$$\geq \sum_{i=1}^n \alpha(n)\left(\frac{diam(RR^*C_i)}{2}\right)^n - \epsilon \geq \mathcal{H}^n_\delta(RR^*T) - \epsilon.$$

Since ϵ is arbitrary, this shows $\mathcal{H}^n_\delta(T) \geq \mathcal{H}^n_\delta(RR^*T)$. Now letting $\delta \to 0$ yields the conclusion of the lemma.

This lemma has the following obvious corollary.

Corollary 20.8 *If P is a mapping which decreases distances, then*

$$\mathcal{H}^n(T) \geq \mathcal{H}^n(PT).$$

The arguments which follow are motivated by the earlier approach in which the classical change of variables formula was presented. We will define a measure and use the Radon Nikodym theorem to obtain a function which is of interest to us. Then we will identify this function. In order to do this, we need some technical lemmas first.

Lemma 20.9 *Let $\mathbf{x} \in \Omega$ be a point where $D\bar{\mathbf{h}}(\mathbf{x})$ exists and $U(\mathbf{x})^{-1}$ exists. Then if $\epsilon \in (0, 1)$ the following hold for all r small enough.*

$$R^*(\mathbf{x})\,\bar{\mathbf{h}}\left(\overline{B(\mathbf{x}, r)}\right) \supseteq R^*(\mathbf{x})\,\bar{\mathbf{h}}(\mathbf{x}) + U(\mathbf{x})\,B(0, r(1 - \epsilon)), \qquad (5)$$

$$\mathcal{H}^n\left(\bar{\mathbf{h}}\left(\overline{B(\mathbf{x}, r)}\right)\right) =$$

$$\mathcal{H}^n\left(\bar{\mathbf{h}}(B(\mathbf{x}, r))\right) \geq m_n\left(U(\mathbf{x})\,B(0, r(1 - \epsilon))\right), \qquad (6)$$

$$\bar{\mathbf{h}}(B(\mathbf{x}, r)) \subseteq \bar{\mathbf{h}}(\mathbf{x}) + R(\mathbf{x})\,U(\mathbf{x})\,B(0, r(1 + \epsilon)), \qquad (7)$$

$$\mathcal{H}^n\left(\bar{\mathbf{h}}(B(\mathbf{x}, r))\right) \leq m_n\left(U(\mathbf{x})\,B(0, r(1 + \epsilon))\right). \qquad (8)$$

If $\mathbf{x}$ is also a point of density of Ω, then

$$\lim_{r \to 0} \frac{\mathcal{H}^n\left(\bar{\mathbf{h}}(B(\mathbf{x}, r) \cap \Omega)\right)}{\mathcal{H}^n\left(\bar{\mathbf{h}}(B(\mathbf{x}, r))\right)} = 1. \qquad (9)$$

Proof: Since $D\bar{\mathbf{h}}(\mathbf{x})$ exists,

$$\bar{\mathbf{h}}(\mathbf{x} + \mathbf{v}) = \bar{\mathbf{h}}(\mathbf{x}) + D\bar{\mathbf{h}}(\mathbf{x})\,\mathbf{v} + o(|\mathbf{v}|). \qquad (10)$$

Consequently, when r is small enough, Formula 7 holds. Using the fact that $R(\mathbf{x})$ preserves all distances, and Theorem 19.15 of Chapter 19,

$$\mathcal{H}^n\left(\bar{\mathbf{h}}(B(\mathbf{x}, r))\right) \leq \mathcal{H}^n\left(R(\mathbf{x})\,U(\mathbf{x})\,B(0, r(1 + \epsilon))\right)$$

$$= \mathcal{H}^n\left(U(\mathbf{x})\,B(0, r(1 + \epsilon))\right) = m_n\left(U(\mathbf{x})\,B(0, r(1 + \epsilon))\right)$$

which shows Formula 8. From Formula 10,

$$R^*(\mathbf{x})\,\bar{\mathbf{h}}(\mathbf{x} + \mathbf{v}) = R^*(\mathbf{x})\,\bar{\mathbf{h}}(\mathbf{x}) + U(\mathbf{x})\,(\mathbf{v} + o(|\mathbf{v}|)).$$

Thus, from the assumption that $U(\mathbf{x})^{-1}$ exists,

$$U(\mathbf{x})^{-1}\,R^*(\mathbf{x})\,\bar{\mathbf{h}}(\mathbf{x} + \mathbf{v}) - U(\mathbf{x})^{-1}\,R^*(\mathbf{x})\,\bar{\mathbf{h}}(\mathbf{x}) - \mathbf{v} = o(|\mathbf{v}|). \qquad (11)$$

Letting

$$\mathbf{F}(\mathbf{v}) = U(\mathbf{x})^{-1}\,R^*(\mathbf{x})\,\bar{\mathbf{h}}(\mathbf{x} + \mathbf{v}) - U(\mathbf{x})^{-1}\,R^*(\mathbf{x})\,\bar{\mathbf{h}}(\mathbf{x}),$$

we can apply Lemma 20.5 in 11 to conclude that for r small enough,

$$U\left(\mathbf{x}\right)^{-1} R^{*}\left(\mathbf{x}\right) \bar{\mathbf{h}}\left(\mathbf{x}+\overline{B\left(0,r\right)}\right) - U\left(\mathbf{x}\right)^{-1} R^{*}\left(\mathbf{x}\right) \bar{\mathbf{h}}\left(\mathbf{x}\right) \supseteq B\left(0,\left(1-\epsilon\right)r\right).$$

Therefore,

$$R^{*}\left(\mathbf{x}\right) \bar{\mathbf{h}}\left(\overline{B\left(\mathbf{x},r\right)}\right) \supseteq R^{*}\left(\mathbf{x}\right) \bar{\mathbf{h}}\left(\mathbf{x}\right) + U\left(\mathbf{x}\right) B\left(0,\left(1-\epsilon\right)r\right)$$

which proves Formula 5. Therefore,

$$R\left(\mathbf{x}\right) R^{*}\left(\mathbf{x}\right) \bar{\mathbf{h}}\left(\overline{B\left(\mathbf{x},r\right)}\right) \supseteq$$

$$R\left(\mathbf{x}\right) R^{*}\left(\mathbf{x}\right) \bar{\mathbf{h}}\left(\mathbf{x}\right) + R\left(\mathbf{x}\right) U\left(\mathbf{x}\right) B\left(0,r\left(1-\epsilon\right)\right).$$

From Lemma 20.7, this implies

$$\mathcal{H}^{n}\left(\bar{\mathbf{h}}\left(\overline{B\left(\mathbf{x},r\right)}\right)\right) \geq \mathcal{H}^{n}\left(R\left(\mathbf{x}\right) R^{*}\left(\mathbf{x}\right) \bar{\mathbf{h}}\left(\overline{B\left(\mathbf{x},r\right)}\right)\right)$$

$$\geq \mathcal{H}^{n}\left(R\left(\mathbf{x}\right) U\left(\mathbf{x}\right) B\left(0,r\left(1-\epsilon\right)\right)\right)$$

$$= \mathcal{H}^{n}\left(U\left(\mathbf{x}\right) B\left(0,r\left(1-\epsilon\right)\right)\right) = m_{n}\left(U\left(\mathbf{x}\right) B\left(0,r\left(1-\epsilon\right)\right)\right)$$

which shows Formula 6.

Now suppose that $\mathbf{x}$ is also a point of density of Ω. Then whenever r is small enough,

$$m_{n}\left(B\left(\mathbf{x},r\right) \setminus \Omega\right) < \epsilon \alpha\left(n\right) r^{n}. \tag{12}$$

Then for such r we use Lemma 20.4, Formula 12, and Formula 5 to write

$$1 \geq \frac{\mathcal{H}^{n}\left(\bar{\mathbf{h}}\left(B\left(\mathbf{x},r\right) \cap \Omega\right)\right)}{\mathcal{H}^{n}\left(\bar{\mathbf{h}}\left(B\left(\mathbf{x},r\right)\right)\right)}$$

$$\geq \frac{\mathcal{H}^{n}\left(\bar{\mathbf{h}}\left(B\left(\mathbf{x},r\right)\right)\right) - \mathcal{H}^{n}\left(\bar{\mathbf{h}}\left(B\left(\mathbf{x},r\right) \setminus \Omega\right)\right)}{\mathcal{H}^{n}\left(\bar{\mathbf{h}}\left(B\left(\mathbf{x},r\right)\right)\right)}.$$

From Lemma 20.4, and Formula 6, this is no larger than

$$1 - \frac{Lip\left(\bar{\mathbf{h}}\right)^{n} \epsilon \alpha\left(n\right) r^{n}}{m_{n}\left(U\left(\mathbf{x}\right) B\left(0,r\left(1-\epsilon\right)\right)\right)}.$$

By the theorem on the change of variables for a linear map, this expression equals

$$1 - \frac{Lip\left(\bar{\mathbf{h}}\right)^{n} \epsilon \alpha\left(n\right) r^{n}}{J\left(\mathbf{x}\right) r^{n} \alpha\left(n\right)\left(1-\epsilon\right)^{n}} \equiv 1 - g\left(\epsilon\right)$$

where $\lim_{\epsilon \to 0} g\left(\epsilon\right) = 0$. Here $J\left(\mathbf{x}\right) \equiv \det U\left(\mathbf{x}\right)$ as in Chapter 19. Then for all r small enough,

$$1 \geq \frac{\mathcal{H}^{n}\left(\bar{\mathbf{h}}\left(B\left(\mathbf{x},r\right) \cap \Omega\right)\right)}{\mathcal{H}^{n}\left(\bar{\mathbf{h}}\left(B\left(\mathbf{x},r\right)\right)\right)} \geq 1 - g\left(\epsilon\right)$$

which proves the lemma since ϵ is arbitrary.

Theorem 20.10 *Let* $N \equiv \{\mathbf{x} \in \Omega : D\bar{\mathbf{h}}(\mathbf{x}) \text{ does not exist }\}$. *Then* N *has measure zero and if* $\mathbf{x} \notin N$ *then*

$$J(\mathbf{x}) = \lim_{r \to 0} \frac{\mathcal{H}^n \left(\bar{\mathbf{h}}\left(B\left(\mathbf{x},r\right)\right)\right)}{m_n \left(B\left(\mathbf{x},r\right)\right)}, \tag{13}$$

where $J(\mathbf{x}) \equiv \det\left(U\left(\mathbf{x}\right)\right) = \det\left(D\bar{\mathbf{h}}\left(\mathbf{x}\right)^* D\bar{\mathbf{h}}\left(\mathbf{x}\right)\right)^{1/2}$.

Proof: Suppose first that $U(\mathbf{x})^{-1}$ exists. Using Formula 6, Formula 8 and the change of variables formula for linear maps,

$$
\begin{aligned}
J(\mathbf{x})(1-\epsilon)^n &= \frac{m_n\left(U\left(\mathbf{x}\right)B\left(0, r\left(1-\epsilon\right)\right)\right)}{m_n\left(B\left(\mathbf{x},r\right)\right)} \leq \frac{\mathcal{H}^n\left(\bar{\mathbf{h}}(B\left(\mathbf{x},r\right))\right)}{m_n\left(B\left(\mathbf{x},r\right)\right)} \\
&\leq \frac{m_n\left(U\left(\mathbf{x}\right)B\left(0, r\left(1+\epsilon\right)\right)\right)}{m_n\left(B\left(\mathbf{x},r\right)\right)} = J(\mathbf{x})(1+\epsilon)^n
\end{aligned}
$$

whenever r is small enough. It follows that since $\epsilon > 0$ is arbitrary, Formula 13 holds.

Now suppose $U(\mathbf{x})^{-1}$ does not exist. We just showed that the conclusion of the theorem holds when $J(\mathbf{x}) \neq 0$. We will apply this that was just shown to a modified function. Let

$$\mathbf{k} : \mathbb{R}^n \to \mathbb{R}^m \times \mathbb{R}^n$$

be defined as

$$\mathbf{k}(\mathbf{x}) \equiv \left(\begin{array}{c} \bar{\mathbf{h}}(\mathbf{x}) \\ \epsilon \mathbf{x} \end{array}\right).$$

Then

$$D\mathbf{k}(\mathbf{x})^* D\mathbf{k}(\mathbf{x}) = D\bar{\mathbf{h}}(\mathbf{x})^* D\bar{\mathbf{h}}(\mathbf{x}) + \epsilon^2 I_n$$

and so

$$
\begin{aligned}
J\mathbf{k}(\mathbf{x})^2 &\equiv \det\left(D\bar{\mathbf{h}}(\mathbf{x})^* D\bar{\mathbf{h}}(\mathbf{x}) + \epsilon^2 I_n\right) \\
&= \det\left(Q^* D Q + \epsilon^2 I_n\right)
\end{aligned}
$$

where D is a diagonal matrix having the squares of the eigenvalues of $D\bar{\mathbf{h}}(\mathbf{x})^* D\bar{\mathbf{h}}(\mathbf{x})$ down the main diagonal. Thus, since one of these eigenvalues equals 0,

$$0 < J\mathbf{k}(\mathbf{x})^2 = \prod_{i=1}^{n} \left(\lambda_i^2 + \epsilon^2\right) \leq C^2 \epsilon^2. \tag{14}$$

Therefore, what was just shown applies to $\mathbf{k}$.

If
$$T \equiv \left\{ \left(\overline{\mathbf{h}}\left(\mathbf{w}\right), \mathbf{0}\right)^T : \mathbf{w} \in B\left(\mathbf{x},r\right) \right\},$$

$$T_\epsilon \equiv \left\{ \left(\overline{\mathbf{h}}\left(\mathbf{w}\right), \epsilon\mathbf{w}\right)^T : \mathbf{w} \in B\left(\mathbf{x},r\right) \right\} \equiv \mathbf{k}\left(B\left(\mathbf{x},r\right)\right),$$

then
$$T = PT_\epsilon$$

where P is the projection map defined by

$$P\left(\begin{matrix} \mathbf{x} \\ \mathbf{y} \end{matrix} \right) \equiv \left(\begin{matrix} \mathbf{x} \\ 0 \end{matrix} \right).$$

Since P decreases distances,

$$\mathcal{H}^n\left(\overline{\mathbf{h}}\left(B\left(\mathbf{x},r\right)\right)\right) = \mathcal{H}^n\left(T\right) = \mathcal{H}^n\left(PT_\epsilon\right) \leq \mathcal{H}^n\left(T_\epsilon\right) = \mathcal{H}^n\left(\mathbf{k}\left(B\left(\mathbf{x},r\right)\right)\right).$$

It follows from Formula 14 and the first part of the proof applied to $\mathbf{k}$ that

$$C\epsilon \geq J\mathbf{k}\left(\mathbf{x}\right) = \lim_{r \to 0} \frac{\mathcal{H}^n\left(\mathbf{k}\left(B\left(\mathbf{x},r\right)\right)\right)}{m_n\left(B\left(\mathbf{x},r\right)\right)} \geq \limsup_{r \to 0} \frac{\mathcal{H}^n\left(\overline{\mathbf{h}}\left(B\left(\mathbf{x},r\right)\right)\right)}{m_n\left(B\left(\mathbf{x},r\right)\right)}.$$

Since ϵ is arbitrary, this establishes Formula 13 in the case where $U\left(\mathbf{x}\right)^{-1}$ does not exist and completes the proof of the theorem.

For another setting in which this argument works, see Problem 15.

We define the following two sets for future reference

$$S \equiv \{\mathbf{x} \in \Omega : D\overline{\mathbf{h}}\left(\mathbf{x}\right) \text{ exists but } U\left(\mathbf{x}\right)^{-1} \text{ does not exist}\} \tag{15}$$

$$N \equiv \{\mathbf{x} \in \Omega : D\overline{\mathbf{h}}\left(\mathbf{x}\right) \text{ does not exist}\}. \tag{16}$$

20.2 The area formula for one to one Lipschitz mappings

In this section we assume $\mathbf{h} : \Omega \to \mathbb{R}^m$ is one to one and Lipschitz. Since $\mathbf{h}$ is one to one, Lemma 20.4 implies we can define a measure, ν, on the $\sigma-$ algebra of Lebesgue measurable sets as follows:

$$\nu\left(E\right) \equiv \mathcal{H}^n\left(\mathbf{h}\left(E \cap \Omega\right)\right).$$

By Lemma 20.4, we see this is a measure and $\nu << m$. Therefore by the corollary to the Radon Nikodym theorem, Corollary 13.3, there exists $f \in L^1_{loc}\left(\mathbb{R}^n\right)$, $f \geq 0$, $f\left(\mathbf{x}\right) = 0$ if $\mathbf{x} \notin \Omega$, and

$$\nu\left(E\right) = \int_E f\,dm = \int_{\Omega \cap E} f\,dm.$$

We want to identify f. Define

$$E \equiv \{\mathbf{x} \in \Omega : \mathbf{x} \text{ is not a point of density of } \Omega\} \cup N \cup$$

$$\{\mathbf{x} \in \Omega : \mathbf{x} \text{ is not a Lebesgue point of } f\}.$$

Then E is a set of measure zero and if $\mathbf{x} \in (\Omega \setminus E) \cap S^C$, Lemma 20.9 and Theorem 20.10 imply

$$
\begin{aligned}
f(\mathbf{x}) &= \lim_{r \to 0} \frac{1}{m_n(B(\mathbf{x},r))} \int_{B(\mathbf{x},r)} f(\mathbf{y}) \, dm = \lim_{r \to 0} \frac{\mathcal{H}^n(\overline{\mathbf{h}}(B(\mathbf{x},r) \cap \Omega))}{m_n(B(\mathbf{x},r))} \\
&= \lim_{r \to 0} \frac{\mathcal{H}^n(\overline{\mathbf{h}}(B(\mathbf{x},r) \cap \Omega))}{\mathcal{H}^n(\overline{\mathbf{h}}(B(\mathbf{x},r)))} \frac{\mathcal{H}^n(\overline{\mathbf{h}}(B(\mathbf{x},r)))}{m_n(B(\mathbf{x},r))} = J(\mathbf{x}).
\end{aligned}
$$

On the other hand, if $\mathbf{x} \in (\Omega \setminus E) \cap S$, then by Theorem 20.10,

$$
\begin{aligned}
f(\mathbf{x}) &= \lim_{r \to 0} \frac{1}{m_n(B(\mathbf{x},r))} \int_{B(\mathbf{x},r)} f(\mathbf{y}) \, dm = \lim_{r \to 0} \frac{\mathcal{H}^n(\overline{\mathbf{h}}(B(\mathbf{x},r) \cap \Omega))}{m_n(B(\mathbf{x},r))} \\
&\leq \lim_{r \to 0} \frac{\mathcal{H}^n(\overline{\mathbf{h}}(B(\mathbf{x},r)))}{m_n(B(\mathbf{x},r))} = J(\mathbf{x}) = 0.
\end{aligned}
$$

Therefore, $f(\mathbf{x}) = J(\mathbf{x})$ a.e., whenever $\mathbf{x} \in \Omega \setminus E$.

Now let F be a Borel measurable set in $\mathbb{R}^m$. Recall this implies F is $\mathcal{H}^n$ measurable. Then

$$\int_{\mathbf{h}(\Omega)} \mathcal{X}_F(\mathbf{y}) \, d\mathcal{H}^n = \int \mathcal{X}_{F \cap \mathbf{h}(\Omega)}(\mathbf{y}) \, d\mathcal{H}^n = \mathcal{H}^n(\mathbf{h}(\mathbf{h}^{-1}(F) \cap \Omega))$$

$$= \nu(\mathbf{h}^{-1}(F)) = \int \mathcal{X}_{\Omega \cap \mathbf{h}^{-1}(F)}(\mathbf{x}) J(\mathbf{x}) \, dm = \int_\Omega \mathcal{X}_F(\mathbf{h}(\mathbf{x})) J(\mathbf{x}) \, dm.$$
$$(17)$$

Can we write a similar formula for F only $\mathcal{H}^n$ measurable? Note that there are no measurability questions in the above formula because $\mathbf{h}^{-1}(F)$ is a Borel set due to the continuity of $\mathbf{h}$.

First consider the case where E is only $\mathcal{H}^n$ measurable but

$$\mathcal{H}^n(E \cap \mathbf{h}(\Omega)) = 0.$$

By Theorem 19.11 of Chapter 19, there exists a Borel set $F \supseteq E \cap \mathbf{h}(\Omega)$ such that

$$\mathcal{H}^n(F) = \mathcal{H}^n(E \cap \mathbf{h}(\Omega)) = 0.$$

Then from Formula 17,

$$\mathcal{X}_{\Omega \cap \mathbf{h}^{-1}(F)}(\mathbf{x}) J(\mathbf{x}) = 0 \text{ a.e.}$$

But

$$0 \leq \mathcal{X}_{\Omega \cap \mathbf{h}^{-1}(E)}(\mathbf{x}) J(\mathbf{x}) \leq \mathcal{X}_{\Omega \cap \mathbf{h}^{-1}(F)}(\mathbf{x}) J(\mathbf{x}) \qquad (18)$$

The area formula for one to one Lipschitz mappings

which shows the two functions in Formula 18 are equal a.e. Therefore

$$\mathcal{X}_{\Omega \cap \mathbf{h}^{-1}(E)}\left(\mathbf{x}\right) J\left(\mathbf{x}\right)$$

is Lebesgue measurable and so from Formula 17,

$$0 = \int \mathcal{X}_{E \cap \mathbf{h}(\Omega)}\left(\mathbf{y}\right) d\mathcal{H}^n = \int \mathcal{X}_{F \cap \mathbf{h}(\Omega)}\left(\mathbf{y}\right) d\mathcal{H}^n$$

$$= \int \mathcal{X}_{\Omega \cap \mathbf{h}^{-1}(F)}\left(\mathbf{x}\right) J\left(\mathbf{x}\right) dm = \int \mathcal{X}_{\Omega \cap \mathbf{h}^{-1}(E)}\left(\mathbf{x}\right) J\left(\mathbf{x}\right) dm, \qquad (19)$$

which shows Formula 17 holds in this case where

$$\mathcal{H}^n\left(E \cap \mathbf{h}\left(\Omega\right)\right) = 0.$$

Now let $\Omega_R \equiv \Omega \cap B\left(\mathbf{0}, R\right)$ where R is large enough that $\Omega_R \neq \emptyset$ and let E be $\mathcal{H}^n$ measurable. By Theorem 19.11 of Chapter 19, there exists $F \supseteq E \cap \mathbf{h}\left(\Omega_R\right)$ such that F is Borel and

$$\mathcal{H}^n\left(F \setminus \left(E \cap \mathbf{h}\left(\Omega_R\right)\right)\right) = 0. \qquad (20)$$

Now

$$\left(E \cap \mathbf{h}\left(\Omega_R\right)\right) \cup \left(F \setminus \left(E \cap \mathbf{h}\left(\Omega_R\right)\right) \cap \mathbf{h}\left(\Omega_R\right)\right) = F \cap \mathbf{h}\left(\Omega_R\right)$$

and so

$$\mathcal{X}_{\Omega_R \cap \mathbf{h}^{-1}(F)} J = \mathcal{X}_{\Omega_R \cap \mathbf{h}^{-1}(E)} J + \mathcal{X}_{\Omega_R \cap \mathbf{h}^{-1}(F \setminus (E \cap \mathbf{h}(\Omega_R)))} J$$

where from Formula 20 and Formula 19, the second function on the right of the equal sign is Lebesgue measurable and equals zero a.e. Therefore, the first function on the right of the equal sign is also Lebesgue measurable and equals the function on the left a.e. Thus,

$$\int \mathcal{X}_{E \cap \mathbf{h}(\Omega_R)}\left(\mathbf{y}\right) d\mathcal{H}^n = \int \mathcal{X}_{F \cap \mathbf{h}(\Omega_R)}\left(\mathbf{y}\right) d\mathcal{H}^n$$

$$= \int \mathcal{X}_{\Omega_R \cap \mathbf{h}^{-1}(F)}\left(\mathbf{x}\right) J\left(\mathbf{x}\right) dm = \int \mathcal{X}_{\Omega_R \cap \mathbf{h}^{-1}(E)}\left(\mathbf{x}\right) J\left(\mathbf{x}\right) dm. \qquad (21)$$

Letting $R \to \infty$ we obtain Formula 21 with Ω replacing Ω_R and the function

$$\mathbf{x} \to \mathcal{X}_{\Omega_R \cap \mathbf{h}^{-1}(E)}\left(\mathbf{x}\right) J\left(\mathbf{x}\right)$$

is Lebesgue measurable. Writing this in a more familiar form yields

$$\int_{\mathbf{h}(\Omega)} \mathcal{X}_E\left(\mathbf{y}\right) d\mathcal{H}^n = \int_{\Omega} \mathcal{X}_E\left(\mathbf{h}\left(\mathbf{x}\right)\right) J\left(\mathbf{x}\right) dm. \qquad (22)$$

From this, it follows that if s is a nonnegative $\mathcal{H}^n$ measurable simple function, Formula 22 continues to be valid with s in place of $\mathcal{X}_E$. Then approximating an arbitrary nonnegative $\mathcal{H}^n$ measurable function, g, by an increasing sequence of simple functions, it follows that Formula 22 holds with g in place of $\mathcal{X}_E$ and there are no measurability problems because $\mathbf{x} \to g(\mathbf{h}(\mathbf{x})) J(\mathbf{x})$ is Lebesgue measurable. This proves the main result which is known as the area formula.

Theorem 20.11 *Let $g : \mathbf{h}(\Omega) \to [0, \infty]$ be $\mathcal{H}^n$ measurable where $\mathbf{h}$ is one to one and Lipschitz on Ω, and Ω is a Lebesgue measurable set. Then if $J(\mathbf{x}) \equiv 0$ for $\mathbf{x} \in N$,*

$$\mathbf{x} \to (g \circ \mathbf{h})(\mathbf{x}) J(\mathbf{x})$$

is Lebesgue measurable and

$$\int_{\mathbf{h}(\Omega)} g(\mathbf{y}) \, d\mathcal{H}^n = \int_{\Omega} g(\mathbf{h}(\mathbf{x})) J(\mathbf{x}) \, dm.$$

For another version of this theorem based on the same arguments given here, in which the function $\mathbf{h}$ is not assumed to be Lipschitz, see Problems 13 - 16.

20.3 Mappings that are not one to one

In this section, $\mathbf{h} : \Omega \to \mathbb{R}^m$ will only be Lipschitz. We drop the requirement that $\mathbf{h}$ be one to one. Let S and N be given in Formula 15 and Formula 16. The following lemma is a version of Sard's theorem.

Lemma 20.12 *For S defined above, $\mathcal{H}^n(\mathbf{h}(S)) = 0$.*

Proof: From Theorem 20.10, whenever $\mathbf{x} \in S$ and r is small enough,

$$\frac{\mathcal{H}^n(\overline{\mathbf{h}}(B(\mathbf{x},r)))}{m_n(B(\mathbf{x},r))} < \epsilon.$$

Therefore, whenever $\mathbf{x} \in S$ and r small enough,

$$\mathcal{H}^n(\overline{\mathbf{h}}(B(\mathbf{x},r))) \leq \epsilon \alpha(n) r^n. \tag{23}$$

Let $S_k = S \cap B(\mathbf{0},k)$ and for each $\mathbf{x} \in S_k$, let $r_{\mathbf{x}}$ be such that Formula 23 holds with r replaced by $5r_{\mathbf{x}}$ and

$$B(\mathbf{x},r_{\mathbf{x}}) \subseteq B(\mathbf{0},k).$$

By the Vitali covering theorem, there is a disjoint subsequence of these balls, $\{B(\mathbf{x}_i, r_i)\}$, with the property that $\{B(\mathbf{x}_i, 5r_i)\} \equiv \{\widehat{B}_i\}$ covers

S_k. Then by the way these balls were defined, with Formula 23 holding for $r = 5r_i$,

$$\mathcal{H}^n\left(\overline{\mathbf{h}}\left(S_k\right)\right) \leq \sum_{i=1}^{\infty} \mathcal{H}^n\left(\overline{\mathbf{h}}\left(\widehat{B}_i\right)\right) \leq 5^n \epsilon \sum_{i=1}^{\infty} \alpha\left(n\right) r_i^n$$

$$= 5^n \epsilon \sum_{i=1}^{\infty} m_n\left(B\left(\mathbf{x}_i, r_i\right)\right) \leq 5^n \epsilon m_n\left(B\left(\mathbf{0}, k\right)\right).$$

Since ϵ is arbitrary, this shows $\mathcal{H}^n\left(\overline{\mathbf{h}}\left(S_k\right)\right) = 0$. Now letting $k \to \infty$, this shows $\mathcal{H}^n\left(\overline{\mathbf{h}}\left(S\right)\right) = 0$ which proves the lemma.

Thus $m_n\left(N\right) = 0$ and $\mathcal{H}^n\left(\mathbf{h}\left(S\right)\right) = 0$ and so by Lemma 20.4

$$\mathcal{H}^n\left(\mathbf{h}\left(S \cup N\right)\right) \leq \mathcal{H}^n\left(\mathbf{h}\left(S\right)\right) + \mathcal{H}^n\left(\mathbf{h}\left(N\right)\right) = 0. \tag{24}$$

Let $B \equiv \Omega \setminus \left(S \cup N\right)$.

A similar lemma to the following was proved in the section on the change of variables formula for a C^1 map. There the proof was based on the inverse function theorem. However, this is no longer possible; so, a slightly more technical argument is required.

Lemma 20.13 *There exists a sequence of disjoint measurable sets, $\{F_i\}$, such that*

$$\cup_{i=1}^{\infty} F_i = B$$

and $\mathbf{h}$ is one to one on F_i.

Proof: $\mathcal{L}\left(\mathbb{R}^n, \mathbb{R}^n\right)$ is a finite dimensional normed linear space. In fact,

$$\{\mathbf{e}_i \otimes \mathbf{e}_j : i, j \in \{1, \cdots, n\}\}$$

is easily seen to be a basis. Let $\mathcal{I}$ be the elements of $\mathcal{L}\left(\mathbb{R}^n, \mathbb{R}^n\right)$ which are invertible and let $\mathcal{F}$ be a countable dense subset of $\mathcal{I}$. Also let C be a countable dense subset of B. For $\mathbf{c} \in C$ and $T \in \mathcal{F}$,

$$E\left(\mathbf{c}, T, i\right) \equiv \{\mathbf{b} \in B\left(\mathbf{c}, i^{-1}\right) \cap B \text{ such that } (a.) \text{ and } (b.) \text{ hold}\}$$

where the conditions *(a.)* and *(b.)* are as follows.

$$\frac{1}{1+\epsilon}|T\mathbf{v}| \leq |U\left(\mathbf{b}\right)\mathbf{v}| \text{ for all } \mathbf{v} \tag{a.}$$

$$|\mathbf{h}\left(\mathbf{a}\right) - \mathbf{h}\left(\mathbf{b}\right) - D\mathbf{h}\left(\mathbf{b}\right)\left(\mathbf{a} - \mathbf{b}\right)| \leq \epsilon|T\left(\mathbf{a} - \mathbf{b}\right)| \tag{b.}$$

for all $\mathbf{a} \in B\left(\mathbf{b}, 2i^{-1}\right)$. Here $0 < \epsilon < 1/2$.

Obviously, there are countably many $E\left(\mathbf{c}, T, i\right)$. Now suppose $\mathbf{a}, \mathbf{b} \in E\left(\mathbf{c}, T, i\right)$ and $\mathbf{h}\left(\mathbf{a}\right) = \mathbf{h}\left(\mathbf{b}\right)$. Then

$$|\mathbf{a} - \mathbf{b}| \leq |\mathbf{a} - \mathbf{c}| + |\mathbf{c} - \mathbf{b}| < \frac{2}{i}.$$

Therefore, from $(a.)$ and $(b.)$,

$$\frac{1}{1+\epsilon}\left|T\left(\mathbf{a}-\mathbf{b}\right)\right| \leq \left|U\left(\mathbf{b}\right)\left(\mathbf{a}-\mathbf{b}\right)\right| = \left|Dh\left(\mathbf{b}\right)\left(\mathbf{a}-\mathbf{b}\right)\right|$$
$$= \left|\mathbf{h}\left(\mathbf{a}\right)-\mathbf{h}\left(\mathbf{b}\right)-Dh\left(\mathbf{b}\right)\left(\mathbf{a}-\mathbf{b}\right)\right| \leq \epsilon\left|T\left(\mathbf{a}-\mathbf{b}\right)\right|.$$

Since T is one to one, this shows that $\mathbf{a} = \mathbf{b}$. Thus $\mathbf{h}$ is one to one on $E\left(\mathbf{c}, T, i\right)$.

Now let $\mathbf{b} \in B$. Choose $T \in \mathcal{F}$ such that

$$\left\| U\left(\mathbf{b}\right) - T \right\| < \epsilon \left\| U\left(\mathbf{b}\right)^{-1} \right\|^{-1}.$$

Then for all $\mathbf{v} \in \mathbb{R}^n$,

$$\left| T\mathbf{v} - U\left(\mathbf{b}\right)\mathbf{v} \right| \leq \epsilon \left\| U\left(\mathbf{b}\right)^{-1} \right\|^{-1} \left| \mathbf{v} \right| \leq \epsilon \left| U\left(\mathbf{b}\right)\mathbf{v} \right|$$

and so

$$\left| T\mathbf{v} \right| \leq \left(1 + \epsilon\right)\left| U\left(\mathbf{b}\right)\mathbf{v} \right|$$

which yields $(a.)$. Now choose i large enough that for $\left|\mathbf{a}-\mathbf{b}\right| < 2i^{-1}$,

$$\left|\mathbf{h}\left(\mathbf{a}\right)-\mathbf{h}\left(\mathbf{b}\right)-Dh\left(\mathbf{b}\right)\left(\mathbf{a}-\mathbf{b}\right)\right| < \frac{\epsilon}{\left\|T^{-1}\right\|}\left|\mathbf{a}-\mathbf{b}\right|$$
$$\leq \epsilon\left|T\left(\mathbf{a}-\mathbf{b}\right)\right|$$

and pick $\mathbf{c} \in C \cap B\left(\mathbf{b}, i^{-1}\right)$. Then $\mathbf{b} \in E\left(\mathbf{c}, T, i\right)$ and this shows that B equals the union of these sets.

Let $\{E_i\}$ be an enumeration of these sets and define $F_1 \equiv E_1$, and if $F_1, \cdots, F_n$ have been chosen, $F_{n+1} \equiv E_{n+1} \setminus \cup_{i=1}^{n} F_i$. Then $\{F_i\}$ satisfies the conditions of the lemma and this proves the lemma.

The following corollary will not be needed in this chapter but it is of interest.

Corollary 20.14 *For each E_i in Lemma 20.13, $\mathbf{h}^{-1}$ is Lipschitz on $\mathbf{h}\left(E_i\right)$.*

Proof: Pick $\mathbf{a}, \mathbf{b} \in E_i$. Then by condition $a.$ and $b.$,

$$\left|\mathbf{h}\left(\mathbf{a}\right)-\mathbf{h}\left(\mathbf{b}\right)\right| \geq \left|Dh\left(\mathbf{b}\right)\left(\mathbf{a}-\mathbf{b}\right)\right| - \epsilon\left|T\left(\mathbf{a}-\mathbf{b}\right)\right|$$

$$\geq \left(\frac{1}{1+\epsilon} - \epsilon\right)\left|T\left(\mathbf{a}-\mathbf{b}\right)\right| \geq r\left|\mathbf{a}-\mathbf{b}\right|$$

for some $r > 0$ by the equivalence of all norms on a finite dimensional space. Therefore,

$$\left|\mathbf{h}^{-1}\left(\mathbf{h}\left(\mathbf{a}\right)\right)-\mathbf{h}^{-1}\left(\mathbf{h}\left(\mathbf{b}\right)\right)\right| \leq \frac{1}{r}\left|\mathbf{h}\left(\mathbf{a}\right)-\mathbf{h}\left(\mathbf{b}\right)\right|$$

and this proves the corollary.

Now let $g : \mathbf{h}(\Omega) \to [0, \infty]$ be $\mathcal{H}^n$ measurable. By Theorem 20.11,

$$\int_{\mathbf{h}(\Omega)} \mathcal{X}_{\mathbf{h}(F_i)}(\mathbf{y}) g(\mathbf{y}) d\mathcal{H}^n = \int_{F_i} g(\mathbf{h}(\mathbf{x})) J(\mathbf{x}) dm. \tag{25}$$

Now define

$$\mathfrak{n}(\mathbf{y}) = \sum_{i=1}^{\infty} \mathcal{X}_{\mathbf{h}(F_i)}(\mathbf{y}).$$

By Lemma 20.4, $\mathbf{h}(F_i)$ is $\mathcal{H}^n$ measurable and so $\mathfrak{n}$ is a $\mathcal{H}^n$ measurable function. For each $\mathbf{y} \in B$, $\mathfrak{n}(\mathbf{y})$ gives the number of elements in $\mathbf{h}^{-1}(\mathbf{y}) \cap B$. From Formula 25,

$$\int_{\mathbf{h}(\Omega)} \mathfrak{n}(\mathbf{y}) g(\mathbf{y}) d\mathcal{H}^n = \int_B g(\mathbf{h}(\mathbf{x})) J(\mathbf{x}) dm. \tag{26}$$

Now define

$$\#(\mathbf{y}) \equiv \text{ number of elements in } \mathbf{h}^{-1}(\mathbf{y}).$$

Theorem 20.15 *The function* $\mathbf{y} \to \#(\mathbf{y})$ *is* $\mathcal{H}^n$ *measurable and if*

$$g : \mathbf{h}(\Omega) \to [0, \infty]$$

is $\mathcal{H}^n$ *measurable, then*

$$\int_{\mathbf{h}(\Omega)} g(\mathbf{y}) \#(\mathbf{y}) d\mathcal{H}^n = \int_\Omega g(\mathbf{h}(\mathbf{x})) J(\mathbf{x}) dm.$$

Proof: If $\mathbf{y} \notin \mathbf{h}(S \cup N)$, then $\mathfrak{n}(\mathbf{y}) = \#(\mathbf{y})$. By Formula 24

$$\mathcal{H}^n(\mathbf{h}(S \cup N)) = 0$$

and so $\mathfrak{n}(\mathbf{y}) = \#(\mathbf{y})$ a.e. Since $\mathcal{H}^n$ is a complete measure, $\#(\cdot)$ is $\mathcal{H}^n$ measurable. Letting

$$G \equiv \mathbf{h}(\Omega) \setminus \mathbf{h}(S \cup N),$$

26 implies

$$\begin{aligned} \int_{\mathbf{h}(\Omega)} g(\mathbf{y}) \#(\mathbf{y}) d\mathcal{H}^n &= \int_G g(\mathbf{y}) \mathfrak{n}(\mathbf{y}) d\mathcal{H}^n \\ &= \int_B g(\mathbf{h}(\mathbf{x})) J(\mathbf{x}) dm \\ &= \int_\Omega g(\mathbf{h}(\mathbf{x})) J(\mathbf{x}) dm. \end{aligned}$$

This proves the theorem.

20.4 Surface measure

The area formula makes possible a short and elegant treatment of surface measure. There are really two aspects of the subject, one being the definition of the surface measure and the other being the computation of surface measure through some parametrization. Without the proper preparation the presentation can be quite lengthy and involve partitions of unity, and many intricate definitions along with questions of whether the definition is well defined. Here we will simply define the surface measure to be an appropriate Hausdorff measure and use the area formula to see how to deal with the computations.

Definition 20.16 Let M be a closed subset of $\mathbb{R}^m$. We will refer to M as an n-dimensional Lipschitz surface if there exist measurable subsets of $\mathbb{R}^n$, $\{B_i\}$, with $m_n(B_i) < \infty$, and mappings $\alpha_i : B_i \to \mathbb{R}^m$ such that

$$\alpha_i(B_i) \cap \alpha_j(B_j) = \emptyset, \ i \neq j$$

$$M = \cup_{i=1}^{\infty} \alpha_i(B_i)$$

α_i is one to one and Lipschitz

$\{\alpha_i(B_i)\}_{i=1}^{\infty}$ is locally finite.

By locally finite we mean that if $\mathbf{y} \in \mathbb{R}^m$, there exists $r > 0$ such that $B(\mathbf{y}, r)$ intersects only finitely many sets of $\{\alpha_i(B_i)\}_{i=1}^{\infty}$.

We will follow the notation of the earlier sections in which $\bar{\mathbf{f}}$ is a Lipschitz extension of $\mathbf{f}$ to all of $\mathbb{R}^n$.

Definition 20.17 Define $\sigma \equiv \mathcal{H}^n|_M$ on the σ algebra of sets of the form

$$S(\sigma) \equiv \{E \cap M : E \text{ is } \mathcal{H}^n \text{ measurable}\}. \tag{27}$$

The next goal is to show that σ is a Radon measure.

Theorem 20.18 σ *is a Radon measure on the σ algebra 27.*

Proof: First note that M is a locally compact Hausdorff space with respect to the relative topology in which the open sets are of the form $U \cap M$ where U is an open set in $\mathbb{R}^m$. It is clear that $S(\sigma)$ contains the Borel sets and that $(\sigma, S(\sigma))$ is a complete measure space.

Claim: σ is σ finite and $\sigma(K) < \infty$ whenever K is compact.
Proof: By the area formula,

$$\sigma(\alpha_i(B_i)) = \int_{\alpha(B_i)} d\mathcal{H}^n = \int_{B_i} J_i(\mathbf{x}) \, dm_n < \infty.$$

This shows σ is σ finite. Now if $K \subseteq M$ is compact, then since $\{\alpha_i(B_i)\}$ are locally finite, for each $k \in K$, there exists an open set U_k containing k such that U_k intersects only finitely many of the sets $\alpha_i(B_i)$. Finitely many of these sets, U_k, cover K, $U_{k_1}, \cdots, U_{k_r}$ and each of these intersects only finitely many sets of $\{\alpha_i(B_i)\}$. Therefore, K is contained in the union of finitely many sets,

$$\alpha_1(B_1), \cdots, \alpha_p(B_p).$$

It follows that

$$\sigma(K) \leq \sum_{i=1}^{p} \sigma(\alpha_i(B_i)) < \infty.$$

By Theorem 9.26, $\overline{\sigma}$ is a Radon measure on a σ algebra, $\overline{\mathcal{S}}$, containing the Borel sets. By Lemma 9.6, $\overline{\mathcal{S}} = \mathcal{S}(\sigma)$ and $\overline{\sigma} = \sigma$ on this σ algebra. This proves the theorem.

We want to find $J(\mathbf{x})$ in the case where

$$\alpha(\mathbf{x}) = (\mathbf{x}, g(\mathbf{x}))^T \tag{28}$$

for $\mathbf{x} \in B \subseteq \mathbb{R}^n$ and $g : B \to \mathbb{R}$ is Lipschitz. This is the case where the $n+1$ st variable is given as a function of the other n variables. Thus, in this case $D\alpha(\mathbf{x})$ has the $(n+1) \times n$ matrix given as follows.

$$D\alpha(\mathbf{x}) = \begin{pmatrix} 1 & & 0 \\ & \ddots & \\ 0 & & 1 \\ g_{,x_1} & \cdots & g_{,x_n} \end{pmatrix}$$

Also from Lemma 19.16 of Chapter 19,

$$J(\mathbf{x}) = \left(\det\left(D\bar{\alpha}(\mathbf{x})^* D\bar{\alpha}(\mathbf{x})\right)\right)^{1/2}.$$

Therefore, $J(\mathbf{x})$ is the determinant of the following $n \times n$ matrix.

$$\begin{pmatrix} 1 + (g_{,x_1})^2 & g_{,x_1}g_{,x_2} & \cdots & g_{,x_1}g_{,x_n} \\ g_{,x_2}g_{,x_1} & 1 + (g_{,x_2})^2 & \cdots & g_{,x_2}g_{,x_n} \\ \vdots & & \ddots & \vdots \\ g_{,x_n}g_{,x_1} & g_{,x_n}g_{,x_2} & \cdots & 1 + (g_{,x_n})^2 \end{pmatrix} \tag{29}$$

The following theorem is the main result. It is a generalization of the familiar formula found in most calculus books for the surface area of a surface given in the form $z = f(x, y)$.

Theorem 20.19 *If α is given by Formula 28 then $J(\mathbf{x})$ is given by*

$$J(\mathbf{x}) = \left(1 + \sum_{i=1}^{n}(g_{,x_i}(\mathbf{x}))^2\right)^{1/2}$$

This theorem follows from the following lemma and Formula 29.

Lemma 20.20 *Let $a_1, \cdots, a_n$ be numbers and let $A(a_1, \cdots, a_n)$ be the matrix which has $1 + a_i^2$ in the ii th slot and $a_i a_j$ in the ij th slot when $i \neq j$. Then*

$$\det A = 1 + \sum_{i=1}^{n} a_i^2.$$

Proof: The lemma is obviously true if $n = 1$. Suppose it is true for $n - 1$, $n > 1$ and consider

$$A(a_1, \cdots, a_n).$$

If any of the a_i equal 0, then the i th column and i th row vanish except for a 1 in the ii th position. Expanding along the i th column we see that we have to find

$$\det(A(b_1, \cdots, b_{n-1}))$$

where $b_j = a_j$ if $j < i$ and $b_j = a_{j+1}$ if $j \geq i$. By induction, the determinant equals

$$1 + \sum_{i=1}^{n-1} b_i^2 = 1 + \sum_{i=1}^{n} a_i^2.$$

Now suppose none of the a_i equal 0. The last two rows are

$$\begin{matrix} a_{n-1}a_1 & a_{n-1}a_2 & \cdots & 1 + a_{n-1}^2 & a_{n-1}a_n \\ a_n a_1 & a_n a_2 & \cdots & a_n a_{n-1} & 1 + a_n^2 \end{matrix}$$

Multiply the $n - 1$ st row by $-a_n(a_{n-1})^{-1}$ and add to the n th row. This yields

$$\begin{matrix} a_{n-1}a_1 & a_{n-1}a_2 & \cdots & 1 + a_{n-1}^2 & a_{n-1}a_n \\ 0 & 0 & \cdots & \frac{-a_n}{a_{n-1}} & 1 \end{matrix}$$

for the bottom two rows. Now the last two columns of the resulting modified matrix are

$$\begin{matrix} a_1 a_{n-1} & a_1 a_n \\ a_2 a_{n-1} & a_2 a_n \\ \vdots & \vdots \\ 1 + a_{n-1}^2 & a_{n-1}a_n \\ \frac{-a_n}{a_{n-1}} & 1 \end{matrix}$$

Multiply the $n-1$ st column by $-a_n (a_{n-1})^{-1}$ and add to the n th column. This yields

$$
\begin{matrix}
a_1 a_{n-1} & 0 \\
a_2 a_{n-1} & 0 \\
\vdots & \vdots \\
1 + a_{n-1}^2 & \frac{-a_n}{a_{n-1}} \\
\frac{-a_n}{a_{n-1}} & 1 + \frac{a_n^2}{a_{n-1}^2}
\end{matrix}
$$

Therefore, this sequence of row operations yields a modified matrix whose determinant is the same as that of the original matrix but which is of the form

$$
\begin{pmatrix}
A\left(a_1, \cdots, a_{n-1}\right) & \mathbf{d}_{n-1} \\
\mathbf{d}_{n-1}^T & 1 + \frac{a_n^2}{a_{n-1}^2}
\end{pmatrix}
$$

where $\mathbf{d}_{n-1}^T = \left(0, , \cdots, 0, \frac{-a_n}{a_{n-1}}\right)$. Therefore, by induction, expanding along the last column yields

$$
\det\left(A\left(a_1, \cdots, a_n\right)\right)
$$

$$
= \left(1 + \frac{a_n^2}{a_{n-1}^2}\right)\left(1 + \sum_{i=1}^{n-1} a_i^2\right) - \left(\frac{a_n}{a_{n-1}}\right)^2 \left(1 + \sum_{i=1}^{n-2} a_i^2\right)
$$

$$
= 1 + \sum_{i=1}^{n} a_i^2.
$$

and this proves the lemma.

20.5   The divergence theorem

In this section we give a proof of a general version of the divergence theorem. We say a bounded open subset, U, of $\mathbb{R}^n$ has Lipschitz boundary and lies locally on one side of its boundary if it satisfies the following conditions.

For each $p \in \partial U \equiv \overline{U} \setminus U$, there exists an open set, Q, containing p, an open interval (a, b), an open box $B \subseteq \mathbb{R}^{n-1}$, and an orthogonal transformation R such that

$$
RQ = B \times (a, b), \tag{30}
$$

$$
R\left(Q \cap U\right) = \{\mathbf{y} \in \mathbb{R}^n : \widehat{\mathbf{y}} \in B,\ a < y_n < g\left(\widehat{\mathbf{y}}\right)\} \tag{31}
$$

where g is Lipschitz continuous and

$$
\widehat{\mathbf{y}} \equiv \left(y_1, \cdots, y_{n-1}\right).
$$

Note that finitely many of these sets Q cover ∂U because ∂U is compact. Letting $W = Q \cap U$ the following picture describes the situation.

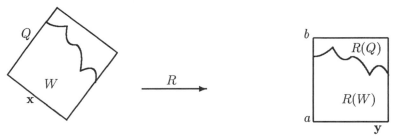

Lemma 20.21 *Let V be a bounded open set and let X be the closed subspace of $C\left(\overline{V}\right)$, the space of continuous functions defined on $\overline{V}$, which is given by the following.*

$$X = \{u \in C\left(\overline{V}\right) : u\left(\mathbf{x}\right) = 0 \text{ on } \partial V\}.$$

Then $C_c^\infty\left(V\right)$ is dense in X.

Proof: Let $O \subseteq \overline{O} \subseteq W \subseteq \overline{W} \subseteq V$ be such that $dist\left(\overline{O}, V^C\right) < \eta$ and let $\psi_\delta\left(\cdot\right)$ be a mollifier. Let $u \in X$ and consider $\mathcal{X}_W u * \psi_\delta$. Let $\epsilon > 0$ be given and let η be small enough that $|u\left(\mathbf{x}\right)| < \epsilon/2$ whenever $\mathbf{x} \in V \setminus \overline{O}$. Then if δ is small enough $|\mathcal{X}_W u * \psi_\delta\left(\mathbf{x}\right) - u\left(\mathbf{x}\right)| < \epsilon$ for all $\mathbf{x} \in \overline{O}$ and $\mathcal{X}_W u * \psi_\delta$ is in $C_c^\infty\left(V\right)$. For $\mathbf{x} \in V \setminus \overline{O}$, $|\mathcal{X}_W u * \psi_\delta\left(\mathbf{x}\right)| \leq \epsilon/2$ and so for such $\mathbf{x}$,

$$|\mathcal{X}_W u * \psi_\delta\left(\mathbf{x}\right) - u\left(\mathbf{x}\right)| \leq \epsilon.$$

This proves the lemma since ϵ was arbitrary.

The next lemma is the main result in proving the divergence theorem. The existence part of the proof is really only a computation involving Fubini's theorem, and the chain rule. In the argument we use the notation $C_c^1\left(Q\right)$ to denote functions which have one continuous derivative and whose support is contained in the open set Q.

Lemma 20.22 *Let Q be an open set satisfying Formula 30 and Formula 31. Then there exists a vector field $\mathbf{n}\left(\mathbf{x}\right)$ for $\mathbf{x} \in \partial U \cap Q$ such that*

$$n_i\left(\cdot\right) \in L^\infty\left(Q \cap \partial U\right), \ i = 1, 2, \cdots, n \tag{32}$$

and for each $|\mathbf{w}| = 1$,

$$\lim_{t \to 0} \frac{1}{t} \int_{U \cap Q} [f\left(\mathbf{x} + t\mathbf{w}\right) - f\left(\mathbf{x}\right)]dx = \int_{\partial U \cap Q} f\left(\mathbf{n} \cdot \mathbf{w}\right) d\sigma \tag{33}$$

for all $f \in C_c^1\left(Q\right)$. If $\widehat{\mathbf{n}}$ is any other vector field satisfying Formula 32 and Formula 33, then

$$\mathbf{n}\left(\mathbf{x}\right) = \widehat{\mathbf{n}}\left(\mathbf{x}\right) \ \sigma \ a.e.$$

Furthermore, $\mathbf{n}(\mathbf{x})$ *can be computed a.e. by*

$$\mathbf{n}(\mathbf{x}) = R^* \mathbf{N}(R(\mathbf{x})), \tag{34}$$

where

$$\mathbf{N}(\mathbf{y}) \equiv [-\sum_{j=1}^{n-1} g_{,j}(\hat{\mathbf{y}}) e_j(\mathbf{y}) + e_n(\mathbf{y})]\cdot$$

$$\left(1 + \sum_{j=1}^{n-1} (g_{,j}(\hat{\mathbf{y}}))^2\right)^{-1/2}. \tag{35}$$

Proof:

$$\frac{1}{t}\int_{U\cap Q}[f(\mathbf{x}+t\mathbf{w}) - f(\mathbf{x})]dx = \frac{1}{t}\int_B\int_{-\infty}^{g(\mathbf{y})}(f\circ R^*)(\mathbf{y}+tR\mathbf{w})$$

$$- (f\circ R^*)(\mathbf{y})\,dy_n d\hat{y}$$

$$= \frac{1}{t}\int_B\int_{-\infty}^0 F(\hat{\mathbf{y}}+tR\mathbf{w}, z + g(\hat{\mathbf{y}}) + t(R\mathbf{w})_n) - F(\hat{\mathbf{y}}, z + g(\hat{\mathbf{y}}))\,dzd\hat{y}$$

where $F = f\circ R^*$. Taking the limit as $t\to 0$ yields

$$\int_B\int_{-\infty}^0\sum_{j=1}^{n-1} F_{,j}(\hat{\mathbf{y}}, z + g(\hat{\mathbf{y}}))(R\mathbf{w})_j + F_{,n}(\hat{\mathbf{y}}, z + g(\hat{\mathbf{y}}))(R\mathbf{w})_n\,dzd\hat{y}.$$

This equals

$$\int_B\int_{-\infty}^0\sum_{j=1}^{n-1}[\frac{\partial}{\partial y_j}F(\hat{\mathbf{y}}, z + g(\hat{\mathbf{y}})) - \frac{\partial}{\partial y_n}F(\hat{\mathbf{y}}, z + g(\hat{\mathbf{y}}))g_{,j}(\hat{\mathbf{y}})](R\mathbf{w})_j$$

$$+F_{,n}(\hat{\mathbf{y}}, z + g(\hat{\mathbf{y}}))(R\mathbf{w})_n\,dzd\hat{y}.$$

Now since $f\in C_c^1(Q)$, $F\in C_c^1(B\times(a,b))$. Therefore this reduces to

$$\int_B -\sum_{j=1}^{n-1} F(\hat{\mathbf{y}}, g(\hat{\mathbf{y}}))g_{,j}(\hat{\mathbf{y}})(R\mathbf{w})_j + F(\hat{\mathbf{y}}, g(\hat{\mathbf{y}}))(R\mathbf{w})_n\,d\hat{y}. \tag{36}$$

Now by Theorem 20.19,

$$d\sigma = \det\left(D\alpha(\hat{\mathbf{y}})^T RR^T D\alpha(\hat{\mathbf{y}})\right)d\hat{y} = \det\left(D\alpha(\hat{\mathbf{y}})^T D\alpha(\hat{\mathbf{y}})\right)d\hat{y}$$

$$= \left(1 + \sum_{j=1}^{n-1}(g_{,j}(\hat{\mathbf{y}}))^2\right)^{1/2}d\hat{y}$$

where $\alpha\left(\widehat{\mathbf{y}}\right) = \left(\widehat{\mathbf{y}}, g\left(\widehat{\mathbf{y}}\right)\right)^T$. Therefore, Formula 36 equals

$$\int_{\partial U \cap Q} f\left(\mathbf{x}\right)\left(\mathbf{n}\left(\mathbf{x}\right) \cdot \mathbf{w}\right) d\sigma.$$

$\left(\left(\mathbf{n}\left(\mathbf{x}\right) \cdot \mathbf{w}\right) = \left(R^*\mathbf{N}\left(R\left(\mathbf{x}\right)\right) \cdot \mathbf{w}\right) = \left(\mathbf{N}\left(R\mathbf{x}\right) \cdot R\mathbf{w}\right) = \left(\mathbf{N}\left(\mathbf{y}\right) \cdot R\mathbf{w}\right)\right).$

Now we verify the uniqueness assertion. Let

$$X \equiv \{u \in C\left(\overline{Q}\right) : u = 0 \text{ on } \partial Q\}.$$

By Lemma 20.21, $C_c^1\left(Q\right)$ is dense in the Banach space X. Define on $C_c^1\left(Q\right)$

$$\Lambda f \equiv \lim_{t \to 0} \frac{1}{t} \int_{Q \cap U} [f\left(\mathbf{x}+t\mathbf{w}\right) - f\left(\mathbf{x}\right)]dx$$

$$= \int_{\partial U \cap Q} f\left(\mathbf{n} \cdot \mathbf{w}\right) d\sigma.$$

Thus $|\Lambda f| \underset{\approx}{\leq} ||f||_X \sigma\left(\partial U \cap Q\right)$. It follows that Λ has a unique continuous extension, $\widetilde{\Lambda}$, to all of X satisfying

$$\widetilde{\Lambda} f = \int_{\partial U \cap Q} f\left(\mathbf{n} \cdot \mathbf{w}\right) d\sigma \tag{37}$$

for all $f \in X$. If $\widehat{\mathbf{n}}\left(\mathbf{x}\right)$ satisfies Formula 32 and Formula 33, then

$$\widetilde{\Lambda} f = \int_{\partial U \cap Q} f\left(\widehat{\mathbf{n}} \cdot \mathbf{w}\right) d\sigma \tag{38}$$

for all $f \in X$. Consequently, whenever $h \in C_c\left(\partial U \cap Q\right)$,

$$\int_{\partial U \cap Q} h\left(\widehat{\mathbf{n}} \cdot \mathbf{w}\right) d\sigma = \int_{\partial U \cap Q} h\left(\mathbf{n} \cdot \mathbf{w}\right) d\sigma.$$

To see this, let $f \in X$ be such that $f|_{\partial U \cap Q} = h$. Then apply Formulas 37 and 38. By regularity of σ, $C_c\left(\partial U \cap Q\right)$ is dense in $L^1\left(\partial U \cap Q, \sigma\right)$. Hence

$$\left(\widehat{\mathbf{n}}\left(\mathbf{x}\right) \cdot \mathbf{w}\right) = \left(\mathbf{n}\left(\mathbf{x}\right) \cdot \mathbf{w}\right)$$

for σ a.e. x. Let $\{\mathbf{w}_k\}_{k=1}^{\infty}$ be a dense subset of S^{n-1}, the unit sphere, and let

$$\widehat{\mathbf{n}}\left(\mathbf{x}\right) \cdot \mathbf{w}_k = \mathbf{n}\left(\mathbf{x}\right) \cdot \mathbf{w}_k, \ \mathbf{x} \notin E_k$$

where $\sigma\left(E_k\right) = 0$. Then let

$$E = \cup_{k=1}^{\infty} E_k.$$

If $\mathbf{x} \notin E$, it follows that

$$\left(\widehat{\mathbf{n}}\left(\mathbf{x}\right) - \mathbf{n}\left(\mathbf{x}\right)\right) \cdot \mathbf{w} = 0$$

whenever $|\mathbf{w}| = 1$. Thus $\hat{\mathbf{n}}(\mathbf{x}) = \mathbf{n}(\mathbf{x}) \ \sigma$ a.e. and this proves the lemma. Next we extend this lemma, which is a local result, to all of U.

Lemma 20.23 *There exists a vector field, $\mathbf{n}(\mathbf{x})$, for σ a.e. $\mathbf{x} \in \partial U$ such that*

$$n_i(\cdot) \in L^\infty(\partial U) \ i = 1, 2, \cdots, n$$

and for each $|\mathbf{w}| = 1$,

$$\lim_{t \to 0} \frac{1}{t} \int_U [f(\mathbf{x}+t\mathbf{w}) - f(\mathbf{x})]dx = \int_{\partial U} f(\mathbf{n} \cdot \mathbf{w}) \, d\sigma$$

for all $f \in C_c^1(\mathbb{R}^n)$.

Proof: Let $Q_1, \cdots, Q_p$ be open sets satisfying Formula 30 and Formula 31 and such that

$$\partial U \subseteq \cup_{i=1}^p Q_i.$$

Let Q_0 be an open set satisfying

$$Q_0 \subseteq \overline{Q_0} \subseteq U, \ U \subseteq \cup_{i=0}^p Q_i$$

and let H be an open set such that

$$\overline{U} \subseteq H \subseteq \overline{H} \subseteq \cup_{i=0}^p Q_i$$

and let $\psi_0, \psi_1, \cdots, \psi_p$ be a C_c^∞ partition of unity on $\overline{H}$ subordinate to $Q_0, \cdots, Q_p$. Thus

$$\psi_i \in C_c^\infty(Q_i), \ \psi_i \geq 0, \ \sum_{i=0}^p \psi_i(\mathbf{x}) = 1$$

for $\mathbf{x} \in \overline{H}$. Such a partition of unity exists by the same argument used to prove Theorem 9.11 of Chapter 9 applied to the smoother functions of the sort described in Theorem 12.21 of Chapter 12. Then whenever $|t|$ is small enough,

$$\frac{1}{t} \int_U [f(\mathbf{x}+t\mathbf{w}) - f(\mathbf{x})]dx = \sum_{i=1}^p \frac{1}{t} \int_{Q_i \cap U} [\psi_i f(\mathbf{x}+t\mathbf{w}) - \psi_i f(\mathbf{x})]dx. \quad (39)$$

Note the sum in Formula 39 starts at $i = 1$. By Lemma 20.22,

$$\lim_{t \to 0} \frac{1}{t} \int_U [f(\mathbf{x}+t\mathbf{w}) - f(\mathbf{x})]dx = \sum_{i=1}^p \int_{\partial U \cap Q_i} \psi_i(\mathbf{x}) f(\mathbf{x})(\mathbf{n}^i(\mathbf{x}) \cdot \mathbf{w}) \, d\sigma \quad (40)$$

where $\mathbf{n}^i(\mathbf{x})$ is the vector of Lemma 20.22. The last expression in Formula 40 equals

$$\sum_{i=1}^{p} \int_{\partial U} f(\mathbf{x}) \left(\psi_i(\mathbf{x}) \mathbf{n}^i(\mathbf{x}) \cdot \mathbf{w} \right) d\sigma.$$

Hence

$$\lim_{t \to 0} \frac{1}{t} \int_{U} [f(\mathbf{x}+t\mathbf{w}) - f(\mathbf{x})] dx = \int_{\partial U} f(\mathbf{x}) \left(\sum_{i=1}^{p} \psi_i(\mathbf{x}) \mathbf{n}^i(\mathbf{x}) \cdot \mathbf{w} \right) d\sigma.$$

Letting

$$\mathbf{n}(\mathbf{x}) = \sum_{i=1}^{p} \psi_i(\mathbf{x}) \mathbf{n}^i(\mathbf{x}),$$

this proves the lemma.

Theorem 20.24 *There exists a σ a.e. unique vector field $\mathbf{n}(\mathbf{x})$ defined for a.e. $\mathbf{x} \in \partial U$ such that*

$$n_i(\cdot) \in L^\infty(\partial U) \ i = 1, \cdots, n \tag{41}$$

and for each $|\mathbf{w}| = 1$,

$$\lim_{t \to 0} \frac{1}{t} \int_{U} [f(\mathbf{x}+t\mathbf{w}) - f(\mathbf{x})] dx = \int_{\partial U} f(\mathbf{n} \cdot \mathbf{w}) d\sigma \tag{42}$$

for all $f \in C_c^1(\mathbb{R}^n)$.

Proof: The existence part has been done in Lemma 20.23. It remains to show $\mathbf{n}$ is a.e. unique. Let $\hat{\mathbf{n}}$ satisfy Formula 41 and Formula 42. Let D be an open ball containing $\overline{U}$ and define

$$X \equiv \{u \in C(\overline{D}) : u(\mathbf{x}) = 0 \text{ for } \mathbf{x} \in \partial D\}.$$

Then let

$$\Lambda f \equiv \lim_{t \to 0} \frac{1}{t} \int_{U} [f(\mathbf{x}+t\mathbf{w}) - f(\mathbf{x})] dx = \int_{\partial U} f(\mathbf{n} \cdot \mathbf{w}) d\sigma \tag{43}$$

for all $f \in C_c^1(D)$. Then

$$|\Lambda f| \le \sigma(\partial U) \, \|f\|_X$$

where X is defined in Lemma 20.21 with D replacing V. Therefore, Λ has a unique continuous extension, $\widetilde{\Lambda}$, defined on all of X satisfying

$$\widetilde{\Lambda} f = \int_{\partial U} f(\mathbf{n} \cdot \mathbf{w}) d\sigma \tag{44}$$

410

for all $f \in X$. The expression in Formula 43 involving the limit does not depend on $\mathbf{n}$ and so if $\widehat{\mathbf{n}}$ satisfies Formula 42, we can conclude that we also have

$$\widetilde{\Lambda} f = \int_{\partial U} f\left(\widehat{\mathbf{n}} \cdot \mathbf{w}\right) d\sigma \tag{45}$$

for all $f \in X$. Now if $h \in C\left(\partial U\right)$, we can choose $f \in X$ such that $f|_{\partial U} = h$. Then Formula 44 and Formula 45 imply

$$\int_{\partial U} h\left(\widehat{\mathbf{n}} - \mathbf{n} \cdot \mathbf{w}\right) d\sigma = 0$$

for all $h \in C\left(\partial U\right)$, a dense subset of $L^1\left(\partial U, \sigma\right)$. Therefore, as in Lemma 20.22, this implies

$$\mathbf{n}\left(\mathbf{x}\right) = \widehat{\mathbf{n}}\left(\mathbf{x}\right)$$

σ a.e. $\mathbf{x}$. This proves the theorem.

Corollary 20.25 *We can compute $\mathbf{n}\left(\mathbf{x}\right)$ a.e. $\mathbf{x}$ using Formula 34 and Formula 35 for $\mathbf{x} \in Q \cap U$ where Q satisfies Formula 30 and Formula 31.*

Proof: Consider $f \in C_c^1\left(Q\right)$. If $\mathbf{n}_0\left(\mathbf{x}\right)$ is the vector field of Formula 34 and Formula 35, we have by Lemma 20.22 and Theorem 20.24,

$$\int_{\partial U \cap Q} f\left(\mathbf{n} \cdot \mathbf{w}\right) d\sigma = \int_{\partial U \cap Q} f\left(\mathbf{n}_0 \cdot \mathbf{w}\right) d\sigma$$

for all $f \in C_c^1\left(Q\right)$. By Lemma 20.22, $\mathbf{n}_0\left(\mathbf{x}\right) = \mathbf{n}\left(\mathbf{x}\right)$ a.e. This proves the corollary.

We call $\mathbf{n}\left(\mathbf{x}\right)$ the unit outer normal. Note that from Formula 35, $\mathbf{N}\left(\mathbf{y}\right)$ is the unit outward normal to $R\left(U \cap Q\right)$. Thus $\mathbf{n}\left(\mathbf{x}\right) \equiv R^* \mathbf{N}\left(R\mathbf{x}\right)$ is the unit outward normal to $U \cap Q$.

The next theorem is the main result.

Theorem 20.26 *(Integration by parts and divergence theorem) Let $f \in C_c^1\left(\mathbb{R}^n\right)$. Then*

$$\int_U f_{,i}\left(\mathbf{x}\right) dx = \int_{\partial U} f n_i d\sigma, \tag{46}$$

and if $\mathbf{F} = \sum_i f_i\left(\mathbf{x}\right) \mathbf{e}_i$ is a vector field defined on $\mathbb{R}^n$ such that $f_i \in C_c^1\left(\mathbb{R}^n\right)$, then

$$\int_U div\left(\mathbf{F}\right) dx = \int_{\partial U} \mathbf{F} \cdot \mathbf{n} d\sigma. \tag{47}$$

Proof: Formula 46 follows from Formula 42, the dominated convergence theorem, and choosing $\mathbf{w} = \mathbf{e}_i$. Formula 47 follows immediately from this.

We conclude with a simple corollary.

Corollary 20.27 *Both formulas in Theorem 20.26 are valid if f is only assumed to be Lipschitz on $\overline{U}$.*

Proof: To show Formula 46, let $\psi \in C_c^1(\mathbb{R}^n)$, $\psi = 1$ on an open set containing $\overline{U}$. Let $\bar{f}$ be a Lipschitz extension of f to all of $\mathbb{R}^n$ and let η_ϵ be a mollifier. Let

$$f_\epsilon \equiv \psi \bar{f} * \eta_\epsilon.$$

Then it is routine to show that

$$f_{\epsilon,i} = (\psi \bar{f})_{,i} * \eta_\epsilon$$

and f_ϵ converges uniformly to f on $\overline{U}$. Now $(\psi \bar{f})_{,i}$ has compact support and is bounded. Therefore it is in $L^2(\mathbb{R}^n)$. It follows that $f_{\epsilon,i} \to (\psi \bar{f})_{,i}$ in $L^2(\mathbb{R}^n)$. Select a subsequence, still denoted by ϵ, such that convergence also takes place a.e. Then by the dominated convergence theorem,

$$
\begin{aligned}
\int_U f_{,i}\,dx &= \int_U (\psi\bar{f})_{,i}\,dx = \lim_{\epsilon \to 0} \int_U f_{\epsilon,i}\,dx \\
&= \lim_{\epsilon \to 0} \int_{\partial U} f_\epsilon n_i\,d\sigma = \int_{\partial U} f n_i\,d\sigma.
\end{aligned}
$$

As in Theorem 20.26, Formula 47 follows from this.

20.6 Exercises

1. Suppose $f \in C(\mathbb{R}^n)$ and $f_{,i} \in L_{loc}^1(\mathbb{R}^n)$ where $f_{,i}$ is the weak partial derivative defined by

$$f_{,i}(\phi) \equiv -\int_{\mathbb{R}^n} f(\mathbf{x})\phi_{,i}(\mathbf{x})\,dx$$

for all $\phi \in C_c^\infty(\mathbb{R}^n)$. Show that under these conditions the formula, Formula 46, is valid. **Hint:** Let $f_m \equiv f * \phi_m$ where ϕ_m is a mollifier. Show $f_m \in C_c^\infty(\mathbb{R}^n)$ and $f_{m,i} = f_{,i} * \phi_m$. Then apply the divergence theorem for f_m and let $m \to \infty$.

2. Let $\{a_n\}$ be an increasing sequence of points of $(0,1)$ with

$$\lim_{n\to\infty} a_n = 1$$

and let g_n be a nonnegative continuous function whose support is in (a_n, a_{n+1}) and whose integral equals n^{-1}. Let

$$g(x) \equiv \sum_{n=1}^{\infty} (-1)^{n+1} g_n(x)$$

and let

$$G(x) \equiv \int_0^x g(t)\, dt, \quad G(1) \equiv \sum_{n=1}^{\infty} (-1)^{n+1} \frac{1}{n} = \ln(2).$$

Then show $G'(x) = g(x)$, $G \in C([0,1])$, $G' \in C((0,1))$, but $G' \notin L^1(0,1)$. What does this show about the divergence theorem in the case where $f \in C(\overline{\Omega})$ and $f_{,i} \in C(\Omega)$? Can this difficulty be removed by defining the integral in another way so that the formula

$$\int_{\Omega} f_{,i}\, dx = \int_{\partial\Omega} f n_i\, d\sigma$$

is still valid when we only assume $f \in C(\overline{\Omega})$ and $f_{,i} \in C(\Omega)$?

3. Let Q be one of the open sets described in the section on the divergence theorem and let $f \in C_c^1(Q)$. Let $\nu \equiv R^* \mathbf{e}_n$ where

$$\mathbf{n}(\mathbf{x}) \equiv R^* N(R(\mathbf{x}))$$

$$= R^* \left(-\sum_{i=1}^{n-1} g_{,i}(R(\mathbf{x}))\, \mathbf{e}_i + \mathbf{e}_n \right) \left(1 + \sum_{i=1}^{n-1} (g_{,i}(R(\mathbf{x})))^2 \right)^{-1/2}$$

Then show

$$\nu \cdot \mathbf{n} = \left(1 + \sum_{i=1}^{n-1} (g_{,i}(R(\mathbf{x})))^2 \right)^{-1/2}$$

and argue that $\nu \cdot \mathbf{n} \geq \left(1 + n\, Lip(g)^2 \right)^{-1/2} \geq \delta > 0$ on $\partial U \cap Q$. Then show there exists a constant, C, independent of Q such that for $f \in C_c^1(Q)$,

$$\int_{\partial U} |f|\, d\sigma \leq C \int_U |\nabla f|\, dx.$$

Hint: Use Theorem 20.24 to argue

$$\delta \int_{\partial U} \phi_\epsilon(f)\, d\sigma \leq \int_{\partial U} \phi_\epsilon(f)(\nu \cdot \mathbf{n})\, d\sigma = \int_U \nabla \phi_\epsilon(f) \cdot \nu\, d\sigma$$

where $\phi_\epsilon(t) \equiv (t^2 + \epsilon^2)^{1/2} - \epsilon$. Then let $\epsilon \to 0$.

4. ↑Now let $f \in C_c^1(\mathbb{R}^n)$ and let $Q_0, Q_1, \cdots, Q_p$ be as in the proof of the divergence theorem and, as there, let $\{\psi_i\}$ be a C_c^∞ partition of unity on $\overline{U}$ subordinate to $Q_0, \cdots, Q_p$. Then

$$\int_{\partial U} |f\psi_i| \, d\sigma \leq C \int_U |\nabla (f\psi_i)| \, dx.$$

Show there is a constant, C, such that

$$\int_{\partial U} |f| \, d\sigma \leq C \left(\int_U |f| + \sum_{j=1}^n \left| \frac{\partial f}{\partial x_j} \right| dx \right)$$

and if $p > 1$,

$$\int_{\partial U} |f|^p \, d\sigma \leq C \left(\int_U |f|^p \, dx + \int_U \sum_{j=1}^n \left| \frac{\partial f}{\partial x_j} \right| |f|^{p-1} \right)$$

$$\leq C \left(\int_U |f|^p + |\nabla f|^p \, dx \right).$$

5. Let $u \in L^p(U)$ where U is an open set in $\mathbb{R}^n$. We say $u_{,i} \in L^p(U)$ if there exists a function $g \in L^p(U)$ such that

$$-\int_U u\phi_{,i} dx = \int_U g\phi dx.$$

Then we say $u_{,i} = g$. Show this is well defined.

6. Show that if U is an open set which has Lipschitz boundary and lies locally on one side of its boundary, then $C_c^1(\overline{U}) \subseteq X^{1p}(U)$ where this space is defined in Problem 2 of Chapter 18 and that if $u \in C_c^1(\overline{U})$, then the weak partial derivatives coincide with the usual partial derivatives. Here

$$C_c^1(\overline{U}) \equiv \text{ restrictions of functions in } C_c^1(\mathbb{R}^n) \text{ to } \overline{U}.$$

Does this result also hold for arbitrary open sets for which the divergence theorem has not been proved?

7. ↑ Let U be an open set. Then $W^{1p}(U)$ is defined as the closure in $X^{1p}(U)$ of $C^\infty(\overline{U})$. (It can be shown that $X^{1p}(U) = W^{1p}(U)$ for the Lipschitz bounded open sets U but we will not use this fact here.) Establish the following for such Lipschitz bounded open sets. There exists $\gamma \in \mathcal{L}(W^{1p}(U), L^p(\partial U))$ such that $\gamma u(\mathbf{x}) = u(\mathbf{x})$ if $u \in C^1(\overline{U})$. Show that for $u \in W^{1p}(U)$,

$$\int_U u_{,i} dx = \int_{\partial U} \gamma u n_i \, d\sigma,$$

414

where $u_{,i}$ is the weak partial derivative discussed in Problem 5. This mapping is called the trace map. **Hint:** See Problems 3 and 4.

8. ↑ Let $B \times (a, b)$ be as described in the section on the divergence theorem and let $\psi \in C_c^\infty (B \times (a, b))$. Let $h : S \to \mathbb{R}$ be continuous where

$$S \equiv \{(\hat{\mathbf{y}}, g(\hat{\mathbf{y}})) : \hat{\mathbf{y}} \in B\}$$

and let

$$H(\hat{\mathbf{y}}, y_n) \equiv h(\hat{\mathbf{y}}, g(\hat{\mathbf{y}})) \psi(\mathbf{y}).$$

Then let $H_k \equiv H * \phi_k$ where ϕ_k is a mollifier. Show γH_k converges uniformly to $\gamma(h\psi)$. Now recall that $C(\partial U)$ is dense in $L^p(\partial U)$. Show $\gamma(C^\infty(\overline{U}))$ is dense in $L^p(\partial U)$.

9. Let $B \times (a, b)$ be as described in the section on the divergence theorem and let

$$V^- \equiv \{(\hat{\mathbf{y}}, y_n) : y_n < g(\hat{\mathbf{y}})\}, \ V^+ \equiv \{(\hat{\mathbf{y}}, y_n) : y_n > g(\hat{\mathbf{y}})\},$$

for g a Lipschitz function. Suppose u^+ and u^- are Lipschitz functions defined on $\overline{V^+}$ and $\overline{V^-}$ respectively and suppose that $u^+(\hat{\mathbf{y}}, g(\hat{\mathbf{y}})) = u^-(\hat{\mathbf{y}}, g(\hat{\mathbf{y}}))$ for all $\hat{\mathbf{y}} \in B$. Let

$$u(\hat{\mathbf{y}}, y_n) \equiv \begin{cases} u^+(\hat{\mathbf{y}}, y_n) & \text{if } (\hat{\mathbf{y}}, y_n) \in V^+ \\ u^-(\hat{\mathbf{y}}, y_n) & \text{if } (\hat{\mathbf{y}}, y_n) \in V^- \end{cases}$$

and suppose $spt(u) \subseteq B \times (a, b)$. Then if we extend u to be 0 off of $B \times (a, b)$, show u is continuous and that the weak partial derivatives, $u_{,i}$, are all in $L^\infty(\mathbb{R}^n) \cap L^p(\mathbb{R}^n)$ for all $p > 1$ and that $u_{,i} = u_{,i}^+$ on V^+ and $u_{,i} = u_{,i}^-$ on V^-.

10. ↑ In the situation of 9, let $u \in C_c^1(B \times (a, b))$ and define

$$w(\hat{\mathbf{y}}, y_n) \equiv \begin{cases} u^-(\hat{\mathbf{y}}, y_n) & \text{if } (\hat{\mathbf{y}}, y_n) \in V^-, \\ u^+(\hat{\mathbf{y}}, y_n) & \text{if } (\hat{\mathbf{y}}, y_n) \in V^+, \end{cases}$$

while

$$w(\mathbf{y}) \equiv 0 \text{ if } \mathbf{y} \notin B \times (a, b).$$

where on V^+,

$$u^+(\hat{\mathbf{y}}, y_n) \equiv u(\hat{\mathbf{y}}, 2g(\hat{\mathbf{y}}) - y_n).$$

Show there exists a constant, C, such that

$$\|w\|_{W^{1,p}(\mathbb{R}^n)} \leq C \|u\|_{W^{1,p}(V^-)}$$

and that in fact $w \in W^{1p}(\mathbb{R}^n)$. To verify this last assertion, you must show that if $w, w_{,i} \in L^p(\mathbb{R}^n)$, for $i = 1, 2, \cdots, n$, then there exists a sequence of functions in $C_c^\infty(\mathbb{R}^n)$, w_k, such that $w_k \to w$ and $w_{k,i} \to w_{,i}$ in $L^p(\mathbb{R}^n)$. **Hint:** Let $\psi \in C_c^\infty(B(0,2))$ and $\psi = 1$ on $\overline{B(0,1)}$. Then let $\psi_k(\mathbf{x}) \equiv \psi\left(\frac{\mathbf{x}}{k}\right)$ and consider $(w\psi_k)$. Show that for k large enough

$$\left\|(w\psi_k)_{,i} - w_{,i}\right\|_p, \ \|(w\psi_k) - w\|_p$$

are both small. Argue that it suffices to let w have compact support. Then consider $w_k \equiv w * \phi_k$ where ϕ_k is a mollifier. Show $w_{k,i} = w_{,i} * \phi_k$.

11. ↑ Let U be a bounded open set with Lipschitz boundary locally on one side of its boundary as described. Show there exists $E \in \mathcal{L}(W^{1p}(U), \ W^{1p}(\mathbb{R}^n))$ such that $Eu = u$ a.e. on $\overline{U}$. This is an example of an extension theorem for Sobolev space.

12. Let $\mathbf{h}: \mathbb{R}^n \to \mathbb{R}^m$ where $m \geq n > 0$ and $\mathbf{h}$ is Lipschitz. Let

$$A = \{\mathbf{x} : \mathbf{h}(\mathbf{x}) = \mathbf{c}\}$$

where $\mathbf{c}$ is a constant vector. Show $J(\mathbf{x}) = 0$ a.e. on A. **Hint:** Use Theorem 20.15.

13. Let U be an open subset of $\mathbb{R}^n$ and let $\mathbf{h}: U \to \mathbb{R}^m$ be differentiable on $A \subseteq U$ for some A a Lebesgue measurable set. Show that if $T \subseteq A$ and $m_n(T) = 0$, then $\mathcal{H}^n(\mathbf{h}(T)) = 0$. **Hint:** Let

$$T_k \equiv \{\mathbf{x} \in T : \|D\mathbf{h}(\mathbf{x})\| < k\}$$

and let $\epsilon > 0$ be given. Now let V be an open set containing T_k which is contained in U such that $m_n(V) < \frac{\epsilon}{k^n 5^n}$ and let $\delta > 0$ be given. Using differentiability of $\mathbf{h}$, for each $\mathbf{x} \in T_k$ there exists $r_\mathbf{x} < \delta$ such that $B(\mathbf{x}, 5r_\mathbf{x}) \subseteq V$ and

$$\mathbf{h}(B(\mathbf{x}, r_\mathbf{x})) \subseteq B(\mathbf{h}(\mathbf{x}), 5kr_\mathbf{x}).$$

Use the same argument found in Lemma 20.3 to conclude

$$\mathcal{H}^n(\mathbf{h}(T_k)) = 0.$$

Now

$$\mathcal{H}^n(\mathbf{h}(T)) = \lim_{k \to \infty} \mathcal{H}^n(\mathbf{h}(T_k)) = 0.$$

14. ↑In the context of 13 show that if S is a Lebesgue measurable subset of A, then $\mathbf{h}(S)$ is $\mathcal{H}^n$ measurable. **Hint: Use the same argument found in Lemma 20.4.**

15. ↑ Suppose also that $\mathbf{h}$ is differentiable on U. Show the following holds. Let $\mathbf{x} \in A$ be a point where $U(\mathbf{x})^{-1}$ exists $(D\mathbf{h}(\mathbf{x}) \equiv R(\mathbf{x})U(\mathbf{x}))$. Then if $\epsilon \in (0,1)$ the following hold for all r small enough.

$$R^*(\mathbf{x})\mathbf{h}\left(\overline{B(\mathbf{x},r)}\right) \supseteq R^*(\mathbf{x})\mathbf{h}(\mathbf{x}) + U(\mathbf{x})B(\mathbf{0}, r(1-\epsilon)), \qquad (48)$$

$$\mathcal{H}^n\left(\mathbf{h}\left(\overline{B(\mathbf{x},r)}\right)\right) = \mathcal{H}^n\left(\mathbf{h}(B(\mathbf{x},r))\right) \geq m_n\left(U(\mathbf{x})B(\mathbf{0}, r(1-\epsilon))\right), \tag{49}$$

$$\mathbf{h}(B(\mathbf{x},r)) \subseteq \mathbf{h}(\mathbf{x}) + R(\mathbf{x})U(\mathbf{x})B(\mathbf{0}, r(1+\epsilon)), \tag{50}$$

$$\mathcal{H}^n\left(\mathbf{h}(B(\mathbf{x},r))\right) \leq m_n\left(U(\mathbf{x})B(\mathbf{0}, r(1+\epsilon))\right). \tag{51}$$

If $U \setminus A$ has measure 0, then for each $\mathbf{x} \in A$,

$$\lim_{r \to 0} \frac{\mathcal{H}^n\left(\mathbf{h}(B(\mathbf{x},r) \cap A)\right)}{\mathcal{H}^n\left(\mathbf{h}(B(\mathbf{x},r))\right)} = 1. \tag{52}$$

Also show that for $\mathbf{x} \in A$,

$$J(\mathbf{x}) = \lim_{r \to 0} \frac{\mathcal{H}^n\left(\mathbf{h}(B(\mathbf{x},r))\right)}{m_n(B(\mathbf{x},r))}, \tag{53}$$

where $J(\mathbf{x}) \equiv \det(U(\mathbf{x})) = \det\left(D\mathbf{h}(\mathbf{x})^* D\mathbf{h}(\mathbf{x})\right)^{1/2}$.

16. ↑ Assuming the context of 13 - 15 let $\mathbf{h}$ be one to one on A and establish that for F Borel measurable in $\mathbb{R}^m$

$$\int_{\mathbf{h}(A)} \mathcal{X}_F(\mathbf{y})\, d\mathcal{H}^n = \int_A \mathcal{X}_F(\mathbf{h}(\mathbf{x}))\, J(\mathbf{x})\, dm.$$

This is like 17. Next show, using the arguments of 18 - 22, that an area formula of the form

$$\int_{\mathbf{h}(A)} g(\mathbf{y})\, d\mathcal{H}^n = \int_A g(\mathbf{h}(\mathbf{x}))\, J(\mathbf{x})\, dm$$

holds whenever $g : \mathbf{h}(A) \to [0, \infty]$ is $\mathcal{H}^n$ measurable.

Chapter 21
The Coarea Formula

In Chapter 20 the area formula was discussed. This formula implies that for E a measurable set

$$\mathcal{H}^n\left(\mathbf{f}\left(E\right)\right) = \int \mathcal{X}_E\left(\mathbf{x}\right) J\left(\mathbf{x}\right) dm$$

where $\mathbf{f} : \mathbb{R}^n \to \mathbb{R}^m$ for $\mathbf{f}$ a Lipschitz mapping and $m \geq n$. It is a version of the change of variables formula for multiple integrals. The coarea formula is a statement about the Hausdorff measure of a set which involves the inverse image of $\mathbf{f}$. It is an extension of Fubini's theorem. Recall that if $n > m$ and $\mathbb{R}^n = \mathbb{R}^m \times \mathbb{R}^{n-m}$, we may take a product measurable set, $E \subseteq \mathbb{R}^n$, and obtain its Lebesgue measure by the formula

$$
\begin{aligned}
m_n\left(E\right) &= \int_{\mathbb{R}^m} \int_{\mathbb{R}^{n-m}} \mathcal{X}_E\left(\mathbf{y}, \mathbf{x}\right) dm_{n-m} dm_m \\
&= \int_{\mathbb{R}^m} m_{n-m}\left(E^{\mathbf{y}}\right) dm_m = \int_{\mathbb{R}^m} \mathcal{H}^{n-m}\left(E^{\mathbf{y}}\right) dm_m.
\end{aligned}
$$

Let π_1 and π_2 be defined by

$$\pi_2\left(\mathbf{y}, \mathbf{x}\right) = \mathbf{x}, \; \pi_1\left(\mathbf{y}, \mathbf{x}\right) = \mathbf{y}.$$

Then

$$E^{\mathbf{y}} = \pi_2\left(\pi_1^{-1}\left(\mathbf{y}\right) \cap E\right)$$

and so

$$
\begin{aligned}
m_n\left(E\right) &= \int_{\mathbb{R}^m} \mathcal{H}^{n-m}\left(\pi_2\left(\pi_1^{-1}\left(\mathbf{y}\right) \cap E\right)\right) dm_m \\
&= \int_{\mathbb{R}^m} \mathcal{H}^{n-m}\left(\pi_1^{-1}\left(\mathbf{y}\right) \cap E\right) dm_m. \tag{1}
\end{aligned}
$$

Thus, the notion of product measure yields a formula for the measure of a set in terms of the inverse image of one of the projection maps onto a smaller dimensional subspace. The coarea formula gives a generalization of Formula 1 in the case where π_1 is replaced by an arbitrary Lipschitz function mapping $\mathbb{R}^n$ to $\mathbb{R}^m$.

It is possible to obtain the coarea formula as a computation involving the area formula and some simple linear algebra and this is the approach taken here.

21.1   A determinant identity

This section is devoted to proving the following theorem and its corollary which will serve as the basis for the derivation of the coarea formula.

Theorem 21.1 *Let $p \le q$ and let A be a $p \times q$ matrix. Let the positive eigenvectors of AA^* be $\{\lambda_1, \cdots, \lambda_r\}$ and let Q be an orthogonal matrix such that*

$$Q^* AA^* Q = \begin{pmatrix} D & 0 \\ 0 & 0 \end{pmatrix}_{p \times p}$$

where

$$D = \begin{pmatrix} \lambda_1 & & \\ & \ddots & \\ & & \lambda_r \end{pmatrix}$$

is a diagonal matrix. Then there exists an orthogonal matrix R, such that

$$R^* A^* AR = \begin{pmatrix} D & 0 \\ 0 & 0 \end{pmatrix}_{q \times q}.$$

Proof: Let the columns of Q be $(\mathbf{v}_1, \cdots, \mathbf{v}_r, \mathbf{v}_{r+1}, \cdots, \mathbf{v}_p)$. Thus $AA^* \mathbf{v}_i = \lambda_i \mathbf{v}_i$ for each $\mathbf{v}_i$ with $i \le r$. Now define for each $i \le r$, $\mathbf{v}_i^1 \equiv A^* \mathbf{v}_i / |A^* \mathbf{v}_i|$. This is well defined because $|A^* \mathbf{v}_i| > 0$ due to the assumption that $\lambda_i > 0$. Also $(\mathbf{v}_i^1, \mathbf{v}_j^1) = \delta_{ij}$ and $A^* A (A^* \mathbf{v}_i) = \lambda_i (A^* \mathbf{v}_i)$ so $\mathbf{v}_i^1$ is an eigenvector of $A^* A$ with eigenvalues λ_i. Enlarge $\{\mathbf{v}_1^1, \cdots, \mathbf{v}_r^1\}$ to an orthonormal basis of eigenvectors, $\{\mathbf{v}_1^1, \cdots, \mathbf{v}_r^1, \mathbf{w}_1, \cdots, \mathbf{w}_{q-r}\}$.

Claim: $A^* A \mathbf{w}_i = \mathbf{0}$ for each eigenvector, $\mathbf{w}_i$.

Proof: If not, then $A^* A \mathbf{w}_k = \mu_k \mathbf{w}_k$ for some $\mu_k \ne 0$. Hence

$$AA^* (A \mathbf{w}_k) = \mu_k (A \mathbf{w}_k)$$

and so $\mu_k = \lambda_i$ for some $i \le r$ and $A \mathbf{w}_k = \sum_{i=1}^r \alpha_i \mathbf{v}_i$ where not all the α_i are equal to 0. Hence

$$A \mathbf{w}_k = \sum_{i=1}^r \frac{\alpha_i}{\lambda_i} AA^* \mathbf{v}_i$$

which implies, after factoring out A and multiplying by A^*, that

$$A^* A \Big(\mathbf{w}_k - \sum_{i=1}^r \frac{\alpha_i}{\lambda_i} A^* \mathbf{v}_i \Big) = 0.$$

Therefore the expression in the parentheses is an eigenvector of $A^* A$ corresponding to a zero eigenvalue. It follows the vector in the parentheses must be a linear combination of the $\mathbf{w}_k$, those eigenvectors having zero eigenvalue. Therefore, using the definition of the $\mathbf{v}_i^1$,

$$\mathbf{w}_k - \sum_{i=1}^r \frac{\alpha_i}{\lambda_i} A^* \mathbf{v}_i = \mathbf{w}_k - \sum_{i=1}^r \frac{\alpha_i}{\lambda_i} |A^* \mathbf{v}_i| \mathbf{v}_i^1 = \sum_{j=1}^{q-r} a_j \mathbf{w}_j.$$

420

The Coarea formula

This implies

$$\mathbf{w}_k - \sum_{j=1}^{q-r} a_j \mathbf{w}_j = \sum_{i=1}^{r} \frac{\alpha_i}{\lambda_i} |A^* \mathbf{v}_i| \mathbf{v}_i^1 = 0,$$

a contradiction because not all the α_i equal zero.

To conclude the proof, let R be the matrix whose columns are

$$(\mathbf{v}_1^1, \cdots, \mathbf{v}_r^1, \mathbf{w}_1, \cdots, \mathbf{w}_{q-r}).$$

Then the above shows

$$R^* A^* A R = \begin{pmatrix} D & 0 \\ 0 & 0 \end{pmatrix}_{q \times q}$$

as claimed. This proves the theorem.

The following corollary is what will be used in this chapter. It follows immediately from the above theorem.

Corollary 21.2 *Let A be an $m \times n$ matrix. Then*

$$\det (I + A A^*) = \det (I + A^* A).$$

21.2 The Coarea formula

We need to use the chain rule and the next theorem presents a version of this rule.

Theorem 21.3 *Let $\mathbf{f}$ and $\mathbf{g}$ be Lipschitz mappings from $\mathbb{R}^n$ to $\mathbb{R}^n$ with $\mathbf{f}(\mathbf{g}(\mathbf{x})) = \mathbf{x}$ on A, a measurable set. Then for a.e. $\mathbf{x} \in A$, $D\mathbf{g}(\mathbf{f}(\mathbf{x}))$, $D\mathbf{f}(\mathbf{x})$, and $D(\mathbf{f} \circ \mathbf{g})(\mathbf{x})$ all exist and*

$$I = D(\mathbf{g} \circ \mathbf{f})(\mathbf{x}) = D\mathbf{g}(\mathbf{f}(\mathbf{x})) D\mathbf{f}(\mathbf{x}).$$

The proof of this theorem is based on the following lemma.

Lemma 21.4 *If $\mathbf{h} : \mathbb{R}^n \to \mathbb{R}^n$ is Lipschitz, then if $\mathbf{h}(\mathbf{x}) = 0$ for all $\mathbf{x} \in A$, then $\det (D\mathbf{h}(\mathbf{x})) = 0$ a.e.*

Proof: By the Area formula, $0 = \int_{\{0\}} \#(\mathbf{y}) \, dy = \int_A |\det (D\mathbf{h}(\mathbf{x}))| \, dx$ and so $\det (D\mathbf{h}(\mathbf{x})) = 0$ a.e.

Proof of the theorem: On A, $\mathbf{g}(\mathbf{f}(\mathbf{x})) - \mathbf{x} = 0$ and so by the lemma, there exists a set of measure zero, N_1 such that if $\mathbf{x} \notin N_1$, $D(\mathbf{g} \circ \mathbf{f})(\mathbf{x}) - I = 0$. Let M be the set of points in $\mathbf{f}(\mathbb{R}^n)$ where $\mathbf{g}$ fails to be differentiable and let $N_2 \equiv \mathbf{g}(M) \cap A$, also a set of measure zero. Finally let N_3 be the set of points where $\mathbf{f}$ fails to be differentiable. Then if $\mathbf{x} \notin N_1 \cup N_2 \cup N_3$, the chain rule implies $I = D\left(\mathbf{g} \circ \mathbf{f}\right)(\mathbf{x}) = D\mathbf{g}\left(\mathbf{f}\left(\mathbf{x}\right)\right) D\mathbf{f}\left(\mathbf{x}\right)$. This proves the theorem.

It is convenient to define the following for a measure space $(\Omega, \mathcal{S}, \mu)$ and $f : \Omega \to [0, \infty]$, an arbitrary function, maybe not measurable.

$$\int^* f d\mu \equiv \int_\Omega^* f d\mu \equiv \inf\{\int_\Omega g d\mu : g \geq f, \text{ and } g \text{ measurable}\}$$

Lemma 21.5 *Suppose* $f_n \geq 0$ *and*

$$\lim_{n \to \infty} \sup \int^* f_n d\mu = 0.$$

Then there is a subsequence f_{n_k} *such that* $f_{n_k}(x) \to 0$ *a.e.* x.

Proof: For n large enough, $\int^* f_n d\mu < \infty$. Let n be this large and pick $g_n \geq f_n$, g_n measurable, such that

$$\int^* f_n d\mu + n^{-1} > \int g_n d\mu.$$

Thus

$$\limsup \int g_n d\mu = \liminf \int g_n d\mu = \lim \int g_n d\mu = 0.$$

If $n = 1, 2, \cdots$, let $k_n > \max(k_{n-1}, n)$ be such that

$$\int g_{k_n} d\mu < 2^{-n}.$$

Then

$$\mu\left(\left[g_{k_n} \geq n^{-1}\right]\right) \leq 2^{-n} n.$$

Thus

$$\sum_{n=1}^{\infty} \mu\left(\left[g_{k_n} \geq n^{-1}\right]\right) < \infty$$

and so for all N,

$$\mu\left(\cap_{n=1}^{\infty} \cup_{m \geq n} \left[g_{k_m} \geq m^{-1}\right]\right) \leq \sum_{n=N}^{\infty} \mu\left(\left[g_{k_m} \geq m^{-1}\right]\right)$$

$$\leq \sum_{n=N}^{\infty} n 2^{-n}.$$

Thus

$$\mu\left(\cap_{n=1}^{\infty} \cup_{m \geq n} \left[g_{k_m} \geq m^{-1}\right]\right) = 0.$$

If $g_{k_m}(x)$ fails to converge to 0, then x is in $\cap_{n=1}^{\infty} \cup_{m \geq n} \left[g_{k_m} \geq m^{-1}\right]$ and so $g_{k_m}(x) \to 0$ a.e. x. Since $f_{k_m}(x) \leq g_{k_m}(x)$, this proves the lemma.

It might help a little before proceeding further to recall the concept of a level surface of a function of n variables. If $f : U \subseteq \mathbb{R}^n \to \mathbb{R}$, such a level surface is of the form $f^{-1}(y)$ and we would expect it to be an $n-1$ dimensional thing in some sense. In the next lemma, we consider a more general construction in which the function has values in $\mathbb{R}^m$. In this more general case, we would expect $\mathbf{f}^{-1}(\mathbf{y})$ to be something which is an $n-m$ dimensional thing.

Lemma 21.6 *Let $A \subseteq \mathbb{R}^p$ and let $\mathbf{f} : \mathbb{R}^p \to \mathbb{R}^m$ be Lipschitz. Then*

$$\int_{\mathbb{R}^m}^* \mathcal{H}^s \left(A \cap \mathbf{f}^{-1}(\mathbf{y}) \right) d\mathcal{H}^m$$

$$\leq \frac{\alpha(s)\,\alpha(m)}{\alpha(s+m)} \left(Lip\,(\mathbf{f}) \right)^m \mathcal{H}^{s+m}(A).$$

Proof: The formula is obvious if $\mathcal{H}^{s+m}(A) = \infty$ so assume

$$\mathcal{H}^{s+m}(A) < \infty.$$

$$A \subseteq \cup_{i=1}^\infty B_i^j, \; diam\left(B_i^j \right) \leq j^{-1}, \; B_i^j \text{ is closed,}$$

and

$$\mathcal{H}_{j-1}^{s+m}(A) + j^{-1} \geq \sum_{i=1}^\infty \alpha(s+m) \left(\frac{diam\left(B_i^j \right)}{2} \right)^{s+m} \tag{2}$$

Now define

$$g_i^j(\mathbf{y}) \equiv \alpha(s) \left(\frac{diam\left(B_i^j \right)}{2} \right)^s \mathcal{X}_{\mathbf{f}(B_i^j)}(\mathbf{y}).$$

Thus $g_i^j(\mathbf{y}) \neq 0$ when $\mathbf{f}^{-1}(\mathbf{y}) \in B_i^j$. Since each B_i^j is closed, $\mathbf{f}\left(B_i^j \right)$ is a Borel set. Therefore g_i^j is a Borel measurable function. The set, $A \cap \mathbf{f}^{-1}(\mathbf{y})$ is covered by some subsequence, $\{B_{i_k}^j\}_{k=1}^\infty$ with the property that for each of these sets, $\mathbf{y} \in \mathbf{f}(B_{i_k}^j)$. Thus, since $g_i^j(\mathbf{y}) = 0$ if $\mathbf{y} \notin \mathbf{f}\left(B_i^j \right)$,

$$\mathcal{H}_{j-1}^s \left(A \cap \mathbf{f}^{-1}(\mathbf{y}) \right) \leq \sum_{k=1}^\infty \alpha(s) \left(\frac{diam\left(B_{i_k}^j \right)}{2} \right)^s = \sum_{i=1}^\infty g_i^j(\mathbf{y}),$$

a Borel measurable function. It follows,

$$\int_{\mathbb{R}^m}^* \mathcal{H}^s \left(A \cap \mathbf{f}^{-1}(\mathbf{y}) \right) d\mathcal{H}^m \quad = \quad \int_{\mathbb{R}^m}^* \lim_{j \to \infty} \mathcal{H}_{j-1}^s \left(A \cap \mathbf{f}^{-1}(\mathbf{y}) \right) d\mathcal{H}^m$$

$$\leq \quad \int_{\mathbb{R}^m}^* \lim_{j \to \infty} \inf \sum_{i=1}^\infty g_i^j(\mathbf{y}) \, d\mathcal{H}^m.$$

By Borel measurability of the integrand,

$$\leq \int_{\mathbb{R}^m} \lim_{j \to \infty} \inf \sum_{i=1}^{\infty} g_i^j(\mathbf{y}) \, d\mathcal{H}^m.$$

By Fatou's lemma,

$$\leq \lim_{j \to \infty} \inf \int_{\mathbb{R}^m} \sum_{i=1}^{\infty} g_i^j(\mathbf{y}) \, d\mathcal{H}^m$$

$$= \lim_{j \to \infty} \inf \sum_{i=1}^{\infty} \alpha(s) \left(\frac{diam \left(B_i^j \right)}{2} \right)^s \mathcal{H}^m \left(\mathbf{f} \left(B_i^j \right) \right).$$

By the isodiametric inequality and equality of $\mathcal{H}^m$ and Lebesgue measure on $\mathbb{R}^m$,

$$\leq \lim_{j \to \infty} \inf \sum_{i=1}^{\infty} \alpha(s) \left(\frac{diam \left(B_i^j \right)}{2} \right)^s \left(\frac{diam \left(\mathbf{f}(B_i^j) \right)}{2} \right)^m \alpha(m)$$

$$\leq \alpha(s) \alpha(m) Lip(\mathbf{f})^m \lim_{j \to \infty} \inf \sum_{i=1}^{\infty} \left(\frac{diam \left(B_i^j \right)}{2} \right)^{s+m}$$

$$\leq \frac{\alpha(s) \alpha(m) Lip(\mathbf{f})^m}{\alpha(s+m)} \lim_{j \to \infty} \inf \sum_{i=1}^{\infty} \alpha(s+m) \left(\frac{diam \left(B_i^j \right)}{2} \right)^{s+m}$$

$$\leq \frac{\alpha(s) \alpha(m) Lip(\mathbf{f})^m}{\alpha(s+m)} \mathcal{H}^{s+m}(A)$$

by Formula 2. This proves the lemma.

We want to get away from always having to use $\int^*$ rather than $\int$. The next lemma will enable the change in notation.

Lemma 21.7 *Let A be a Lebesgue measurable subset of $\mathbb{R}^n$ and let $\mathbf{f} : \mathbb{R}^n \to \mathbb{R}^m$ be Lipschitz. Then*

$$\mathbf{y} \to \mathcal{H}^{n-m} \left(A \cap \mathbf{f}^{-1}(\mathbf{y}) \right)$$

is Lebesgue measurable. If A is compact, this function is Borel measurable.

Proof: Suppose first that A is compact. Then $A \cap \mathbf{f}^{-1}(\mathbf{y})$ is also and so it is $\mathcal{H}^{n-m}$ measurable. Suppose $\mathcal{H}^{n-m} \left(A \cap \mathbf{f}^{-1}(\mathbf{y}) \right) < t$. Then for all $\delta > 0$,

$$\mathcal{H}_\delta^{n-m} \left(A \cap \mathbf{f}^{-1}(\mathbf{y}) \right) < t$$

and so there exist open sets S_i, satisfying

$$diam\,(S_i) < \delta,\ A \cap \mathbf{f}^{-1}(\mathbf{y}) \subseteq \cup_{i=1}^{\infty} S_i,$$

and

$$\sum_{i=1}^{\infty} \alpha\,(n-m) \left(\frac{diam\,(S_i)}{2} \right)^{n-m} < t.$$

Claim: If $\mathbf{z}$ is close enough to $\mathbf{y}$, then $A \cap \mathbf{f}^{-1}(\mathbf{z}) \subseteq \cup_{i=1}^{\infty} S_i$.
Proof: If not, then there exists a sequence $\{\mathbf{z}_k\}$ such that

$$\mathbf{z}_k \to \mathbf{y},$$

and

$$\mathbf{x}_k \in (A \cap \mathbf{f}^{-1}(\mathbf{z}_k)) \setminus \cup_{i=1}^{\infty} S_i.$$

Taking a subsequence still denoted by k we can have

$$\mathbf{z}_k \to \mathbf{y},\ \mathbf{x}_k \to \mathbf{x} \in A \setminus \cup_{i=1}^{\infty} S_i.$$

Hence

$$\mathbf{f}\,(\mathbf{x}) = \lim_{k \to \infty} \mathbf{f}\,(\mathbf{x}_k) = \lim_{k \to \infty} \mathbf{z}_k = \mathbf{y}.$$

But $\mathbf{x} \notin \cup_{i=1}^{\infty} S_i$ contrary to the assumption that $A \cap \mathbf{f}^{-1}(\mathbf{y}) \subseteq \cup_{i=1}^{\infty} S_i$.
It follows from this claim that whenever $\mathbf{z}$ is close enough to $\mathbf{y}$,

$$\mathcal{H}_{\delta}^{n-m}\left(A \cap \mathbf{f}^{-1}(\mathbf{z}) \right) < t.$$

Thus, if

$$U_{\delta} \equiv \{\mathbf{z} : \mathcal{H}_{\delta}^{n-m}\left(A \cap \mathbf{f}^{-1}(\mathbf{z}) \right) < t + \delta\},$$

then U_{δ} is open. Hence, letting $\delta_i \to 0+$,

$$\{\mathbf{z} : \mathcal{H}^{n-m}\left(A \cap \mathbf{f}^{-1}(\mathbf{z}) \right) \leq t\} = \cap_{i=1}^{\infty} U_{\delta_i} = \text{Borel set.}$$

Thus, if A is compact, then for each $\mathbf{y} \in \mathbb{R}^m$, $A \cap \mathbf{f}^{-1}(\mathbf{y})$ is $\mathcal{H}^{n-m}$ measurable and also the function

$$\mathbf{y} \to \mathcal{H}^{n-m}\left(A \cap \mathbf{f}^{-1}(\mathbf{y}) \right)$$

is a Borel measurable function, hence Lebesgue measurable. Now let V be an open set and let

$$A_k \uparrow V,\ A_k \text{ compact.}$$

Then

$$\mathcal{H}^{n-m}\left(V \cap \mathbf{f}^{-1}(\mathbf{y}) \right) = \lim_{k \to \infty} \mathcal{H}^{n-m}\left(A_k \cap \mathbf{f}^{-1}(\mathbf{y}) \right)$$

so $\mathbf{y} \to \mathcal{H}^{n-m} \left(V \cap \mathbf{f}^{-1} (\mathbf{y}) \right)$ is measurable for all V open. Now let A be only Lebesgue measurable with $m(A) < \infty$ and let

$$V_k \supseteq A, \ V_k \text{ open }, \ \cdots, V_k \supseteq V_{k+1} \cdots, \ m(V_1) < \infty, \ m(V_k) \downarrow m(A).$$

Then $m(V_k \setminus A) \to 0$. Now if $\mathcal{H}^{n-m} \left(A \cap \mathbf{f}^{-1} (\mathbf{y}) \right)$ is finite,

$$\mathcal{H}^{n-m} \left(V_k \cap \mathbf{f}^{-1} (\mathbf{y}) \right) - \mathcal{H}^{n-m} \left(A \cap \mathbf{f}^{-1} (\mathbf{y}) \right)$$

$$\leq \mathcal{H}^{n-m} \left((V_k \setminus A) \cap \mathbf{f}^{-1} (\mathbf{y}) \right)$$

also by Lemma 21.6,

$$\int_{\mathbb{R}^m}^{*} \mathcal{H}^{n-m} \left(A \cap \mathbf{f}^{-1} (\mathbf{y}) \right) d\mathcal{H}^m$$

$$\leq \frac{\alpha(m) \alpha(n-m)}{\alpha(n)} \left(Lip \left(\mathbf{f} \right) \right)^m m(A) < \infty$$

and so $\mathcal{H}^{n-m} \left(A \cap \mathbf{f}^{-1} (\mathbf{y}) \right) < \infty$ a.e. $\mathbf{y}$. Hence for $\mathcal{H}^m$ a.e. $\mathbf{y}$,

$$\left| \mathcal{H}^{n-m} \left(V_k \cap \mathbf{f}^{-1} (\mathbf{y}) \right) - \mathcal{H}^{n-m} \left(A \cap \mathbf{f}^{-1} (\mathbf{y}) \right) \right|$$

$$\leq \mathcal{H}^{n-m} \left((V_k \setminus A) \cap \mathbf{f}^{-1} (\mathbf{y}) \right).$$

Thus,

$$\limsup_{k \to \infty} \int_{\mathbb{R}^m}^{*} \left| \mathcal{H}^{n-m} \left(V_k \cap \mathbf{f}^{-1} (\mathbf{y}) \right) - \mathcal{H}^{n-m} \left(A \cap \mathbf{f}^{-1} (\mathbf{y}) \right) \right| dy \leq$$

$$\limsup_{k \to \infty} \int_{\mathbb{R}^m}^{*} \mathcal{H}^{n-m} \left((V_k \setminus A) \cap \mathbf{f}^{-1} (\mathbf{y}) \right) dy \leq$$

$$\limsup_{k \to \infty} \frac{\alpha(m) \alpha(n-m)}{\alpha(n)} \left(Lip \left(\mathbf{f} \right) \right)^m m(V_k \setminus A) = 0.$$

By Lemma 21.5 there is a subsequence k_i such that

$$\mathcal{H}^{n-m} \left(V_{k_i} \cap \mathbf{f}^{-1} (\mathbf{y}) \right) \to \mathcal{H}^{n-m} \left(A \cap \mathbf{f}^{-1} (\mathbf{y}) \right)$$

for $\mathcal{H}^m$ a.e. $\mathbf{y}$. Hence $\mathbf{y} \to \mathcal{H}^{n-m} \left(A \cap \mathbf{f}^{-1} (\mathbf{y}) \right)$ is measurable as claimed if $m(A) < \infty$. If A is an arbitrary Lebesgue measurable set, then

$$\mathcal{H}^{n-m} \left(A \cap \mathbf{f}^{-1} (\mathbf{y}) \right) = \lim_{k \to \infty} \mathcal{H}^{n-m} \left(A \cap B(0,k) \cap \mathbf{f}^{-1} (\mathbf{y}) \right)$$

and so the function, $\mathbf{y} \to \mathcal{H}^{n-m} \left(A \cap \mathbf{f}^{-1} (\mathbf{y}) \right)$, is measurable. This proves the lemma.

With this lemma proved, it is possible to obtain the following useful inequality which will be used repeatedly.

Lemma 21.8 *If $A \subseteq \mathbb{R}^n$ is Lebesgue measurable, then*

$$\int_{\mathbb{R}^m} \mathcal{H}^{n-m} \left(A \cap \mathbf{f}^{-1}\left(\mathbf{y}\right) \right) dy$$

$$\leq \frac{\alpha\left(n-m\right)\alpha\left(m\right)}{\alpha\left(n\right)} \left(Lip\left(\mathbf{f}\right) \right)^m m_n\left(A\right).$$

Proof: This follows from Lemma 21.6 and Lemma 21.7. Since

$$\mathbf{y} \to \mathcal{H}^{n-m}\left(A \cap \mathbf{f}^{-1}\left(\mathbf{y}\right)\right)$$

is measurable,

$$\int_{\mathbb{R}^m} \mathcal{H}^{n-m}\left(A \cap \mathbf{f}^{-1}\left(\mathbf{y}\right)\right) dy = \int_{\mathbb{R}^m}^* \mathcal{H}^{n-m}\left(A \cap \mathbf{f}^{-1}\left(\mathbf{y}\right)\right) dy.$$

Now let $p = n$, and $s = n - m$ in Lemma 21.6. This proves the lemma.

With these lemmas it is now possible to establish the coarea formula. First we define $\Lambda\left(n, m\right)$ as all possible ordered lists of m numbers taken from $\{1, 2, \cdots, n\}$.

Theorem 21.9 *Let A be a measurable set in $\mathbb{R}^n$ and let $\mathbf{f} : \mathbb{R}^n \to \mathbb{R}^m$ be a Lipschitz map. Then the following formula holds along with all measurability assertions needed for it to make sense.*

$$\int_{\mathbb{R}^m} \mathcal{H}^{n-m}\left(A \cap \mathbf{f}^{-1}\left(\mathbf{y}\right)\right) dy = \int_A J\mathbf{f}\left(\mathbf{x}\right) dx \tag{3}$$

where

$$J\mathbf{f}\left(\mathbf{x}\right) \equiv \det\left(D\mathbf{f}\left(\mathbf{x}\right) D\mathbf{f}\left(\mathbf{x}\right)^*\right)^{1/2}.$$

Proof: For $\mathbf{x} \in \mathbb{R}^n$, and $\mathbf{i} \in \Lambda\left(n, m\right)$, with $\mathbf{i} = \left(i_1, \cdots, i_m\right)$, we define $\mathbf{x_i} \equiv \left(x_{i_1}, \cdots, x_{i_m}\right)$, and $\pi_{\mathbf{i}}\mathbf{x} \equiv \mathbf{x_i}$. Also for $\mathbf{i} \in \Lambda\left(n, m\right)$, let $\mathbf{i}_c \in \Lambda\left(n, n - m\right)$ consist of the remaining indices taken in order. For $\mathbf{f} : \mathbb{R}^n \to \mathbb{R}^m$ where $m \leq n$, we define $J\mathbf{f}\left(\mathbf{x}\right) \equiv \det\left(D\mathbf{f}\left(\mathbf{x}\right) D\mathbf{f}\left(\mathbf{x}\right)^*\right)^{1/2}$ and let $S \equiv \{\mathbf{x} : J\mathbf{f}\left(\mathbf{x}\right) = 0\}$ and $N \equiv \{\mathbf{x} : D\mathbf{f}\left(\mathbf{x}\right)$ does not exist$\}$. Thus $m\left(N\right) = 0$. Let A be a closed subset of $\mathbb{R}^n \setminus \{S \cup N\}$. For each $\mathbf{i} \in \Lambda\left(n, m\right)$, define

$$\mathbf{f^i}\left(\mathbf{x}\right) \equiv \left(\begin{array}{c} \mathbf{f}\left(\mathbf{x}\right) \\ \mathbf{x_{i_c}} \end{array} \right).$$

By Lemma 20.13 and Corollary 20.14, there exist disjoint measurable sets $\{F_j^{\mathbf{i}}\}_{j=1}^{\infty}$ such that $\mathbf{f^i}$ is one to one on $F_j^{\mathbf{i}}$, $\left(\mathbf{f^i}\right)^{-1}$ is Lipschitz on $\mathbf{f^i}\left(F_j^{\mathbf{i}}\right)$, and $\cup_{j=1}^{\infty} F_j^{\mathbf{i}} = \{\mathbf{x} : D\mathbf{f^i}\left(\mathbf{x}\right)$ exists and $\det D\mathbf{f^i}\left(\mathbf{x}\right) \neq 0\}$. If $\mathbf{x} \in \mathbb{R}^n \setminus \{S \cup N\}$, it follows for some $\mathbf{i} \in \Lambda\left(n, m\right)$ that $\det\left(D_{\mathbf{x_i}}\mathbf{f}\left(\mathbf{x}\right)\right) \neq 0$. But $\det\left(D_{\mathbf{x_i}}\mathbf{f}\left(\mathbf{x}\right)\right) = \det\left(D\mathbf{f^i}\left(\mathbf{x}\right)\right)$ and so $\mathbf{x} \in F_j^{\mathbf{i}}$ for some $\mathbf{i}$ and j. Hence

$$\cup_{\mathbf{i},j} F_j^{\mathbf{i}} \supseteq A.$$

Now let $\{E_j^i\}$ be measurable sets such that $E_j^i \subseteq F_k^i$ for some k, the sets are disjoint, and their union coincides with $\cup_{i,j} F_j^i$. Then

$$\int_A Jf(x)\, dx = \sum_{i \, \in \Lambda(n,m)} \sum_{j=1}^{\infty} \int_{E_j^i \cap A} \det\left(Df(x)\, Df(x)^*\right)^{1/2} dx. \quad (4)$$

Let $g : \mathbb{R}^n \to \mathbb{R}^n$ be a Lipschitz extension of $(f^i)^{-1}$ so $g \circ f^i(x) = x$ for all $x \in E_j^i$. First, using Theorem 21.3, and the fact that Lipschitz mappings take sets of measure zero to sets of measure zero, we replace E_j^i with $\widetilde{E}_j^i \subseteq E_j^i$ such that $E_j^i \setminus \widetilde{E}_j^i$ has measure zero and

$$Df^i(g(y))\, Dg(y) = I$$

on $f^i\left(\widetilde{E}_j^i\right)$. Changing the variables using the area formula and making this change, the expression in Formula 4 equals

$$\int_A Jf(x)\, dx = \sum_{i \, \in \Lambda(n,m)} \sum_{j=1}^{\infty} \int_{f^i\left(\widetilde{E}_j^i \cap A\right)} \cdot$$

$$\det\left(Df(g(y))\, Df(g(y))^*\right)^{1/2} \left|\det Df^i(g(y))\right|^{-1} dy. \quad (5)$$

Now to avoid technicalities, we assume a Borel measurable representative of the integrand on the right has been chosen and we also replace $\widetilde{E}_j^i$ with a compact set, K_j^i contained in it. Thus,

$$\int_{K_j^i \cap A} \det\left(Df(x)\, Df(x)^*\right)^{1/2} dx =$$

$$\int_{\mathbb{R}^n} \mathcal{X}_{f^i\left(K_j^i \cap A\right)}(y) \det\left(Df(g(y))\, Df(g(y))^*\right)^{1/2} \cdot$$

$$\left|\det Df^i(g(y))\right|^{-1} dy.$$

$$= \int_{\mathbb{R}^m} \int_{\pi_{i_c}\left(f^{-1}(y_1) \cap K_j^i \cap A\right)} \det\left(Df(g(y))\, Df(g(y))^*\right)^{1/2} \cdot$$

$$\left|\det Df_{x_i}(g(y))\right|^{-1} dy_2 dy_1 \quad (6)$$

where $y_1 = f(x)$ and $y_2 = x_{i_c}$. Thus

$$y_2 = \pi_{i_c} g(y) = \pi_{i_c} g\left(f^i(x)\right) = x_{i_c}. \quad (7)$$

Now consider the inner integral in Formula 6 in which y_1 is fixed. The integrand equals

$$\det\left[\left(D_{x_i}f(g(y))\, D_{x_{i_c}}f(g(y))\right)\begin{pmatrix} D_{x_i}f(g(y))^* \\ D_{x_{i_c}}f(g(y))^* \end{pmatrix}\right]^{1/2}.$$

$$|\det D\mathbf{f}_{\mathbf{x_i}}\,(\mathbf{g}\,(\mathbf{y}))|^{-1}. \tag{8}$$

Since $\mathbf{y}_1$ is fixed, and $\mathbf{y}_1 = \mathbf{f}\,(\pi_{\mathbf{i}}\mathbf{g}\,(\mathbf{y}),\pi_{\mathbf{i}_c}\mathbf{g}\,(\mathbf{y}))$, it follows

$$
\begin{aligned}
0 &= D_{\mathbf{x_i}}\mathbf{f}\,(\mathbf{g}\,(\mathbf{y}))\,D_{\mathbf{y}_2}\pi_{\mathbf{i}}\mathbf{g}\,(\mathbf{y}) + D_{\mathbf{x}_{i_c}}\mathbf{f}\,(\mathbf{g}\,(\mathbf{y}))\,D_{\mathbf{y}_2}\pi_{\mathbf{i}_c}\mathbf{g}\,(\mathbf{y})\\
&= D_{\mathbf{x_i}}\mathbf{f}\,(\mathbf{g}\,(\mathbf{y}))\,D_{\mathbf{y}_2}\pi_{\mathbf{i}}\mathbf{g}\,(\mathbf{y}) + D_{\mathbf{x}_{i_c}}\mathbf{f}\,(\mathbf{g}\,(\mathbf{y})).
\end{aligned}
$$

Letting $A \equiv D_{\mathbf{x_i}}\mathbf{f}\,(\mathbf{g}\,(\mathbf{y}))$ and $B \equiv D_{\mathbf{y}_2}\pi_{\mathbf{i}}\mathbf{g}\,(\mathbf{y})$ and using the above formula, 8 is of the form

$$\det\left[(\ A\ \ -AB\ )\left(\begin{array}{c} A^* \\ -B^*A^* \end{array}\right)\right]^{1/2}|\det A|^{-1} = \det\,(I + BB^*)^{1/2},$$

which, by Corollary 21.2, equals $\det\,(I + B^*B)^{1/2}$. (Note the size of the identity changes in these two expressions, the first being an $m \times m$ matrix and the second being a $n-m \times n-m$ matrix.) By Formula 7 $\pi_{\mathbf{i}_c}\mathbf{g}\,(\mathbf{y}) = \mathbf{y}_2$ and so,

$$\det\,(I + B^*B)^{1/2} = \det\left[(\ B^*\ \ I\ )\left(\begin{array}{c} B \\ I \end{array}\right)\right]^{1/2}$$

$$= \det\left[(\ D_{\mathbf{y}_2}\pi_{\mathbf{i}}\mathbf{g}\,(\mathbf{y})^*\ \ D_{\mathbf{y}_2}\pi_{\mathbf{i}_c}\mathbf{g}\,(\mathbf{y})^*\ )\left(\begin{array}{c} D_{\mathbf{y}_2}\pi_{\mathbf{i}}\mathbf{g}\,(\mathbf{y}) \\ D_{\mathbf{y}_2}\pi_{\mathbf{i}_c}\mathbf{g}\,(\mathbf{y}) \end{array}\right)\right]^{1/2}$$

$$= \det\,(D_{\mathbf{y}_2}\mathbf{g}\,(\mathbf{y})^*\,D_{\mathbf{y}_2}\mathbf{g}\,(\mathbf{y}))^{1/2}.$$

Therefore, Formula 6 reduces to

$$\int_{K_j^{\mathbf{i}}\cap A}\det\,(D\mathbf{f}\,(\mathbf{x})\,D\mathbf{f}\,(\mathbf{x})^*)^{1/2}\,d\mathbf{x} = \int_{\mathbb{R}^m}\int_{\pi_{\mathbf{i}_c}\left(\mathbf{f}^{-1}(\mathbf{y}_1)\cap K_j^{\mathbf{i}}\cap A\right)}$$

$$\det\,(D_{\mathbf{y}_2}\mathbf{g}\,(\mathbf{y})^*\,D_{\mathbf{y}_2}\mathbf{g}\,(\mathbf{y}))^{1/2}\,dy_2 dy_1. \tag{9}$$

If $\mathbf{y}_2 = \mathbf{x}_{i_c}$ where $(\mathbf{x_i},\mathbf{x}_{i_c}) \in \mathbf{f}^{-1}\,(\mathbf{y}_1) \cap K_j^{\mathbf{i}} \cap A$, then by Formula 7,

$$\mathbf{f}\,(\mathbf{x_i},\pi_{\mathbf{i}_c}\mathbf{g}\,(\mathbf{y})) = \mathbf{y}_1,\ \mathbf{f}\,(\pi_{\mathbf{i}}\mathbf{g}\,(\mathbf{y}),\pi_{\mathbf{i}_c}\mathbf{g}\,(\mathbf{y})) = \mathbf{y}_1$$

and so $\mathbf{x_i} = \pi_{\mathbf{i}}\mathbf{g}\,(\mathbf{y})$ which means

$$\mathbf{g}\,(\mathbf{y}) = (\pi_{\mathbf{i}}\mathbf{g}\,(\mathbf{y}),\pi_{\mathbf{i}_c}\mathbf{g}\,(\mathbf{y})) = (\mathbf{x_i},\mathbf{x}_{i_c})$$

$$\in \mathbf{f}^{-1}\,(\mathbf{y}_1) \cap K_j^{\mathbf{i}} \cap A.$$

Thus $\mathbf{y}_2 \rightarrow \mathbf{g}\,(\mathbf{y})$ where

$$\mathbf{y}_2 \in \pi_{\mathbf{i}_c}\left(\mathbf{f}^{-1}\,(\mathbf{y}_1) \cap K_j^{\mathbf{i}} \cap A\right)$$

is a parametrization of

$$\mathbf{f}^{-1}\,(\mathbf{y}_1) \cap K_j^{\mathbf{i}} \cap A.$$

The Coarea Formula

By Theorem 20.18, the equation in Formula 9 implies

$$\int_{K_j^i \cap A} \det \left(Df\left(\mathbf{x}\right) Df\left(\mathbf{x}\right)^*\right)^{1/2} dx$$

$$= \int_{\mathbb{R}^m} \mathcal{H}^{n-m} \left(\mathbf{f}^{-1}\left(\mathbf{y}_1\right) \cap K_j^i \cap A\right) dy_1.$$

Using Lemmas 21.8 and 21.7, along with the inner regularity of Lebesgue measure, K_j^i can be replaced with E_j^i. Therefore, summing the terms over all i and j,

$$\int_A \det \left(Df\left(\mathbf{x}\right) Df\left(\mathbf{x}\right)^*\right)^{1/2} dx = \int_{\mathbb{R}^m} \mathcal{H}^{n-m} \left(\mathbf{f}^{-1}\left(\mathbf{y}\right) \cap A\right) dy$$

which verifies the coarea formula whenever A is a closed subset of $\mathbb{R}^n \setminus \{S \cup N\}$. By Lemma 21.8 again, this formula is true for all A a closed subset of $\mathbb{R}^n \setminus S$. Using the same two lemmas again, we see this coarea formula holds for all A a measurable subset of $\mathbb{R}^n \setminus S$. It remains to verify the formula for all measurable sets, A, whether or not they intersect S.

Next we consider the case where $A \subseteq S \equiv \{\mathbf{x} : J\left(Df\left(\mathbf{x}\right)\right) = 0\}$. Let A be compact so that by Lemma 21.7, $\mathbf{y} \to \mathcal{H}^{n-m} \left(A \cap \mathbf{f}^{-1}\left(\mathbf{y}\right)\right)$ is Borel. For $\epsilon > 0$, define $\mathbf{k}, \mathbf{p} : \mathbb{R}^n \times \mathbb{R}^m \to \mathbb{R}^m$ by

$$\mathbf{k}\left(\mathbf{x}, \mathbf{y}\right) \equiv \mathbf{f}\left(\mathbf{x}\right) + \epsilon \mathbf{y}, \quad \mathbf{p}\left(\mathbf{x}, \mathbf{y}\right) \equiv \mathbf{y}.$$

Then

$$D\mathbf{k}\left(\mathbf{x}, \mathbf{y}\right) = \left(Df\left(\mathbf{x}\right), \epsilon I\right) = \left(UR, \epsilon I\right)$$

where the dependence of U and R on $\mathbf{x}$ has been suppressed. Thus

$$J\mathbf{k}^2 = \det\left(UR, \epsilon I\right) \begin{pmatrix} R^*U \\ \epsilon I \end{pmatrix} = \det\left(U^2 + \epsilon^2 I\right)$$

$$= \det\left(Q^* DQ Q^* DQ + \epsilon^2 I\right) = \det\left(D^2 + \epsilon^2 I\right)$$

$$= \prod_{i=1}^{m} \left(\lambda_i^2 + \epsilon^2\right) \in [\epsilon^{2m}, C^2 \epsilon^2] \tag{10}$$

since one of the λ_i equals 0. All the eigenvalues must be bounded independent of $\mathbf{x}$, since $\|Df\left(\mathbf{x}\right)\|$ is bounded independent of $\mathbf{x}$ due to the assumption that $\mathbf{f}$ is Lipschitz. Since $J\mathbf{k} \neq 0$, the first part of the argument implies

$$\epsilon C m_{n+m} \left(A \times \overline{B\left(0,1\right)}\right) \geq \int_{A \times \overline{B(0,1)}} |J\mathbf{k}| \, dm_{n+m}$$

$$= \int_{\mathbb{R}^m} \mathcal{H}^n \left(\mathbf{k}^{-1}\left(\mathbf{y}\right) \cap A \times \overline{B\left(0,1\right)}\right) dy$$

Which by Lemma 21.6, is no smaller than

$$C_{nm} \int_{\mathbb{R}^m} \int_{\mathbb{R}^m} \mathcal{H}^{n-m} \left(\mathbf{k}^{-1}(\mathbf{y}) \cap \mathbf{p}^{-1}(\mathbf{w}) \cap A \times \overline{B(0,1)} \right) dw dy$$

where $C_{nm} = \frac{\alpha(n)}{\alpha(n-m)\alpha(m)}$. This is no smaller than

$$C_{nm} \int_{\mathbb{R}^m} \int_{B(0,1)} \mathcal{H}^{n-m} \left(\mathbf{f}^{-1}(\mathbf{y} - \epsilon\mathbf{w}) \cap A \right) dw dy \qquad (*)$$

which by the Borel measurability of the integrand,

$$= C_{nm} \int_{B(0,1)} \int_{\mathbb{R}^m} \mathcal{H}^{n-m} \left(\mathbf{f}^{-1}(\mathbf{y} - \epsilon\mathbf{w}) \cap A \right) dy dw$$

$$= \frac{\alpha(n)}{\alpha(n-m)} \int_{\mathbb{R}^m} \mathcal{H}^{n-m} \left(\mathbf{f}^{-1}(\mathbf{y}) \cap A \right) dy. \qquad (**)$$

Here formula $*$ is implied by the following claim.

Claim:
$$\mathcal{H}^{n-m} \left(\mathbf{k}^{-1}(\mathbf{y}) \cap \mathbf{p}^{-1}(\mathbf{w}) \cap A \times \overline{B(0,1)} \right)$$
$$\geq \mathcal{X}_{\overline{B(0,1)}}(\mathbf{w}) \, \mathcal{H}^{n-m} \left(\mathbf{f}^{-1}(\mathbf{y} - \epsilon\mathbf{w}) \cap A \right).$$

Proof of the claim: If $\mathbf{w} \notin \overline{B(0,1)}$, there is nothing to prove so assume $\mathbf{w} \in \overline{B(0,1)}$. For such $\mathbf{w}$,

$$(\mathbf{x}, \mathbf{w}_1) \in \mathbf{k}^{-1}(\mathbf{y}) \cap \mathbf{p}^{-1}(\mathbf{w}) \cap A \times \overline{B(0,1)}$$

if and only if $\mathbf{f}(\mathbf{x}) + \epsilon\mathbf{w}_1 = \mathbf{y}$, $\mathbf{w}_1 = \mathbf{w}$, and $\mathbf{x} \in A$, if and only if

$$(\mathbf{x}, \mathbf{w}_1) \in \mathbf{f}^{-1}(\mathbf{y} - \epsilon\mathbf{w}) \cap A \times \{\mathbf{w}\}.$$

Therefore for $\mathbf{w} \in \overline{B(0,1)}$,

$$\mathcal{H}^{n-m} \left(\mathbf{k}^{-1}(\mathbf{y}) \cap \mathbf{p}^{-1}(\mathbf{w}) \cap A \times \overline{B(0,1)} \right)$$

$$\geq \mathcal{H}^{n-m} \left(\mathbf{f}^{-1}(\mathbf{y} - \epsilon\mathbf{w}) \cap A \times \{\mathbf{w}\} \right) = \mathcal{H}^{n-m} \left(\mathbf{f}^{-1}(\mathbf{y} - \epsilon\mathbf{w}) \cap A \right).$$

(Actually equality holds in the claim.)

Now by $**$, it follows since $\epsilon > 0$ is arbitrary,

$$\int_{\mathbb{R}^m} \mathcal{H}^{n-m} \left(A \cap \mathbf{f}^{-1}(\mathbf{y}) \right) dy = 0 = \int_A J\mathbf{f}(\mathbf{x}) \, dx.$$

Since this holds for arbitrary compact sets in S, it follows from Lemma 21.8 and inner regularity of Lebesgue measure that the equation holds for all measurable subsets of S. This completes the proof of the coarea formula.

There is a simple corollary to this theorem in the case of locally Lipschitz maps.

Corollary 21.10 *Let* $\mathbf{f} : \mathbb{R}^n \to \mathbb{R}^m$ *where* $m \leq n$ *and* $\mathbf{f}$ *is locally Lipschitz. This means that for each* $r > 0$, $\mathbf{f}$ *is Lipschitz on* $B(0,r)$. *Then the Coarea formula, Formula 3, holds for* $\mathbf{f}$.

Proof: Let $A \subseteq B(0,r)$ and let $\mathbf{f}_r$ be Lipschitz with

$$\mathbf{f}(\mathbf{x}) = \mathbf{f}_r(\mathbf{x})$$

for $\mathbf{x} \in B(0,r+1)$. Then

$$\int_A J(D\mathbf{f}(\mathbf{x}))\, dx = \int_A J(D\mathbf{f}_r(\mathbf{x}))dx = \int_{\mathbb{R}^m} \mathcal{H}^{n-m}\left(A \cap \mathbf{f}_r^{-1}(\mathbf{y})\right) dy$$

$$= \int_{\mathbf{f}_r(A)} \mathcal{H}^{n-m}\left(A \cap \mathbf{f}_r^{-1}(\mathbf{y})\right) dy = \int_{\mathbf{f}(A)} \mathcal{H}^{n-m}\left(A \cap \mathbf{f}^{-1}(\mathbf{y})\right) dy$$

$$= \int_{\mathbb{R}^m} \mathcal{H}^{n-m}\left(A \cap \mathbf{f}^{-1}(\mathbf{y})\right) dy$$

Now for arbitrary measurable A the above shows for $k = 1, 2, \cdots$

$$\int_{A \cap B(0,k)} J(D\mathbf{f}(\mathbf{x}))\, dx = \int_{\mathbb{R}^m} \mathcal{H}^{n-m}\left(A \cap B(0,k) \cap \mathbf{f}^{-1}(\mathbf{y})\right) dy.$$

Use the monotone convergence theorem to obtain Formula 3.

From the definition of Hausdorff measure in Chapter 19, it is easy to verify that $\mathcal{H}^0(E)$ equals the number of elements in E. Thus, if $n = m$, the Coarea formula implies

$$\int_A J(D\mathbf{f}(\mathbf{x}))\, dx = \int_{\mathbf{f}(A)} \mathcal{H}^0\left(A \cap \mathbf{f}^{-1}(\mathbf{y})\right) dy = \int_{\mathbf{f}(A)} \#(y)\, dy$$

in the notation of Chapter 20. Note also that this gives a version of Sard's theorem by letting $S = A$.

21.3   Change of variables

We say that the Coarea formula holds for $\mathbf{f} : \mathbb{R}^n \to \mathbb{R}^m, n \geq m$ if whenever A is a Lebesgue measurable subset of $\mathbb{R}^n$, Formula 3 holds. Note this is the same as

$$\int_A J(D\mathbf{f}(\mathbf{x}))dx = \int_{\mathbf{f}(A)} \mathcal{H}^{n-m}\left(A \cap \mathbf{f}^{-1}(\mathbf{y})\right) dy.$$

Now let

$$s\left(\mathbf{x}\right) = \sum_{i=1}^{p} c_i \mathcal{X}_{E_i}\left(\mathbf{x}\right)$$

where E_i is measurable and $c_i \geq 0$. Then

$$
\begin{aligned}
\int_{\mathbb{R}^n} s\left(\mathbf{x}\right) J\left(\left(D\mathbf{f}\left(\mathbf{x}\right)\right)\right) dx &= \sum_{i=1}^{p} c_i \int_{E_i} J\left(D\mathbf{f}\left(\mathbf{x}\right)\right) dx \\
&= \sum_{i=1}^{p} c_i \int_{\mathbf{f}(E_i)} \mathcal{H}^{n-m}\left(E_i \cap \mathbf{f}^{-1}(\mathbf{y})\right) dy \\
&= \int_{\mathbf{f}(\mathbb{R}^n)} \sum_{i=1}^{p} c_i \mathcal{H}^{n-m}\left(E_i \cap \mathbf{f}^{-1}(\mathbf{y})\right) dy \\
&= \int_{\mathbf{f}(\mathbb{R}^n)} \left[\int_{\mathbf{f}^{-1}(\mathbf{y})} s \, d\mathcal{H}^{n-m}\right] dy
\end{aligned}
$$

$$= \int_{\mathbf{f}(\mathbb{R}^n)} \left[\int_{\mathbf{f}^{-1}(\mathbf{y})} s \, d\mathcal{H}^{n-m}\right] dy. \tag{11}$$

Theorem 21.11 *Let $g \geq 0$ be Lebesgue measurable and let*

$$\mathbf{f} : \mathbb{R}^n \to \mathbb{R}^m, \ n \geq m$$

satisfy the Coarea formula. Then

$$\int_{\mathbb{R}^n} g\left(\mathbf{x}\right) J\left(\left(D\mathbf{f}\left(\mathbf{x}\right)\right)\right) dx = \int_{\mathbf{f}(\mathbb{R}^n)} \left[\int_{\mathbf{f}^{-1}(\mathbf{y})} g \, d\mathcal{H}^{n-m}\right] dy.$$

Proof: Let $s_i \uparrow g$ where s_i is a simple function satisfying Formula 11. Then let $i \to \infty$ and use the monotone convergence theorem to replace s_i with g. This proves the change of variables formula.

Note that this formula is a nonlinear version of Fubini's theorem. The "$n - m$ dimensional surface", $\mathbf{f}^{-1}(\mathbf{y})$, plays the role of $\mathbb{R}^{n-m}$ and $\mathcal{H}^{n-m}$ is like $n - m$ dimensional Lebesgue measure. The term, $J\left(\left(D\mathbf{f}\left(\mathbf{x}\right)\right)\right)$, corrects for the error occurring because of the lack of flatness of $\mathbf{f}^{-1}(\mathbf{y})$.

21.4   Exercises

1. Suppose U is an open subset of $\mathbb{R}^n$ and let $\mathbf{f} : U \to \mathbb{R}^m$ be C^1. Does the Coarea formula hold for all A Lebesgue measurable and a subset of U?

2. Let $f(x) = |x|$. What does the Coarea formula give for this example?

3. Let $f : \mathbb{R}^n \to [0,\infty)$ be Lipschitz and let S be a Lebesgue measurable set. Also let $R_t \equiv \{x \in \mathbb{R}^n : f(x) \leq t\}$. Show that for a.e. $t \in \mathbb{R}$,

$$\frac{d}{dt}\left(\int_{R_t \cap S} J(Df(x))\,dx\right) = \mathcal{H}^{n-1}\left(S \cap R_t \cap f^{-1}(t)\right).$$

4. Let U be an open set and let $\mathbf{f} : U \subseteq \mathbb{R}^n \to \mathbb{R}^n$ be continuous and one to one. Suppose $D\mathbf{f}(x)$ exists and has an inverse for some $x \in U$. Show there exists $\delta > 0$ such that $B(\mathbf{f}(x),\delta) \subseteq \mathbf{f}(U)$ and that $(D\mathbf{f}^{-1})(\mathbf{f}(x)) = D\mathbf{f}(x)^{-1}$. **Hint:** Let $\mathbf{h}(v) \equiv \mathbf{f}(x+v) - \mathbf{f}(x)$ and show

$$\left| D\mathbf{f}(x)^{-1}\mathbf{h}(v) - v \right| < \epsilon |v|$$

whenever $|v|$ is sufficiently small. Then apply Lemma 20.5 to get an inequality of the following sort.

$$D\mathbf{f}(x)^{-1}\mathbf{h}(B(0,r)) \supseteq B(0,r(1-\epsilon)).$$

Thus, if $D\mathbf{f}(x)^{-1}\mathbf{w} \in B(0,r(1-\epsilon))$, there exists small $\mathbf{v}$ such that

$$D\mathbf{f}(x)^{-1}(\mathbf{f}(x+v) - \mathbf{f}(x)) = D\mathbf{f}(x)^{-1}\mathbf{w}.$$

Chapter 22
Fourier Analysis in $\mathbb{R}^n$

The purpose of this chapter is to present some of the most important theorems on Fourier analysis in $\mathbb{R}^n$. These theorems are the Marcinkiewicz interpolation theorem, the Calderon Zygmund decomposition, Mihlin's theorem, and the Calderon Zygmund theory of singular integrals. They are all fundamental results whose proofs depend on the methods of real analysis. Our purpose is to present proofs of these theorems, showing how they follow from these methods. We leave their application to other works.

22.1 The Marcinkiewicz interpolation theorem

Let $(\Omega, \mu, \mathcal{S})$ be a measure space. Then we make the following definition.

Definition 22.1 $L^p(\Omega) + L^1(\Omega)$ will denote the space of measurable functions, f, such that f is the sum of a function in $L^p(\Omega)$ and $L^1(\Omega)$. Also, if $T : L^p(\Omega) + L^1(\Omega) \rightarrow$ space of measurable functions, we say that T is subadditive if

$$|T(f + g)(x)| \leq |Tf(x)| + |Tg(x)|.$$

We say T is of type (p, p) if there exists a constant independent of $f \in L^p(\Omega)$ such that

$$\|Tf\|_p \leq A \|f\|_p, \ f \in L^p(\Omega).$$

We say T is weak type (p, p) if there exists a constant A independent of f such that

$$\mu[x : |Tf(x)| > \alpha] \leq \left(\frac{A}{\alpha} \|f\|_p\right)^p, \ f \in L^p(\Omega).$$

Lemma 22.2 If $p \in [1, r]$, then $L^p(\Omega) \subseteq L^1(\Omega) + L^r(\Omega)$.

Proof: Let $\lambda > 0$ and let $f \in L^p(\Omega)$

$$f_1(x) \equiv \begin{cases} f(x) \text{ if } |f(x)| \leq \lambda \\ 0 \text{ if } |f(x)| > \lambda \end{cases}, \ f_2(x) \equiv \begin{cases} f(x) \text{ if } |f(x)| > \lambda \\ 0 \text{ if } |f(x)| \leq \lambda \end{cases}.$$

Then $f(x) = f_1(x) + f_2(x)$.

$$\int |f_1(x)|^r \, d\mu = \int_{[|f| \leq \lambda]} |f(x)|^r \, d\mu \leq \lambda^{r-p} \int_{[|f| \leq \lambda]} |f(x)|^p \, d\mu < \infty.$$

Therefore, $f_1 \in L^r(\Omega)$.

$$\int |f_2(x)| \, d\mu = \int_{[|f|>\lambda]} |f(x)| \, d\mu \leq \mu \left[|f| > \lambda\right]^{1/p'} \left(\int |f|^p \, d\mu\right)^{1/p}.$$

This proves the lemma since $f = f_1 + f_2$, $f_1 \in L^r$ and $f_2 \in L^1$.

The next lemma is very useful and for a proof in the σ finite case based on Fubini's theorem see Problem 9 of Chapter 11 or Problem 2 of Chapter 14. Here we give a slightly more general argument.

Lemma 22.3 *Let $\phi(0) = 0$, ϕ is strictly increasing, and C^1. Let $f : \Omega \to [0, \infty)$ be measurable. Then*

$$\int_\Omega \phi(f(x)) \, d\mu = \int_0^\infty \phi'(\alpha) \mu[f > \alpha] \, d\alpha. \tag{1}$$

Proof: First suppose

$$f = \sum_{i=1}^m a_i \mathcal{X}_{E_i}$$

where $a_i > 0$ and the a_i are all distinct nonzero values of f, the sets being disjoint. Thus,

$$\int_\Omega \phi(f(x)) \, d\mu = \sum_{i=1}^m \phi(a_i) \mu(E_i).$$

We can suppose without loss of generality that $a_1 < a_2 < \cdots < a_m$. We also observe that

$$\alpha \to \mu([|f| > \alpha])$$

is constant on the intervals $[0, a_1), [a_1, a_2), \cdots$. For example, on $[a_i, a_{i+1})$, this function has the value

$$\sum_{j=i+1}^m \mu(E_j).$$

The function equals zero on $[a_m, \infty)$. Therefore,

$$\alpha \to \phi'(\alpha) \mu([|f| > \alpha])$$

is Lebesgue measurable and the second integral in Formula 1 equals

$$\phi(a_1) \sum_{i=1}^m \mu(E_i) + (\phi(a_2) - \phi(a_1)) \sum_{i=2}^m \mu(E_i) + \cdots$$

$$+ (\phi(a_k) - \phi(a_{k-1})) \sum_{i=k}^m \mu(E_i) + \cdots + (\phi(a_m) - \phi(a_{m-1})) \mu(E_m).$$

This equals

$$\sum_{i=1}^{m} \phi(a_i) \mu(E_i),$$

and so this establishes Formula 1 in the case when f is a nonnegative simple function. Since every measurable nonnegative function may be written as the pointwise limit of such simple functions, the desired result will follow by the Monotone convergence theorem and the next claim.

Claim: If $f_n \uparrow f$, then for each $\alpha > 0$,

$$\mu([f > \alpha]) = \lim_{n \to \infty} \mu([f_n > \alpha]).$$

Proof of the claim: $[f_n > \alpha] \uparrow [f > \alpha]$ and so

$$\mu([f_n > \alpha]) \uparrow \mu([f > \alpha]).$$

This proves the lemma. (Note the importance of the strict inequality in $[f > \alpha]$ in proving the claim.)

The next theorem is the main result in this section. It is called the Marcinkiewicz interpolation theorem.

Theorem 22.4 *Let* $(\Omega, \mu, \mathcal{S})$ *be a σ finite measure space, $1 < r < \infty$, and let*

$$T : L^1(\Omega) + L^r(\Omega) \to \quad space \ of \ measurable \ functions$$

be subadditive, weak (r, r), and weak $(1, 1)$. Then T is of type (p, p) for every $p \in (1, r)$ and

$$\|Tf\|_p \leq A_p \|f\|_p$$

where the constant A_p depends only on p and the constants in the definition of weak $(1, 1)$ and weak (r, r).

Proof: Let $\alpha > 0$ and let f_1 and f_2 be defined as in Lemma 22.2,

$$f_1(x) \equiv \begin{cases} f(x) & \text{if } |f(x)| \leq \alpha \\ 0 & \text{if } |f(x)| > \alpha \end{cases}, \quad f_2(x) \equiv \begin{cases} f(x) & \text{if } |f(x)| > \alpha \\ 0 & \text{if } |f(x)| \leq \alpha \end{cases}.$$

Thus $f = f_1 + f_2$ where $f_1 \in L^r$ and $f_2 \in L^1$. Since T is subadditive ,

$$[|Tf| > \alpha] \subseteq [|Tf_1| > \alpha/2] \cup [|Tf_2| > \alpha/2].$$

Let $p \in (1, r)$. By Lemma 22.3,

$$\int |Tf|^p \, d\mu \leq p \int_0^\infty \alpha^{p-1} \mu([|Tf_1| > \alpha/2]) \, d\alpha +$$

$$+ p \int_0^\infty \alpha^{p-1} \mu([|Tf_2| > \alpha/2]) \, d\alpha.$$

437

Therefore, since T is weak $(1,1)$ and weak (r,r),

$$\int |Tf|^p \, d\mu \le p \int_0^\infty \alpha^{p-1} \left(\frac{2A_r}{\alpha} \|f_1\|_r \right)^r d\alpha + p \int_0^\infty \alpha^{p-1} \frac{2A_1}{\alpha} \|f_2\|_1 \, d\alpha. \tag{2}$$

Therefore, the right side of Formula 2 equals

$$p \, (2A_r)^r \int_0^\infty \alpha^{p-1-r} \int_\Omega |f_1|^r \, d\mu d\alpha + 2A_1 p \int_0^\infty \alpha^{p-2} \int_\Omega |f_2| \, d\mu d\alpha.$$

From Fubini's theorem this equals

$$p \, (2A_r)^r \int_\Omega \int_0^\infty \alpha^{p-1-r} |f_1|^r \, d\alpha d\mu + 2A_1 p \int_\Omega \int_0^\infty \alpha^{p-2} |f_2| \, d\alpha d\mu.$$

Now from the definition of f_1 and f_2, this equals

$$p \, (2A_r)^r \int_\Omega |f(x)|^r \int_{|f(x)|}^\infty \alpha^{p-1-r} \, d\alpha d\mu + 2A_1 p \int_\Omega |f(x)| \int_0^{|f(x)|} \alpha^{p-2} \, d\alpha d\mu$$

which equals

$$\frac{2^r A_r^r p}{r-p} \int_\Omega |f(x)|^p \, d\mu + \frac{2pA_1}{p-1} \int_\Omega |f(x)|^p \, d\mu$$

and this proves the theorem.

22.2   The Calderon Zygmund decomposition

In this section, we will see that for a given function, $\mathbb{R}^n$ can be decomposed into a set where the function is small and a set which is the union of disjoint cubes on which the average of the function is under some control. The measure in this section will always be Lebesgue measure on $\mathbb{R}^n$. This theorem depends on the Lebesgue theory of differentiation.

Theorem 22.5 *Let $f \ge 0$, $\int f dx < \infty$, and let α be a positive constant. Then there exist sets F and Ω such that*

$$\mathbb{R}^n = F \cup \Omega, \ F \cap \Omega = \emptyset \tag{a.}$$

$$f(x) \le \alpha \ a.e. \ on \ F \tag{b.}$$

$\Omega = \cup_{k=1}^\infty Q_k$ where the interiors of the cubes are disjoint and for each cube, Q_k,

$$\alpha < \frac{1}{m(Q_k)} \int_{Q_k} f(x) \, dx \le 2^n \alpha. \tag{c.}$$

The Calderon Zygmund decomposition

Proof: Let S_0 consist of all cubes of the form $\prod_{i=1}^{n} [a_i, b_i]$ where $a_i = iM$, i an integer, and $b_i = a_i + M$. Here M is so large that if Q is one of these cubes, then

$$\frac{1}{m(Q)} \int_Q f \, dm \leq \alpha. \tag{3}$$

Suppose $S_0, \cdots, S_m$ have been chosen. To get S_{m+1}, replace each cube of S_m by the 2^n cubes obtained by bisecting the sides. Then retain in S_{m+1} exactly those cubes for which Formula 3 holds. Now define

$$F \equiv \{ \mathbf{x} : \mathbf{x} \text{ is contained in some cube from } S_m \text{ for all } m \},$$

$$\Omega \equiv \mathbb{R}^n \setminus F.$$

Now a.e. point of F is a Lebesgue point of f. Let $\mathbf{x}$ be such a point of F and suppose $\mathbf{x} \in Q_k$ for $Q_k \in S_k$. Let $d_k \equiv$ diameter of Q_k. Thus $d_k \to 0$.

$$\frac{1}{m(Q_k)} \int_{Q_k} |f(\mathbf{y}) - f(\mathbf{x})| \, dy \leq \frac{1}{m(Q_k)} \int_{B(\mathbf{x}, d_k)} |f(\mathbf{y}) - f(\mathbf{x})| \, dy$$

$$= \frac{m(B(\mathbf{x}, d_k))}{m(Q_k)} \frac{1}{m(B(\mathbf{x}, d_k))} \int_{B(\mathbf{x}, d_k)} |f(\mathbf{x}) - f(\mathbf{y})| \, dy$$

$$\leq K_n \frac{1}{m(B(\mathbf{x}, d_k))} \int_{B(\mathbf{x}, d_k)} |f(\mathbf{x}) - f(\mathbf{y})| \, dy$$

where K_n is a constant which depends on n and measures the ratio of the volume of a ball and a cube with the same diameter. The last expression converges to 0 because $\mathbf{x}$ is a Lebesgue point. Hence

$$f(\mathbf{x}) = \lim_{k \to \infty} \frac{1}{m(Q_k)} \int_{Q_k} f(\mathbf{y}) \, dy \leq \alpha$$

and this shows $f(\mathbf{x}) \leq \alpha$ a.e. on F. If $\mathbf{x} \notin F$, then let m be the first index where $\mathbf{x} \notin \cup S_m$. Let Q be the cube in S_{m-1} containing $\mathbf{x}$ and let Q^* be the cube in the bisection of Q which contains $\mathbf{x}$. Therefore Formula 3 does not hold for Q^*. Thus $\mathbf{x} \in Q^*$ and

$$\alpha < \frac{1}{m(Q^*)} \int_{Q^*} f \, dx \leq \frac{m(Q)}{m(Q^*)} \frac{1}{m(Q)} \int_Q f \, dx = 2^n \alpha.$$

This proves the theorem.

22.3 Mihlin's theorem

In this section, the Marcinkiewicz interpolation theorem and Calderon Zygmund decomposition will be used to establish a remarkable theorem of Mihlin, a generalization of Plancherel's theorem to the L^p spaces. It is of fundamental importance in the study of elliptic partial differential equations and later it will be used to give proofs of the theorems about singular integrals.

Recall that by Corollary 16.14, if $f \in L^2(\mathbb{R}^n)$ and if $\phi \in \mathfrak{S}$, then $f * \phi \in L^2(\mathbb{R}^n)$ and

$$F(f * \phi)(\mathbf{x}) = (2\pi)^{n/2} F\phi(\mathbf{x}) Ff(\mathbf{x}).$$

The next lemma is essentially a weak $(1, 1)$ estimate.

Lemma 22.6 *Suppose $\rho \in L^\infty(\mathbb{R}^n) \cap L^2(\mathbb{R}^n)$ and suppose also that there exists a constant C_1 such that*

$$\int_{|\mathbf{x}| \geq 2|\mathbf{y}|} |F^{-1}\rho(\mathbf{x} - \mathbf{y}) - F^{-1}\rho(\mathbf{x})| \, dx \leq C_1. \tag{4}$$

Then there exists a constant A depending only on $C_1, \|\rho\|_\infty$, and n such that

$$m\left[\mathbf{x} : |F^{-1}\rho * \phi(\mathbf{x})| > \alpha\right] \leq \frac{A}{\alpha} \|\phi\|_1$$

for all $\phi \in \mathfrak{S}$.

Proof: Let $\phi \in \mathfrak{S}$ and use the Calderon decomposition to write $\mathbb{R}^n = E \cup \Omega$ where Ω is a union of cubes with disjoint interiors $\{Q_i\}$ such that

$$\alpha m(Q_i) \leq \int_{Q_i} |\phi(\mathbf{x})| \, dx \leq 2^n \alpha m(Q_i), \quad |\phi(\mathbf{x})| \leq \alpha \text{ a.e. on } E. \tag{5}$$

Now define a good function, g, and a bad function, b, as follows.

$$g(\mathbf{x}) = \begin{cases} \phi(\mathbf{x}) & \text{if } \mathbf{x} \in E \\ \frac{1}{m(Q_i)} \int_{Q_i} \phi(\mathbf{x}) \, dx & \text{if } \mathbf{x} \in Q_i \subseteq \Omega \end{cases}, \quad g(\mathbf{x}) + b(\mathbf{x}) = \phi(\mathbf{x}). \tag{6}$$

Thus

$$\int_{Q_i} b(\mathbf{x}) \, dx = 0, \quad b(\mathbf{x}) = 0 \text{ if } \mathbf{x} \notin \Omega. \tag{7}$$

Using Formula 5 and Formula 6, we verify that

$$\|g\|_2^2 \leq \alpha(1 + 4^n) \|\phi\|_1, \quad \|g\|_1 \leq 2 \|\phi\|_1. \tag{8}$$

To see the first of these inequalities,

$$\|g\|_2^2 = \|g\|_{L^2(E)}^2 + \|g\|_{L^2(\Omega)}^2.$$

Now,

$$
\begin{aligned}
\|g\|^2_{L^2(\Omega)} &= \sum_i \int_{Q_i} |g(x)|^2 \, dx \\
&\leq \sum_i \int_{Q_i} \left(\frac{1}{m(Q_i)} \int_{Q_i} |\phi(y)| \, dy \right)^2 \, dx \\
&\leq \sum_i \int_{Q_i} (2^n \alpha)^2 \, dx \leq 4^n \alpha^2 \sum_i m(Q_i) \\
&\leq 4^n \alpha^2 \frac{1}{\alpha} \sum_i \int_{Q_i} |\phi(x)| \, dx \leq 4^n \alpha \|\phi\|_1.
\end{aligned}
$$

$$
\|g\|^2_{L^2(E)} = \int_E |\phi(x)|^2 \, dx \leq \alpha \int_E |\phi(x)| \, dx = \alpha \|\phi\|_1.
$$

The proof of the second inequality is easier and is left to the reader. Because of Formula 8, $g \in L^1(\mathbb{R}^n)$ and so $F^{-1}\rho * g \in L^2(\mathbb{R}^n)$. By Plancherel's theorem,

$$
\|F^{-1}\rho * g\|_2 = \|F(F^{-1}\rho * g)\|_2.
$$

By Corollary 16.14 of Chapter 16, the expression on the right equals

$$
(2\pi)^{n/2} \|\rho F g\|_2
$$

and so

$$
\|F^{-1}\rho * g\|_2 = (2\pi)^{n/2} \|\rho F g\|_2 \leq C_n \|\rho\|_\infty \|g\|_2.
$$

It follows from this and Formula 8 that

$$
m\left[|F^{-1}\rho * g(\mathbf{x})| \geq \alpha/2 \right]
$$

$$
\leq \frac{C_n \|\rho\|^2_\infty}{\alpha^2} \alpha (1 + 4^n) \|\phi\|_1 = C_n \alpha^{-1} \|\phi\|_1. \tag{9}
$$

Now if Q is one of the cubes whose union is Ω, let Q^* be the cube with the same center as Q but whose sides are $2\sqrt{n}$ times as long.

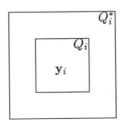

Let

$$
\Omega^* \equiv \cup_{i=1}^\infty Q_i^*
$$

441

and let

$$E^* \equiv \mathbb{R}^n \setminus \Omega^*.$$

Thus $E^* \subseteq E$. Let $\mathbf{x} \in E^*$. Then because of Formula 7,

$$\int_{Q_i} F^{-1}\rho\,(\mathbf{x} - \mathbf{y})\,b\,(\mathbf{y})\,dy$$

$$= \int_{Q_i} \left[F^{-1}\rho\,(\mathbf{x} - \mathbf{y}) - F^{-1}\rho\,(\mathbf{x} - \mathbf{y}_i) \right] b\,(\mathbf{y})\,dy, \qquad (10)$$

where $\mathbf{y}_i$ is the center of Q_i. Consequently if the sides of Q_i have length $2t/\sqrt{n}$, Formula 10 implies

$$\int_{E^*} \left| \int_{Q_i} F^{-1}\rho\,(\mathbf{x} - \mathbf{y})\,b\,(\mathbf{y})\,dy \right| dx \leq$$

$$\int_{E^*} \int_{Q_i} \left| F^{-1}\rho\,(\mathbf{x} - \mathbf{y}) - F^{-1}\rho\,(\mathbf{x} - \mathbf{y}_i) \right| |b\,(\mathbf{y})|\,dydx$$

$$= \int_{Q_i} \int_{E^*} \left| F^{-1}\rho\,(\mathbf{x} - \mathbf{y}) - F^{-1}\rho\,(\mathbf{x} - \mathbf{y}_i) \right| dx\,|b\,(\mathbf{y})|\,dy \qquad (11)$$

$$\leq \int_{Q_i} \int_{|\mathbf{x} - \mathbf{y}_i| \geq 2t} \left| F^{-1}\rho\,(\mathbf{x} - \mathbf{y}) - F^{-1}\rho\,(\mathbf{x} - \mathbf{y}_i) \right| dx\,|b\,(\mathbf{y})|\,dy$$

since if $\mathbf{x} \in E^*$, then $|\mathbf{x} - \mathbf{y}_i| \geq 2t$. Now for $\mathbf{y} \in Q_i$,

$$|\mathbf{y} - \mathbf{y}_i| \leq \left(\sum_{j=1}^{n} \left(\frac{t}{\sqrt{n}} \right)^2 \right)^{1/2} = t.$$

Therefore we apply Formula 4 and the change of variables $\mathbf{u} = \mathbf{x} - \mathbf{y}_i$ to conclude from Formula 11,

$$\int_{E^*} \left| \int_{Q_i} F^{-1}\rho\,(\mathbf{x} - \mathbf{y})\,b\,(\mathbf{y})\,dy \right| dx \leq C_1 \int_{Q_i} |b\,(\mathbf{y})|\,dy. \qquad (12)$$

Now from Formula 12,

$$\int_{E^*} \sum_{i=1}^{\infty} \left| \int_{Q_i} F^{-1}\rho\,(\mathbf{x} - \mathbf{y})\,b\,(\mathbf{y})\,dy \right| dx$$

$$= \sum_{i=1}^{\infty} \int_{E^*} \left| \int_{Q_i} F^{-1}\rho\,(\mathbf{x} - \mathbf{y})\,b\,(\mathbf{y})\,dy \right| dx$$

$$\leq \sum_{i=1}^{\infty} C_1 \int_{Q_i} |b\,(\mathbf{y})|\,dy = C_1\,||b||_1. \qquad (13)$$

The first term in Formula 13 is no smaller than

$$\int_{E^*} \left| \sum_{i=1}^{\infty} \int_{Q_i} F^{-1}\rho\left(\mathbf{x} - \mathbf{y}\right) b\left(\mathbf{y}\right) dy \right| dx$$

which, since $b\left(\mathbf{y}\right) = 0$ for $\mathbf{y} \notin \Omega$, equals

$$\int_{E^*} \left| \int_{\mathbb{R}^n} F^{-1}\rho\left(\mathbf{x} - \mathbf{y}\right) b\left(\mathbf{y}\right) dy \right| dx = \int_{E^*} \left| F^{-1}\rho * b\left(\mathbf{x}\right) \right| dx.$$

Thus, by Formula 8,

$$
\begin{aligned}
\int_{E^*} \left| F^{-1}\rho * b\left(\mathbf{x}\right) \right| dx
&\leq C_1 \left\| b \right\|_1 \\
&\leq C_1 \left[\left\| \phi \right\|_1 + \left\| g \right\|_1 \right] \\
&\leq C_1 \left[\left\| \phi \right\|_1 + 2 \left\| \phi \right\|_1 \right] \\
&\leq 3C_1 \left\| \phi \right\|_1 .
\end{aligned}
$$

Consequently,

$$m\left(\left[\left| F^{-1}\rho * b \right| \geq \frac{\alpha}{2} \right] \cap E^* \right) \leq \frac{6C_1}{\alpha} \left\| \phi \right\|_1 .$$

From Formulas 6, 9, and 5,

$$m\left[\left| F^{-1}\rho * \phi \right| > \alpha \right] \leq m\left[\left| F^{-1}\rho * g \right| \geq \frac{\alpha}{2} \right] + m\left[\left| F^{-1}\rho * b \right| \geq \frac{\alpha}{2} \right]$$

$$\leq \frac{C_n}{\alpha} \left\| \phi \right\|_1 + m\left(\left[\left| F^{-1}\rho * b \right| \geq \frac{\alpha}{2} \right] \cap E^* \right) + m\left(\Omega^* \right)$$

$$\leq \frac{C_n}{\alpha} \left\| \phi \right\|_1 + \frac{6C_1}{\alpha} \left\| \phi \right\|_1 + C_n m\left(\Omega \right) \leq \frac{A}{\alpha} \left\| \phi \right\|_1$$

because

$$m\left(\Omega \right) \leq \alpha^{-1} \left\| \phi \right\|_1$$

by Formula 5. This proves the lemma.

The next lemma extends this lemma by giving a weak $(2, 2)$ estimate and a $(2, 2)$ estimate.

Lemma 22.7 *Suppose* $\rho \in L^{\infty}\left(\mathbb{R}^n\right) \cap L^2\left(\mathbb{R}^n\right)$ *and suppose also that there exists a constant* C_1 *such that*

$$\int_{\left| \mathbf{x} \right| > 2\left| \mathbf{y} \right|} \left| F^{-1}\rho\left(\mathbf{x} - \mathbf{y}\right) - F^{-1}\rho\left(\mathbf{x}\right) \right| dx \leq C_1. \tag{14}$$

Then $F^{-1}\rho *$ *maps* $L^1\left(\mathbb{R}^n\right) + L^2\left(\mathbb{R}^n\right)$ *to measurable functions and there exists a constant* A *depending only on* $C_1, n, \left\| \rho \right\|_{\infty}$ *such that*

$$m\left[\left| F^{-1}\rho * f \right| > \alpha \right] \leq A \frac{\left\| f \right\|_1}{\alpha} \ \text{if} \ f \in L^1\left(\mathbb{R}^n\right), \tag{15}$$

$$m\left[\left|F^{-1}\rho * f\right| > \alpha\right] \le \left(A\frac{||f||_2}{\alpha}\right)^2 \ \text{if } f \in L^2\left(\mathbb{R}^n\right). \tag{16}$$

Thus, $F^{-1}\rho *$ is weak type $(1,1)$ and weak type $(2,2)$. Also we have

$$\left|\left|F^{-1}\rho * f\right|\right|_2 \le A\left|\left|f\right|\right|_2 \ \text{if } f \in L^2\left(\mathbb{R}^n\right). \tag{17}$$

Proof: We know by Plancherel's theorem that $F^{-1}\rho$ is in $L^2\left(\mathbb{R}^n\right)$. If $f \in L^1\left(\mathbb{R}^n\right)$, then by Minkowski's inequality, just as in Corollary 16.14,

$$F^{-1}\rho * f \in L^2\left(\mathbb{R}^n\right).$$

Now let $g \in L^2\left(\mathbb{R}^n\right)$. By Holder's inequality, the following is well defined a.e.

$$F^{-1}\rho * g\left(\mathbf{x}\right) \equiv \int F^{-1}\rho\left(\mathbf{x} - \mathbf{y}\right)g\left(\mathbf{y}\right)dy$$

and by continuity of translation in $L^2\left(\mathbb{R}^n\right)$, this shows $\mathbf{x} \to F^{-1}\rho * g\left(\mathbf{x}\right)$ is continuous. Therefore, $F^{-1}\rho *$ maps $L^1\left(\mathbb{R}^n\right) + L^2\left(\mathbb{R}^n\right)$ to the space of measurable functions. It is clear that $F^{-1}\rho *$ is subadditive.

If $\phi \in \mathfrak{S}$, we can apply Plancherel's theorem to conclude

$$\left|\left|F^{-1}\rho * \phi\right|\right|_2 = \left|\left|F\left(F^{-1}\rho * \phi\right)\right|\right|_2 =$$

$$(2\pi)^{n/2}\left|\left|\rho F\phi\right|\right|_2 \le (2\pi)^{n/2}\left|\left|\rho\right|\right|_\infty\left|\left|\phi\right|\right|_2. \tag{18}$$

Now let $f \in L^2\left(\mathbb{R}^n\right)$ and let $\phi_k \in \mathfrak{S}$, with

$$\left|\left|\phi_k - f\right|\right|_2 \to 0.$$

Then by Holder's inequality,

$$\left|\int F^{-1}\rho\left(\mathbf{x} - \mathbf{y}\right)f\left(\mathbf{y}\right)dy\right| = \lim_{k \to \infty}\left|\int F^{-1}\rho\left(\mathbf{x} - \mathbf{y}\right)\phi_k\left(\mathbf{y}\right)dy\right|$$

and so by Fatou's lemma, Plancherel's theorem, and Formula 18,

$$\left|\left|F^{-1}\rho * f\right|\right|_2 = \left(\int\left|\int F^{-1}\rho\left(\mathbf{x} - \mathbf{y}\right)f\left(\mathbf{y}\right)dy\right|^2 dx\right)^{1/2}$$

$$\lim_{k \to \infty}\inf\left(\int\left|\int F^{-1}\rho\left(\mathbf{x} - \mathbf{y}\right)\phi_k\left(\mathbf{y}\right)dy\right|^2 dx\right)^{1/2} = \lim_{k \to \infty}\inf\left|\left|F^{-1}\rho * \phi_k\right|\right|_2$$

$$\le \left|\left|\rho\right|\right|_\infty (2\pi)^{n/2}\lim_{k \to \infty}\inf\left|\left|\phi_k\right|\right|_2 = \left|\left|\rho\right|\right|_\infty (2\pi)^{n/2}\left|\left|f\right|\right|_2.$$

Thus, Formula 17 holds with $A = \left|\left|\rho\right|\right|_\infty (2\pi)^{n/2}$. Consequently,

$$A\left|\left|f\right|\right|_2 \ge \left(\int_{\left[\left|F^{-1}\rho * f\right| > \alpha\right]}\left|F^{-1}\rho * f\left(\mathbf{x}\right)\right|^2 dx\right)^{1/2}$$

$$\geq \alpha m \left[\left| F^{-1} \rho * f \right| > \alpha \right]^{1/2}$$

and so Formula 16 follows.

It remains to prove Formula 15 which holds for all $f \in \mathfrak{S}$ by Lemma 22.6. Let $f \in L^1(\mathbb{R}^n)$ and let $\phi_k \to f$ in $L^1(\mathbb{R}^n)$, $\phi_k \in \mathfrak{S}$. Without loss of generality, we may assume that both f and $F^{-1}\rho$ are Borel measurable. Therefore, by Minkowski's inequality, and Plancherel's theorem,

$$\left\| F^{-1}\rho * \phi_k - F^{-1}\rho * f \right\|_2$$

$$\leq \left(\int \left| \int F^{-1}\rho(\mathbf{x} - \mathbf{y})(\phi_k(\mathbf{y}) - f(\mathbf{y})) \, dy \right|^2 dx \right)^{1/2}$$

$$\leq \left\| \phi_k - f \right\|_1 \left\| \rho \right\|_2$$

which shows that $F^{-1}\rho * \phi_k$ converges to $F^{-1}\rho * f$ in $L^2(\mathbb{R}^n)$. Therefore, we may take a subsequence such that the convergence is pointwise a.e. Then, denoting the subsequence by k,

$$\mathcal{X}_{[|F^{-1}\rho*f|>\alpha]}(\mathbf{x}) \leq \lim_{k \to \infty} \inf \mathcal{X}_{[|F^{-1}\rho*\phi_k|>\alpha]}(\mathbf{x}) \quad \text{a.e. } \mathbf{x}.$$

Thus by Lemma 22.6 and Fatou's lemma, there exists a constant, A, depending on C_1, n, and $\|\rho\|_\infty$ such that

$$m\left[\left| F^{-1}\rho * f \right| > \alpha \right] \leq \lim_{k \to \infty} \inf m\left[\left| F^{-1}\rho * \phi_k \right| > \alpha \right]$$

$$\leq \lim_{k \to \infty} \inf A \frac{\|\phi_k\|_1}{\alpha} = A \frac{\|f\|_1}{\alpha}.$$

This shows Formula 15 and proves the lemma.

Theorem 22.8 *Let $\rho \in L^2(\mathbb{R}^n) \cap L^\infty(\mathbb{R}^n)$ and suppose*

$$\int_{|\mathbf{x}| \geq 2|\mathbf{y}|} \left| F^{-1}\rho(\mathbf{x} - \mathbf{y}) - F^{-1}\rho(\mathbf{x}) \right| dx \leq C_1.$$

Then for each $p \in (1, \infty)$, there exists a constant, A_p, depending only on

$$p, n, \|\rho\|_\infty,$$

and C_1 such that for all $\phi \in \mathfrak{S}$,

$$\left\| F^{-1}\rho * \phi \right\|_p \leq A_p \|\phi\|_p.$$

Proof: From Lemma 22.7, $F^{-1}\rho*$ is weak $(1,1)$, weak $(2,2)$, and maps

$$L^1(\mathbb{R}^n) + L^2(\mathbb{R}^n)$$

445

to measurable functions. Therefore, by the Marcinkiewicz interpolation theorem, there exists a constant A_p depending only on p, C_1, n, and $||\rho||_\infty$ for $p \in (1, 2]$, such that for $f \in L^p(\mathbb{R}^n)$, and $p \in (1, 2]$,

$$||F^{-1}\rho * f||_p \leq A_p ||f||_p.$$

Now suppose $p > 2$. Then $p' < 2$ where

$$\frac{1}{p} + \frac{1}{p'} = 1.$$

Now by Plancherel's theorem and Theorem 16.23,

$$
\begin{aligned}
\int F^{-1}\rho * \phi(\mathbf{x}) \psi(\mathbf{x}) \, dx &= (2\pi)^{n/2} \int \rho(\mathbf{x}) F\phi(\mathbf{x}) F\psi(\mathbf{x}) \, dx \\
&= \int F(F^{-1}\rho * \psi) F\phi \, dx \\
&= \int (F^{-1}\rho * \psi)(\phi) \, dx.
\end{aligned}
$$

Thus by the case for $p \in (1, 2)$ and Holder's inequality,

$$
\begin{aligned}
\left| \int F^{-1}\rho * \phi(\mathbf{x}) \psi(\mathbf{x}) \, dx \right| &= \left| \int (F^{-1}\rho * \psi)(\phi) \, dx \right| \\
&\leq ||F^{-1}\rho * \psi||_{p'} ||\phi||_p \\
&\leq A_{p'} ||\psi||_{p'} ||\phi||_p.
\end{aligned}
$$

Letting $L\psi \equiv \int F^{-1}\rho * \phi(\mathbf{x}) \psi(\mathbf{x}) \, dx$, this shows $L \in L^{p'}(\mathbb{R}^n)'$ which implies by the Riesz representation theorem that $F^{-1}\rho * \phi$ represents L and

$$||F^{-1}\rho * \phi||_p \leq A_{p'} ||\phi||_p.$$

Since $p' = p/(p-1)$, this proves the theorem.

It is possible to give verifiable conditions on ρ which imply Formula 14. The conditions on ρ which we consider here are

$$C_0 \geq \sup\{|\mathbf{x}|^{|\alpha|} |D^\alpha \rho(\mathbf{x})| : |\alpha| \leq L, \mathbf{x} \in \mathbb{R}^n \setminus \{\mathbf{0}\}\}, L > n/2. \quad (19)$$

$$\rho \in C^L(\mathbb{R}^n \setminus \{\mathbf{0}\}) \text{ where } L \text{ is an integer.}$$

Here α is a multi-index and $|\alpha| = \sum_{i=1}^n \alpha_i$. Also recall the notation

$$\mathbf{x}^\alpha \equiv x_1^{\alpha_1} \cdots x_n^{\alpha_n}$$

where $\alpha = (\alpha_1 \cdots \alpha_n)$. For more general conditions, see [31].

Lemma 22.9 *Let Formula 19 hold and suppose* $\psi \in C_c^\infty\left(\mathbb{R}^n \setminus \{0\}\right)$. *Then for each* α, $|\alpha| \leq L$, *there exists a constant* $C \equiv C\left(\alpha, n, \psi\right)$ *such that*

$$\sup_{\mathbf{x} \in \mathbb{R}^n} |\mathbf{x}|^{|\alpha|} \left|D^\alpha \left(\rho\left(\mathbf{x}\right) \psi\left(2^k \mathbf{x}\right)\right)\right| \leq C C_0.$$

Proof:

$$|\mathbf{x}|^{|\alpha|} \left|D^\alpha \rho\left(\mathbf{x}\right) \psi\left(2^k \mathbf{x}\right)\right| \leq |\mathbf{x}|^{|\alpha|} \sum_{\beta + \gamma = \alpha} \left|D^\beta \rho\left(\mathbf{x}\right)\right| 2^{k|\gamma|} \left|D^\gamma \psi\left(2^k \mathbf{x}\right)\right|$$

$$= \sum_{\beta + \gamma = \alpha} |\mathbf{x}|^{|\beta|} \left|D^\beta \dot\rho\left(\mathbf{x}\right)\right| \left|2^k \mathbf{x}\right|^{|\gamma|} \left|D^\gamma \psi\left(2^k \mathbf{x}\right)\right|$$

$$\leq C_0 C\left(\alpha, n\right) \sum_{|\gamma| \leq |\alpha|} \sup\{|\mathbf{z}|^{|\gamma|} \left|D^\gamma \psi\left(\mathbf{z}\right)\right| : \mathbf{z} \in \mathbb{R}^n\} = C_0 C\left(\alpha, n, \psi\right)$$

and this proves the lemma.

Lemma 22.10 *There exists*

$$\phi \in C_c^\infty\left(\left[\mathbf{x} : 4^{-1} < |\mathbf{x}| < 4\right]\right), \ \phi\left(\mathbf{x}\right) \geq 0,$$

and

$$\sum_{k=-\infty}^{\infty} \phi\left(2^k \mathbf{x}\right) = 1$$

for each $\mathbf{x} \neq 0$.

Proof: Let

$$\psi \geq 0, \ \psi = 1 \ \text{on} \ \left[2^{-1} \leq |\mathbf{x}| \leq 2\right],$$

$$spt\left(\psi\right) \subseteq \left[4^{-1} < |\mathbf{x}| < 4\right].$$

Consider

$$g\left(\mathbf{x}\right) = \sum_{k=-\infty}^{\infty} \psi\left(2^k \mathbf{x}\right).$$

Then for each $\mathbf{x}$, only finitely many terms are not equal to 0. Also, $g\left(\mathbf{x}\right) > 0$ for all $\mathbf{x} \neq 0$. To verify this last claim, suppose $g\left(\mathbf{x}\right) = 0$ for some $\mathbf{x}$. Then for every k, either $2^k |\mathbf{x}| < 2^{-1}$ or $2^k |\mathbf{x}| > 2$. But if the first alternative holds, then let l be such that $2^{k+l} |\mathbf{x}| \geq 2^{-1}$ but $2^{k+l-1} |\mathbf{x}| < 2^{-1}$. Then $2^{k+l} |\mathbf{x}| < 1$ and so $\psi\left(2^{k+l} \mathbf{x}\right) \neq 0$ a contradiction. Similar reasoning yields the claim if the second alternative holds.

Now notice that

$$g\left(2^r \mathbf{x}\right) = \sum_{k=-\infty}^{\infty} \psi\left(2^k 2^r \mathbf{x}\right) = \sum_{k=-\infty}^{\infty} \psi\left(2^k \mathbf{x}\right) = g\left(\mathbf{x}\right).$$

Let $\phi(\mathbf{x}) \equiv \psi(\mathbf{x}) g(\mathbf{x})^{-1}$. Then

$$\sum_{k=-\infty}^{\infty} \phi(2^k\mathbf{x}) = \sum_{k=-\infty}^{\infty} \frac{\psi(2^k\mathbf{x})}{g(2^k\mathbf{x})} = g(\mathbf{x})^{-1} \sum_{k=-\infty}^{\infty} \psi(2^k\mathbf{x}) = 1$$

for each $\mathbf{x} \neq \mathbf{0}$. This proves the lemma.

Now define

$$\rho_m(\mathbf{x}) \equiv \sum_{k=-m}^{m} \rho(\mathbf{x}) \phi(2^k\mathbf{x}), \ \ \gamma_k(\mathbf{x}) \equiv \rho(\mathbf{x}) \phi(2^k\mathbf{x}).$$

Let $t > 0$ and let $|\mathbf{y}| \leq t$. We want to estimate

$$\int_{|\mathbf{x}| \geq 2t} \left| F^{-1}\gamma_k(\mathbf{x} - \mathbf{y}) - F^{-1}\gamma_k(\mathbf{x}) \right| dx \tag{20}$$

and we will estimate it two ways. In the following estimates, $C(a, b, \cdots, d)$ will denote a generic constant depending only on the indicated objects, $a, b, \cdots, d$. For the first estimate, note that, since $|\mathbf{y}| \leq t$, Formula 20 is no larger than

$$2 \int_{|\mathbf{x}| \geq t} \left| F^{-1}\gamma_k(\mathbf{x}) \right| dx = 2 \int_{|\mathbf{x}| \geq t} \left| F^{-1}\gamma_k(\mathbf{x}) \right| |\mathbf{x}|^{-L} |\mathbf{x}|^L dx$$

$$\leq 2 \left(\int_{|\mathbf{x}| \geq t} |\mathbf{x}|^{-2L} dx \right)^{1/2} \left(\int_{|\mathbf{x}| \geq t} |\mathbf{x}|^{2L} \left| F^{-1}\gamma_k(\mathbf{x}) \right|^2 dx \right)^{1/2}$$

$$\leq C(n, L) t^{n/2 - L} \left(\int |\mathbf{x}|^{2L} \left| F^{-1}\gamma_k(\mathbf{x}) \right|^2 dx \right)^{1/2} \tag{21}$$

$$\leq C(n, L) t^{n/2 - L} \left(\int \sum_{j=1}^{n} |x_j|^{2L} \left| F^{-1}\gamma_k(\mathbf{x}) \right|^2 dx \right)^{1/2}$$

$$\leq C(n, L) t^{n/2 - L} \left(\sum_{j=1}^{n} \int \left| F^{-1}D_j^L\gamma_k(\mathbf{x}) \right|^2 dx \right)^{1/2}$$

$$= C(n, L) t^{n/2 - L} \left(\sum_{j=1}^{n} \int_{S_k} \left| D_j^L\gamma_k(\mathbf{x}) \right|^2 dx \right)^{1/2}$$

by Plancherel's theorem with the definition that

$$S_k \equiv [\mathbf{x} : 2^{-2-k} < |\mathbf{x}| < 2^{2-k}], \tag{22}$$

a set containing the support of γ_k. Now from the definition of γ_k,

$$\left| D_j^L \gamma_k \left(\mathbf{z} \right) \right| = \left| D_j^L \left(\rho \left(\mathbf{z} \right) \phi \left(2^k \mathbf{z} \right) \right) \right|.$$

By Lemma 22.9, this is no larger than

$$C \left(L, n, \phi \right) C_0 \left| \mathbf{z} \right|^{-L}.$$

It follows, using polar coordinates, that the last expression in Formula 21 is no larger than

$$C \left(n, L, \phi, C_0 \right) t^{n/2 - L} \left(\int_{S_k} \left| \mathbf{z} \right|^{-2L} dz \right)^{1/2} \leq C \left(n, L, \phi, C_0 \right) t^{n/2 - L}. \quad (23)$$

$$\left(\int_{2^{-2-k}} \rho^{n-1-2L} d\rho \right)^{1/2} \leq C \left(n, L, \phi, C_0 \right) t^{n/2 - L} 2^{k(L-n/2)}.$$

Now we estimate Formula 20 in another way. The support of γ_k is in S_k, a bounded set, and so $F^{-1}\gamma_k$ is differentiable. Therefore,

$$\int_{|\mathbf{x}| \geq 2t} \left| F^{-1}\gamma_k \left(\mathbf{x} - \mathbf{y} \right) - F^{-1}\gamma_k \left(\mathbf{x} \right) \right| dx =$$

$$\int_{|\mathbf{x}| \geq 2t} \left| \int_0^1 \sum_{j=1}^n D_j F^{-1}\gamma_k \left(\mathbf{x} - s\mathbf{y} \right) y_j ds \right| dx$$

$$\leq t \int_{|\mathbf{x}| \geq 2t} \int_0^1 \sum_{j=1}^n \left| D_j F^{-1}\gamma_k \left(\mathbf{x} - s\mathbf{y} \right) \right| ds dx \leq t \int \sum_{j=1}^n \left| D_j F^{-1}\gamma_k \left(\mathbf{x} \right) \right| dx$$

$$\leq t \sum_{j=1}^n \left(\int \left(1 + \left| 2^{-k} \mathbf{x} \right|^2 \right)^{-L} dx \right)^{1/2}.$$

$$\left(\int \left(1 + \left| 2^{-k} \mathbf{x} \right|^2 \right)^L \left| D_j F^{-1}\gamma_k \left(\mathbf{x} \right) \right|^2 dx \right)^{1/2}$$

$$\leq C \left(n, L \right) t 2^{kn/2} \sum_{j=1}^n \left(\int \left(1 + \left| 2^{-k} \mathbf{x} \right|^2 \right)^L \left| D_j F^{-1}\gamma_k \left(\mathbf{x} \right) \right|^2 dx \right)^{1/2}. \quad (24)$$

Now consider the jth term in the last sum in Formula 24.

$$\int \left(1 + \left| 2^{-k} \mathbf{x} \right|^2 \right)^L \left| D_j F^{-1}\gamma_k \left(\mathbf{x} \right) \right|^2 dx \leq$$

$$C \left(n, L \right) \int \sum_{|\alpha| \leq L} 2^{-2k|\alpha|} \mathbf{x}^{2\alpha} \left| D_j F^{-1}\gamma_k \left(\mathbf{x} \right) \right|^2 dx$$

$$= C(n, L) \sum_{|\alpha| \leq L} 2^{-2k|\alpha|} \int \mathbf{x}^{2\alpha} \left| F^{-1} (\pi_j \gamma_k) (\mathbf{x}) \right|^2 dx$$

where $\pi_j (\mathbf{z}) \equiv z_j$.

$$= C(n, L) \sum_{|\alpha| \leq L} 2^{-2k|\alpha|} \int \left| F^{-1} D^{\alpha} (\pi_j \gamma_k) (\mathbf{x}) \right|^2 dx$$

$$= C(n, L) \sum_{|\alpha| \leq L} 2^{-2k|\alpha|} \int_{S_k} \left| D^{\alpha} (z_j \gamma_k (\mathbf{z})) \right|^2 dz \tag{25}$$

where S_k is given in Formula 22 and we used Plancherel's theorem along with the fact that the support of γ_k is contained in S_k. Now

$$\begin{aligned}
\left| D^{\alpha} (z_j \gamma_k (\mathbf{z})) \right| &= 2^{-k} \left| D^{\alpha} \left(\rho (\mathbf{z}) z_j 2^k \phi \left(2^k \mathbf{z} \right) \right) \right| \\
&= 2^{-k} \left| D^{\alpha} \left(\rho (\mathbf{z}) \psi_j \left(2^k \mathbf{z} \right) \right) \right|
\end{aligned}$$

where $\psi_j (\mathbf{z}) \equiv z_j \phi (\mathbf{z})$. By Lemma 22.9, this is dominated by

$$2^{-k} C(\alpha, n, \phi, j, C_0) |\mathbf{z}|^{-|\alpha|}.$$

Therefore, Formula 25 is dominated by

$$\begin{aligned}
&C(L, n, \phi, j, C_0) \sum_{|\alpha| \leq L} 2^{-2k|\alpha|} \int_{S_k} 2^{-2k} |\mathbf{z}|^{-2|\alpha|} dz \\
\leq \ &C(L, n, \phi, j, C_0) \sum_{|\alpha| \leq L} 2^{-2k|\alpha|} 2^{-2k} \left(2^{-2-k} \right)^{(-2|\alpha|)} \left(2^{2-k} \right)^n \\
\leq \ &C(L, n, \phi, j, C_0) \sum_{|\alpha| \leq L} 2^{-kn-2k}
\end{aligned}$$

$$\leq C(L, n, \phi, j, C_0) \, 2^{-kn} 2^{-2k}.$$

It follows that Formula 24 is no larger than

$$C(L, n, \phi, C_0) \, t 2^{kn/2} 2^{-kn/2} 2^{-k} = C(L, n, \phi, C_0) \, t 2^{-k}. \tag{26}$$

It follows from Formula 26 and Formula 23 that if $|\mathbf{y}| \leq t$,

$$\int_{|\mathbf{x}| \geq 2t} \left| F^{-1} \gamma_k (\mathbf{x} - \mathbf{y}) - F^{-1} \gamma_k (\mathbf{x}) \right| dx \leq$$

$$C(L, n, \phi, C_0) \min \left(t 2^{-k}, \left(2^{-k} t \right)^{n/2 - L} \right).$$

With this inequality, we are ready to prove the next lemma which is the desired result.

Lemma 22.11 *There exists a constant depending only on the indicated objects, $C_1 = C(L, n, \phi, C_0)$ such that when $|\mathbf{y}| \leq t$,*

$$\int_{|\mathbf{x}| \geq 2t} \left| F^{-1} \rho(\mathbf{x} - \mathbf{y}) - F^{-1} \rho(\mathbf{x}) \right| dx \leq C_1$$

$$\int_{|\mathbf{x}| \geq 2t} \left| F^{-1} \rho_m(\mathbf{x} - \mathbf{y}) - F^{-1} \rho_m(\mathbf{x}) \right| dx \leq C_1. \tag{27}$$

Proof: $F^{-1}\rho = \lim_{m \to \infty} F^{-1}\rho_m$ in $L^2(\mathbb{R}^n)$. Let $m_k \to \infty$ be such that convergence is pointwise a.e. Then if $|\mathbf{y}| \leq t$, Fatou's lemma implies

$$\int_{|\mathbf{x}| \geq 2t} \left| F^{-1} \rho(\mathbf{x} - \mathbf{y}) - F^{-1} \rho(\mathbf{x}) \right| dx \leq$$

$$\liminf_{l \to \infty} \int_{|\mathbf{x}| \geq 2t} \left| F^{-1} \rho_{m_l}(\mathbf{x} - \mathbf{y}) - F^{-1} \rho_{m_l}(\mathbf{x}) \right| dx$$

$$\leq \liminf_{l \to \infty} \sum_{k=-m_l}^{m_l} \int_{|\mathbf{x}| \geq 2t} \left| F^{-1} \gamma_k(\mathbf{x} - \mathbf{y}) - F^{-1} \gamma_k(\mathbf{x}) \right| dx$$

$$\leq C(L, n, \phi, C_0) \sum_{k=-\infty}^{\infty} \min\left(t2^{-k}, \left(2^{-k} t\right)^{n/2 - L} \right). \tag{28}$$

Now consider the sum in Formula 28,

$$\sum_{k=-\infty}^{\infty} \min\left(t2^{-k}, \left(2^{-k} t\right)^{n/2 - L} \right). \tag{29}$$

$t2^j = \min\left(t2^j, \left(2^j t\right)^{n/2 - L} \right)$ exactly when $t2^j \leq 1$. This occurs if and only if $j \leq -\ln(t)/\ln(2)$. Therefore Formula 29 is no larger than

$$\sum_{j \leq -\ln(t)/\ln(2)} 2^j t + \sum_{j \geq -\ln(t)/\ln(2)} \left(2^j t\right)^{n/2 - L}.$$

Letting $a = L - n/2$, this equals

$$t \sum_{k \geq \ln(t)/\ln(2)} 2^{-k} + t^{-\alpha} \sum_{j \geq -\ln(t)/\ln(2)} \left(2^{-a}\right)^j$$

$$\leq 2t \left(\frac{1}{2}\right)^{\ln(t)/\ln(2)} + t^{-a} \left(\frac{1}{2^a}\right)^{-\ln(t)/\ln(2)} \leq 2 + 1 = 3.$$

Similarly, Formula 27 holds. This proves the lemma.

With this lemma we are ready to prove Mihlin's theorem.

Theorem 22.12 *(Mihlin's theorem) Suppose ρ satisfies*

$$C_0 \geq \sup\{|\mathbf{x}|^{|\alpha|} |D^\alpha \rho(\mathbf{x})| : |\alpha| \leq L, \, \mathbf{x} \in \mathbb{R}^n \setminus \{\mathbf{0}\}\},$$

where L is an integer greater than $n/2$ and $\rho \in C^L(\mathbb{R}^n \setminus \{\mathbf{0}\})$. Then for every $p > 1$, there exists a constant A_p depending only on p, C_0, ϕ, n, and L, such that for all $\psi \in \mathfrak{S}$,

$$\left\| F^{-1}\rho * \psi \right\|_p \leq A_p \left\| \psi \right\|_p.$$

Proof: Since ρ_m satisfies Formula 27, and is obviously in $L^2(\mathbb{R}^n) \cap L^\infty(\mathbb{R}^n)$, we may apply Theorem 22.8 to obtain a constant A_p depending only on $p, n, \|\rho_m\|_\infty$, and C_1 such that for all $\psi \in \mathfrak{S}$ and $p \in (1, \infty)$,

$$\left\| F^{-1}\rho_m * \psi \right\|_p \leq A_p \left\| \psi \right\|_p.$$

Now $\|\rho_m\|_\infty \leq \|\rho\|_\infty$ because

$$|\rho_m(\mathbf{x})| \leq |\rho(\mathbf{x})| \sum_{k=-m}^{m} \phi(2^k \mathbf{x}) \leq |\rho(\mathbf{x})|. \tag{30}$$

Therefore, since $C_1 = C_1(L, n, \phi, C_0)$ and $C_0 \geq \|\rho\|_\infty$,

$$\left\| F^{-1}\rho_m * \psi \right\|_p \leq A_p(L, n, \phi, C_0, p) \left\| \psi \right\|_p.$$

In particular, A_p does not depend on m. Now, by Formula 30, and the observation that $\rho \in L^\infty(\mathbb{R}^n)$, we may use

$$\lim_{m \to \infty} \rho_m(\mathbf{y}) = \rho(\mathbf{y})$$

and the dominated convergence theorem to conclude the following for $\theta \in \mathfrak{S}$.

$$\left| (F^{-1}\rho * \psi)(\theta) \right| \equiv \left| (2\pi)^{n/2} \int \rho(\mathbf{x}) F\psi(\mathbf{x}) F^{-1}\theta(\mathbf{x}) \, d\mathbf{x} \right|$$

$$= \lim_{m \to \infty} \left| (F^{-1}\rho_m * \psi)(\theta) \right| \leq \lim_{m \to \infty} \sup \left\| F^{-1}\rho_m * \psi \right\|_p \|\theta\|_{p'}$$

$$\leq A_p(L, n, \phi, C_0, p) \|\psi\|_p \|\theta\|_{p'}.$$

Hence $F^{-1}\rho * \psi \in L^p(\mathbb{R}^n)$ and $\left\| F^{-1}\rho * \psi \right\|_p \leq A_p \|\psi\|_p$. This proves the theorem.

22.4   Singular integrals

We know that if $K \in L^1 (\mathbb{R}^n)$ then when $p > 1$,

$$\|K * f\|_p \leq \|f\|_p .$$

It turns out that some meaning can be assigned to $K * f$ for some functions K which are not in L^1. This involves assuming a certain form for K and exploiting cancellation. The resulting theory of singular integrals is very useful. To illustrate, an application will be given to the Helmholtz decomposition of vector fields in the next section. Like Mihlin's theorem, the theory presented here rests on Theorem 22.8.

Lemma 22.13 *Suppose*

$$K \in L^2 (\mathbb{R}^n) , \ \|FK\|_\infty \leq B < \infty,$$

and

$$\int_{|\mathbf{x}|>2|\mathbf{y}|} |K (\mathbf{x} - \mathbf{y}) - K (\mathbf{x})| \, dx \leq B.$$

Then for all $p > 1$, there exists a constant, $A (p, n, B)$, depending only on the indicated quantities such that

$$\|K * f\|_p \leq A (p, n, B) \|f\|_p$$

for all $f \in \mathfrak{S}$.

Proof: Let $FK = \rho$. Then $\rho \in L^2 (\mathbb{R}^n) \cap L^\infty (\mathbb{R}^n)$ and $K = F^{-1}\rho$. By Theorem 22.8 of Chapter 22,

$$\|K * f\|_p \leq A (p, n, B) \|f\|_p$$

for all $f \in \mathfrak{S}$. This proves the lemma.

The next lemma provides a situation in which we will be able to apply the previous lemma.

Lemma 22.14 *Suppose*

$$|K (\mathbf{x})| \leq B |\mathbf{x}|^{-n} , \tag{31}$$

$$\int_{a<|\mathbf{x}|<b} K (\mathbf{x}) \, dx = 0, \tag{32}$$

$$\int_{|\mathbf{x}|>2|\mathbf{y}|} |K (\mathbf{x} - \mathbf{y}) - K (\mathbf{x})| \, dx \leq B. \tag{33}$$

Define

$$K_\epsilon (\mathbf{x}) = \begin{cases} K (\mathbf{x}) & \text{if } |\mathbf{x}| \geq \epsilon, \\ 0 & \text{if } |\mathbf{x}| < \epsilon. \end{cases} \tag{34}$$

Then there exists a constant $C(n)$ such that

$$\int_{|\mathbf{x}|>2|\mathbf{y}|} |K_\epsilon(\mathbf{x}-\mathbf{y}) - K_\epsilon(\mathbf{x})| \, dx \le C(n) B \qquad (35)$$

and

$$\|FK_\epsilon\|_\infty \le C(n) B. \qquad (36)$$

Proof: In the argument, $C(n)$ will denote a generic constant depending only on n. We verify Formula 35 first.

$$\int_{|\mathbf{x}|\ge 2|\mathbf{y}|} |K_\epsilon(\mathbf{x}-\mathbf{y}) - K_\epsilon(\mathbf{x})| \, dx$$

$$= \int_{|\mathbf{x}|\ge 2|\mathbf{y}|, |\mathbf{x}-\mathbf{y}|>\epsilon, |\mathbf{x}|<\epsilon} |K_\epsilon(\mathbf{x}-\mathbf{y}) - K_\epsilon(\mathbf{x})| \, dx +$$

$$\int_{|\mathbf{x}|\ge 2|\mathbf{y}|, |\mathbf{x}-\mathbf{y}|<\epsilon, |\mathbf{x}|\ge\epsilon} |K_\epsilon(\mathbf{x}-\mathbf{y}) - K_\epsilon(\mathbf{x})| \, dx$$

$$+ \int_{|\mathbf{x}|\ge 2|\mathbf{y}|, |\mathbf{x}-\mathbf{y}|>\epsilon, |\mathbf{x}|>\epsilon} |K_\epsilon(\mathbf{x}-\mathbf{y}) - K_\epsilon(\mathbf{x})| \, dx$$

$$+ \int_{|\mathbf{x}|\ge 2|\mathbf{y}|, |\mathbf{x}-\mathbf{y}|<\epsilon, |\mathbf{x}|<\epsilon} |K_\epsilon(\mathbf{x}-\mathbf{y}) - K_\epsilon(\mathbf{x})| \, dx. \qquad (37)$$

Now we consider the terms in the above expression. The last integral in Formula 37 equals 0. The third integral on the right is no larger than B by the definition of K_ϵ and assumption Formula 33. Consider the second integral on the right. This integral is no larger than

$$\int_{|\mathbf{x}|\ge 2|\mathbf{y}|, |\mathbf{x}|\ge\epsilon, |\mathbf{x}-\mathbf{y}|<\epsilon} B |\mathbf{x}|^{-n} \, dx.$$

Now $|\mathbf{x}| \le |\mathbf{y}| + \epsilon \le |\mathbf{x}|/2 + \epsilon$ and so $|\mathbf{x}| < 2\epsilon$. Thus this is no larger than

$$\int_{\epsilon\le|\mathbf{x}|\le 2\epsilon} B |\mathbf{x}|^{-n} \, dx \le BC(n) \ln 2 = C(n) B.$$

It remains to estimate the first integral on the right in Formula 37. This integral is bounded by

$$\int_{|\mathbf{x}|\ge 2|\mathbf{y}|, |\mathbf{x}-\mathbf{y}|>\epsilon, |\mathbf{x}|<\epsilon} B |\mathbf{x}-\mathbf{y}|^{-n} \, dx = \int_{|\mathbf{z}+\mathbf{y}|\ge 2|\mathbf{y}|, |\mathbf{z}|>\epsilon, |\mathbf{z}+\mathbf{y}|<\epsilon} B |\mathbf{z}|^{-n} \, dz.$$

In the integral above,

$$\epsilon \leq |\mathbf{z}|, |\mathbf{z}| \leq |\mathbf{y}| + \epsilon, \text{ and } |\mathbf{y}| \leq |\mathbf{z} + \mathbf{y}|/2 \leq \epsilon/2.$$

Thus,

$$\epsilon \leq |\mathbf{z}| \leq (3/2)\,\epsilon$$

and so the integral above is no larger than $BC(n)\ln(3/2)$. This establishes Formula 35.

Now it remains to show Formula 36, a statement about the Fourier transforms of K_ϵ. Fix ϵ and let

$$K_{\epsilon R}(\mathbf{y}) \equiv \begin{cases} K_\epsilon(\mathbf{y}) \text{ if } |\mathbf{y}| < R, \\ 0 \text{ if } |\mathbf{y}| \geq R. \end{cases}$$

Then

$$|F K_{\epsilon R}(\mathbf{y})| \leq \left| \int_{0<|\mathbf{x}|<3\pi|\mathbf{y}|^{-1}} K_\epsilon(\mathbf{x})\, e^{-i\mathbf{x}\cdot\mathbf{y}} dx \right|$$

$$+ \left| \int_{3\pi|\mathbf{y}|^{-1}<|\mathbf{x}|\leq R} K_\epsilon(\mathbf{x})\, e^{-i\mathbf{x}\cdot\mathbf{y}} dx \right|$$

$$= \mathbf{A} + \mathbf{B}. \tag{38}$$

First we estimate $\mathbf{A}$. By Formula 32 this equals

$$\left| \int_{\epsilon<|\mathbf{x}|<3\pi|\mathbf{y}|^{-1}} K_\epsilon(\mathbf{x})\left(e^{-i\mathbf{x}\cdot\mathbf{y}} - 1\right) dx \right|.$$

Now $\left| e^{-i\mathbf{x}\cdot\mathbf{y}} - 1 \right| = |2 - 2\cos(\mathbf{x}\cdot\mathbf{y})|^{1/2} \leq 2|\mathbf{x}||\mathbf{y}|$ so, using polar coordinates, this expression is no larger than

$$2B \int_{\epsilon<|\mathbf{x}|<3\pi|\mathbf{y}|^{-1}} |\mathbf{x}|^{-n} |\mathbf{x}||\mathbf{y}| dx \leq C(n) B |\mathbf{y}| \int_\epsilon^{3\pi/|\mathbf{y}|} d\rho \leq BC(n).$$

Next, we must estimate $\mathbf{B}$. This estimate is based on the trick which follows. Let

$$\mathbf{z} \equiv \mathbf{y}\pi/|\mathbf{y}|^2$$

so that

$$|\mathbf{z}| = \pi/|\mathbf{y}|, \ \mathbf{z}\cdot\mathbf{y} = \pi.$$

Then

$$\int_{3\pi|\mathbf{y}|^{-1}<|\mathbf{x}|\leq R} K_\epsilon(\mathbf{x})\, e^{-i\mathbf{x}\cdot\mathbf{y}} dx = \frac{1}{2} \int_{3\pi|\mathbf{y}|^{-1}<|\mathbf{x}|\leq R} K_\epsilon(\mathbf{x})\, e^{-i\mathbf{x}\cdot\mathbf{y}} dx$$

$$-\frac{1}{2} \int_{3\pi|\mathbf{y}|^{-1}<|\mathbf{x}|\leq R} K_\epsilon(\mathbf{x})\, e^{-i(\mathbf{x}+\mathbf{z})\cdot\mathbf{y}}\, dx$$

$$=\frac{1}{2} \int_{3\pi|\mathbf{y}|^{-1}<|\mathbf{x}|\leq R} K_\epsilon(\mathbf{x})\, e^{-i\mathbf{x}\cdot\mathbf{y}}\, dx -\frac{1}{2} \int_{3\pi|\mathbf{y}|^{-1}<|\mathbf{x}-\mathbf{z}|\leq R} K_\epsilon(\mathbf{x}-\mathbf{z})\, e^{-i\mathbf{x}\cdot\mathbf{y}}\, dx.$$

Thus

$$\int_{3\pi|\mathbf{y}|^{-1}<|\mathbf{x}|\leq R} K_\epsilon(\mathbf{x})\, e^{-i\mathbf{x}\cdot\mathbf{y}}\, dx =$$

$$\frac{1}{2} \int_{|\mathbf{x}|\leq R} K_\epsilon(\mathbf{x})\, e^{-i\mathbf{x}\cdot\mathbf{y}}\, dx -\frac{1}{2} \int_{|\mathbf{x}-\mathbf{z}|\leq R} K_\epsilon(\mathbf{x}-\mathbf{z})\, e^{-i\mathbf{x}\cdot\mathbf{y}}\, dx$$

$$+\frac{1}{2} \int_{|\mathbf{x}-\mathbf{z}|\leq 3\pi|\mathbf{y}|^{-1}} K_\epsilon(\mathbf{x}-\mathbf{z})\, e^{-i\mathbf{x}\cdot\mathbf{y}}\, dx -\frac{1}{2} \int_{|\mathbf{x}|\leq 3\pi|\mathbf{y}|^{-1}} K_\epsilon(\mathbf{x})\, e^{-i\mathbf{x}\cdot\mathbf{y}}\, dx. \tag{39}$$

Having obtained this, we can now estimate the terms. Since $|\mathbf{z}| = \pi/|\mathbf{y}|$, the following picture describes the situation. In this picture, r equals either R or $3\pi|\mathbf{y}|^{-1}$ and each integral above is taken over one of the two balls in the picture, either the one centered at $\mathbf{0}$ or the one centered at $\mathbf{z}$.

To begin with, we consider the integrals which involve $K_\epsilon(\mathbf{x}-\mathbf{z})$.

$$\int_{|\mathbf{x}-\mathbf{z}|\leq R} K_\epsilon(\mathbf{x}-\mathbf{z})\, e^{-i\mathbf{x}\cdot\mathbf{y}}\, dx$$

$$=\int_{|\mathbf{x}|\leq R} K_\epsilon(\mathbf{x}-\mathbf{z})\, e^{-i\mathbf{x}\cdot\mathbf{y}}\, dx -\int_{|\mathbf{x}-\mathbf{z}|>R,|\mathbf{x}|<R} K_\epsilon(\mathbf{x}-\mathbf{z})\, e^{-i\mathbf{x}\cdot\mathbf{y}}\, dx$$

$$+\int_{|\mathbf{x}-\mathbf{z}|<R,|\mathbf{x}|>R} K_\epsilon(\mathbf{x}-\mathbf{z})\, e^{-i\mathbf{x}\cdot\mathbf{y}}\, dx. \tag{40}$$

Similarly,

$$\int_{|\mathbf{x}-\mathbf{z}|\leq 3\pi|\mathbf{y}|^{-1}} K_\epsilon(\mathbf{x}-\mathbf{z})\, e^{-i\mathbf{x}\cdot\mathbf{y}}\, dx =\int_{|\mathbf{x}|\leq 3\pi|\mathbf{y}|^{-1}} K_\epsilon(\mathbf{x}-\mathbf{z})\, e^{-i\mathbf{x}\cdot\mathbf{y}}\, dx$$

$$-\int_{|\mathbf{x}-\mathbf{z}|>3\pi|\mathbf{y}|^{-1},|\mathbf{x}|<3\pi|\mathbf{y}|^{-1}} K_\epsilon(\mathbf{x}-\mathbf{z})\, e^{-i\mathbf{x}\cdot\mathbf{y}}\, dx+$$

$$\int\limits_{|x-z|<3\pi|y|^{-1},|x|>3\pi|y|^{-1}} K_\epsilon\left(x-z\right)e^{-ix\cdot y}\,dx. \tag{41}$$

The last integral in Formula 40 is taken over a set that is contained in

$$B\left(0,R+|z|\right)\setminus B\left(0,R\right)$$

and so this integral is dominated by

$$B\left(\frac{1}{(R-|z|)^n}\right)\alpha\left(n\right)\left((R+|z|)^n-R^n\right),$$

an expression which converges to 0 as $R\to\infty$. Similarly, the second integral on the right in Formula 40 converges to zero as $R\to\infty$. Now consider the last two integrals in Formula 41. Letting $3\pi\left|y\right|^{-1}$ play the role of R, these are each dominated by an expression of the form

$$B\left(\frac{1}{\left(3\pi\left|y\right|^{-1}-|z|\right)^n}\right)\alpha\left(n\right)\left(\left(3\pi\left|y\right|^{-1}+|z|\right)^n-\left(3\pi\left|y\right|^{-1}\right)^n\right)$$

$$=\alpha\left(n\right)B\frac{|y|^n}{(2\pi)^n}\frac{1}{|y|^n}\left((4\pi)^n-(3\pi)^n\right)=C\left(n\right)B$$

because $|z|=\pi/\left|y\right|$.

Returning to Formula 39, and collecting the terms which have not yet been estimated along with those that have,

$$\mathbf{B}=\left|\int\limits_{3\pi|y|^{-1}<|x|\le R} K_\epsilon\left(x\right)e^{-ix\cdot y}\,dx\right|$$

$$\le\frac{1}{2}\left|\int\limits_{|x|<R} K_\epsilon\left(x\right)e^{-ix\cdot y}\,dx-\int\limits_{|x|<R} K_\epsilon\left(x-z\right)e^{-ix\cdot y}\,dx\right.$$

$$+\int\limits_{|x|<3\pi|y|^{-1}} K_\epsilon\left(x-z\right)e^{-ix\cdot y}\,dx-\left.\int\limits_{|x|<3\pi|y|^{-1}} K_\epsilon\left(x\right)e^{-ix\cdot y}\,dx\right|$$

$$+C\left(n\right)B+g\left(R\right)$$

where $g\left(R\right)\to0$ as $R\to\infty$. Thus, using $|z|=\pi/\left|y\right|$ again,

$$\mathbf{B}\le\frac{1}{2}\int\limits_{3|z|<|x|<R}\left|K_\epsilon\left(x\right)-K_\epsilon\left(x-z\right)\right|\,dx+C\left(n\right)B+g\left(R\right).$$

But the integral in the above is dominated by $C(n)B$ by Formula 35 which was established earlier. Therefore, from Formula 38,

$$|FK_{\epsilon R}| \leq C(n)B + g(R)$$

where $g(R) \to 0$.

Now $K_{\epsilon R} \to K_\epsilon$ in $L^2(\mathbb{R}^n)$ and so $FK_{\epsilon R} \to FK_\epsilon$ in $L^2(\mathbb{R}^n)$ by Plancherel's theorem. Therefore, by taking a subsequence, still denoted by R, $FK_{\epsilon R}(\mathbf{y}) \to FK_\epsilon(\mathbf{y})$ a.e. which shows

$$|FK_\epsilon(\mathbf{y})| \leq C(n)B \text{ a.e.}$$

This proves the lemma.

Corollary 22.15 *Suppose Formula 31- Formula 33 hold. Then if $g \in C_c^1(\mathbb{R}^n)$, $K_\epsilon * g$ converges uniformly and in $L^p(\mathbb{R}^n)$ as $\epsilon \to 0$.*

Proof:

$$K_\epsilon * g(\mathbf{x}) \equiv \int K_\epsilon(\mathbf{y}) g(\mathbf{x} - \mathbf{y}) \, dy.$$

When $\epsilon < 1$, this equals

$$\int_{|\mathbf{y}| \geq 1} K(\mathbf{y}) g(\mathbf{x} - \mathbf{y}) \, dy + \int_{1 \geq |\mathbf{y}| \geq \epsilon} K(\mathbf{y}) [g(\mathbf{x} - \mathbf{y}) - g(\mathbf{x})] \, dy.$$

Now there exists a constant C such that

$$|g(\mathbf{x} - \mathbf{y}) - g(\mathbf{x})| \leq C|\mathbf{y}|$$

and so it follows from the dominated convergence theorem that the second integral converges uniformly for $\mathbf{x} \in \mathbb{R}$ as $\epsilon \to 0$. Hence the convergence also takes place in $L^p(\mathbb{R}^n)$ because

$$\mathbf{x} \to \int_{1 \geq |\mathbf{y}| \geq \epsilon} K(\mathbf{y}) [g(\mathbf{x} - \mathbf{y}) - g(\mathbf{x})] \, dy$$

has compact support independent of ϵ. This proves the corollary.

Theorem 22.16 *Suppose Formula 31- Formula 33. Then for K_ϵ given by Formula 34 and $p > 1$, there exists a constant $A(p, n, B)$ such that for all $f \in L^p(\mathbb{R}^n)$,*

$$\|K_\epsilon * f\|_p \leq A(p, n, B) \|f\|_p. \tag{42}$$

Also, for each $f \in L^p(\mathbb{R}^n)$,

$$Tf \equiv \lim_{\epsilon \to 0} K_\epsilon * f \tag{43}$$

exists in $L^p(\mathbb{R}^n)$ and for all $f \in L^p(\mathbb{R}^n)$,

$$\|Tf\|_p \le A(p,n,B)\|f\|_p. \tag{44}$$

Thus T is a linear and continuous map defined on $L^p(\mathbb{R}^n)$ for each $p > 1$.

Proof: From Formula 31 it follows $K_\epsilon \in L^{p'}(\mathbb{R}^n)$ where, as usual, $1/p + 1/p' = 1$. By continuity of translation in $L^{p'}(\mathbb{R}^n)$, $x \to K_\epsilon * f(x)$ is a continuous function. Also by Formula 31, $K_\epsilon \in L^2(\mathbb{R}^n)$ and by Lemma 22.14, $\|FK_\epsilon\|_\infty \le C(n)B$ for all ϵ. Therefore, by Lemma 22.13,

$$\|K_\epsilon * g\|_p \le A(p,n,B)\|g\|_p$$

for all $g \in \mathfrak{S}$. Now let $f \in L^p(\mathbb{R}^n)$ and $g_k \to f$ in $L^p(\mathbb{R}^n)$ where $g_k \in \mathfrak{S}$. Then

$$
\begin{aligned}
|K_\epsilon * f(\mathbf{x}) - K_\epsilon * g_k(\mathbf{x})| &\le \int |K_\epsilon(\mathbf{x}-\mathbf{y})||g_k(\mathbf{y}) - f(\mathbf{y})|\,dy \\
&\le \|K_\epsilon\|_{p'}\|g_k - f\|_p
\end{aligned}
$$

and so by Fatou's lemma,

$$
\begin{aligned}
\|K_\epsilon * f\|_p &\le \lim_{k \to \infty} \inf \|K_\epsilon * g_k\|_p \le \lim_{k \to \infty} \inf A(p,n,B)\|g_k\|_p \\
&= A(p,n,B)\|f\|_p.
\end{aligned}
$$

This verifies Formula 42.

To verify Formula 43, let $\delta > 0$ be given and let

$$f \in L^p(\mathbb{R}^n), g \in C_c^\infty(\mathbb{R}^n).$$

$$
\begin{aligned}
\|K_\epsilon * f - K_\eta * f\|_p &\le \|K_\epsilon * (f-g)\|_p + \|K_\epsilon * g - K_\eta * g\|_p \\
&\quad + \|K_\eta * (f-g)\|_p
\end{aligned}
$$

$$\le 2A(p,n,B)\|f-g\|_p + \|K_\epsilon * g - K_\eta * g\|_p.$$

Choose g such that $2A(p,n,B)\|f-g\|_p \le \delta/2$. Then if ϵ, η are small enough, Corollary 22.15 implies the last term is also less than $\delta/2$. Thus, $\lim_{\epsilon \to 0} K_\epsilon * f$ exists in $L^p(\mathbb{R}^n)$. Let Tf be the element of $L^p(\mathbb{R}^n)$ to which it converges. Then Formula 44 follows and T is obviously linear. This proves the theorem.

When do conditions 31-33 hold? We will show next that if K is given by the following, then Formula 31- Formula 33 are satisfied.

$$K(\mathbf{x}) \equiv \frac{\Omega(\mathbf{x})}{|\mathbf{x}|^n}, \tag{45}$$

where

$$\Omega(\lambda \mathbf{x}) = \Omega(\mathbf{x}) \text{ for all } \lambda > 0, \tag{46}$$

$$\Omega \text{ is Lipschitz on } S^{n-1},$$

$$\int_{S^{n-1}} \Omega(\mathbf{x}) \, d\sigma = 0. \tag{47}$$

Theorem 22.17 *For K given by Formula 45- Formula 47, it follows that there exists a constant B such that*

$$|K(\mathbf{x})| \le B |\mathbf{x}|^{-n}, \tag{48}$$

$$\int_{a<|\mathbf{x}|<b} K(\mathbf{x}) \, dx = 0, \tag{49}$$

$$\int_{|\mathbf{x}|>2|\mathbf{y}|} |K(\mathbf{x}-\mathbf{y}) - K(\mathbf{x})| \, dx \le B. \tag{50}$$

Consequently, the conclusions of Theorem 22.16 hold also.

Proof: Formula 48 is obvious. To verify Formula 49,

$$
\begin{aligned}
\int_{a<|\mathbf{x}|<b} K(\mathbf{x}) \, dx &= \int_a^b \int_{S^{n-1}} \frac{\Omega(\rho \mathbf{w})}{\rho^n} \rho^{n-1} \, d\sigma d\rho \\
&= \int_a^b \frac{1}{\rho} \int_{S^{n-1}} \Omega(\mathbf{w}) \, d\sigma d\rho = 0.
\end{aligned}
$$

It remains to show Formula 50.

$$
\begin{aligned}
K(\mathbf{x}-\mathbf{y}) - K(\mathbf{x}) &= |\mathbf{x}-\mathbf{y}|^{-n} \left(\Omega\left(\frac{\mathbf{x}-\mathbf{y}}{|\mathbf{x}-\mathbf{y}|}\right) - \Omega\left(\frac{\mathbf{x}}{|\mathbf{x}|}\right) \right) \\
&\quad + \Omega(\mathbf{x}) \left(\frac{1}{|\mathbf{x}-\mathbf{y}|^n} - \frac{1}{|\mathbf{x}|^n} \right)
\end{aligned}
\tag{51}
$$

where we used Formula 46 to write $\Omega\left(\frac{\mathbf{z}}{|\mathbf{z}|}\right) = \Omega(\mathbf{z})$. The first group of terms in Formula 51 is dominated by

$$|\mathbf{x}-\mathbf{y}|^{-n} \, Lip(\Omega) \left| \frac{\mathbf{x}-\mathbf{y}}{|\mathbf{x}-\mathbf{y}|} - \frac{\mathbf{x}}{|\mathbf{x}|} \right|$$

and we need an estimate when $|\mathbf{x}| > 2|\mathbf{y}|$. Since $|\mathbf{x}| > 2|\mathbf{y}|$,

$$|\mathbf{x}-\mathbf{y}|^{-n} \le (|\mathbf{x}|-|\mathbf{y}|)^{-n} \le \frac{2^n}{|\mathbf{x}|^n}.$$

Also

$$\left| \frac{\mathbf{x}-\mathbf{y}}{|\mathbf{x}-\mathbf{y}|} - \frac{\mathbf{x}}{|\mathbf{x}|} \right| = \left| \frac{(\mathbf{x}-\mathbf{y})|\mathbf{x}| - \mathbf{x}|\mathbf{x}-\mathbf{y}|}{|\mathbf{x}||\mathbf{x}-\mathbf{y}|} \right|$$

$$\leq \frac{2}{|\mathbf{x}|^2} |\mathbf{x} |\mathbf{x}| - \mathbf{y} |\mathbf{x}| - \mathbf{x} |\mathbf{x} - \mathbf{y}||$$

$$= \frac{2}{|\mathbf{x}|^2} |\mathbf{x} (|\mathbf{x}| - |\mathbf{x} - \mathbf{y}|) - \mathbf{y} |\mathbf{x}||$$

$$\leq \frac{4}{|\mathbf{x}|^2} |\mathbf{x}| |\mathbf{y}| = 4 \frac{|\mathbf{y}|}{|\mathbf{x}|}.$$

Therefore,

$$\int_{|\mathbf{x}|>2|\mathbf{y}|} |\mathbf{x} - \mathbf{y}|^{-n} \left| \Omega \left(\frac{\mathbf{x} - \mathbf{y}}{|\mathbf{x} - \mathbf{y}|} \right) - \Omega \left(\frac{\mathbf{x}}{|\mathbf{x}|} \right) \right| dx$$

$$\leq 4 \left(2^n \right) \int_{|\mathbf{x}|>2|\mathbf{y}|} \frac{1}{|\mathbf{x}|^n} \frac{|\mathbf{y}|}{|\mathbf{x}|} dx \, Lip \left(\Omega \right)$$

$$= C \left(n, Lip\Omega \right) \int_{|\mathbf{x}|>2|\mathbf{y}|} \frac{|\mathbf{y}|}{|\mathbf{x}|^{n+1}} dx$$

$$= C \left(n, Lip\Omega \right) \int_{|\mathbf{u}|>2} \frac{1}{|\mathbf{u}|^{n+1}} du. \tag{52}$$

It remains to consider the second group of terms in Formula 51 when $|\mathbf{x}| > 2 |\mathbf{y}|$.

$$\left| \frac{1}{|\mathbf{x} - \mathbf{y}|^n} - \frac{1}{|\mathbf{x}|^n} \right| = \left| \frac{|\mathbf{x}|^n - |\mathbf{x} - \mathbf{y}|^n}{|\mathbf{x} - \mathbf{y}|^n |\mathbf{x}|^n} \right|$$

$$\leq \frac{2^n}{|\mathbf{x}|^{2n}} ||\mathbf{x}|^n - |\mathbf{x} - \mathbf{y}|^n|$$

$$\leq \frac{2^n}{|\mathbf{x}|^{2n}} |\mathbf{y}| \left[|\mathbf{x}|^{n-1} + |\mathbf{x}|^{n-2} |\mathbf{x} - \mathbf{y}| + \right.$$

$$\left. \cdots + |\mathbf{x}| |\mathbf{x} - \mathbf{y}|^{n-2} + |\mathbf{x} - \mathbf{y}|^{n-1} \right]$$

$$\leq \frac{2^n |\mathbf{y}| C \left(n \right) |\mathbf{x}|^{n-1}}{|\mathbf{x}|^{2n}} = \frac{C \left(n \right) 2^n |\mathbf{y}|}{|\mathbf{x}|^{n+1}}.$$

Thus

$$\int_{|\mathbf{x}|>2|\mathbf{y}|} \left| \Omega \left(\mathbf{x} \right) \left(\frac{1}{|\mathbf{x} - \mathbf{y}|^n} - \frac{1}{|\mathbf{x}|^n} \right) \right| dx \leq C \left(n \right) \int_{|\mathbf{x}|>2|\mathbf{y}|} \frac{|\mathbf{y}|}{|\mathbf{x}|^{n+1}} dx$$

$$\leq C \left(n \right) \int_{|\mathbf{u}|>2} \frac{1}{|\mathbf{u}|^{n+1}} du. \tag{53}$$

From Formula 52 and Formula 53,

$$\int_{|\mathbf{x}|>2|\mathbf{y}|} |K \left(\mathbf{x} - \mathbf{y} \right) - K \left(\mathbf{x} \right)| \, dx \leq C \left(n, Lip\Omega \right).$$

This proves the theorem.

22.5 The Helmholtz decomposition of vector fields

In this section we give an application of singular integrals to the Helmholtz decomposition of a vector field. It turns out that every vector field which has its components in L^p can be written as a sum of a gradient and a vector field which has zero divergence. This is a very remarkable result when applied to vector fields which are only in L^p, although it is not too hard to see if enough smoothness is assumed.

Definition 22.18 Define

$$\Phi\left(\mathbf{y}\right) \equiv \begin{cases} -\frac{1}{2\alpha(2)}\ln|\mathbf{y}|\,, & \text{if } n = 2, \\ \frac{1}{(n-2)n\alpha(n)}|\mathbf{y}|^{2-n}\,, & \text{if } n > 2. \end{cases}$$

Then it is easy to verify that $\Delta\Phi = 0$ away from 0. Also if $n > 2$,

$$\Phi_{,ii}\left(\mathbf{y}\right) = C_n\left[\frac{1}{|\mathbf{y}|^n} - n\frac{y_i^2}{|\mathbf{y}|^{n+2}}\right], \ \Phi_{,ij}\left(\mathbf{y}\right) = C_n\frac{y_i y_j}{|\mathbf{y}|^{n+2}}, \qquad (54)$$

while if $n = 2$,

$$\Phi_{,22}\left(\mathbf{y}\right) = C_2\frac{y_1^2 - y_2^2}{\left(y_1^2 + y_2^2\right)^2}, \ \Phi_{,11}\left(\mathbf{y}\right) = C_2\frac{y_2^2 - y_1^2}{\left(y_1^2 + y_2^2\right)^2},$$

$$\Phi_{,ij}\left(\mathbf{y}\right) = C_2\frac{y_1 y_2}{\left(y_1^2 + y_2^2\right)^2}.$$

In the above the subscripts following a comma denote partial derivatives.

Lemma 22.19 *For $n \geq 2$*

$$\Phi_{,ij}\left(\mathbf{y}\right) = \frac{\Omega_{ij}\left(\mathbf{y}\right)}{|\mathbf{y}|^n}$$

where

$$\Omega_{ij} \text{ is Lipschitz continuous on } S^{n-1}, \qquad (55)$$

$$\Omega_{ij}\left(\lambda\mathbf{y}\right) = \Omega_{ij}\left(\mathbf{y}\right), \qquad (56)$$

for all $\lambda > 0$, and

$$\int_{S^{n-1}} \Omega_{ij}\left(\mathbf{y}\right) d\sigma = 0. \qquad (57)$$

Proof: The case $n = 2$ is left to the reader. Formulas 55 and 56 are obvious from the above descriptions. It remains to verify Formula 57. If $n \geq 3$ and $i \neq j$, then this formula is also clear from Formula 54. Thus, we consider the case when $n \geq 3$ and $i = j$.

$$\int_{S^{n-1}} 1 - ny_i^2\, d\sigma = \int_{S^{n-1}} 1 - ny_n^2\, d\sigma =$$

$$2 \int_{\sum_{i=1}^{n-1} y_i^2 \leq 1} \left[1 - n \left(1 - \sum_{i=1}^{n-1} y_i^2 \right) \right] \left(1 - \sum_{i=1}^{n-1} y_i^2 \right)^{-1/2} dy_1 \cdots dy_{n-1}$$

$$= 2 \int_{\sum_{i=1}^{n-1} y_i^2 \leq 1} \left(1 - \sum_{i=1}^{n-1} y_i^2 \right)^{-1/2} - n \left(1 - \sum_{i=1}^{n-1} y_i^2 \right)^{1/2} dy_1 \cdots dy_{n-1}$$

$$= 2 \int_{\sum_{i=1}^{n-2} y_i^2 \leq 1} \int_{-r}^{r} \left(r^2 - z^2 \right)^{-1/2} - n \left(r^2 - z^2 \right)^{1/2} dz dy_1 \cdots dy_{n-2}$$

where $r^2 = 1 - \sum_{i=1}^{n-2} y_i^2$. Thus this reduces to

$$2 \int_{\sum_{i=1}^{n-2} y_i^2 \leq 1} \left[\pi - \frac{\pi}{2} n \left(1 - \sum_{i=1}^{n-2} y_i^2 \right) \right] dy_1 \cdots dy_{n-2}$$

$$= 2\pi \int_{S^{n-3}} \int_0^1 \left[1 - \frac{n}{2} \left(1 - \rho^2 \right) \right] \rho^{n-3} d\rho d\sigma$$

$$= 2\pi \int_{S^{n-3}} \frac{\rho^{n-2}}{n-2} + \frac{-2n\rho^n}{2n(n-2)} \Big|_0^1 d\sigma = 0.$$

This proves the lemma.

Let $B = B(\mathbf{0}, R)$ where

$$B \supseteq U - U \equiv \{ \mathbf{x} - \mathbf{y} : \mathbf{x} \in U, \mathbf{y} \in U \}$$

for U a bounded open set with Lipschitz boundary. Let $f \in C_c^\infty(U)$ and define for $\mathbf{x} \in U$,

$$u(\mathbf{x}) \equiv \int_B \Phi(\mathbf{y}) f(\mathbf{x} - \mathbf{y}) dy = \int_U \Phi(\mathbf{x} - \mathbf{y}) f(\mathbf{y}) dy.$$

Let $h(\mathbf{y}) = f(\mathbf{x} - \mathbf{y})$. Then since Φ is in $L^1(B)$,

$$\Delta u(\mathbf{x}) = \int_B \Phi(\mathbf{y}) \Delta f(\mathbf{x} - \mathbf{y}) dy = \int_B \Phi(\mathbf{y}) \Delta h(\mathbf{y}) dy$$

$$= \int_{B \setminus B(\mathbf{0}, \epsilon)} \nabla \cdot (\nabla h(\mathbf{y}) \Phi(\mathbf{y})) - \nabla \Phi(\mathbf{y}) \cdot \nabla h(\mathbf{y}) dy$$

$$+ \int_{B(\mathbf{0}, \epsilon)} \Phi(\mathbf{y}) \Delta h(\mathbf{y}) dy.$$

The last term converges to 0 as $\epsilon \to 0$. Since $spt(h) \subseteq B$, the divergence theorem implies

$$\Delta u(\mathbf{x}) = - \int_{\partial B(\mathbf{0}, \epsilon)} \Phi(\mathbf{y}) \nabla h(\mathbf{y}) \cdot \mathbf{n} d\sigma$$

$$-\int_{B\backslash B(0,\epsilon)} \nabla\Phi\left(\mathbf{y}\right)\cdot\nabla h\left(\mathbf{y}\right)dy + e\left(\epsilon\right)$$

where $e\left(\epsilon\right)\to 0$ as $\epsilon\to 0$. The first term converges to 0 as $\epsilon\to 0$ and since $\Delta\Phi\left(\mathbf{y}\right)=0$,

$$\nabla\Phi\left(\mathbf{y}\right)\cdot\nabla h\left(\mathbf{y}\right) = \nabla\cdot\left(\nabla\Phi\left(\mathbf{y}\right)h\left(\mathbf{y}\right)\right).$$

Consequently

$$\Delta u\left(\mathbf{x}\right) = -\int_{B\backslash B(0,\epsilon)} \nabla\cdot\left(\nabla\Phi\left(\mathbf{y}\right)h\left(\mathbf{y}\right)\right)dy + e\left(\epsilon\right).$$

Thus, by the divergence theorem and the definition of h above,

$$\begin{aligned}
\Delta u\left(\mathbf{x}\right) &= \int_{\partial B(0,\epsilon)} f\left(\mathbf{x}-\mathbf{y}\right)\nabla\Phi\left(\mathbf{y}\right)\cdot\mathbf{n}d\sigma + e\left(\epsilon\right) \\
&= -\left(\int_{\partial B(0,\epsilon)} f\left(\mathbf{x}-\mathbf{y}\right)d\sigma\left(y\right)\right)\frac{1}{n\alpha\left(n\right)\epsilon^{n-1}} + e\left(\epsilon\right).
\end{aligned}$$

Letting $\epsilon\to 0$, and using the fact that $\sigma\left(\partial B\left(0,\epsilon\right)\right)=n\alpha\left(n\right)\epsilon^{n-1}$, we see

$$-\Delta u\left(\mathbf{x}\right) = f\left(\mathbf{x}\right).$$

This proves the following lemma.

Lemma 22.20 *Let U be a bounded open set with Lipschitz boundary locally on one side of U and let $B\supseteq U-U$ where $B=B\left(0,R\right)$. Let $f\in C_c^\infty\left(U\right)$. Then*

$$\int_B \Phi\left(\mathbf{y}\right)f\left(\mathbf{x}-\mathbf{y}\right)dy = \int_U \Phi\left(\mathbf{x}-\mathbf{y}\right)f\left(\mathbf{y}\right)dy,$$

and if

$$u\left(\mathbf{x}\right)\equiv\int_B \Phi\left(\mathbf{y}\right)f\left(\mathbf{x}-\mathbf{y}\right)dy = \int_U \Phi\left(\mathbf{x}-\mathbf{y}\right)f\left(\mathbf{y}\right)dy,$$

it follows that for all $x\in U$,

$$-\Delta u\left(\mathbf{x}\right) = f\left(\mathbf{x}\right).$$

Theorem 22.21 *Let $f\in L^p\left(U\right)$. Then there exists $u\in L^p\left(U\right)$ whose weak derivatives are also in $L^p\left(U\right)$ such that in the sense of weak derivatives,*

$$-\Delta u = f.$$

It is given by

$$u\left(\mathbf{x}\right) = \int_B \Phi\left(\mathbf{y}\right)\widetilde{f}\left(\mathbf{x}-\mathbf{y}\right)dy = \int_U \Phi\left(\mathbf{x}-\mathbf{y}\right)f\left(\mathbf{y}\right)dy \qquad (58)$$

The Helmholtz decomposition of vector fields

where $\widetilde{f}$ denotes the zero extension of f off of U.

Proof: Let $f \in L^p(U)$ and let $f_k \in C_c^\infty(U)$, $\|f_k - f\|_{L^p(U)} \to 0$, and let u_k be given by Formula 58 with f_k in place of f. Then by Minkowski's inequality,

$$\|u - u_k\|_{L^p(U)} \leq \int_B |\Phi(\mathbf{y})| \, dy \, \|f - f_k\|_{L^p(U)} = C(B) \, \|f - f_k\|_{L^p(U)}$$

and so $u_k \to u$ in L^p. Also

$$u_{k,i}(\mathbf{x}) = \int_U \Phi_{,i}(\mathbf{x} - \mathbf{y}) f_k(\mathbf{y}) \, dy = \int_B f_k(\mathbf{x} - \mathbf{y}) \Phi_{,i}(\mathbf{y}) \, dy$$

and since $\Phi_{,i} \in L^1(B)$, it follows from Minkowski's inequality that

$$\|u_{k,i} - w_i\|_{L^p(U)} \to 0$$

where

$$w_i \equiv \int_B f(\mathbf{x} - \mathbf{y}) \Phi_{,i}(\mathbf{y}) \, dy.$$

Now let $\phi \in C_c^\infty(U)$. Then

$$\int_U w_i \phi \, dx = -\lim_{k \to \infty} \int_U u_k \phi_{,i} \, dx = -\int_U u \phi_{,i} \, dx.$$

Thus $u_{,i} = w_i \in L^p(\mathbb{R}^n)$ and so if $\phi \in C_c^\infty(U)$,

$$\int_U f\phi \, dx = \lim_{k \to \infty} \int_U f_k \phi \, dx = \lim_{k \to \infty} \int_U \nabla u_k \cdot \nabla \phi \, dx = \int_U \nabla u \cdot \nabla \phi \, dx$$

and so $-\Delta u = f$ as claimed. This proves the theorem.

One could also ask whether the second weak partial derivatives of u are in $L^p(U)$. To answer this question, we use the theory of singular integrals.

Lemma 22.22 *Let $f \in L^p(U)$ and let*

$$w(\mathbf{x}) \equiv \int_U \Phi_{,i}(\mathbf{x} - \mathbf{y}) f(\mathbf{y}) \, dy.$$

Then $w_{,j} \in L^p(U)$ for each $j = 1 \cdots n$ and the map $f \to w_{,j}$ is continuous and linear on $L^p(U)$.

Proof: First let $f \in C_c^\infty(U)$. For such f,

$$
\begin{aligned}
w(\mathbf{x}) &= \int_U \Phi_{,i}(\mathbf{x} - \mathbf{y}) f(\mathbf{y}) \, dy = \int_{\mathbb{R}^n} \Phi_{,i}(\mathbf{x} - \mathbf{y}) f(\mathbf{y}) \, dy \\
&= \int_{\mathbb{R}^n} \Phi_{,i}(\mathbf{y}) f(\mathbf{x} - \mathbf{y}) \, dy = \int_B \Phi_{,i}(\mathbf{y}) f(\mathbf{x} - \mathbf{y}) \, dy
\end{aligned}
$$

465

and

$$w_{,j}(\mathbf{x}) = \int_B \Phi_{,i}(\mathbf{y}) f_{,j}(\mathbf{x}-\mathbf{y})\, dy$$

$$= \int_{B \backslash B(0,\epsilon)} \Phi_{,i}(\mathbf{y}) f_{,j}(\mathbf{x}-\mathbf{y})\, dy + \int_{B(0,\epsilon)} \Phi_{,i}(\mathbf{y}) f_{,j}(\mathbf{x}-\mathbf{y})\, dy.$$

The second term converges to 0 because $f_{,j}$ is bounded and $\Phi_{,i} \in L^1_{loc}$. Thus

$$w_{,j}(\mathbf{x}) = \int_{B \backslash B(0,\epsilon)} \Phi_{,i}(\mathbf{y}) f_{,j}(\mathbf{x}-\mathbf{y})\, dy + e(\epsilon)$$

$$= \int_{B \backslash B(0,\epsilon)} -(\Phi_{,i}(\mathbf{y}) f(\mathbf{x}-\mathbf{y}))_{,j} - \Phi_{,ij}(\mathbf{y}) f(\mathbf{x}-\mathbf{y})\, dy + e(\epsilon)$$

where $e(\epsilon) \to 0$ as $\epsilon \to 0$. Using the divergence theorem, this yields

$$w_{,j}(\mathbf{x}) = \int_{\partial B(0,\epsilon)} \Phi_{,i}(\mathbf{y}) f(\mathbf{x}-\mathbf{y}) n_j\, d\sigma$$

$$- \int_{B \backslash B(0,\epsilon)} \Phi_{,ij}(\mathbf{y}) f(\mathbf{x}-\mathbf{y})\, dy + e(\epsilon).$$

Consider the first term on the right. This term equals, after letting $\mathbf{y} = \epsilon \mathbf{z}$,

$$\epsilon^{n-1} \int_{\partial B(0,1)} \Phi_{,i}(\epsilon \mathbf{z}) f(\mathbf{x}-\epsilon \mathbf{z}) n_j\, d\sigma$$

$$= C_n \epsilon^{n-1} \int_{\partial B(0,1)} \epsilon^{1-n} z_i z_j f(\mathbf{x}-\epsilon \mathbf{z})\, d\sigma(z)$$

$$= C_n \int_{\partial B(0,1)} z_i z_j f(\mathbf{x}-\epsilon \mathbf{z})\, d\sigma(z)$$

and this converges to 0 if $i \neq j$ and it converges to

$$C_n f(\mathbf{x}) \int_{\partial B(0,1)} z_i^2\, d\sigma(z)$$

if $i = j$. Thus

$$w_{,j}(\mathbf{x}) = C_n \delta_{ij} f(\mathbf{x}) - \int_{B \backslash B(0,\epsilon)} \Phi_{,ij}(\mathbf{y}) f(\mathbf{x}-\mathbf{y})\, dy + e(\epsilon).$$

Letting

$$\Phi_{ij}^\epsilon \equiv \begin{cases} 0 & \text{if } |\mathbf{y}| < \epsilon, \\ \Phi_{,ij}(\mathbf{y}) & \text{if } |\mathbf{y}| \geq \epsilon, \end{cases}$$

it follows

$$w_{,j}(\mathbf{x}) = C_n \delta_{ij} f(\mathbf{x}) - \Phi_{ij}^\epsilon * \tilde{f}(\mathbf{x}) + e(\epsilon).$$

By the theory of singular integrals, we may define

$$K_{ij} f \equiv \lim_{\epsilon \to 0} \Phi_{ij}^{\epsilon} * f$$

and $K_{ij} \in \mathcal{L}\left(L^p\left(\mathbb{R}^n\right), L^p\left(\mathbb{R}^n\right)\right)$. Therefore, letting $\epsilon \to 0$,

$$w_{,j} = C_n \delta_{ij} f - K_{ij} \widetilde{f}$$

whenever $f \in C_c^{\infty}(U)$. Now let $f \in L^p(U)$, let

$$\|f_k - f\|_{L^p(U)} \to 0,$$

where $f_k \in C_c^{\infty}(U)$, and let

$$w_k(\mathbf{x}) = \int_U \Phi_{,i}(\mathbf{x} - \mathbf{y}) f_k(\mathbf{y}) \, dy.$$

Then it follows that $w_k \to w$ in $L^p(U)$ and

$$w_{k,j} = C_n \delta_{ij} f_k - K_{ij} \widetilde{f_k}.$$

Now let $\phi \in C_c^{\infty}(U)$.

$$
\begin{aligned}
w_{,j}(\phi) &\equiv -\int_U w \phi_{,j} \, dx = -\lim_{k \to \infty} \int_U w_k \phi_{,j} \, dx \\
&= \lim_{k \to \infty} \int_U w_{k,j} \phi \, dx = \lim_{k \to \infty} \int_U \left(C_n \delta_{ij} \widetilde{f_k} - K_{ij} \widetilde{f_k}\right) \phi \, dx \\
&= \int_U \left(C_n \delta_{ij} \widetilde{f} - K_{ij} \widetilde{f}\right) \phi \, dx.
\end{aligned}
$$

It follows

$$w_{,j} = C_n \delta_{ij} \widetilde{f} - K_{ij} \widetilde{f}$$

and this proves the lemma.

Corollary 22.23 *In the situation of Theorem 22.21, all weak derivatives of u of order 2 are in $L^p(U)$ and also $f \to u_{,ij}$ is a continuous map.*

Proof:

$$u_{,i}(\mathbf{x}) = \int_U \Phi_{,i}(\mathbf{x} - \mathbf{y}) f(\mathbf{y}) \, dy$$

and so $u_{,ij} \in L^p(U)$ and $f \to u_{,ij}$ is continuous by Lemma 22.22.

Next we discuss the Helmholtz decomposition of a vector field. Let $\mathbf{F} \in L^p(U; \mathbb{R}^n)$ and define

$$\phi(\mathbf{x}) \equiv \int_U \nabla \Phi(\mathbf{x} - \mathbf{y}) \cdot \mathbf{F}(\mathbf{y}) \, dy. \tag{59}$$

Then by Lemma 22.22,

$$\phi_{,j} = C_n \widetilde{F}_j - \sum_i K_{ij} \widetilde{F}_i \in L^p(\mathbb{R}^n)$$

and the mapping $\mathbf{F} \to \nabla\phi$ is continuous from $L^p(U;\mathbb{R}^n)$ to $L^p(U;\mathbb{R}^n)$.
Now suppose $\mathbf{F} \in C_c^\infty(U;\mathbb{R}^n)$. Then

$$
\begin{aligned}
\phi(\mathbf{x}) &= \int_U \sum_{i=1}^n -\frac{\partial}{\partial y^i}\left(\Phi(\mathbf{x}-\mathbf{y})F_i(\mathbf{y})\right) + \Phi(\mathbf{x}-\mathbf{y})\nabla\cdot\mathbf{F}(\mathbf{y})\,dy \\
&= \int_U \Phi(\mathbf{x}-\mathbf{y})\nabla\cdot\mathbf{F}(\mathbf{y})\,dy
\end{aligned}
$$

and so by Lemma 22.20,

$$\nabla\cdot\nabla\phi = \Delta\phi = -\nabla\cdot\mathbf{F}.$$

This continues to hold in the sense of weak derivatives if $\mathbf{F}$ is only in $L^p(U;\mathbb{R}^n)$ because by Minkowski's inequality and Formula 59 the map $\mathbf{F} \to \phi$ is continuous. Also note that for $\mathbf{F} \in C_c^\infty(U;\mathbb{R}^n)$,

$$\phi(\mathbf{x}) = \int_B \Phi(\mathbf{y})\nabla\cdot\mathbf{F}(\mathbf{x}-\mathbf{y})\,dy.$$

Next define $\pi : L^p(U;\mathbb{R}^n) \to L^p(U;\mathbb{R}^n)$ by

$$\pi\mathbf{F} = -\nabla\phi, \quad \phi(\mathbf{x}) = \int_U \nabla\Phi(\mathbf{x}-\mathbf{y})\cdot\mathbf{F}(\mathbf{y})\,dy.$$

We have observed that π is continuous, linear, and $\nabla\cdot\pi\mathbf{F} = \nabla\cdot\mathbf{F}$. It is also true that π is a projection. To see this, let $\mathbf{F} \in C_c^\infty(U;\mathbb{R}^n)$.

$$
\begin{aligned}
\pi^2\mathbf{F}(\mathbf{x}) &= -\nabla\int_B \Phi(\mathbf{z})\nabla\cdot\pi\mathbf{F}(\mathbf{x}-\mathbf{z})\,dz \\
&= -\nabla\int_B \Phi(\mathbf{z})\nabla\cdot\nabla\int_B \Phi(\mathbf{w})\nabla\cdot\mathbf{F}(\mathbf{x}-\mathbf{z}-\mathbf{w})\,dw\,dz \\
&= -\nabla\int_B \Phi(\mathbf{z})\nabla\cdot\mathbf{F}(\mathbf{x}-\mathbf{z})\,dz = \pi\mathbf{F}(\mathbf{x}).
\end{aligned}
$$

Since π is continuous and $C_c^\infty(U;\mathbb{R}^n)$ is dense in $L^p(U;\mathbb{R}^n)$, $\pi^2\mathbf{F} = \pi\mathbf{F}$ for all $\mathbf{F} \in L^p(U;\mathbb{R}^n)$. This proves the following theorem.

Theorem 22.24 *There exists a continuous projection*

$$\pi : L^p(U;\mathbb{R}^n) \to L^p(U;\mathbb{R}^n)$$

such that $\pi\mathbf{F}$ is a gradient and

$$\nabla\cdot(\mathbf{F}-\pi\mathbf{F}) = 0$$

in the sense of weak derivatives.

Note this theorem shows that any L^p vector field is the sum of a gradient and a part which is divergence free. $\mathbf{F} = \mathbf{F}-\pi\mathbf{F}+\pi\mathbf{F}$.

22.6 Exercises

1. This problem is about a general concept called a quotient space. Let X be a Banach space and let Y be a closed subspace of X. Define an equivalence relation, $\sim$, as $x_1 \sim x_2$ if and only if $x_1 - x_2 \in Y$. Let X/Y denote the collection of equivalence classes and let $[x]$ denote one of these equivalence classes. Define

$$[x_1] + [x] \equiv [x + x_1], \quad c[x] = [cx]$$

$$||[x]|| \equiv \inf \{||x + y|| : y \in Y\}.$$

   Show that these definitions are all well defined and that $(X/Y, ||\cdot||)$ is a Banach space. The next sequence of exercises outlines a real variable method of interpolation in Banach space.

2. ↑ Let A_0 and A_1 be two Banach spaces and let X be a topological vector space such that $A_i \subseteq X$ for $i = 1, 2$, and the identity map from A_i to X is continuous. Let

$$Y \subseteq A_0 \times A_1$$

   be defined as $\{(a_0, a_1) : a_0 + a_1 = 0\}$. Show Y is a closed subspace of $A_0 \times A_1$ where we define $||(a_0, a_1)|| \equiv \left(||a_0||_0^2 + ||a_1||_1^2 \right)^{1/2}$.

3. ↑ Define a norm on $A_0 + A_1$ by

$$||a|| \equiv ||[(a_0, a_1)]||$$

   where $a_0 + a_1 = a$ and the second norm refers to the norm in the quotient space of Problem 1. Thus

$$||a|| \equiv \inf \left\{ \left(||a_0||_0^2 + ||a_1||_1^2 \right)^{1/2} : a_0 + a_1 = a \right\}$$

   and by the earlier problems, $A_0 + A_1$ is a Banach space with this norm. Now let

$$||a||_t \equiv \inf \left\{ \left(||a_0||_0^2 + t^2 ||a_1||_1^2 \right)^{1/2} : a_0 + a_1 = a \right\}.$$

   Show $(A_0 + A_1, ||\cdot||_t)$ is a Banach space and that all the norms are equivalent.

469

4. ↑ Let $1 \leq q < \infty, 0 < \theta < 1$. Define $(A_0, A_1)_{\theta, q}$ to be those elements of $A_0 + A_1, a$, such that

$$\|a\|_{\theta, q} \equiv \left[\int_0^\infty \left(t^{-\theta} \|a\|_t \right)^q \frac{dt}{t} \right]^{1/q} < \infty.$$

Show $(A_0, A_1)_{\theta, q}$ is a normed linear space satisfying

$$A_0 \cap A_1 \subseteq (A_0, A_1)_{\theta, q} \subseteq A_0 + A_1, \tag{60}$$

with the inclusion maps continuous, and

$$\left((A_0, A_1)_{\theta, q}, \|\cdot\|_{\theta, q} \right) \text{ is a Banach space.} \tag{61}$$

If $a \in A_0 \cap A_1$, then

$$\|a\|_{\theta, q} \leq \left(\frac{1}{q\theta(1 - \theta)} \right)^{1/q} \|a\|_1^\theta \|a\|_0^{1 - \theta}. \tag{62}$$

If $A_0 \subseteq A_1$ with $\|\cdot\|_0 \geq \|\cdot\|_1$, then

$$A_0 \cap A_1 \subseteq A_0 \subseteq (A_0, A_1)_{\theta, q} \subseteq A_1 = A_0 + A_1.$$

Also, if bounded sets in A_0 have compact closures in A_1 then the same is true if A_1 is replaced with $(A_0, A_1)_{\theta, q}$. Finally, if

$$T \in \mathcal{L}(A_0, B_0), T \in \mathcal{L}(A_1, B_1),$$

and T is a linear map from $A_0 + A_1$ to $B_0 + B_1$ where the A_i and B_i are Banach spaces with the properties described above, then it follows that

$$T \in \mathcal{L}\left((A_0, A_1)_{\theta, q}, (B_0, B_1)_{\theta, q} \right)$$

and if M is its norm, and M_0 and M_1 are the norms of T as a map in $\mathcal{L}(A_0, B_0)$ and $\mathcal{L}(A_1, B_1)$ respectively, then

$$M \leq M_0^{1 - \theta} M_1^\theta.$$

Hint: When Formula 60 is shown, get Formula 61 as follows. If $\{a_n\}$ is Cauchy in $(A_0, A_1)_{\theta, q}$, then $a_n \to a$ in $A_0 + A_1$. Now show $a_n \to a$ in $(A_0, A_1)_{\theta, q}$ using Fatou's lemma. To obtain Formula 62,

$$\|a\|_{\theta, q} \leq \left(\int_0^r \left(t^{1-\theta} \|a\|_1 \right)^q \frac{dt}{t} + \int_r^\infty \left(t^{-\theta} \|a\|_0 \right)^q \frac{dt}{t} \right)^{1/q}$$

$$\leq \left(\|a\|_1^q \frac{r^{q - q\theta}}{q - q\theta} + \|a\|_0^q \frac{r^{-\theta q}}{\theta q} \right)^{1/q}$$

Now use the value of r which minimizes this last expression, $r = \frac{\|a\|_0}{\|a\|_1}$. To verify the second to the last assertion, use Formula 62 to write

$$\|a\|_{\theta,q} \leq \left(\left(\frac{1}{\epsilon} \right)^{1/\theta} \|a\|_1 + \epsilon^{1/(1-\theta)} \|a\|_0 \right) \frac{1}{q\theta(1-\theta)}.$$

This will be of use in showing the claimed result. For the final assertion,

$$\|Ta\|_{\theta,q} = \left(\int_0^\infty \left(t^{-\theta} \|Ta\|_t \right)^q \frac{dt}{t} \right)^{1/q} \leq$$

$$\left(\int_0^\infty \left(t^{-\theta} M_0 \inf \left\{ \|a_0\|_0 + t \frac{M_1}{M_0} \|a_1\|_1 : a_0 + a_1 = a \right\} \right)^q \frac{dt}{t} \right)^{1/q}.$$

Now let $s = \frac{M_1}{M_0} t$.

5. Let H be a Hilbert space and let N be a closed subspace of H. Then the Banach space H/N is a Hilbert space with the inner product given by $([x],[y]) \equiv (Qx, Qy)$ where P is the projection onto N and $Q = I - P$.

6. ↑ Suppose, in the context of Problem 1 - 4, that A_0 and A_1 are both Hilbert spaces. For each $t > 0$, define an inner product, $(\cdot, \cdot)_t$, on $A_0 \times A_1$ by

$$((a_0, a_1), (b_0, b_1))_t \equiv (a_0, b_0)_0 + t^2 (a_1, b_1)_1$$

where $(\cdot, \cdot)_i$ denotes the inner product in $A_i, i = 0, 1$. This determines a norm on $A_0 \times A_1$ defined as

$$|(a_0, a_1)|_t \equiv \left(a_0^2 + t^2 |a_1|_1^2 \right)^{1/2}.$$

By Problem 5, $(A_0 \times A_1, |\cdot|_t) / \ker(+)$ is a Hilbert space whose inner product is

$$([(a_0, a_1)], [(b_0, b_1)])_t = (Q_t(a_0, a_1), Q_t(b_0, b_1))$$

where P_t is the projection of $(A_0 \times A_1, |\cdot|_t)$ onto $\ker(+)$ and $Q_t = I - P_t$. Define $\alpha : (A_0 \times A_1, |\cdot|_t) / \ker(+) \to (A_0 + A_1, \|\cdot\|_t)$ by

$$\alpha([(a_0, a_1)]) \equiv a_0 + a_1.$$

Show α is one to one, onto, and preserves the norms. Conclude

$$(A_0 + A_1, \|\cdot\|_t)$$

is a Hilbert space.

7. ↑ For $a, b \in (A_0, A_1)_{\theta, 2}$ define

$$(a, b)_{\theta, 2} \equiv \int_0^\infty t^{-2\theta} ((a, b))_t \frac{dt}{t}$$

where $((\cdot, \cdot))_t$ is the inner product of Problem 6. Show the norm determined by this inner product is $||\cdot||_{\theta, 2}$ and that, therefore, $(A_0, A_1)_{\theta, 2}$ is a Hilbert space.

8. Suppose ρ satisfies the conditions of Mihlin's theorem. Show there exists a constant A_p depending on $p > 1$ but not on ϕ such that for all $\phi \in \mathfrak{S}$,

$$||F^{-1} (\rho F (\phi))||_p \le A_p ||\phi||_p .$$

Show $\phi \rightarrow F^{-1} (\rho F (\phi))$ can be extended uniquely to a continuous linear map on $L^p (\mathbb{R}^n)$.

9. ↑ For $u \in \mathfrak{S}$,
$$||u||_{H_p^s} \equiv ||F^{-1} (e^s Fu)||_p$$

where $p > 1$ and $e (\mathbf{y}) \equiv (1 + |\mathbf{y}|^2)^{1/2}$. Let $\pi_j (\mathbf{y}) \equiv y_j, \pi^\alpha \equiv \mathbf{y}^\alpha$. Recall that for $u \in \mathfrak{S}$,

$$||u||_{m, p} \equiv \sum_{|\alpha| \le m} ||D^\alpha u||_p .$$

Show that for $s = m$ an integer or if $p = 2$, the two norms $||\cdot||_{H_p^m}$ and $||\cdot||_{m, p}$ are equivalent. Note how easy it is to define $||\cdot||_{H_p^s}$ for any $s \ge 0$. The completion of $\mathfrak{S}$ with respect to this norm is called the space of Bessel potentials and is denoted by H_p^s. **Hint:** For the case when $p = 2$ see Problem 10 of Chapter 16. For the case when p is arbitrary but $s = m$, justify the following steps using Mihlin's theorem.
$$||D^\alpha \phi||_p = ||F^{-1} (F D^\alpha \phi)||_p$$
$$= ||F^{-1} (\pi^\alpha F \phi)||_p = ||F^{-1} (\pi^\alpha e^{-m} e^m F \phi)||_p$$
$$= ||F^{-1} (\pi^\alpha e^{-m} F [f^{-1} (e^m F \phi)])||_p \le C_p ||F^{-1} (e^m F \phi)||_p .$$
Now let $\psi \in C^\infty (\mathbb{R}^n), \psi = 1$ if $|\mathbf{x}| \ge 2, \psi = 0$ if $|\mathbf{x}| \le 1$. Show the two functions

$$e^m \left(1 + \sum_{j=1}^n (\psi \circ \pi_j) |\pi_j|^m \right)^{-1} , \quad \psi \circ \pi_j |\pi_j|^m |\pi_j|^{-m}$$

both satisfy the conditions for ρ in Mihlin's theorem. If $\phi \in \mathfrak{S}$,

$$\left|\left|F^{-1}\left(e^m F\phi\right)\right|\right|_p =$$

$$\left|\left|F^{-1}\left\{e^m\left(1+\sum_{j=1}^{n}(\psi\circ\pi_j)\,|\pi_j|^m\right)^{-1}\right.\right.$$

$$\left.\left.\left(1+\sum_{j=1}^{n}(\psi\circ\pi_j)\,|\pi_j|^m\right)F\phi\right\}\right|\right|_p$$

$$=\left|\left|F^{-1}\left\{e^m\left(1+\sum_{j=1}^{n}(\psi\circ\pi_j)\,|\pi_j|^m\right)^{-1}\right.\right.$$

$$\left.\left.F\left[F^{-1}\left(\left(1+\sum_{j=1}^{n}(\psi\circ\pi_j)\,|\pi_j|^m\right)F\phi\right)\right]\right\}\right|\right|_p$$

$$\leq C_p\left|\left|F^{-1}\left(\left(1+\sum_{j=1}^{n}(\psi\circ\pi_j)\,|\pi_j|^m\right)F\phi\right)\right|\right|_p$$

$$\leq C_p\left(||\phi||_p+\sum_{j=1}^{n}\left|\left|F^{-1}\left((\psi\circ\pi_j)\,|\pi_j|^m\,F\phi\right)\right|\right|_p\right)$$

$$\leq C_p[||\phi||_p$$

$$+\sum_{j=1}^{n}\left|\left|F^{-1}\left((\psi\circ\pi_j)\,|\pi_j|^m\,|\pi_j|^{-m}\,F[F^{-1}(\pi_j^m\,F\phi)]\right)\right|\right|_p]$$

$$\leq C_p\left(||\phi||_p+\sum_{j=1}^{n}\left|\left|F^{-1}\left(\pi_j^m\,F\phi\right)\right|\right|_p\right)$$

$$\leq C_p\left(||\phi||_p+\sum_{j=1}^{n}\left|\left|D_j^m\,(\phi)\right|\right|_p\right)\leq C_p\,||\phi||_{m,p}\,.$$

10. ↑ For $\phi \in \mathfrak{S}$, let $|||\phi||| \equiv ||\phi||_p + \sum_{j=1}^{n}\left|\left|D_j^m\,(\phi)\right|\right|_p$. Show $|||\cdot|||$ is equivalent to $||\cdot||_{m,p}$ on $\mathfrak{S}$.

Chapter 23
Integration for Vector Valued Functions

23.1 Strong and weak measurability

In this chapter $(\Omega, \mathcal{S}, \mu)$ will be a σ finite measure space and X will be a Banach space which contains the values of either a function or a measure. The Banach space will be either a real or a complex Banach space but the field of scalars does not matter and so we will denote it by $\mathbb{F}$ with the understanding that $\mathbb{F} = \mathbb{C}$ unless otherwise stated. We note that the theory presented here includes the case where $X = \mathbb{R}^n$ or $\mathbb{C}^n$ but it does not include all of Chapter 7 as a special case because it omits spaces such as $[0, \infty]$. To begin with we give the following definition.

Definition 23.1 We say a function, $x : \Omega \to X$, for X a Banach space, is a simple function if it is of the form

$$x(s) = \sum_{i=1}^{n} a_i \mathcal{X}_{B_i}(s)$$

where $B_i \in \mathcal{S}$ and $\mu(B_i) < \infty$ for each i. A function x from Ω to X is said to be strongly measurable if there exists a sequence of simple functions $\{x_n\}$ converging pointwise to x. The function x is said to be weakly measurable if, for each $f \in X'$,

$$f \circ x$$

is a scalar valued measurable function.

The first task is to verify that this definition of measurability is often the same as the one given earlier which involves inverse images of open sets.

Theorem 23.2 x is strongly measurable if and only if $x^{-1}(U)$ is measurable for all U open in X and $x(\Omega)$ is separable.

Proof: Suppose first $x^{-1}(U)$ is measurable for all U open in X and $x(\Omega)$ is separable. Let $\{a_n\}_{n=1}^{\infty}$ be the dense subset of $x(\Omega)$. We know $x^{-1}(B)$ is measurable for all B Borel because

$$\{B : x^{-1}(B) \text{ is measurable}\}$$

is a σ algebra containing the open sets. Let

$$U_k^n \equiv \{z \in X : \|z - a_k\| \leq \min\{\|z - a_l\|_{l=1}^n\}\},$$

$$B_k^n \equiv x^{-1}(U_k^n), \ D_k^n \equiv B_k^n \setminus \left(\cup_{i=1}^{k-1} B_i^n\right), \ D_1^n \equiv B_1^n,$$

and

$$x_n(s) \equiv \sum_{k=1}^{n} a_k \mathcal{X}_{D_k^n}(s).$$

Thus $x_n(s)$ is a closest approximation to $x(s)$ from $\{a_k\}_{k=1}^n$ and so $x_n(s) \to x(s)$ because $\{a_n\}_{n=1}^\infty$ is dense in $x(\Omega)$. Since $(\Omega, \mathcal{S}, \mu)$ is σ finite, there exists $\Omega_n \uparrow \Omega$ with $\mu(\Omega_n) < \infty$. Let

$$y_n(s) \equiv \mathcal{X}_{\Omega_n}(s) x_n(s).$$

Then $y_n(s) \to x(s)$ for each s because for any s, $s \in \Omega_n$ if n is large enough. Also y_n is a simple function because it equals 0 off a set of finite measure.

Now suppose that x is strongly measurable. Then some sequence of simple functions, $\{x_n\}$, converges pointwise to x. We have $x_n^{-1}(W)$ is measurable for every open set W because it is just a finite union of measurable sets. Thus, $x_n^{-1}(W)$ is measurable for every Borel set W. Let U be an open set in X and let $\{V_n\}$ be a sequence of open sets satisfying

$$\overline{V}_n \subseteq U, \ \overline{V}_n \subseteq V_{n+1}, \ U = \cup_{n=1}^\infty V_n.$$

Then

$$x^{-1}(V_m) \subseteq \bigcup_{n<\infty} \bigcap_{k\geq n} x_k^{-1}(V_m) \subseteq x^{-1}(\overline{V}_m).$$

This implies

$$x^{-1}(U) = \bigcup_{m<\infty} x^{-1}(V_m)$$

$$\subseteq \bigcup_{m<\infty} \bigcup_{n<\infty} \bigcap_{k\geq n} x_k^{-1}(V_m) \subseteq \bigcup_{m<\infty} x^{-1}(\overline{V}_m) \subseteq x^{-1}(U).$$

Since

$$x^{-1}(U) = \bigcup_{m<\infty} \bigcup_{n<\infty} \bigcap_{k\geq n} x_k^{-1}(V_m),$$

it follows that $x^{-1}(U)$ is measurable for every open U. It remains to show $x(\Omega)$ is separable. Let

$$D \equiv \text{ all values of the simple functions } x_n$$

which converge to x pointwise. Then D is clearly countable and dense in $\overline{D}$, a set which contains $x(\Omega)$. Therefore $x(\Omega)$ is separable by Problem 24 of Chapter 1. This proves the theorem.

Corollary 23.3 *If X is a separable Banach space then x is strongly measurable if and only if $x^{-1}(U)$ is measurable for all U open in X.*

The next lemma is interesting for its own sake. Roughly it says that if a Banach space is separable, then the unit ball in the dual space is weak $*$ separable. This will be used to prove Pettis's theorem, one of the major theorems in this subject which relates weak measurability to strong measurability.

Lemma 23.4 *If X is a separable Banach space with B' the unit ball in X', then there exists a sequence $\{f_n\}_{n=1}^{\infty} \subseteq B'$ with the property that for every $f_0 \in B'$, there exists a subsequence $\{f_{n'}\}$ of $\{f_n\}_{n=1}^{\infty}$ such that*

$$f_0(x) = \lim_{n' \to \infty} f_{n'}(x)$$

for every $x \in X$.

Proof: Let $\{a_k\}$ be a countable dense set in X and consider the mapping

$$\phi_n : B' \to \mathbb{F}^n$$

given by

$$\phi_n(f) \equiv (f(a_1), \cdots, f(a_n)).$$

$\mathbb{F}^n$ is separable and so $\phi_n(B')$ is separable. That is, there exists

$$\{f_k^n\}_{k=1}^{\infty} \subseteq B'$$

such that

$$\{\phi_n(f_k^n)\}_{k=1}^{\infty}$$

is dense in $\phi_n(B')$. Let

$$\{f_m\}_{m=1}^{\infty} = \{f_k^n\}_{n,k=1}^{\infty}.$$

This is the promised sequence for if $f_0 \in B'$, choose $f_{m_l} \in \{f_m\}$ such that

$$|f_0(a_i) - f_{m_l}(a_i)| \le l^{-1} \text{ for } i = 1, \cdots, l.$$

Therefore

$$f_{m_l}(a_i) \to f_0(a_i)$$

for every a_i. It follows that $f_{m_l}(x) \to f_0(x)$ for all $x \in X$ because $\{a_i\}$ is dense and

$$|f_{m_l}(x) - f_0(x)|$$

$$\le \quad |f_{m_l}(x) - f_{m_l}(a_i)| + |f_{m_l}(a_i) - f_0(a_i)| + |f_0(a_i) - f_0(x)|$$
$$\le \quad 2|x - a_i| + |f_{m_l}(a_i) - f_0(a_i)|.$$

This proves the lemma.

The next theorem is one of the most important results in the subject. It is due to Pettis and appeared in 1938.

Theorem 23.5 *If x has values in a separable Banach space, X, and if x is weakly measurable, then x is strongly measurable.*

Proof: First we show that the weak measurability of x implies $||x||$ is measurable. Therefore, suppose x is weakly measurable. Let

$$A \equiv \{s : ||x(s)|| \le a\}, \ A_f \equiv \{s : |f(x(s))| \le a\}.$$

Then

$$A \subseteq \bigcap_{||f|| \le 1} A_f.$$

If

$$s \in \bigcap_{||f|| \le 1} A_f$$

then $|f(x(s))| \le a$ for all $f \in B'$. But by the Hahn Banach theorem, there exists f_s such that

$$||f_s|| = 1 \text{ and } |f_s(x(s))| = ||x(s)||.$$

Therefore, $||x(s)|| \le a$ and so

$$\bigcap_{||f|| \le 1} A_f = A.$$

By Lemma 23.4,

$$\bigcap_{||f|| \le 1} A_f = \bigcap_{m=1}^{\infty} A_{f_m}$$

where $\{f_m\}$ is the sequence of that lemma because if $s \in \bigcap_{m=1}^{\infty} A_{f_m}$, then $|f_m(x(s))| \le a$ for all m. Therefore if $f \in B'$,

$$|f(x(s))| = \lim_{l \to \infty} |f_{m_l}(x(s))| \le a.$$

Thus,

$$A = \bigcap_{m=1}^{\infty} A_{f_m},$$

a countable intersection of measurable sets. Hence A is measurable. This shows that $||x||$ is measurable whenever x is weakly measurable. Now if x is weakly measurable, so is $x - a$. Therefore,

$$x^{-1}(B(a,r)) = \{s \in \Omega : ||x(s) - a|| < r\}$$

478

is measurable for every ball $B(a, r)$. Since X has a countable basis of such balls, $x^{-1}(U)$ is measurable for every open U. By Theorem 23.2, x is strongly measurable. This proves the theorem.

Corollary 23.6 x *is strongly measurable if and only if* $x(\Omega)$ *is separable and* x *is weakly measurable.*

Proof: Strong measurability clearly implies weak measurability because if $x_n(s) \to x(s)$ where x_n is simple, then $f(x_n(s)) \to f(x(s))$ for all $f \in X'$. Hence $f \circ x$ is measurable because it is the limit of a sequence of measurable functions. Let D denote the set of all values of x_n. Then $\overline{D}$ is a separable set containing $x(\Omega)$. Therefore $x(\Omega)$ is separable also by Problem 24 of Chapter 1.

Now suppose D is a countable dense subset of $x(\Omega)$ and x is weakly measurable. Let Z be the subset consisting of all finite linear combinations of D with the scalars coming from the set of rational points of $\mathbb{F}$. Thus, Z is countable. Letting $Y = \overline{Z}$, Y is a separable Banach space containing $x(\Omega)$. If $f \in Y'$, f can be extended to an element of X' by the Hahn Banach theorem. Therefore, x is a weakly measurable Y valued function. Now use Theorem 23.5 to conclude that x is strongly measurable.

One can also talk about weak $*$ measurability and prove a theorem just like the Pettis theorem above. The next lemma is the analogue of Lemma 23.4.

Lemma 23.7 *Let B be the unit ball in X. If X' is separable, there exists a sequence $\{x_m\}_{m=1}^{\infty} \subseteq B$ with the property that for all $x \in B$ there exists a subsequence $\{x_{m'}\}$ of $\{x_m\}$ with*

$$x^*(x_{m'}) \to x^*(x)$$

for all $x^* \in X'$.

Proof: Let

$$D \equiv \{x_k^*\}_{k=1}^{\infty}$$

be the dense subspace of X'. Define $\phi_n : B \to \mathbb{F}^n$ by

$$\phi_n(x) \equiv (x_1^*(x), \cdots, x_n^*(x)).$$

$\mathbb{F}^n$ is separable; so, there exists $\{x_k^n\}_{k=1}^{\infty} \subseteq B$ with $\{\phi_n(x_k^n)\}_{k=1}^{\infty}$ dense in $\phi_n(B)$. Let

$$\{x_m\} \equiv \{x_k^n\}_{k,n=1}^{\infty}.$$

Thus if $x \in B$, there exists a sequence $\{x_{m'}\} \subseteq \{x_m\}$ satisfying

$$x^*(x_{m'}) \to x^*(x)$$

479

for every $x^* \in D$. Since D is dense, it follows just as in the proof of Lemma 23.4 that this holds for all $x^* \in X'$.

The next theorem is another version of the Pettis theorem. First we make the following definition.

Definition 23.8 A function y having values in X' is weak $*$ measurable, when for each $x \in X$, $y(\cdot)(x)$ is a measurable scalar valued function.

Theorem 23.9 *If X' is separable and $y : \Omega \to X'$ is weak $*$ measurable, then y is strongly measurable.*

Proof: As in the proof of the Pettis theorem, we show that y is weak $*$ measurable implies that $||y||$ is measurable. Letting $x \in B$,

$$A \equiv \{s \in \Omega : ||y(s)|| \leq a\}, \ A_x \equiv \{s \in \Omega : |y(s)(x)| \leq a\}.$$

Clearly

$$A \subseteq \bigcap_{x \in B} A_x.$$

But if $|y(s)(x)| \leq a$ for all $x \in B$, then $||y(s)|| \leq a$ and so

$$A = \bigcap_{x \in B} A_x.$$

By using Lemma 23.7,

$$\bigcap_{x \in B} A_x = \bigcap_{m=1}^{\infty} A_{x_m}$$

where $\{x_m\}$ is the sequence promised there. This shows that

$$A = \bigcap_{m=1}^{\infty} A_{x_m},$$

a countable intersection of measurable sets. Thus, $||y||$ is measurable. If $a^* \in X'$ and y is weak $*$ measurable, so is $y - a^*$. Therefore,

$$y^{-1}(B(a^*, r)) = \{s \in \Omega : ||y(s) - a^*|| \leq r\}$$

is measurable for every ball $B(a^*, r)$ and since X' has a countable basis of balls, this shows that

$$y^{-1}(U)$$

is measurable for any open U. By Theorem 23.2, y is strongly measurable. This proves the theorem.

The following are interesting consequences of the theory developed so far and are of interest independent of the theory of integration of vector valued functions.

Theorem 23.10 *If X' is separable, then so is X.*

Proof: Let $\{x_m\} \subseteq B$, the unit ball of X, be the sequence promised by Lemma 23.7. Let V be all finite linear combinations of elements of $\{x_m\}$ with rational scalars. Thus $\overline{V}$ is a separable subspace of X. The claim is that $\overline{V} = X$. If not, there exists

$$x_0 \in X \setminus \overline{V}.$$

But by the Hahn Banach theorem there exists $x_0^* \in X'$ satisfying $x_0^*(x_0) \neq 0$, but $x_0^*(v) = 0$ for every $v \in \overline{V}$. Hence

$$x_0^* \left(\frac{x_0}{\|x_0\|} \right) = \lim_{m' \to \infty} x_0^*(x_{m'})$$

for some $\{x_{m'}\} \subseteq \{x_m\}$. But $x_0^*(x_{m'}) = 0$ and so $x_0^* \left(\frac{x_0}{\|x_0\|} \right) = 0$ also, a contradiction. This proves the theorem.

Corollary 23.11 *If X is reflexive, then X is separable if and only if X' is separable.*

23.2   The Bochner integral

Definition 23.12 Let

$$x(s) = \sum_{k=1}^{n} a_k \mathcal{X}_{E_k}(s) \tag{1}$$

where for each k, E_k is measurable and $\mu(E_k) < \infty$. Then we define

$$\int_\Omega x(s) \, d\mu \equiv \sum_{k=1}^{n} a_k \mu(E_k).$$

Proposition 23.13 *Definition 23.12 is well defined.*

Proof: It suffices to verify that if

$$\sum_{k=1}^{n} a_k \mathcal{X}_{E_k}(s) = 0,$$

then

$$\sum_{k=1}^{n} a_k \mu(E_k) = 0.$$

Let $f \in X'$. Then

$$f\left(\sum_{k=1}^{n} a_k \mathcal{X}_{E_k}(s)\right) = \sum_{k=1}^{n} f(a_k) \mathcal{X}_{E_k}(s) = 0$$

and, therefore,

$$0 = \int_{\Omega}\left(\sum_{k=1}^{n} f(a_k) \mathcal{X}_{E_k}(s)\right) d\mu = \sum_{k=1}^{n} f(a_k) \mu(E_k) = f\left(\sum_{k=1}^{n} a_k \mu(E_k)\right).$$

Since $f \in X'$ is arbitrary, and X' separates the points of X, it follows that

$$\sum_{k=1}^{n} a_k \mu(E_k) = 0$$

as claimed. This proves the proposition.

It follows easily from this proposition that $\int_{\Omega} d\mu$ is well defined and linear on simple functions.

Definition 23.14 A strongly measurable function x is Bochner integrable if there exists a sequence of simple functions x_n converging to x pointwise and satisfying

$$\int_{\Omega} \|x_n(s) - x_m(s)\| d\mu \to 0 \text{ as } m, n \to \infty. \tag{2}$$

If x is Bochner integrable, we define

$$\int_{\Omega} x(s) d\mu \equiv \lim_{n \to \infty} \int_{\Omega} x_n(s) d\mu. \tag{3}$$

Theorem 23.15 *The Bochner integral is well defined and if x is Bochner integrable and $f \in X'$,*

$$f\left(\int_{\Omega} x(s) d\mu\right) = \int_{\Omega} f(x(s)) d\mu \tag{4}$$

and

$$\left\|\int_{\Omega} x(s) d\mu\right\| \le \int_{\Omega} \|x(s)\| d\mu. \tag{5}$$

Proof: The limit in Formula 3 exists because of the definition of the integral on simple functions. Thus, if x is given by Formula 1 with the E_k disjoint,

$$\left\|\int_{\Omega} x(s) d\mu\right\|$$

$$= \left\| \int_\Omega \sum_{k=1}^n a_k \mathcal{X}_{E_k}(s) \, d\mu \right\| = \left\| \sum_{k=1}^n a_k \mu(E_k) \right\|$$

$$\leq \sum_{k=1}^n \|a_k\| \mu(E_k) = \int_\Omega \sum_{k=1}^n \|a_k\| \mathcal{X}_{E_k}(s) \, d\mu = \int_\Omega \|x(s)\| \, d\mu$$

which shows the triangle inequality holds on simple functions. This implies

$$\left\| \int_\Omega x_n(s) \, d\mu - \int_\Omega x_m(s) \, d\mu \right\| = \left\| \int_\Omega (x_n(s) - x_m(s)) \, d\mu \right\|$$

$$\leq \int_\Omega \|x_n(s) - x_m(s)\| \, d\mu$$

which verifies the existence of the limit. We need to show that the integral does not depend on the choice of the sequence satisfying Formula 2. Suppose y_n, x_n both satisfy Formula 2 and converge to x pointwise. By Fatou's lemma,

$$\left\| \int_\Omega y_n \, d\mu - \int_\Omega x_m \, d\mu \right\| \leq \int_\Omega \|y_n - x\| \, d\mu + \int_\Omega \|x - x_m\| \, d\mu$$

$$\leq \lim_{k \to \infty} \inf \int_\Omega \|y_n - y_k\| \, d\mu + \lim_{k \to \infty} \inf \int_\Omega \|x_k - x_m\|$$

$$\leq \epsilon/2 + \epsilon/2$$

if m and n are chosen large enough. Since ϵ is arbitrary, this shows that the limit is the same for both sequences and demonstrates the Bochner integral is well defined.

Now let x be Bochner integrable and let x_n be a sequence which satisfies the conditions of the definition. Define

$$y_n(s) = \begin{cases} x_n(s) & \text{if } \|x_n(s)\| \leq 2\|x(s)\|, \\ 0 & \text{if } \|x_n(s)\| > 2\|x(s)\|. \end{cases} \tag{6}$$

If $x(s) = 0$ then $y_n(s) = 0$ for all n. If $x(s) > 0$ then for all n large enough,

$$y_n(s) = x_n(s).$$

Thus, $y_n(s) \to x(s)$ and

$$\|y_n(s)\| \leq 2\|x(s)\|. \tag{7}$$

By Fatou's lemma,

$$\int_\Omega \|x\| \, d\mu \leq \lim_{n \to \infty} \inf \int_\Omega \|x_n\| \, d\mu < \infty \tag{8}$$

483

thanks to Formula 2. Therefore, by Formula 7, Formula 8, and the dominated convergence theorem,

$$0 = \lim_{n,m\to\infty} \int_\Omega \|y_n - y_m\| \, d\mu \tag{9}$$

and it follows we can use y_n in place of x_n in Definition 23.14.

From Definition 23.12,

$$f\left(\int_\Omega y_n \, d\mu\right) = \int_\Omega f(y_n) \, d\mu.$$

Thus,

$$f\left(\int_\Omega x \, d\mu\right) = \lim_{n\to\infty} f\left(\int_\Omega y_n \, d\mu\right) = \lim_{n\to\infty} \int_\Omega f(y_n) \, d\mu = \int_\Omega f(x) \, d\mu,$$

the last equation holding from the dominated convergence theorem and Formula 7, Formula 8. This shows Formula 4. To verify Formula 5,

$$\left\| \int_\Omega x(s) \, d\mu \right\| = \lim_{n\to\infty} \left\| \int_\Omega y_n(s) \, d\mu \right\|$$

$$\leq \lim_{n\to\infty} \int_\Omega \|y_n(s)\| \, d\mu = \int_\Omega \|x(s)\| \, d\mu$$

where the last equation follows from the dominated convergence theorem and 7, 8. This proves the theorem.

Theorem 23.16 *An X valued function, x, is Bochner integrable if and only if x is strongly measurable and*

$$\int_\Omega \|x(s)\| \, d\mu < \infty. \tag{10}$$

In this case there exists a sequence of simple functions $\{y_n\}$ satisfying Formula 2, $y_n(s)$ converges pointwise to $x(s)$,

$$\|y_n(s)\| \leq 2\|x(s)\|$$

and

$$\lim_{n\to\infty} \int_\Omega \|x(s) - y_n(s)\| \, ds = 0. \tag{11}$$

Proof: The "if" part of the theorem follows from the definition of what it means to be Bochner integrable and Fatou's lemma as in 8. Suppose x is strongly measurable and Formula 10 holds. Then there exists a sequence of simple functions, $\{x_n\}$, converging pointwise to x. Let y_n be defined as in Formula 6. Then just as in the proof of Theorem 23.15, Formula 7 and Formula 9 hold and y_n converges to x pointwise. This shows

484

the conditions of Definition 23.14 are satisfied and verifies x is Bochner integrable. Equation 11 follows from 7 and the dominated convergence theorem. This proves the theorem.

Definition 23.17 $x \in L^p (\Omega; X)$ for $p \in [0, \infty)$ if x is strongly measurable and

$$\int_\Omega ||x (s)||^p \, d\mu < \infty$$

Also

$$||x||_{L^p(\Omega;X)} \equiv ||x||_p \equiv \left(\int_\Omega ||x (s)||^p \, d\mu \right)^{1/p}. \tag{12}$$

As in the case of scalar valued functions, we agree to identify two functions in $L^p (\Omega; X)$ if they are equal a.e. With this convention, and using the same arguments found in Chapter 12, it is clear that $L^p (\Omega; X)$ is a normed linear space with the norm given by Formula 12. In fact, $L^p (\Omega; X)$ is a Banach space. This is the main contribution of the next theorem. First we give a lemma whose proof is very similar to the corresponding proof in Chapter 12.

Lemma 23.18 *If x_n is a Cauchy sequence in $L^p (\Omega; X)$ satisfying*

$$\sum_{n=1}^\infty ||x_{n+1} - x_n||_p < \infty,$$

then there exists $x \in L^p (\Omega; X)$ such that $x_n (s) \to x (s)$ a.e. and

$$||x - x_n||_p \to 0.$$

Proof: Let

$$g_N (s) \equiv \sum_{n=1}^N ||x_{n+1} (s) - x_n (s)||_X .$$

Then by the triangle inequality,

$$\left(\int_\Omega g_N (s)^p \, d\mu \right)^{1/p} \leq \sum_{n=1}^N \left(\int_\Omega ||x_{n+1} (s) - x_n (s)||^p \, d\mu \right)^{1/p}$$

$$\leq \sum_{n=1}^\infty ||x_{n+1} - x_n||_p < \infty.$$

Let

$$g (s) = \lim_{N \to \infty} g_N (s) = \sum_{n=1}^\infty ||x_{n+1} (s) - x_n (s)||_X .$$

485

By the monotone convergence theorem,

$$\left(\int_\Omega g\left(s\right)^p d\mu \right)^{1/p} = \lim_{N\to\infty} \left(\int_\Omega g_N\left(s\right)^p d\mu \right)^{1/p} < \infty.$$

Therefore, there exists a set of measure 0, E, such that for $s \notin E$, $g\left(s\right) < \infty$. Hence, for $s \notin E$,

$$\lim_{N\to\infty} x_{N+1}\left(s\right)$$

exists because

$$x_{N+1}\left(s\right) = x_{N+1}\left(s\right) - x_1\left(s\right) + x_1\left(s\right) = \sum_{n=1}^{N}\left(x_{n+1}\left(s\right) - x_n\left(s\right)\right) + x_1\left(s\right).$$

Thus, if $N > M$,

$$\|x_{N+1}\left(s\right) - x_{M+1}\left(s\right)\|_X \leq \sum_{n=M+1}^{N} \|x_{n+1}\left(s\right) - x_n\left(s\right)\|_X$$

$$\leq \sum_{n=M+1}^{\infty} \|x_{n+1}\left(s\right) - x_n\left(s\right)\|_X$$

which shows that $\{x_{N+1}\left(s\right)\}_{N=1}^{\infty}$ is a Cauchy sequence. Now let

$$x\left(s\right) \equiv \begin{cases} \lim_{N\to\infty} x_N\left(s\right) \text{ if } s \notin E, \\ 0 \text{ if } s \in E. \end{cases}$$

By Theorem 23.2, $x_n\left(\Omega\right)$ is separable for each n. Therefore, $x\left(\Omega\right)$ is also separable. Also, if

$$f \in X',$$

then

$$f\left(x\left(s\right)\right) = \lim_{N\to\infty} f\left(x_N\left(s\right)\right)$$

if $s \notin E$ and $f\left(x\left(s\right)\right) = 0$ if $s \in E$. Therefore, $f \circ x$ is measurable because it is the limit of the measurable functions,

$$f \circ x_N \mathcal{X}_{E^c}.$$

Since x is weakly measurable and $x\left(\Omega\right)$ is separable, Corollary 23.6 shows that x is strongly measurable. By Fatou's lemma,

$$\int_\Omega \|x\left(s\right) - x_N\left(s\right)\|^p d\mu \leq \lim_{M\to\infty} \inf \int_\Omega \|x_M\left(s\right) - x_N\left(s\right)\|^p d\mu.$$

But if N and M are large enough with $M > N$,

$$\left(\int_\Omega \|x_M\left(s\right) - x_N\left(s\right)\|^p d\mu \right)^{1/p} \leq \sum_{n=N}^{M} \|x_{n+1} - x_n\|_p$$

$$\leq \sum_{n=N}^{\infty} \|x_{n+1} - x_n\|_p < \epsilon$$

and this shows, since ϵ is arbitrary, that

$$\lim_{N\to\infty} \int_\Omega ||x(s) - x_N(s)||^p \, d\mu = 0.$$

It remains to show $x \in L^p(\Omega; X)$. This follows from the above and the triangle inequality. Thus, for N large enough,

$$\left(\int_\Omega ||x(s)||^p \, d\mu \right)^{1/p}$$

$$\leq \left(\int_\Omega ||x_N(s)||^p \, d\mu \right)^{1/p} + \left(\int_\Omega ||x(s) - x_N(s)||^p \, d\mu \right)^{1/p}$$

$$\leq \left(\int_\Omega ||x_N(s)||^p \, d\mu \right)^{1/p} + \epsilon < \infty.$$

This proves the lemma.

Theorem 23.19 $L^p(\Omega; X)$ *is complete. Also every Cauchy sequence has a subsequence which converges pointwise.*

Proof: If $\{x_n\}$ is Cauchy in $L^p(\Omega; X)$, extract a subsequence $\{x_{n_k}\}$ satisfying

$$||x_{n_{k+1}} - x_{n_k}||_p \leq 2^{-k}$$

and apply Lemma 23.18. The pointwise convergence of this subsequence was established in the proof of this lemma. This proves the theorem.

Clearly Fatou's lemma and the monotone convergence theorem make no sense for functions with values in a Banach space but the dominated convergence theorem holds in this setting.

Theorem 23.20 *If x is strongly measurable and $x_n(s) \to x(s)$ a.e. with*

$$||x_n(s)|| \leq g(s) \quad a.e.$$

where $g \in L^1(\Omega)$, then x is Bochner integrable and

$$\int_\Omega x(s) \, d\mu = \lim_{n\to\infty} \int_\Omega x_n(s) \, d\mu.$$

Proof: $||x_n(s) - x(s)|| \leq 2g(s)$ a.e. so by the usual dominated convergence theorem,

$$0 = \lim_{n\to\infty} \int_\Omega ||x_n(s) - x(s)|| \, d\mu.$$

Also,

$$\int_\Omega \|x_n(s) - x_m(s)\| \, d\mu$$

$$\leq \int_\Omega \|x_n(s) - x(s)\| \, d\mu + \int_\Omega \|x_m(s) - x(s)\| \, d\mu,$$

and so $\{x_n\}$ is a Cauchy sequence in $L^1(\Omega; X)$. Therefore, by Theorem 23.19, there exists $y \in L^1(\Omega; X)$ and a subsequence $x_{n'}$ satisfying

$$x_{n'}(s) \to y(s) \text{ a.e. and in } L^1(\Omega; X).$$

But $x(s) = \lim_{n' \to \infty} x_{n'}(s)$ a.e. and so $x(s) = y(s)$ a.e. Hence

$$\int_\Omega \|x(s)\| \, d\mu = \int_\Omega \|y(s)\| \, d\mu < \infty$$

which shows that x is Bochner integrable. Finally,

$$\left\| \int_\Omega x(s) \, d\mu - \int_\Omega x_n(s) \, ds \right\| \leq \int_\Omega \|x_n(s) - x(s)\| \, d\mu,$$

and this last integral converges to 0. This proves the theorem.

23.3   Measurable representatives

In this section we consider the special case where $X = L^1(B, \nu)$ where $(B, \mathcal{F}, \nu)$ is a σ finite measure space and $x \in L^1(\Omega; X)$. Thus for each $s \in \Omega$, $x(s) \in L^1(B, \nu)$. In general, the map

$$(s, t) \to x(s)(t)$$

will not be product measurable, but one can obtain a product measurable representative. This is important because it allows the use of Fubini's theorem on the measurable representative.

By Theorem 23.16, there exists a sequence of simple functions, $\{x_n\}$, of the form

$$x_n(s) = \sum_{k=1}^m a_k \mathcal{X}_{E_k}(s) \tag{13}$$

where $a_k \in L^1(B, \nu)$ which satisfy the conditions of Definition 23.14 and

$$\|x_n - x\|_1 \to 0. \tag{14}$$

Because of the form of x_n given in Formula 13, if we define

$$x_n(s, t) \equiv x_n(s)(t),$$

then x_n is product measurable.

$$\int_\Omega \int_B |x_n(s,t) - x_m(s,t)| \, d\nu d\mu \leq \int_\Omega \int_B |x_n(s,t) - x(s)(t)| \, d\nu d\mu \quad (15)$$

$$+ \int_\Omega \int_B |x_m(s,t) - x(s)(t)| \, d\nu d\mu. \quad (16)$$

It follows from Formula 16 and Formula 14 that $\{x_n\}$ is a Cauchy sequence in $L^1(\Omega \times B)$. Therefore, there exists $y \in L^1(\Omega \times B)$ and a subsequence of $\{x_n\}$, still denoted by $\{x_n\}$, such that

$$\lim_{n \to \infty} x_n(s,t) = y(s,t) \text{ a.e.}$$

and

$$\lim_{n \to \infty} \|x_n - y\|_1 = 0.$$

It follows that

$$\int_\Omega \int_B |y(s,t) - x(s)(t)| \, d\nu d\mu \leq \int_\Omega \int_B |y(s,t) - x_n(s,t)| \, d\nu d\mu \quad (17)$$

$$+ \int_\Omega \int_B |x(s)(t) - x_n(s,t)| \, d\nu d\mu.$$

Since $\lim_{n \to \infty} \|x_n - x\|_1 = 0$, it follows from Formula 17 that $y = x$ in $L^1(\Omega; X)$. Thus, for a.e. s,

$$y(s, \cdot) = x(s) \text{ in } X = L^1(B).$$

Now

$$\int_\Omega x(s) \, d\mu \in X$$

so it makes sense to ask for

$$\left(\int_\Omega x(s) \, d\mu \right)(t),$$

at least a.e. To find what this is, note

$$\left\| \int_\Omega x_n(s) \, d\mu - \int_\Omega x(s) \, d\mu \right\|_X \leq \int_\Omega \|x_n(s) - x(s)\|_X \, d\mu.$$

Therefore, since the right side converges to 0,

$$\lim_{n \to \infty} \int_B \left| \left(\int_\Omega x_n(s) \, d\mu \right)(t) - \left(\int_\Omega x(s) \, d\mu \right)(t) \right| d\nu = 0.$$

But

$$\left(\int_\Omega x_n(s) \, d\mu \right)(t) = \int_\Omega x_n(s,t) \, d\mu \text{ a.e. } t.$$

Therefore

$$\lim_{n\to\infty} \int_B \left| \int_\Omega x_n(s,t)\, d\mu - \left(\int_\Omega x(s)\, d\mu \right)(t) \right| d\nu = 0. \qquad (18)$$

Also, since $x_n \to y$ in $L^1(\Omega \times B)$,

$$\lim_{n\to\infty} \int_B \left| \int_\Omega x_n(s,t)\, d\mu - \int_\Omega y(s,t)\, d\mu \right| d\nu = 0. \qquad (19)$$

From Formula 18 and Formula 19 we conclude

$$\int_\Omega y(s,t)\, d\mu = \left(\int_\Omega x(s)\, d\mu \right)(t) \quad \text{a.e. } t.$$

This proves the following theorem.

Theorem 23.21 *Let $X = L^1(B)$ where $(B, \mathcal{F}, \nu)$ is a σ finite measure space and let $x \in L^1(\Omega; X)$. Then there exists a product measurable representative, $y \in L^1(\Omega \times B)$, such that*

$$x(s) = y(s, \cdot) \quad \text{a.e. } s \text{ in } \Omega,$$

and

$$\int_\Omega y(s,t)\, d\mu = \left(\int_\Omega x(s)\, d\mu \right)(t) \quad \text{a.e. } t.$$

23.4   Vector measures

Just as in Chapter 13, there is a concept of a vector measure.

Definition 23.22 Let $(\Omega, \mathcal{S})$ be a set and a σ algebra of subsets of Ω. A mapping

$$F : \mathcal{S} \to X$$

is said to be a vector measure if

$$F\left(\cup_{i=1}^\infty E_i \right) = \sum_{i=1}^\infty F(E_i)$$

whenever $\{E_i\}_{i=1}^\infty$ is a sequence of disjoint elements of $\mathcal{S}$. For F a vector measure,

$$|F|(A) \equiv \sup\{ \sum_{F \in \pi(A)} \|\mu(F)\| : \pi(A) \text{ is a partition of } A\}.$$

This is the same definition that was given in Chapter 13, the only difference being the fact that now we have in mind a general Banach space X as the vector space of values of the vector measure. The same theorem about $|F|$ proved there holds in this context with the same proof.

Theorem 23.23 *If $|F|(\Omega) < \infty$, then $|F|$ is a measure on $\mathcal{S}$.*

Definition 23.24 A Banach space is said to have the Radon Nikodym property if whenever

$$(\Omega, \mathcal{S}, \mu) \text{ is a finite measure space}$$

$$F : \mathcal{S} \to X \text{ is a vector measure with } |F|(\Omega) < \infty$$

$$F << \mu$$

then one may conclude that there exists $g \in L^1(\Omega; X)$ such that

$$F(E) = \int_E g(s)\, d\mu$$

for all $E \in \mathcal{S}$.

Some Banach spaces have the Radon Nikodym property and some don't. We shall not attempt to give a complete answer to the question of which Banach spaces have this property but the next theorem gives examples of many spaces which do.

Theorem 23.25 *Let X' be a separable dual space. Then X' has the Radon Nikodym property.*

Proof: Let $F << \mu$ and let $|F|(\Omega) < \infty$ for $F : \mathcal{S} \to X'$, a vector measure. Pick $x \in X$ and consider the map

$$E \to F(E)(x)$$

for $E \in \mathcal{S}$. It is routine to verify that this defines a complex measure which is absolutely continuous with respect to $|F|$. Therefore, by the Radon Nikodym theorem, there exists $f_x \in L^1(\Omega, |F|)$ such that

$$F(E)(x) = \int_E f_x(s)\, d|F|. \tag{20}$$

It is easy to see from Formula 20 that

$$|\operatorname{Re} f_x(s)| \leq ||x||,\ |\operatorname{Im} f_x(s)| \leq ||x|| \text{ for } |F| \text{ a.e. } s. \tag{21}$$

Therefore,

$$|f_x(s)| \leq \sqrt{2}\,||x|| \text{ for } |F| \text{ a.e. } s. \tag{22}$$

Denote the exceptional set of measure zero by N_x. By Theorem 23.10, X is separable. Letting D be the dense, countable subset of X, define

$$N_1 \equiv \cup_{x \in D} N_x.$$

Thus

$$|F|(N_1) = 0.$$

For any $E \in \mathcal{S}$, $x, y \in D$, and $a, b \in \mathbb{F}$,

$$\int_E f_{ax+by}(s) \, d\,|F| = F(E)(ax + by) = aF(E)(x) + bF(E)(y)$$

$$= \int_E a f_x(s) + b f_y(s) \, d\,|F|.$$

Since Formula 21 holds for all $E \in \mathcal{S}$, it follows that

$$f_{ax+by}(s) = a f_x(s) + b f_y(s)$$

for $|F|$ a.e. s. Let $\tilde{D}$ consist of all finite linear combinations of the form

$$\sum_{i=1}^{m} a_i x_i$$

where a_i is a rational point of $\mathbb{F}$ and $x_i \in D$. If

$$\sum_{i=1}^{m} a_i x_i \in \tilde{D},$$

the above argument implies

$$f_{\sum_{i=1}^{m} a_i x_i}(s) = \sum_{i=1}^{m} a_i f_{x_i}(s) \text{ a.e.}$$

Since $\tilde{D}$ is countable, there exists a set, N_2, with

$$|F|(N_2) = 0$$

such that for $s \notin N_2$,

$$f_{\sum_{i=1}^{m} a_i x_i}(s) = \sum_{i=1}^{m} a_i f_{x_i}(s)$$

whenever $\sum_{i=1}^{m} a_i x_i \in \tilde{D}$. Let

$$N = N_1 \cup N_2$$

and let

$$\tilde{h}_x(s) \equiv \mathcal{X}_{N^C}(s) f_x(s)$$

for all $x \in \tilde{D}$. Now let

$$h_x(s) \equiv \lim_{x' \to x} \{\tilde{h}_{x'}(s) : x' \in \tilde{D}\}.$$

This is well defined because if x' and y' are elements of $\tilde{D}$, Formula 22 implies

$$\left|\tilde{h}_{x'}(s) - \tilde{h}_{y'}(s)\right| = \left|\tilde{h}_{(x'-y')}(s)\right| \leq \sqrt{2}\,\|x' - y'\|.$$

By Formula 22, the dominated convergence theorem may be applied to conclude that for $x_n \to x$, with $x_n \in \tilde{D}$,

$$\int_E h_x(s)\,d\,|F| = \lim_{n \to \infty} \int_E \tilde{h}_{x_n}(s)\,d\,|F| = \lim_{n \to \infty} F(E)(x_n) = F(E)(x).$$
$$(23)$$

It follows from the density of $\tilde{D}$ that for all $x, y \in X$ and $a, b \in \mathbb{F}$,

$$|h_x(s)| \leq \sqrt{2}\,\|x\|, \quad h_{ax+by}(s) = ah_x(s) + bh_y(s), \qquad (24)$$

for all s. If $s \in N$, both sides of the equation in Formula 24 equal 0.

Let $\theta(s)$ be given by

$$\theta(s)(x) = h_x(s).$$

By Formula 24 it follows that $\theta(s) \in X'$ for each s. Also

$$\theta(s)(x) = h_x(s) \in L^1(\Omega)$$

so $\theta(\cdot)$ is weak $*$ measurable. Since X' is separable, Theorem 23.9 implies that θ is strongly measurable. Furthermore, by Formula 24,

$$\|\theta(s)\| \equiv \sup_{\|x\| \leq 1} |\theta(s)(x)| \leq \sup_{\|x\| \leq 1} |h_x(s)| \leq \sqrt{2}.$$

Therefore,

$$\int_\Omega \|\theta(s)\|\,d\,|F| < \infty$$

so $\theta \in L^1(\Omega; X')$. By Formula 4, if $E \in \mathcal{S}$,

$$\int_E h_x(s)\,d\,|F| = \int_E \theta(s)(x)\,d\,|F| = \left(\int_E \theta(s)\,d\,|F|\right)(x). \qquad (25)$$

From Formula 23 and Formula 25,

$$\left(\int_E \theta(s)\,d\,|F|\right)(x) = F(E)(x)$$

for all $x \in X$ and therefore,

$$\int_E \theta(s) \, d\,|F| = F(E).$$

Finally, since $F << \mu, |F| << \mu$ also and so there exists $k \in L^1(\Omega)$ such that

$$|F|(E) = \int_E k(s) \, d\mu$$

for all $E \in \mathcal{S}$, by the Radon Nikodym Theorem. It follows easily that

$$F(E) = \int_E \theta(s) \, d\,|F| = \int_E \theta(s) \, k(s) \, d\mu.$$

Letting $g(s) = \theta(s) \, k(s)$, this has proved the theorem.

Corollary 23.26 *Any separable reflexive Banach space has the Radon Nikodym property.*

It is not necessary to assume separability in the above corollary. For the proof of a more general result, consult *Vector Measures* by Diestal and Uhl, [13].

23.5   The Riesz representation theorem

In this section we give the Riesz representation theorem for the spaces $L^p(\Omega; X)$ under certain conditions. The proof follows the earlier proofs in Chapter 13.

Definition 23.27 *If X and Y are two Banach spaces, we say X is isometric to Y if there exists $\theta \in \mathcal{L}(X, Y)$ such that*

$$||\theta x||_Y = ||x||_X .$$

We write $X \cong Y$ to indicate that X is isometric to Y. The map θ is called an isometry.

The next theorem says that $L^{p'}(\Omega; X')$ is always isometric to a subspace of $(L^p(\Omega; X))'$ for any Banach space, X.

Theorem 23.28 *Let X be any Banach space and let $(\Omega, \mathcal{S}, \mu)$ be a finite measure space. Let $p \geq 1$ and let $1/p + 1/p' = 1$.(If $p = 1$, $p' \equiv \infty$.) Then $L^{p'}(\Omega; X')$ is isometric to a subspace of $(L^p(\Omega; X))'$. Also, for $g \in L^{p'}(\Omega; X')$,*

$$\sup_{||f||_p \leq 1} \left| \int_\Omega g(s)(f(s)) \, d\mu \right| = ||g||_{p'} .$$

Proof: First observe that for $f \in L^p(\Omega; X)$ and $g \in L^{p'}(\Omega; X')$,

$$s \to g(s)(f(s))$$

is a function in $L^1(\Omega)$. (To obtain measurability, write f as a limit of simple functions. Holder's inequality then yields the function is in $L^1(\Omega)$.) Define

$$\theta : L^{p'}(\Omega; X') \to (L^p(\Omega; X))'$$

by

$$\theta g(f) \equiv \int_\Omega g(s)(f(s)) \, d\mu.$$

Holder's inequality implies

$$\|\theta g\| \le \|g\|_{p'} \qquad (26)$$

and it is also clear that θ is linear. We need to show

$$\|\theta g\| = \|g\|.$$

This will first be verified for simple functions. Let

$$g(s) = \sum_{i=1}^m c_i \mathcal{X}_{E_i}(s)$$

where $c_i \in X'$, the E_i are disjoint and

$$\cup_{i=1}^m E_i = \Omega.$$

Then $\|g\| \in L^{p'}(\Omega)$. Let $\epsilon > 0$ be given. By the scalar Riesz representation theorem, there exists $h \in L^p(\Omega)$ such that $\|h\|_p = 1$ and

$$\int_\Omega \|g(s)\|_{X'} \, h(s) \, d\mu \ge \|g\|_{L^{p'}(\Omega; X')} - \epsilon.$$

Now let d_i be chosen such that

$$c_i(d_i) \ge \|c_i\|_{X'} - \epsilon / \|h\|_{L^1(\Omega)}$$

and $\|d_i\|_X \le 1$. Let

$$f(s) \equiv \sum_{i=1}^m d_i h(s) \mathcal{X}_{E_i}(s).$$

Thus $f \in L^p(\Omega; X)$ and $\|f\|_{L^p(\Omega; X)} \le 1$. This follows from

$$\|f\|_p^p = \int_\Omega \sum_{i=1}^m \|d_i\|^p \, |h(s)|^p \, \mathcal{X}_{E_i}(s)$$

$$= \sum_{i=1}^m \left(\int_{E_i} |h(s)^p| \, d\mu \right) \|d_i\|^p \le \int_\Omega |h|^p \, d\mu = 1.$$

Also

$$\|\theta g\| \geq |\theta g\,(f)| = \left| \int_\Omega g\,(s)\,(f\,(s))\,d\mu \right| \geq$$

$$\left| \int_\Omega \sum_{i=1}^{m} \left(\|c_i\|_{X'} - \epsilon/\|h\|_{L^1(\Omega)} \right) h\,(s)\,\mathcal{X}_{E_i}\,(s)\,d\mu \right|$$

$$\geq \left| \int_\Omega \|g\,(s)\|_{X'}\,h\,(s)\,d\mu \right| - \epsilon \left| \int_\Omega h\,(s)\,/\,\|h\|_{L^1(\Omega)}\,d\mu \right|$$

$$\geq \|g\|_{L^{p'}(\Omega;X')} - 2\epsilon.$$

Since ϵ was arbitrary,

$$\|\theta g\| \geq \|g\| \qquad (27)$$

and from Formula 26 this shows equality holds in Formula 27 whenever g is a simple function.

In general, let $g \in L^{p'}(\Omega; X')$ and let g_n be a sequence of simple functions converging to g in $L^{p'}(\Omega; X')$. Then

$$\|\theta g\| = \lim_{n\to\infty} \|\theta g_n\| = \lim_{n\to\infty} \|g_n\| = \|g\|.$$

This proves the theorem and shows θ is the desired isometry.

Theorem 23.29 *If X is a Banach space and X' has the Radon Nikodym property, then if $(\Omega, \mathcal{S}, \mu)$ is a finite measure space,*

$$(L^p(\Omega; X))' \cong L^{p'}(\Omega; X')$$

and in fact the mapping θ of Theorem 23.28 is onto.

Proof: Let $l \in (L^p(\Omega; X))'$ and define $F(E) \in X'$ by

$$F(E)(x) \equiv l(\mathcal{X}_E(\cdot)x).$$

Lemma 23.30 *F defined above is a vector measure with values in X' and $|F|(\Omega) < \infty$.*

Proof of the lemma: Clearly $F(E)$ is linear. Also

$$\|F(E)\| = \sup_{\|x\|\leq 1} \|F(E)(x)\|$$

$$\leq \|l\| \sup_{\|x\|\leq 1} \|\mathcal{X}_E(\cdot)x\|_{L^p(\Omega;X)} \leq \|l\|\,\mu(E)^{1/p}.$$

Let $\{E_i\}_{i=1}^{\infty}$ be a sequence of disjoint elements of $\mathcal{S}$ and let $E = \cup_{n<\infty} E_n$.

$$
\left| F(E)(x) - \sum_{k=1}^{n} F(E_k)(x) \right| = \left| l(\mathcal{X}_E(\cdot)x) - \sum_{i=1}^{n} l(\mathcal{X}_{E_i}(\cdot)x) \right| \quad (28)
$$

$$
\leq \|l\| \left\| \mathcal{X}_E(\cdot)x - \sum_{i=1}^{n} \mathcal{X}_{E_i}(\cdot)x \right\|_{L^p(\Omega;X)}
$$

$$
\leq \|l\| \mu \left(\bigcup_{k>n} E_k \right)^{1/p} \|x\|.
$$

Since $\mu(\Omega) < \infty$,

$$
\lim_{n \to \infty} \mu \left(\bigcup_{k>n} E_k \right)^{1/p} = 0
$$

and so inequality 28 shows that

$$
\lim_{n \to \infty} \left\| F(E) - \sum_{k=1}^{n} F(E_k) \right\|_{X'} = 0.
$$

To show $|F|(\Omega) < \infty$, let $\epsilon > 0$ be given and let $\|x_i\| \leq 1$ be chosen in such a way that

$$
F(H_i)(x_i) > \|F(H_i)\| - \epsilon/n.
$$

Thus

$$
-\epsilon + \sum_{i=1}^{n} \|F(H_i)\| < \sum_{i=1}^{n} l(\mathcal{X}_{H_i}(\cdot)x_i) \leq \|l\| \left\| \sum_{i=1}^{n} \mathcal{X}_{H_i}(\cdot)x_i \right\|_{L^p(\Omega;X)}
$$

$$
\leq \|l\| \left(\int_{\Omega} \sum_{i=1}^{n} \mathcal{X}_{H_i}(s)\,d\mu \right)^{1/p} = \|l\| \mu(\Omega)^{1/p}.
$$

Since $\epsilon > 0$ was arbitrary,

$$
\sum_{i=1}^{n} \|F(H_i)\| < \|l\| \mu(\Omega)^{1/p}.
$$

This implies that $|F|(\Omega) < \infty$ because $\{H_1, \cdots, H_n\}$ was an arbitrary partition of Ω. This proves the lemma.

Continuing with the proof of Theorem 23.29, note that

$$
F << \mu.
$$

Since X' has the Radon Nikodym property, there exists $g \in L^1(\Omega; X')$ such that

$$F(E) = \int_E g(s)\, d\mu.$$

Also, we have from the definition of $F(E)$

$$l\left(\sum_{i=1}^n x_i \mathcal{X}_{E_i}(\cdot)\right) = \sum_{i=1}^n l(\mathcal{X}_{E_i}(\cdot) x_i)$$

$$= \sum_{i=1}^n F(E_i)(x_i) = \sum_{i=1}^n \int_{E_i} g(s)(x_i)\, d\mu. \tag{29}$$

It follows from Formula 29 that whenever h is a simple function,

$$l(h) = \int_\Omega g(s)(h(s))\, d\mu. \tag{30}$$

Let

$$G_n \equiv \{s : \|g(s)\|_{X'} \le n\}$$

and let

$$j : L^p(G_n; X) \to L^p(\Omega; X)$$

be given by

$$jh(s) = \begin{cases} h(s) & \text{if } s \in G_n, \\ 0 & \text{if } s \notin G_n. \end{cases}$$

Letting h be a simple function in $L^p(G_n; X)$, we have

$$j^* l(h) = l(jh) = \int_{G_n} g(s)(h(s))\, d\mu. \tag{31}$$

Since the simple functions are dense in $L^p(G_n; X)$, and $g \in L^{p'}(G_n; X')$, it follows that Formula 31 holds for all $h \in L^p(G_n; X)$. By Theorem 23.28,

$$\|g\|_{L^{p'}(G_n; X')} = \|j^* l\|_{(L^p(G_n; X))'} \le \|l\|_{(L^p(\Omega; X))'}.$$

By the monotone convergence theorem or the definition of $\|\cdot\|_\infty$,

$$\|g\|_{L^{p'}(\Omega; X')} = \lim_{n \to \infty} \|g\|_{L^{p'}(G_n; X')} \le \|l\|_{(L^p(\Omega; X))'}.$$

Therefore $g \in L^{p'}(\Omega; X')$ and since simple functions are dense in $L^p(\Omega; X)$, Formula 30 holds for all

$$h \in L^p(\Omega; X).$$

Thus $l = \theta g$ and the theorem is proved because, by Theorem 23.28, $\|l\| = \|g\|$ and the mapping θ is onto because l was arbitrary.

Corollary 23.31 *If X' is separable, then*

$$(L^p(\Omega; X))' \cong L^{p'}(\Omega; X').$$

Corollary 23.32 *If X is separable and reflexive, then*

$$(L^p(\Omega; X))' \cong L^{p'}(\Omega; X').$$

23.6 Exercises

1. Let X be a Banach space. Show that if $\int_a^b g\phi dt = 0$ for all $\phi \in C_c^\infty(a, b)$ where $g \in L^1([a, b]; X)$, then $g(t) = 0$ a.e.

2. ↑ Let X be a Banach space and suppose $f \in L^p([a, b]; X)$ for some $p \geq 1$. We define f' as a linear function mapping $C_c^\infty(a, b)$ to X by

$$f'(\phi) \equiv -\int_a^b f\phi' dt.$$

If there exists $g \in L^r([a, b]; X)$ for some $r \geq 1$ with

$$\int_a^b g\phi dt = f'(\phi)$$

then we say $f' = g$ and that $f' \in L^r([a, b]; X)$. Show this definition is well defined and that if $f, f' \in L^1([a, b]; X)$, then f equals a continuous function a.e. Denoting this function by f, we have the formula

$$f(t) = f(a) + \int_a^t f'(s)\, ds.$$

3. Let $E \subseteq W \subseteq X$ where the injection map is continuous from W to X and compact from E to W. By this we mean that bounded sets in E have compact closures in the topology of W. In other words, if a sequence is bounded in E it has a convergent subsequence converging in W. Show that in this situation, for all $\epsilon > 0$ there exists a constant C_ϵ such that for all $u \in E$,

$$\|u\|_W \leq \epsilon \|u\|_E + C_\epsilon \|u\|_X.$$

Hint: If not, there exists $\epsilon > 0$ and a sequence u_n such that

$$\|u_n\|_W > \epsilon \|u_n\|_E + n \|u_n\|_X.$$

Letting $v_n = u_n / ||u_n||_E$, we may obtain the same inequality under the assumption that $||v_n||_E = 1$. Now use the compactness of the embedding to get a contradiction.

4. ↑ This problem gives a version of the Ascoli Arzela theorem generalized to the case where the functions have values in infinite dimensional space. Let the Banach spaces be as just described, let $q > 1$, and let S be defined by

$$\left\{ u \text{ such that } ||u(t)||_E + ||u'||_{L^q([a,b];X)} \leq R \text{ for all } t \in [0, T] \right\}.$$

Then S is precompact in $C(a, b; W)$ in the sense that the closure of this set taken in the topology of $C(a, b; W)$ is a compact set in this topology. Here the norm in $C(a, b; W)$ is defined as

$$||f|| \equiv \max \{ f(t) : t \in [a, b] \}.$$

Hint: Let $\{u_n\}$ be any sequence in S and show this is equicontinuous using Problem 3 as follows.

$$||u_n(t) - u_n(s)||_W$$

$$\leq \frac{\epsilon}{4R} ||u_n(t) - u_n(s)||_E + C_\epsilon ||u_n(t) - u_n(s)||_X$$

$$\leq \frac{\epsilon}{2} + C_\epsilon R |t - s|^{1/q'}.$$

Now use the Cantor diagonal process and compactness of the embedding of E into W to select a subsequence which converges at each point of a countable dense subset of $[a, b]$. Using equicontinuity, show this is Cauchy in $C(a, b; W)$. Alternatively, adapt the proof of the Ascoli Arzela theorem given in Chapter 2.

5. This problem presents a result which is very useful in partial differential equations. Let the Banach spaces be as described in problem 3, let $p \geq 1$, let $q > 1$, and define

$$S \equiv \{ u \in L^p([a, b]; E) : u' \in L^q([a, b]; X)$$

$$\text{and } ||u||_{L^p([a,b];E)} + ||u'||_{L^q([a,b];X)} \leq R \}.$$

Then S is precompact in $L^p([a, b]; W)$. **Hint:** Show S has an η net for each $\eta > 0$ as follows. If not, there exists a sequence $\{u_n\} \subseteq S$, such that

$$||u_n - u_m|| \geq \eta \tag{32}$$

for all $n \neq m$ and the norm refers to $L^p([a,b];W)$. Let

$$a = t_0 < t_1 < \cdots < t_n = b, \ t_k - t_{k-1} = T/k.$$

Now define

$$\overline{u}_n(t) \equiv \sum_{i=1}^{k} \overline{u}_{n_i} \mathcal{X}_{[t_{i-1},t_i)}(t), \ \overline{u}_{n_i} \equiv \frac{1}{t_i - t_{i-1}} \int_{t_{i-1}}^{t_i} u_n(s)\,ds.$$

Next show

$$\int_a^b \|u_n(t) - \overline{u}_n(t)\|_W^p\,dt$$

$$\leq 2^{p-1} \sum_{i=1}^{k} \int_{t_{i-1}}^{t_i} \frac{1}{t_i - t_{i-1}} \int_{t_{i-1}}^{t_i} \epsilon^p \|u_n(s) - u_n(t)\|_E^p$$
$$+ C_\epsilon^p \|u_n(s) - u_n(t)\|_X^p\,ds\,dt,$$

and

$$\|u_n(s) - u_n(t)\|_X \leq R|t - s|^{1/q'}.$$

Therefore this is no larger than

$$2^{p-1} \sum_{i=1}^{k} \int_{t_{i-1}}^{t_i} \frac{1}{t_i - t_{i-1}} \int_{t_{i-1}}^{t_i}$$

$$\left[\epsilon^p \|u_n(s) - u_n(t)\|_E^p + C_\epsilon^p R^p \left(\frac{b-a}{k} \right)^{p/q'} \right] ds\,dt,$$

$$\leq 4^{p-1} \sum_{i=1}^{k} \int_{t_{i-1}}^{t_i} \frac{1}{t_i - t_{i-1}} \int_{t_{i-1}}^{t_i} \epsilon^p \left(\|u_n(s)\|_E^p + \|u_n(t)\|_E^p \right) ds\,dt$$
$$+ 2^{p-1} C_\epsilon^p R^p (b-a)^{1+p/q'} k^{-(p/q')}$$

$$\leq 2 \cdot 4^{p-1} \cdot \epsilon^p R^p + 2^{p-1} C_\epsilon^p R^p (b-a)^{1+p/q'} k^{-(p/q')}.$$

Choose ϵ small enough and then choose k large enough that

$$\|u_n - \overline{u}_n\|_{L^p([a,b];W)} < \frac{\eta}{4}.$$

Now use compactness of the embedding of E into W to obtain a subsequence such that $\{\overline{u}_n\}$ is Cauchy in $L^p([a,b];W)$ and use this to contradict 32.

6. Show $L^1(\mathbb{R})$ is not reflexive. **Hint:** $L^1(\mathbb{R})$ is separable. What about $L^\infty(\mathbb{R})$?

Chapter 24
Convex Functions

In this chapter some of the basic theorems on convex functions are presented. It will draw on material presented in Chapters 6, 18, and 23 and the notation used will be from these chapters. Throughout this chapter X will be a real locally convex topological vector space unless specified otherwise.

Definition 24.1 We say a function $\phi : X \to (-\infty, \infty]$ is convex if whenever $\lambda \in [0, 1]$,

$$\phi (\lambda x + (1 - \lambda) y) \leq \lambda \phi (x) + (1 - \lambda) \phi (y).$$

We also define

$$epi (\phi) \equiv \{(x, a) : a \geq \phi (x)\}$$

and say ϕ is lower semi-continuous, $l.s.c.$, if $epi (\phi)$ is a closed subset of $X \times (-\infty, \infty]$. If in addition to this, $\phi (x) < \infty$ for some $x \in X$, we say ϕ is proper. The effective domain of ϕ is defined as

$$dom (\phi) \equiv \{x \in X \text{ such that } \phi (x) < \infty\}.$$

In this chapter it will always be the case that ϕ is proper.

24.1   Continuity properties of convex functions

Here we consider continuity properties of convex functions. Recall that if $\phi : \mathbb{R} \to \mathbb{R}$, is convex, then ϕ is continuous. This was Problem 3 of Chapter 12. More general results will be presented here in the case where the convex function is defined on a topological vector space.

Lemma 24.2 *Suppose ϕ is convex, l.s.c., and proper. Also suppose*

$$\phi (y) < a$$

for all $y \in U$ where U is some open set containing x. Then ϕ is continuous at x. In fact, if $B_A (x, r) \subseteq U$ where A is a finite subset of Ψ, the set of seminorms, then for $\epsilon \in (0, 1/2)$, and $y \in B_A (x, \epsilon r)$,

$$2\epsilon \phi (x) - 2\epsilon a \leq \phi (y) - \phi (x) \leq 2\epsilon a + 2\epsilon \phi (x). \tag{1}$$

Proof: Let the open set be U and let $x \in B_A (x, r) \subseteq U$ where A is a finite subset of Ψ. Let $\epsilon \in (0, 1/2)$. If $y \in B_A (x, \epsilon r)$, then

$$\frac{y - x}{2\epsilon} + x \in B_A (x, r)$$

and

$$y = 2\epsilon \left(\frac{y - x}{2\epsilon} + x \right) + (1 - 2\epsilon) \, x.$$

Therefore, if $y \in \overline{B_A (x, \epsilon r)}$,

$$\begin{aligned}
\phi \, (y) \; &\leq \; 2\epsilon \phi \left(\frac{y - x}{2\epsilon} + x \right) + (1 - 2\epsilon) \, \phi \, (x) \quad (2) \\
&\leq \; 2\epsilon a + (1 - 2\epsilon) \, \phi \, (x).
\end{aligned}$$

Now it is also true that

$$y = (1 + 2\epsilon) \, x - 2\epsilon \left(x + \frac{x - y}{2\epsilon} \right)$$

and so

$$x = \frac{y}{1 + 2\epsilon} + \frac{2\epsilon}{1 + 2\epsilon} \left(x + \frac{x - y}{2\epsilon} \right)$$

and, therefore,

$$\begin{aligned}
\phi \, (x) \; &\leq \; \frac{\phi \, (y)}{1 + 2\epsilon} + \frac{2\epsilon}{1 + 2\epsilon} \phi \left(x + \frac{x - y}{2\epsilon} \right) \quad (3) \\
&\leq \; \frac{\phi \, (y)}{1 + 2\epsilon} + \frac{2\epsilon a}{1 + 2\epsilon}.
\end{aligned}$$

By Formula 2 and Formula 3,

$$(1 + 2\epsilon) \, \phi \, (x) - 2a\epsilon \leq \phi \, (y) \leq (1 - 2\epsilon) \, \phi \, (x) + 2\epsilon a$$

and so

$$2\epsilon \phi \, (x) - 2a\epsilon < \phi \, (y) - \phi \, (x) \leq 2a\epsilon - 2\epsilon \phi \, (x) \, .$$

Since ϵ is arbitrary, this proves Formula 1 and the continuity of ϕ at x. The following corollary generalizes the earlier homework Problem.

Corollary 24.3 *If ϕ is convex and proper, defined on $\mathbb{R}^n$, then ϕ is continuous on the interior of dom (ϕ).*

Proof: Let $x \in$ interior of $(dom \, (\phi))$. Then $x \in B \, (x, r) \in dom \, (\phi)$ for some $r > 0$. Let $\epsilon \sqrt{n} < r$ and define $v_i \equiv \epsilon e_i$ for $i = 1, \cdots, n$, and $v_0 = 0$. Let

$$S \equiv \left\{ \sum_{i=0}^{n} \lambda_i v_i : \sum_{i=0}^{n} \lambda_i = 1 \text{ and } \lambda_i > 0 \text{ for all } i \right\}.$$

Then S is an open set of diameter less than $\epsilon \sqrt{n}$ which contains the point

$$\sum_{i=0}^{n} \frac{1}{1 + n} v_i \equiv p.$$

Continuity properties of convex functions

Let $U \equiv \mathbf{x} - \mathbf{p} + S$. Then $\mathbf{x} \in U \subseteq \mathbf{B}(\mathbf{x}, r)$ and

$$\phi\left(\mathbf{x} - \mathbf{p} + \sum_{i=0}^{n} \lambda_i \mathbf{v}_i\right) \leq \sum_{i=0}^{n} \lambda_i \phi\left(\mathbf{x} - \mathbf{p} + \mathbf{v}_i\right) \leq \max\left\{\phi\left(\mathbf{x} - \mathbf{p} + \mathbf{v}_i\right)\right\}_{i=0}^{n}.$$

Therefore, ϕ is continuous by Lemma 24.2.

More can be said about the continuity of ϕ near a point where ϕ is continuous. In fact the convex function is Lipschitz continuous near such a point of continuity.

Theorem 24.4 *Let X be a normed linear space and $\phi : X \to (-\infty, \infty]$ is convex and proper. If $x \in \text{dom}(\phi)$ and ϕ is continuous at x, then ϕ is Lipschitz continuous on some open set containing x.*

Proof: There exists $r_0 > 0$ such that for $y \in B(x, r_0)$,

$$|\phi(x) - \phi(y)| < M.$$

Consider u_0 and $u_1 \in B(x, r_0/2)$ and let $v_1, \cdots, v_m$ be points on the line segment joining u_0 and u_1 such that

$$\|v_{i+1} - v_i\| < \frac{r_0}{4}, \quad u_0 = v_1, \ u_1 = v_m, \text{ and}$$

$$\|u_0 - u_1\| = \sum_{i=1}^{m-1} \|v_{i+1} - v_i\|.$$

For v_i, consider the function $v \to \phi(v) - \phi(v_i)$, a convex function bounded above by $2M$ for all $v \in B\left(v_i, \frac{r_0}{2}\right) \subseteq B(x, r_0)$. Therefore, letting $\epsilon = \frac{2}{r_0}\|v_{i+1} - v_i\|$, it follows

$$v_{i+1} \in \overline{B\left(v_i, \epsilon \frac{r_0}{2}\right)} \subseteq B(x, r_0)$$

and so by Lemma 24.2,

$$|\phi(v_{i+1}) - \phi(v_i)| \leq 2\epsilon(2M) = \frac{8M}{r_0}\|v_{i+1} - v_i\|.$$

Thus,

$$|\phi(u_0) - \phi(u_1)| \leq \sum_{i=1}^{m-1} |\phi(v_{i+1}) - \phi(v_i)| \leq$$

$$\frac{8M}{r_0} \sum_{i=1}^{m-1} \|v_{i+1} - v_i\| = \frac{8M}{r_0}\|u_0 - u_1\|.$$

This shows ϕ is Lipschitz on $B\left(x, \frac{r_0}{2}\right)$ as claimed.

505

Corollary 24.5 *Let $\phi : \mathbb{R}^n \to (-\infty, \infty]$ be convex and proper. Then ϕ is locally Lipschitz on $dom\,(\phi)$.*

The next theorem extends these results slightly. It turns out that on a Banach space, as opposed to a normed linear space, a convex function is locally Lipschitz continuous on the interior of its domain. Thus on a Banach space one gets the continuity at a point in the interior of the domain for free.

Theorem 24.6 *Suppose X is a Banach space and $\phi : X \to (-\infty, \infty]$ is l.s.c. and proper. Then ϕ is locally Lipschitz on the interior of $dom\,(\phi)$.*

Proof: Let $x_0 \in$ interior of $dom\,(\phi)$ and define $\psi(x) \equiv \phi(x_0 + x)$. Let $a > \psi(0) = \phi(x_0)$. Thus $0 \in$ interior of $dom\,(\psi)$. Define

$$E \equiv \{x \in X : \psi(x) \le a\}.$$

Then E is convex and contains 0.
Claim: $X = \cup_{n=1}^{\infty} n\,(E \cap (-E))$.
Proof of the claim: Let $y \in X$ be arbitrary and let $g(t) \equiv \psi(ty)$ for $t \in \mathbb{R}$. Thus $0 \in$ interior of $dom\,(g)$ and so g is continuous at 0 by Corollary 24.3 applied to $\mathbb{R}$. Therefore, for $|t|$ small enough, $ty \in E$ and $-ty \in E$. Thus $ty \in E \cap (-E)$ and this proves the claim.

It follows from the completeness of X and the Baire category theorem that the interior of $n\,(E \cap (-E))$ is nonempty for some n. Therefore, the interior of $E \cap (-E)$ is also nonempty and for some p and $r > 0$, $B(p, r) \subseteq E \cap (-E)$. Then $B(-p, r) \subseteq E \cap (-E)$ also. By the identity

$$z = \frac{1}{2}((z - p) + (z + p))$$

and the convexity of $E \cap (-E)$,

$$B(0, r) \subseteq \left\{ \frac{1}{2}(x + y) : x \in B(p, r) \text{ and } y \in B(-p, r) \right\} \subseteq E \cap (-E).$$

Therefore $\psi(x) = \phi(x + x_0) \le a$ for all $x \in B(0, r)$ and so $\phi(y) \le a$ for all $y \in B(x, r)$. By Lemma 24.2, this proves ϕ is continuous at x. Theorem 24.4 implies ϕ is Lipschitz on some open set containing x.

24.2 Separation properties

The separation properties in locally convex topological vector space imply corresponding separation properties for the epigraph of a convex function. These separation properties are essential in much of what follows.

Theorem 24.7 *Let* $\phi : X \to (-\infty, \infty]$ *be convex l.s.c. and proper. Then if* $(x_0, a_0) \notin epi(\phi)$, *there exists* $x_0^* \in X'$ *and* $c \in \mathbb{R}$ *such that for all* $x \in X$,

$$\phi(x) + x_0^*(x) > c > x_0^*(x_0) + a_0.$$

The proof involves a few lemmas.

Lemma 24.8 $epi(\phi) \cap (X \times \mathbb{R})$ *is closed in* $X \times \mathbb{R}$.

Proof: If $(x, a) \notin epi(\phi) \cap (X \times \mathbb{R})$, then $a < \phi(x)$ so $a \neq \infty$. Therefore, $(x, a) \in X \times \mathbb{R}$. There exists an open set U in $X \times (-\infty, \infty]$ such that $(x, a) \in U$ but $U \cap epi(\phi) = \emptyset$. Then $U \cap (X \times \mathbb{R})$ is an open set containing (x, a) which does not intersect $epi(\phi) \cap (X \times \mathbb{R})$.

Definition 24.9 $X \times \mathbb{R}$ is a topological vector space relative to the seminorms

$$\bar{\rho}((x, a)) \equiv \rho(x) + |a|$$

where $\rho \in \Psi$, the seminorms defined on X.

Lemma 24.10 *If* $f \in (X \times \mathbb{R})'$ *then there exists* $x^* \in X'$ *and* $\beta \in \mathbb{R}$ *such that*

$$f((x, a)) = x^*(x) + \beta a$$

for all $(x, a) \in X \times \mathbb{R}$.

Proof: If $f \in (X \times \mathbb{R})'$, Theorem 6.4 implies there exists a finite subset, A, of Ψ such that

$$f((x, a)) \leq \bar{\rho}_A((x, a))$$

where

$$\bar{\rho}_A(x, a) \equiv \max(\rho(x) : \rho \in A) + |a| = \max(\bar{\rho}((x, a)) : \rho \in A).$$

Now define $x^*(x) \equiv f((x, 0))$. Thus $|x^*(x)| \leq \rho_A(x)$ so $x^* \in X'$. Similarly, use the Riesz representation theorem for Hilbert space to define $\beta \in \mathbb{R}$ such that $f((0, a)) = \beta a$. Thus

$$f((x, a)) = f((x, 0)) + f((0, a)) = x^*(x) + \beta a.$$

Proof of the theorem: Let $(x_0, a_0) \notin epi(\phi)$. Then $a_0 < \phi(x_0)$ and so $(x_0, a_0) \in X \times \mathbb{R}$. By Lemmas 24.8 and 24.10 and the separation theorems in topological vector space, there exists $x_0^* \in X'$, $c \in \mathbb{R}$, and $\beta \in \mathbb{R}$ such that for all x,

$$x_0^*(x_0) + \beta a_0 < c < \beta\phi(x) + x_0^*(x).$$

If $\beta \leq 0$, a contradiction results from the assumption that $a_0 < \phi(x_0)$. It follows we may divide by β and, replacing x_0^* with x_0^*/β, the conclusion is obtained. This proves the theorem.

Recall that in locally convex topological vector spaces, if a set is closed and convex, it follows it is weakly closed also. The corresponding theorem in this context is the following corollary.

Corollary 24.11 *If $\phi : X \to (-\infty, \infty]$ is convex, l.s.c., and proper, then the same is true if X is given the weak topology.*

Proof: Let $(x_0, a_0) \notin epi(\phi)$. Then by Theorem 24.7 there exists $x^* \in X'$ and $c \in \mathbb{R}$ such that

$$x^*(x) + a > c > x^*(x_0) + a_0$$

whenever $a \geq \phi(x)$. It follows there exists U, a weakly open set containing 0 and $\eta > 0$ such that

$$c > x^*(x_0 + z) + a_0 + b$$

whenever $z \in U$ and $|b| < \eta$. Thus the set, $(U + x_0) \times (a_0 - \eta, a_0 + \eta)$, open in the topology of $(X, \text{weak topology}) \times \mathbb{R}$ does not intersect $epi(\phi)$ and so $epi(\phi)$ is closed. This proves the corollary.

Definition 24.12 We say a convex, proper, *l.s.c.* function, ϕ, defined on a Banach space, X, is coercive if

$$\lim_{||x|| \to \infty} \phi(x) = +\infty.$$

With Corollary 24.11, it is easy to establish the following existence theorem. This is easy because we can work with the weak topology and use the Eberlein Smulian theorem to extract a weakly convergent subsequence once an estimate has been established from the assumption of coercivity. This type of argument is quite common in the study of weak solutions to various partial differential equations.

Theorem 24.13 *Let $\phi : X \to (-\infty, \infty]$ be convex, proper, l.s.c., and coercive where X is a reflexive Banach space. Then ϕ has a global minimum.*

Proof: Let $\phi(x_0) < \infty$. By the assumption that ϕ is coercive, there exists $R > 0$ such that $x_0 \in B(0, R)$ and if $||x|| > R$, then $\phi(x) > \phi(x_0)$.

Claim: ϕ is bounded below.

Proof: If not, we can define $A_n \equiv \{x \in X : \phi(x) \leq -n\}$. Then A_n is weakly closed by Corollary 24.11 because if $x \notin A_n$, then $\phi(x) >$

$-n$ which implies $(x, -n) \notin epi(\phi)$ and so there is a weakly open set containing x, U, such that $U \times \{-n\} \subseteq X \times \mathbb{R} \setminus epi(\phi)$ and so for all $y \in U$, $\phi(y) > -n$. Since the complement of A_n is open, it follows A_n is closed.

Let $K_n \equiv A_n \cap \overline{B(0, R)}$, a weakly compact set. For n large enough, K_n must be empty because if not, it would follow there exists a point in all the K_n, x, and therefore, $\phi(x) = -\infty$ contrary to the assumption that ϕ maps into $(-\infty, \infty]$. Thus ϕ must be bounded below.

Now let $\lambda \equiv \inf \{\phi(x) : x \in X\} > -\infty$. Let $x_n \in \overline{B(0, R)}$ be such that $\phi(x_n) \to \lambda$. By taking a subsequence, if necessary, we may also assume $x_n \to x \in B(0, R)$ weakly. Then since $epi(\phi)$ is closed, it follows $(x, \lambda) \in epi(\phi)$. Thus $\lambda \geq \phi(x) \geq \lambda$, the second inequality following because of the definition of λ. This proves the theorem.

24.3 Conjugate functions

Definition 24.14 Let $\phi : X \to (-\infty, \infty]$ be some function, not necessarily convex but satisfying $\phi(y) < \infty$ for some $y \in X$. We define $\phi^* : X' \to (-\infty, \infty]$ by

$$\phi^*(x^*) \equiv \sup \{x^*(y) - \phi(y) : y \in X\}.$$

This function, ϕ^*, defined above, is called the conjugate function of ϕ or the polar of ϕ. Note $\phi^*(x^*) \neq -\infty$ because $\phi(y) < \infty$ for some y.

Theorem 24.15 *Let X be a real Banach space. Then ϕ^* is convex and l.s.c.*

Proof: Let $\lambda \in [0, 1]$. Then

$$\phi^*(\lambda x^* + (1 - \lambda) y^*) = \sup \{(\lambda x^* + (1 - \lambda) y^*)(y) - \phi(y) : y \in X\}$$

$$\sup \{\lambda(x^*(y) - \phi(y)) + (1 - \lambda)(y^*(y) - \phi(y)) : y \in X\}$$

$$\leq \lambda \phi^*(x^*) + (1 - \lambda) \phi^*(y^*).$$

It remains to show the function is *l.s.c.* Consider $f_y(x^*) \equiv x^*(y) - \phi(y)$. Then f_y is obviously convex. Also

$$epi(\phi^*) = \cap_{y \in X} epi(f_y).$$

Therefore, if we can show $epi(f_y)$ is closed, we are done. If $(x^*, a) \notin epi(f_y)$, then $a < x^*(y) - \phi(y)$ and, by continuity, this condition will persist for all y^* near x^*. Thus $epi(f_y)$ is closed and this proves the theorem.

509

Note this theorem holds with no change in the proof if X is only a locally convex topological vector space and X' is given the weak $*$ topology.

Definition 24.16 We define ϕ^{**} on X by

$$\phi^{**}(x) \equiv \sup\{x^*(x) - \phi^*(x^*), x^* \in X'\}.$$

Theorem 24.17 $\phi^{**}(x) \leq \phi(x)$ for all x and if ϕ is convex and l.s.c., $\phi^{**}(x) = \phi(x)$ for all $x \in X$.

Proof:

$$\phi^{**}(x) \equiv \sup\{x^*(x) - \sup\{x^*(y) - \phi(y) : y \in X\} : x^* \in X'\}$$

$$\leq \sup\{x^*(x) - (x^*(x) - \phi(x))\} = \phi(x).$$

Next suppose ϕ is convex and l.s.c. If $\phi^{**}(x_0) < \phi(x_0)$, then by Theorem 24.7 there exists x_0^* such that for all $y \in X$,

$$x_0^*(y) + \phi(y) > c > x_0^*(x_0) + \phi^{**}(x_0). \tag{4}$$

It follows from Formula 4 that

$$\phi^{**}(x_0) \geq (-x_0^*)(x_0) - \phi^*(-x_0^*) =$$

$$(-x_0^*)(x_0) - \sup\{(-x_0^*)(y) - \phi(y) : y \in X\}$$

$$= (-x_0^*)(x_0) + \inf\{x_0^*(y) + \phi(y) : y \in X\}$$

$$\geq c - x_0^*(x_0) > \phi^{**}(x_0),$$

a contradiction. Hence, if ϕ is convex and l.s.c., $\phi^{**}(x) = \phi(x)$ for all x and this proves the theorem.

The following corollary is descriptive of the situation just discussed. It says that to find $epi(\phi^{**})$ it suffices to take the intersection of all closed convex sets which contain $epi(\phi)$.

Corollary 24.18 $epi(\phi^{**})$ is the smallest closed convex set containing $epi(\phi)$.

Proof: $epi(\phi^{**}) \supseteq epi(\phi)$ from Theorem 24.17. Also $epi(\phi^{**})$ is closed by the proof of Theorem 24.15. Suppose $epi(\phi) \subseteq K \subseteq epi(\phi^{**})$ and K is convex and closed. Let

$$\psi(x) \equiv \min\{a : (x, a) \in K\}.$$

($\{a : (x, a) \in K\}$ is a closed subset of $(-\infty, \infty]$ so the minimum exists.) ψ is also a convex function with $epi\,(\psi) = K$. To see ψ is convex, let $\lambda \in [0, 1]$. Then, by the convexity of K,

$$\lambda\,(x, \psi\,(x)) + (1 - \lambda)\,(y, \psi\,(y))$$

$$= (\lambda x + (1 - \lambda)\,y, \lambda\psi\,(x) + (1 - \lambda)\,\psi\,(y)) \in K.$$

It follows from the definition of ψ that

$$\psi\,(\lambda x + (1 - \lambda)\,y) \leq \lambda\psi\,(x) + (1 - \lambda)\,\psi\,(y).$$

Then

$$\phi^{**} \leq \psi \leq \phi$$

and so from the definitions,

$$\phi^{***} \geq \psi^* \geq \phi^*$$

which implies from the definitions and Theorem 24.17 that

$$\phi^{**} = \phi^{****} \leq \psi^{**} = \psi \leq \phi^{**}.$$

Therefore, $\psi = \phi^{**}$ and $epi\,(\phi^{**})$ is the smallest closed convex set containing $epi\,(\phi)$ as claimed.

24.4   Subgradients

Definition 24.19 Let X be a real locally convex topological vector space. For $x \in X$, $\delta\phi\,(x) \subseteq X'$, possibly $\emptyset$. This subset of X' is defined by $y^* \in \delta\phi\,(x)$ means for all $z \in X$,

$$y^*\,(z - x) \leq \phi\,(z) - \phi\,(x).$$

Also $x \in \delta\phi^*\,(y^*)$ means that for all $z^* \in X'$,

$$(z^* - y^*)\,(x) \leq \phi^*\,(z^*) - \phi^*\,(y^*).$$

We define $dom\,(\delta\phi) \equiv \{x : \delta\phi\,(x) \neq \emptyset\}$.

The subgradient is an attempt to generalize the derivative. For example, a function may have a subgradient but fail to be differentiable at some point. A good example is $f\,(x) = |x|$. At $x = 0$, this function fails to have a derivative but it does have a subgradient. In fact, $\delta f\,(0) = [-1, 1]$.

To begin with we consider the question of existence of the subgradient of a convex function. There is a very simple criterion for existence. It is essentially that the subgradient is nonempty at every point of the interior of the domain of ϕ.

Theorem 24.20 *Let $\phi : X \to (-\infty, \infty]$ be convex and suppose for some $u \in dom(\phi)$, ϕ is continuous. Then $\delta\phi(x) \neq \emptyset$ for all $x \in int(dom(\phi))$. Thus*

$$dom(\delta\phi) \supseteq int(dom(\phi)).$$

Proof: Let $x_0 \in int(dom(\phi))$ and let

$$A \equiv \{(x_0, \phi(x_0))\}, B \equiv epi(\phi) \cap X \times \mathbb{R}.$$

Then A and B are both nonempty and convex. Since ϕ is continuous at $u \in dom(\phi)$,

$$(u, \phi(u) + 1) \in int(epi\phi \cap X \times \mathbb{R}).$$

Thus $int(B) \neq \emptyset$ and also $int(B) \cap A = \emptyset$. Therefore, by Lemma 24.10 and Corollary 6.11, there exists $x^* \in X'$ and $\beta \in \mathbb{R}$ such that

$$(x^*, \beta) \neq (0, 0) \tag{5}$$

and for all $(x, a) \in B$,

$$x^*(x) + \beta a \geq x^*(x_0) + \beta\phi(x_0). \tag{6}$$

In particular, whenever $x \in dom(\phi)$,

$$x^*(x) + \beta\phi(x) \geq x^*(x_0) + \beta\phi(x_0).$$

If $\beta = 0$, this would mean $x^*(x - x_0) \geq 0$ for all $x \in dom(\phi)$. Since $x_0 \in int(dom(\phi))$, this implies $x^* = 0$, contradicting Formula 5. If $\beta < 0$, we could apply Formula 6 to the case when $a = \phi(x_0) + 1$ and $x = x_0$ to obtain a contradiction. It follows $\beta > 0$ and so $-x^*/\beta \in \delta\phi(x_0)$. This proves the theorem.

There is an interesting symmetry which relates $\delta\phi, \delta\phi^*, \phi$, and ϕ^*.

Theorem 24.21 *Suppose ϕ is convex, l.s.c., and proper. Then*

$$y^* \in \delta\phi(x) \text{ if and only if } x \in \delta\phi^*(y^*)$$

and, in this case,

$$y^*(x) = \phi^*(y^*) + \phi(x).$$

Proof: If $y^* \in \delta\phi(x)$ then $y^*(z - x) \leq \phi(z) - \phi(x)$ and so

$$y^*(z) - \phi(z) \leq y^*(x) - \phi(x)$$

for all $z \in X$. Therefore,

$$\phi^* (y^*) \leq y^* (x) - \phi (x) \leq \phi^* (y^*).$$

Hence

$$y^* (x) = \phi^* (y^*) + \phi (x). \qquad (7)$$

Now if $z^* \in X'$ is arbitrary, Formula 7 shows

$$(z^* - y^*) (x) = z^* (x) - \phi^* (y^*) - \phi (x) \leq \phi^* (z^*) - \phi^* (y^*)$$

and this shows $x \in \delta\phi^* (y^*)$.

Now suppose $x \in \delta\phi^* (y^*)$. Then for $z^* \in X'$,

$$(z^* - y^*) (x) \leq \phi^* (z^*) - \phi^* (y^*)$$

and so, taking sup over all z^*, and using Theorem 24.17,

$$\phi^{**} (x) = \phi (x) \leq y^* (x) - \phi^* (y^*) \leq \phi^{**} (x).$$

Thus

$$y^* (x) = \phi^* (y^*) + \phi^{**} (x) = \phi^* (y^*) + \phi (x) \geq y^* (z) - \phi (z) + \phi (x)$$

for all $z \in X$ and this implies for all $z \in X$,

$$\phi (z) - \phi (x) \geq y^* (z - x)$$

so $y^* \in \delta\phi (x)$ and this proves the theorem.

The following theorem is a form of the chain rule in which the derivative is replaced by the subgradient.

Definition 24.22 If X is a Banach space, $u \in H^1 (0, T; X)$ if there exists $g \in L^2 (0, T; X)$ such that

$$u (t) = u (0) + \int_0^t g (s) \, ds$$

and we define $u' (\cdot) \equiv g (\cdot)$.

Theorem 24.23 *Suppose $u \in H^1 (0, T; X), z \in L^2 (0, T; X')$, and $z (t) \in \delta\phi (u (t))$ a.e. Then $t \to \phi (u (t))$ is in $W^{1,1} (0, T)$, the space of functions in $L^1 (0, T)$ whose weak derivatives are in $L^1 (0, T)$, and*

$$\frac{d}{dt} \phi (u (\cdot)) = z (\cdot) (\phi (u (\cdot))).$$

Proof: First we show $t \to \phi(u(t))$ is in $L^1(0,T)$. Pick $z(t_0) \in \delta\phi(u(t_0))$. Then for a.e. $t \in [0,T]$,

$$z(t_0)(u(t) - u(t_0)) + \phi(u(t_0)) \leq \phi(u(t)) \leq z(t)(u(t) - u(t_0)) + \phi(u(t_0))$$

since $z(t)(u(t_0) - u(t)) \leq \phi(u(t_0)) - \phi(u(t))$ a.e. t. This inequality shows $t \to \phi(u(t))$ is in $L^1(0,T)$ since $z \in L^2(0,T;X')$ and $u \in L^2(0,T;X)$. Also, for a.e. $t \in [0, T-h]$,

$$z(t)\left(\frac{u(t+h) - u(t)}{h}\right) \leq \frac{\phi(u(t+h)) - \phi(u(t))}{h}$$

$$\leq z(t+h)\left(\frac{u(t+h) - u(t)}{h}\right)$$

for a.e. h. Now $z(\cdot + h) \to z(\cdot)$ in $L^2(0,T;X')$ by continuity of translation in $L^2(0,T;X')$. Also it is routine to show that

$$\mathcal{X}_{[0,T-h]}(\cdot)\frac{u(\cdot + h) - u(\cdot)}{h} \to u'(\cdot)$$

in $L^2(0,T;X)$ and so

$$\frac{\phi(u(\cdot + h)) - \phi(u(\cdot))}{h} \to z(u')$$

in $L^1(0,T)$.

It follows from the definition of weak derivatives that in the sense of weak derivatives,

$$\frac{d}{dt}(\phi \circ u) = z(u') \in L^1(0,T).$$

Note that by Theorem 18.2 this implies that for a.e. $t \in [0,T]$, $\phi(u(t))$ is equal to an absolutely continuous function, $\phi \circ u$, and

$$(\phi \circ u)(t) - (\phi \circ u)(0) = \int_0^t z(u')\, ds.$$

There are other rules of calculus which have a generalization to subgradients. The following theorem is on such a generalization. It generalizes the theorem which states that the derivative of a sum equals the sum of the derivatives.

Theorem 24.24 Let ϕ_1 and ϕ_2 be convex, l.s.c. and proper. Then

$$\delta(\lambda\phi_i)(x) = \lambda\delta\phi_i(x), \quad \delta(\phi_1 + \phi_2)(x) \supseteq \delta\phi_1(x) + \delta\phi_2(x) \qquad (8)$$

if $\lambda > 0$. If there exists $\overline{x} \in dom\,(\phi_1) \cap dom\,(\phi_2)$ and ϕ_1 is continuous at $\overline{x}$ then for all $x \in X$,

$$\delta\,(\phi_1 + \phi_2)\,(x) = \delta\phi_1\,(x) + \delta\phi_2\,(x). \tag{9}$$

Proof: Formula 8 is obvious so we only need to show Formula 9. Suppose $\overline{x}$ is as described. It is clear Formula 9 holds whenever $x \notin dom\,(\phi_1) \cap dom\,(\phi_2)$ since then both sides equal $\emptyset$. Therefore, we will assume $x \in dom\,(\phi_1) \cap dom\,(\phi_2)$ in what follows. Let $x^* \in \delta\,(\phi_1 + \phi_2)\,(x)$. We need to show x^* is the sum of an element of $\delta\phi_1\,(x)$ and $\delta\phi_2\,(x)$. Define

$$C_1 \equiv \{(y, a) \in X \times \mathbb{R} : \phi_1\,(y) - x^*\,(y - x) - \phi_1\,(x) \le a\},$$

$$C_2 \equiv \{(y, a) \in X \times \mathbb{R} : a \le \phi_2\,(x) - \phi_2\,(y)\}.$$

Both C_1 and C_2 are convex and nonempty. In addition to this,

$$(\overline{x}, \phi_1\,(\overline{x}) - x^*\,(\overline{x} - x) - \phi_1\,(x) + 1) \in int\,(C_1)$$

due to the assumed continuity of ϕ_1 at $\overline{x}$. If $(y, a) \in int\,(C_1)$ then

$$\phi_1\,(y) - x^*\,(y - x) - \phi_1\,(x) \le a - \epsilon$$

whenever ϵ is small enough. Therefore, if (y, a) is also in C_2, the assumption that $x^* \in \delta\,(\phi_1 + \phi_2)\,(x)$ implies

$$a - \epsilon \ge \phi_1\,(y) - x^*\,(y - x) - \phi_1\,(x) \ge \phi_2\,(x) - \phi_2\,(y) \ge a,$$

a contradiction. Therefore $int\,(C_1) \cap C_2 = \emptyset$ and so by Corollary 6.11 and Lemma 24.10, there exists $(w^*, \beta) \in X' \times \mathbb{R}$ with

$$(w^*, \beta) \neq (0, 0), \tag{10}$$

and

$$w^*\,(y) + \beta a \ge w^*\,(y_1) + \beta a_1, \tag{11}$$

whenever $(y, a) \in C_1$ and $(y_1, a_1) \in C_2$.

Claim: $\beta > 0$.

Proof of claim: If $\beta < 0$ let

$$a = \phi_1\,(\overline{x}) - x^*\,(\overline{x} - x) - \phi_1\,(x) + 1,$$

$$a_1 = \phi_2\,(x) - \phi_2\,(\overline{x}), \text{ and } y = y_1 = \overline{x}.$$

Then

$$\beta\,(\phi_1\,(\overline{x}) - x^*\,(\overline{x} - x) - \phi_1\,(x) + 1) \ge \beta\,(\phi_2\,(x) - \phi_2\,(\overline{x})).$$

Dividing by β yields

$$\phi_1\left(\overline{x}\right) - x^*\left(\overline{x} - x\right) - \phi_1\left(x\right) + 1 \leq \phi_2\left(x\right) - \phi_2\left(\overline{x}\right)$$

and so

$$\phi_1\left(\overline{x}\right) + \phi_2\left(\overline{x}\right) - \left(\phi_1\left(x\right) + \phi_2\left(x\right)\right) + 1 \leq x^*\left(\overline{x} - x\right)$$

$$\leq \phi_1\left(\overline{x}\right) + \phi_2\left(\overline{x}\right) - \left(\phi_1\left(x\right) + \phi_2\left(x\right)\right),$$

a contradiction. Therefore, $\beta \geq 0$.

Now suppose $\beta = 0$. Letting

$$a = \phi_1\left(\overline{x}\right) - x^*\left(\overline{x} - x\right) - \phi_1\left(x\right) + 1,$$

$$\left(\overline{x}, a\right) \in int\left(C_1\right),$$

and so there exists an open set U containing 0 and $\eta > 0$ such that

$$\overline{x} + U \times \left(a - \eta, a + \eta\right) \subseteq C_1.$$

Therefore, Formula 11 applied to $\left(\overline{x} + z, a\right) \in C_1$ and $\left(\overline{x}, \phi_2\left(x\right) - \phi_2\left(\overline{x}\right)\right) \in C_2$ for $z \in U$ yields

$$w^*\left(\overline{x} + z\right) \geq w^*\left(\overline{x}\right)$$

for all $z \in U$. Hence $w^*\left(z\right) = 0$ on U which implies $w^* = 0$, contradicting Formula 10. This proves the claim.

Now with the claim, it follows $\beta > 0$ and so, letting $z^* = w^*/\beta$, Formula 11 implies

$$z^*\left(y\right) + a \geq z^*\left(y_1\right) + a_1$$

whenever $\left(y, a\right) \in C_1$ and $\left(y_1, a_1\right) \in C_2$. In particular,

$$\left(y, \phi_1\left(y\right) - x^*\left(y - x\right) - \phi_1\left(x\right)\right) \in C_1 \text{ and}$$

$$\left(y_1, \phi_2\left(x\right) - \phi_2\left(y_1\right)\right) \in C_2. \tag{12}$$

So letting $y = x$,

$$z^*\left(x\right) + \left(\phi_1\left(x\right) - x^*\left(x - x\right) - \phi_1\left(x\right)\right) \geq z^*\left(y_1\right) + \phi_2\left(x\right) - \phi_2\left(y_1\right).$$

Therefore,

$$z^*\left(y_1 - x\right) \leq \phi_2\left(y_1\right) - \phi_2\left(x\right)$$

for all y_1 and so $z^* \in \delta\phi_2\left(x\right)$. Now let $y_1 = x$ in Formula 12. Then

$$z^*\left(y\right) + \phi_1\left(y\right) - x^*\left(y - x\right) - \phi_1\left(x\right) \geq z^*\left(x\right)$$

and so $x^* - z^* \in \delta\phi_1\left(x\right)$ so $x^* = z^* + \left(x^* - z^*\right) \in \delta\phi_2\left(x\right) + \delta\phi_1\left(x\right)$ and this proves the theorem.

Next we discuss a very important example known as the duality map from a Banach space to its dual space. Before doing so, consider a Hilbert space H. We can define a map R from H to H', called the Riesz map, by the rule

$$R(x)(y) \equiv (y, x).$$

By the Riesz representation theorem, we know this map is onto and one to one with the properties

$$R(x)(x) = ||x||^2, \text{ and } ||Rx||^2 = ||x||^2.$$

The duality map from a Banach space to its dual is an attempt to generalize this notion of Riesz map to an arbitrary Banach space.

Definition 24.25 For X a Banach space define $F : X \to \mathcal{P}(X')$ by

$$F(x) \equiv \left\{ x^* \in X' : x^*(x) = ||x||^2, ||x^*|| \leq ||x|| \right\}. \tag{13}$$

Lemma 24.26 With $F(x)$ defined as above, it follows that

$$F(x) = \left\{ x^* \in X' : x^*(x) = ||x||^2, ||x^*|| = ||x|| \right\}$$

and $F(x)$ is a closed, nonempty, convex subset of X'.

Proof: If x^* is in the set described in Formula 13,

$$x^* \left(\frac{x}{||x||} \right) = ||x||$$

and so $||x^*|| \geq ||x||$. Therefore

$$x^* \in \left\{ x^* \in X' : x^*(x) = ||x||^2, ||x^*|| = ||x|| \right\}.$$

This shows this set and the set of Formula 13 are equal. It is also clear the set of Formula 13 is closed and convex. It only remains to show this set is nonempty.

Define $f : \mathbb{R}x \to \mathbb{R}$ by $f(\alpha x) = \alpha ||x||^2$. Then the norm of f on $\mathbb{R}x$ is $||x||$ and $f(x) = ||x||^2$. By the Hahn Banach theorem, f has an extension to all of X, x^*, and this extension is in the set of Formula 13, showing this set is nonempty as required.

The next theorem shows this duality map is the subgradient of $\frac{1}{2} ||x||^2$.

Theorem 24.27 For X a real Banach space, let $\phi(x) \equiv \frac{1}{2} ||x||^2$. Then $F(x) = \delta\phi(x)$.

Proof: Let $x^* \in F(x)$. Then

$$
\begin{aligned}
x^*(y-x) &= x^*(y) - x^*(x) \\
&\leq ||x|| \, ||y|| - ||x||^2 \leq \frac{1}{2} ||y||^2 - \frac{1}{2} ||x||^2.
\end{aligned}
$$

This shows $F(x) \subseteq \delta\phi(x)$. Now let $x^* \in \delta\phi(x)$. Then for all $t \in \mathbb{R}$,

$$
x^*(ty) = x^*(ty + x - x) \leq \frac{1}{2}\left(||x + ty||^2 - ||x||^2 \right). \tag{14}
$$

Now if $t > 0$, divide both sides by t. This yields

$$
x^*(y) \leq \frac{1}{2t}\left(2t \, ||x|| \, ||y|| + t^2 \, ||x||^2 \right).
$$

Letting $t \to 0$, we see that

$$
x^*(y) \leq ||x|| \, ||y||. \tag{15}
$$

Next suppose $t = -s$, where $s > 0$ in Formula 14. Then

$$
x^*(y) \geq \frac{1}{2s}\left[||x||^2 - ||x - sy||^2 \right] \geq
$$

$$
\frac{1}{2s}\left[||x - sy||^2 - 2 ||x - sy|| \, ||sy|| + ||sy||^2 - ||x - sy|| \right]^2. \tag{16}
$$

Taking a limit as $s \to 0$ yields

$$
x^*(y) \geq - ||x|| \, ||y||. \tag{17}
$$

It follows from Formula 17 and Formula 15 that

$$
|x^*(y)| \leq ||x|| \, ||y||
$$

and that, therefore, $||x^*|| \leq ||x||$ and $|x^*(x)| \leq ||x||^2$. Now return to Formula 16 and let $y = x$. Then

$$
x^*(x) \geq \frac{1}{2s}\left[||x||^2 \left(1 - (1 - s)^2 \right) \right].
$$

Letting $s \to 0$,

$$
x^*(x) \geq ||x||^2.
$$

Since it was already shown that $|x^*(x)| \leq ||x||^2$, this shows $x^* \in F(x)$ and proves the theorem.

The next result is an easy corollary of Theorem 24.13 and gives conditions under which the subgradient is onto. By this we mean that if $y^* \in X'$, then there exists $x \in X$ such that $y^* \in \delta\phi(x)$.

Theorem 24.28 *Suppose X is a reflexive Banach space and suppose $\phi : X \to (-\infty, \infty]$ is convex, proper, l.s.c., and for all $y^* \in X'$, $x \to \phi(x) - y^*(x)$ is coercive. Then $\delta\phi$ is onto.*

Proof: The function $x \to \phi(x) - y^*(x) \equiv \psi(x)$ is convex, proper, l.s.c., and coercive. Therefore, by Theorem 24.13, there exists x which minimizes this function. Therefore,

$$0 \in \delta\psi(x)$$

by Theorem 24.24, $0 \in \delta\psi(x) = \delta\phi(x) - y^*$ and this proves the theorem.

Corollary 24.29 *Suppose X is a reflexive Banach space and $\phi : X \to (-\infty, \infty]$ is convex, proper, and l.s.c. Then for each $y^* \in X'$ there exist $x \in X$, $x_1^* \in F(x)$, and $x_2^* \in \delta\phi(x)$ such that*

$$y^* = x_1^* + x_2^*.$$

Proof: Apply Theorem 24.28 to the convex function $\frac{1}{2}\|x\|^2 + \phi(x)$ and use Theorems 24.24 and 24.27.

24.5   Hilbert space

In this section we consider subgradients of a slightly different form defined on a subset of H, a real Hilbert space. In Hilbert space the duality map is just the Riesz map defined earlier by

$$Rx(y) \equiv (y, x).$$

Definition 24.30 *$dom(\partial\phi) \equiv dom(\delta\phi)$ and for $x \in dom(\partial\phi)$,*

$$\partial\phi(x) \equiv R^{-1}\delta\phi(x).$$

Thus $y \in \partial\phi(x)$ if and only if for all $z \in H$,

$$Ry(z - x) = (y, z - x) \le \phi(z) - \phi(x).$$

Definition 24.31 *A mapping $A : D(A) \subseteq H \to \mathcal{P}(H)$ is called monotone if whenever $y_i \in Ax_i$,*

$$(y_1 - y_2, x_1 - x_2) \ge 0.$$

A monotone map is called maximal monotone if whenever $z \in H$, there exists $x \in D(A)$ and $y \in A(x)$ such that $z = y + x$. Put more simply, $I + A$ maps $D(A)$ onto H.

The following lemma states, among other things, that when ϕ is a convex, proper, *l.s.c.* function defined on a Hilbert space, $\partial\phi$ is maximal monotone.

Lemma 24.32 *If ϕ is a convex, proper, l.s.c. function defined on a Hilbert space, then $\partial\phi$ is maximal monotone and $(I + \partial\phi)^{-1}$ is a Lipschitz continuous map from H to $dom\,(\partial\phi)$ having Lipschitz constant 1.*

Proof: Let $y \in H$. Then $Ry \in H'$ and by Corollary 24.29, there exists $x \in dom\,(\delta\phi)$ such that $Rx + \delta\phi(x) \ni Ry$. Multiplying by R^{-1} we see $y \in x + \partial\phi(x)$. This shows $I + \partial\phi$ is onto. If $y_i \in \partial\phi(x_i)$, then $Ry_i \in \delta\phi(x_i)$ and so by the definition of subgradients,

$$
\begin{aligned}
(y_1 - y_2, x_1 - x_2) &= R(y_1 - y_2)(x_1 - x_2) \\
&= Ry_1(x_1 - x_2) - Ry_2(x_1 - x_2) \\
&\geq \phi(x_1) - \phi(x_2) - (\phi(x_1) - \phi(x_2)) = 0
\end{aligned}
$$

showing $\partial\phi$ is monotone. Now suppose $x_i \in (I + \partial\phi)^{-1}(y)$. Then $y - x_i \in \partial\phi(x_i)$ and by monotonicity of $\partial\phi$,

$$
-|x_1 - x_2|^2 = (y - x_1 - (y - x_2), x_1 - x_2) \geq 0
$$

and so $x_1 = x_2$. Thus $(I + \partial\phi)^{-1}$ is well defined. If $x_i = (I + \partial\phi)^{-1}(y_i)$, then by the monotonicity of $\partial\phi$,

$$
(y_1 - x_1 - (y_1 - x_2), x_1 - x_2) \geq 0
$$

and so

$$
|y_1 - y_2||x_1 - x_2| \geq |x_1 - x_2|^2
$$

which shows

$$
\left|(I + \partial\phi)^{-1}(y_1) - (I + \partial\phi)^{-1}(y_2)\right| \leq |y_1 - y_2|.
$$

This proves the lemma.

The next theorem is a famous result called Alexandrov's theorem. It states that any convex function defined on $\mathbb{R}^n$ has a second derivative a.e. First we define the gradient by

$$
\nabla\phi(\mathbf{y}) \cdot \mathbf{v} \equiv D\phi(\mathbf{y})(\mathbf{v}).
$$

Thus the $\nabla\phi(\mathbf{y})$ is just $R^{-1}D\phi(\mathbf{y})$.

Theorem 24.33 *Let $\phi : \mathbb{R}^n \to \mathbb{R}$ be convex. Then there exists a set, $F \subseteq \mathbb{R}^n$, whose complement has measure 0 such that if $\mathbf{y} \in F$, there exists $L(\mathbf{y}) \in \mathcal{L}(\mathbb{R}^n, \mathbb{R}^n)$ such that for a.e. $\mathbf{v}$,*

$$
\nabla\phi(\mathbf{y} + \mathbf{v}) = \nabla\phi(\mathbf{y}) + L(\mathbf{y})\mathbf{v} + o(|\mathbf{v}|).
$$

Proof: Let $J(\mathbf{x}) \equiv (I + \partial\phi)^{-1}(\mathbf{x})$. By Theorem 24.20 and Lemma 24.32, J maps $\mathbb{R}^n$ onto $\mathbb{R}^n$. Also ϕ is locally Lipschitz by Corollary 24.5 and so $\nabla\phi(\mathbf{y})$ exists for a.e. $\mathbf{y} \in \mathbb{R}^n$. It is easy to see that $\partial\phi(\mathbf{y}) = \nabla\phi(\mathbf{y})$ whenever $\nabla\phi(\mathbf{y})$ exists. Now let

$$F \equiv \{\mathbf{y} \in \mathbb{R}^n : \nabla\phi(\mathbf{y}) \text{ exists}, \mathbf{y} = J(\mathbf{x})$$

where $DJ(\mathbf{x})$ exists and is nonsingular$\}$.

Then $\mathbb{R}^n \setminus F$ is a set of measure zero by Rademacher's theorem and Lemma 20.12. The proof will proceed using the following lemma.

Lemma 24.34 *If $J(\mathbf{x}) \in F$ and $0 < \epsilon < \min\left(\frac{3}{4}, \left\|J(\mathbf{x})^{-1}\right\|^{-1}\right)$, there exist constants, $r, \eta > 0$ such that for all $|\mathbf{v}| < r$, there exists $\mathbf{u}$ such that $|\mathbf{u}| \leq \eta$ and*

$$J(\mathbf{x}) + \mathbf{v} = J(\mathbf{x} + \mathbf{u}) \tag{18}$$

and

$$|\mathbf{v}| \leq |\mathbf{u}| \leq K|\mathbf{u}|, \tag{19}$$

where K depends on $\left\|J(\mathbf{x})^{-1}\right\|$ and ϵ.

Proof: By assumption, $DJ(\mathbf{x})$ exists and is nonsingular. Thus,

$$DJ(\mathbf{x})^{-1}(J(\mathbf{x} + \mathbf{u}) - J(\mathbf{x})) - \mathbf{u} = o(|\mathbf{u}|).$$

Picking $\epsilon \in \left(0, \min\left(\frac{3}{4}, \left\|J(\mathbf{x})^{-1}\right\|^{-1}\right)\right)$, there exists $\eta > 0$ such that $|\mathbf{u}| < \eta$ implies

$$|o(|\mathbf{u}|)| < \epsilon|\mathbf{u}|. \tag{20}$$

By Lemma 20.5, it follows that for each $|\mathbf{w}| < (1 - \epsilon)\eta$, there exists $\mathbf{u}$ with $|\mathbf{u}| \leq \eta$ such that

$$DJ(\mathbf{x})^{-1}(J(\mathbf{x} + \mathbf{u}) - J(\mathbf{x})) = \mathbf{w}.$$

Now let

$$|\mathbf{v}| < \frac{(1 - \epsilon)\eta}{\left\|DJ(\mathbf{x})^{-1}\right\|}.$$

If $\mathbf{w} \equiv DJ(\mathbf{x})^{-1}\mathbf{v}$, it follows $|\mathbf{w}| < (1 - \epsilon)\eta$ and so there exists $\mathbf{u}$ with $|\mathbf{u}| \leq \eta$ such that

$$DJ(\mathbf{x})^{-1}(J(\mathbf{x} + \mathbf{u}) - J(\mathbf{x})) = DJ(\mathbf{x})^{-1}\mathbf{v}$$

and so

$$J(\mathbf{x} + \mathbf{u}) = J(\mathbf{x}) + \mathbf{v}.$$

Let $r \equiv \frac{(1-\epsilon)\eta}{\left\|DJ(\mathbf{x})^{-1}\right\|}$. This establishes Formula 18. It remains to verify Formula 19.

$$\mathbf{v} + J(\mathbf{x}) = J(\mathbf{x} + \mathbf{u}) \tag{21}$$

and by Lemma 24.32, $|\mathbf{v}| \leq |\mathbf{u}|$. Now by the assumption $J(\mathbf{x}) \in F$,

$$J(\mathbf{x} + \mathbf{u}) = J(\mathbf{x}) + DJ(\mathbf{x})\mathbf{u} + o(|\mathbf{u}|)$$

and so from Formula 21,

$$\mathbf{v} = DJ(\mathbf{x})\mathbf{u} + o(|\mathbf{u}|) \tag{22}$$

and by Formula 20, and the choice of ϵ,

$$\left| DJ(\mathbf{x})^{-1}\mathbf{v} \right| \geq |\mathbf{u}| - \epsilon|\mathbf{u}|$$

$$\geq \left(1 - \min\left(\frac{3}{4}, \left\|J(\mathbf{x})^{-1}\right\|^{-1}\right)\right)|\mathbf{u}|.$$

This verifies Formula 19 with

$$K = \left\|DJ(\mathbf{x})^{-1}\right\| \left(1 - \min\left(\frac{3}{4}, \left\|J(\mathbf{x})^{-1}\right\|^{-1}\right)\right)^{-1}.$$

This proves the lemma.

Now we continue with the proof of the theorem. Let $J(\mathbf{x}) \in F$ and let $\mathbf{v}$ be such that $D\phi(J(\mathbf{x}) + \mathbf{v})$ exists. Since $\nabla\phi = \partial\phi$ at such points, $\mathbf{x} - J(\mathbf{x}) = \nabla\phi(J(\mathbf{x}))$. Now Formula 22 and Formula 19 imply

$$\begin{aligned}
\nabla\phi(J(\mathbf{x}) + \mathbf{v}) &= \nabla\phi(J(\mathbf{x} + \mathbf{u})) = \mathbf{x} + \mathbf{u} - J(\mathbf{x} + \mathbf{u}) \\
&= \mathbf{x} + \mathbf{u} - J(\mathbf{x}) - DJ(\mathbf{x})(\mathbf{u}) + o(|\mathbf{u}|) \\
&= \nabla\phi(J(\mathbf{x})) + (I - DJ(\mathbf{x}))\mathbf{u} + o(|\mathbf{v}|) \\
&= \nabla\phi(J(\mathbf{x})) + (I - DJ(\mathbf{x}))DJ(\mathbf{x})^{-1}(\mathbf{v}) + o(|\mathbf{v}|)
\end{aligned}$$

and this shows that for $J(\mathbf{x}) \in F$ then for a.e. $\mathbf{v}$,

$$\nabla\phi(J(\mathbf{x}) + \mathbf{v}) = \nabla\phi(J(\mathbf{x})) + \left(DJ(\mathbf{x})^{-1} - I\right)(\mathbf{v}) + o(|\mathbf{v}|)$$

which proves the theorem with $L(J(\mathbf{x})) \equiv \left(DJ(\mathbf{x})^{-1} - I\right)$.

The following theorem is of interest in this context. It relates the existence of the second derivative of a Lipschitz function to an equation involving the function. Note that the assumptions in this theorem are weaker than usual, because we do not know anything about the continuity of the second derivative. Therefore, the usual approach must be modified.

Theorem 24.35 *Suppose $\phi : U \to \mathbb{R}$ is Lipschitz where U is an open subset of $\mathbb{R}^n$, and let $\mathbf{y} \in U$. Suppose there exists $r > 0$ such that for a.e. $\mathbf{v} \in B(\mathbf{0}, r)$,*

$$\nabla\phi(\mathbf{y} + \mathbf{v}) = \nabla\phi(\mathbf{y}) + L(\mathbf{y})\mathbf{v} + o(|\mathbf{v}|).$$

Then for a.e. $\mathbf{v} \in B(\mathbf{0}, r)$

$$\phi(\mathbf{y} + \mathbf{v}) = \phi(\mathbf{y}) + \nabla\phi(\mathbf{y}) \cdot \mathbf{v} + \frac{1}{2}L(\mathbf{y})\mathbf{v} \cdot \mathbf{v} + o\left(|\mathbf{v}|^2\right).$$

Proof: Let $N \equiv \{\mathbf{v} \in B(\mathbf{0}, r) : \nabla\phi(\mathbf{y} + \mathbf{v}) \text{ does not exist}\}$. Thus N has measure zero by assumption. We enlarge N if necessary, so that we can assume N is also a Borel set. Then using polar coordinates,

$$\int_{S^{n-1}} \int_0^r \mathcal{X}_N(\rho\omega)\rho^{n-1}d\rho d\sigma = 0,$$

and so for σ a.e. $\omega \in S^{n-1}$,

$$\int_0^r \mathcal{X}_N(\rho\omega)\rho^{n-1}d\rho = 0$$

and so $\rho\omega \notin N$ for a.e. ρ. Let

$$G \equiv \left\{\omega \in S^{n-1} \text{ such that } \int_0^r \mathcal{X}_N(\rho\omega)\rho^{n-1}d\rho = 0\right\}$$

and let

$$F \equiv \{\rho\omega : \rho < r, \omega \in G\}.$$

Then $m(B(\mathbf{0}, r) \setminus F) = 0$ and for each $\mathbf{v} \in F$, $\mathbf{y} + t\mathbf{v}$ is a point where $\nabla\phi$ exists for a.e. $t \in [0, 1]$.

Let $\mathbf{v} \in F$. Then let $f(t) \equiv \phi(\mathbf{y} + t\mathbf{v}) - \phi(\mathbf{y})$. Thus, since f is Lipschitz, $f(1) = \int_0^1 f'(t)\,dt$ and

$$f'(t) = \nabla\phi(\mathbf{y} + t\mathbf{v}) \cdot \mathbf{v}$$

for a.e. t. Hence

$$
\begin{aligned}
f(1) &= \phi(\mathbf{y} + \mathbf{v}) - \phi(\mathbf{y}) = \int_0^1 \nabla\phi(\mathbf{y} + t\mathbf{v}) \cdot \mathbf{v}\,dt \\
&= \int_0^1 \nabla\phi(\mathbf{y}) \cdot \mathbf{v} + tL(\mathbf{y})\mathbf{v} \cdot \mathbf{v} + o(|t\mathbf{v}|) \cdot \mathbf{v}\,dt \\
&= \nabla\phi(\mathbf{y}) \cdot \mathbf{v} + \frac{1}{2}L(\mathbf{y})\mathbf{v} \cdot \mathbf{v} + o\left(|\mathbf{v}|^2\right)
\end{aligned}
$$

which proves the theorem.

24.6 Exercises

1. For A a maximal monotone operator defined on a Hilbert space H, let
$$G(A) \equiv \{[x, y] : x \in D(A) \text{ and } y \in Ax\}.$$
Show that for $\lambda > 0$, λA is also maximal monotone. We define $J_\lambda(A)$, written as J_λ for short, by
$$J_\lambda(A)(x) \equiv (I + \lambda A)^{-1} x.$$
Show
$$|J_\lambda x - J_\lambda y| \leq |x - y|.$$
Hint: For $r \in (-1, 1)$ and $f \in H$, show there exists a solution, u, to the equation,
$$(1 + r) u + Au = (1 + r) f,$$
as follows. Let $J_1 = (I + A)^{-1}$ and show J_1 is Lipschitz continuous with Lipschitz constant 1. This equation has a solution if and only if
$$u = J_1 ((1 + r) f - ru) = Tu.$$
Show T is a contraction map.

2. ↑ Define for A maximal monotone,
$$A_\lambda x \equiv \frac{1}{\lambda} x - \frac{1}{\lambda} J_\lambda x.$$
Show A_λ is Lipschitz continuous with Lipschitz constant no more than $\frac{2}{\lambda}$. Also verify that
$$A_\lambda x \in A J_\lambda x,$$
and
$$|A_\lambda x| \leq |y|$$
for all $y \in Ax$ if $x \in D(A)$. This operator, A_λ, is called the Yosida approximation to A.

3. ↑ Suppose $(y_1 - y, x_1 - x) \geq 0$ for all $[x, y] \in G(A)$ where A is maximal monotone. Show that this implies $x_1 \in D(A)$ and $y_1 \in Ax_1$. **Hint:** Try to show
$$J_\lambda(x_1 + \lambda y_1) = x_1$$

because then it will follow $x_1 \in D(A)$ and $y_1 \in Ax_1$. To verify this, use the assumption and Problem 2 to conclude

$$0 \leq (y_1 - A_\lambda (x_1 + \lambda y_1), x_1 - J_\lambda (x_1 + \lambda y_1)).$$

Then simplify to find

$$0 \leq -\frac{1}{\lambda} (x_1 - J_\lambda (x_1 + \lambda y_1), x_1 - J_\lambda (x_1 + \lambda y_1)).$$

The problem shows the graphs of these operators are maximal with respect to also being monotone and this is the reason for the name, maximal monotone.

4. ↑ Suppose $[x_k, y_k] \in G(A)$ and $x_k \rightarrow x, y_k \rightharpoonup y$ where the half arrow denotes weak convergence. Show that then $[x, y] \in G(A)$.

5. ↑ Let A be maximal monotone and let B be Lipschitz and monotone. Then $A + B$ is maximal monotone. **Hint:** First suppose B has Lipschitz constant less than one. Then consider

$$Tx \equiv (I + A)^{-1} (y - Bx).$$

Show T is a contraction map and consequently has a fixed point x which satisfies
$$y \in x + Ax + Bx.$$

Next let $A + B$ play the role of A to conclude that $A + B + B$ is maximal monotone. Continuing in this way, show that any Lipschitz constant is all right.

6. ↑ Let A and B be maximal monotone, let

$$y \in x_\lambda + B_\lambda x_\lambda + Ax_\lambda,$$

and suppose $B_\lambda x_\lambda$ is bounded independent of λ. Show there exists $x_1 \in D(A) \cap D(B)$ such that

$$y \in x_1 + Bx_1 + Ax_1.$$

Hint: $y - x_\lambda - B_\lambda x_\lambda \in Ax_\lambda$ and so

$$|x_\lambda - x_\mu|^2 \leq (B_\mu x_\mu - B_\lambda x_\lambda, x_\lambda - x_\mu)$$
$$= -(B_\lambda x_\lambda - B_\mu x_\mu, x_\lambda - x_\mu)$$
$$= -(B_\lambda x_\lambda - B_\mu x_\mu, J_\lambda (B) x_\lambda - J_\mu (B) x_\mu) +$$
$$(B_\lambda x_\lambda - B_\mu x_\mu, \lambda B_\lambda x_\lambda - \mu B_\mu x_\mu)$$

$$\leq (B_\lambda x_\lambda - B_\mu x_\mu, \lambda B_\lambda x_\lambda - \mu B_\mu x_\mu).$$

Conclude $\{x_\lambda\}$ is Cauchy as $\lambda \to 0$. Select a subsequence

$$x_\lambda \to x_1, B_\lambda x_\lambda \rightharpoonup z_1, \text{ and } y - x_\lambda - B_\lambda x_\lambda \rightharpoonup z_2.$$

Use Problem 4 and the observation that $J_\lambda(B) x_\lambda - x_\lambda \to 0$ to conclude $z_1 \in Bx_1$, $z_2 \in Ax_1$, and $y = x + z_1 + z_2$.

7. ↑ Let A be maximal monotone and let $B = \partial\phi$ where ϕ is proper, lower semicontinuous, and convex. Suppose

$$\phi(J_\lambda(A)x) \leq \phi(x) + C\lambda$$

and there exists $\xi \in D(A) \cap D(\phi)$. Then $A + \partial\phi$ is maximal monotone. **Hint:** Let $y \in H$ be arbitrary and let x_λ be given by

$$y = x_\lambda + \partial\phi(x_\lambda) + A_\lambda x_\lambda$$

and show $A_\lambda x_\lambda$ is bounded. Using Problem 6 it will follow $A + \partial\phi$ is maximal monotone. To do this, note

$$(y - x_\lambda - A_\lambda x_\lambda, J_\lambda(x_\lambda) - x_\lambda) \leq C\lambda$$

because $y - x_\lambda - A_\lambda x_\lambda \in \partial\phi(x_\lambda)$. Thus,

$$-(y - x_\lambda - A_\lambda x_\lambda, A_\lambda x_\lambda) \leq C. \tag{23}$$

Also since $A_\lambda \xi$ is bounded independent of λ, (Problem 2), and A_λ is monotone,

$$\phi(\xi) - \phi(x_\lambda) \geq (y - x_\lambda - A_\lambda x_\lambda, \xi - x_\lambda)$$

$$\geq (y - x_\lambda, \xi - x_\lambda) - (A_\lambda x_\lambda, \xi - x_\lambda)$$

$$\geq (y - x_\lambda, \xi - x_\lambda) - (A_\lambda \xi, \xi - x_\lambda) \geq |x_\lambda|^2 - C|x_\lambda|$$

for some C independent of λ. Hence $C \geq \phi(x_\lambda) + |x_\lambda|^2$. By Theorem 24.7 we can find $|x_\lambda|$ is bounded and then Formula 23 shows $A_\lambda x_\lambda$ is bounded.

8. Let ϕ be a proper convex function defined on a normed linear space. Show ϕ is lower semicontinuous if and only if whenever $u_n \to u$, $\phi(u) \leq \liminf_{n\to\infty} \phi(u_n)$.

9. Let $L : D(L) \subseteq L^2(\Omega) \to L^2(\Omega; \mathbb{R}^n)$ be given by $Lu \equiv \nabla u$ where $D(L)$ is defined to be the space of functions in $L^2(\Omega)$ whose weak

derivatives are also in $L^2(\Omega)$. Show L is a closed operator. Now define

$$\phi(u) \equiv \begin{cases} \frac{1}{2}\int_\Omega |\nabla u|^2 + a(\mathbf{x})\, u^2 dx & \text{if } u \in H^1(\Omega), \\ +\infty & \text{otherwise.} \end{cases}$$

where $a(\cdot) \in L^\infty(\Omega)$. Show ϕ is proper, lower semicontinuous, and convex on $L^2(\Omega)$. What is $\partial\phi(u)$?

10. We say a maximal monotone operator is coercive if for all m, there exists n such that if $|u| \geq n$ and $u \in D(A)$, then for any $z \in Au$,

$$\frac{|(z, u)|}{|u|} \geq m.$$

Note that if $D(A)$ is bounded, then A is automatically coercive. Show that if A is maximal monotone and coercive, then A is onto. **Hint:** Let

$$\epsilon u_\epsilon + z_\epsilon = f$$

where $z_\epsilon \in Au_\epsilon$. Use coercivity to show the u_ϵ are bounded. Then use the Eberlein Smulian theorem to obtain a subsequence converging to 0 such that $u_\epsilon \rightharpoonup u$, $z_\epsilon \rightharpoonup z$ where the arrow denotes weak convergence. Then $z = f$ and because u_ϵ is bounded,

$$(z_\epsilon, u_\epsilon - x) \to (f, u - x) = (z, u - x).$$

Now let $[x, y] \in G(A)$ and argue

$$(z - y, u - x) = \lim_{\epsilon \to 0} (z_\epsilon - y, u_\epsilon - x) \geq 0.$$

Use Problem 3.

11. ↑ This problem uses Problem 7 to obtain an existence theorem for an evolution equation. Let ϕ be a nonnegative, convex, proper, lower semicontinuous function defined on a Hilbert space H and let $\mathcal{H} \equiv L^2(0, T; H)$. Let Φ be defined on $\mathcal{H}$ by

$$\Phi(u) \equiv \int_0^T \phi(u(s))\, ds.$$

(Since ϕ is lower semicontinuous, $\phi^{-1}(a, \infty]$ is open. From this, ϕ is Borel measurable on H and, therefore, $\phi \circ u$ is measurable.) Show Φ is also nonnegative, convex, proper, and lower semicontinuous. Let

$$D(L) \equiv \{u \in L^2(0, T; H) : u' \in L^2(0, T; H) \text{ and}$$

$$u(0) = u_0 \in D(\phi)\}$$

and for $u \in D(L)$, let $Lu \equiv u'$. Show L is maximal monotone on $\mathcal{H}$. Next verify the conditions of Problem 7 hold for Φ and $J_\lambda(L)$. Show that if $\partial \Phi$ is coercive, there exists a unique solution to the differential inclusion with initial condition,

$$u' + \partial \Phi(u) \ni f, \ u(0) = u_0$$

where $f \in \mathcal{H}$. If $z \in \partial \Phi(u)$, show $z(t) \in \partial \phi(u(t))$ a.e.

Appendix 1: The Hausdorff Maximal theorem

The purpose of this appendix is to prove the equivalence between the axiom of choice, the Hausdorff maximal theorem, and the well-ordering principle. The Hausdorff maximal theorem and the well-ordering principle are very useful but a little hard to believe; so, it may be surprising that they are equivalent to the axiom of choice. First we give a proof that the axiom of choice implies the Hausdorff maximal theorem, a remarkable theorem about partially ordered sets.

We say that a nonempty set is partially ordered if there exists a partial order, $\prec$, satisfying

$$x \prec x$$

and

$$\text{if } x \prec y \text{ and } y \prec z \text{ then } x \prec z.$$

An example of a partially ordered set is the set of all subsets of a given set and $\prec \equiv \subseteq$. Note that we can not conclude that any two elements in a partially ordered set are related. In other words, just because x, y are in the partially ordered set, it does not follow that either $x \prec y$ or $y \prec x$. We call a subset of a partially ordered set, C, a chain if x, $y \in C$ implies that either $x \prec y$ or $y \prec x$. Sometimes this is called a totally ordered set. We say C is a maximal chain if whenever $\tilde{C}$ is a chain containing C, it follows the two chains are equal. In other words C is a maximal chain if there is no strictly larger chain.

Lemma 24.36 *Let $\mathcal{F}$ be a nonempty partially ordered set with partial order $\prec$. Then assuming the axiom of choice, there exists a maximal chain in $\mathcal{F}$.*

Proof: Let $\mathcal{X}$ be the set of all chains from $\mathcal{F}$. For $C \in \mathcal{X}$, let

$$S_C = \{x \in \mathcal{F} \text{ such that } C \cup \{x\} \text{ is a chain strictly larger than } C\}.$$

If $S_C = \emptyset$ for any C, then C is maximal and we are done. Thus, assume $S_C \neq \emptyset$ for all $C \in \mathcal{X}$. Let $f(S_C) \in S_C$. (This is where the axiom of choice is being used.) Let

$$g(C) = C \cup \{f(S_C)\}.$$

Thus $g(C) \supsetneq C$ and $g(C) \setminus C = \{f(S_C)\} = \{\text{a single element of } \mathcal{F}\}$. We call a subset T of $\mathcal{X}$ a tower if

$$\emptyset \in T,$$

$$C \in T \text{ implies } g(C) \in T,$$

Appendix 1: The Hausdorff Maximal theorem

and if $S \subseteq T$ is totally ordered with respect to set inclusion, then

$$\cup S \in T.$$

Note that $\mathcal{X}$ is a tower. Let T_0 be the intersection of all towers. Thus, T_0 is a tower, the smallest tower. Let $C_0 \in T_0$ and let

$$B \equiv \{\mathcal{D} \in T_0 \text{ such that } \mathcal{D} \supsetneq C_0 \text{ but } f(C_0) \notin \mathcal{D}\}.$$

Now define

$$\widetilde{T}_0 = T_0 \setminus B.$$

Claim: $\widetilde{T}_0$ is a tower.

Proof of the claim: It is clear that $\emptyset \notin B$ and so $\emptyset \in \widetilde{T}_0$. Now suppose $C \in \widetilde{T}_0$. Then $g(C) \in T_0$. We need to show $g(C) \notin B$. If $g(C) \in B$, then

$$g(C) = C \cup \{f(C)\} \supsetneq C_0 \text{ but } f(C_0) \notin g(C). \tag{*}$$

If $f(C) \notin C_0$, then $C \supseteq C_0$. If $C = C_0$, then $f(C_0) = f(C) \in g(C)$, a contradiction to $*$. Thus $C \supsetneq C_0$. But, since $C \notin B$, this requires $f(C_0) \in C \subseteq g(C)$, a contradiction to $*$. The other alternative is that $f(C) \in C_0$. This requires that $C \supsetneq C_0$ but then since $C \notin B$, $f(C_0) \in C \subseteq g(C)$, contradicting $*$ again. Thus $g(C) \in \widetilde{T}_0$.

If $S \subseteq \widetilde{T}_0$ is totally ordered, suppose $\cup S \in B$. Then

$$\cup S \supsetneq C_0 \text{ but } f(C_0) \notin \cup S.$$

Therefore, $C \supsetneq C_0$ for some $C \in S$ and so $f(C_0) \notin C$ because $C \subseteq \cup S$. But $f(C_0) \in C$ because $C \in \widetilde{T}_0$. Hence $\cup S \notin B$. This shows $\widetilde{T}_0$ is a tower which proves the claim.

Thus $\widetilde{T}_0$ is a tower which is a subset of the smallest tower, T_0, and so $\widetilde{T}_0 = T_0$. Now define

$$T_1 = \{C \in T_0 \text{ such that } C \text{ is comparable to all elements of } T_0\}.$$

(By C "is comparable" we mean that if $\mathcal{D} \in T_0$, either $\mathcal{D} \subseteq C$ or $\mathcal{D} \supseteq C$.) Such sets exist because $\emptyset$ is an example. We will use the following claim.

Claim: T_1 is a tower.

Proof of the claim: It is clear that $\emptyset \in T_1$. Suppose $C \in T_1$ and let $\mathcal{D} \in T_0$. If $\mathcal{D} \subseteq C$, then $\mathcal{D} \subseteq g(C)$. If $\mathcal{D} \supsetneq C$, then from what was just shown, $f(C) \in \mathcal{D}$ and so $g(C) \subseteq \mathcal{D}$.

If $S \subseteq T_1$ is totally ordered, and $\mathcal{D} \in T_0$, then either $\mathcal{D}$ contains all elements of S or some element of S contains $\mathcal{D}$. Either way, $\mathcal{D}$ is comparable to $\cup S$. This shows T_1 is a tower and proves the claim.

Therefore, $T_1 = T_0$ which shows that T_0 is totally ordered. Since T_0 is a tower,

$$\cup T_0 \in T_0, \; g(\cup T_0) \in T_0.$$

Since $\mathcal{T}_0$ is a totally ordered set of chains, this requires

$$g\left(\cup\mathcal{T}_0\right) \subseteq \cup\mathcal{T}_0.$$

But also from the definition of g,

$$g\left(\cup\mathcal{T}_0\right) \supsetneq \cup\mathcal{T}_0,$$

a contradiction. Thus $S_\mathcal{C} = \emptyset$ for some $\mathcal{C}$ and the Hausdorff maximal principle is proved.

If X is a nonempty set, we say $\leq$ is an order on X if

$$x \leq x,$$

and if $x,\ y \in X$, then

$$\text{either } x \leq y \text{ or } y \leq x$$

and

$$\text{if } x \leq y \text{ and } y \leq z \text{ then } x \leq z.$$

We say that $\leq$ is a well order and say that $(X, \leq)$ is a well-ordered set if every nonempty subset of X has a smallest element. More precisely, if $S \neq \emptyset$ and $S \subseteq X$ then there exists an $x \in S$ such that $x \leq y$ for all $y \in S$. A familiar example of a well-ordered set is the natural numbers.

Lemma 24.37 *The Hausdorff maximal principle implies every nonempty set can be well-ordered.*

Proof: Let X be a nonempty set and let $a \in X$. Then $\{a\}$ is a well-ordered subset of X. Let

$$\mathcal{F} = \{S \subseteq X : \text{there exists a well order for } S\}.$$

Thus $\mathcal{F} \neq \emptyset$. We will say that for $S_1,\ S_2 \in \mathcal{F}$, $S_1 \prec S_2$ if $S_1 \subseteq S_2$ and there exists a well order for S_2, $\leq_2$ such that

$$(S_2, \leq_2) \text{ is well-ordered}$$

and if

$$y \in S_2 \setminus S_1 \text{ then } x \leq_2 y \text{ for all } x \in S_1,$$

and if $\leq_1$ is the well order of S_1 then the two orders are consistent on S_1. Then we observe that $\prec$ is a partial order on $\mathcal{F}$. By the Hausdorff maximal principle, we let $\mathcal{C}$ be a maximal chain in $\mathcal{F}$ and let

$$X_\infty = \cup\mathcal{C}.$$

We also define an order, $\leq$, on X_∞ as follows. If x, y are elements of X_∞, we pick $S \in \mathcal{C}$ such that x, y are both in S. Then if $\leq_S$ is the order on S, we let $x \leq y$ if and only if $x \leq_S y$. This definition is well defined because of the definition of the order, $\prec$. Now let U be any nonempty subset of X_∞. Then $S \cap U \neq \emptyset$ for some $S \in \mathcal{C}$. Because of the definition of $\leq$, if $y \in S_2 \setminus S_1$, $S_i \in \mathcal{C}$, then $x \leq y$ for all $x \in S_1$. Thus, if $y \in X_\infty \setminus S$ then $x \leq y$ for all $x \in S$ and so the smallest element of $S \cap U$ exists and is the smallest element in U. Therefore X_∞ is well-ordered. Now suppose there exists $z \in X \setminus X_\infty$. Define the following order, $\leq_1$, on $X_\infty \cup \{z\}$.

$$x \leq_1 y \text{ if and only if } x \leq y \text{ whenever } x, y \in X_\infty$$

$$x \leq_1 z \text{ whenever } x \in X_\infty.$$

Then let

$$\widetilde{\mathcal{C}} = \{S \in \mathcal{C} \text{ or } X_\infty \cup \{z\}\}.$$

Then $\widetilde{\mathcal{C}}$ is a strictly larger chain than $\mathcal{C}$ contradicting maximality of $\mathcal{C}$. Thus $X \setminus X_\infty = \emptyset$ and this shows X is well-ordered by $\leq$. This proves the lemma.

With these two lemmas we can now state the main result.

Theorem 24.38 *The following are equivalent.*

The axiom of choice

The Hausdorff maximal principle

The well-ordering principle.

Proof: It only remains to prove that the well-ordering principle implies the axiom of choice. Let I be a nonempty set and let X_i be a nonempty set for each $i \in I$. Let $X = \cup \{X_i : i \in I\}$ and well order X. Let $f(i)$ be the smallest element of X_i. Then

$$f \in \prod_{i \in I} X_i.$$

24.1 Exercises

1. Zorn's lemma states that in a nonempty partially ordered set, if every chain has an upper bound, there exists a maximal element, x in the partially ordered set. When we say x is maximal, we mean that if

$x \prec y$, it follows $y = x$. Show Zorn's lemma is equivalent to the Hausdorff maximal theorem.

2. Let X be a vector space. We say $Y \subseteq X$ is a Hamel basis if every element of X can be written in a unique way as a finite linear combination of elements in Y. Show that every vector space has a Hamel basis and that if Y, Y_1 are two Hamel bases of X, then there exists a one to one and onto map from Y to Y_1.

3. ↑ Using the Baire category theorem of Chapter 3 show that any Hamel basis of a Banach space is either finite or uncountable.

Appendix 2: Stone's Theorem and Partitions of Unity

Let S be a topological space. We say that a collection of sets $\mathfrak{D}$ is a refinement of an open cover, $\mathfrak{G}$, if every set of $\mathfrak{D}$ is contained in some set of $\mathfrak{G}$. An open refinement would be one in which all sets are open, with a similar convention holding for the term "closed refinement".

Definition 24.39 We say that a collection of sets, $\mathfrak{D}$, is locally finite if for all $p \in S$, there exists V an open set containing p such that V has nonempty intersection with only finitely many sets of $\mathfrak{D}$.

Definition 24.40 We say S is paracompact if it is Hausdorff and for every open cover $\mathfrak{G}$, there exists an open refinement $\mathfrak{D}$ such that $\mathfrak{D}$ is locally finite and $\mathfrak{D}$ covers S.

Theorem 24.41 *If $\mathfrak{D}$ is locally finite then*

$$\cup\{\overline{D} : D \in \mathfrak{D}\} = \overline{\cup\{D : D \in \mathfrak{D}\}}.$$

Proof: It is clear the left side is a subset of the right. Let p be a limit point of

$$\cup\{D : D \in \mathfrak{D}\}$$

and let $p \in V$, an open set intersecting only finitely many sets of $\mathfrak{D}$, $D_1...D_n$. If p is not in any of $\overline{D_i}$ then $p \in W$ where W is some open set which contains no points of $\cup_{i=1}^n D_i$. Then $V \cap W$ contains no points of any set of $\mathfrak{D}$ and this contradicts the assumption that p is a limit point of

$$\cup\{D : D \in \mathfrak{D}\}.$$

Thus $p \in \overline{D_i}$ for some i and this proves the theorem.

We say $\mathfrak{G} \subseteq \mathcal{P}(S)$ is countably locally finite if

$$\mathfrak{G} = \cup_{n=1}^{\infty} \mathfrak{G}_n$$

and each $\mathfrak{G}_n$ is locally finite. The following theorem appeared in the 1950's. It will be used to prove Stone's theorem.

Theorem 24.42 *Let S be a regular topological space. The following are equivalent*

1.) Every open covering of S has a refinement that is open, covers S and is countably locally finite.

2.) Every open covering of S has a refinement that is locally finite and covers S.

3.) Every open covering of S has a refinement that is closed, locally finite, and covers S.

4.) Every open covering of S has a refinement that is open, locally finite, and covers S.

Proof:

1.)$\Rightarrow$ 2.)

Let $\mathfrak{S}$ be an open cover of S and let $\mathfrak{B}$ be an open countably locally finite refinement

$$\mathfrak{B} = \cup_{n=1}^{\infty} \mathfrak{B}_n$$

where $\mathfrak{B}_n$ is an open refinement of $\mathfrak{S}$ and $\mathfrak{B}_n$ is locally finite. For $B \in \mathfrak{B}_n$, let

$$E_n(B) = B \setminus \bigcup_{k<n} (\cup\{B : B \in \mathfrak{B}_k\}).$$

Claim: $\{E_n(B) : n \in \mathbb{N}, B \in \mathfrak{B}_n\}$ is locally finite.

Proof of the claim: Let $p \in S$. Then $p \in B_0 \in \mathfrak{B}_n$ for some n. Let V be open, $p \in V$, and V intersects only finitely many sets of $\mathfrak{B}_1 \cup ... \cup \mathfrak{B}_n$. Then consider $B_0 \cap V$. If $m > n$,

$$(B_0 \cap V) \cap E_m(B) \subseteq [\bigcup_{k<m} (\cup\{B : B \in \mathfrak{B}_k\})]^C \subseteq B_0^C.$$

Thus $(B_0 \cap V) \cap E_m(B) = \emptyset$ for $m > n$. This establishes the claim.

Claim: $\{E_n(B) : n \in \mathbb{N}, B \in \mathfrak{B}_n\}$ covers S.

Proof: Let $p \in S$ and let $n = \min\{k \in \mathbb{N} : p \in B \text{ for some } B \in \mathfrak{B}_k\}$. Let $p \in B \in \mathfrak{B}_n$. Then $p \in E_n(B)$.

The two claims show that 1.)$\Rightarrow$ 2.).

2.)$\Rightarrow$ 3.)

Let $\mathfrak{S}$ be an open cover and let

$$\mathcal{G} \equiv \{U : U \text{ is open and } \overline{U} \subseteq V \in \mathfrak{S} \text{ for some } V \in \mathfrak{S}\}.$$

Then since S is regular, $\mathcal{G}$ covers S. By 2.), $\mathcal{G}$ has a locally finite refinement, $\mathfrak{C}$, covering S. Consider

$$\{\overline{E} : E \in \mathfrak{C}\}.$$

This covers S and is locally finite because if $p \in S$, there exists V, $p \in V$, and V has nonempty intersections with only finitely many elements of $\mathfrak{C}$, say $E_1, \cdots, E_n$. If $\overline{E} \cap V \neq \emptyset$, then $E \cap V \neq \emptyset$ and so V intersects only $\overline{E_1}, \cdots, \overline{E_n}$. This shows 2.)$\Rightarrow$ 3.).

3.)$\Rightarrow$ 4.)

Let $\mathfrak{S}$ be an open cover and let $\mathfrak{B}$ be a locally finite refinement which covers S. By 3.) we can take $\mathfrak{B}$ to be a closed refinement but this is not important here. Let

$$\mathfrak{F} \equiv \{U : U \text{ is open and } U \text{ intersects only finitely many sets of } \mathfrak{B}\}.$$

Then $\mathfrak{F}$ covers S because $\mathfrak{B}$ is locally finite. By 3.), $\mathfrak{F}$ has a locally finite closed refinement, $\mathfrak{C}$, which covers S. Define for $B \in \mathfrak{B}$

$$\mathfrak{C}(B) \equiv \{C \in \mathfrak{C} : C \cap B = \emptyset\}$$

We use $\mathfrak{C}(B)$ to fatten up B. Let

$$E(B) \equiv (\cup\{C : C \in \mathfrak{C}(B)\})^C.$$

Then since $\mathfrak{C}(B)$ is locally finite, $E(B)$ is an open set by Theorem 24.41. Now let

$$B \subseteq F(B) \in \mathfrak{S}$$

and let

$$\mathfrak{L} = \{E(B) \cap F(B) : B \in \mathfrak{B}\}$$

Claim: $\mathfrak{L}$ covers S.

This claim is obvious because if $p \in S$ then $p \in B$ for some $B \in \mathfrak{B}$. Hence

$$p \in E(B) \cap F(B) \in \mathfrak{L}.$$

Claim: $\mathfrak{L}$ is locally finite and a refinement of $\mathfrak{S}$.

Proof: It is clear $\mathfrak{L}$ is a refinement of $\mathfrak{S}$ because every set of $\mathfrak{L}$ is a subset of a set of $\mathfrak{S}$, $F(B)$. Let $p \in S$. There exists an open set, W, such that $p \in W$ and W intersects only $C_1, \cdots, C_n$, elements of $\mathfrak{C}$. Hence $W \subseteq C_1 \cup, \cdots, \cup C_n$ since $\mathfrak{C}$ covers S. If $C_i \cap E(B) \neq \emptyset$, then $C_i \cap B \neq \emptyset$. If not, $C_i \in \mathfrak{C}(B)$ and $C_i \cap E(B) = \emptyset$. But C_i is contained in a set $U_i \in \mathfrak{F}$ which intersects only finitely many sets of $\mathfrak{B}$. Thus each C_i intersects only finitely many $B \in \mathfrak{B}$ and so each C_i intersects only finitely many of the sets, $E(B)$. Thus W intersects only finitely many of the $E(B)$. Hence $\mathfrak{L}$ is locally finite.

It is obvious that 4.)$\Rightarrow$ 1.) and so this proves the theorem.

The following theorem is Stone's theorem.

Theorem 24.43 *If S is a metric space then S is paracompact.*

Proof: Let $\mathfrak{S}$ be an open cover. Well order $\mathfrak{S}$. For $B \in \mathfrak{S}$,

$$F_n(B) \equiv \{x \in B : B\left(x, 2^{-n}\right) \subseteq B\}, \ n = 1, 2, \cdots.$$

Let

$$E_n(B) = F_n(B) \setminus \cup\{D : D \prec B \text{ and } D \neq B\}$$

where $\prec$ denotes the well order. If $B, D \in \mathfrak{S}$, then one is first in the well order. Let $D \prec B$. Then

$$dist\left(E_n(D), D^C\right) = 2^{-n}$$

and $E_n(B) \subseteq D^C$. Hence

$$dist(E_n(B), E_n(D)) \geq 2^{-n}$$

for all $B, D \in \mathfrak{S}$. Let

$$\widetilde{E_n(B)} \equiv \cup\{B(x, 8^{-n}) : x \in E_n(B)\}.$$

Thus $\widetilde{E_n(B)} \subseteq B$ and

$$dist\left(\widetilde{E_n(B)}, \widetilde{E_n(D)}\right) \geq 2^{-n} - 2^{1-3n} > 0.$$

It follows that the collection

$$\{\widetilde{E_n(B)} : B \in \mathfrak{S}\} \equiv \mathfrak{B}_n$$

is locally finite. In addition to this,

$$S \subseteq \cup\{\widetilde{E_n(B)} : n \in \mathbb{N}, B \in \mathfrak{S}\}$$

because if $p \in S$, let B be the first set in $\mathfrak{S}$ to contain p. Then $p \in E_n(B)$ for n large enough. This proves the theorem.

Lemma 24.44 *Suppose W and U are open sets with $\overline{W} \subseteq U$. Then there exists a continuous function $f : S \to [0,1]$ such that $f(x) = 1$ on $\overline{W}$ and $f(x) = 0$ on U^C.*

Proof: Let

$$f(x) = 1 - \frac{dist(x, \overline{W})}{dist(x, \overline{W}) + dist(x, U^C)}.$$

Theorem 24.45 *Let S be a metric space and let $\mathfrak{S}$ be any open cover of S. Then there exists a set, $\mathfrak{F}$, and functions $\{\phi_F : F \in \mathfrak{F}\}$ such that*

$$\phi_F : S \to [0,1]$$

ϕ_F is continuous

$\phi_F(x)$ equals 0 for all but finitely many $F \in \mathfrak{F}$

ϕ_F equals 0 on the complement of some $U \in \mathfrak{S}$

$$\sum\{\phi_F(x) : F \in \mathfrak{F}\} = 1 \text{ for all } x \in S.$$

Proof: Let

$$\mathcal{G} \equiv \{U : U \text{ is open and } \overline{U} \text{ is a subset of some set of } \mathfrak{S}\}.$$

Let $\mathfrak{S}_1$ be a locally finite open cover which is a refinement of $\mathcal{G}$. Thus if $U \in \mathfrak{S}_1$, it follows that $\overline{U}$ is a subset of some set of $\mathfrak{S}$. Now repeat this argument to get $\mathfrak{F}$, a locally finite open cover which is a refinement of $\mathfrak{S}_1$ and satisfies $\overline{F}$ is a subset of some set of $\mathfrak{S}_1$ for each set $F \in \mathfrak{F}$. For each $F \in \mathfrak{F}$, pick $U_F \in \mathfrak{S}_1$ such that $\overline{F} \subseteq U_F$ and note that

$$\{U_F : F \in \mathfrak{F}\}$$

yields a locally finite open cover of S which is a refinement of $\mathfrak{S}$. For $F \in \mathfrak{F}$, define a continuous function g_F satisfying

$$g_F(x) \equiv \begin{cases} 1 \text{ if } x \in \overline{F}, \\ 0 \text{ if } x \notin U_F. \end{cases}$$

Let

$$\phi_F(x) \equiv \left(\sum\{g_F(x) : F \in \mathfrak{F}\}\right)^{-1} g_F(x).$$

Now

$$\sum\{g_F(x) : F \in \mathfrak{F}\}$$

is a continuous function because if $x \in S$, then there exists an open set W with $x \in W$ and W has nonempty intersection with only finitely many sets of

$$\{U_F : F \in \mathfrak{F}\},$$

say $U_{F_1}, \cdots, U_{F_n}$. Then for $y \in W$,

$$\sum\{g_F(y) : F \in \mathfrak{F}\} = \sum_{i=1}^{n} g_{F_i}(y).$$

Since $\mathfrak{F}$ is a cover of S,

$$\sum\{g_F(x) : F \in \mathfrak{F}\} \neq 0$$

for any $x \in S$. Hence ϕ_F is continuous. This also shows $\phi_F(x) = 0$ for all but finitely many $F \in \mathfrak{F}$. Since each U_F is a subset of some set of $\mathfrak{S}$, this shows $\phi_F = 0$ on the complement of some set of $\mathfrak{S}$. It is obvious that

$$\sum\{\phi_F(x) : F \in \mathfrak{F}\} = 1$$

from the definition. Thus all the conclusions of the theorem have been shown and this proves the theorem.

The functions described above are called a partition of unity subordinate to the open cover $\mathfrak{S}$.

Appendix 2: Stone's theorem and Partitions of Unity

24.1 General partitions of unity

Theorem 24.46 *If (X, τ) is paracompact, then (X, τ) is regular and normal.*

Proof: Let p be a point and let H be a closed set not containing p. For each $x \in H$, there exists U_x, an open set containing x such that $p \notin \overline{U}_x$. Let

$$\mathfrak{C} = \{X \setminus H, \ U_x : x \in H\}.$$

Let $\mathcal{G}$ be an open refinement of $\mathfrak{C}$ which is locally finite and covers X. Let $\mathcal{G}_1$ be those sets of $\mathcal{G}$ which intersect H. Thus, if $G \in \mathcal{G}_1$, then $p \notin \overline{G}$ and also

$$\cup \{G : G \in \mathcal{G}_1\} \supseteq H.$$

Then letting

$$W \equiv \cup \{G : G \in \mathcal{G}_1\}$$

we see that

$$\overline{W} = \cup \{\overline{G} : G \in \mathcal{G}_1\}$$

because $\mathcal{G}_1$ is locally finite. Hence $p \in \overline{W}^C$, $H \subseteq W$. This proves X is regular.

Now let $H \cap K = \emptyset$, H and K closed. Since (X, τ) is regular, there exists, for each $x \in H$, an open set U_x such that $x \in U_x$ and $\overline{U}_x \cap K = \emptyset$. Let

$$\mathfrak{C} = \{X \setminus H, U_x : x \in H\}$$

and let $\mathcal{G}$ be an open locally finite refinement of $\mathfrak{C}$ which covers X. Let $\mathcal{G}_1$ be those sets of $\mathcal{G}$ which intersect H. Let

$$W \equiv \cup \{G : G \in \mathcal{G}_1\}.$$

Thus $W \supseteq H$ and

$$\overline{W} = \cup \{\overline{G} : G \in \mathcal{G}_1\}.$$

Hence

$$H \subseteq W, \ K \subseteq \overline{W}^C$$

and $W, \overline{W}^C$ are disjoint open sets. This proves the theorem.

The next theorem is a generalization of Theorem 24.45.

Theorem 24.47 *Let (S, τ) be paracompact and let $\mathfrak{S}$ be any open cover of S. Then there exists a set, $\mathfrak{F}$, and functions $\{\phi_F : f \in \mathfrak{F}\}$ such that*

$$\phi_F : S \to [0, 1]$$

ϕ_F is continuous

$\phi_F(x)$ *equals* 0 *for all but finitely many* $F \in \mathfrak{F}$

ϕ_F *equals* 0 *on the complement of some* $U \in \mathfrak{S}$

$$\sum \{\phi_F(x) : F \in \mathfrak{F}\} = 1 \ \textit{for all } x \in S.$$

Proof: The proof is just like that of Theorem 24.45 but instead of using Lemma 24.44, we use Urysohn's lemma.

24.2 A general metrization theorem

With these theorems on paracompactness and normal spaces, it is not too hard to present a general metrization theorem, the Nagata Smirnov theorem. To begin with, we say that a basis for a topology, $\mathfrak{B}$ is σ locally finite if

$$\mathfrak{B} = \bigcup_{n=1}^{\infty} \mathfrak{B}_n$$

where $\mathfrak{B}_n$ is locally finite. We say a topological space (X, τ) is metrizable if there exists a metric d defined on X such that the topology induced by the metric, for which a basis is the collection of balls, coincides with the original topology.

Theorem 24.48 *Suppose* (X, τ) *is regular, Hausdorff, and has a* σ *locally finite basis. Then* (X, τ) *is metrizable.*

Proof: The basis $\mathfrak{B}$ is of the form

$$\mathfrak{B} = \bigcup_{n=1}^{\infty} \mathfrak{B}_n$$

where $\mathfrak{B}_n$ is locally finite. Without loss of generality, we may assume that

$$\mathfrak{B}_n \subseteq \mathfrak{B}_{n+1}.$$

If $U \in \mathfrak{B}_n$ let $F(U)$ be defined by

$$F(U) = \cup\{\overline{V} : \overline{V} \subseteq U \text{ and } V \in \mathfrak{B}_n\}.$$

Thus $F(U)$ is a closed subset of U because $\mathfrak{B}_n$ is locally finite. Let

$$\mathcal{D}_n = \{(F(U), U) : U \in \mathfrak{B}_n \text{ and } F(U) \neq \emptyset\}.$$

Then for n large enough, $\mathcal{D}_n \neq \emptyset$. Furthermore, if $p \in W$, with W open, there exists for some n large enough, $(F(U), U) \in \mathcal{D}_n$ such that

$$p \in F(U) \subseteq U \subseteq W.$$

Appendix 2: Stone's theorem and Partitions of Unity

Next we observe that (X, τ) is paracompact. This is easy to see because if $\mathfrak{C}$ is an open cover, we may take

$$\widetilde{\mathfrak{B}} \equiv \{B \in \mathfrak{B} : B \text{ is a subset of some set of } \mathfrak{C}\}.$$

Then $\widetilde{\mathfrak{B}}$ is a σ locally finite open refinement covering X. Since every open cover has a σ locally finite open refinement covering X, it follows from Theorem 24.42 that every open cover has a locally finite open refinement covering X. Thus (X, τ) is paracompact. Consequently, (X, τ) is normal. Therefore for each

$$D \equiv (F(U), U) \in \mathcal{D}_n,$$

define a continuous function f_D mapping X to $[0,1]$ such that

$$f_D(x) = \begin{cases} 1 \text{ if } x \in F(U), \\ 0 \text{ if } x \notin U. \end{cases}$$

Now define

$$d_n(x, y) \equiv \sum \{|f_D(x) - f_D(y)| : D \in \mathcal{D}_n\}.$$

Claim 1 $y \to d_n(x, y)$ is continuous.

Proof: Let $p \in X$ and let $p \in W_0 \in \tau$ where W_0 intersects only finitely many sets of $\mathfrak{B}_n$. Let $x \in W_1 \in \tau$ where W_1 intersects only finitely many sets of $\mathfrak{B}_n$ also. Let

$$\{U_1, \cdots, U_l\}$$

be those sets of $\mathfrak{B}_n$ which are intersected by either W_0 or W_1. Let

$$\{U_1, \cdots, U_r\}$$

be those sets of $\{U_1, \cdots, U_l\}$ for which $F(U_i) \neq \emptyset$ and let $D_i = (F(U_i), U_i)$ for these sets. Thus if

$$D = (F(U), U) \in \mathcal{D}_n$$

and $U \notin \{U_1, \cdots, U_r\}$, then $f_D(y) = 0$ for $y \in W_0 \cup W_1$. Consequently, for $y \in W_0$,

$$d_n(x, y) = \sum_{i=1}^{r} |f_{D_i}(y) - f_{D_i}(x)|$$

and therefore $y \to d_n(x, y)$ is continuous at p. It follows $y \to d_n(x, y)$ is continuous because p is arbitrary.

Now define

$$d(x, y) \equiv \sum_{n=1}^{\infty} 2^{-n} \left(\frac{d_n(x, y)}{1 + d_n(x, y)} \right).$$

Claim 2 d is a metric on X.

Proof: All axioms are obvious except for $d(x, y) = 0$ if and only if $x = y$. Suppose $x \neq y$. Then there exists n such that $(F(U), U) \in \mathcal{D}_n$ and such that

$$x \in F(U), \ y \notin U.$$

Thus $d(x, y) > 0$ if $x \neq y$.

It remains to show the two topologies, τ and τ_d, coincide. Here τ_d is the topology induced by the metric and τ is the original topology. Let $W \in \tau$ and let $p \in W$. Then

$$p \in F(U) \subseteq U \subseteq W, \ U \in \mathcal{B}_n$$

for some n. Choose $r > 0$ small enough that

$$\frac{r}{2^{-n} - r} < \frac{1}{2}$$

and suppose $d(p, y) < r$. Then

$$\frac{2^{-n} d_n(p, y)}{1 + d_n(p, y)} < r$$

and so for $D = (F(U), U)$,

$$|f_D(p) - f_D(y)| \leq d_n(p, y) < \frac{r}{2^{-n} - r} < \frac{1}{2}.$$

Thus $d(p, y) < r$ implies that $y \in U \subseteq W$ because of the above inequality. (If $y \notin U$, then $|f_D(p) - f_D(y)| = 1$.) In other words,

$$B_d(p, r) \subseteq W$$

and, therefore, $\tau \subseteq \tau_d$.

Next let $p \in B(q, r)$ and choose $r_1 > 0$ such that

$$p \in B(p, r_1) \subseteq B(q, r).$$

Choose N large enough that

$$\sum_{n=N}^{\infty} 2^{-n} < \frac{r_1}{2}.$$

By **Claim 1** there exists an open set, $V \in \tau$, containing p such that if $y \in V$,

$$\sum_{n=1}^{N} \frac{d_n(p, y)}{1 + d_n(p, y)} 2^{-n} < \frac{r_1}{2}.$$

Thus for $y \in V \in \tau$,

$$d(y, p) < \sum_{n=1}^{N} \frac{d_n(p, y)}{1 + d_n(p, y)} 2^{-n} + \sum_{n=N}^{\infty} 2^{-n} < r_1$$

and so

$$p \in V \subseteq B_d(q, r).$$

This shows that $B_d(q, r)$ is in τ and consequently $\tau_d \subseteq \tau$. This proves the theorem.

Theorem 24.49 *If X is a metric space, then X has a σ locally finite basis.*

Proof: From Stone's theorem, X is paracompact. Let

$$\mathfrak{C}_n = \{B(p, r) : r \leq 2^{-n}\}.$$

Let $\mathfrak{B}_n$ be a locally finite refinement of $\mathfrak{C}_n$ which covers X and consider

$$\mathfrak{B} \equiv \bigcup_{n=1}^{\infty} \mathfrak{B}_n.$$

This is a basis because if $p \in W$, an open set, then

$$B(p, 3 \cdot 2^{-n}) \subseteq W$$

for some n. Then $p \in B \in \mathfrak{B}_n$. Therefore,

$$dist(p, W^C) \geq 3 \cdot 2^{-n}$$

and if $y \in B$, then

$$d(p, y) < 2 \cdot 2^{-n}.$$

Thus $p \in B \subseteq W$. This shows $\mathfrak{B}$ is a basis as claimed and proves the theorem.

Summarizing these two theorems we can state the following corollary.

Corollary 24.50 *Let (X, τ) be Hausdorff and regular. Then X is metrizable if and only if there exists a σ locally finite basis for τ.*

Appendix 3: Taylor Series and Analytic Functions

24.1 Taylor's formula

First we recall the usual Taylor formula with the Lagrange form of the remainder which is encountered in calculus classes. Since we will only need it on a specific interval, we will state it for this interval. When this has been done, the generalization to Taylor polynomials in the context of Banach spaces will be given.

Theorem 24.51 *Let* $h : (-\delta, 1 + \delta) \to \mathbb{R}$ *have* $m + 1$ *derivatives. Then there exists* $t \in [0, 1]$ *such that*

$$h(1) = h(0) + \sum_{k=1}^{m} \frac{h^{(k)}(0)}{k!} + \frac{h^{(m+1)}(t)}{(m+1)!}.$$

Now let $f : U \to Y$ where U is an open subset of a Banach space X, and suppose $f \in C^m(U)$. Let $x \in U$ and let $r > 0$ be such that

$$B(x, r) \subseteq U.$$

Then for $\|v\| < r$ and $y^* \in Y'$ we consider

$$\operatorname{Re} y^* (f(x+tv) - f(x)) \equiv h(t)$$

for $t \in [0, 1]$. Then

$$h'(t) = \operatorname{Re} y^* (Df(x+tv)(v)), \; h''(t) = \operatorname{Re} y^* (D^2 f(x+tv)(v)(v))$$

and continuing in this way, we see that

$$h^{(k)}(t) = \operatorname{Re} y^* \left(D^{(k)} f(x+tv)(v)(v) \cdots (v) \right) \equiv \operatorname{Re} y^* \left(D^{(k)} f(x+tv) v^k \right).$$

It follows from Taylor's formula for a function of one variable that

$$\operatorname{Re} y^* f(x+v) = \operatorname{Re} y^* f(x) + \sum_{k=1}^{m} \operatorname{Re} y^* \left(\frac{D^{(k)} f(x) v^k}{k!} \right)$$

$$+ \operatorname{Re} y^* \left(\frac{D^{(m+1)} f(x+tv) v^{m+1}}{(m+1)!} \right).$$

Since this holds for all $y^* \in Y'$, it follows the equation holds still if $\operatorname{Re} y^*$ is removed from both sides. This proves the following theorem.

Theorem 24.52 *Let $f : U \to Y$ and let $f \in C^{m+1}(U)$. Then if*

$$B(x,r) \subseteq U,$$

and $\|v\| < r$, there exists $t \in (0,1)$ such that

$$f(x+v) - f(x)$$

$$= \sum_{k=1}^{m} \left(\frac{D^{(k)}f(x)v^k}{k!} \right) + \left(\frac{D^{(m+1)}f(x+tv)v^{m+1}}{(m+1)!} \right).$$

24.2   Analytic functions

In calculus, there was a difference between functions of a real variable and functions of a complex variable. In the latter case the existence of a single derivative implied the existence of all derivatives and in fact the Taylor series converged to the function. It is reasonable to ask if a similar phenomenon occurs in the case of complex Banach spaces versus real Banach spaces. This section presents a quick introduction to this topic based on the assumption that the reader has had some exposure to complex analysis. Some of the details involving questions of convergence and term by term differentiation are left to the reader who will find these questions routine if he or she has read through Chapter 11. Also if h maps an open subset of $\mathbb{C}$ to a complex Banach space, X, and has a first derivative, then the usual Cauchy integral formula,

$$h(z) = \frac{1}{2\pi i} \int_C \frac{h(w)}{w - z} dw,$$

holds if C is a circle contained, together with its interior, in the open set on which h has a derivative. The integral can be defined as the ordinary Riemann integral using Riemann sums or it can be defined in terms of a Bochner integral. These details are routine and are left to the reader. There are several equivalent definitions of an analytic function defined on a complex Banach space. The following is the one we will use since it resembles the familiar definition encountered in undergraduate complex variable courses.

Definition 24.53 Let X and Y be complex Banach spaces and let $U \subseteq X$ be an open set. We say $f : U \to Y$ is analytic and bounded on U if

$$\lim_{z \to 0} \frac{f(x + zh) - f(x)}{z}$$

exists for all $x \in U$ and also $\|f(x)\| \leq M < \infty$ for all $x \in U$.

Let $x \in U$ and let $\delta(x) \equiv dist(x, U^C)$. Let $\mathbf{h} \in X^l$ and consider all $\mathbf{z} \in \mathbb{C}^l$ small enough that

$$\sum_{m=1}^{l} |z_m| + |z_m| \, \|h_m\| < \frac{\delta(x)}{2l+1}$$

and let C_1 be a circle with radius $\frac{\delta}{2l}$ centered at 0. Then consider

$$z_m \to f\left(x + \sum_{m=1}^{l} z_m h_m\right)$$

which is analytic on and inside C_1. Thus

$$f\left(x + z_1 h_1 + \sum_{m=2}^{l} z_m h_m\right)$$

$$= \frac{1}{2\pi i} \int_{C_1} \frac{f\left(x + w_1 h_1 + \sum_{m=2}^{l} z_m h_m\right)}{(w_1 - z_1)} dw_1$$

$$= \left(\frac{1}{2\pi i}\right)^2 \int_{C_1} \frac{1}{w_1 - z_1} \cdot$$

$$\int_{C_1} \frac{f\left(x + w_1 h_1 + w_2 h_2 + \sum_{m=3}^{l} z_m h_m\right)}{(w_2 - z_2)} dw_2 dw_1 =$$

$$\left(\frac{1}{2\pi i}\right)^l \int_{C_1} \cdots \int_{C_1} \frac{f(x + w_1 h_1 + w_2 h_2 + \cdots + w_l h_l)}{\prod_{m=1}^{l}(w_m - z_m)} dw_l \cdots dw_1.$$

Consider the case when $l = 2$.

$$\left(\frac{1}{2\pi i}\right)^2 \int_{C_1} \int_{C_1} \frac{f(x + w_1 h_1 + w_2 h_2)}{(w_1 - z_1)(w_2 - z_2)} dw_2 dw_1 =$$

$$\left(\frac{1}{2\pi i}\right)^2 \int_{C_1} \int_{C_1} f(x + w_1 h_1 + w_2 h_2) \cdot$$

$$\sum_{k_2=0}^{\infty} \frac{z_2^{k_2}}{w_2^{k_2+1}} \sum_{k_1=0}^{\infty} \frac{z_1^{k_1}}{w_1^{k_1+1}} dw_2 dw_1 =$$

$$\left(\frac{1}{2\pi i}\right)^2 \sum_{k_2=0}^{\infty} \sum_{k_1=0}^{\infty} \left(\int_{C_1} \int_{C_1} \frac{f(x + w_1 h_1 + w_2 h_2)}{w_2^{k_2+1} w_1^{k_1+1}} dw_2 dw_1\right) z_2^{k_2} z_1^{k_1}.$$

Similarly, for arbitrary l, and letting C be any circle centered at 0 with radius smaller than $\frac{\delta}{l}$,

$$f\left(x + \sum_{m=1}^{l} z_m h_m\right) = \sum_{k_l=0}^{\infty} \cdots \sum_{k_1=0}^{\infty} a_{k_1 \cdots k_l}(x, h_l \cdots h_1) z_1^{k_1} \cdots z_l^{k_k} \quad (1)$$

where

$$a_{k_1 \cdots k_l} (x, h_l \cdots h_1)$$

$$= \left(\frac{1}{2\pi i} \right)^l \int_C \cdots \int_C \frac{f \left(x + \sum_{m=1}^{l} w_m h_m \right)}{\prod_{m=1}^{l} w_m^{k_m + 1}} dw_1 \cdots dw_l. \qquad (2)$$

Lemma 24.54 *Let $l \geq 1$ and let $t_m \in \mathbb{C}$. Then if $\mathbf{h} \in X^l$, then whenever $|z|$ is small enough, Formula 1 holds. Also the coefficients satisfy*

$$a_{k_1 \cdots k_l} (x, t_l h_l \cdots t_1 h_1) = \left(\prod_{m=1}^{l} t_m^{k_m} \right) a_{k_1 \cdots k_l} (x, h_l \cdots h_1) \qquad (3)$$

and

$$\| a_{k_1 \cdots k_l} (x, h_l \cdots h_1) \| \leq C \prod_{m=1}^{l} \| h_m \| \qquad (4)$$

for some constant C.

Proof: Let C be small enough that $t_m C$ for all $m = 1, \cdots, l$ and C have radius less than $\frac{\delta}{l}$. First assume $t_m \neq 0$ for all m. Then

$$a_{k_1 \cdots k_l} (x, t_l h_l \cdots t_1 h_1)$$

$$= \left(\frac{1}{2\pi i} \right)^l \int_C \cdots \int_C \frac{f \left(x + \sum_{m=1}^{l} w_m t_m h_m \right)}{\prod_{m=1}^{l} (w_m t_m)^{k_m + 1}} \cdot$$

$$\prod_{m=1}^{l} t_m^{k_m + 1} dw_1 \cdots dw_l$$

$$= \left(\frac{1}{2\pi i} \right)^l \int_{t_l C} \cdots \int_{t_1 C} \frac{f \left(x + \sum_{m=1}^{l} u_m h_m \right)}{\prod_{m=1}^{l} (u_m)^{k_m + 1}} du_1 \cdots du_l \prod_{m=1}^{l} t_m^{k_m}$$

$$= a_{k_1 \cdots k_l} (x, h_l \cdots h_1) \prod_{m=1}^{l} t_m^{k_m}.$$

If $t_m = 0$ for any m, the result of both sides in the above equals zero due to the fact that

$$\int_C \frac{1}{w_m^{k_m + 1}} dw_m = 0$$

whenever $k_m \geq 1$.

To verify Formula 4, use Formula 3 to conclude

$$\| a_{k_1 \cdots k_l} (x, h_l \cdots h_1) \| \leq$$

$$\left\| a_{k_1 \cdots k_l} \left(x, \frac{h_l}{\|h_l\|} \cdots \frac{h_1}{\|h_1\|} \right) \right\| \prod_{m=1}^{l} \|h_m\|^{k_m}$$

and $\left\| a_{k_1 \cdots k_l} \left(x, \frac{h_l}{\|h_l\|} \cdots \frac{h_1}{\|h_1\|} \right) \right\|$ is bounded by

$$\frac{M}{(2\pi)^l} \int_C \cdots \int_C \frac{1}{\prod_{m=1}^{l} |w_m|^{k_m+1}} d\,|w_1| \cdots d\,|w_l| \equiv C.$$

This proves the lemma.

Lemma 24.55 *Suppose*

$$g(x + zh) = g(x) + \sum_{m=1}^{\infty} b_m(x, h) z^m$$

for all z small enough. Then

$$b_1(x, h_1 + h_2) = b_1(x, h_1) + b_1(x, h_2).$$

Proof:

$$g(x + z_1 h_1 + z_2 h_2) = \sum_{m=0}^{\infty} \sum_{n=0}^{\infty} g_{mn}(x, h_1, h_2) z_1^m z_2^n.$$

Thus,

$$g(x + z_1 h_1) = \sum_{m=0}^{\infty} g_{m0}(x, h_1 h_2) z_1^m,$$

$$g(x + z_2 h_2) = \sum_{n=0}^{\infty} g_{0n}(x, h_1 h_2) z_2^n$$

which implies

$$g_{m0}(x, h_1, h_2) = b_m(x, h_1), \quad g_{0n}(x, h_1, h_2) = b_n(x, h_2).$$

Now let $z_1 = z_2 = z$. Then

$$g(x + z(h_1 + h_2)) = g(x) + \sum_{n=0}^{\infty} b_n(x, h_1 + h_2) z^n$$

$$= g(x) + z(g_{10}(x, h_1, h_2) + g_{01}(x, h_1 h_2)) + \text{higher order terms in } z.$$

Therefore,

$$\begin{aligned} b_1(x, h_1 + h_2) &= g_{10}(x, h_1, h_2) + g_{01}(x, h_1 h_2) \\ &= b_1(x, h_1) + b_1(x, h_2) \end{aligned}$$

and this proves the lemma.

Lemma 24.56 *Suppose $a(x, h_l \cdots h_1)$ is multilinear, $(h_i \to a(x, h_l \cdots h_1)$ is linear),*

$$\|a(x, h_l \cdots h_1)\| \le C \prod_{m=1}^{l} \|h_m\|,$$

and

$$D^{l-1} f(x + h_l)(h_{l-1}) \cdots (h_1) - D^{l-1} f(x)(h_{l-1}) \cdots (h_1)$$

$$-a(x, h_l \cdots h_1) = o(\|h_n\|).$$

Then $D^l f(x)$ exists and

$$D^l f(x)(h_l)(h_{l-1}) \cdots (h_1) = a(x, h_l \cdots h_1).$$

Proof: If $n = 1$, the conclusion is obvious and is nothing more than the definition of the derivative. Next let $n = 2$.

$$Df(x + h)(h_1) - Df(x)(h_1) - a(x, h, h_1) = o(\|h\|).$$

Let $L(x)$ be defined by

$$L(x)(h)(h_1) \equiv a(x, h, h_1).$$

Then $L(x) \in \mathcal{L}(U, \mathcal{L}(X, Y))$ because

$$\|L(x)\| \equiv \sup_{\|h\| \le 1} \|L(x)(h)\| \equiv \sup_{\|h\| \le 1} \sup_{\|h_1\| \le 1} \|L(x)(h)(h_1)\| \le C.$$

Also

$$\|Df(x + h) - Df(x) - L(x)h\|$$

$$\equiv \sup_{\|h_1\| \le 1} \|Df(x + h)(h_1) - Df(x)(h_1) - L(x)(h)(h_1)\|$$

$$= \sup_{\|h_1\| \le 1} \|Df(x + h)(h_1) - Df(x)(h_1) - a(x, h, h_1)\| = o(\|h\|)$$

and so $L(x) = D^2 f(x)$. Continuing in this way, we verify the conclusion of the lemma.

Lemma 24.57 *If f is analytic on U, then $f \in C^\infty(U)$.*

Proof: By Lemma 24.55 applied to $g = f$ and Lemma 24.54, $Df(x)$ exists and

$$Df(x)(h) = a_1(x, h).$$

Analytic functions

Suppose $D^{l-1} f(x)$ exists for $l \geq 2$.

$$f\left(x + \sum_{m=1}^{l} z_m h_m\right) = \sum_{n_l=0}^{\infty} \cdots \sum_{n_1=0}^{\infty} a_{n_1 \cdots n_l}(x, h_l \cdots h_1) z_1^{n_1} \cdots z_l^{n_l}.$$

Differentiate with respect to $z_1, \cdots, z_{l-1}$ to obtain

$$D^{l-1} f\left(x + \sum_{m=1}^{l-1} z_m h_m + z_l h_l\right)(h_{l-1}) \cdots (h_1) =$$

$$\sum_{n_l=0}^{\infty} \sum_{n_{l-1}=1}^{\infty} \cdots \sum_{n_1=1}^{\infty} a_{n_l n_{l-1} \cdots n_1}(x, h_1 \cdots h_l)\left(\prod_{m=1}^{l-1} n_m\right) z_1^{n_1-1} \cdots z_{l-1}^{n_{l-1}-1} z_l^{n_l}.$$

Take $z_i = 0$ for $i = 1, \cdots, l-1$. Then

$$D^{l-1} f(x + z_l h_l)(h_{l-1}) \cdots (h_1) = \sum_{n_l=0}^{\infty} a_{n_l 1 \cdots 1}(x, h_l \cdots h_1) z_l^{n_l}. \qquad (5)$$

Now we apply Lemma 24.55 to the function

$$z_l \rightarrow D^{l-1} f(x + z_l h_l)(h_{l-1}) \cdots (h_1)$$

and conclude

$$h_l \rightarrow a_{1 \cdots 1}(x, h_l \cdots h_1)$$

is linear. Thus from Formula 5,

$$D^{l-1} f(x + z_l h_l)(h_{l-1}) \cdots (h_1) - D^{l-1} f(x)(h_{l-1}) \cdots (h_1)$$

$$= a_{1 \cdots 1}(x, h_l \cdots h_1) z_l + o(z_l h_l). \qquad (6)$$

From this equation, it follows that

$$a_{1 \cdots 1}\left(x, h_l \cdots h_i + \widehat{h_i} \cdots h_1\right) z_l - a_{1 \cdots 1}(x, h_l \cdots h_i \cdots h_1) z_l$$

$$- a_{1 \cdots 1}\left(x, h_l \cdots \widehat{h_i} \cdots h_1\right) z_l = o(z_l h_l)$$

because for each z_l, the left side of Formula 6 is linear in h_i for each $i \leq l-1$. Dividing both sides of the above by z_l and then letting $z_l \rightarrow 0$, we see that $a_{1 \cdots 1}$ is linear in each of the h_i. Denoting $z_l h_l$ by h_l,

$$D^{l-1} f(x + h_l)(h_{l-1}) \cdots (h_1) - D^{l-1} f(x)(h_{l-1}) \cdots (h_1)$$

$$= a_{1 \cdots 1}(x, h_l \cdots h_1) + o(\|h_l\|)$$

and so by Lemma 24.56, $D^l f(x)$ exists and

$$D^l f(x)(h_l) \cdots (h_1) = a_{1 \cdots 1}(x, h_l \cdots h_1).$$

This proves the lemma.

With these lemmas, the main result can be established. This is the generalizaton of the well known result for analytic functions.

Theorem 24.58 *Let X and Y be two complex Banach spaces and let U be an open set in X. Then $f : U \to Y$ is analytic on U if and only if $Df(x)$ exists for each $x \in U$ and in this case, $f \in C^\infty(U)$, and if $h \in X$, then whenever z is small enough,*

$$f(x + zh) = f(x) + \sum_{n=1}^{\infty} \frac{D^n f(x) h^n z^n}{n!}.$$

Proof: We know

$$f(x + zh) = f(x) + \sum_{n=1}^{\infty} a_n(x, h) z^n.$$

Differentiating, we obtain

$$D^k f(x + zh) h^k = k! a_k(x, h) + \sum_{n=k+1}^{\infty} n(n-1) \cdots (n-k+1) z^{n-k}.$$

Letting $z = 0$ this shows

$$D^k f(x) h^k = k! a_k(x, h)$$

and this proves half the theorem.

Conversely, if $Df(x)$ exists on U, it is clear that for each $f \in U$, f is analytic on some ball, $B(x, r) \subseteq U$. Therefore the formula involving the series follows. This proves the theorem.

24.3   Ordinary differential equations

In this section we give an application to ordinary differential equations. To begin with, here are two Banach spaces which will be of use. Let Z be a complex Banach space and let X be the space of functions mapping $\overline{B(0,1)} \equiv D_1$ to Z such that the functions are continuous on D_1 and analytic on $B_1 \equiv B(0, 1)$, the derivative is the restriction to B_1 of a continuous function defined on D_1, and the function equals 0 at 0. The norm on X will be $||\phi||_X \equiv ||\phi||_\infty + ||\phi'||_\infty$ where

$$||\phi||_\infty \equiv \sup\{||\phi(t)||_Z : t \in B_1\}.$$

(Note that for a function continuous on D_1 it does not matter in the above definition of $||\cdot||_\infty$ whether we use B_1 or D_1 in the definition.) We define Y to be the space of continuous functions which are defined on D_1 having values in Z which are also analytic on B_1. The norm on Y is defined as

$$||\phi||_\infty \equiv ||\phi||_Y .$$

Lemma 24.59 *The spaces X and Y with the given norms are Banach spaces and if $L : X \to Y$ is defined as $L\phi(t) = \phi'(t)$ for all $t \in B_1$, then L is one to one, onto and continuous.*

Proof: It is clear that X and Y are both normed linear spaces. It remains to show they are Banach spaces. Suppose $\{\phi_n\}$ is a Cauchy sequence in X. Then $\phi_n \to \phi$ uniformly and $\phi'_n \to \psi$ uniformly where ψ and ϕ are continuous on D_1. We need to verify that $\psi = \phi'$ on B_1. Letting C_1 be the unit circle, the Cauchy integral formula implies for $t \in B_1$,

$$\phi(t) = \lim_{n \to \infty} \phi_n(t) = \lim_{n \to \infty} \frac{1}{2\pi i} \int_{C_1} \frac{\phi_n(w)}{w - t} dw = \frac{1}{2\pi i} \int_{C_1} \frac{\phi(w)}{w - t} dw$$

which shows $\phi'(t)$ exists on B_1. Also for $t \in B_1$,

$$\psi(t) = \lim_{n \to \infty} \phi'_n(t) = \lim_{n \to \infty} \frac{1}{2\pi i} \int_{C_1} \frac{\phi_n(w)}{(w - t)^2} dw$$

$$= \frac{1}{2\pi i} \int_{C_1} \frac{\phi(w)}{(w - t)^2} dw = \phi'(t).$$

This shows X is a Banach space. A similar argument using the Cauchy integral theorem shows Y is a Banach space also. It is obvious that L is continuous. It remains to show L is one to one and onto.

Let $\phi \in Y$. We need to show $\phi = L\psi$ for some $\psi \in X$. Let

$$\psi(t) \equiv \int_\Gamma \phi(w) \, dw$$

where Γ is any piecewise smooth curve from 0 to t. By the Cauchy integral theorem, this definition is well defined and it is clear that $\psi(0) = 0$, $\psi'(t) = \phi(t)$, and ψ is continuous on D_1. This shows L is onto. It only remains to show L is one to one. Suppose $L\phi = 0$. Since $\phi(0) = 0$,

$$\phi(t) = \int_0^1 \phi'(ts) \, t \, ds = 0$$

if $t \neq 0$. But $\phi(0)$ is given to equal zero. Thus L is one to one as claimed. This proves the lemma.

Appendix 3: Taylor Series and Analytic Functions

Theorem 24.60 *Let Λ and Z be complex Banach spaces and let W be an open subset of $\mathbb{C} \times Z \times \Lambda$ containing $(0, y_0, \lambda)$. Also let $f : W \to Z$ be analytic. Then there exists a unique $y = y(y_0, \lambda)$ solving*

$$y' = f(t, y, \lambda), y(0) = y_0 \tag{7}$$

valid for $t \in D_\alpha \equiv \overline{B(0, |\alpha|)}$ where $\alpha = \alpha(y_0, \lambda)$. Furthermore, the map

$$(t, y_0, \lambda) \to y(y_0, \lambda)(t)$$

is analytic.

Proof: Let $\alpha s = t$ and define $\phi(s) \equiv y(t) - y_0$. Then y is a solution to Formula 7 for $t \in D_\alpha$ if and only if ϕ is a solution for $s \in D_1$ to the equations

$$\phi'(s) = \alpha f(\alpha s, \phi(s) + y_0, \lambda), \ \phi(0) = 0.$$

Let X, Y, and L be given above and define

$$\widetilde{W} \equiv \{(\alpha, \widehat{y_0}, \mu, \phi) \in \mathbb{C} \times Z \times \Lambda \times X :$$

$$\text{for } s \in D_1, (s\alpha, \widehat{y_0} + \phi(s), \mu) \in W\}.$$

For a given $(\alpha, \widehat{y_0}, \mu, \phi) \in \widetilde{W}$,

$$\{(s\alpha, \widehat{y_0} + \phi(s), \mu) : s \in D_1\}$$

is a compact subset of W. Consequently, the distance from this set to W^C is positive and so if $(\beta, y_0, \lambda, \psi)$ is sufficiently close to $(\alpha, \widehat{y_0}, \mu, \phi)$ in $\mathbb{C} \times Z \times \Lambda \times X$ it follows $(\beta, y_0, \lambda, \psi)$ is also in $\widetilde{W}$. This shows $\widetilde{W}$ is an open subset of $\mathbb{C} \times Z \times \Lambda \times X$. Now define $F : \widetilde{W} \to Y$ by

$$F(\alpha, \widehat{y_0}, \mu, \phi)(s) \equiv L\phi(s) - \alpha f(\alpha s, \phi(s) + \widehat{y_0}, \mu).$$

Then

$$F(0, y_0, \lambda, 0) = 0,$$

and F is analytic in $\widetilde{W}$. Also

$$D_4 F(0, y_0, \lambda, 0) \psi = L\psi = \psi'$$

and so $D_4 F(0, y_0, \lambda, 0) \in \mathcal{L}(X, Y)$, is one to one, onto and continuous by Lemma 24.59. By the open mapping theorem, its inverse is also continuous. Therefore, the conditions of the implicit function theorem are satisfied and so there exists $r > 0$ such that if

$$|\alpha| + ||\mu - \lambda|| + ||\widehat{y_0} - y_0|| < r,$$

then there exists a unique $\phi \in X$ such that

$$F\left(\alpha, \widehat{y}_0, \mu, \phi\right) = 0,$$

and ϕ is an analytic function of $(\alpha, \widehat{y}_0, \mu)$. Fixing $0 < \alpha < r$, it follows

$$\left(\widehat{y}_0, \mu\right) \to y\left(\widehat{y}_0, \mu\right)$$

is analytic on an open subset of $Z \times \Lambda$. Also $t \to y\left(\widehat{y}_0, \mu\right)(t)$ is an analytic function because of the definition of y in terms of ϕ. It follows that for $t \in B\left(0, |\alpha|\right)$,

$$\left(\widehat{y}_0, \mu\right) \to y\left(\widehat{y}_0, \mu\right)(t) \ \text{ and } t \to y\left(\widehat{y}_0, \mu\right)(t)$$

are both analytic. This implies the conclusion of the theorem.

Appendix 4: The Brouwer Fixed Point theorem

Here we present a short proof of one of the most important of all fixed point theorems. There are many approaches to this. To see a different proof, one may consult the book by Deimling [14], the one by Dunford and Schwartz [16], the book on fixed point theorems by Smart [51], or the book on homology theory by Vick [55]. The proof given here is from Even's book on partial differential equations [19]. It is quite elementary and is based on the following lemma.

Lemma 24.61 *Let $p : U \to V$ be C^2 where U and V are open subsets of $\mathbb{R}^n$. Then*

$$\sum_{j=1}^{n} (cof \nabla p)_{ij,j} = 0,$$

where here $(\nabla p)_{ij} \equiv p_{i,j} \equiv \frac{\partial p_i}{\partial x_j}$.

Proof: $\det (\nabla p) = \sum_{i=1}^{n} p_{i,j} cof (\nabla p)_{ij}$ and so

$$\frac{\partial \det (\nabla p)}{\partial p_{i,j}} = cof (\nabla p)_{ij}. \tag{1}$$

Also

$$\delta_{kj} \det (\nabla p) = \sum_{i} p_{i,k} (cof \nabla p)_{ij}. \tag{2}$$

The reason for this is that if $k \neq j$ this is just the expansion of a determinant of a matrix in which the k th and j th columns are equal. Differentiate Formula 2 with respect to x_j and sum on j. This yields

$$\sum_{r,s,j} \delta_{kj} \frac{\partial (\det \nabla p)}{\partial p_{r,s}} p_{r,sj} = \sum_{ij} p_{i,kj} (cof \nabla p)_{ij} + \sum_{ij} p_{i,k} cof (\nabla p)_{ij,j}.$$

Hence, using $\delta_{kj} = 0$ if $j \neq k$ and Formula 1,

$$\sum_{rs} (cof \nabla p)_{rs} p_{r,sk} = \sum_{rs} p_{r,ks} (cof \nabla p)_{rs} + \sum_{ij} p_{i,k} cof (\nabla p)_{ij,j}.$$

Subtracting the first sum on the right from both sides and using the equality of mixed partials,

$$\sum_{i} p_{i,k} \left(\sum_{j} (cof (\nabla p))_{ij,j} \right) = 0.$$

If $\det(p_{i,k}) \neq 0$ so that $(p_{i,k})$ is invertible, this shows $\sum_j (cof\nabla p)_{ij,j} = 0$. If $\det(\nabla p) = 0$, let

$$p_k = p + \epsilon_k I$$

where $\epsilon_k \to 0$ and $\det(\nabla p + \epsilon_k I) \equiv \det(\nabla p_k) \neq 0$. Then

$$\sum_j (cof\nabla p)_{ij,j} = \lim_{k \to \infty} \sum_j (cof\nabla p_k)_{ij,j} = 0$$

and this proves the lemma.

A function w is said to be a retraction of B onto ∂B if $w(x) = x$ for $x \in \partial B$ and $w(x) \in \partial B$ for all $x \in B$.

Lemma 24.62 Let $B = \overline{B(0,r)}$. Then there does not exist a $C^2(\mathbb{R}^n)$ retraction of B onto ∂B.

Proof: Suppose w is such a smooth retraction. Let $id(x) = x$.

$$I(\tau) \equiv \int_B \det \nabla (id + \tau(w - id)) \, dx, \ \tau \in [0, 1]. \tag{3}$$

$$I'(\tau) = \int_B \sum_{ij} \frac{\partial (\det \nabla (id + \tau(w - id)))}{\partial p_{i,j}} \frac{\partial p_{i,j}}{\partial \tau}$$

where $p_i = x_i + \tau(w_i - x_i)$ for $\mathbf{x} = (x_1, \cdots, x_n)$. Thus

$$
\begin{aligned}
I'(\tau) &= \int_B \sum_i \sum_j cof\nabla (id + \tau(w - id))_{ij} (w_i - id_i)_{,j} \\
&= \sum_i \int_B \sum_j cof\nabla (id + \tau(w - id))_{ij} (w_i - id_i)_{,j} \, dx.
\end{aligned}
$$

Now integrate by parts and use the assumption that $w = id$ on ∂B. Then the above equals

$$-\sum_i \int_B \sum_j cof\nabla (id + \tau(w - id))_{ij,j} (w_i - id_i) \, dx = 0$$

by Lemma 24.61. This shows that $i'(\tau) = 0$ and so

$$I(0) = I(1). \tag{4}$$

But $I(0) = $ volume of $B > 0$ while $I(1) = 0$, a contradiction to Formula 4. The reason that $I(1) = 0$ is as follows.

$$I(1) = \int_B \det(\nabla w) \, dx$$

and

$$\sum_i w_i w_i = |w|^2 = r^2$$

so

$$2 \sum_i w_{i,j} w_i = 0.$$

Hence $\det(w_{i,j}) = 0$. This proves the lemma.

Lemma 24.63 *Let $B = \overline{B(0,r)}$. Then there does not exist a continuous retraction of B onto ∂B.*

Proof: Suppose $w : B \to \partial B$ is continuous, $w(x) = x$ on ∂B. Then extend w to all of $\mathbb{R}^n$ by

$$\overline{w}(x) = \begin{cases} x \text{ if } |x| \geq r, \\ w(x) \text{ if } |x| \leq r. \end{cases}$$

Thus, $\overline{w}$ is continuous on $\mathbb{R}^n$. Let η_ϵ be a mollifier satisfying

$$\eta_\epsilon(x) = \eta_\epsilon(|x|)$$

so that

$$\int_{\mathbb{R}^n} \eta_\epsilon(x) \, x \, dx = 0. \tag{5}$$

Let $spt(\eta_\epsilon) \subseteq B(0, \epsilon)$. Then $\eta_\epsilon * \overline{w}$ converges uniformly to $\overline{w}$ for all x. Also $|\overline{w}(x)| \geq r$ for all x. Let $\epsilon < r$ and also be small enough that

$$(\eta_\epsilon * \overline{w})(x) \neq 0$$

for all x. Then if $|x| \geq 2r$, we use Formula 5 to obtain

$$
\begin{aligned}
(\eta_\epsilon * \overline{w})(x) &= \int_{B(0,\epsilon)} y \eta_\epsilon(x-y) \, dy \tag{6} \\
&= \int_{B(0,\epsilon)} (y-x) \eta_\epsilon(x-y) \, dy + x \int_{B(0,\epsilon)} \eta_\epsilon(x-y) \, dy \\
&= x.
\end{aligned}
$$

Consider

$$u(x) \equiv \frac{2r \eta_\epsilon * \overline{w}(x)}{|\eta_\epsilon * \overline{w}(x)|}.$$

Then $u \in C^2(\mathbb{R}^n)$, $u(x) = x$ if $|x| = 2r$, and $u : \mathbb{R}^n \to \partial(B(0, 2r))$. This contradicts Lemma 2.

Theorem 24.64 *(Brouwer fixed point theorem) Let $B = \overline{B(0,r)}$ and let $f : B \to B$ be continuous. Then f has a fixed point.*

Proof: Suppose there is no fixed point for f. Then define $h(x)$ as shown in the following picture.

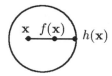

Then $h : B \to \partial B$ is a continuous retraction contradicting Lemma 3. This proves the theorem.

Corollary 24.65 *Let K be any compact convex subset of $\mathbb{R}^n$ and let $f : K \to K$ be continuous. Then f has a fixed point.*

Proof: Let $K \subseteq B(0, r)$ and define $g : \overline{B(0,r)} \to \overline{B(0,r)}$ by

$$g(x) \equiv f \circ proj_K(x).$$

Then g is continuous so $g(x) = x$ for some $x \in B(0, r)$. Thus

$$f(proj_K(x)) = x.$$

Since $f(K) \subseteq K$, it follows that $x \in K$ and so $proj_K(x) = x$. Hence $f(x) = x$. This proves the corollary.

References

[1] **Abraham R., Marsden J.**, and **Ratiu T.** *Manifolds, Tensor Analysis and Applications,* Addison-Wesley 1983.

[2] **Adams R.** *Sobolev Spaces,* Academic Press, New York, San Francisco, London, 1975.

[3] **Ash R.B.** *Real Analysis and Probability,* Academic Press, 1972.

[4] **Bartle R.** *The Elements of Real Analysis,* Wiley, 1976.

[5] **Bartle R.** and **Sherbert D.** *Introduction to Real Analysis,* Wiley, 1992.

[6] **Billingsley P.**, *Probability and Measure,* Wiley, 1979.

[7] **Brezis H.** *Opérateurs maximaux monotones et semigroupes de contractions dans les espaces de Hilbert,* Math Studies, 5, North Holland, 1973.

[8] **Calderón A.P.** and **Zygmund A.** On the existence of certain singular integrals, *Acta Math.* 88 (1952), 85-139.

[9] **Chow S.N.** and **Hale J.K.** *Methods of Bifurcation Theory,* Springer-Verlag, New York 1982.

[10] **Chung K.L.** *A Course in Probability Theory,* Harcourt Brace and World, 1968.

[11] **Crandall M.G., Ishii, H.**, and **Lions P.L.** User's Guide to Viscosity Solutions, *Bulletin of A.M.S.,* July 1992.

[12] **Davis H.** and **Snider A.** *Introduction to Vector Analysis,* Wm.C. Brown, 1995.

[13] **Diestal J.** and **Uhl J.** *Vector Measures,* American Math. Society, Providence, R.I., 1977.

[14] **Deimling K.** *Nonlinear Functional Analysis,* Springer-Verlag, 1985.

[15] **Dontchev A.L.** The Graves theorem Revisited, *Journal of Convex Analysis,* Vol. 3, 1996, No.1, 45-53.

References

[16] **Dunford N.** and **Schwartz J.T.** *Linear Operators,* Interscience Publishers, a division of John Wiley and Sons, New York, part 1 1958, part 2 1963, part 3 1971.

[17] **Ekeland I.** and **Temam R.** *Convex Analysis and Variational Problems,* North Holland Publishing Company, 1976.

[18] **Evans L.C.** and **Gariepy,** *Measure Theory and Fine Properties of Functions,* CRC Press, 1992.

[19] **Evans L.C.** *Partial Differential Equations,* Berkeley Mathematics Lecture Notes. 1993.

[20] **Evans L. C.** *Weak convergence methods for nonlinear partial differential equations,* regional conference series in Mathematics, number 74, A.M.S.

[21] **Federer H.,** *Geometric Measure Theory,* Springer-Verlag, New York, 1969.

[22] **de Guzman M.** *Differentiation of Integrals in* $\mathbb{R}^n$. Lecture notes in Mathematics, 481, Springer-Verlag, Berlin, 1975.

[23] **Halmos P.R.** *Naive Set Theory,* D. Van Nostrand Co. Inc. Princeton, N.J., 1960.

[24] **Halmos P.R.** *Measure Theory,* D. Van Nostrand Co. Inc., Princeton N.J., 1950.

[25] **Hardt R.,** *An Introduction to Geometric Measure Theory,* Lecture Notes, Melbourne University, 1979.

[26] **Hardy G.H.** and **Littlewood J.E.** A maximal theorem with function theoretic applications, *Acta Math.* 54 (1930), 81-116.

[27] **Hewitt E.** and **Stromberg K.** *Real and Abstract Analysis,* Springer-Verlag, New York, 1965.

[28] **Hewitt E.** and **Ross K.** *Abstract Harmonic Analysis I,* Springer-Verlag, Heidelberg, 1963.

[29] **Hoffman K.** and **Kunze R.** *Linear Algebra,* Prentice-Hall, 1971.

[30] **Hogg R.** and **Craig A.** *Introduction to Mathematical Statistics*, Macmillan, 1970.

[31] **Hörmander L.** Estimates for translation invariant operators in L^p spaces, *Acta Math.* 104 1960, 93-139.

[32] **Klambauer G.** *Real Analysis*, American Elsevier Publishing company, 1973.

[33] **Kolmogorov A.N.** and **Fomin S.V.** *Introductory Real Analysis*, Dover, 1975.

[34] **Lions J.L.** *Quelques Methods de Resolution des Problemes aux Limites Non Lineaires*, Dunod, Paris, 1969.

[35] **Marcinkiewicz J.** Sur l'interpolation d'opérations, *C.R. Acad. Sci. Paris* 208 1939, 1272-1273.

[36] **Marcus M.** and **Minc H.** *Introduction to Linear Algebra*, Dover, 1965.

[37] **McShane E. J.** *Integration*, Princeton University Press, Princeton, N.J. 1944.

[38] **Munkres J.R.** *Topology: A First Course*, Prentice-Hall, Englewood Cliffs, N. J., 1975.

[39] **Naniewicz Z.** and **Panagiotopoulos P.D.** *Mathematical Theory of Hemivariational Inequalities and Applications*, Marcel Dekker, 1995.

[40] **Patty C.** *Foundations of Topology*, PWS-KENT, 1993.

[41] **Ray W.O.** *Real Analysis*, Prentice-Hall, 1988.

[42] **Robbin J. W.** On the Existence theorem for Differential Equations, *Proc. Am. Math. Soc.* 19(1968), 1005-1006.

[43] **Rohatgi V.K.** *An introduction to probability and Mathematical Statistics*, John Wiley and Sons, 1976.

[44] **Royden H. L.** *Real Analysis*, Macmillan, New York, 1963.

References

[45] **Rudin W.** *Principles of Mathematical Analysis,* McGraw-Hill, 1976.

[46] **Rudin W.** *Real and Complex Analysis,* third edition, McGraw-Hill, 1987.

[47] **Rudin W.** *Functional Analysis,* second edition, McGraw-Hill, 1991.

[48] **Saks S.** *Theory of the Integral,* second edition, Monografie Matematycyne, vol. 28, Warsaw, 1952.

[49] **Saks S.** Remark on the differentiability of the Lebesgue indefinite integral, *Fund. Math.* 22 1934, 257-261.

[50] **Simon L.** *Lectures on Geometric Measure Theory,* Centre for Mathematical Analysis, Australian National University, 1984.

[51] **Smart D.R.** *Fixed point theorems* Cambridge University Press, 1974.

[52] **Stein E.** *Singular Integrals and Differentiability Properties of Functions.* Princeton University Press, Princeton, N. J., 1970.

[53] **Stein E.** and **Weiss G.** *Introduction to Fourier Analysis on Euclidean Spaces,* Princeton University Press, Princeton, N.J. 1971.

[54] **Strook D. W.** *A Concise Introduction to the Theory of Integration,* Birkhauser, 1994.

[55] **Vick J.W.** *Homology theory; an introduction to algebraic topology,* Academic Press, New York, 1973.

[56] **Yosida K.** *Functional Analysis,* Springer-Verlag, New York, 1978.

Chapter Notes

Chapter 1: This chapter is based on material found in Halmos [23], and Patty [40]. Another good source for the set theory is Hewitt and Stromberg [27].
Chapter 2: Theorems on compactness in metric space and the Ascoli Arzela theorem are proved in Patty [40] and most good advanced calculus books. The proof of the Ascoli Arzela theorem given here follows that given in Patty [40]. A different proof of the Stone Weierstrass theorem which is not based on the Bernstein polynomials and which is closer to

the original ideas of Weierstrass is in Rudin [45].

Chapters 3 and **4:** This material is standard introductory functional analysis and may be read in any functional analysis book. A good source for more on these theorems is Rudin [47], where all of them are presented in greater generality. See also Dunford and Schwartz [16] and Yosida [56].

Chapter 5: This is advanced calculus in the context of Banach space and as such it follows any good advanced calculus book. A good source for this material is Chow and Hale [9] and the references listed there or Abraham, Marsden and Ratiu [1]. Some of the generalizations of the inverse function theorem which are alluded to in the exercises may be found in Patty [40]. The very interesting existence and uniqueness theorem for ordinary differential equations is found in [9] and according to this reference is due to Robbin [42]. Graves theorem in the exercises follows the presentation in [15].

Chapter 6: The treatment of locally convex topological vector space given here is a more concrete version of that presented in Rudin [47]. The proof of the Tychonoff fixed point theorem is based on Deimling [14]. For more on set valued pseudomonotone operators, see [39], a recent monograph on the subject.

Chapters 7 and **8:** The approach to the Lebesgue integral and measure theory taken here follows Rudin [46]. To see it done differently, see Evans and Gariepy [18], or Dunford and Schwartz [16] part one. The Daniell integral approach is used or discussed in many other books. See Federer [21], Ray [41], Royden [44], Strook [54], or McShane [37]. This other approach involves defining the integral first and then getting the measure from it.

Chapters 9 and **10:** This may be found in Hewitt and Stromberg [27] and in Rudin [46]. The treatment of nonmeasurable sets and sets which are Lebesgue measurable but not Borel measurable which is found in the exercises is from McShane [37]. A very different approach which is much more set theoretic is in Hewitt and Stromberg [27]. The problem about the Borel set which has small measure yet its intersection with every interval has positive measure is from Rudin [46]. The proof of the Riesz representation theorem for positive linear functionals is a modification of that presented in Rudin [46] and uses the notation of this book. The version of Fubini's theorem presented in this chapter is a modification of an approach outlined in Hewitt and Stromberg [27] and is discussed more in Hewitt and Ross [28].

Chapter 11: The version of product measure presented here follows Hewitt and Stromberg [27] and Rudin [46]. A different version based more directly on the outer measure approach of Caratheodory which gives a complete measure more quickly is found in Evans and Gariepy [18] and Federer [21]. For much more on product measure, including product measure for infinite products, see Hewitt and Stromberg [27] or [3].

Chapter 12: This chapter includes all the standard material on the L^p spaces. See most books on measure and integration for similar presentations. The theorems about density of smooth functions are useful in studying function spaces such as the Sobolev spaces. The technique of using a mollifier to smooth out a function is used extensively in books on this subject. See Adams [2] for example.

Chapter 13: The treatment of the Radon Nikodym theorem is due to Von Neumann and is based on the Riesz representation theorem for the dual of a Hilbert space. This is used to establish the Riesz representation theorem for L^p as in Rudin [46]. The other approach to the Riesz representation theorem for L^p is based on Clarkson inequalities and may be found in Hewitt and Stromberg [27]. Another very good source for this approach is Ray [41]. This way of presenting these theorems has the advantage of emphasizing the uniform convexity of the norm and it does not require σ finite measure spaces. The disadvantage of this approach is the technical complexity of the Clarkson inequalities. In the second half of the chapter I have given a proof based on the easy Clarkson inequality which handles the case $p \geq 2$ and the case where $p < 2$ is obtained from a duality argument rather than the hard Clarkson inequalities. The representation theorems for $C(X)$ and $C_0(X)$ follow Evans and Gariepy [18] and were taken by them from Simon [50].

Chapters 14 and **15:** The elegant proof of the Vitali covering theorem is from Evans and Gariepy [18] and they got it from Simon [50]. The Hardy Littlewood Maximal function is discussed at length in Stein [52] which contains more on the subject than has been presented in this chapter. The change of variables formula for multiple integrals based on the Radon Nikodym theorem is a simpler version of that presented in Rudin [46] who does it for differentiable transformations, not C^1 transformations as done here. The Besicovitch covering theorem is a slight simplification of that presented in Evans and Gariepy [18] and they used the treatment in Hardt [25]. The differentiation theory for arbitrary Radon measures is based on a maximal function just as it is for Lebesgue measure. A very different approach to the differentiation theorems is in Evans and Gariepy [18]. The applications of the Lebesgue Besicovitch differentiation theorem are discussed in Evans [20]. I am grateful to Dan Hayden whose master's degree project helped me to collect the ideas which appear here. Several topics would have been much longer and much less elegant without his help.

Chapter 16: The treatment of Fourier transforms found in this chapter is based on the Schwartz class as it is in Rudin [47]. The L^1 theory is much easier and is left for the exercises.

Chapter 17: This chapter contains basic material on probability and random vectors. The first proof of the central limit theorem is based on [43] and avoids a direct use of the continuity theorem for characteristic functions. For different presentations of conditional probability and expectation, and for more on probability, consult [3], [10], or [6]. Some of the problems follow the development in [30] and this is a good reference for much more on statistics. I am grateful to Prof. Pinelis who found several errors and omissions in this chapter which needed to be corrected.

Chapter 18: I have avoided the topological aspect of distributions. This is discussed in Yosida [56] and Rudin [47]. Rademacher's theorem is proved like it is in Evans and Gariepy [18] and Evans [19] and is obtained as a corollary of a more general theorem. Evans and Gariepy [18] also give a more direct proof of this important theorem.

Chapters 19 and 20: The material on Hausdorff measures is standard and may be found in Evans and Gariepy [18] with somewhat different proofs. The proof of the area formula is a modification of the argument given in Rudin [46] for the change of variables formula and is based on the Brouwer fixed point theorem as it is there. It yields a simple and elegant proof. The divergence theorem is given for general Lipschitz domains but there are more general versions available in Evans and Gariepy [18]. The argument I have used is a modified version of the argument given in Strook [54].

Chapter 21: A different approach to the coarea formula may be found in Evans and Gariepy [18].

Chapters 22: The proof of Mihlin's theorem given here follows the paper by Hörmander [31]. Another approach is in Stein [52]. The material on the Helmholtz decomposition of an L^p vector field follows Davis and Snider [12].

Chapter 23: A well-known reference containing the topics in this chapter as well as many others is Diestal and Uhl [13]. I have tried to give an introduction to this subject which includes the topics which are of the most use in partial differential equations. Some of the problems follow [34] which makes extensive use of the material in this chapter. I am grateful to Ivo Dinov for the work he did as part of his master's project which assisted me in preparing this chapter.

Chapter 24: This is a brief introduction to the subject of convex functions which includes some of the most important theorems which are closely related to earlier topics in the book. For much more on this subject, consult Ekeland and Temam [17]. The proof of Alexandrov's theorem is from a paper in the Bulletin of the A.M.S. [11]. Another very different proof may be found in Evans and Gariepy [18]. The exercises on maximal monotone operators are all specializations of theorems presented in Brezis [7].

The Appendices: The proof of the equivalence of the well ordering principle, the axiom of choice and the Hausdorff maximal theorem is a modification of the treatment in Halmos [23]. Stone's theorem and general partitions of unity follow Munkres [38]. The theorems on analytic

References

functions and Taylor series are stated in Chow and Hale [9]. The proof of the Brouwer fixed point theorem is from Evans [19].

Index